OCEANOGRAPHY

Readings from

OCEANOGRAPHY

with introductions by
J. Robert Moore,
The University of Wisconsin

W. H. Freeman and Company
San Francisco

Most of the SCIENTIFIC AMERICAN articles in *Oceanography* are available as separate Offprints. For a complete list of more than 800 articles now available as Offprints, write to W. H. Freeman and Company, 660 Market Street, San Francisco, California 94104.

Printed in the United States of America

Library of Congress Catalog Card Number: 70-172644

Standard Book Number: 0-7167-0981-3 (cloth)
0-7167-0980-5 (paper)

9 8 7 6 5 4 3 2 1

PREFACE

The date December 23, 1972, is an important one to all who devote their professional lives to the scientific study of the sea—it marks the centenary of the sailing of Her Majesty's Ship *Challenger* on the first major expedition of oceanographic research. Indeed, in its lasting historical significance, the *Challenger* cruise was an event of importance to all mankind, most certainly as important as the opening of the Panama Canal, the laying of the first transoceanic telegraph cable, and the invention of self-contained underwater breathing apparatus—all three of which are not only ocean-oriented, but also trace portions of their lineage to developments that resulted from the *Challenger* cruise. Moreover, modern oceanography—as a distinct, albeit interdisciplinary, field of study and research—can be said to have originated with that cruise. Even today, the scientists engaged in research on sea-floor sediments, ocean water, and marine life continue to cite the observations made and data recorded a century ago by scientists aboard that British ship. Surely, few of the research monographs published today can hope to achieve such scientific longevity. Thus, the *Challenger* holds a special place in the hearts of oceanographers; for all scholars, the centenary of the cruise and the decade of the 1970s provide a time of unparalleled opportunity to assess the developments of ocean science during the intervening one hundred years, to review the present status of man's effort in understanding and using the sea, and to prognosticate on the state of the science in the years ahead, all in the faint but promising light of a new century dawning only three short decades hence. For whatever challenges and conflicts have confronted man in his past endeavors to answer his questions about the sea, there is a basic premise for the future: man must turn to the sea for his survival.

Although early oceanographers directed their attention primarily to purely scientific matters, the oceanographer of the future—if he regards himself as one of charity and concern for all men—must devote his research in large measure to resolving crucial difficulties, including finding new food and physical resources, abating pollution, aiding in conservation, and solving a host of special problems—problems that relate to what we call, in the political jargon of our times, *marine affairs*—including naval defense, trade and commerce, ocean law, and international politics. Each year, an increasing world population demands more of the sea, not only as a source of additional food and minerals, but as a site for dumping waste, as a place for recreation, and as a vast highway to serve expanding international commerce. It is naive, indeed, to assume that men can demand so much of the sea without causing it permanent damage and without creating temporal conflicts and misunderstandings with their neighbors. It must be remembered that the ocean is accessible to all nations, friend and foe alike. To ensure the priceless heritage of the freedom to transit and to use the ocean, citizens of each nation must understand, at least in terms of fundamental processes, the necessary limitations of, as well as the expectations for, man's exploitation of the common ocean.

With this in mind, I have assembled forty-one articles that originally appeared in *Scientific American*, and I present them here as instructive reading for those interested in oceanography and ocean-related problems. The increase in university and high school courses in oceanography, the reporting of ocean-oriented news by the press, and particularly the many programs about the ocean, both entertaining and enlightening, that appear on television, have combined to make oceanography a subject of popular interest. Furthermore, a clear understanding of the nature, the causes, and the effects of many important national and international incidents of recent times—including the Santa Barbara oil spill, the capture of the U.S.S. *Pueblo*, the loss of the U.S.S. *Thresher*, the seizure of American fishing vessels off South American coasts, and the historic voyage of the tanker *Manhattan*—requires some understanding of the science of the sea. Aggregately, the essays presented in this volume provide the reader with an overview of oceanography, including simplified explanations, examinations of processes at work in the ocean, examples of scholarly research, and commentaries on man's relations to the sea.

In considering the selection of essays presented here, the reader should recognize that, although a careful survey of *Scientific American* articles bearing on marine science was a necessary part of the preparation of this book, I have not attempted to provide a complete coverage of ocean-related topics. Indeed, the subject area is so vast that only a representative coverage is possible. Thus, the selection of essays was made on the basis of two criteria: the relevance of the articles to man's use of the sea and the instructive value of their scientific content.

June 1971 J. Robert Moore

CONTENTS

I OCEANOGRAPHY—SOME PERSPECTIVES

II THE SEA IN MOTION

III THE FLOOR OF THE OCEAN

IV OCEANIC LIFE

V MAN AND THE SEA

Note on cross-references: References to articles included in this book are noted by the title of the article and the page on which it begins; references to articles that are available as Offprints, but are not included here, are noted by the article's title and Offprint number; references to articles published by SCIENTIFIC AMERICAN, but which are not available as Offprints, are noted by the title of the article and the month and year of its publication.

OCEANOGRAPHY

I

OCEANOGRAPHY — SOME PERSPECTIVES

I

OCEANOGRAPHY—SOME PERSPECTIVES

INTRODUCTION

Oceanography began with a controversy.

In 1831, Charles Darwin, who had failed as a student at Cambridge University and was then an idle youth without prospects of a gentlemanly career, met another young man, Captain Robert Fitzroy, an experienced mariner of twenty-six and only four years Darwin's senior. Fitzroy was soon to master Her Majesty's Ship *Beagle* on a cruise to make navigational measurements at sea and to explore, in general, the coast of South America. With foresight, Fitzroy realized that the voyage would provide great opportunities for studying the natural science of distant waters and shores, and he offered Darwin the post of naturalist on the expedition. Strangely—for Charles Darwin had grown up accustomed to the amenities and pleasures available to the sons of well-to-do parents—he accepted the offer, and thus began the chain of events that led to the dispatch of the *Challenger* some four decades later.

Although many are familiar with Charles Darwin's most heralded contribution, his classic *Origin of Species*—a monumental synthesis that argued convincingly the idea of natural selection as a mechanism of organic evolution—few are cognizant of Darwin's lucid, instructive writings recording his observations of fossils, island fauna and flora, coral reefs, and other features of nature that he observed while on the *Beagle* cruise. The influence of Darwin's observations of oceanic phenomena on the development of the theories he presented in his *Origin of Species* cannot be overemphasized. Oceanographers, marine biologists, geologists, and paleontologists continue to cite Darwin's *Coral Reefs*, particularly because of his careful observations of certain remote atolls and his discussions of reef formation. The ocean was very much a part of many of Darwin's contributions, including his explanation of the isolated, insular characteristics of the fauna of the Galapagos, his theory on the origin of coral reefs, and his discussion of problems related to changes of sea level, modification of coast lines, and the interpretation of marine strata.

The *Beagle* initiated a series of cruises that brought the attention of the scientific community to the sea. As vehicles of exploration, ships provided built-in rewards for the investigator, as we shall see.

Subsequent to the voyage of the *Beagle*, Edward Forbes, a young British zoologist, did extensive dredging in European waters. On the basis of his dredging cruises in the Aegean Sea, he reported that natural groupings of marine animals occur in definite depth zones from the shore seaward. He also reported that the number of animals diminishes with each successively deeper life zone until a point is reached beyond which there is no life. The depth limit for marine life on the floor of the Aegean Sea was determined, by Forbes, to be about three hundred fathoms, or eighteen hundred feet.

The characteristics of the academic community being what they were (the work of each scientist at that time was generally well known to the others), it was inevitable that someone would challenge Forbes' controversial postulation of a deep lifeless zone on the sea floor. There followed a succession of dredging cruises, all in search of deep-sea life, led by distinguished men, each of them receiving some logistical support from the navy of his country. The most important of these men, and the one most critical to our story, was Charles Wyville Thompson, a professor of zoology at Belfast. In his book *The Depths of the Sea*, Thompson expressed his respect for Forbes, while at the same time recommending a direction for future marine investigations.

> I will give a brief sketch of the general results to which Forbes was led by his labours, and I shall have to point out hereafter, that although we are now inclined to look somewhat differently on certain very fundamental points, and although recent investigations with better appliances and more extended experience have invalidated many of his conclusions, to Forbes is due the credit of having been the first to treat these questions in a broad philosophical sense, and to point out that the only means of acquiring a true knowledge of the *rationale* of the distribution of our present fauna, is to make ourselves acquainted with its history, to connect the present with the past. This is the direction which must be taken by future inquiry. Forbes, as a pioneer in this line of research, was scarcely in a position to appreciate the full value of his work. Every year adds enormously to our stock of data, and every new fact indicates more clearly the brilliant results which are to be obtained by following his his methods, and by emulating his enthusiasm and his indefatigable industry.[1]

Thompson could not have been more charitable in his praise for the man who, caught up in the bloom of scientific cruises resulting from the *Beagle's* success, presented the thesis of a lifeless zone in the depths of the sea and, thus, promulgated the argument that ultimately led to the *Challenger* cruise. Indeed, Thompson's praise was both compassionate and wise.

Responding to the appeal for vessels by Thompson and his colleague William B. Carpenter, the British Admiralty provided two ships, H.M.S. *Lightning* and H.M.S. *Porcupine*, with which dredging expeditions in the North Atlantic and Mediterranean seas were conducted during the summers of 1868, 1869, and 1870. The dredging operations were carried out primarily for the benefit of biologists and to gain evidence with which to refute Forbes's idea of a lifeless zone. Thompson succeeded admirably in obtaining evidence of life at great depths. He had had but little to say about marine deposits, but his brief report on the sediment samples is critically important.

The dredging of a part of the North Atlantic floored with *Globigerina* ooze—a mud composed of microscopic shells—proved to be most fortuitous: the resulting sample material was identified as the modern

1. Charles Wyville Thompson, *The Depths of the Sea*, 2nd ed. (London: Macmillan, 1874), p. 6.

depositional counterpart of an ancient sediment, the Cretaceous chalk of England. Thompson wrote:

> Very speedily after the first sample of the bottom of the mid-Atlantic had been brought up by the sounding line, and submitted to chemical analysis and to microscopical examination, many observers were struck with the great similarity between its composition and structure and that of the ancient chalk. . . . Altogether two slides—one of washed down white chalk, and the other of Atlantic ooze—resemble one another so clearly, that it is not always easy for even an accomplished microscopist to distinguish them.[2]

Thompson also reported that the Cretaceous chalk and the deposits of *Globigerina* ooze possessed a common fauna, an argument both superfluous and unproven. The important point to remember is that Thompson was the first marine-oriented scientist to put forth the idea that modern sea floor sediments—not necessarily sedimentation processes—could be used in comparative studies of ancient sedimentary strata by direct sample-to-sample comparison. In his vividly descriptive and technically detailed narrative, including instructions for proper dredging technique, thermometry, and sample handling, Thompson provided the profession with its first general reference on oceanographic observations. His graphs of temperature plotted against water depth are so carefully drawn that any modern student of physical oceanography will immediately recognize the thermoclines. His book, though dated, is still useful.

There is no doubt that Charles Wyville Thompson was the father of modern oceanography. The following excerpt from the preface to his book shows his enthusiasm and joy at the time just prior to the *Challenger's* departure:

> On the return of the 'Porcupine' from her last cruise, so much interest was felt in the bearings of the new discoveries upon important biological, geological, and physical problems, that a representation was made to Government by the Council of the Royal Society, urging the despatch of an expedition to traverse the great ocean basins, and take an outline survey of the vast new field of research—the bottom of the sea.
>
> Rear-Admiral Richards, C.B., F.R.S., the Hydrographer to the Navy, warmly supported the proposal, and while I am writing a noble ship is lying at Sheerness equipped for scientific research, under his wise and liberal directions, as no ship of any nation was ever equipped before.[3]

The world awaited the *Challenger*.

---

2. *Ibid.*, pp. 467–468.

---

3. *Ibid.*, p. xi.

Before going further, let us remember the importance of Thompson and his pre-*Challenger* contributions. The influence of the previous expeditions by the *Lightning* and the *Porcupine* on the subsequent decision by the Admiralty to dispatch the *Challenger* is unquestioned. Indeed, the great ship sailed under the scientific leadership of Thompson himself, a man not only highly respected as a senior scientist by both the Royal Society and the Admiralty, but an experienced oceanographer with some three previous cruises—all successful—to his credit: when Thompson took the *Challenger* to sea, he was, quite simply, transferring the British flag of leadership to a larger ship on a longer cruise. Details of the *Challenger* cruise, and of the impact of the many contributions that resulted from the numerous studies of the collections that the ship brought back to England, are given in "The Voyage of the *Challenger*," by Herbert S. Bailey, Jr. (page 20).

Subsequent to the cruise, numerous expeditions were dispatched to observe and collect at sea. Some of these were organized solely to enhance national prestige. Others chose more pragmatic objectives, such as surveying deep-sea routes for telegraph cables and charting safe navigation routes through coral-island passages. Collectively, the numerous expeditions between that of the *Challenger* and the beginning of World War I added a wealth of descriptive details to the scientific literature. In fact, someone browsing in the stacks of any large university library might well be surprised at the many contributions—most in magnificently ornate volumes—that were published before the turn of the century.

On the American side of the Atlantic, there was also an awakening interest in marine science during the middle of the nineteenth century, although the American effort was more a response to the practical needs of the mariner than a response to scientific challenge. The man most directly responsible for the entry of the United States into oceanographic investigations was Matthew Fontaine Maury, an American naval officer.

In 1855—some two decades prior to the *Challenger* expedition—Maury published his classic book *Physical Geography of the Sea,* a volume that became an immediate best-seller and remained so for many years after. Interestingly, had it not been for an injury that prevented him from active sea duty, there is doubt that Lt. Maury would have made such a significant contribution.

Assigned to custodianship of numerous ships' logs, charts, and reports on observations of wind, waves, and currents, Maury undertook the herculean task of synthesizing the scattered data. The results were published as a semitechnical monograph that included charts to provide the shipmaster with the information necessary for plotting the quickest course on a long voyage. Basically, Maury recommended courses that took advantage of surface currents to provide additional speed and that avoided areas of adverse or inadequate winds. His recommendations appealed to the practical-minded shipmasters and shipowners—so much so that, on one occasion, he was feted with a banquet and the gift of a fancy tea service. (Such an honor today would surely bring the officer before an investigating committee!) Maury's contributions (particularly his current charts), combined with the subsequent *Challenger* observations, also provided the scientific community with a broad, but reasonably accurate, picture of the main systems of oceanic circulation.

Around the turn of the century, the principal leader of oceanographic endeavor in the United States was Alexander Agassiz at Harvard. Often using his own money, Agassiz undertook surveys in the Atlantic, the Gulf of Mexico, the Pacific, and even the Indian Ocean, frequently arranging for a collier to meet and refuel his ship at a distant rendezvous, in order that no time be lost that might be used for trawling and dredging. Alexander Agassiz also had the support of the old United States Fish Commission, and sailed aboard the Commission's steamer *Albatross* on several research cruises.

Some of the ships that added to the large reservoir of data and descriptions were the *Enterprise, Medusa, Hirondelle, Princess Alice, Fram, Michael Sars, Ingolf,* and *Valdivia.* As impressive as a listing of these ships may seem, they provided no data significantly different from that contained in the *Challenger* reports.

With the onset of World War I, there began a period of scientific isolation and lessening activity in oceanographic research. This was particularly true for the United States, where oceanographic research remained in the doldrums until it was revitalized in the late 1920s by the late Henry Bryant Bigelow. The only major oceanographic efforts in the period between the wars were the cruises of the *Atlantis* of the Woods Hole Oceanographic Institution (largely in the North Atlantic), the British John Murray expedition to the Indian Ocean, the *E. W. Scripps* cruises in the Pacific, and the highly successful surveys of the South Atlantic by the German research vessel *Meteor.*

One of the most important steps taken to awaken and encourage oceanographic research in America following World War I was the appointment, in 1927, of the Committee on Oceanography by the National Academy of Sciences. The Committee was composed of some of the most distinguished scientists of the period—William Bowie, E. G. Conklin, B. M. Duggar, John Merriam, T. Wayland Vaughan, and Frank Lillie—and chose as its secretary an equally famous scientist, Henry Bigelow of Harvard University.

The committee was charged with fulfilling the United States' role in a world-wide program of oceanographic research. The selection of Bigelow as secretary was a wise choice, for it was Bigelow who most clearly saw the dangers inherent in the decline of marine research in the United States. Moreover, Bigelow had the ability to prepare a report that clearly and forcefully stressed priorities: When Bigelow's report was submitted in 1929, it was given serious attention by the National Academy, and a vigorous campaign was undertaken to find the resources necessary for implementing the committee's recommendations—chiefly, the acquisition of new research vessels, seashore laboratories, and other facilities and capital equipment. In the end, it was Bigelow who, through his personal appeals to industrial executives, philanthropic foundations, and government, secured the necessary financial support to rekindle the flickering flame of oceanography in America. The young reader should bear in mind that Bigelow's appeals for help came at a most crucial time in American history —the beginning of the Great Depression of the 1930s. Nevertheless, funds were secured, as were various other forms of assistance, including the use of private yachts by oceanographers.

Timely generosity that developed in the early 1930s was exemplified by Eldridge R. Johnson, who not only placed his yacht *Caroline* at the disposal of the Smithsonian Institution for an expedition to explore the Puerto Rico Deep, but also financed the entire expedition. Such acts of generosity were commonplace, and the response of the private and institutional sectors to Bigelow's appeals was to place oceanography on a firm basis for growth. During this era, the Woods Hole Oceanographic Institution was created, the research vessel *Atlantis* was purchased, and financial support was given to the Scripps Institution of Oceanography at La Jolla, California, the University of Washington at Seattle, and several smaller centers of marine science.

After World War II, many scientific advances, including improvements in navigation aids and methods, radar, sonar, acoustic precision depth recording, and rapid instrumental analysis in the laboratory, provided a major boost to oceanography. Also, the acute awareness of submarine warfare as the inevitable replacement for battleship-oriented surface warfare brought considerable financial support to oceanographic institutions and to individual scientists for their research.

Great strides in oceanography have not been made—and will not be made in the future—without overcoming, in large measure, several difficult problems, chiefly those of financial support, personnel, and the law. Because these problems relate to the taxpayer and to the college student as much as they do to the professional oceanographer, it behooves us to consider them here, at least briefly.

By far the most critical problem in oceanographic research is that of obtaining adequate funding for periods of time sufficient to complete a major investigation. All aspects of oceanographic research are expensive, and ships are particularly costly. In fact, the most expensive part of any ocean survey is the operation of the vessel. Even Thompson, it may be noted, had to seek the Admiralty's help to obtain an adequate ship. Most of the newer oceanographic vessels (for example, the *Atlantis II* of the Woods Hole Oceanographic Institution) represent an overall expense of several thousand dollars a day, whether the ship is at the dock or on a cruise in distant waters.

One alternative to new oceanographic ships has been the use of "ships of opportunity"—that is, government and private vessels, normally employed in other activities, that can be used for research for short periods of time. Because the problem of the availability of vessels is common to all nations engaged in ocean-oriented research, exploration, and resource exploitation, the best solution may be a greater international sharing of ships, or, alternatively, the use of naval vessels, many of which are already available and presumably satisfactory for oceanographic purposes.

Financial support for oceanographic research in the United States has increased enormously in the years since World War II. Although considerable funds are still available from private foundations, industries, and state governments, by far the largest amount of money to support marine research has come from such federal agencies as the Office of Naval Research, the National Science Foundation, and the National Oceanic and Atmospheric Administration, particularly NOAA's office of Sea Grant Programs. Although this infusion of capital has encouraged oceanographers and pushed marine science to the fore, such support has caused many research institutions to expand beyond the point of prudence. Consequently, owing to the shifting political climate in Washington, many oceanographers are faced with feast-or-famine budgeting and its disturbing effect on the continuity of their research. This situation obtains in academic, governmental, and industrial sectors alike.

At a recent scientific meeting, I overheard a distinguished oceanographer say: "Our group seems to have no difficulty in finding money; what we need, though, is experienced staff—we are people-limited." This brings up the second major problem in oceanography today: personnel.

At the time of the *Challenge* cruise, there were no formal university courses in oceanography, and the scientist who chose to pursue marine research did so by, as the British say, "getting on with it." In other words, the scientist simply applied whatever basic talent he already had, as a zoologist, chemist, or physicist, to interpreting the data, collections, and observations made at sea. Surprisingly, this same approach is still used a century after the *Challenger.* With the expansion of formal courses and definite degree programs in oceanography after World War II, the profession began to enlist additional capable, imaginative young scientists into its ranks. Nevertheless, many of the famous oceanographers of today were not formerly educated in oceanography, and there has not been a pronounced correlation between the degree objectives of many marine scientists and the research topics in which they have made their most important contributions, although the two frequently overlap. Examples are manifold among the authors of papers included in this reader: Willard Bascom, author of "Ocean Waves" and "Beaches," was a mining engineer in the western United States prior to becoming a leader in industrial oceanography; the writer of "Salt and Rain," A. H. Woodcock, one of the senior staff scientists at the prestigious Woods Hole Oceanographic Institution, began his career as a farmer; and the author of "The Nature of Oceanic Life," John Isaacs, a scientist of exceptionally high stature, began his college training in engineering, subsequently worked as a merchant seaman, fire lookout, and commercial fisherman, and finally became a marine scientist at the Scripps Institution of Oceanography. There are others, to be sure, who were formally trained in oceanography and have continued to be active researchers in the field. Bruce Heezen, who wrote "The Origin of Submarine Canyons," is an example of the latter.

These few examples suggest, then, that professional oceanographers of today have not, as a group, had a formal background of university-level preparation in oceanography. Although there is a definite trend to pursue specialized degree programs as an entrée to the professional body, it appears that much future research will be undertaken by young scientists educated in other fields who will simply apply their special skills to solving marine problems, as many before them have done.

Why then do we also speak of the lack of experienced scientists? The key word is *experienced*. Frankly, the life of the practicing oceanographer is not as romantic as it is made to seem on television. For the most part, the life of an oceanographer consists of long cruises, hard physical work (often in the broiling sun of the tropics or in the icy winds of the high latitudes), frequent and prolonged absences from friends and family, and unbelievable demands on one's time, both at sea and ashore. Some researchers never become accustomed to a rolling ship, and all are subject to a lack of the amenities that are taken for granted by their professional peers who live in suburbia. Yet, for those who do pass the tests of life at sea and do enjoy the rewards of research, oceanography can be a fascinating career.

These observations suggest that we should perhaps look first for those who have a natural desire for the kind of life an oceanographer must live before we consider their specific technical training. As we have noted, training in oceanography, per se, is not an absolute prerequisite: curiosity and enthusiasm coupled with basic knowledge in *any* field of science are often sufficient.

The third problem of concern to those who explore and exploit the sea is the increasing uncertainty over the legal aspects of scientific exploration, commercial fishing, undersea dredging and mining, emplacement of structures on the sea floor, and military use of the sea.

Unlike the freedom experienced by the men of the *Challenger*, who could go wherever they so desired and take samples in any and all waters—even unlike the relative freedom I experienced on the cruises that I made as a university student only twenty years ago—the freedom of the oceanographer of the 1970s is limited: he is faced with many legal limitations as to where he may explore, at what depths, at what distance from shore, and with what kinds of equipment. Some academic investigators have been threatened by foreign gunboats, as were those aboard the University of Miami's *Pillsbury* off the South American coast recently. For the industrial explorer who wishes to survey the sea floor for minerals of economic importance, the legal problems are still more difficult to resolve. Which governmental agency has the authority to issue exploration permits, much less to regulate exploitation and to establish conservation and antipollution measures? Which of the various international, national, state, and local agencies have the *legal* authority to sanction and regulate ocean-oriented enterprises? Although some headway has been made through international agreements and the enactment of statutes, there are still many legal problems to be resolved. Furthermore, legal problems have the nasty habit of becoming political problems when injustices, real or imagined, are somehow connected with them. That the entire marine community needs guidance and counsel there is no doubt: the field of ocean law, particularly as it relates to mineral and food resources, holds promising rewards for the young lawyer in search of a challenging career.

Before concluding this introduction, we should compare and—in some cases—contrast the early oceanographic cruises, as exemplified by the cruise of the *Challenger,* with their counterparts in the 1970s.

There is no doubt that we do have much larger and more sophisticated vessels and facilities now than we had a century ago. The vessels of today are designed to be effective at sea for prolonged periods. There is instant, worldwide radio-telephone and radio-telegraph communication and, on some of the newest ships, equipment for sophisticated navigation using the earth-circling satellites. Even underwater television is available. Like the oceanographers of the mid-1800s, those of today still benefit both from government financial support and from the assistance of the navy. Probably the greatest single advantage that the modern oceanographer has over his predecessor is access to a wide variety of computers and equipment for instrumental analysis, which make it easy for him to model his hypotheses mathematically prior to collecting data at sea for empirically testing those hypotheses. The computer has also revolutionized navigation, particularly satellite navigation, by providing the mariner with accurate fixes at sea, thus allowing him to make measurements of the global ocean that are infinitely more precise. Determinations of mineral components and chemical elements, and measurements of a host of other physical and chemical parameters of sea water, organisms, sea floor sediments, rocks, and even the air over the sea, are now readily obtained through rapid instrumental analysis in the laboratory using X-ray diffraction, atomic absorption, thermal-neutron activation, and other modern analytical tools. Of course, much of the laboratory study of materials collected at sea is still done using the microscope, but microscopy has been improved by television-scanning and phase-contrast devices. Many studies, especially those of very small plant and animal remains, have gone beyond optical microscopy to the electron microscope, a tool unimagined in the time of the *Challenger.*

Some aspects of oceanography have not changed. To one who reads between the lines of the following articles on Darwin and the *Challenger,* it is apparent that the progress of science is largely the result of imagination and hard work. I might also go so far as to speculate that many of the successful marine scientists of the middle 1800s would be classified as "mavericks," or "promoters," or both, were they alive and working today. Nevertheless, the oceanographic frontier has been pushed forward for the past century by a rather special breed of men, and I judge that this may continue to be true in the century ahead.

We should encourage young scientists and engineers, regardless of their particular background, to grapple with some of the technical problems that confront oceanographers today. Likewise, for students interested in law, political science, or business, there are enough problems in the world of marine affairs to enable each to choose his own discipline of concentration without overcrowding the field. From the *Challenger* to the present, the number of those who were and who are devoted to being marine scientists for a lifetime has been remarkably small.

In bringing these introductory remarks to a close, I would ask the reader to keep in mind the following questions: What were the motivations of the principal scientists? What were the obstacles that each had to overcome? What did each contribute to the science of oceanography?

1

CHARLES DARWIN

LOREN C. EISELEY
February 1956

In the autumn of 1831 the past and the future met and dined in London—in the guise of two young men who little realized where the years ahead would take them. One, Robert Fitzroy, was a sea captain who at 26 had already charted the remote, sea-beaten edges of the world and now proposed another long voyage. A religious man with a strong animosity toward the new-fangled geology, Captain Fitzroy wanted a naturalist who would share his experience of wild lands and refute those who used rocks to promote heretical whisperings. The young man who faced him across the table hesitated. Charles Darwin, four years Fitzroy's junior, was a gentleman idler after hounds who had failed at medicine and whose family, in desperation, hoped he might still succeed as a country parson. His mind shifted uncertainly from fox hunting in Shropshire to the thought of shooting llamas in South America. Did he really want to go? While he fumbled for a decision and the future hung irresolute, Captain Fitzroy took command.

"Fitzroy," wrote Darwin later to his sister Susan, "says the stormy sea is exaggerated; that if I do not choose to remain with them, I can at any time get home to England; and that if I like, I shall be left in some healthy, safe and nice country; that I shall always have assistance; that he has many books, all instruments, guns, at my service. . . . There is indeed a tide in the affairs of men, and I have experienced it. Dearest Susan, Goodbye."

They sailed from Devonport December 27, 1831, in H.M.S. *Beagle*, a 10-gun brig. Their plan was to survey the South American coastline and to carry a string of chronometrical measurements around the world. The voyage almost ended before it began, for they at once encountered a violent storm. "The sea ran very high," young Darwin recorded in his diary, "and the vessel pitched bows under and suffered most dreadfully; such a night I never passed, on every side nothing but misery; such a whistling of the wind and roar of the sea, the hoarse screams of the officers and shouts of the men, made a concert that I shall not soon forget." Captain Fitzroy and his officers held the ship on the sea by the grace of God and the cat-o'-nine-tails. With an almost irrational stubbornness Darwin decided, in spite of his uncomfortable discovery of his susceptibility to seasickness, that "I did right to accept the offer." When the *Beagle* was buffeted back into Plymouth Harbor, Darwin did not resign. His mind was made up. "If it is desirable to see the world," he wrote in his journal, "what a rare and excellent opportunity this is. Perhaps I may have the same opportunity of drilling my mind that I threw away at Cambridge."

So began the journey in which a great mind untouched by an old-fashioned classical education was to feed its hunger upon rocks and broken bits of bone at the world's end, and eventually was to shape from such diverse things as bird beaks and the fused wing-cases of island beetles a theory that would shake the foundations of scientific thought in all the countries of the earth.

The Intellectual Setting

The intellectual climate from which Darwin set forth on his historic voyage was predominantly conservative. Insular England had been horrified by the excesses of the French Revolution and was extremely wary of emerging new ideas which it attributed to "French atheists." Religious dogma still held its powerful influence over natural science. True, the 17th-century notion that the world had been created in 4004 B.C. was beginning to weaken in the face of naturalists' studies of the rocks and their succession of life forms. But the conception of a truly ancient and evolving planet was still unformed. No one could dream that the age of the earth was as vast as we now know it to be. And the notion of a continuity of events—of one animal changing by degrees into another—seemed to fly in the face not only of religious beliefs but also of common sense. Many of the greatest biologists of the time—men like Louis Agassiz and Richard Owen—tended to the belief that the successive forms of life in the geological record were all separate creations, some of which had simply been extinguished by historic accidents.

Yet Darwin did not compose the theory of evolution out of thin air. Like so many great scientific generalizations, the theory with which his name is associated had already had premonitory beginnings. All of the elements which were to enter into the theory were in men's minds and were being widely discussed during Darwin's college years. His own grandfather, Erasmus Darwin, who died seven years before Charles was born, had boldly proposed a theory of the "transmutation" of living forms. Jean Baptiste Lamarck had glimpsed a vision of evolutionary continuity. And Sir Charles Lyell—later to be Darwin's lifelong confidant—had opened the way for the evolutionary point of view by demonstrating that the planet must be very old—old enough to allow extremely slow organic change. Lyell dismissed the notion of catastrophic extinction of animal forms on a world-wide scale as impossible, and he made plain that natural forces—the work of wind and frost and water—were sufficient to explain most of the phenomena found in the rocks, provided these forces were seen as operating over enormous periods. Without Lyell's gift of time in immense quantities, Darwin would not have been able to devise the theory of natural selection.

If all the essential elements of the Darwinian scheme of nature were known prior to Darwin, why is he accorded so important a place in biological history? The answer is simple: Almost every great scientific generalization is a supreme act of creative synthesis. There comes a time when an accumulation of smaller discoveries and observations can be combined in some great and comprehensive view of nature. At this point the need is not so much for increased numbers of facts as for a mind of great insight capable of taking the assembled information and rendering it intelligible. Such a synthesis represents the scientific mind at its highest point of achievement. The stature of the discoverer is not diminished by the fact that he has slid into place the last piece of a tremendous puzzle on which many others have worked. To finish the task he must see correctly over a vast and diverse array of data.

Still it must be recognized that Darwin came at a fortunate time. The fact that another man, Alfred Russel Wallace, conceived the Darwinian theory independently before Darwin published it shows clearly that the principle which came to be called natural selection was in the air—was in a sense demanding to be born. Darwin himself pointed out in his autobiography that "innumerable well-observed facts were stored in the minds of naturalists ready to take their proper places as soon as any theory which would receive them was sufficiently explained."

The Voyage

Darwin, then, set out on his voyage with a mind both inquisitive to see and receptive to what he saw. No detail was too small to be fascinating and provocative. Sailing down the South American coast, he notes the octopus changing its color angrily in the waters of a cove. In the dry arroyos of the pampas he observes great bones and shrewdly seeks to relate them to animals of the present. The local inhabitants insist that the fossil bones grew after death, and also that certain rivers have the power of "changing small bones into large." Everywhere men wonder, but they are deceived through their thirst for easy explanations. Darwin, by contrast, is a working dreamer. He rides, climbs, spends long days on the Indian-haunted pampas in constant peril of his life. Asking at a house whether robbers are numerous, he receives the cryptic reply: "The thistles are not up yet." The huge thistles, high as a horse's back at their full growth, provide ecological cover for bandits. Darwin notes the fact and rides on. The thistles are overrunning the pampas; the whole aspect of the vegetation is altering under the impact of man. Wild dogs howl in the brakes; the common cat, run wild, has grown large and fierce. All is struggle, mutability, change. Staring into the face of an evil relative of the rattlesnake, he observes a fact "which appears to me very curious and instructive, as showing how every character, even though it may be in some degree independent of structure . . . has a tendency to vary by slow degrees."

He pays great attention to strange animals existing in difficult environments. A queer little toad with a scarlet belly he whimsically nicknames ***diabolicus*** because it is "a fit toad to preach in the ear of Eve." He notes it lives among sand dunes under the burning sun, and unlike its brethren, cannot swim. From toads to grasshoppers, from pebbles to mountain ranges, nothing escapes his attention. The wearing away of stone, the downstream travel of rock fragments and boulders, the great crevices and upthrusts of the Andes, an earthquake—all confirm the dynamic character of the earth and its great age.

Captain Fitzroy by now is anxious to voyage on. The sails are set. With the towering Andes on their right flank they run north for the Galápagos Islands, lying directly on the Equator 600 miles off the west coast of South America. A one-time refuge of buccaneers, these islands are essentially chimneys of burned-out volcanoes. Darwin remarks that they

PHOTOGRAPHIC PORTRAIT of Darwin was made some years after the appearance of *Origin of Species*. It is from the collection of George Eastman House in Rochester, N. Y.

THREE IMPORTANT FIGURES in the life of Darwin are shown here and on the following page. They appear in *Portraits of Men of Eminence*, three volumes of which were published between 1863 and 1865. This book is also from George Eastman House. At left is Robert Fitzroy, Captain of the *Beagle;* at right, Charles Lyell, the geologist who was Darwin's lifelong confidant.

remind him of huge iron foundries surrounded by piles of waste. "A little world in itself," he marvels, "with inhabitants such as are found nowhere else." Giant armored tortoises clank through the undergrowth like prehistoric monsters, feeding upon the cacti. Birds in this tiny Eden do not fear men: "One day a mocking bird alighted on the edge of a pitcher which I held in my hand. It began very quietly to sip the water, and allowed me to lift it with the vessel from the ground." Big sea lizards three feet long drowse on the beaches, and feed, fantastically, upon the seaweed. Surveying these "imps of darkness, black as the porous rocks over which they crawl," Darwin is led to comment that "there is no other quarter of the world, where this order replaces the herbivorous mammalia in so extraordinary a manner."

Yet only by degrees did Darwin awake to the fact that he had stumbled by chance into one of the most marvelous evolutionary laboratories on the planet. Here in the Galápagos was a wealth of variations from island to island—among the big tortoises, among plants and especially among the famous finches with remarkably diverse beaks. Dwellers on the islands, notably Vice Governor Lawson, called Darwin's attention to these strange variations, but as he confessed later, with typical Darwinian lack of pretense, "I did not for some time pay sufficient attention to this statement." Whether his visit to the Galápagos was the single event that mainly led Darwin to the central conceptions of his evolutionary mechanism—hereditary change within the organism coupled with external selective factors which might cause plants and animals a few miles apart in the same climate to diverge—is a moot point upon which Darwin himself in later years shed no clear light. Perhaps, like many great men, nagged long after the event for a precise account of the dawn of a great discovery, Darwin no longer clearly remembered the beginning of the intellectual journey which had paralleled so dramatically his passage on the seven seas. Perhaps there had never been a clear beginning at all—only a slowly widening comprehension until what had been seen at first mistily and through a veil grew magnified and clear.

The Invalid and the Book

The paths to greatness are tricky and diverse. Sometimes a man's weaknesses have as much to do with his rise as his virtues. In Darwin's case it proved to be a unique combination of both. He had gathered his material by a courageous and indefatigable pursuit of knowledge that took him through the long vicissitudes of a voyage around the world. But his great work was written in sickness and seclusion. When Darwin reached home after the voyage of the *Beagle*, he was an ailing man, and he remained so to the end of his life. Today we know that this illness was in some degree psychosomatic, that he was anxiety-ridden, subject to mysterious headaches and nausea. Shortly after his voyage Darwin married his cousin Emma Wedgwood, granddaughter of the founder of the great pottery works, and isolated himself and his family in a little village in Kent. He avoided travel like the plague,

THE THIRD IMPORTANT FIGURE, shown above, is Thomas Huxley, who defended Darwin in debate.

save for brief trips to watering places for his health. His seclusion became his strength and protected him; his very fears and doubts of himself resulted in the organization of that enormous battery of facts which documented the theory of evolution as it had never been documented before.

Let us examine the way in which Darwin developed his great theory. The nature of his observations has already been indicated—the bird beaks, the recognition of variation and so on. But it is an easier thing to perceive that evolution has come about than to identify the mechanism involved in it. For a long time this problem frustrated Darwin. He was not satisfied with vague references to climatic influence or the inheritance of acquired characters. Finally he reached the conclusion that since variation in individual characteristics existed among the members of any species, selection of some individuals and elimination of others must be the key to organic change.

This idea he got from the common recognition of the importance of selective breeding in the improvement of domestic plants and livestock. He still did not understand, however, what selective force could be at work in wild nature. Then in 1838 he chanced to read Thomas Malthus, and the solution came to him. Malthus had written in 1798 a widely read population study in which he pointed out that the human population tended to increase faster than its food supply, precipitating in consequence a struggle for existence.

Darwin applied this principle to the whole world of organic life and argued that the struggle for existence under changing environmental conditions was what induced alterations in the physical structure of organisms. To put it in other words, fortuitous and random variations occurred in living things. The struggle for life perpetuated advantageous variations by means of heredity. The weak and unfit were eliminated and those with the best heredity for any given environment were "selected" to be the parents of the next generation. Since neither life nor climate nor geology ever ceased changing, evolution was perpetual. No ogran and no animal was ever in complete equilibrium with its surroundings.

This, briefly stated, is the crux of the Darwinian argument. Facts which had been known before Darwin but had not been recognized as parts of a single scheme—variation, inheritance of variation, selective breeding of domestic plants and animals, the struggle for existence—all suddenly fell into place as "natural selection," as "Darwinism."

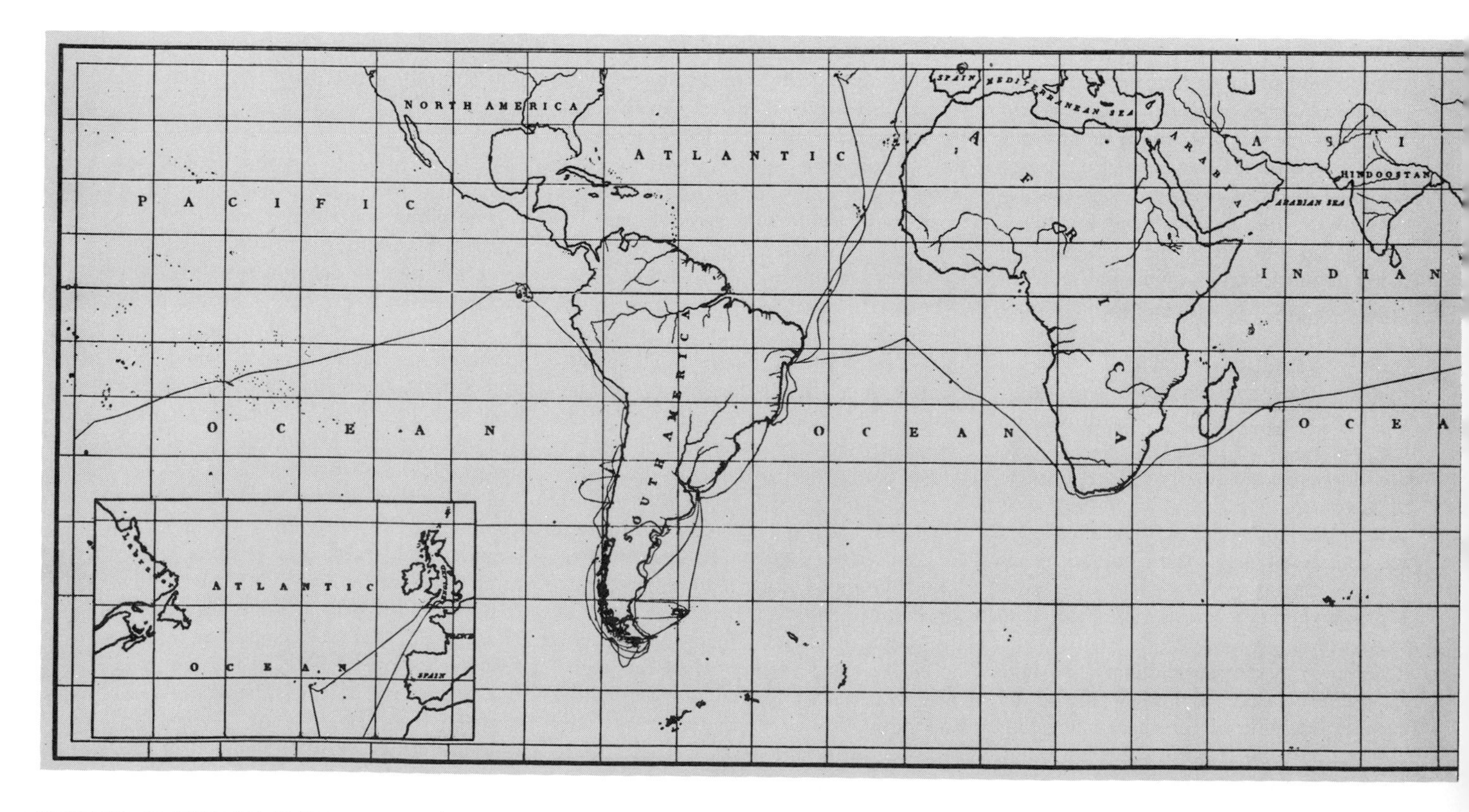

VOYAGE OF THE BEAGLE is traced in this map from Fitzroy's *Narrative of the Surveying Voyage of His Majesty's Ships Adventure and Beagle.* The *Beagle*'s course on her departure from and return to England is at lower left. The ship made frequent stops at

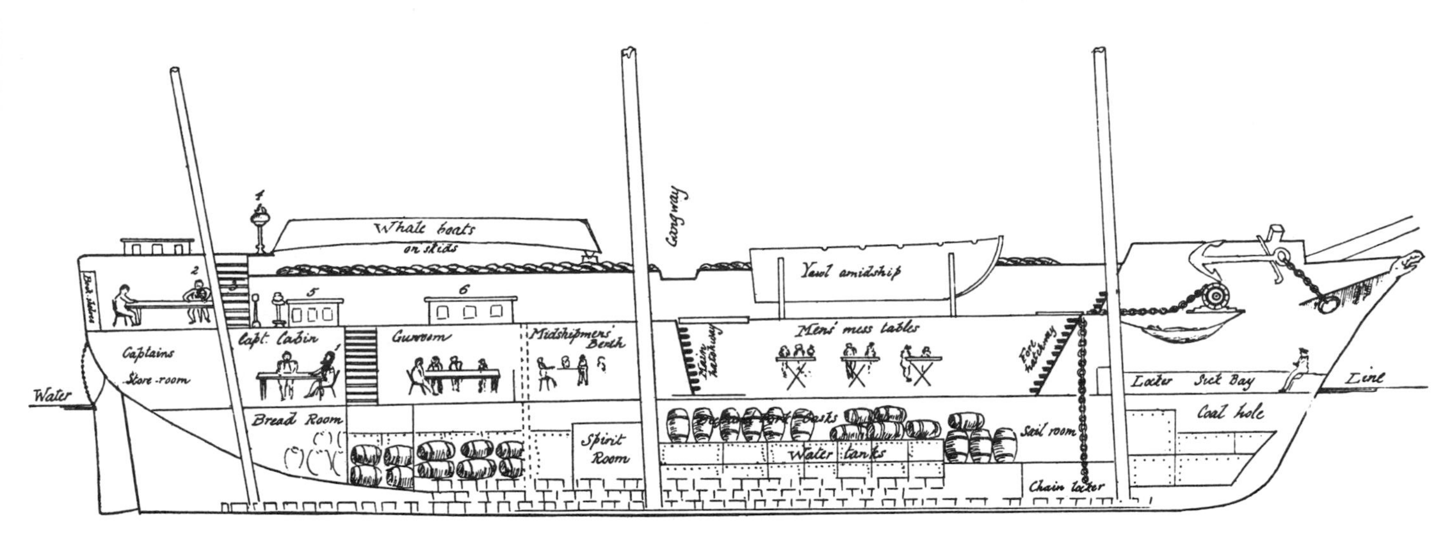

H. M. S. BEAGLE was drawn in cross section many years after the voyage by Philip Gidley King, who accompanied Darwin when he was ashore during the voyage. Darwin is shown in two places: the captain's cabin (*small figure 1 at upper left*) and poop cabin (*2*).

Procrastination

While he developed his theory and marshaled his data, Darwin remained in seclusion and retreat, hoarding the secret of his discovery. For 22 years after the *Beagle*'s return he published not one word beyond the bare journal of his trip (later titled *A Naturalist's Voyage around the World*) and technical monographs on his observations.

Let us not be misled, however, by Darwin's seclusiveness and illness. No more lovable or sweet-tempered invalid ever lived. Visitors, however beloved, always aggravated his illness, but instead of the surly misanthropy which afflicts most people under similar circumstances, the result in Darwin's case was merely nights of sleeplessness. Throughout the long night hours his restless mind went on working with deep concentration; more than once, walking alone in the dark hours of winter, he met the foxes trotting home at dawn.

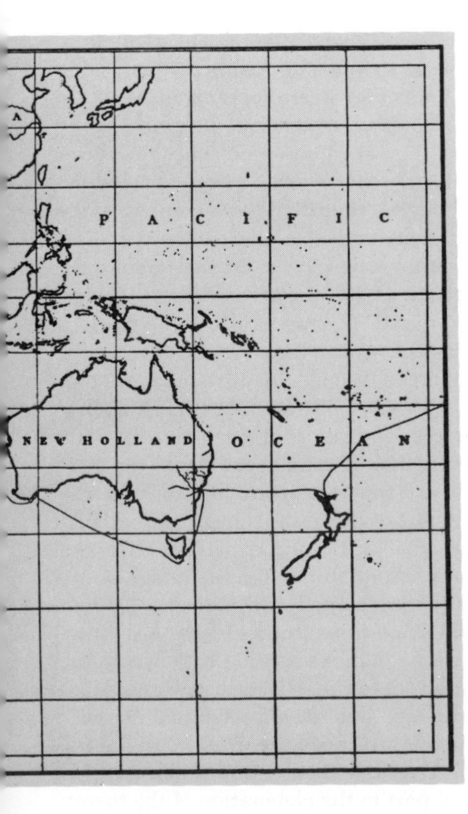

oceanic islands. The Galápagos Islands are on the Equator to the west of South America.

Darwin's gardener is said to have responded once to a visitor who inquired about his master's health: "Poor man, he just stands and stares at a yellow flower for minutes at a time. He would be better off with something to do." Darwin's work was of an intangible nature which eluded people around him. Much of it consisted in just such standing and staring as his gardener reported. It was a kind of magic at which he excelled. On a visit to the Isle of Wight he watched thistle seed wafted about on offshore winds and formulated theories of plant dispersal. Sometimes he engaged in activities which his good wife must surely have struggled to keep from reaching the neighbors. When a friend sent him a half ounce of locust dung from Africa, Darwin triumphantly grew seven plants from the specimen. "There is no error," he assured Lyell, "for I dissected the seeds out of the middle of the pellets." To discover how plant seeds traveled, Darwin would go all the way down a grasshopper's gullet, or worse, without embarrassment. His eldest son Francis spoke amusedly of his father's botanical experiments: "I think he personified each seed as a small demon trying to elude him by getting into the wrong heap, or jumping away all together; and this gave to the work the excitement of a game."

The point of his game Darwin kept largely to himself, waiting until it should be completely finished. He piled up vast stores of data and dreamed of presenting his evolution theory in a definitive, monumental book, so large that it would certainly have fallen dead and unreadable from the press. In the meantime, Robert Chambers, a bookseller and journalist, wrote and brought out anonymously a modified version of Lamarckian evolution, under the title *Vestiges of the Natural History of Creation*. Amateurish in some degree, the book drew savage onslaughts from the critics, including Thomas Huxley, but it caught the public fancy and was widely read. It passed through numerous editions both in England and America—evidence that *sub rosa* there was a good deal more interest on the part of the public in the "development hypothesis," as evolution was then called, than the fulminations of critics would have suggested.

Throughout this period Darwin remained stonily silent. Many explanations of his silence have been ventured by his biographers: that he was busy accumulating materials; that he did not wish to affront Fitzroy; that the attack on the *Vestiges* had intimidated him; that he thought it wise not to write upon so controversial a subject until he had first acquired a reputation as a professional naturalist of the first rank. Primarily, however, the basic reason lay in his personality—a nature reluctant to face the storm that publication would bring about his ears. It was pleasanter to procrastinate, to talk of the secret to a few chosen companions such as Lyell

QUARTER-DECK of the *Beagle* is depicted in this drawing by King. In the center is the wheel, the circumference of which is inscribed: "England expects every man to do his duty."

and the great botanist Joseph Hooker.

The Darwin family had been well-to-do since the time of grandfather Erasmus. Charles was independent, in a position to devote all his energies to research and under no academic pressure to publish in haste.

"You will be anticipated," Lyell warned him. "You had better publish." That was in the spring of 1856. Darwin promised, but again delayed. We know that he left instructions for his wife to see to the publication of his notes in the event of his death. It was almost as if present fame or notoriety were more than he could bear. At all events he continued to delay, and this situation might very well have continued to the end of his life, had not Lyell's warning suddenly come true and broken his pleasant dream.

Alfred Russel Wallace, a comparatively unknown, youthful naturalist, had divined Darwin's great secret in a moment of fever-ridden insight while on a collecting trip in Indonesia. He, too, had put together the pieces and gained a clear conception of the scheme of evolution. Ironically enough, it was to Darwin, in all innocence, that he sent his manuscript for criticism in June of 1858. He sensed in Darwin a sympathetic and traveled listener.

Darwin was understandably shaken. The work which had been so close to his heart, the dream to which he had devoted 20 years, was a private secret no longer. A newcomer threatened his priority. Yet Darwin, wanting to do what was decent and ethical, had been placed in an awkward position by the communication. His first impulse was to withdraw totally in favor of Wallace. "I would far rather burn my whole book," he insisted, "than that he or any other man should think that I had behaved in a paltry spirit." It is fortunate for science that before pursuing this quixotic course Darwin turned to his friends Lyell and Hooker, who knew the many years he had been laboring upon his *magnum opus*. The two distinguished scientists arranged for the delivery of a short summary by Darwin to accompany Wallace's paper before the Linnaean Society. Thus the theory of the two men was announced simultaneously.

Publication

The papers drew little comment at the meeting but set in motion a mild undercurrent of excitement. Darwin, though upset by the death of his son Charles, went to work to explain his views more fully in a book. Ironically he called it *An Abstract of an Essay on the Origin of Species* and insisted it would be only a kind of preview of a much larger work. Anxiety and devotion to his great hoard of data still possessed him. He did not like to put all his hopes in this volume, which must now be written at top speed. He bolstered himself by references to the "real" book—that Utopian volume in which all that could not be made clear in his abstract would be clarified.

His timidity and his fears were totally groundless. When the *Origin of Species* (the title distilled by his astute publisher from Darwin's cumbersome and half-hearted one) was published in the fall of 1859, the first edition was sold in a single day. The book which Darwin had so apologetically bowed into existence was, of course, soon to be recognized as one of the great books of all time. It would not be long before its author would sigh happily and think no more of that huge, ideal volume which he had imagined would be necessary to convince the public. The public and his brother scientists would find the *Origin* quite heavy going enough. His book to end all books would never be written. It did not need to be. The world of science in the end could only agree with the sharp-minded Huxley, whose immediate reaction upon reading the *Origin* was: "How extremely stupid not to have thought of that!" And so it frequently seems in science, once the great synthesizer has done his work. The ideas were not new, but the synthesis was. Men would never again look upon the world in the same manner as before.

No great philosophic conception ever entered the world more fortunately. Though it is customary to emphasize the religious and scientific storm the book aroused—epitomized by the famous debate at Oxford between Bishop Wilberforce and Thomas Huxley—the truth is that Darwinism found relatively easy acceptance among scientists and most of the public. The way had been prepared by the long labors of Lyell and the wide popularity of Chambers' book, the *Vestiges*. Moreover, Darwin had won the support of the great Hooker and of Huxley, the most formidable scientific debater of all time. Lyell, though more cautious, helped to publicize Darwin and at no time attacked him. Asa Gray, one of America's leading botanists, came to his defense. His codiscoverer, Wallace, as generous-hearted as Darwin, himself advanced the word "Darwinism" for Darwin's theory, and minimized his own part in the elaboration of the theory as "one week to 20 years."

This sturdy band of converts assumed

the defense of Darwin before the public, while Charles remained aloof. Sequestered in his estate at Down, he calmly answered letters and listened, but not too much, to the tumult over the horizon. "It is something unintelligible to me how anyone can argue in public like orators do," he confessed to Hooker, though he was deeply grateful for the verbal swordplay of his cohorts. Hewett Watson, another botanist of note, wrote to him shortly after the publication of the *Origin:* "Your leading idea will assuredly become recognized as an established truth in science, *i.e.*, 'Natural Selection.' It has the characteristics of all great natural truths, clarifying what was obscure, simplifying what was intricate, adding greatly to previous knowledge. You are the greatest revolutionist in natural history of this century, if not of all centuries."

Watson's statement was clairvoyant. Not a line of his appraisal would need to be altered today. Within 10 years the *Origin* and its author were known all over the globe, and evolution had become the guiding motif in all biological studies.

Summing up the achievement of this book, we may say today, first, that Darwin had proved the reality of evolutionary change beyond any reasonable doubt, and secondly, that he had demonstrated, in natural selection, a principle capable of wide, if not universal, application. Natural selection dispelled the confusions that had been introduced into biology by the notion of individual creation of species. The lad who in 1832 had noted with excited interest "that there are three sorts of birds which use their wings for more purposes than flying; the Steamer [duck] as paddles, the Penguin as fins, and the Ostrich (*Rhea*) spreads its plumes like sails" now had his answer—"descent with modification." "If you go any considerable lengths in the admission of modification," warned Darwin, "I can see no possible means of drawing the line, and saying here you must stop." Rung by rung, was his plain implication, one was forced to descend down the full length of life's mysterious ladder until one stood in the brewing vats where the thing was made. And similarly, rung by rung, from mudfish to reptile and mammal, the process ascended to man.

A Small Place for Man

Darwin had cautiously avoided direct references to man in the *Origin of Species.* But 12 years later, after its triumph was assured, he published a study of human evolution entitled *The Descent of Man.* He had been preceded in this field by Huxley's *Evidences as to Man's Place in Nature* (1863). Huxley's brief work was written with wonderful clarity and directness. By contrast, the *Descent of Man* has some of the labored and inchoate quality of Darwin's overfull folios of data. It is contradictory in spots, as

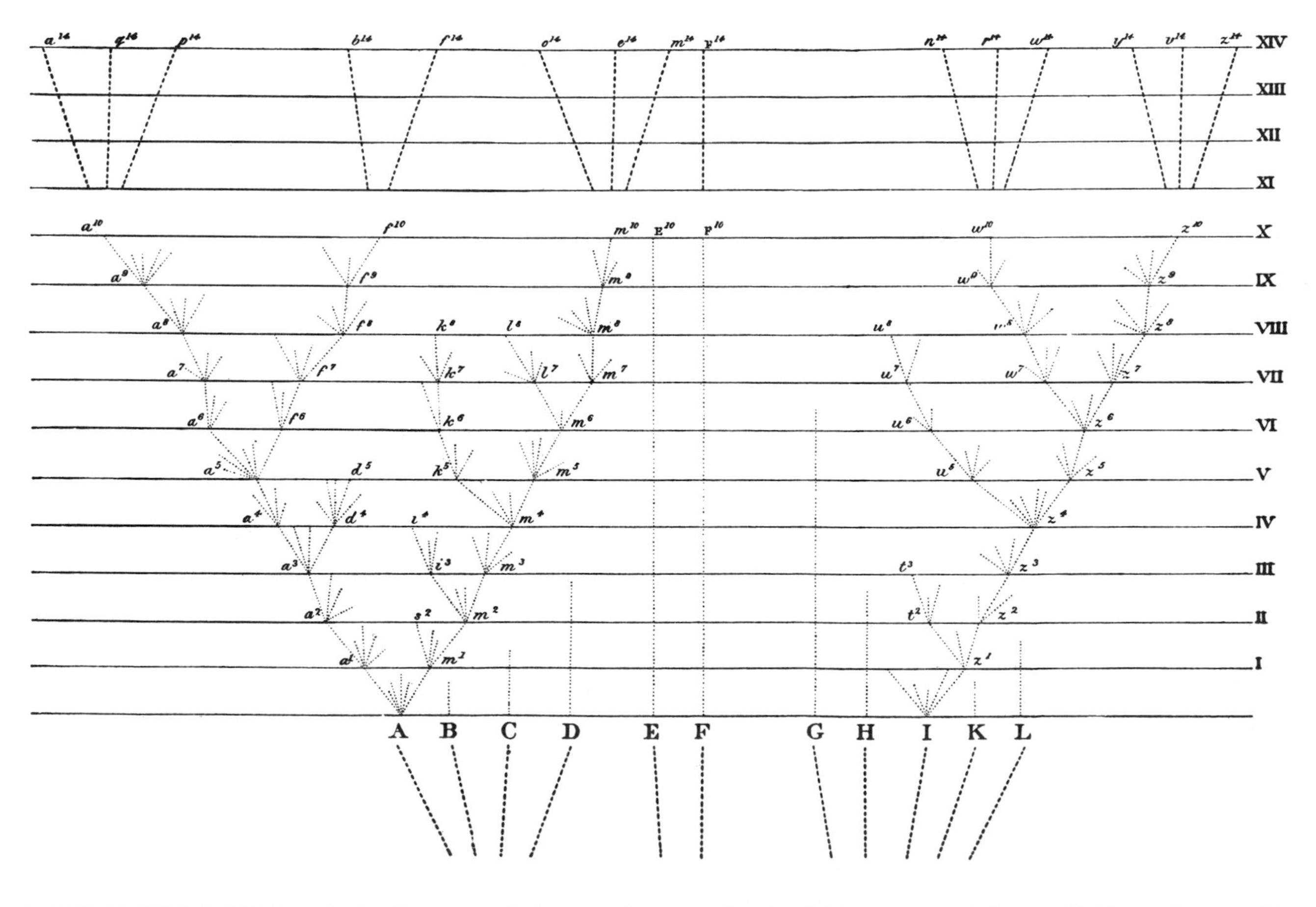

NATURAL SELECTION through the divergence of characters is illustrated in *Origin of Species.* The capital letters at the bottom of the illustration represent different species of the same genus. Each horizontal line, labeled with a Roman numeral at the right, represents 1,000 or more generations. Darwin believed that some of the original species, such as A, would diverge more than others. After many generations they would give rise to new varieties, such as a^1 and m^1. These new varieties would diverge in turn. After thousands of generations the new varieties would give rise to entirely new species, such as a^{14}, q^{14}, p^{14} and so on. The original species would meantime have died out. Darwin thought that only some species of the original genus would diverge sufficiently to give rise to new species. Some of the species, such as F, would remain much the same. Others, such as B, C and D, would die out.

though the author simply poured his notes together and never fully read the completed manuscript to make sure it was an organic whole.

One of its defects is Darwin's failure to distinguish consistently between biological inheritance and cultural influences upon the behavior and evolution of human beings. In this, of course, Darwin was making a mistake common to biologists of the time. Anthropology was then in its infancy. In the biological realm, the *Descent of Man* did make plain in a general way that man was related to the rest of the primate order, though the precise relationship was left ambiguous. After all, we must remember that no one had yet unearthed any clear fossils of early man. A student of evolution had to content himself largely with tracing morphological similarities between living man and the great apes. This left considerable room for speculation as to the precise nature of the human ancestors. It is not surprising that they were occasionally visualized as gorilloid beasts with huge canine teeth, nor that Darwin wavered between this and gentler interpretations.

An honest biographer must record the fact that man was not Darwin's best subject. In the words of a 19th-century critic, his "was a world of insects and pigeons, apes and curious plants, but man as he exists, had no place in it." Allowing for the hyperbole of this religious opponent, it is nonetheless probable that Darwin did derive more sheer delight from writing his book on earthworms than from any amount of contemplation of a creature who could talk back and who was apt stubbornly to hold ill-founded opinions. In any case, no man afflicted with a weak stomach and insomnia has any business investigating his own kind. At least it is best to wait until they have undergone the petrification incident to becoming part of a geological stratum.

Darwin knew this. He had fled London to work in peace. When he dealt with the timid gropings of climbing plants, the intricacies of orchids or the calculated malice of the carnivorous sundew, he was not bedeviled by metaphysicians, by talk of ethics, morals or the nature of religion. Darwin did not wish to leave man an exception to his system, but he was content to consider man simply as a part of that vast, sprawling, endlessly ramifying ferment called "life." The rest of him could be left to the philosophers. "I have often," he once complained to a friend, "been made wroth (even by Lyell) at the confidence with which people speak of the introduction of man, as if they had seen him walk on the stage and as if in a geological sense it was more important than the entry of any other mammifer."

Darwin's fame as the author of the theory of evolution has tended to obscure the fact that he was, without doubt, one of the great field naturalists of all time. His capacity to see deep problems in simple objects is nowhere better illustrated than in his study of movement in plants, published some two years before his death. He subjected twining plants, previously little studied, to a series of ingenious investigations of pioneer importance in experimental botany. Perhaps Darwin's intuitive comparison of plants to animals accounted for much of his success in this field. There is an entertaining story that illustrates how much more perceptive than his contemporaries he was here. To Huxley and another visitor, Darwin was trying to explain the remarkable behavior of *Drosera*, the sundew plant, which catches insects with grasping, sticky hairs. The two visitors listened to Darwin as one might listen politely to a friend who is slightly "touched." But as they watched the plant, their tolerant poise suddenly vanished and Huxley cried out in amazement: "Look, it *is* moving!"

The Islands

As one surveys the long and tangled course that led to Darwin's great discovery, one cannot but be struck by the part played in it by oceanic islands. It is a part little considered by the general public today. The word "evolution" is commonly supposed to stand for something that occurred in the past, something involving fossil apes and dinosaurs, something pecked out of the rocks of eroding mountains—a history of the world largely demonstrated and proved by the bone hunter. Yet, paradoxically, in Darwin's time it was this very history that most cogently challenged the evolutionary point of view. Paleontology was not nearly so extensively developed as today, and the record was notable mainly for its gaps. "Where are the links?" the critics used to rail at Darwin. "Where are the links between man and ape—between your lost land animal and the whale? Show us the fossils; prove your case." Darwin could only repeat: "This is the most obvious and gravest objection which can be urged against my theory. The explanation lies, as I believe, in the extreme imperfection of the geological record." The evidence for the continuity of life must be found elsewhere. And it was the oceanic islands that finally supplied the clue.

Until Darwin turned his attention to them, it appears to have been generally assumed that island plants and animals were simply marooned evidences of a past connection with the nearest continent. Darwin, however, noted that whole classes of continental life were absent from the islands; that certain plants which were herbaceous (nonwoody) on the mainland had developed into trees on the islands; that island animals often differed from their counterparts on the mainland.

Above all, the fantastically varied finches of the Galápagos particularly amazed and puzzled him. The finches diverged mainly in their beaks. There were parrot-beaks, curved beaks for probing flowers, straight beaks, small beaks—beaks for every conceivable purpose. These beak variations existed nowhere but on the islands; they must have evolved there. Darwin had early observed: "One might really fancy that, from an original paucity of birds in this archipelago, one species had been taken and modified for different ends." The birds had become transformed, through the struggle for existence on their little islets, into a series of types suited to particular environmental niches where, properly adapted, they could obtain food and survive. As the ornithologist David Lack has remarked: "Darwin's finches form a little world of their own, but one which intimately reflects the world as a whole" [see "Darwin's Finches," by David Lack; SCIENTIFIC AMERICAN Offprint 22].

Darwin's recognition of the significance of this miniature world, where the forces operating to create new beings could be plainly seen, was indispensable to his discovery of the origin of species. The island worlds reduced the confusion of continental life to more simple proportions; one could separate the factors involved with greater success. Over and over Darwin emphasized the importance of islands in his thinking. Nothing would aid natural history more, he contended to Lyell, "than careful collecting and investigating of *all the productions* of the most isolated islands. . . . Every sea shell and insect and plant is of value from such spots."

Darwin was born in precisely the right age even in terms of the great scientific voyages. A little earlier the story the islands had to tell could not have been read; a little later much of it began to be erased. Today all over the globe the populations of these little worlds are vanishing, many without ever having

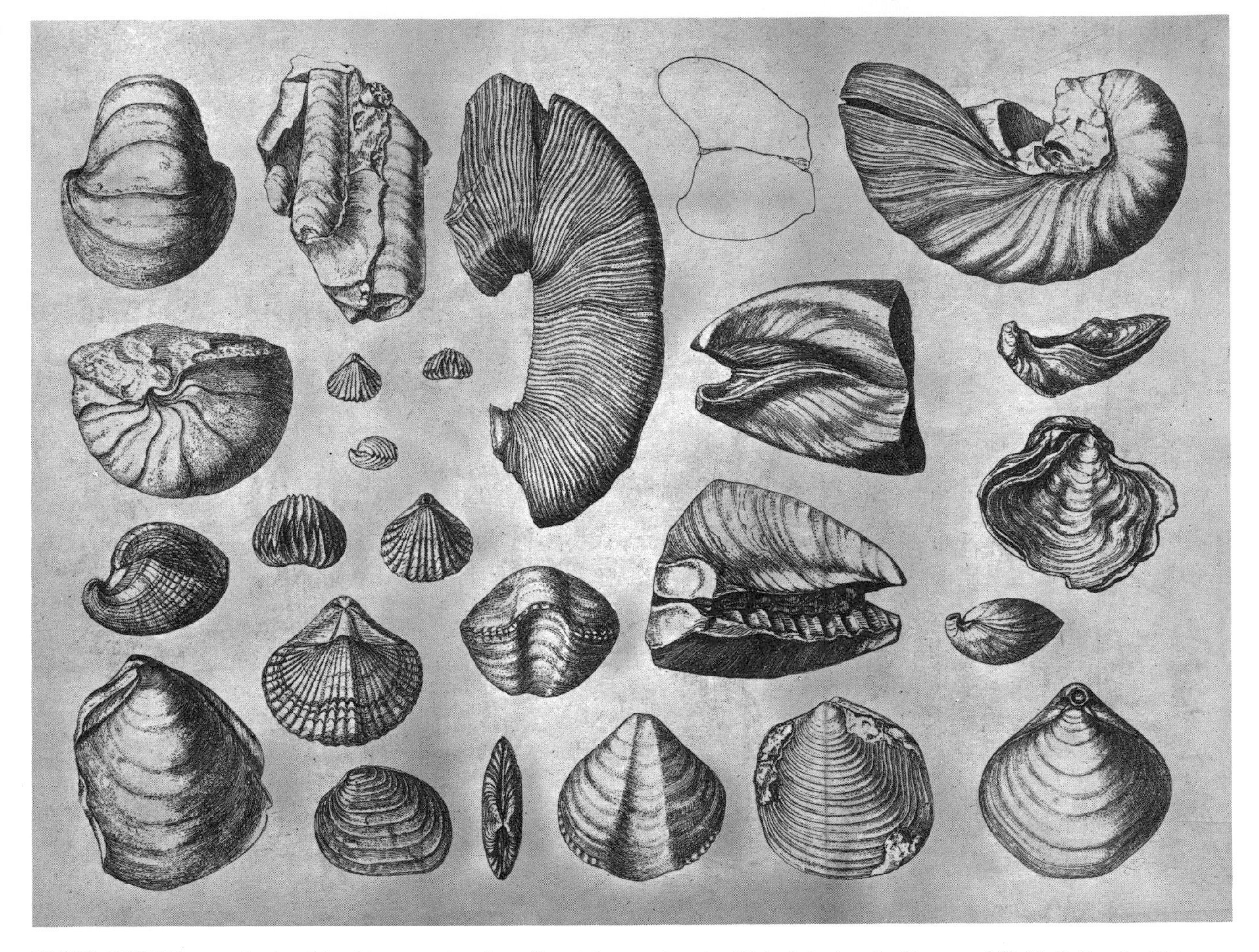

FOSSIL SHELLS were depicted in this engraving from Darwin's *Geological Observations on the Volcanic Islands and Parts of South America Visited during the Voyage of H. M. S. Beagle.* This was a technical work published by Darwin before *Origin of Species.*

been seriously investigated. Man, breaking into their isolation, has brought with him cats, rats, pigs, goats, weeds and insects from the continents. In the face of these hardier, tougher, more aggressive competitors, the island faunas—the rare, the antique, the strange, the beautiful—are vanishing without a trace. The giant Galápagos tortoises are almost extinct, as is the land lizard with which Darwin played. Some of the odd little finches and rare plants have gone or will go. On the island of Madagascar our own remote relatives, the lemurs, which have radiated into many curious forms, are now being exterminated through the destruction of the forests. Even that continental island Australia is suffering from the decimation wrought by man. The Robinson Crusoe worlds where small castaways could create existences idyllically remote from the ravening slaughter of man and his associates are about to pass away forever. Every such spot is now a potential air base where the cries of birds are drowned in the roar of jets, and the crevices once frequented by bird life are flattened into the long runways of the bombers. All this would not have surprised Darwin, one would guess.

Of Darwin's final thoughts in the last hours of his life in 1882, when he struggled with a weakening heart, no record remains. One cannot but wonder whether this man who had no faith in Paradise may not have seen rising on his dying sight the pounding surf and black slag heaps of the Galápagos, those islands called by Fitzroy "a fit shore for Pandemonium." None would ever see them again as Darwin had seen them—smoldering sullenly under the equatorial sun and crawling with uncertain black reptiles lost from some earlier creation. Once he had cried out suddenly in anguish: "What a book a devil's chaplain might write on the clumsy, wasteful, blundering, low and horribly cruel works of nature!" He never spoke or wrote in quite that way again. It was more characteristic of his mind to dwell on such memories as that Eden-like bird drinking softly from the pitcher held in his hand. When the end came, he remarked with simple dignity, "I am not in the least afraid of death."

It was in that spirit he had ventured upon a great voyage in his youth. It would suffice him for one more journey.

2

THE VOYAGE OF THE *CHALLENGER*

HERBERT S. BAILEY, Jr.
May 1953

JUST 77 years ago this month a spar-decked little ship of 2,300 tons sailed into the harbor of Spithead, England. She was home from a voyage of three and a half years and 68,890 miles over the seven seas. Her expedition had been a bold attack upon the unknown in the tradition of the great sea explorations of the 15th and 16th centuries. The unknown she had explored was the sea bottom. When she had left England, the ocean deeps were an almost unfathomed mystery. When she returned, she had sounded the depths of every ocean except the Arctic and laid the foundation for the modern science of oceanography.

The ship was called the *Challenger*. Her name and voyage are already covered with the dust of time, but her story is worth reviving today, when far more handsomely outfitted expeditions are once more exploring the sea deeps. They are filling in details and retouching parts of a picture which in its broad outlines has remained essentially unchanged since that pioneering voyage. It was the *Challenger*, rigged with crude but ingenious sounding equipment, that charted what is still our basic map of the world under the oceans.

Before the *Challenger*, only a few isolated soundings had been taken in the deep seas. Magellan is believed to have made the first. During his voyage around the globe in 1521 he lowered hand lines to a depth of perhaps 200 fathoms (1,200 feet) in the Pacific; failing to reach bottom, he concluded that he was over the deepest part of the ocean. (Actually the water where he took his soundings is 12,000 feet deep, far from the deepest bottom in the Pacific.) After Magellan no deep-sea soundings were taken for about 300 years. In the 19th century a few sea captains and layers of telegraph cables began to plumb deep waters, some of them getting their lines down as deep as two miles or more.

One of the first men to take a scientific interest in the ocean depths was Edward Forbes, professor of natural philosophy at the University of Edinburgh. He did some dredging in the Aegean Sea, studying the distribution of flora and fauna and their relation to depths, temperatures and other factors. Forbes never dredged deeper than about 1,200 feet, and he acquired some curious notions, including a belief that nothing lived in the sea below 1,500 feet. But his pioneering work led the way for the *Challenger* expedition.

H.M.S. Challenger, *as depicted in the official* Challenger *report*

THE MAN WHO organized the expedition was Charles Wyville Thomson, Forbes's successor as professor of natural philosophy at Edinburgh. Thomson first made some summer dredging cruises in ships borrowed from the British Admiralty, and the results were so interesting that they prompted Thomson and the Royal Society to approach the Admiralty with a much more ambitious project. They asked for a vessel that could carry out an investigation of the "conditions of the Deep Sea throughout all the Great Oceanic Basins." The naval authorities, now fully awake to the importance of oceanic research, provided H.M.S. *Challenger*, a corvette fitted with auxiliary steam power in addition to her sails. A naval crew under Captain George S. Nares was assigned to the mission, and Thomson selected a staff of scientists and other civilians to assist him.

They proceeded to adapt or improvise the necessary scientific equipment and to fit out laboratories on the ship. To make room for their gear they removed all but two of the warship's 18 guns. Their equipment included instruments for taking soundings, bottom samples and undersea temperatures; winches and a donkey engine; 144 miles of sounding rope and 12.5 miles of sounding wire; sinkers, nets, dredges, a small library, hundreds of miscellaneous scientific instruments and "spirits of wine" for preserving specimens.

The expedition, coming after Charles Darwin's famous voyage in the *Beagle* and in the midst of the great uproar over his new theory of evolution, naturally attracted public attention. Even

Dredging and sounding apparatus on the deck of the Challenger

Punch gave the *Challenger* a send-off:

Her task's to sound Ocean, smooth humours or rough in,
To examine old Nep's deep-sea bed. . . .
In a word, all her secrets from Nature to wheedle,
And the great freight of facts homeward bear.

The *Challenger* sailed from Portsmouth on December 21, 1872. She immediately ran into a howling storm at sea. Thomson found this no evil omen, pointing out that the gale "brought all our weak points to light" and increased confidence in the arrangements. His staff spent the first leg of the voyage, as far as Bermuda, in training and practice on their work: sounding, dredging, trawling and making measurements. Holding the ship steady with her steam engines, the civil and naval crews each time took a standard series of observations: the total depth of water, the temperatures at various depths, the atmospheric and meteorological conditions, the direction and rate of the current on the ocean surface and occasionally of the currents at different depths. They also dredged up samples of the bottom, including its plant and animal life, and dipped up samples of the water and of the sea life at various levels. They found they had to make their soundings with the hemp rope, because the wire tended to kink and break. Attached to the line were sinkers, thermometers and water bottles; when the sinkers hit the bottom they were automatically detached. By the time they had finished their voyage, they had made such observations at 360 stations scattered over the 140 million square miles of the ocean floor.

The routine was long and laborious. In really deep water it took more than an hour and a half to reach bottom—and much longer to haul the line back. "Dredging," wrote one of the naval officers, "was our *bête noire*. The romance of deep-water trawling or dredging in the *Challenger*, when repeated several hundred times, was regarded from two points of view: the one was the naval officer's, who had to stand for 10 or 12 hours at a stretch carrying on the work . . . the other was the naturalist's . . . to whom some new worm, coral, or echinoderm is a joy forever, who retires to a comfortable cabin to describe with enthusiasm this new animal, which we, without much enthusiasm, and with much weariness of spirit, to the rumbling tune of the donkey engine only, had dragged up for him from the bottom of the sea."

THE TRIP was not, however, all drudgery. There was romance and adventure enough to inspire the officers and scientists, almost to a man, to produce memoirs, logs and other accounts for an eagerly waiting public at the end of the voyage. One of the first diversions occurred in the South Atlantic. Putting in at the tiny colony of Tristan Island, the *Challenger's* crew learned from the inhabitants that two brothers named Stoltenhoff, seeking their fortune at seal-hunting, had marooned themselves nearly two years earlier on aptly named Inaccessible Island. The ship diverted its course to rescue the brothers. They had kept themselves alive on a diet of penguins' eggs and wild pigs, but had had no luck catching seals. The ship also stopped at nearby Nightingale Island, a rookery for hundreds of thousands of penguins, and found it so covered with eggs that the shore party could hardly walk without stepping on them. The penguins defended their nests furiously and pecked one of the ship's dogs to death.

Beyond Cape Horn, on Marion Island, they saw multitudes of white albatross, but, heeding the warning of the Ancient Mariner, they killed none. This was a rare exception, for it was their practice to collect specimens of indigenous flora and fauna wherever they touched land.

Exploring in the southernmost Indian Ocean, the *Challenger* became the first steamship to cross the Antarctic Circle. The scientists were tremendously interested in the icebergs and even fired a cannon at one to break off a chunk. In an unsuccessful attempt to find the "Termination Land" reported by the U. S. explorer Charles Wilkes, the *Challenger* ran into a sudden antarctic storm while traversing a pack of icebergs. With the wind at 42 miles per hour and night coming on, the ship took refuge in the lee of a large berg, holding position close beside it with the steam engine. During an unexpected lull in the wind,

The zoological laboratory of the Challenger (above), *and the principle of its deep-sea dredges* (below)

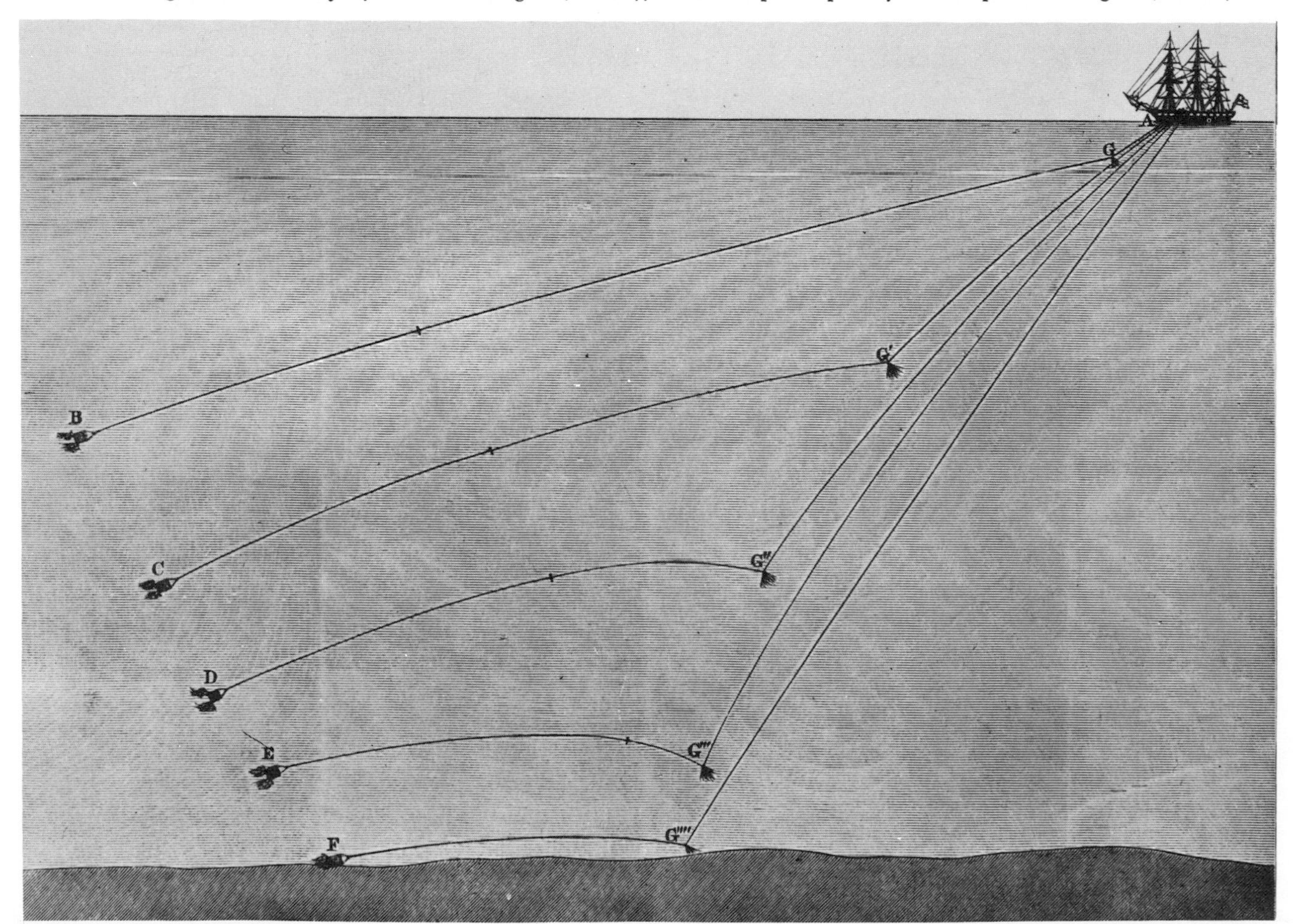

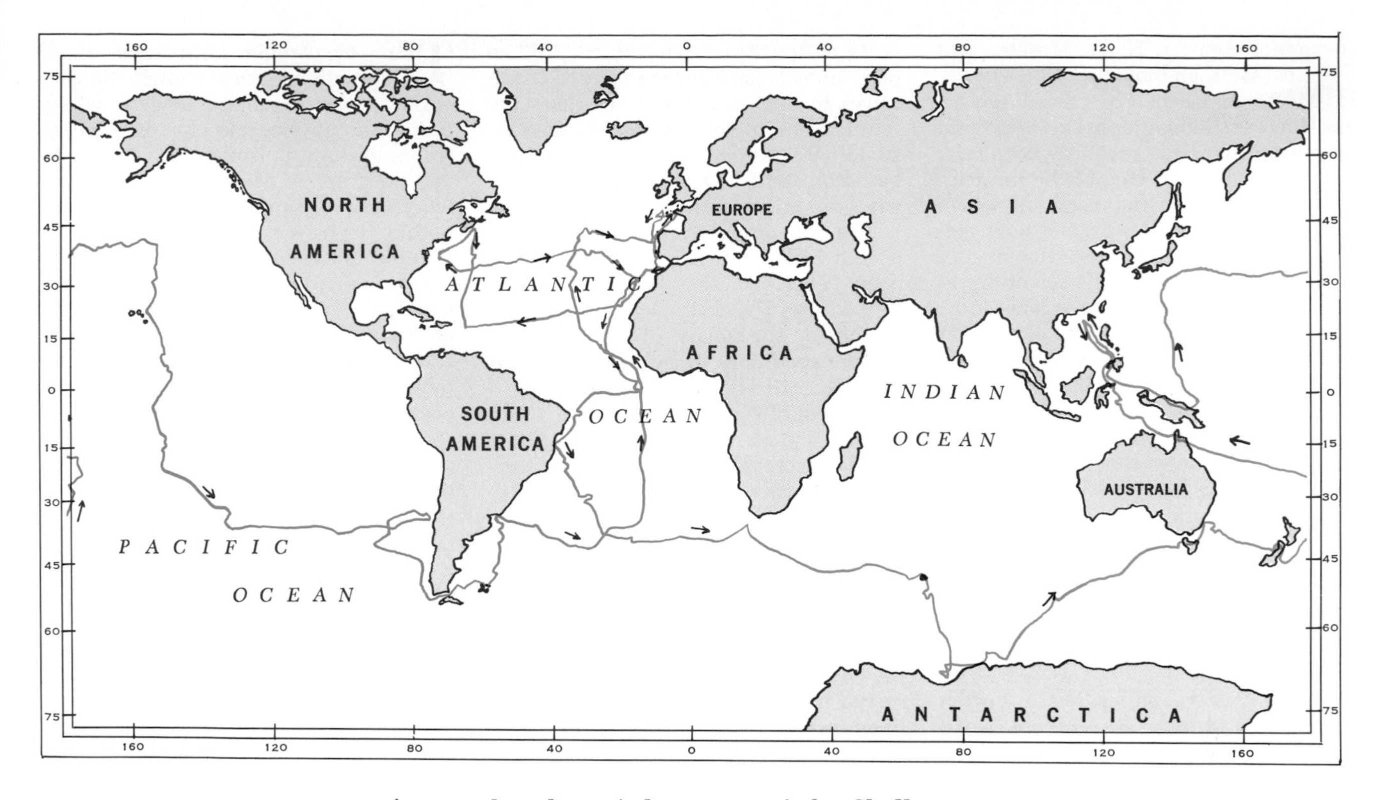

A general outline of the voyage of the Challenger

the ship rammed the berg before the engine could be reversed and lost its jib boom and other rigging. The damage was not serious and the rigging was recovered, but the company spent an anxious night steaming back and forth in a dense snowstorm between two large icebergs.

The ship next went on to Australia, New Zealand and the Pacific islands. In the Fijis they interviewed King Thackombau, a converted Christian, who had earlier cut out a prisoner's tongue and eaten it in his sight—before eating the prisoner himself. The ship called at the Philippines, Japan, China and the Admiralties. On March 23, 1875, off the Marianas Islands, the explorers hit their deepest sounding—26,850 feet. This was not very far from the deepest of all time: the record to date is a sounding of 34,440 feet, made in the Mindanao Trench off the Philippines by a U. S. Navy vessel in 1950.

The *Challenger* zigzagged across the Pacific, stopping at the Hawaiian Islands and Tahiti, and then rounded South America through the Strait of Magellan. It swung north through the South and North Atlantic and finally arrived home in England on May 24, 1876.

Of the *Challenger's* crew of some 240 men, seven died during the trip: two by drowning, one of yellow fever, the others of accidents and miscellaneous causes. Several of the crew jumped ship in Australia. The remainder returned to a joyful welcome—and to the long, hard task of organizing the vast amount of data accumulated on the voyage.

A COMMISSION was set up in Edinburgh to assess the results of the voyage, which were eventually published in an official report of 50 volumes. Two volumes contain a "summary of scientific results"; two a "narrative of the voyage." The other 46 are monographs written by some of the leading scientists of the day, among them T. H. Huxley,

Cladodactyla crocea, *taken in the Falkland Islands*

Japetella prismatica

Alexander Agassiz, H. N. Moseley and the great German biologist Ernst Haeckel. Most famous of the official reports are Haeckel's monographs on certain sea organisms that had previously been relatively little known. One of the most interesting is on the radiolaria, of which the expedition collected 3,508 new species to add to the 600 then known.

To see the *Challenger's* scientific results in proper perspective one must remember that the voyage took place at a time when every new discovery was an exciting prize to be fitted into the evolutionary table. The *Challenger* discovered 715 new genera and 4,417 species of living things, thus demonstrating that the oceans were teeming with unknown life waiting to be classified. It proved beyond question that life existed at great depths in the sea. The voyage opened the great descriptive era of oceanography, which was followed by the analytic oceanography of our own century.

The summary volumes were written by Sir John Murray, who became head of the commission after Thomson, exhausted by the voyage, died in 1882. Murray's comments and theories have had an important influence on oceanography. He strongly put forward the view, for example, that at equivalent latitudes both the Arctic and the Antarctic have similar marine organisms, and that these are not to be found in the more temperate zones. This "bipolarity" theory has now been discarded, but for a time it stimulated much investigation. Murray also asserted that, contrary to what had been hoped and expected, the deep sea did not yield a widespread fauna of great antiquity, though some very ancient species were found. He believed that under about 600 feet below sea level the bottom deposits and fauna become more uniform with increasing depth until a point is reached at which conditions are almost the same in all parts of the world. He added, "When once animals have accommodated themselves to deep-sea conditions there are few barriers to further vertical or horizontal migration." Such suggestions, based on the *Challenger's* observations, gave direction to further investigation.

The expedition washed out of existence a form of living matter that had been "observed" by Huxley and described by Haeckel. On the basis of preserved specimens dredged during earlier expeditions, Haeckel had decided that the entire ocean floor, or at least a major part of it, was covered with a thin layer of almost structureless living slime which he named "Bathybius." At a time when Darwin's theories were still under severe debate, "Bathybius" had been hailed as a living example of the primordial protoplasm. The scientists of the *Challenger* looked for it in vain, and finally discovered the answer to the puzzle. The alcohol and sea water in which the sea-bottom specimens were preserved had combined to form an amorphous precipitate of sulfate of lime. This was Haeckel's "Bathybius."

THE *Challenger* expedition made thousands of other contributions to various sciences—meteorology, hydrography, the physics and chemistry of sea water, geology, petrology, botany, zoology, geography. Murray's map of the world-wide sampling of oozes and other bottom deposits collected by the expedition has not been changed much by the many subsequent explorations. The *Challenger's* crew perfected the method of "swinging the compass" to get accurate magnetic readings. The voyage established the main contour lines of the ocean basins and disproved the myth of the lost continent Atlantis. It yielded the first systematic plot of currents and temperatures in the oceans, and showed that the temperature in each zone was fairly constant in all seasons.

The achievement of the *Challenger* was tremendous: a barrier had been broken and the world of the depths explored. In a sense the *Challenger* had answered the question that had echoed down the ages in the words of Ecclesiastes: "That which is far off and exceeding deep, who can find it out?"

II

THE SEA IN MOTION

OCEAN AND ATMOSPHERE, the two thin fluid films in which life is sustained and whose nature and motion determine the environment, dominate this color photograph of the watery hemisphere of the earth. The picture was made on January 21, 1968, by a spin-scan camera on NASA's Applications Technology Satellite 3, in synchronous orbit 22,300 miles above the Pacific Ocean. The camera experiment was proposed and developed by Verner E. Suomi and Robert J. Parent of the University of Wisconsin's Space Science and Engineering Center. As the camera spins it scans a 2.2-mile-wide strip across the earth, then steps down in latitude and scans another strip; in about 25 minutes a 2,400-strip picture is completed. What the camera transmits to the earth is an electrical signal representing the amount of green, red and blue light in each successive picture element, and from these signals a color negative is built up at the receiving station. Such photographs yield information on the interrelation of atmospheric and oceanic phenomena. In this picture, for example, the convective pattern over the southeastern Pacific indicates that air heated by the sea is rising.

II

THE SEA IN MOTION

INTRODUCTION

Some years ago, as a student at Harvard, I enrolled in a physical oceanography course taught by the great oceanographer Columbus O'Donnel Iselin. By way of introducing the course, Dr. Iselin made a few remarks on its scope and content, and then, to preface the lecture, said, "The ocean is motion." At the time, I considered his euphonious phrase a bit lacking in what I then considered professorial erudition. With the passing of years, however, that simple phrase—a jingle, almost—has remained as my constant reminder of the activities of an oceanographer living and working aboard ship: sampling, dredging, navigating, and all the other activities encountered by a mariner, the most important of which, for the oceanographer, is the proper interpretation of his data. Whether we are interested in the movement of sand on a beach, the design of a ship, or the navigation of the green turtle, we must consider the motion of the waves and currents. As the reader will soon perceive, an understanding of all other branches of oceanography requires first an understanding of the physical phenomena of the sea.

The motion of the sea is basically the result of the interaction of two thin films that cover our planet: the atmosphere and the ocean. Thus, the physical oceanographer must not only understand the laws and theories of physics as they relate to the sea, but he must likewise be familiar with the major processes at work in the atmosphere. So closely linked are the two major fluids that it is impossible to separate meteorology from physical oceanography in dealing with problems of oceanic movement.

R. W. Stewart, in his essay "The Atmosphere and the Ocean," provides an excellent overview of the fundamental links between the atmosphere and the motion of the sea, particularly the large current patterns. Although it would be redundant to summarize his explanations here, Stewart does mention one noteworthy subject that should be expanded upon, chiefly because it calls attention to the need for further research, both theoretical and empirical, and because it relates to international relations.

This subject is the nature of the Ekman-layer flow, as originally conceived by Walfrid Ekman at the turn of this century. The Ekman spiral predicts that current flows in the northern hemisphere are induced to the right of winds passing over the surface of the sea. Although the concept has been used by oceanographers for many years to interpret observations made at sea, there is, as yet, neither a satisfactory explanation of the turbulent processes involved in the Ekman model nor a method for reliably predicting changes in upwelling currents resulting from Ekman-layer flow. An example of the international importance of upwelling is found off the coast of Peru. Here, in the Southern Hemisphere, the Ekman-flow water movement is offshore, and the surface waters are replaced by colder, nutrient-rich waters from below. The additional nutrients are introduced into the food chain and provide for exceptionally rich fishing grounds. So much are the Peruvians dependent upon the large fishery that this upwelling provides (a major part of their national economy is based upon fish) that they have *unilaterally* extended their national sovereignty over the waters of the Pacific for 200 miles seaward. Their claim, like those of certain other South American countries, is enforced by armed patrol vessels.

Occasional changes in upwelling have caused serious effects on the fisheries and, subsequently, on the national economy. Early warning of economic peril would allow Peru and other states to take steps, in time, to minimize hardships for this sector of their economy and, importantly, for their people.

Although the international political problems associated with Peru's claim of 200 miles of oceanic sovereignty cannot be readily resolved—as our own State Department has learned—we could, perhaps, use the knowledge that we have gained through synoptic oceanographic surveys of upwelling waters off Peru to predict changes and, thus, to foster a better international understanding. Furthermore, as the world population expands, it would behoove all nations with upwelling waters off their coasts, including the United States, to pursue a vigorous research program aimed at defining the maximum tonnage of fish that may be taken from such waters without decreasing the fishery as a renewable food resource.

The aforementioned problem is only one of the several complex problems that require the combined talents of physical oceanographers and meteorologists. Clearly, there are others: the coupling of wind energy to the sea, periodic climatic changes, shifting of major currents, and precise measurements of the heat budget of the global ocean.

Willard Bascom, in "Ocean Waves," draws our attention to the fundamentals of wave theory. His semitechnical discourse on the important parameters of wave length, height, amplitude, velocity, and period provides the information necessary to comprehend simple wave motion, including reflection, refraction, and the nature of breakers on different types of shores. The reader should not be misled by the apparent simplicity of ocean waves, for the subject is monstrously complex; to pursue research in wave theory requires a firm grasp of advanced physics, mathematics, and computer science. For the oceanographer who enjoys a balance between laboratory and shipboard research, the study of waves is an ideal career. We shall address ourselves to some of man's problems with waves shortly, but first let us consider two other types of waves. Both are usually misunderstood.

From time to time, the press and television news services report "tidal wave" destruction of coastal village, fishing ports, and ships in remote parts of the Pacific. The so-called "tidal waves" are actually caused by major geological disturbances on the sea floor—earthquakes, submarine volcano eruptions, and giant mud slides—and have no direct relation to tides. In fact, by their very nature, waves that are truly tidal are periodic and, thus, predictable; the destructive waves, which are called *tsunamis* by oceanographers, are aperiodic, and their occurrence cannot be predicted, at least not yet. Inasmuch as tsunamis can, and have, destroyed entire coastal communities, not to mention pleasure piers, beaches, and lighthouses, there is need to know the basic character of tsunamis and the distribution of their likely sites of origin. Joseph Bernstein, in his essay "Tsunamis," provides this information and reports on the effort that man has made in recent years to establish an early-warning system.

To all who live along the coast, the rhythmic rise and fall of the tide and the local currents caused by the tides are well known. For centuries, the daily routine and commerce of port cities have been singularly controlled, as it were, by the rising and falling of the tides. Yet, few laymen are cognizant of the fact that tides are no more than solitary, long-period waves formed in response to the attraction of the sea by the moon and, to a lesser degree, by the sun. It is interesting that mathematical treatment of a simple wave may also be applied to the tidal wave.

Waves are manifestly important in the activities of man. Let us consider some examples that are encountered by property owners, civic officials, government planners, and businessmen.

When purchasing beach-front property, for any use, the buyer should consider both the "typical" waves and the storm waves along the shore. Waves are active agents of erosion—particularly storm waves—and a piece of property facing a large lake or an ocean and subject to high waves is surely to be eroded, particularly where the coast is composed of unconsolidated sands or other loose sediments.

The construction of pleasure beaches is becoming an increasingly important activity of municipalities, especially where the population is expanding rapidly. Unfortunately, the simple requisites of shore ownership and the dumping of tons of sand will not ensure a desirable bathing beach, nor guarantee its lifetime. Aside from considerations of water quality, the proposed beach must not be subject to waves of such magnitude that they will place the sand in suspension and remove it by longshore drift, nor should the winter waves be such that they will damage piers, boathouses, bathhouses, and other structures during the off season, necessitating expensive repairs each spring. Frequently, the selection of new beach sites and the obligation of tax revenue are based more on political and aesthetic concerns than on scientific evaluation of the environmental factors, including waves. Surely, any citizen can alert his local government to such natural phenomena and their consequences; it is hoped that a simple understanding of waves would provide him with the necessary—albeit, limited—knowledge to do so.

Achieving an understanding of the major ocean currents requires consideration of the Coriolis effect. James E. McDonald's opening statement in "The Coriolis Effect" is worthy of repeating: "It is a curious fact that all things which move over the surface of the earth tend to sidle from their appointed paths—to the right in the Northern Hemisphere, to the left in the Southern Hemisphere." In short, McDonald is telling us that a given point on the earth beneath a moving object simply rotates away from that object during its period of movement. The explanation for the Coriolis effect (sometimes termed the Coriolis *force*, and treated as a force mathematically) serves not only the oceanographer who seeks to explain the circulation of oceanic currents, but also the airplane pilot who, unless corrections for the Coriolis effect were made in long flights, would miss his destination. Even the relatively short-trajectoried projectiles fired from naval guns would consistently miss their targets, were prefixing corrections of the guns' horizontal positions (azimuth) not made to account for the Coriolis effect. Likewise, the same phenomena must be considered in rocket trajectories, bird flights, and ocean currents.

Let us consider the role of the Coriolis effect in nature. An excellent example is the migration of a warbler, the lesser whitethroat, a summer visitor to Germany and Scandinavia that winters in southern Africa. This tiny bird, weighing less than an ounce, makes its annual migration solely at night, always alone, over land and sea, and covers a hundred miles or more during a single night. How does this tiny creature correct for the Coriolis effect? E. G. F. Sauer, an ornithologist at the University of Freiburg, has suggested that the lesser whitethroat can perceive gradual changes in the apparent position of the stars, an effect of the earth rotating beneath him during flight. Moreover, the whitethroat can, unerringly, correct his flight to account for the Coriolis effect, and he does so on a flight that begins in the Northern Hemisphere and ends in the Southern Hemisphere. Considering the ingenious electronic equipment used by man to compute corrections for the Coriolis effect, it is truly remarkable that such a tiny bird can do the same with a brain the size of a match head! Surely, a firm understanding of the Coriolis effect is not only necessary to understand the currents of the sea, but it is also necessary to understand the flight of birds across the sea, and, indeed, all navigation, and all objects in motion over the ocean.

Although nearly everyone knows something about the early investigation by Benjamin Franklin of the nature and position of the Gulf Stream, few realize how closely related the Gulf Stream is to their daily lives. The general circulation gyres of the major surface currents of the ocean have been known at least since the pioneering work of Matthew F. Maury over a century ago. His early charts, as we previously mentioned, aided shipmasters, during the era of sail, in their selection of profitable routes to take advantage of favorable winds and currents. Surprisingly, modern commerce takes advantage of known circulation patterns today. An example of this is the routing of petroleum tankers between the oil ports of Texas and the large cities of the East Coast. Inasmuch as the time for a vessel to deliver its cargo and return to its home port is an economic factor in determining the price of a gasoline or heating-oil product, which may be consumed in New York but produced in Texas, then it is axiomatic that the quicker the tanker can deliver its cargo and return, the sooner it can go again. Because salaries, the cost of ship maintenance, and license fees are relatively fixed expenses, profit will be enhanced by making more trips per year. Major petroleum companies operating tankers between Texas and New York are well aware of the benefits of routing a tanker so that it will gain an extra two to four nautical miles per hour in the

Gulf Stream off the Atlantic Coast. Likewise, returning tankers avoid the main flow of the Gulf Stream current, in order not to waste time and fuel by bucking the current. Similar routings are made throughout the world ocean. To do otherwise would be to create operating losses of millions of dollars per year—not to mention a higher price for a gallon of gasoline at your neighborhood service station.

The pattern of major ocean currents is of importance to people of many occupations. For the historian and the anthropologist, a knowledge of ocean currents may afford a means of determining man's past migration routes. The submarine officer must know his underwater position in relation to major currents, not only to correct his underwater course during passage from one water mass of a given velocity to another, but—more importantly—to keep his craft in proper buoyance as he moves into water of a different temperature and, thus, a different density.

Looking toward the future, we see that much research is needed on the relations of oceanic circulation to weather prediction, to climatology, and even to the dispersal of the wastes man dumps at sea. Such research will require many ships and many people, if we are to measure the currents on the scale needed to define precisely their spatial extent and their turbulent motions. The distinguished oceanographer Henry Stommel points out in "The Circulation of the Abyss" that we do not yet fully understand the deep ocean-bottom currents; nonetheless, these currents may be those most important in effecting climatic changes throughout the entire world.

The citizen concerned about this nation's disposal of radioactive wastes and poisonous gases on the floor of the deep ocean will benefit from even a cursory reading of "The Circulation of the Abyss," "The Atmosphere and the Ocean," and "The Sargasso Sea," all of which are included here.

Of all the oceans, scientists know less about the two at the poles: the Arctic and the Antarctic. Yet, these are of particular interest to oceanographers because of their uniqueness: the Arctic is covered by ice, and the circumpolar Antarctic has a limitless fetch.

In considering the Antarctic, we must bear in mind that it is one of the most biologically productive bodies of water on earth. It has been the scene of wholesale killing of the great whales—an atrocity resulting from international permissiveness—and it is the home of the abundant krill, a shrimp that is present in such vast numbers that it may save man in his quest for additional protein. If for no other reason, the physical oceanography of the Antarctic should be studied as a prelude to future utilization and conservation of the vast food resource located there.

There are several reasons for learning more about the Arctic and its blanket of ice. From a purely scientific point of view, we are interested in learning how the currents are generated and maintained in the absence of an atmospheric wind system in contact with the water. From a political and strategic viewpoint, the Arctic is our only common ocean with Russia, and it is a highly sensitive region in terms of defense. Finally, and from an economic point of view, the Arctic may yet prove to be an important seaway when cargo-carrying submarines are introduced, and it is conceivable that great mineral wealth may lie beneath the Arctic on the continental shelf of Alaska. Prime to all of these interests and their development is the need for detailed knowledge of the physical oceanography of the Arctic.

In this brief commentary on physical oceanography, we have stressed the obvious requirement for extensive, detailed studies of the ocean's motion. While there must be a commitment of ships and other expensive capital equipment, including remote-sensing satellites, we may be able to eliminate parts of the surveys at sea through the use of models. Such devices could help reduce shipboard expenses by first suggesting the most critical locations in the sea to be surveyed. Moreover, laboratory models can be constructed to model both large and small areas, and the results can be made computer-compatible for statistical and other mathematical treatment. Although some success in modeling the patterns of ocean circulation has been achieved—notably, with William Von Arx's rotating hemisphere—oceanographers have been prone to avoid models and to seek their answers at sea. Perhaps, by the end of this decade, the expense of operating ships at sea in order to collect synoptic data will become so great that oceanographers will be forced to pursue part of their research through the use of carefully scaled models.

The physical and chemical aspects of the ocean are considered together in the two papers, "Salt and Rain" and "Why the Sea Is Salt," that conclude this section of the book. Although no less important, the chemistry of the sea, unlike its physics, does not readily lend itself to elementary discourse. Nevertheless, some consideration of the chemical elements present in sea water, and of their relations to sea-surface phenomena, marine life, and the economics of utilizing such elements by man, is essential.

Modern marine chemistry, like most other branches of marine science, began with the investigations carried out on the *Challenger* cruise. Almost ninety years ago, Wilhelm Dittmar, from his careful analysis of seventy-seven water samples from scattered sites in the ocean, reported that most of the major elements in sea water are present in more-or-less constant proportions; thus, by measuring one major element—usually chlorine—and by applying the appropriate ratio, amounts of each of the other major elements could be readily computed. I hasten to add that chemical oceanographers have been seeking exceptions to Dittmar's constancy-of-elements "law" ever since.

In his enlightening essay "Salt and Rain," A. H. Woodcock provides an instructive review of his research relating the sizes of salt particles to the dimensions of resulting raindrops. The source of the tiny salt particles is shown to be the breaking of ocean waves and the liberation and drying of small drops of sea water. Such a process ties together the physics of waves and such atmospheric variables as wind and humidity. These relations are obvious, but not so apparent are the effects of this seemingly simple process and the role it plays in weather, airport location, agriculture, house painting, corrosion, and even the possible fertilization of crops far inland. Let us review these effects as they relate to man.

The source of raindrop nuclei and the changing rate of their supply can be critical factors in tropical and subtropical agriculture. For example, rain falling on the pineapple fields of Hawaii is spatially and temporally variable not only from one island to another, but even over parts of the same island. While some variation is related to wind paths and physiography, some of the variation is due directly to variations in the salt nuclei present in the air.

Along a particularly rough coast with abundant breakers, much salt is carried inland to take its toll in corrosion of metal structures and automobiles and in advancing the deterioration of house paint. In fact, the common practice of washing down old house paint before applying a new coat, especially in coastal areas, is done more to remove the salt crystals, which absorb water and inhibit bonding, than it is to remove the dust and stains.

The fact that a small amount of the organic film covering the sea is removed with the water droplets that dry to become nuclei is of interest both to farmers and to public-health officials. It is now believed, for example, that additional natural fertilization is effected by the deposition of this organic nutrient during rainfall; and along the Gulf Coast, where the poisonous "red tide" occurs from time to time, as the result of a bloom of reddish dinoflagellate plankton in the water offshore, it has been observed that the tender mucous membranes of coastal residents become irritated during "red tide" outbreaks. Here we see an obvious need for cooperation between physicians and oceanographers, not only to understand causes but, more importantly, to predict occurrences.

The last years of the 1960s saw oceanographers begin their long-time dream of drilling holes in the deep ocean floor. Cores of sediment and rock collected by the *Glomar Challenger* have already provided marine scientists with proof of ocean-floor spreading. These same cores are also changing our ideas—very much so—about the origin of salt in the sea. Ferren MacIntyre explains this in his essay "Why the Sea is Salt," a bold and imaginative contribution.

As a college freshman at the end of World War II, I was taught that the salt in the sea was simply the result of millions of years of chemical decay of rocks on land and the delivery of the dissolved elements to the ocean by streams. It is surprising that no one seriously challenged this working hypothesis, even though textbooks of the time carried tabular data on ancient oceanic salt formed several hundred million years ago that compared closely with modern oceanic salt. In fact, it was taught that one could make a fair estimate of the age of the ocean by relating the total salt in the ocean to an "average" amount delivered annually by rivers. For some reason that I fail to remember, that discrepancy of several billion years was dismissed about the same time that the class was!

MacIntyre, in his convincing thesis, a "grand geochemical cycle," suggests an obvious and very reasonable interpretation of the total salt cycle. His explanation involves not only the application of modern physical chemistry but also the role of ocean-floor spreading and its basic recycling of crustal material. A careful reading of his synthesis suggests that mining engineers, economists, and politicians will find its implications relevant to their own fields.

Clues to locating new resources of manganese, copper, cobalt, nickel, zirconium, zinc, and vanadium, as well as such exotic elements as uranium and the rare earths, are more likely to come from the study of oceanic chemical processes than from purely empirical observations and regional surveys. It is already being proposed by some ocean-mining firms that rich metal deposits will be found on the floor of the sea near major rifts. The rift floor of the southern Red Sea is already known to have, *in situ,* millions of dollars' worth of metal deposits, as confirmed by some oceanographic scientists, and the area is soon to be explored by German and American consortia. Searches may soon be directed to other similar rift areas in the Pacific. If these searches are successful, what then will be the rules and laws governing the exploitation of the newly discovered resources? Which nations will most directly benefit from new mineral discoveries? What will be the impact on international trade? These are questions that will need to be answered in the decade ahead.

In concluding our comments on the physics and chemistry of the sea, let us remember that these disciplines are basic to all marine sciences. Their fundamentals must be mastered by all oceanographers; their importance should be apparent to all men.

THE ATMOSPHERE AND THE OCEAN

R. W. STEWART
September 1967

The atmosphere drives the great ocean circulations and strongly affects the properties of seawater; to a large extent the atmosphere in turn owes its nature to and derives its energy from the ocean. Indeed, there are few phenomena of physical oceanography that are not somehow dominated by the atmosphere, and there are few atmospheric phenomena for which the ocean is unimportant. It is therefore hard to know where to start a discussion of the interactions of the atmosphere and the ocean, since in a way everything depends on everything else. One must break into this circle somewhere, and arbitrarily I shall begin by considering some of the effects of wind on ocean water.

When wind blows over water, it exerts a force on the surface in the direction of the wind. The mechanism by which it does so is rather complex and is far from being completely understood, but that it does it is beyond dispute. The ocean's response to this force is immensely complicated by a number of factors. The fact that the earth is rotating is of overriding importance. The presence of continental barriers across the natural directions of flow of the ocean complicates matters further. Finally there is the fact that water is a fluid, not a solid.

To simplify the picture somewhat, let us start by looking at what would happen to a slab of material resting on the surface of the earth. Let us further assume that the slab can move without friction. Consider the result of a sharp, brief impulse that sets the slab moving, say, due north [*see top illustration on page* 38]. Looked at by an observer on a rotating earth, any moving object is subject to a "Coriolis acceleration" directed exactly at a right angle to its motion. The magnitude of the acceleration increases with both the speed of the object's motion and the vertical component of the earth's rotation, and in the Northern Hemisphere it is directed to the right of the motion. An acceleration at right angles to the velocity is just what is required to cause motion in a circle, and in the illustration the center of the circle is due east of the original position of the slab. A circular motion of this kind is called an inertial oscillation, and something of this nature may sometimes happen in the ocean, since inertial oscillations are frequently found when careful observations are made with current meters.

An inertial oscillation requires exactly half a pendulum day for a full circle. (A pendulum day is the time required for a complete revolution of a Foucault pendulum. Like the Coriolis effect, it depends on the vertical component of the earth's rotation and therefore varies with latitude, being just under 24 hours at the poles and increasing to several days close to the Equator. To be precise, it is one sidereal—or star time—day divided by the sine of the latitude.) If there were a small amount of friction, the slab would gradually spiral to the center of the circle. Pushing it toward the north thus causes it to end up displaced to the east [*see bottom illustration on page* 38]. More generally, in the Northern Hemisphere a particle is moved to the right of the direction in which it is impelled, and in the Southern Hemisphere it is moved to the left.

Let us turn to what happens to our frictionless slab if, instead of giving it a short impulse, we give it a steady thrust. Again assume that the force is toward the north [*see upper illustration on page* 39]. Under the influence of this force the slab accelerates toward the north, but as soon as it starts to move it comes under the influence of the Coriolis effect and its motion is deflected (in the Northern Hemisphere) to the right—to the east. As long as the slab has at least some component of velocity toward the north the force will continue to add energy to it and its speed will continue to increase. After a quarter of a pendulum day, however, it will be moving due east. In this position the applied force (which is to the north) is pushing at a right angle to the velocity (east), opposing the influence of the Coriolis effect, which is now trying to turn the slab toward the south.

If there has been no loss of energy because of friction, the slab is moving fast enough so that the Coriolis effect dominates, and it turns toward the south. Now there is a component of velocity opposing the applied force, which acts as a brake and takes energy from the motion. At the end of half a pendulum day the process has gone far enough to bring the slab to a full stop, at which point it is directly east of its starting point. If the force continues, it will again accelerate toward the north and the entire process is repeated, so that the slab performs a series of these looping (cycloidal) motions, each loop taking half a pendulum day to execute. Overall, then, a steady force on a frictionless body resting on a rotating earth causes it to move at right angles to the direction of the force. What is happening is that the force is balanced—on the average—by the Coriolis effect.

Now let us look at the situation when there is a certain amount of friction between the slab and the underlying surface [*see lower illustration on page* 39]. Any frictional drag reduces the speed attained by the slab, reducing the Coriolis effect until it is no longer entirely able to overcome the driving force. As a result if the force is toward the north, the slab will move in a more or less north-

Wind force: 4 Wind speed: 5½ Wave period: 5 Wave height: 1

Wind force: 5 Wind speed: 11½ Wave period: 6 Wave height: 2

Wind force: 6 Wind speed: 13 Wave period: 7 Wave height: 3

Wind force: 8 Wind speed: 18 Wave period: 6 Wave height: 5

Wind force: 9 Wind speed: 21 Wave period: 9 Wave height: 8

Wind force: 10 Wind speed: 27 Wave period: 9 Wave height: 7

easterly direction—more northerly if the friction is large, more easterly if it is small.

A body of water acts much like a set of such slabs, one on top of the other [*see illustration on page 40*]. Each slab is able to move largely independently of the others except for the frictional forces among them. If the top slab is pushed by the wind, it will, in the Northern Hemisphere, move in a direction somewhat to the right of the wind. It will exert a frictional force on the second slab down, which will then be set in motion in a direction still farther to the right. At each successive stage the force is somewhat reduced, so that not only does the direction change but also the speed is a bit less. A succession of such effects produces velocities for which the direction spirals as the depth increases. It is known as the Ekman spiral, after the pioneering Swedish oceanographer V. Walfrid Ekman, who first discussed it soon after the beginning of the century. At a certain depth both the current and the frictional forces associated with it become negligibly small. The entire layer above that depth, in which friction is important, is termed the Ekman layer. Since there is negligible friction between the Ekman layer and the water lying under it, the Ekman layer as a whole behaves like the frictionless slab discussed above: its average velocity must be at a right angle to the wind.

The frictional mechanism, which involves turbulence, has proved to be extraordinarily difficult to study either theoretically or through observations, and surprisingly little is known about it. The surface flow does appear to be somewhat to the right of the wind. Primitive theoretical calculations predict that its direction should be 45 degees from the wind, but this theory is certainly inapplicable in detail. More complicated theoretical models have been attempted, but since almost nothing is known of the nature of turbulence in the presence of a free surface these models rest on weak ground. An educated guess, supported by rather flimsy observational evidence, suggests that the angle is much smaller, perhaps nearer to 10 degrees. All that seems fairly certain is that the average flow in the Ekman layer must be at a right angle to the wind and that there must be some kind of spiral in the current directions. We also believe the bottom of the Ekman layer lies 100 meters or so deep, within a factor of two or three. Of the details of the spiral, and of the turbulent mechanisms that determine its nature, we know very little indeed.

This Ekman-layer flow has some important fairly direct effects in several parts of the world. For example, along the coasts of California and Peru the presence of coastal mountains tends to deflect the low-level winds so that they blow parallel to the coast. Typically, in each case, they blow toward the Equator, and so the average Ekman flow—to the right off California and to the left off Peru—is offshore. As the surface water is swept away deeper water wells up to replace it. The upwelling water is significantly colder than the sun-warmed surface waters, somewhat to the discomfort of swimmers (and, since it is also well fertilized compared with the surface water, to the advantage of fishermen and birds).

EFFECT OF WIND on the surface of the sea is shown in a series of photographs made by the Meteorological Service of Canada. Much of the wind's momentum goes into generating waves rather than directly into making currents. The change in the surface as the wind increases is primarily a change in scale, except for the effect of surface tension: the waves break up more, making more whitecaps. For each photograph the wind force is given according to the Beaufort scale; the wind speed is given in meters per second, the wave period in seconds and the wave height in meters. (In the final photograph the waves are only about half as large as they might become if the force-10 wind, which had blown for less than nine hours, were to continue to blow.)

The total amount of flow in the directly driven Ekman layer rarely exceeds a couple of tons per second across each meter of surface. That represents a substantial flow of water, but it is much less than the flow in major ocean currents. These are driven in a different way—also by the wind, but indirectly. To see how this works let us take a look at the North Atlantic [*see bottom illustration on page 41*]. The winds over this ocean, although they vary a good deal from time to time, have a most persistent characteristic: near 45 degrees north latitude or thereabouts the westerlies blow strongly from west to east, and at about 15 degrees the northeast trades blow, with a marked east-to-west component. The induced Ekman flow is to the right in each case, so that in both cases the water is pushed toward the region known as the Sargasso Sea, with its center at 30 degrees north. This "gathering together of the waters" leads not so much to a piling up (the surface level is only about a meter higher at the center than at the edges) as a pushing down.

(If it were not for the continental boundaries, the piling up would be much more important. Because water tends to seek a level, the piled-up water would push north above 30 degrees and south below; the pushing force, like any other force in the Northern Hemisphere, would cause a flow to its right, so that in the northern part of the ocean a strong eastward flow would develop and in the southern part a strong westward one. On the earth as it now exists, however, these east-west flows are blocked by the continents; only in the Southern Ocean, around the Antarctic Continent, is such a flow somewhat free. In the absence of the continents the oceans, like the atmosphere, would be dominated by east-west motion. As it is, only a residue of such motion is possible, and it is the pushing down rather than the piling up of water that is important.)

The downward thrust of the surface waters presses down on the layers of water underneath [*see illustration on page 42*]. For practical purposes water is incompressible, so that pushing it down from the top forces it out at the sides. It must be remembered that this body of underlying water is rotating with the earth. As it is squeezed out laterally its radius of gyration, and therefore its moment of inertia, increase, and so its rate of rotation must slow. If it slows, however, the rotation no longer "fits" the rotation of the underlying earth. There are two possible consequences: either the water can rotate with respect to the earth or it can move to a different latitude where its newly acquired rotation *will* fit. It usually does the latter. Hence a body of water whose rotation has been slowed by being squashed vertically will usually move toward the Equator, where the vertical component of the earth's rotation is smaller; on the other hand, a body whose rotation has been speeded by being bulged up to replace water that has been swept away from the surface will usually move toward the poles.

In the band of water a couple of thousand miles wide along latitude 30 degrees this indirectly wind-driven flow moves water toward the Equator. Of course the regions of the ocean closer to the poles do not become empty of water; somewhere there must be a return flow. The returning water must also attain a rotation that fits the rotation of the underlying earth. If it flows north, it must gain counterclockwise rotation (or lose clockwise rotation). It does this by running in a strong current on the westward side of the ocean, changing its rotation

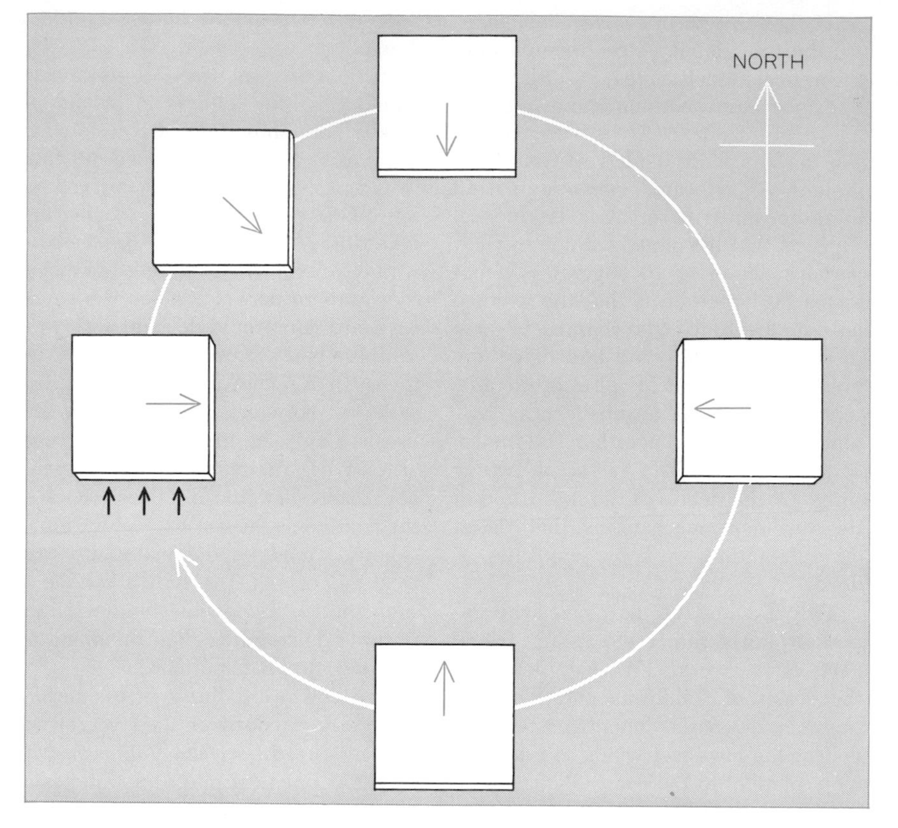

CORIOLIS ACCELERATION, caused by the earth's rotation, affects any object moving on the earth. It is directed at a right angle to the direction of motion (to the right in the Northern Hemisphere). If a frictionless slab is set in motion toward the north by a single impulse (*black arrows*), the Coriolis effect (*colored arrows*) moves the slab in a circle.

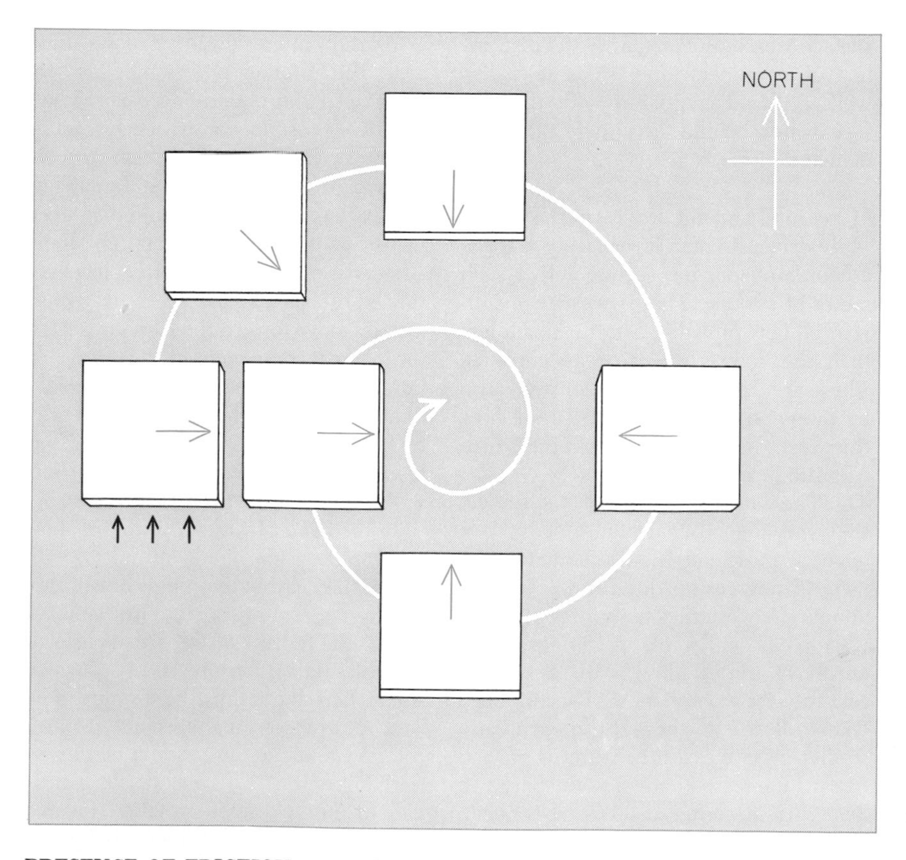

PRESENCE OF FRICTION causes the slab to slow down, spiraling in toward the center of the circle in the top illustration. A push to the north causes a spiral to the east.

by "rubbing its left shoulder" against the shore. The Gulf Stream is such a current; it is the return flow of water that was squeezed south by the wind-driven convergence of surface waters throughout the entire central North Atlantic. Most great ocean currents seem to be indirectly driven in this way.

It is worth noting that these return currents must be on the western side of the oceans (that is, off the eastern coasts of the land) in both hemispheres and regardless of whether the flow is northward or southward. The reason is that the earth's angular velocity of rotation is maximum counterclockwise to an observer looking down at the North Pole and maximum clockwise at the South Pole. Any south-flowing return current in either hemisphere must gain clockwise rotation (or lose counterclockwise rotation) if it is to fit when it arrives. It gains this rotation by friction on its right side, and so it must keep to the right—that is, to the west—of the ocean. On the other hand, a north-flowing return current must keep to the left—again the west!

This description of the general wind-driven circulation accords reasonably well with observations of the long-term characteristics of the ocean circulation. What happens on a shorter term, in response to changes of the atmospheric circulation and the wind-force pattern that results? The characteristic time constant of the Ekman layer is half a pendulum day, and there is every reason to believe this layer adjusts itself within a day or so to changes in the wind field. The indirectly driven flow is much harder to deal with. Its time constant is of the order of years, and we have no clear understanding of how it adjusts; the indirectly driven circulation may still be responding, in ways that are not clear, for years after an atmospheric change.

So far the discussion has been qualitative. To make it quantitative we need to know two things: the nature of the wind over the ocean at each time and place and the amount of force the wind exerts on the surface. Meteorologists are getting better at the first question, although there are some important gaps in our detailed information, notably in the Southern Ocean and in the South Pacific.

Investigation of the second problem, that of the quantitative relation between the wind flow and the force on the surface, is becoming a scientific discipline in its own right. Turbulent flow over a boundary is a complex phenomenon for which there is no really complete theory even in simple laboratory cases. Nevertheless, a great deal of experimental data has been collected on flows over solid surfaces, both in the laboratory and in nature, so that from an engineering point of view at least the situation is fairly well understood. The force exerted on a surface varies with the roughness of that surface and approximately with the square of the wind speed at some fixed height above it. A wind of 10 meters per second (about 20 knots, or 22 miles per hour) measured at a height of 10 meters will produce a force of some 30 tons per square kilometer on a field of mown grass or of about 70 tons per square kilometer on a ripe wheat field. On a really smooth surface such as glass the force is only about 10 tons per square kilometer.

When the wind blows over water, the whole thing is much more complicated. The roughness of the water is not a given characteristic of the surface but depends on the wind itself. Not only that, the elements that constitute the roughness—the waves—themselves move more or less in the direction of the wind. Recent evidence indicates that a large portion of the momentum transferred from the air into the water goes into waves rather than directly into making currents in the water; only as the waves break, or otherwise lose energy, does their momentum become available to generate currents or produce Ekman layers. Waves carry a substantial amount of both energy and momentum (typically about as much as is carried by the wind in a layer about one wavelength thick), and so the wave-generation process is far from negligible. So far we have no theory that accounts in detail for what we observe.

A violently wavy surface belies its appearance by acting, as far as the wind is concerned, as though it were very smooth. At 10 meters per second, recent measurements seem to agree, the force on the surface is quite a lot less than the force over mown grass and scarcely more than it is over glass; some observations in light winds of two or three meters per second indicate that the force on the wavy surface is less than it is on a surface as smooth as glass. In some way the motion of the waves seems to modify the airflow so that air slips over the surface even more freely than it would without the waves. This seems not to be the case at higher wind speeds, above about five meters per second, but the force remains strikingly low compared with that over other natural surfaces.

One serious deficiency is the fact that there are no direct observations at all in those important cases in which the wind speed is greater than about 12 meters per second and has had time and fetch (the distance over water) enough to raise substantial waves. (A wind of even 20 meters per second can raise waves eight or 10 meters high—as high as a three-story building. Making observations under such circumstances with the delicate instruments required is such a formidable task that it is little wonder none have been reported.) Some indirect studies have been made by measuring how water piles up against the shore when driven by the wind, but there are many difficulties and uncertainties in the interpretation of such measurements. Such as they are, they indicate that the apparent roughness of the surface increases somewhat under high-wind conditions, so that the force on the surface increases rather more rapidly than as the square of the wind speed.

Assuming that the force increases at least as the square of the wind speed, it is evident that high-wind conditions produce effects far more important than their frequency of occurrence would suggest. Five hours of 60-knot storm winds will put more momentum into the water than a week of 10-knot breezes. If it should be shown that for high winds the force on the surface increases appreciably more rapidly than as the square of the wind speed, then the transfer of momentum to the ocean will turn out to be dominated by what happens during

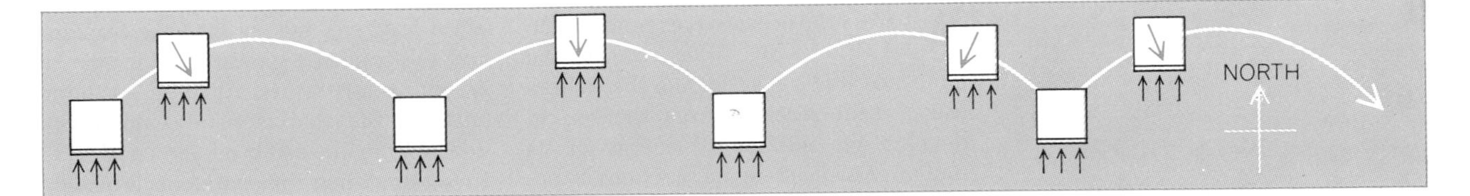

STEADY PUSH (*black arrows*), rather than a single impulse, is balanced, in the absence of friction, by the Coriolis effect (*colored arrows*), causing a series of loops. A steady force on a frictionless slab makes it move at a right angle to the direction of the force.

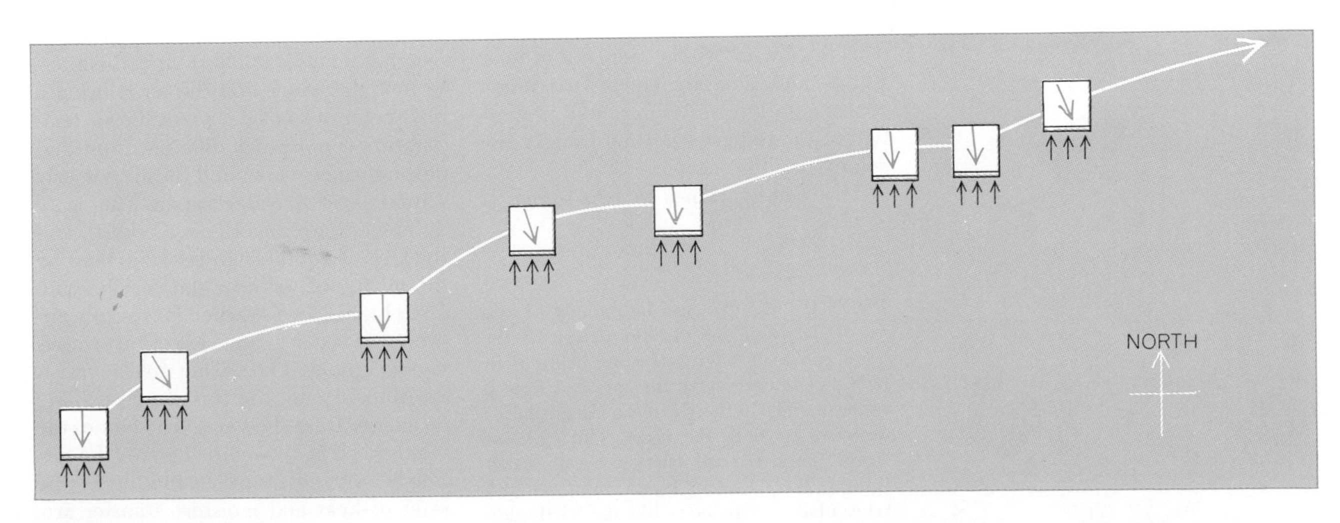

FRICTIONAL DRAG reduces the speed of the slab and thus of the Coriolis effect, which can no longer balance the driving force, and the amplitude of the loops is damped out gradually. A force toward the north therefore moves the slab toward the northeast.

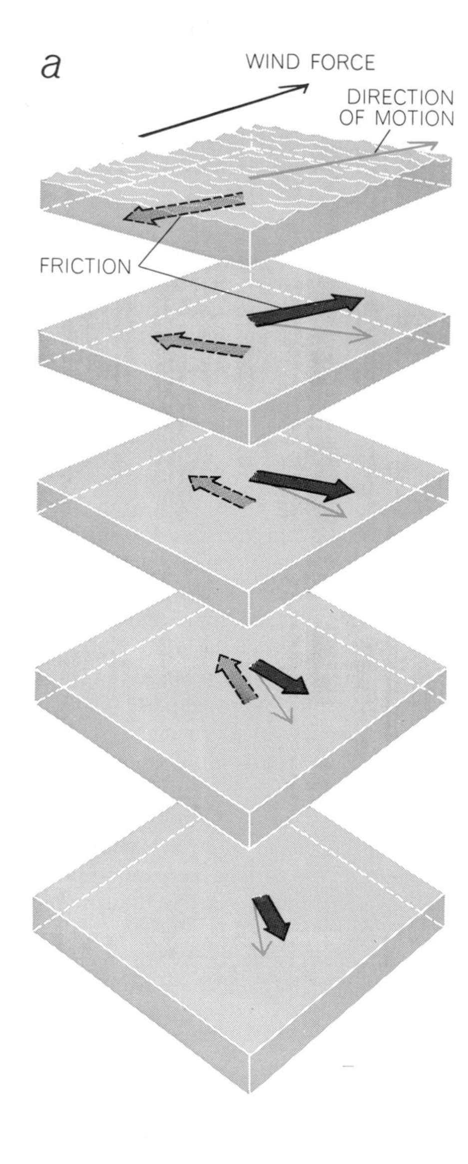

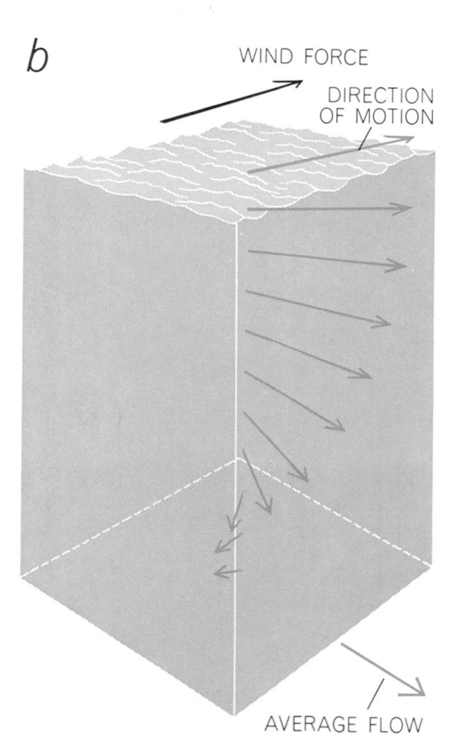

BODY OF WATER can be thought of as a set of slabs (*a*), the top one driven by the wind and each driving the one below it by friction. At each stage the speed of flow is reduced and (in the Northern Hemisphere) directed more to the right. This "Ekman spiral" persists until friction becomes negligible. The "Ekman layer" in which this takes place (*b*) behaves like the frictionless slabs in the preceding illustrations. Its average flow is at right angle to wind driving it.

the occasional storm rather than by the long-term average winds.

It is tempting to try to infer high-wind behavior from what we know about lower wind speeds. Certainly the shapes of wavy surfaces appear nearly the same notwithstanding the size of the waves—as long as one disregards waves less than about five centimeters long, which are strongly affected by surface tension. Yet, curious as it may seem, the only thing that makes one wind-driven wave field different in any fundamental way from another is surface tension, even though it directly affects only these very short waves. Indeed, surface tension is the basis of the entire Beaufort wind scale, which depends on the number and nature of whitecaps; only the fact that the surface tension is better able to hold the surface together at low wind speeds than at high speeds enables us to see a qualitative difference in the nature of the sea surface at different wind speeds [*see illustration on page 36*]. Otherwise the waves would look just the same except for a difference in scale. If we were sure we could ignore surface-tension effects, then we could calculate the force the wind would exert at high wind speeds on the basis of data obtained at lower speeds, but one should be extremely cautious about such calculations, at least until some confirming measurements are available.

Whereas the ocean seems primarily to be driven by surface forces, the atmosphere is a heat engine that makes use of heat received from the sun to develop the mechanical energy of its motion. Any heat engine functions by accepting thermal energy at a comparatively high temperature, discharging some of this thermal energy at a lower temperature and transforming the rest into mechanical energy. The atmosphere does this by absorbing energy at or near its base and radiating it away from much cooler high levels. A substantial proportion of the required heating from below comes from the ocean.

This energy comes in two forms. If cooler air blows over warmer water, there is a direct heat flow into the air. What is usually more important, though, is the evaporation of water from the surface into the air. Evaporation causes cooling, that is, it removes heat, in this case from the surface of the water. When the moisture-laden air is carried to a high altitude, where expansion under reduced atmospheric pressure causes it to cool, the water vapor may recondense into water droplets and the heat that was given up by the surface of the water is transferred to the air. If the cloud that is formed evaporates again, as it sometimes does, the atmosphere gains no net thermal energy. If the water falls to the surface as rain or snow, however, then there has been a net gain and it is available to drive the atmosphere. Typically the heat gained by the atmosphere through this evaporation-condensation process is considerably more than the heat gained by direct thermal transfer through the surface.

Virtually everywhere on the surface of the ocean, averaged over a year, the ocean is a net source of heat to the atmosphere. In some areas the effect is much more marked than in others. For example, some of the most important return currents, such as the Gulf Stream in the western Atlantic and the Kuroshio Current in the western Pacific off Japan, contain very warm water and move so rapidly that the water has not cooled even when it arrives far north of the tropical and subtropical regions where it gained its high temperature. At these northern latitudes the characteristic wind direction is from the west, off the continent. In winter, when the continents are cold, air blowing from them onto this abnormally warm water receives great quantities of heat, both by direct thermal transfer and in the form of water vapor.

The transfer of heat and water vapor depends on a disequilibrium at the interface of the water and the air. Within a millimeter or so of the water the air temperature is not much different from that of the surface water, and the air is nearly saturated with water vapor. The small differences are nevertheless crucial, and the lack of equilibrium is maintained by the mixing of air near the surface with air at higher levels, which is typically appreciably cooler and lower in water-vapor content. The mixing mechanism is a turbulent one, the turbulence gaining its energy from the wind. The highe the wind speed is, the more vigorous the turbulence is and therefore the higher the rates of heat and moisture transfer are. These rates tend to increase linearly with the wind speed, but even less is known about the details of this phenomenon than about the wind force on water. One source of complication is the fact that, as I mentioned above, the wind-to-water transfer of momentum is effected partly by wave-generation mechanisms. When the wind makes waves, it must transfer not only momentum but also important amounts of energy—energy that is not available to provide the turbulence needed to produce the mixing that would effect the transfer of heat and water vapor.

At fairly high wind speeds another phenomenon arises that may be of considerable importance. I mentioned that when surface tension is no longer able to hold the water surface together at high wind speeds, spray droplets blow off the top of the waves. Some of these drops fall back to the surface, but others evaporate and in doing so supply water vapor to the air. They have another important role: The tiny residues of salt that are left over when the droplets of seawater evaporate are small enough and light enough to be carried upward by the turbulent air. They act as nuclei on which condensation may take place, and so they play a role in returning to the atmosphere the heat that is lost in the evaporation process.

The ocean's great effect on climate is illustrated by a comparison of the temperature ranges in three Canadian cities, all at about the same latitude but with very different climates [*see top illustration at right*]. Victoria is a port on the southern tip of Vancouver Island, on the eastern shore of the Pacific Ocean; Winnipeg is in the middle of the North American land mass; St. John's is on the island of Newfoundland, jutting into the western Atlantic. The most striking climatic difference among the three is the enormous temperature range at Winnipeg compared with the two coastal cities. The range at St. John's, although much less, is still greater than at Victoria, probably because at St. John's the air usually blows from the direction of the continent and the effect of the water is somewhat less dominant than at Victoria, which typically receives its air directly from the ocean. St. John's is colder than Victoria because it is surrounded by cold water of the Labrador Current.

	VICTORIA	WINNIPEG	ST. JOHN'S
MEAN JULY MAXIMUM	68	80.1	68.9
MEAN JANUARY MINIMUM	35.6	−8.1	18.5

MODERATING EFFECT of the ocean on climate is illustrated by a comparison of the temperature range (in degrees Fahrenheit) at three Canadian cities. The range between minimums and maximums is much greater at Winnipeg than at coastal Victoria or St. John's.

The influence of the ocean is associated with its enormous thermal capacity. Every day, on the average, the earth absorbs from the sun and reradiates into space enough heat to raise the temperature of the entire atmosphere nearly two degrees Celsius (three degrees Fahrenheit). Yet the thermal capacity of the atmosphere is equivalent to that of only the top three meters of the ocean, or only a few percent of the 100 meters or so of ocean water that is heated in summer and cooled in winter. (The great bulk of ocean water, more than 95 percent of it, is so deep that surface heating does not penetrate, and its temperature is independent of season.) If the ocean lost its entire heat supply for a day but continued to give up heat in a normal way, the temperature of the upper 100 meters would drop by only about a tenth of a degree.

Compared with the land, the ocean heats slowly in summer and cools slowly in winter, so that its temperature is much less variable. Moreover, because air has so much less thermal capacity, when it blows over water it tends to come to the water temperature rather than vice versa. For these reasons maritime climates are much more equable than continental ones.

Although the ocean affects the atmosphere's temperature more than the atmosphere affects the ocean's, the ocean is cooled when it gives up heat to the atmosphere. The density of ocean water is controlled by two factors, temperature and salinity, and evaporative cooling tends to make the water denser by affecting both factors: it lowers the temperature and, since evaporation removes water but comparatively little salt, it also increases the salinity. If surface water becomes denser than the water underlying it, vigorous vertical convective mixing sets in. In a few places in the ocean the cooling at the surface can be so intense that the water will sink and mix to great depths, sometimes right to the bottom. Such occurrences are rare both in space and in time, but once cold water has reached great depths it is heated from above very slowly, and so it tends to stay deep for a long time with little change in temperature; there is some evidence of water that has remained cold and deep in the ocean for more than

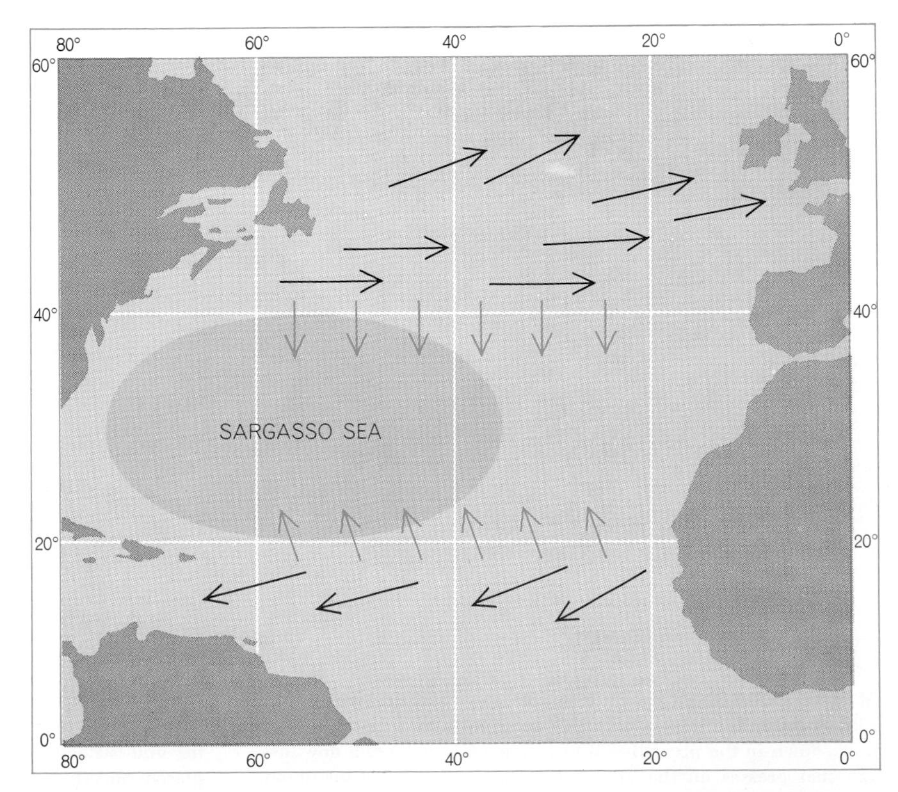

PREVAILING-WIND DIRECTIONS (*black arrows*) and the resulting Ekman-layer flows (*colored arrows*) in the North Atlantic drive water into the region of the Sargasso Sea.

1,000 years. With this length of residence not much of the heavy, cold water needs to be produced every year for it to constitute, as it does, the bulk of the ocean water.

The sinking of water cooled at the surface is one aspect of another important feature of the ocean: the flow induced by differences in density, which is to say the flow induced principally by temperature and salt content. This thermohaline circulation of the ocean is in addition to the wind-driven circulation discussed earlier.

In its thermohaline aspects the ocean itself acts as a heat engine, although it is far less efficient than the atmosphere. Roughly speaking, the ocean can be divided into two layers: a rather thin upper one whose density is comparatively low because it is warmed by the sun, and a thick lower one, a fraction of a percent denser and composed of water only a few degrees above the freezing point that has flowed in from those few areas where it is occasionally created. Somewhere—either distributed over the ocean or perhaps only locally near the shore and in other special places—there is mixing between these layers. The mixing is of such a nature that the cold deep water is mixed into the warm upper water rather than the other way around, that is, the cold water is added to the warm from the bottom [*see upper illustration on page 43*]. Once the water is in the upper layer its motion is largely governed by the wind-driven circulation, although density differences still play a role. In one way or another some of this surface water arrives at a location and time at which it is cooled sufficiently to sink again and thus complete the circulation.

This picture can be rounded out by consideration of the effects of the earth's rotation, which are in some ways quite surprising. The deep water that mixes into the upper layer must have a net upward motion. (The motion is far too

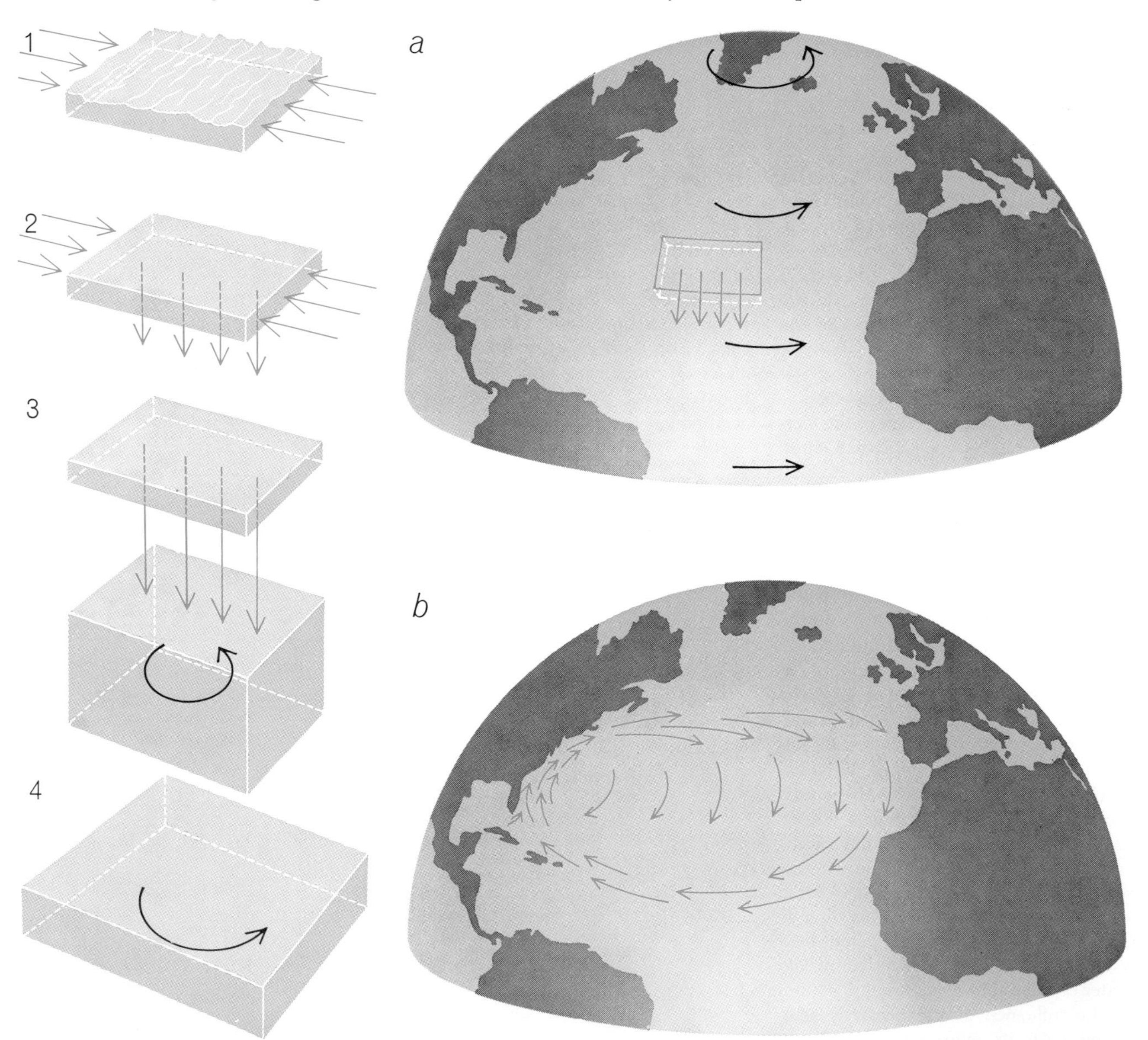

MAJOR CURRENTS are generated by a mechanism involving the Ekman-layer flow and the earth's rotation. The Ekman-layer inflow shown in the preceding illustration (*1*) produces a downflow (*2*) that presses on the underlying water (*3*), squeezing it outward (*4*) and thus reducing its rate of rotation (*curved black arrow*). There is a rate of rotation appropriate to each latitude, and when the rotation of a body of water is reduced, it must move (*colored arrows*) toward the Equator until its new rotation "fits" (*a*). For this reason there is a general movement of water from the mid-latitudes toward the Equator (*b*). That water must be replaced, and the water replacing it must have the proper rotation. This is accomplished by a return flow that runs along the western shore of the ocean, changing its rotation by "rubbing its shoulder" against the coast, as the Gulf Stream does in the Atlantic Ocean.

small to measure, only a few meters per year, but we infer its existence indirectly.) To make possible this upward flow there must be a compensating lateral inflow. Remember that on the rotating earth this lateral inflow results in an increase in speed of rotation, and so for it to continue to fit the rotation of the underlying earth the water must move toward the nearest pole; it must flow away from the equatorial regions. Yet the source of this cold deep water is at high latitudes! How does it get near the Equator to supply the demand?

The answer is similar to the one for the wind-driven circulation: The cold water must flow in a western boundary current, in order to gain the proper rotation as it moves [*see lower illustration on next page*]. There is some direct evidence of the inferred concentrated western boundary current in the North Atlantic, and there are hints of it in the South Pacific, but most of the rest is based on inference. There seems to be no source of cold deep water in the North Pacific, so that the deep water there must come from Antarctic regions.

We have seen that the atmosphere drives the ocean and that heat supplied from the ocean is largely instrumental in releasing energy for the atmo-

THERMOHALINE CIRCULATION, the flow induced by density rather than wind action, begins with the creation of dense, cold water that sinks to great depths. Under certain conditions this deep water mixes upward into the warm surface layer (*color*) as shown here. As it moves up, this water increases its rotation and so it must move generally from the Equator toward the two poles.

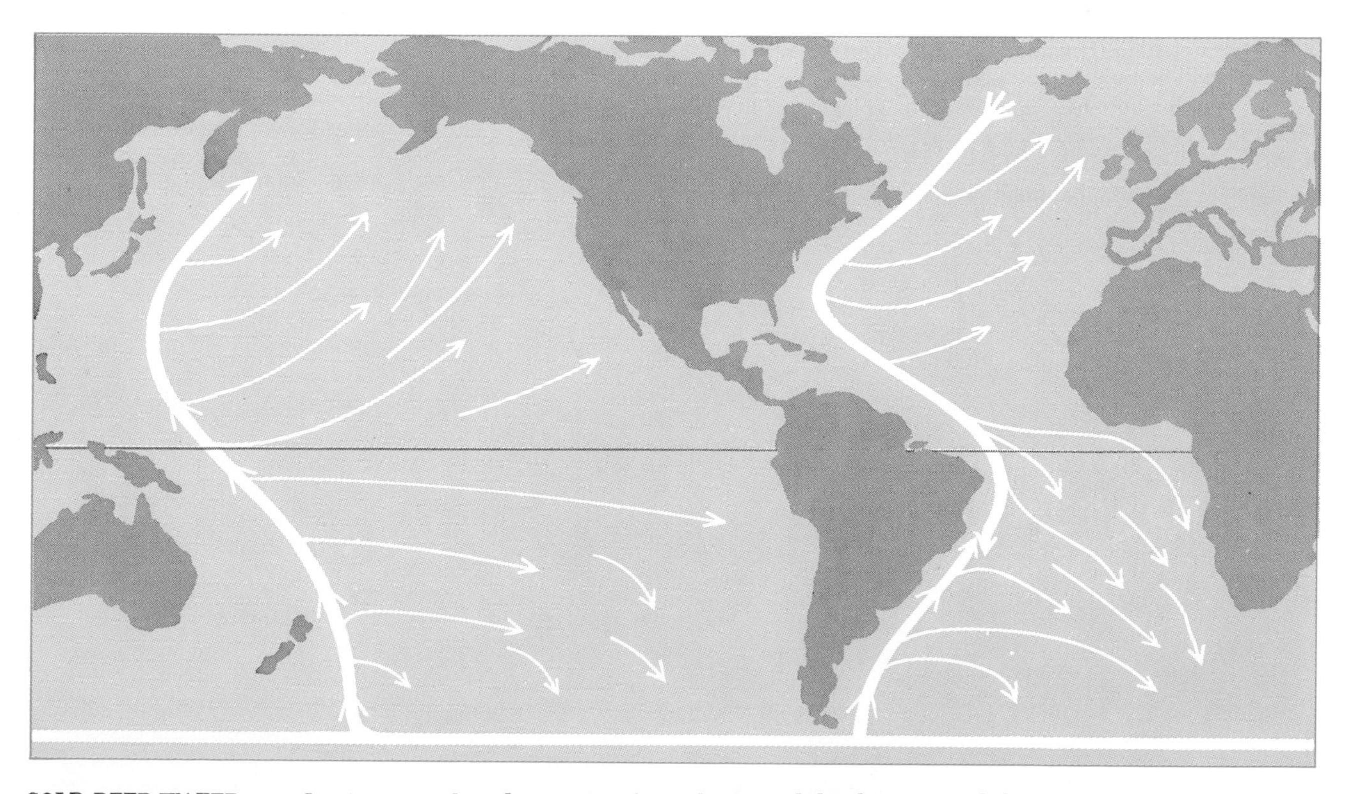

COLD DEEP WATER must flow in western boundary currents in order to arrive at the Equator and thus be able to move poleward as it mixes upward. Details of deep circulation are still almost unknown and the chart is intended only to suggest its approximate directions. There is some evidence of such boundary currents in the North Atlantic and there are some hints in the South Pacific.

INFRARED IMAGERY delineates the temperature structure of bodies of water and is used to study currents and wave patterns. This image of the shoulder of the Gulf Stream is from the Antisubmarine Warfare Environmental Prediction Services Project of the Naval Oceanographic Office. It was made by an airborne scanner at low altitude and shows several hundred yards of the boundary between the warm current and cooler water off Cape Hatteras. The range is from about 13 to 21 degrees Celsius, with the warm water darker.

sphere. There is a great deal of feedback between the two systems. The atmospheric patterns determine the oceanic flows, which in turn influence where—and how much—heat is released to the atmosphere. Further, the atmospheric flow systems determine how much cloud cover there will be over certain parts of the ocean and therefore how much—and where—the ocean will be heated. The system is not a particularly stable one. Every locality has its abnormally cold or mild winters and its abnormally wet or dry summers. The persistence of such anomalies over several months almost certainly involves the ocean, because the characteristic time constants of purely atmospheric phenomena are simply too short. Longer-term climatological variations such as the "little ice age" that lasted for about 40 years near the beginning of the 19th century are even more likely to have involved changes in the ocean's circulation. And then there are the more dramatic events of the great Pleistocene glaciations.

There are any number of theories for these events and, since experts disagree, it is incumbent on the rest of us to refrain from dogmatic statements. Nevertheless, it does not seem impossible that the ocean-atmosphere system has a number of more or less stable configurations. That is, there may be a number of different patterns in which the atmosphere can drive the ocean in such a way that the ocean releases heat to the atmosphere in the right quantity and at the right places to allow the pattern to continue. Of course the atmosphere is extremely turbulent, so that its equilibrium is constantly being disturbed. If the system is stable, then forces must come into play that tend to restore conditions after each such disturbance. If there are a number of different stable patterns, however, it is possible that a particularly large disturbance might tip the system from one stable condition to another.

One can imagine a gambler's die lying on the floor of a truck running over a rough road; the die is stable on any of its six faces, so that in spite of bouncing and vibration the same face usually remains up—until a particularly big bump jars it so that it lands with a different face up, whereupon it is stable in its new position. It seems not at all impossible that the ocean-atmosphere system behaves something like this. Perhaps in recent years we have been bouncing along with, say, a four showing. Perhaps 200 years ago the die flipped over to three for a moment, then flipped back to four. It could one day jounce over to a snake eye and bring a new ice age!

4

OCEAN WAVES

WILLARD BASCOM

August 1959

Man is by nature a wave-watcher. On a ship he finds himself staring vacantly at the constant swell that flexes its muscles just under the sea's surface; on an island he will spend hours leaning against a palm tree absently watching the rhythmic breakers on the beach. He would like to learn the ways of the waves merely by watching them, but he cannot, because they set him dreaming. Try to count a hundred waves sometime and see.

Waves are not always so hypnotic. Sometimes they fill us with terror, for they can be among the most destructive forces in nature, rising up and overwhelming a ship at sea or destroying a town on the shore. Usually we think of waves as being caused by the wind, because these waves are by far the most common. But the most destructive waves are generated by earthquakes and undersea landslides. Other ocean waves, such as those caused by the gravitational attraction of the sun and the moon and by changes in barometric pressure, are much more subtle, often being imperceptible to the eye. Even such passive elements as the contour of the sea bottom, the slope of the beach and the curve of the shoreline play their parts in wave action. A wave becomes a breaker, for example, because as it advances into increasingly shallow water it rises higher and higher until the wave front grows too steep and topples forward into foam and turbulence. Although the causes of this beautiful spectacle are fairly well understood, we cannot say the same of many other aspects of wave activity. The questions asked by the wave-watcher are nonetheless being answered by intensive studies of the sea and by the examination of waves in large experimental tanks. The new knowledge has made it possible to measure the power and to forecast the actions of waves for the welfare of those who live and work on the sea and along its shores.

Toss a pebble into a pond and watch the even train of waves go out. Waves at sea do not look at all like this. They are confused and irregular, with rough diamond-shaped hillocks and crooked valleys. They are so hopelessly complex that 2,000 years of observation by seafarers produced no explanation beyond the obvious one that waves are somehow raised by the wind. The description of the sea surface remained in the province of the poet who found it "troubled, unsettled, restless. Purring with ripples under the caress of a breeze, flying into scattered billows before the torment of a storm and flung as raging surf against the land; heaving with tides breathed by a sleeping giant."

The motions of the oceans were too complex for intuitive understanding. The components had to be sorted out and dealt with one at a time. So the first theoreticians cautiously permitted a perfect train of waves, each exactly alike, to travel endlessly across an infinite ocean. This was an abstraction, but it could at least be dealt with mathematically.

Early observers noticed that passing waves move floating objects back and forth and up and down, but do not transport them horizontally for any great distance. From the motion of seaweeds the motion of the water particles could be deduced. But it was not until 1802 that Franz Gerstner of Germany constructed the first wave theory. He showed that water particles in a wave move in circular orbits. That is, water at the crest moves horizontally in the direction the wave is going, while in the trough it moves in the opposite direction. Thus each water particle at the surface traces a circular orbit, the diameter of which is equal to the height of the wave [*see illustration on next page*]. As each wave passes, the water returns almost to its original position. Gerstner observed that the surface trace of a wave is approximately a trochoid: the curve described by a point on a circle as it rolls along the underside of a line. His work was amplified by Sir George Airy later in the 19th century, by Horace Lamb of England in the present century, and by others.

The first wave experimentalists were Ernst and Wilhelm Weber of Germany, who in 1825 published a book on studies employing a wave tank they had invented. Their tank was five feet long, a foot deep and an inch wide, and it had glass sides. To make waves in the tank they sucked up some of the fluid through a tube at one end of it and allowed the fluid to drop back. Since the Weber brothers experimented not only with water and mercury but also with brandy, their persistence in the face of temptation has been an inspiration to all subsequent investigators. They discovered that waves are reflected without loss of energy, and they determined the shape of the wave surface by quickly plunging in and withdrawing a chalk-dusted slate. By watching particles suspended in the water they confirmed the theory that water particles move in a circular orbit, the size of which diminishes with depth. At the bottom, they observed, these orbits tend to be flattened.

As increasingly bolder workers contributed ideas in the 20th century, many of the complexities of natural waves found their way into equations. However, these gave only a crude, empirical answer to the question of how wind energy is transferred to waves. The necessity for the prediction of waves and surf for amphibious operations in World War II attracted the attention of Harald U.

Sverdrup and Walter Munk of the Scripps Institution of Oceanography. As a result of their wartime studies of the interaction of winds and waves they were the first investigators to give a reasonably complete quantitative description of how wind gets energy into the waves. With this description wave studies seemed to come of age, and a new era of research was launched.

Let us follow waves as they are generated at sea by the wind, travel for perhaps thousands of miles across the ocean and finally break against the shore. The effectiveness of the wind in making waves is due to three factors: its average velocity, the length of time it blows and the extent of the open water across which it blows (called the fetch).

Waves and the Wind

Waves start up when the frictional drag of a breeze on a calm sea creates ripples. As the wind continues to blow, the steep side of each ripple presents a surface against which the moving air can press directly. Because winds are by nature turbulent and gusty, wavelets of all sizes are at first created. The small, steep ones break, forming whitecaps, releasing some of their energy in turbulence and possibly contributing part of it to larger waves that overtake them. Thus as energy is added by the wind the smaller waves continually give way to larger ones which can store the energy better. But more small waves are continually formed, and in the zone where the wind moves faster than the waves there is a wide spectrum of wavelengths. This is the generating area, and in a large storm it may cover thousands of square miles. If storm winds apply more force than a wave can accept, the crest is merely steepened and blown off, forming a breaking wave at sea. This happens when the wave crest becomes a wedge of less than 120 degrees and the height of the wave is about a seventh of its length. Thus a long wave can accept more energy from the wind and rise much higher than a short wave passing under the same wind. When the wind produces waves of many lengths, the shortest ones reach maximum height quickly and then are destroyed, while the longer ones continue to grow.

A simple, regular wave-train can be described by its period (the time it takes two successive crests to pass a point), by its wavelength (the distance between crests) and by its height (the vertical distance between a trough and a succeeding crest). Usually, however, there are several trains of waves with different wavelengths and directions present at the same time, and their intersection creates a random or a short-crested diamond pattern. Under these conditions no meaningful dimensions can be assigned to wave period and length. Height, however, is important, at least to ships; several crests may coincide and add their heights to produce a very large wave. Fortunately crests are much more likely to coincide with troughs and be canceled out. There is no reason to believe that the seventh wave, or some other arbitrarily numbered wave, will be higher than the rest; that is a myth of the sea.

Since waves in a sea are so infinitely variable, statistical methods must be employed to analyze and describe them. A simple way to describe height, for example, is to speak of significant height—the average height of the highest third of the waves. Another method, devised in 1952 by Willard J. Pierson, Jr., of New York University, employs equations like those that describe random noise in information theory to predict the behavior of ocean waves. Pierson superposes the regular wave-trains of classical theory in such a way as to obtain a mathematically irregular pattern. The result is most conveniently described in terms of energy spectra. This scheme assigns a value for the square of the wave height to each

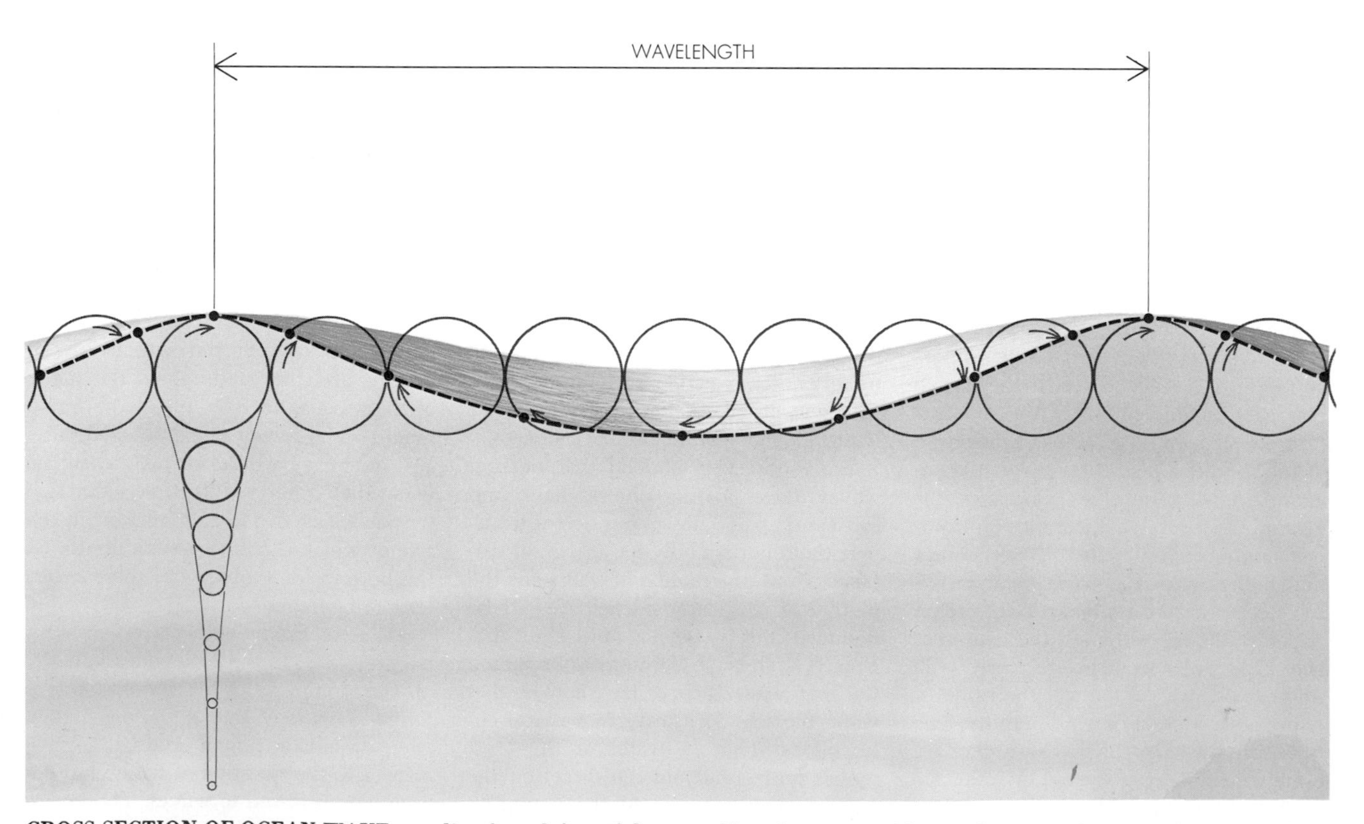

CROSS SECTION OF OCEAN WAVE traveling from left to right shows wavelength as distance between successive crests. The time it takes two crests to pass a point is the wave period. Circles are orbits of water particles in the wave. At the surface their diameter equals the wave height. At a depth of half the wavelength (*left*), orbital diameter is only 4 per cent of that at surface.

frequency and direction. Then, by determining the portion of the spectrum in which most of the energy is concentrated, the average periods and lengths can be obtained for use in wave forecasting.

Over a long fetch, and under a strong, steady wind, the longer waves predominate. It is in such areas of sea that the largest wind waves have been recorded. The height of the waves in a train does not, however, bear any simple relationship to their other two dimensions: the period and the wavelength. The mariner's rule of thumb relates wave height to wind velocity and says that the height ordinarily will not be greater than half the wind speed. This means that an 80-mile-per-hour hurricane would produce waves about 40 feet high.

The question of just how large individual waves at sea can actually be is still unsettled, because observations are difficult to make and substantiate from shipboard in the midst of a violent storm. Vaughan Cornish of England spent half a century collecting data on waves, and concluded that storm waves over 45 feet high are rather common. Much higher waves have been fairly well authenticated on at least two occasions.

In October, 1921, Captain Wilson of the 12,000-ton S.S. *Ascanius* reported an extended storm in which the recording barometer went off the low end of the scale. When the ship was in a trough on an even keel, his observation post on the ship was 60 feet above the water level, and he was certain that some of the waves that obscured the horizon were at least 10 feet higher than he was, accounting for a total height of 70 feet or more. Commodore Hayes of the S.S. *Majestic* reported in February, 1923, that his ship had experienced winds of hurricane force and waves of 80 feet in height. Cornish examined the ship, closely interrogated the officers and concluded that waves 60 to 90 feet high, with an average height of 75 feet, had indeed been witnessed.

A wave reported by Lieutenant Commander R. P. Whitemarsh in the *Proceedings of the U. S. Naval Institute* tops all others. On February 7, 1933, the *U.S.S. Ramapo,* a Navy tanker 478 feet long, was en route from Manila to San Diego when it encountered "a disturbance that was not localized like a typhoon . . . but permitted an unobstructed fetch of thousands of miles." The barometer fell to 29.29 inches and the wind gradually rose from 30 to 60 knots over several days. "We were running directly downwind and with the sea. It would have been disastrous to have steamed on any other course." From among a number of separately determined observations, that of the watch officer on the bridge was selected as the most accurate. He declared that he "saw seas astern at a level above the mainmast crow's-nest and at the moment of observation the horizon was hidden from view by the waves approaching the stern." On working out the geometry of the situation from the ship's plan, Whitemarsh found that this wave must have been at least 112 feet high [*see illustration at the bottom of the next two pages*]. The period of these waves was clocked at 14.8 seconds and their velocity at 55 knots.

As waves move out from under the winds that raise them, their character changes. The crests become lower and more rounded, the form more symmetrical, and they move in trains of similar period and height. They are now called swell, or sometimes ground swell, and in this form they can travel for thousands of miles to distant shores. Happily for mathematicians, swell coincides much more closely with classical theory than do the waves in a rough sea, and this renews their faith in the basic equations.

Curiously enough, although each wave moves forward with a velocity

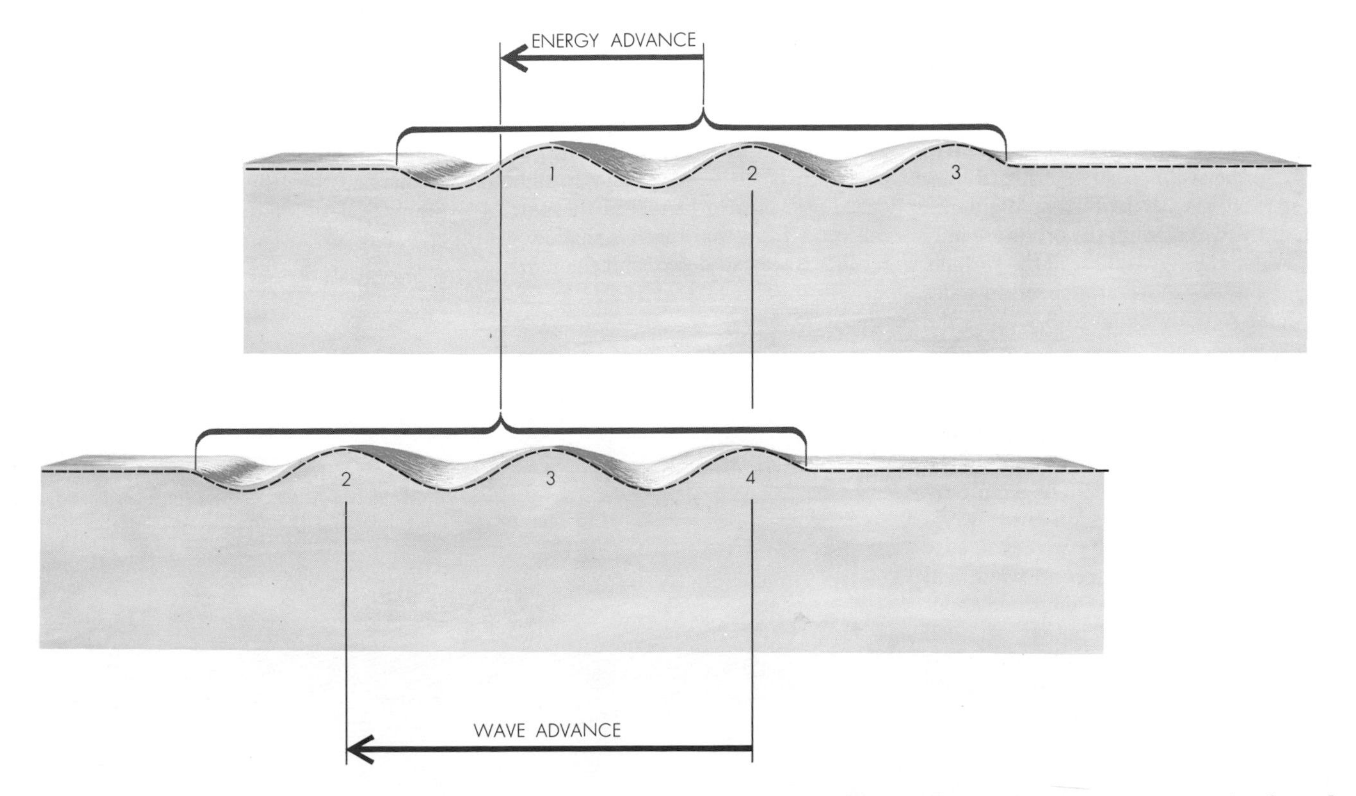

MOVING TRAIN OF WAVES advances at only half the speed of its individual waves. At top is a wave train in its first position. At bottom the train, and its energy, have moved only half as far as wave 2 has. Meanwhile wave 1 has died, but wave 4 has formed at the rear of the train to replace it. Waves arriving at shore are thus remote descendants of waves originally generated.

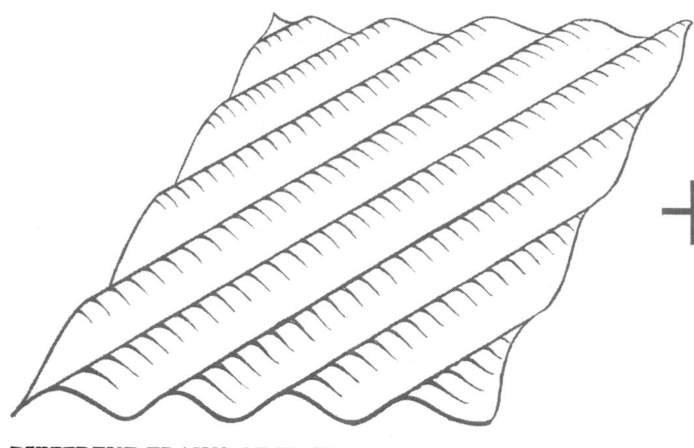

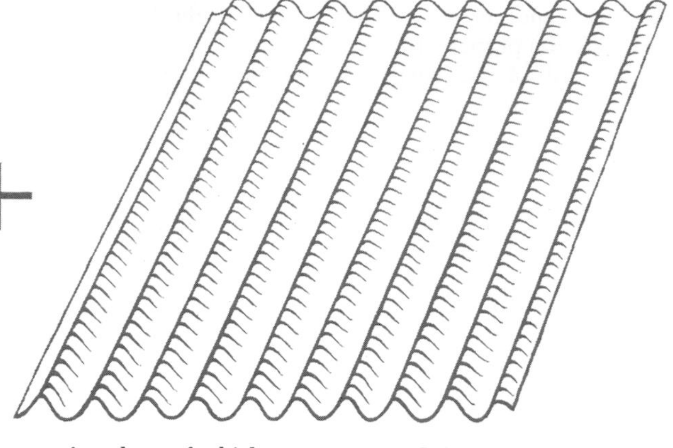

DIFFERENT TRAINS OF WAVES, caused by winds of different directions and strengths, make up the surface of a "sea." The various trains, three of which are represented diagrammatically here, have a wide spectrum of wavelengths, heights and directions. When

that corresponds to its length, the energy of the group moves with a velocity only half that of the individual waves. This is because the waves at the front of a group lose energy to those behind, and gradually disappear while new waves form at the rear of the group. Thus the composition of the group continually changes, and the swells at a distance are but remote descendants of the waves created in the storm [*see illustration on preceding page*]. One can measure the period at the shore and obtain from this a correct value for the wave velocity; however, the energy of the wave train traveled from the storm at only half that speed.

Waves in a swell in the open ocean are called surface waves, which are defined as those moving in water deeper than half the wavelength. Here the bottom has little or no effect on the waves because the water-particle orbits diminish so rapidly with depth that at a depth of half the wavelength the orbits are only 4 per cent as large as those at the surface. Surface waves move at a speed in miles per hour roughly equal to 3.5 times the period in seconds. Thus a wave with a period of 10 seconds will travel about 35 miles per hour. This is the average period of the swell reaching U. S. shores, the period being somewhat longer in the Pacific than the Atlantic. The simple relationship between period and wavelength (length$=5.12T^2$) makes it easy to calculate that a 10-second wave will have a deep-water wavelength of about 512 feet. The longest period of swell ever reported is 22.5 seconds, which corresponds to a wavelength of around 2,600 feet and a speed of 78 miles per hour.

Waves and the Shore

As the waves approach shore they reach water shallower than half their wavelength. Here their velocity is controlled by the depth of the water, and they are now called shallow-water waves. Wavelength decreases, height increases and speed is reduced; only the period is unchanged. The shallow bottom greatly modifies the waves. First, it refracts them, that is, it bends the wave fronts to approximate the shape of the underwater contours. Second, when the water becomes critically shallow, the waves break [*see illustration on page 54*].

Even the most casual observer soon notices the process of refraction. He sees that the larger waves always come in nearly parallel to the shoreline, even though a little way out at sea they seem to be approaching at an angle. This is the result of wave refraction, and it has considerable geological importance because its effect is to distribute wave energy in such a way as to straighten coastlines. Near a headland the part of the wave front that reaches shallow water first is slowed down, and the parts of it in relatively deep water continue to move rapidly. The wave thus bends to converge on the headland from all sides. As it does, the energy is concentrated in less length of crest; consequently the height of the crest is increased. This accounts for the old sailors' saying: "The points draw the waves."

Another segment of the same swell will enter an embayment and the wave front will become elongated so that the height of the waves at any point along the shore is correspondingly low. This is why bays make quiet anchorages and exposed promontories are subject to wave battering and erosion—all by the same waves. One can deal quantitatively with this characteristic of waves and can plot the advance of any wave across waters of known depths. Engineers planning shoreline structures such as jetties or piers customarily draw refraction diagrams to determine in advance the effect of waves of various periods and direction. These diagrams show successive

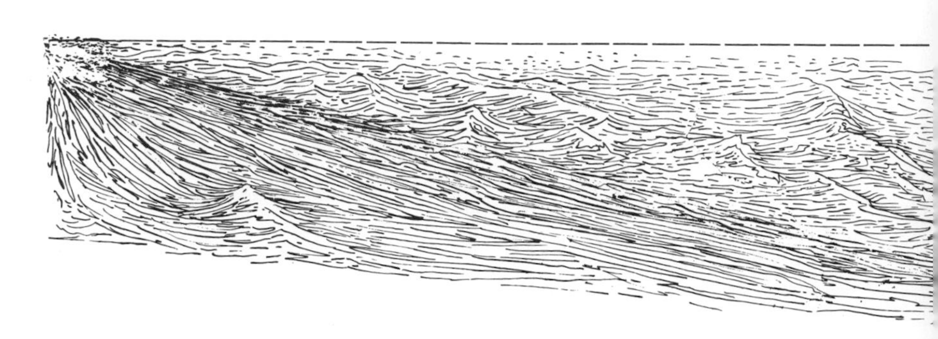

WAVE 112 FEET HIGH, possibly the largest ever measured in the open sea, was encountered in the Pacific in 1933 by the *U.S.S. Ramapo*, a Navy tanker. This diagram shows

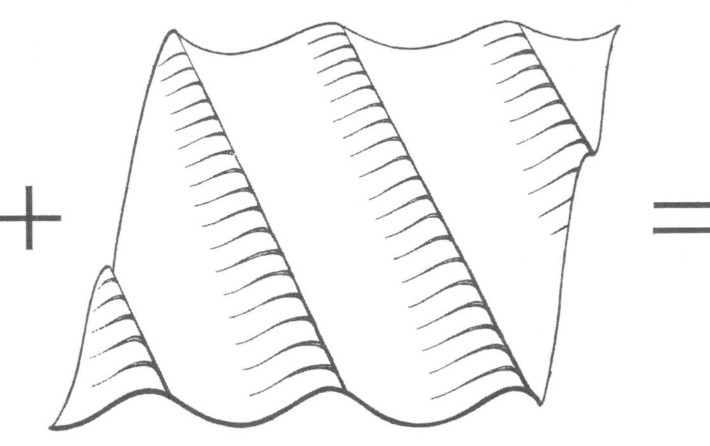
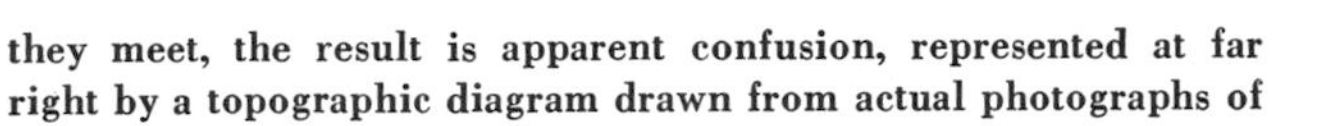
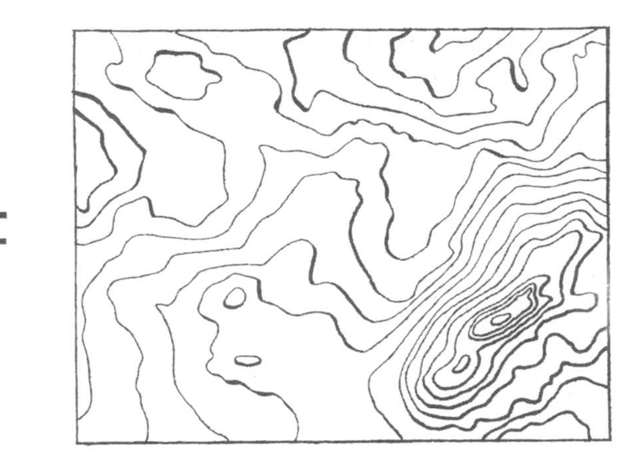

they meet, the result is apparent confusion, represented at far right by a topographic diagram drawn from actual photographs of the sea surface. The pattern becomes so complex that statistical methods must be used to analyze the waves and predict their height.

positions of the wave front, partitioned by orthogonals into zones representing equal wave energy [*see illustration on next page*]. The ratio of the distances between such zones out at sea and at the shore is the refraction coefficient, a convenient means of comparing energy relationships.

Refraction studies must take into account surprisingly small underwater irregularities. For example, after the Long Beach, Calif., breakwater had withstood wave attack for years, a short segment of it was suddenly wrecked by waves from a moderate storm in 1930. The breakwater was repaired, but in 1939 waves breached it again. A refraction study by Paul Horrer of the Scripps Institution of Oceanography revealed that long-period swell from exactly 165 degrees (south-southeast), which was present on only these two occasions, had been focused at the breach by a small hump on the bottom, 250 feet deep and more than seven miles out at sea. The hump had acted as a lens to increase the wave heights to 3.5 times average at the point of damage.

During World War II it was necessary to determine the depth of water off enemy-held beaches against which amphibious landings were planned. Our scientists reversed the normal procedure for refraction studies; by analyzing a carefully timed series of aerial photographs for the changes in length (or velocity) and direction of waves approaching a beach, they were able to map the underwater topography.

The final transformation of normal swell by shoal or shallow water into a breaker is an exciting step. The waves have been shortened and steepened in the final approach because the bottom has squeezed the circular orbital motion of the particles into a tilted ellipse; the particle velocity in the crest increases and the waves peak up as they rush landward. Finally the front of the crest is unsupported and it collapses into the trough. The wave has broken and the orbits exist no more. The result is surf.

If the water continues to get shallower, the broken wave becomes a foam line, a turbulent mass of aerated water. However, if the broken wave passes into deeper water, as it does after breaking on a bar, it can form again with a lesser height that represents the loss of energy in breaking. Then it too will break as it moves into a depth critical to its new height.

The depth of water beneath a breaker, measured down from the still-water level, is at the moment of breaking about 1.3 times the height of the breaker. To estimate the height of a breaker even though it is well offshore, one walks from the top of the beach down until the crest of the breaking wave is seen aligned with the horizon. The vertical distance between the eye and the lowest point to which the water retreats on the face of the beach is then equal to the height of the wave.

The steepness of the bottom influences

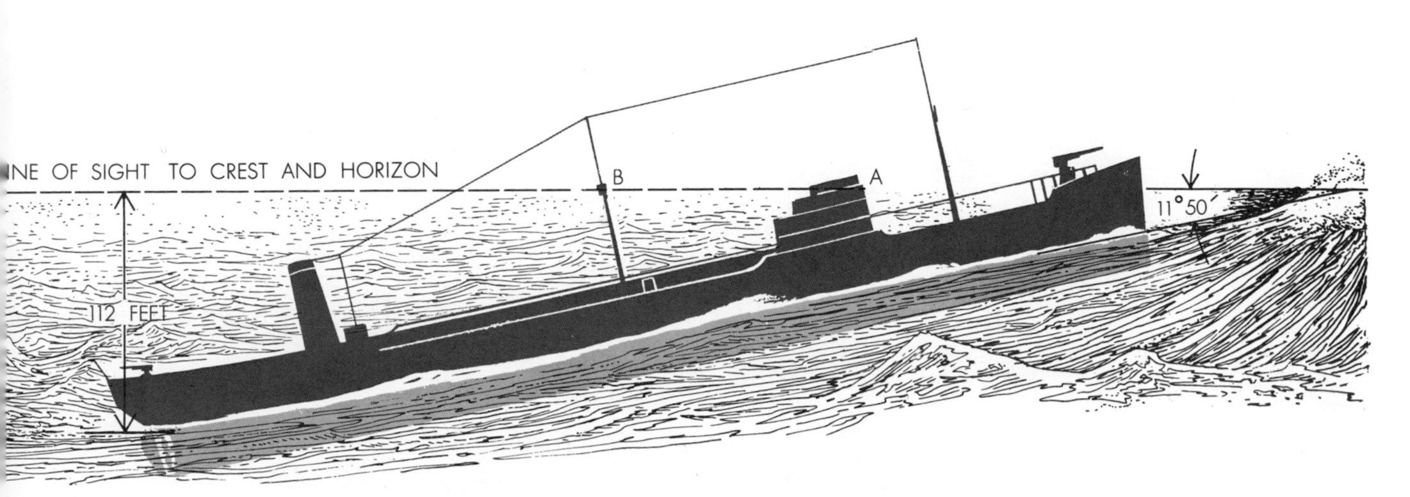

how the great wave was measured. An observer at A on the bridge was looking toward the stern and saw the crow's-nest at B in his line of sight to crest of wave, which had just come in line with horizon. From geometry of situation, wave height was calculated.

the character of the breakers. When a large swell is forced by an abrupt underwater slope to give up its energy rapidly, it forms plunging breakers—violent waves that curl far over, flinging the crest into the trough ahead. Sometimes, the air trapped by the collapsing wave is compressed and explodes with a great roar in a geyser of water [*see illustration on page 52*]. However, if the bottom slope is long and gentle, as at Waikiki in Hawaii, the crest forms a spilling breaker, a line of foam that tumbles down the front of the partly broken wave as it continues to move shoreward.

Since waves are a very effective mechanism for transporting energy against a coast, they are also effective in doing great damage. Captain D. D. Gaillard of the U. S. Army Corps of Engineers devoted his career to studying the forces of waves on engineering structures and in 1904 reported some remarkable examples of their destructive power. At Cherbourg, France, a breakwater was composed of large rocks and capped with a wall 20 feet high. Storm waves hurled 7,000-pound stones over the wall and moved 65-ton concrete blocks 60 feet. At Tillamook Rock Light off the Oregon coast, where severe storms are commonplace, a heavy steel grating now protects the lighthouse beacon, which is 139 feet above low water. This is necessary because rocks hurled up by the waves have broken the beacon several times. On one occasion a rock weighing 135 pounds was thrown well above the lighthouse-keeper's house, the floor of which is 91 feet above the water, and fell back through the roof to wreck the interior.

At Wick, Scotland, the end of the breakwater was capped by an 800-ton block of concrete that was secured to the foundation by iron rods 3.5 inches in diameter. In a great storm in 1872 the designer of the breakwater watched in amazement from a nearby cliff as both cap and foundation, weighing a total of 1,350 tons, were removed as a unit and deposited in the water that the wall was supposed to protect. He rebuilt the structure and added a larger cap weighing 2,600 tons, which was treated similarly by a storm a few years later. There is no record of whether he kept his job

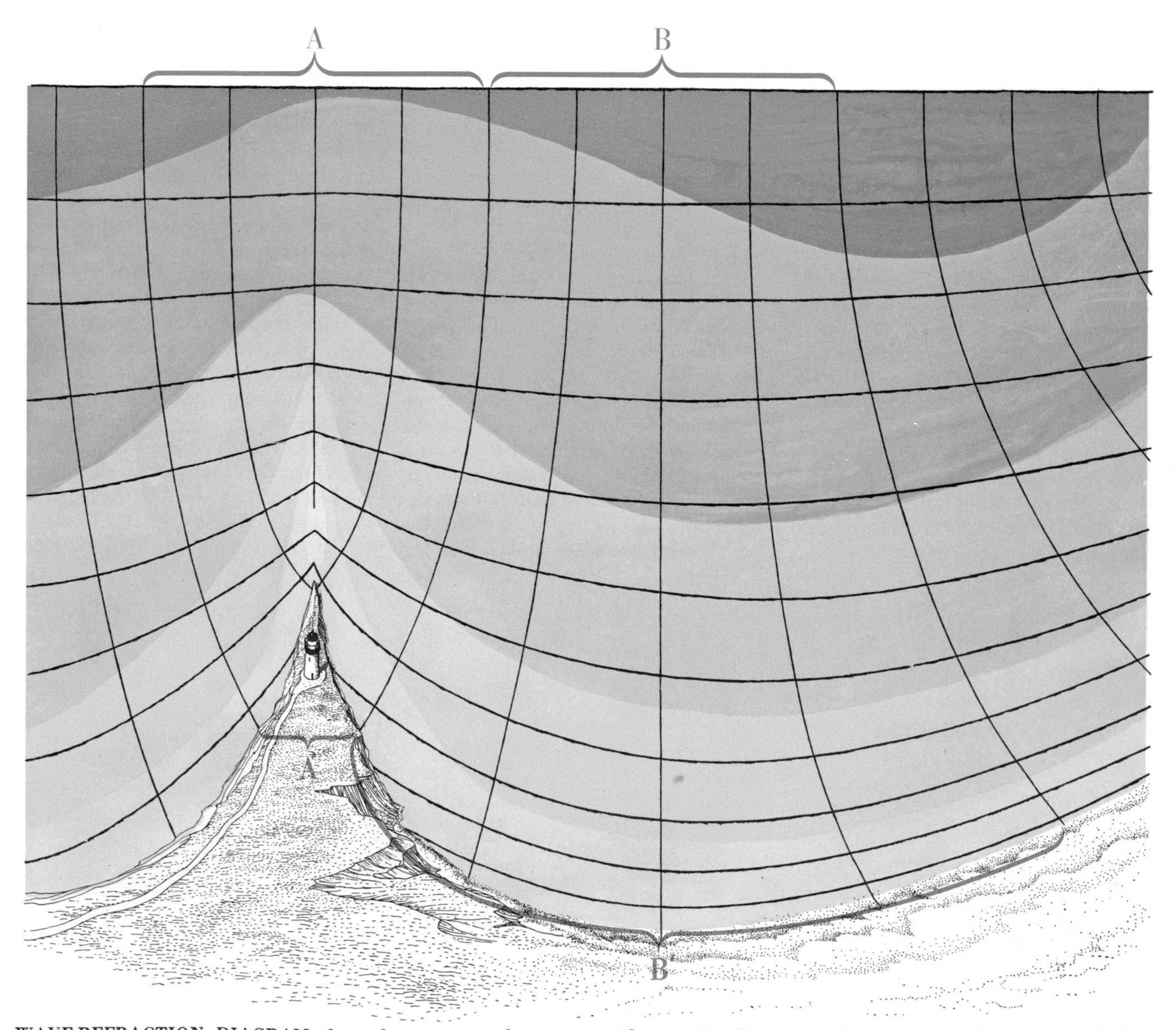

WAVE-REFRACTION DIAGRAM shows how energy of wave front at A is all concentrated by refraction at A′ around small headland area. Same energy at B enters a bay but is spread at beach over wide area B′. Horizontal lines are wave fronts; vertical lines divide energy into equal units for purposes of investigation. Such studies are vital preliminaries to design of shoreline structures.

DIFFRACTION OF OCEAN WAVES is clearly visible in this aerial photograph of Morro Bay, Calif. The waves are diffracted as they pass the end of the lower jetty. Variations in the way the waves break are caused by contours of the shore and the bottom.

and tried again. Gaillard's computations show that the wave forces must have been 6,340 pounds per square foot.

Tsunamis

Even more destructive than wind-generated waves are those generated by a sudden impulse such as an underwater earthquake, landslide or volcano. A man-made variation of the sudden impulse is the explosion of nuclear bombs at the surface of the sea, which in recent years have become large enough to be reckoned with as possible causes of destructive waves.

The public knows such waves as tidal waves, although they are in no way related to the tides and the implication has long irritated oceanographers. It was proposed that the difficulty could be resolved by adopting the Japanese word *tsunami.* Some time later it was discovered that Japanese oceanographers are equally irritated by this word; in literal translation tsunami means tidal wave! However, tsunami has become the favored usage for seismic sea waves.

Like the plunger in a wave channel, the rapid motion or subsidence of a part of the sea bottom can set a train of waves in motion. Once started, these waves travel great distances at high velocity with little loss of energy. Although their height in deep water is only a few feet, on entering shallow water they are able to rise to great heights to smash and inundate shore areas. Their height depends almost entirely on the configuration of the coastline and the nearby underwater contours. Tsunamis have periods of more than 15 minutes and wavelengths of several hundred miles. Since the depth of water is very much less than half the wavelength, they are regarded as long- or shallow-water waves, even in the 13,000-foot average depth of the open ocean, and their velocity is limited by the depth to something like 450 miles per hour.

These fast waves of great destructive potential give no warning except that the disturbance that causes them can be detected by a seismograph. The U. S. Coast Guard operates a tsunami warning network in the Pacific that tracks all earthquakes, and when triangulation indicates that a quake has occurred at sea, it issues alerts. The network also has devices to detect changes in wave period which may indicate that seismic waves are passing [see the article "Tsunamis," by Joseph Bernstein, beginning on page 56]. Curiously the influence of the system may not be entirely beneficial.

WAVE-CREATED "GEYSER" results when large breakers smash into a very steep beach. They curl over and collapse, trapping and compressing air. This compressed air then explodes as shown here, with spray from a 12-foot breaker leaping 50 feet into the air.

Once when an alert was broadcast at Honolulu, thousands of people there dashed down to the beach to see what luckily turned out to be a very small wave.

Certain coasts near zones of unrest in the earth's crust are particularly prone to such destructive waves, especially the shores of the Mediterranean, the Caribbean and the west coast of Asia. On the world-wide scale, they occur more frequently than is generally supposed: nearly once a year.

A well-known seismic sea wave, thoroughly documented by the Royal Society of London, originated with the eruption of the volcano Krakatoa in the East Indies on August 27, 1883. It is not certain whether the waves were caused by the submarine explosion, the violent movements of the sea bottom, the rush of water into the great cavity, or the dropping back into the water of nearly a cubic mile of rock, but the waves were monumental. Their period close to the disturbance was two hours, and at great distances about one hour. Waves at least 100 feet high swept away the town of Merak, 33 miles from the volcano; on the opposite shore the waves carried the man-of-war *Berow* 1.8 miles inland and left it 30 feet above the level of the sea. Some 36,380 people died by the waves in a few hours. Tide gauges in South Africa (4,690 miles from Krakatoa), Cape Horn (7,820 miles) and Panama (11,470 miles) clearly traced the progress of a train of about a dozen waves, and showed that their speed across the Indian Ocean had been between 350 and 450 miles per hour.

A tsunami on April 1, 1946, originating with a landslide in the Aleutian submarine trench, produced similar effects,

HUNDRED-FOOT "TIDAL WAVE," or tsunami, wrought impressive destruction at Scotch Cap, Alaska, in 1946. Reinforced concrete lighthouse that appears in top photograph was demolished, as shown in lower photograph, which was made from a higher angle. Atop the plateau a radio mast, its foundation 103 feet above sea, was also knocked down. Lighthouse debris was on plateau. Same tsunami, started by an Aleutian Island earthquake, hit Hawaiian Islands, South America and islands 4,000 miles away in Oceania.

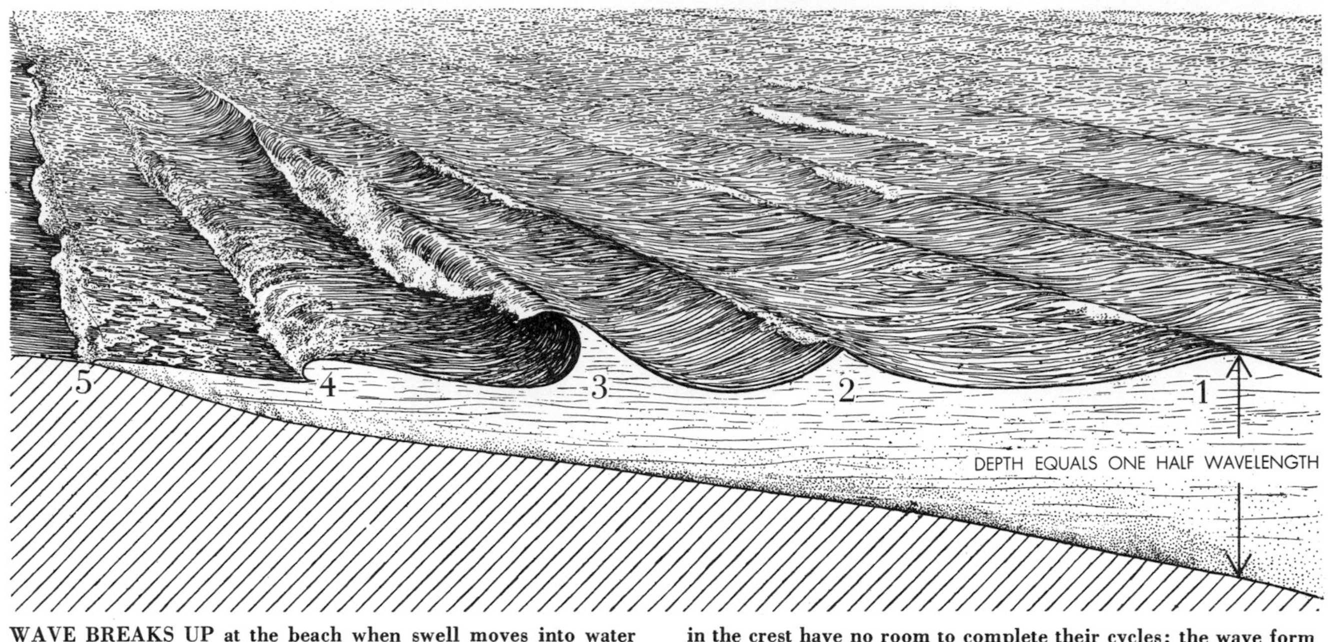

WAVE BREAKS UP at the beach when swell moves into water shallower than half the wavelength (1). The shallow bottom raises wave height and decreases length (2). At a water depth 1.3 times the wave height, water supply is reduced and the particles of water in the crest have no room to complete their cycles; the wave form breaks (3). A foam line forms and water particles, instead of just the wave form, move forward (4). The low remaining wave runs up the face of the beach as a gentle wash called the uprush (5).

fortunately on less-populated shores. It struck hard at the Hawaiian Islands, killing several hundred people and damaging property worth millions of dollars. At Hilo, Hawaii, the tsunami demonstrated that such waves are virtually invisible at sea. The captain of a ship standing off the port was astonished upon looking shoreward to see the harbor and much of the city being demolished by waves he had not noticed passing under his ship. The same waves caused considerable damage throughout the islands of Oceania, 4,000 miles from epicenter, and on the South American coast, but they were most spectacular at Scotch Cap in Alaska. There a two-story reinforced-concrete lighthouse marked a channel through the Aleutian Islands. The building, the base of which was 32 feet above sea level, and a radio mast 100 feet above the sea were reduced to bare foundations by a wave estimated to be more than 100 feet high [*see illustration on preceding page*].

Uncontrollable geologic disturbances will cause many more seismic sea waves in the future, and since the world's coastal population is continuously increasing, the greatest wave disaster is yet to come. Within the next century we can expect that somewhere a wave will at least equal the one that swept the shores of the Bay of Bengal in 1876, leaving 200,000 dead.

Tides and Other Waves

The rhythmic rise and fall of the sea level on a coast indicate the passage of a true wave we call a tide. This wave is driven, as almost everyone knows, by the gravitational influence of the sun and the moon. As these bodies change their relative positions the ocean waters are attracted into a bulge that tends to remain facing the moon as the earth turns under it; a similar bulge travels around the earth on the opposite side. The wave period therefore usually corresponds to half the lunar day.

When the sun and the moon are aligned with the earth, the tides are large (spring tides); when the two bodies are at right angles with respect to the earth, the tides are small (neap tides). By using astronomical data it is possible to predict the tides with considerable accuracy. However, the height and time of the tide at any place not on the open coast are primarily a function of the shape and size of the connection to the ocean.

Still another form of wave is a seiche, a special case of wave reflection. All enclosed bodies of water rock with characteristics related to the size of the basin. The motion is comparable to the sloshing of water in the bathtub when one gets out quickly. In an attempt to return to stability the water sways back and forth with the natural period of the tub (mine has a period of two seconds). Similarly a tsunami or a barometric pressure-change will often set the water in a bay rocking as it passes. In fact, the tsunami itself may reflect back and forth across the ocean as a sort of super-seiche.

In addition to seiches, tides, tsunamis and wind waves there are other waves in the sea. Some travel hundreds of feet beneath the surface along the thermocline, the interface between the cold deep water and the relatively warm surface layer. Of course these waves cannot be seen, but thermometers show that they are there, moving slowly along the boundary between the warm layer and the denser cold water. Their study awaits proper instrumentation. Certain very low waves, with periods of several minutes, issue from storms at sea. These long-period "forerunners" may be caused by the barometric pulsation of the entire storm against the ocean surface. Since they travel at hundreds of miles an hour, they could presumably be used as storm warnings or storm-center locators. Other waves, much longer than tides, with periods of days or weeks and heights of less than an inch, have been discovered by statistical methods and are now an object of study.

The great advances both in wave theory and in the actual measurement of waves at sea have not reduced the need for extensive laboratory studies. The solution of the many complex engineering problems that involve ships, harbors, beaches and shoreline structures requires that waves be simulated under ideal test conditions. Such model studies in advance of expensive construction permit much greater confidence in the designs.

Experimental Tanks

The traditional wave channel in which an endless train of identical small waves is created by an oscillating plunger is still in use, but some of the new wave tanks are much more sophisticated. In some the channel is covered, so that a high velocity draft of air may simulate the wind in making waves. In others, like the large tank at the Stevens Institute of Technology in Hoboken, New Jersey, artificial irregular waves approach the variability of those in the deep ocean. In such tanks proposed ship designs, like those of the America's Cup yacht *Columbia,* are tested at model size to see how they will behave at sea.

The ripple tank, now standard apparatus for teaching physics, has its place in shoreline engineering studies for conveniently modeling diffraction and refraction. Even the fast tsunamis and the very slow waves of the ocean can be modeled in the laboratory. The trick is to use layers of two liquids that do not mix, and create waves on the interface between them. The speeds of the waves can be controlled by adjusting the densities of the liquids.

To reduce the uncertainties in extrapolation from the model to prototype, some of the new wave tanks are very large. The tank of the Beach Erosion Board in Washington, D.C. (630 feet long and 20 feet deep, with a 500-horsepower generator), can subject quarter-scale models of ocean breakwaters to six-foot breakers. The new maneuvering tank now under construction at the David Taylor Model Basin in Carderock, Md., measures 360 by 240 feet, is 35 feet deep along one side and will have wave generators on two sides that can independently produce trains of variable waves. Thus man can almost bring the ocean indoors for study.

The future of wave research seems to lie in refinement of the tools for measuring, statistically examining and reproducing in laboratories the familiar wind waves and swell as well as the more recently discovered varieties. It lies in completing the solution of the problem of wave generation. It lies in the search for forms of ocean waves not yet discovered—some of which may exist only on rare occasions. Nothing less than the complete understanding of all forms of ocean waves must remain the objective of these studies.

5

TSUNAMIS

JOSEPH BERNSTEIN
August 1954

On the morning of April 1, 1946, residents of the Hawaiian Islands awoke to an astonishing scene. In the town of Hilo almost every house on the side of the main street facing Hilo Bay was smashed against the buildings on the other side. At the Wailuku River a steel span of the railroad bridge had been torn from its foundations and tossed 300 yards upstream. Heavy masses of coral, up to four feet wide, were strewn on the beaches. Enormous sections of rock, weighing several tons, had been wrenched from the bottom of the sea and thrown onto reefs. Houses were overturned, railroad tracks ripped from their roadbeds, coastal highways buried, beaches washed away. The waters off the islands were dotted with floating houses, debris and people. The catastrophe, stealing upon Hawaii suddenly and totally unexpectedly, cost the islands 159 lives and $25 million in property damage.

Its cause was the phenomenon commonly known as a "tidal wave," though it has nothing to do with the tidal forces of the moon or sun. More than 2,000 miles from the Hawaiian Islands, somewhere in the Aleutians, the sea bottom had shifted. The disturbance had generated waves which moved swiftly but almost imperceptibly across the ocean and piled up with fantastic force on the Hawaiian coast.

Scientists have generally adopted the name "tsunami," from the Japanese, for the misnamed tidal wave. It ranks among the most terrifying phenomena known

TSUNAMI near the coast of Japan was depicted by Hokusai, great 19th-century Japanese print maker. The title of the print (*upper left*) is approximately translated: "The crest of the great wave off Kanagawa." In the background is the smooth cone of Fujiyama.

to man and has been responsible for some of the worst disasters in human history. What made the 1946 tsunami especially notable was that a number of oceanographers happened to be in the Pacific (in connection with the Bikini atomic bomb test) and were able to observe it at first hand. It became the most thoroughly investigated tsunami in history, and from it came the development of an effective new warning system by the U. S. Coast and Geodetic Survey.

A tsunami may be started by a sea-bottom slide, an earthquake or a volcanic eruption. The most infamous of all was launched by the explosion of the island of Krakatoa in 1883; it raced across the Pacific at 300 miles an hour, devastated the coasts of Java and Sumatra with waves 100 to 130 feet high, and pounded the shore as far away as San Francisco.

The ancient Greeks recorded several catastrophic inundations by huge waves. Whether or not Plato's tale of the lost continent of Atlantis is true, skeptics concede that the myth may have some foundation in a great tsunami of ancient times. Indeed, a tremendously destructive tsunami that arose in the Arabian Sea in 1945 has even revived the interest of geologists and archaeologists in the Biblical story of the Flood.

One of the most damaging tsunamis on record followed the famous Lisbon earthquake of November 1, 1755; its waves persisted for a week and were felt as far away as the English coast. Tsunamis are rare, however, in the Atlantic Ocean; they are far more common in the Pacific. Japan has had 15 destructive ones (eight of them disastrous) since 1596. The Hawaiian Islands are struck severely an average of once every 25 years.

In 1707 an earthquake in Japan generated waves so huge that they piled into the Inland Sea; one wave swamped more than 1,000 ships and boats in Osaka Bay. A tsunami in the Hawaiian Islands in 1869 washed away an entire town (Ponoluu), leaving only two forlorn trees standing where the community had been. In 1896 a Japanese tsunami killed 27,000 people and swept away 10,000 homes.

The dimensions of these waves dwarf all our usual standards of measurement. An ordinary sea wave is rarely more than a few hundred feet long from crest to crest—no longer than 320 feet in the Atlantic or 1,000 feet in the Pacific. But a tsunami often extends more than 100 miles and sometimes as much as 600 miles from crest to crest. While a wind wave never travels at more than about 60 miles per hour, the velocity of a tsunami in the open sea must be reckoned in hundreds of miles per hour. The greater the depth of the water,

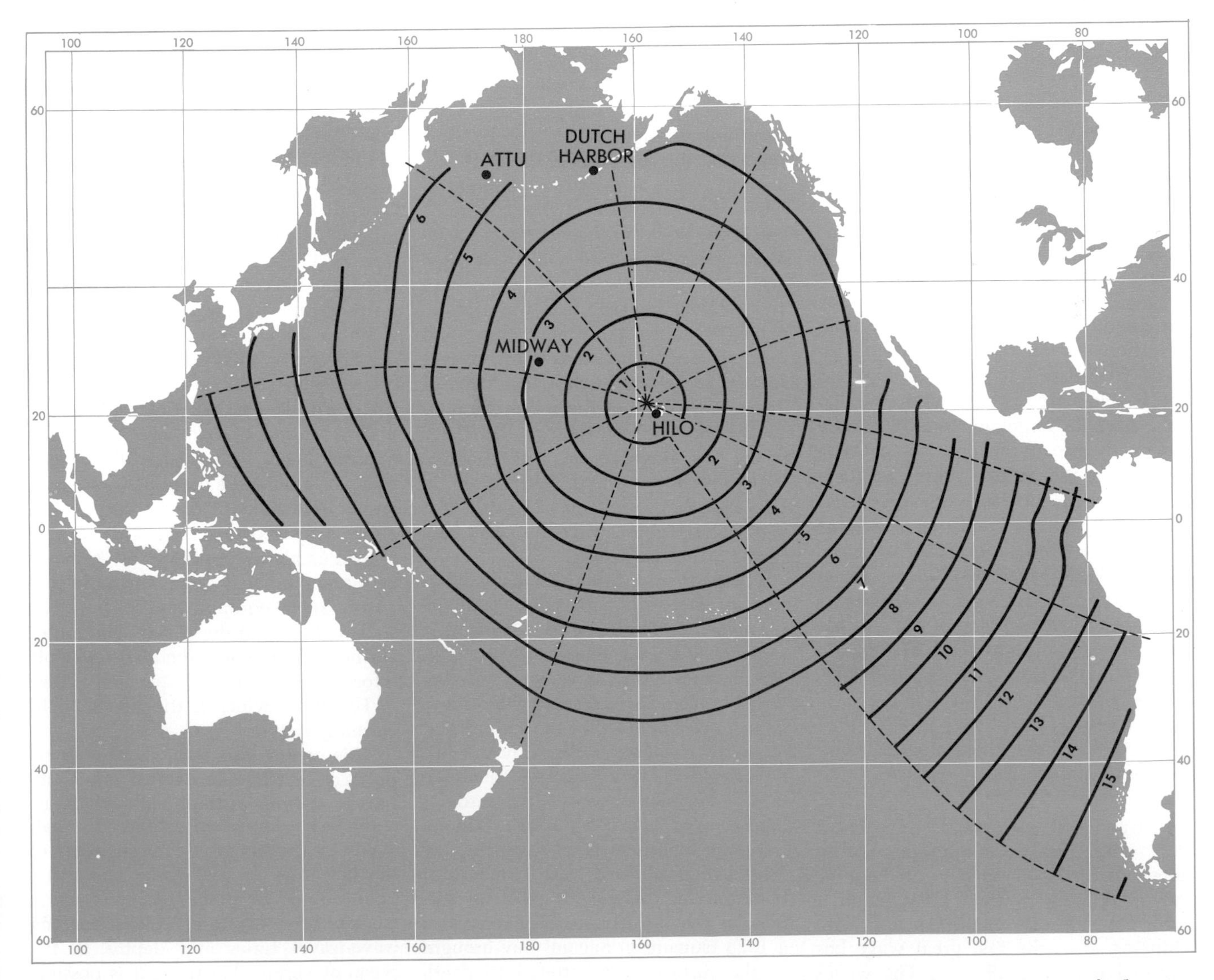

TIME PLOT of tsunamis endangering the Hawaiian Islands may be made from this chart. When the earthquake epicenter is plotted on it, the circles indicate the time in hours for the seismic sea wave to reach Hawaii. Locations of seismic sea-wave detectors in the warning system are named on the chart. Seismographs are located in Japan, Alaska, Guam and along the coast of North and South America.

the greater is the speed of the wave; Lagrange's law says that its velocity is equal to the square root of the product of the depth times the acceleration due to gravity. In the deep waters of the Pacific these waves reach a speed of 500 miles per hour.

Tsunamis are so shallow in comparison with their length that in the open ocean they are hardly detectable. Their amplitude sometimes is as little as two feet from trough to crest. Usually it is only when they approach shallow water or the shore that they build up to their terrifying heights. On the fateful day in 1896 when the great waves approached Japan, fishermen at sea noticed no unusual swells. Not until they sailed home at the end of the day, through a sea strewn with bodies and the wreckage of houses, were they aware of what had happened. The seemingly quiet ocean had crashed a wall of water from 10 to 100 feet high upon beaches crowded with bathers, drowning thousands of them and flattening villages along the shore.

The giant waves are more dangerous on flat shores than on steep ones. They usually range from 20 to 60 feet in height, but when they pour into a V-shaped inlet or harbor they may rise to mountainous proportions.

Generally the first salvo of a tsunami is a rather sharp swell, not different enough from an ordinary wave to alarm casual observers. This is followed by a tremendous suck of water away from the shore as the first great trough arrives. Reefs are left high and dry, and the beaches are covered with stranded fish. At Hilo large numbers of people ran out to inspect the amazing spectacle of the denuded beach. Many of them paid for their curiosity with their lives, for some minutes later the first giant wave roared over the shore. After an earthquake in Japan in 1793 people on the coast at Tugaru were so terrified by the extraordinary ebbing of the sea that they scurried to higher ground. When a second quake came, they dashed back to the beach, fearing that they might be buried under landslides. Just as they reached the shore, the first huge wave crashed upon them.

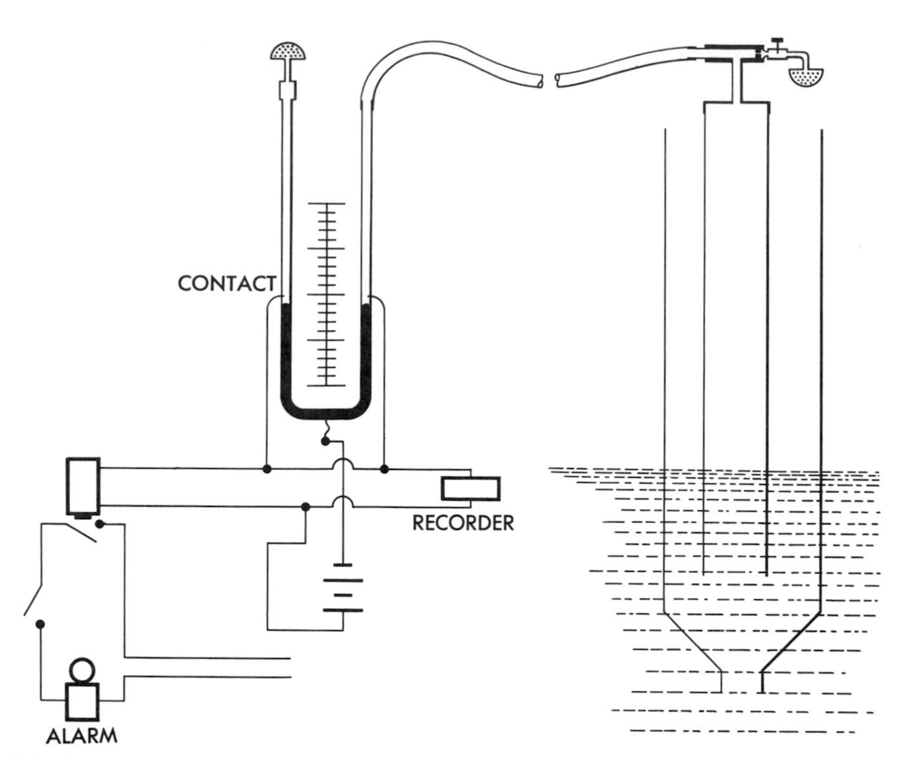

DETECTOR signals the occurrence of a seismic sea wave. An air column in the inner tube at right rises and falls with the tide. A wave of sufficient magnitude increases the air pressure until the mercury in the U-tube is forced around, closing a circuit and sounding an alarm.

A tsunami is not a single wave but a series. The waves are separated by intervals of 15 minutes to an hour or more (because of their great length), and this has often lulled people into thinking after the first great wave has crashed that it is all over. The waves may keep coming for many hours. Usually the third to the eighth waves in the series are the biggest.

Among the observers of the 1946 tsunami at Hilo was Francis P. Shepard of the Scripps Institution of Oceanography, one of the world's foremost marine geologists. He was able to make a detailed inspection of the waves. Their onrush and retreat, he reported, was accompanied by a great hissing, roaring and rattling. The third and fourth waves seemed to be the highest. On some of the islands' beaches the waves came in gently; they were steepest on the shores facing the direction of the seaquake from which the waves had come. In Hilo Bay they were from 21 to 26 feet high. The highest waves, 55 feet, occurred at Pololu Valley.

Scientists and fishermen have occasionally seen strange by-products of the phenomenon. During a 1933 tsunami in Japan the sea glowed brilliantly at night. The luminosity of the water is now believed to have been caused by the stimulation of vast numbers of the luminescent organism *Noctiluca miliaris* by the turbulence of the sea. Japanese fishermen have sometimes observed that sardines hauled up in their nets during a tsunami have enormously swollen stomachs; the fish have swallowed vast numbers of bottom-living diatoms, raised to the surface by the disturbance. The waves of a 1923 tsunami in Sagami Bay brought to the surface and battered to death huge numbers of fishes that normally live at a depth of 3,000 feet. Gratified fishermen hauled them in by the thousands.

The tsunami-warning system developed since the 1946 disaster in Hawaii relies mainly on a simple and ingenious instrument devised by Commander C. K. Green of the Coast and Geodetic Survey staff. It consists of a series of pipes and a pressure-measuring chamber which record the rise and fall of the water surface. Ordinary water movements, such as wind waves and tides, are disregarded. But when waves with a period of between 10 and 40 minutes begin to roll over the ocean, they set in motion a corresponding oscillation in a column of mercury which closes an electric circuit. This in turn sets off an alarm, notifying the observers at the station that a tsunami is in progress. Such equipment has been installed at Hilo, Midway, Attu and Dutch Harbor. The moment the alarm goes off, information is immediately forwarded to Honolulu, which is the center of the warning system.

This center also receives prompt reports on earthquakes from four Coast Survey stations in the Pacific which are equipped with seismographs. Its staff makes a preliminary determination of the epicenter of the quake and alerts tide stations near the epicenter for a tsunami. By means of charts showing wave-travel times and depths in the ocean at various locations, it is possible to estimate the rate of approach and probable time of arrival at Hawaii of a tsunami getting under way at any spot in the Pacific. The civil and military authorities are then advised of the danger, and they issue warnings and take all necessary protective steps. All of these activities are geared to a top-priority communication system, and practice tests have been held to assure that everything will work smoothly.

Since the 1946 disaster there have been 15 tsunamis in the Pacific, but only one was of any consequence. On November 4, 1952, an earthquake occurred under the sea off the Kamchatka Peninsula. At 17:07 that afternoon (Greenwich time) the shock was recorded by the seismograph alarm in Honolulu. The warning system immediately went into action. Within about an hour, with the help of reports from seismic stations in Alaska, Arizona and California, the quake's epicenter was placed at 51 degrees North latitude and 158 degrees East longitude. While accounts of the progress of the tsunami came in from various points in the Pacific (Midway reported it was covered with nine feet of water), the Hawaiian station made its calculations and notified the military services and the police that the first big wave would arrive at Honolulu at 23:30 Greenwich time.

It turned out that the waves were not so high as in 1946. They hurled a cement barge against a freighter in Honolulu Harbor, knocked down telephone lines, marooned automobiles, flooded lawns, killed six cows. But not a single human life was lost, and property damage in the Hawaiian Islands did not exceed $800,000. There is little doubt that the warning system saved lives and reduced the damage.

But it is plain that a warning system, however efficient, is not enough. In the vulnerable areas of the Pacific there should be restrictions against building homes on exposed coasts, or at least a requirement that they be either raised off the ground or anchored strongly against waves.

6

THE CORIOLIS EFFECT

JAMES E. MCDONALD
May 1952

IT IS A curious fact that all things which move over the surface of the earth tend to sidle from their appointed paths—to the right in the Northern Hemisphere, to the left in the Southern Hemisphere. Since man has managed to make himself one of the most mobile of creatures, one might think that so ubiquitous an effect must long have been a matter of common knowledge. It has not been and still is not, even in this era of rapid speeds, which accentuate the sidling tendency. Probably few people realize that as they drive down a straight highway at 60 miles per hour this all-pervading drift would carry them off the road to the right at the rate of some 15 feet per mile were it not for the frictional resistance of the tires to any lateral motion.

This sidewise drifting tendency is called the Coriolis effect, after the 19th-century French mathematician G. G. Coriolis, who made the first complete analysis of it. The effect is due simply to the rotation of the earth, and it appears in all motions as soon as we refer those motions to any coordinate system fixed with respect to the earth (*e.g.*, the latitude-longitude grid).

There is really only one satisfactory way to obtain a vivid impression of the nature of the Coriolis principle. That is to go to a carnival. Every carnival worth the name has a Coriolian coordinate system: *viz.*, the merry-go-round. With only a few balls as laboratory equipment and two assistants, one on the merry-go-round with you and the other on the ground, you can carry out many interesting Coriolian experiments.

When the merry-go-round starts up, you begin a game of catch. Things will probably go very poorly for several throws (which is the reason for your taking the precaution of equipping yourself with several balls). The ball will seem to veer from its thrown direction in the most amazing fashion. Let us say the merry-go-round turns counterclockwise, as does the earth when viewed from above the North Pole. If it makes one complete turn in 10 seconds, and you throw the ball at a speed of 20 feet per second toward an assistant standing 15 feet from you on the merry-go-round, the apparently curving ball will miss the assistant by a little over six feet to the right. When you throw a ball to your other assistant, in the outer world off the merry-go-round, it will again seem to drift rightward. This time, however, by great concentration you may be able to fix your attention on the nonrotating framework of the outer world sufficiently to sense that the ball is really moving as it ought to move, and you may even make proper allowance for the merry-go-round's rotation so that the ball reaches your assistant's hands.

THE APPARENT strangeness of the balls' behavior in these experiments arises from the fact that almost inescapably you take the merry-go-round as your reference system, and in this system the laws of dynamics in their usual form simply do not hold. No such difficulties confront the assistant who stands out on firm ground. He is not so compelled to view these motions with respect to your rotating coordinate system. He will feel certain that the balls have at all times been moving in well-behaved fashion. If he has a little understanding of the problem, he may be able to explain to you that the drifting to the right which you seem to see is really due to the fact that your system is turning out from under the moving balls.

The earth is a spherical merry-go-round, and all of the Coriolis drifts we observe when we use terrestrial coordinate systems are due ultimately to the fact that the earth, like the merry-go-round, is always spinning out from under our dynamical systems. To be sure, there are certain subtleties that enter into some Coriolis effects, but at bottom the whole thing is just the merry-go-round idea. To an observer conscious of Newton's second law of motion, the apparent "acceleration" (deflection from a straight path) of an object moving over the earth suggests that some force is acting on it, and he is strongly tempted to speak of the Coriolis "force." For convenience, meteorologists and others who are concerned with the Coriolis effect do treat it as a force, and their equations work out all right. What they set down as a force in the Newtonian equation is actually a correction for the apparent acceleration. The pure dynamicist looks at it in a different way: he likes to regard these motions as occurring in obedient Newtonian fashion in what he calls "inertial space."

NOW let us look at some interesting examples of the Coriolis effect, as it applies to projectiles, flight, vehicles, ocean currents and even our weather. The Coriolis effect is greatest near the North and South Poles (where the earth turns most rapidly under a moving object) and decreases to zero at the Equator. The magnitude of the effect also depends directly on the speed of the moving object.

In middle latitudes of the Northern Hemisphere a bullet fired with a velocity of 800 feet per second at a target 400 feet away will drift one-tenth of an inch to the right (without considering wind effects or any other interference). That is, in the half-second during which the bullet is in flight, the rotation of the earth has shifted the bull's-eye by about one-tenth of an inch. This is not serious to a pistol marksman, but the effect can make quite a difference to a long-range gunner. A battleship gunner who takes dead aim at the bridge of a destroyer 20 miles away and fires a shell at 2,500 feet per second will miss the destroyer completely, because the lateral Coriolis drift will be more than 200 feet. In World War I the shells of the giant German gun called Big Bertha, which bombarded Paris from a firing site some 70 miles away, took three minutes to reach their destination, and they underwent a Coriolis drift to the right amounting to almost a full mile—an error for which the German ballistics experts carefully allowed.

For a really dramatic effect we can take the case of a rocket fired from the North Pole and aimed at, say, New York City. Assuming, for the sake of simplicity, that the rocket travels at a constant speed of a mile per second, it will be in flight for about 55 minutes. During all of this time the target, New York City, will be traveling at 18 miles per second through solar-system space (the speed of the earth's movement around the sun) and will also be turning with the rotation of the earth at the rate of 15 degrees of longitude per hour. As the result of these motions the rocket, at the end of 55 minutes, will come to earth in some cornfield in northeastern Illinois, not far from Chicago!

The earthbound observers who have been plotting the apparent path of this rocket with their radar network will say that it traced out a graceful curve which started out straight south in the longitude of New York City, but veered steadily westward, arriving in Illinois from a direction about 11 degrees east of north. A less provincial observer out in interplanetary space will see that the effect is entirely a result of the earth itself having turned out from under the moving rocket.

This is an idealized case; in actual situations the Coriolis effect is much less evident, because other forces such as air resistance, neglected in this example, also act on moving objects. Furthermore, the motion of a projectile fired from any place on the earth other than the Poles would be influenced not only by the Coriolis effect but also by the initial impetus from the circumpolar rotation of the launching site.

An airplane experiences Coriolis drifts which would lead to astonishing errors in long flights if no compensation were made for them. A jet fighter that set out on a great-circle heading from Chicago to New York and flew at 600 miles per hour without changing its heading would miss New York by several hundred miles to the south (assuming no allowance for any wind). And if the same pilot tried to fly in a similar way from Seattle to New York, he would find himself down in South America by the time he crossed the meridian through New York! In actual flights a pilot continually banks his plane slightly leftward, in our hemisphere, to compensate for Coriolis drift. It should be noted that, large as these deviations due to the rotation of the earth are, they are still small compared to the effects of cross-winds normally encountered in actual flights. The pilot's Coriolis corrections are thus obscured by the jockeying necessary to compensate for wind drift. To compensate for the Coriolis drift and keep a 20,000-pound jet fighter on a straight terrestrial course at 600 miles per hour requires a leftward force of about 55 pounds in middle latitudes of the Northern Hemisphere. This the pilot manages by manipulation of the plane's wings.

Railroad cars are much more massive, so the Coriolis reaction in their case is greater. A 500-ton locomotive moving at 60 miles per hour develops a lateral

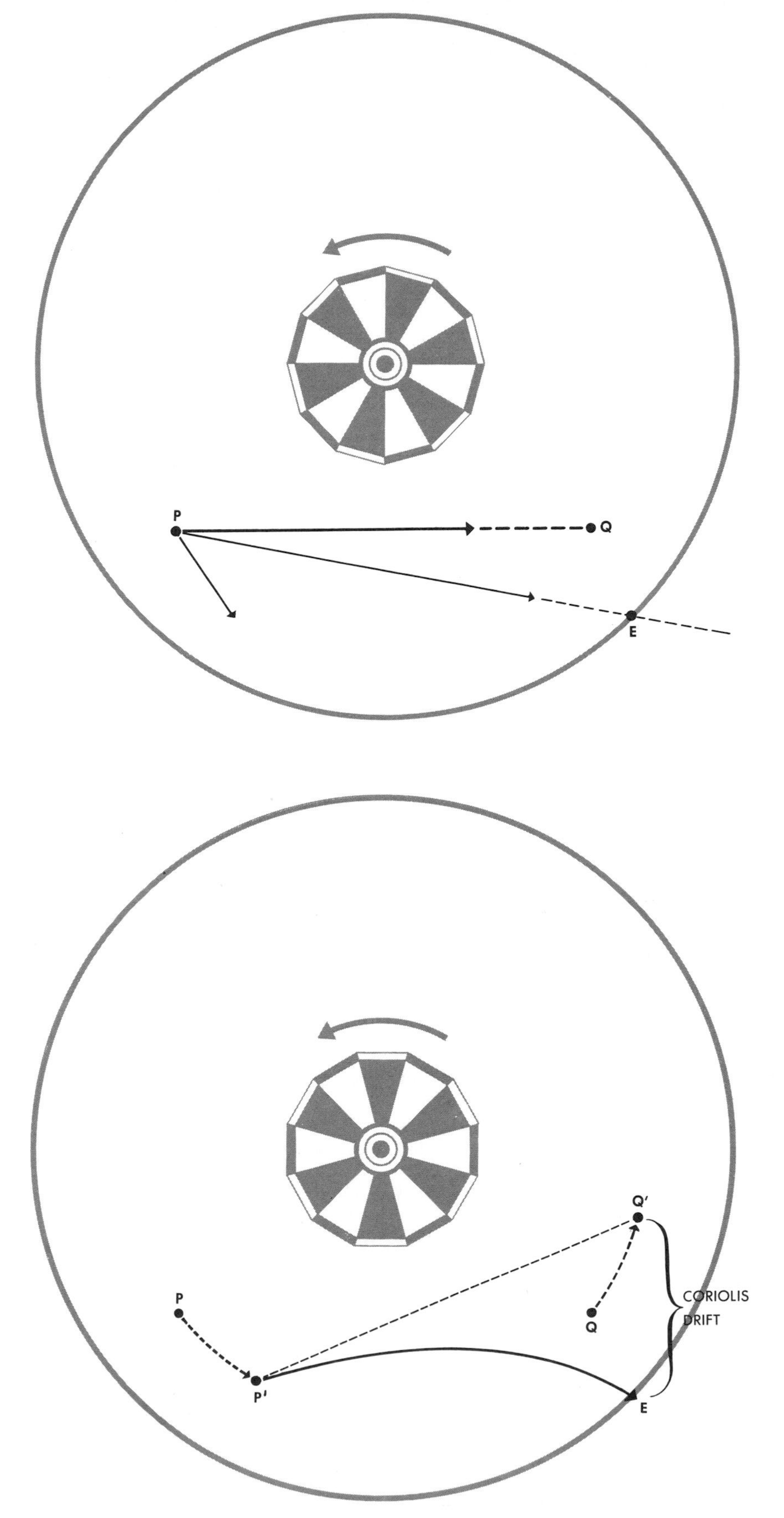

MERRY-GO-ROUND EXPERIMENT demonstrates Coriolis drift. Seen from above, the merry-go-round rotates counterclockwise. In the top drawing a man at P attempts to throw a ball to a man at Q. The rotational motion of the man at P (*short arrow*), however, causes the ball to head in the direction PE. In the bottom drawing the man at P has moved to P′, the man at Q to Q′ and the ball crosses the edge of the merry-go-round at E. To the rotating observers on the merry-go-round the ball appears to have described a curve.

pressure on the rails amounting to about 300 pounds in middle latitudes. This has given rise to the story that the wheels on trains wear unevenly. Such a result could hardly be detected on coaches or freight cars, which for railroading reasons have no definite right or left sides, and the Engineering Department of the Union Pacific Railroad has informed the author that even in the case of locomotive wheels the difference of wear on the flanges of the right and left wheels is too small to be measured.

WHY IS SO universal (one should say, "so terrestrial") an effect not readily apparent in our everyday activities? The answer is that for many moving objects the tendency toward a lateral drift is quite easily counteracted as the motions proceed. Thus in the case of the car speeding down the highway at 60 miles per hour, the potential 15 feet of shift per mile is prevented by the frictional resistance of the tires to lateral motion.

A walking man makes corrections for the Coriolis effect easily and quite unconsciously. On frictionless ice that prevented his making any small lateral corrections (but somehow still permitted him to walk!) a man walking at four miles per hour would drift from his intended straight path by about 250 feet at the end of one mile. Lost polar explorers are reported to have a strong tendency to circle steadily toward the right near the North Pole and to the left near the South Pole; this may very well be due to the Coriolis effect, which is about 50 per cent stronger at the Poles than in middle latitudes. It is said that even the penguins in the Antarctic waddle in arcs to the left, but this the author will have to see to believe.

AMONG ALL the physical phenomena in which the Coriolis effect plays a role, the most striking is the weather. Were it not for the Coriolis effect, winds on the earth would rush directly from higher-pressure areas to lower-pressure ones, and no strong "highs" or "lows" could develop. Hence there would be no opportunity for the build-up of the intense cyclones and the large anticyclones that control and give variability to our weather, and our weather would be much less changeable than it is. This is precisely the situation in the Tropics, where the Coriolis effect is zero or very small. In that almost Coriolis-free belt any atmospheric pressure differences produced by heating of the air at the ground are quickly smoothed out, and the region has well earned the name of "the doldrums." Hurricanes and typhoons never form closer to the Equator than about five degrees of latitude.

Away from the Equator, however, the case is very different. There the Coriolis acceleration causes winds to veer around and blow at right angles to the pressure gradient, instead of parallel with it. The result is the pattern of strong lows and highs and circular movement that is responsible for changes in our weather.

On other planets, where the angular velocity of rotation is different from that of the earth, the Coriolis effect is correspondingly different. Jupiter and Saturn must have very marked Coriolis effects, because each rotates about two and a half times more rapidly than does the earth. Their atmospheres of hydrogen, methane and ammonia must have very steep pressure gradients, if their winds compare in strength with ours. In contrast, the atmosphere of Venus is probably very calm, because Venus rotates much more slowly than the earth—perhaps once in about 30 terrestrial days.

Just as the motions of the atmosphere exhibit the Coriolis effect, so also do the more ponderous movements of the great ocean currents. To simplify the picture a bit, let us assume that the density of the sea is uniform. The oceans are not perfectly level, for the winds shift the waters and give them a gentle relief. Since water flows downhill, the natural tendency of the oceans' water is to flow from regions where the mean sea-level is relatively high to those where it is lower. But as soon as the water tries to move in so forthright a fashion, the Coriolis drift causes the moving water to veer off to the right (in the Northern Hemisphere). Eventually the currents flow steadily along the contour lines, with the water surface sloping upward to the right as one looks in the direction of flow. In practice, of course, internal eddy-stresses within the ocean and the winds blowing across the sea surface modify this trend. But the general rule still holds.

Lest the reader mistakenly conclude that he should have spotted these oceanic hills and valleys on his last sea voyage, it should be mentioned that the total difference of mean height across even the fastest-moving parts of the Gulf Stream system is only about a foot and a half in some 80 or 90 miles. Even this modest slope is only partly due to Coriolis effects, the remainder resulting from the sort of horizontal density gradients we have agreed to overlook. Yet, slight as such surface slopes may be, they constitute a major factor in the dynamics of the ocean currents.

PEOPLE ON the Pacific Coast are well acquainted with certain other consequences of the Coriolis acceleration, though not many realize this is the cause. Coriolis drift is mainly responsible for the notorious California fogs and the coldness of the water on California's beaches. Off the California coast, where the prevailing winds are from the northwest, the wind stress and Coriolis drift generally combine to make the coastal waters sidle off in a southwesterly direction. As water is transported away from the shore toward the southwest, the deficit must be made up somehow. The water moving offshore is replaced by water rising from below. This upwelling brings up water from cold strata lying at depths as great as several

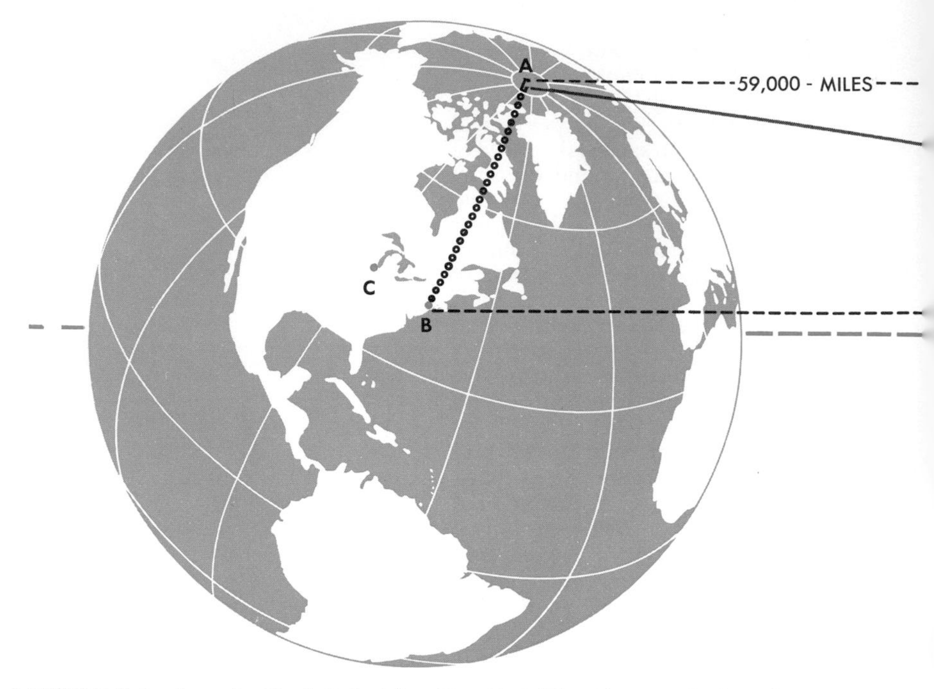

MISSILE flying from the North Pole (A) to New York (B) with a speed of a mile per second would land near Chicago (C) unless the Coriolis effect were taken into account. In the 55 minutes that it would take the missile to reach the latitude of New York the earth

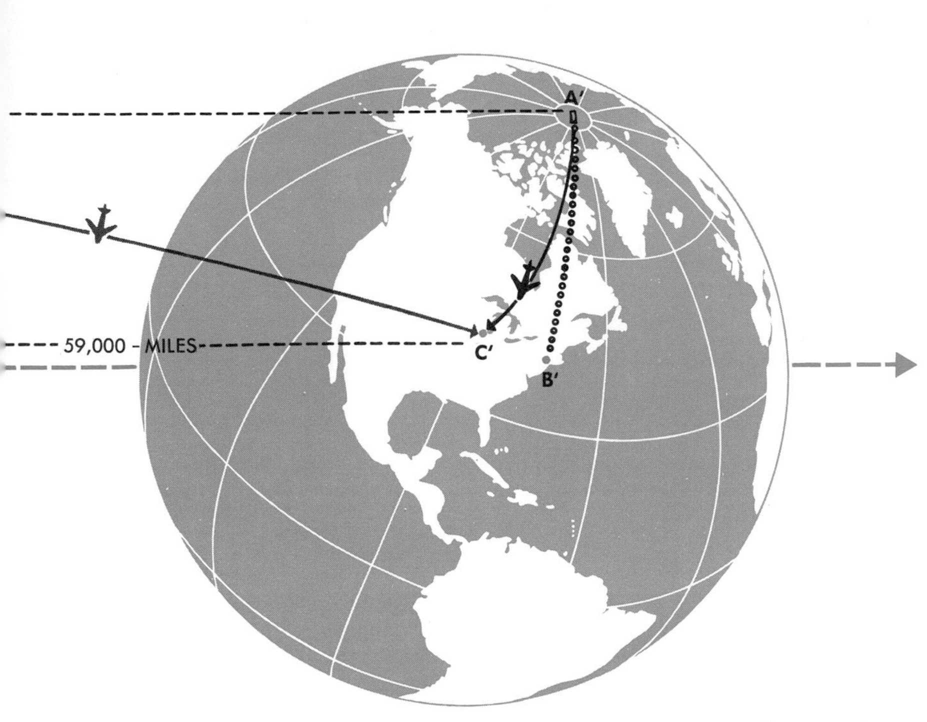

would have rotated on its axis about 15 degrees and traveled along its orbit some 59,000 miles. Within the coordinate system of the earth the missile would appear to describe the trajectory A′C′. Within the coordinate system of the sun and all the planets, it would appear to describe the path AC′ (*gray line running diagonally across these two pages*).

hundred feet. As a result there is a cool strip of water along the California coast, superimposed, in fact, on the already cool California Current flowing down from the north. In summer the warm moist Pacific air streaming in from the northwest is cooled by the coastal water, and this is what forms the fogs for which California regrets to be famous. A similar situation prevails off the coast of Peru and parts of the western coast of Africa.

Some geologists believe the Coriolis effect causes a river to erode one of its banks faster than the other. The Russian scientists P. A. Slavsov and Karl von Baer reported that river valleys in Siberia tend to have steep walls on their right side and gently sloping walls on their left side. Similar asymmetries have been observed in some Alaskan rivers, in the Missouri River and in a number of streams on Long Island. This supposed effect of the Coriolis drift is sometimes called Baer's law. But students of the effect have not all been willing to attribute it to the Coriolis influence. Even in a river a mile wide flowing at the fast rate of five miles per hour in middle latitudes of the Northern Hemisphere, the Coriolis drift to the right would pile up the water only a little more than one inch higher at the right bank than at the left bank. Possibly such a slight difference in height might cause significant differences in erosion over geological periods of time, but the question is still unsettled.

This is as good a place as any to correct the persistent misconception that the Coriolis acceleration causes the water to run out of a washbowl in a clockwise direction in the Northern Hemisphere and counterclockwise in the Southern Hemisphere. The Coriolis influence is so small at the velocity of water in a washbowl, the time involved is so short and other factors are so numerous (hands, noncircular bowl-shapes, and so on) that one may feel sure the Coriolis effect is never in control here. This is regrettable, because if it were, a washbowl would constitute a useful analogue of a cyclone in the atmosphere.

WE SHALL consider one more possible case of a Coriolis effect. There is a theory that some birds may be guided in their migrations by sensitivity to the Coriolis acceleration and to geomagnetic latitude. H. L. Yeagley of Pennsylvania State College has recently studied this amazing theory in an effort to determine the navigating techniques of the homing pigeon.

When a bird flies at constant ground speed in the Northern Hemisphere, its Coriolis acceleration toward the right grows greater the farther north it flies. Yeagley suggests that if, through some delicate sensory organ, the bird can detect slight differences in this acceleration, and can combine this information with an accurate estimate of its ground speed, it may be able to sense its geographical latitude. If, at the same time, another sensory organ with the necessary electrical properties senses differences in the minute electromotive forces generated by virtue of the bird's motion through the earth's magnetic field, then this plus the bird's estimate of its speed would provide a basis for sensing geomagnetic latitude. Now, since the magnetic poles of the earth are displaced some 20 degrees from the geographical poles, the parallels of geomagnetic latitude form a grid with the parallels of geographical latitude, and with this grid it is theoretically possible to navigate.

Most physicists would regard the theory as of very low *a priori* plausibility. Even assuming that a bird's senses are so delicate that it can detect the tiny differences in Coriolis acceleration and magnetic field, these cannot be translated into latitude until the bird has compared each effect with a very precise estimate of its ground speed. Furthermore, the bird must somehow allow for the effect of cross-winds, which is normally much greater than the Coriolis drift. As if this were not enough, the bird would have to defy relativity theory, which says that it could not distinguish the effects of the normal atmospheric electric field from those induced by the bird's motion through the earth's magnetic field. Yet despite these difficulties, certain features of Yeagley's theory seem to have been borne out by his extensive studies with homing pigeons.

If further research should confirm the magnetic-Coriolis theory of bird navigation, the solution of this deep mystery of the animal world will be rather more astonishing than the original mystery. It would certainly be startling to learn that this effect has been used by generations of golden plovers and Arctic terns to hold true to their courses as they fly over thousands of miles of trackless oceans.

Whether the birds are really that clever or not, we may be quite sure that they inexorably tend to drift as they fly. All things that move over the surface of our spinning earth, whether birds, winds, rivers, ocean currents, explorers, cars, trains, bullets or rockets, are inevitably subjected to this effect as we view them in our terrestrial coordinate systems. Even when man gets away from his planetary home and stakes out better-behaved coordinate systems in interplanetary space, he will not be able to omit consideration of the Coriolis effect from his dynamics. For the solar system itself, along with all its near neighbors, is slowly but surely rotating around the hub of our galaxy, some 30,000 light-years away. Undoubtedly a precise analysis of the waddling of Antarctic penguins would show not only Coriolis effects due to the earth's circumpolar rotation, and similar but smaller effects due to our planet's annual circuit around the sun, but also a tiny Coriolis drift due to the stately whirl of our solar system about the center of the galaxy.

Here we find ourselves in somewhat the same situation as Archimedes with his earth-moving lever—all *we* need to demonstrate our point is a suitable coordinate system.

7

THE CIRCULATION OF THE OCEANS

WALTER MUNK
September 1955

Everybody knows the difference between climate and day-to-day weather. It is less known that a similar distinction applies also to the currents of the oceans. Until recently we were aware only of the broad, average features of the ocean movements—the "climatic" circulation. But modern studies have disclosed a fine structure which is superposed on this climate and which shifts from day to day in an unbelievably mercurial manner. If 10 vessels strategically placed in the Gulf Stream were to measure the currents and make a "weather map" of the Stream next Tuesday, the map would differ from the one for Friday. Not long ago we watched a freighter carefully holding a course which according to the climatic chart should have speeded it on its way to Europe by taking advantage of the Gulf Stream. Actually the ship was bucking a two-knot countercurrent; the Gulf Stream was 100 miles off its usual path!

The vagaries of the ocean currents were practically unknown until the last world war, when new techniques and more detailed mapping disclosed that currents in the Atlantic were not as steady or predictable as the earlier climatic maps had suggested. The upshot is that oceanographers have now become interested in two kinds of maps: the climatic map, which shows the average currents over a large area for a year, and the "synoptic" map, which is like a daily or weekly weather report, showing how the currents change from one week to the next. The currents look quite different in the two charts. In the synoptic picture they are narrow, winding and fast; in the climatic picture they are smooth, broad and slow.

Both charts have their uses. If you want to study a long-term phenomenon such as the transport of sediments by currents off the coast of a continent, the climatic chart will be the one you need. On the other hand, if you are piloting a ship or submarine, you will find the synoptic chart much more useful.

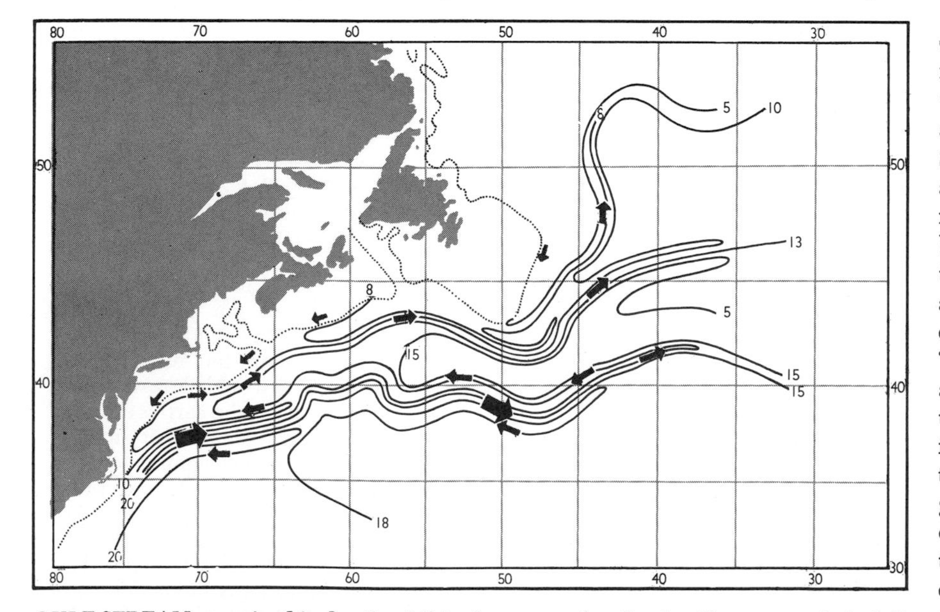

GULF STREAM, examined in detail, exhibits the narrow, fast-flowing filaments typical of all ocean currents. The solid lines mark regions of equal temperature, measured in degrees centigrade. The dotted line indicates the point at which the bottom is deeper than 100 fathoms.

Oceanographers have mapped the general circulation of all the world's oceans, relying mainly on a method which is like that for determining air currents in the atmosphere; that is, the currents are deduced from pressure fields in the sea, which in turn are indicated by measurements of water salinity and temperature [see "The Anatomy of the Atlantic," by Henry Stommel; Scientific American Offprint 810]. The map on facing page summarizes what we know about the climatic circulation of the oceans' surface (the top 1,000 feet).

Is there any system to this complex circulation pattern—any clue to how it may be produced? I think there is, and the chart on page 4 is an attempt to analyze the chief elements of the picture. Suppose we plot the currents that should appear in an idealized rectangular ocean responding to the known winds that blow over the world at the various latitudes. (To simplify things we take into account only the east-west components of the wind system, disregarding "details" such as the winds blowing around the Bermuda high.) The circulation in this schematic ocean then divides into several gyres (rings) corresponding to the wind belts—a counterclockwise gyre in the subpolar region, a clockwise circulation in the subtropical belt above the equator, a narrow gyre on each side of the equator and a counterclockwise gyre in the subtropical region below the equator. In each gyre there is a strong, persistent current on the western side

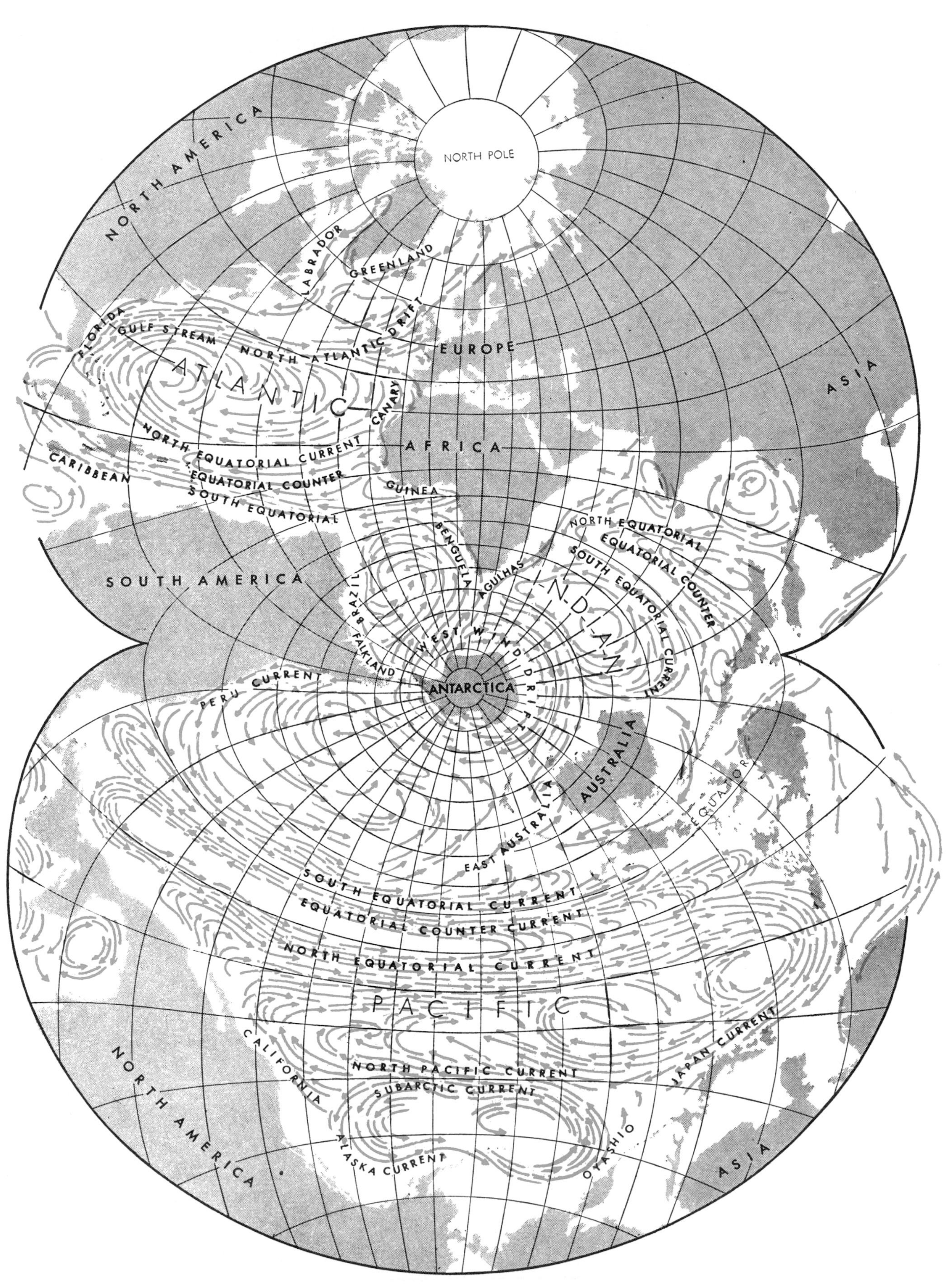

CLIMATIC CIRCULATION in the ocean basins is illustrated on this unusual projection. Colored arrows show the present picture of the average currents. Equatorial countercurrent, which is present in all the oceans, lies five degrees north of the geographic equator.

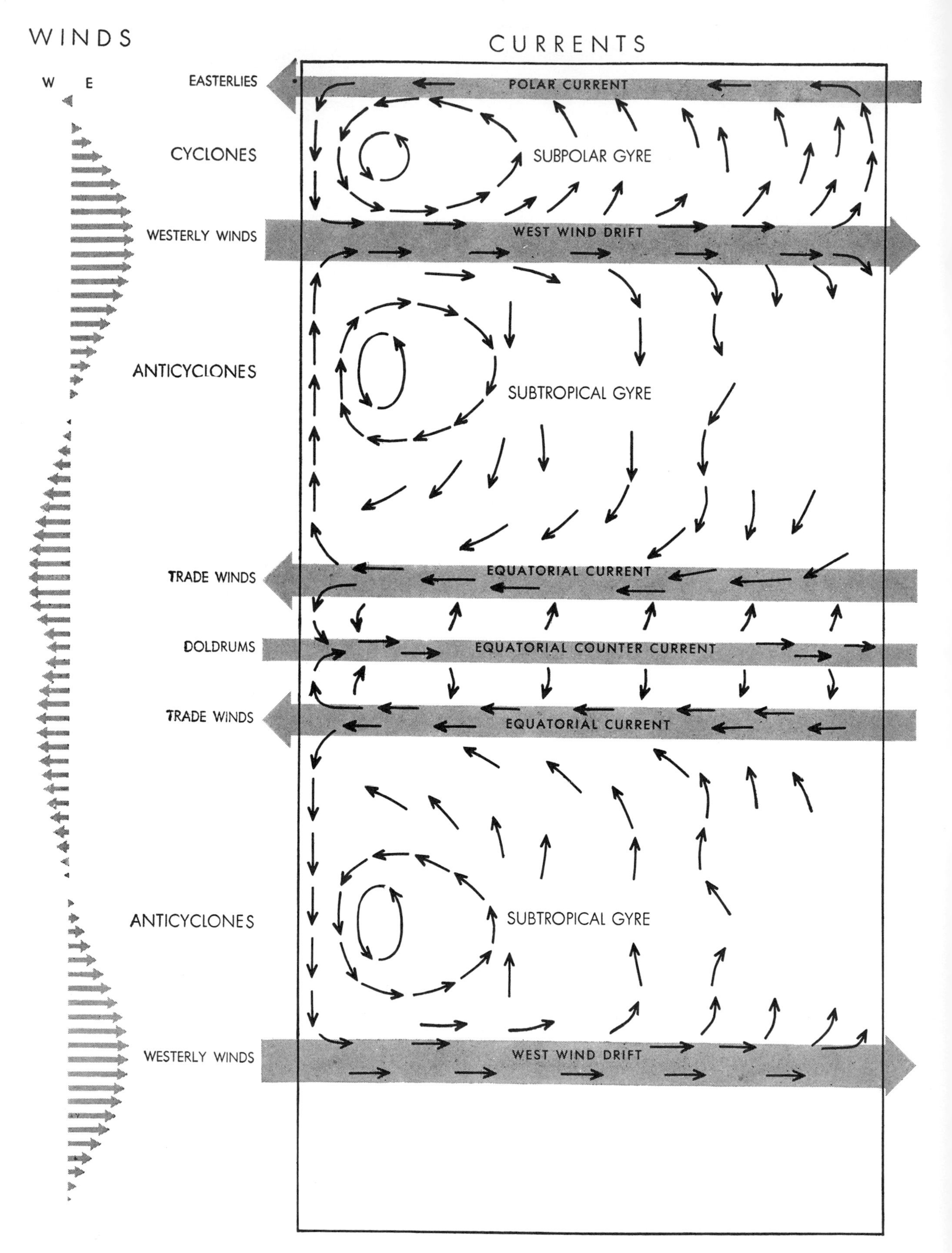

IDEALIZED OCEAN, rectangular in shape and subject only to the horizontal wind forces shown by the broad gray arrows, would have the circulation patterns traced by the black arrows. Approximate relative velocities of surface winds are indicated at left.

(due, as we shall see, to the rotation of the earth) and a compensating drift in the central and eastern portion.

With some imagination we can recognize this pattern in the three major ocean basins of the earth. The strong western current appears as the Gulf Stream in the North Atlantic, the Kuroshio in the North Pacific, the Brazil current in the South Atlantic, the Agulhas in the Indian Ocean, and possibly the East Australia current in the South Pacific. The current driven by the strong west winds in the "roaring forties" of the Southern Hemisphere flows not in a gyre but right around the globe, because no continent stands in its path; this is the mighty Antarctic circumpolar current.

The ocean-current gyres in our picture correspond closely not only to the wind systems but also to chemical and biological properties of the ocean regions. Each subtropical gyre, for example, encloses a sea which is relatively warm, salty, poor in phosphates, low in biological activity and blue in color (blue is the desert color of the sea). At the boundaries of the gyre these conditions change sharply. And the center of each gyre, near the western shore, is an unusually stable environment. The best known such region is the Sargasso Sea in the Atlantic, named after its sargassum, or gulfweed. Very possibly the six other similar regions in the world—the centers of the subtropical gyres in other oceans—will be found to have like populations of floating sea life with narrow environmental tolerances; that remains to be explored.

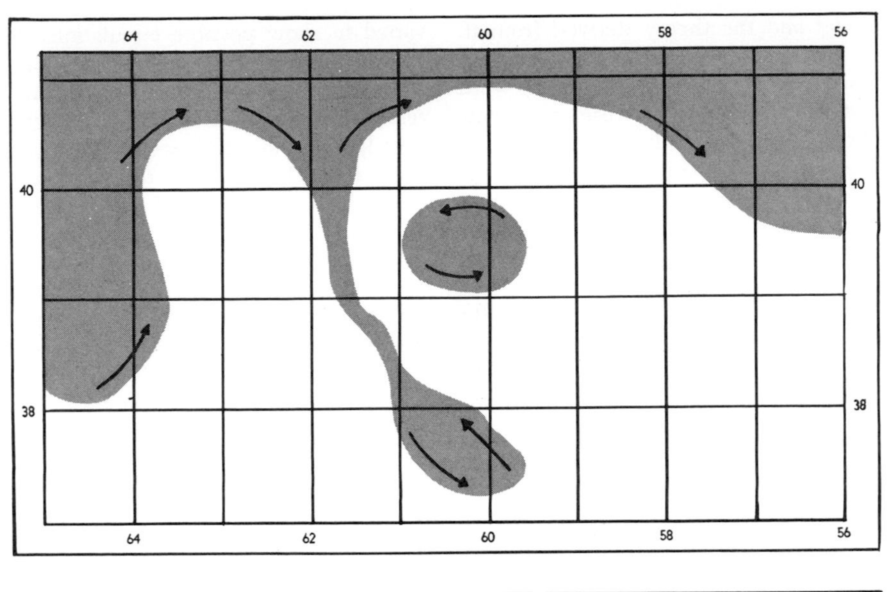

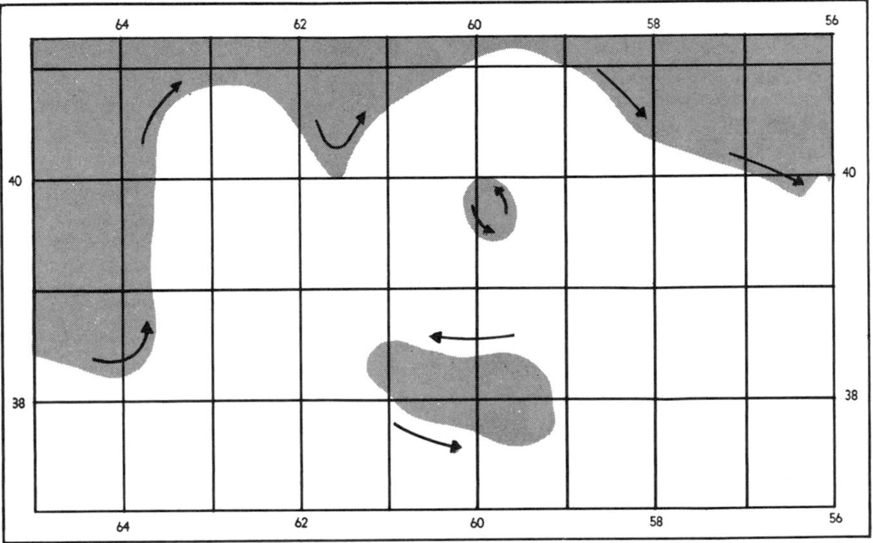

GULF STREAM EDDY begins with the formation of a 250-mile loop in the current (*top*). Four days later, loop has been pinched off (*bottom*). Gray indicates cold subarctic water.

The precise mechanism whereby the winds produce the circulation gyres is complex and not clear. First of all, the action of wind upon water is itself a complicated matter. Wind can move water simply by frictional force as it slides over the surface, even when the surface is smooth. It must also accelerate the motion of water when it picks up spray and throws it down again, particularly during hurricanes, when so much water is pulled up into the air that the "boundary" between the sea and the air is lost. Another important means by which wind drives ocean water is its pressure on the waves as it sweeps over rough water—just as wind blowing over a field bends the blades of grass because pressure is higher on the windward sides than on the lee sides. It turns out that the important elements in the response of water to wind are not the large waves that rock boats and make people seasick, but the tiny bumps, the ripples. If we could cover the North Atlantic with oil and smooth these ripples, the Gulf Stream would lose an appreciable part of its strength. The importance of these tiny waves is surprising. Would any honest seafaring man care to admit that the tiny ripples, to which he paid so little attention, may have been partly responsible for setting him off his course?

How do the driving winds produce the great circulations (gyres) that we see in the oceans? During the last 10 years a theory has been developed. We start with a situation where no land barrier stands in the way of the wind-driven water. The currents will then flow in a great circle around the earth, as they do around the Antarctic Continent [*see page 68*]. Things get more complicated when we introduce land masses. Suppose we erect barriers and make an enclosed sea. Now if winds blow only from the west and have equal force at all latitudes in this sea, there can be no rotary circulation or currents; just as a paddle wheel subject to equal force from the same direction on its opposite blades will not turn. The wind will simply pile up water on the eastern side of the sea. But if the wind is stronger at some latitudes than at others, the stronger will overpower the weaker and the water will begin to circulate. The circulation will be even stronger, of course, if the winds at different latitudes blow in opposite directions. To this effect we must now add the effect of the earth's rotation. The turning of the earth toward the east exerts a torque on the ocean circulation, with the result that the center is displaced toward the west and the currents are intensified on the western side [*see lower drawing on page 69*].

In general the great wind-driven currents in the world's oceans do fit this model and the theory derived from it. The boundaries of the major currents are where they should be in relation to wind systems, and the strong western currents also appear where they should. Moreover, the theory has received some support from a laboratory model simulating ocean circulations. William von Arx, of the Woods Hole Oceanographic Institution, performed these experiments with a rotating basin shaped like a roulette wheel—essentially a hemisphere turned inside out. His "oceans" consist of a thin film of water clinging in an equilibrium distribution over the surface of the whirling basin, and winds are blown on the water from nozzles on vacuum cleaners. Von Arx projects the Northern Hemisphere into this basin, with the North Pole at the low point in the center. Potassium permanganate crystals are placed in the center, and when ink is introduced into the water, it reacts with the chemical to show the flow patterns in different colors. Von Arx's model faithfully reproduces the gyres of the North Atlantic and the North Pacific, including the intense western currents. The model is especially interesting because the topography and winds can be varied to show possible circulations of the oceans in the past, when conditions were different; for instance, one can investigate how the Gulf Stream might have behaved at a time when there was a separation between North and South America in the place of the present Isthmus of Panama.

It must not be supposed that the theory about how the ocean circulations are produced is fully confirmed by these observations and experiments. There are many inconsistencies; in particular, some of the circulation in the oceans of the Southern Hemisphere refuses to fit into the pattern pictured by the theory.

This is where we stand, then, on the climatic circulation. The era of measurement of the synoptic circulation, or day-to-day ocean weather, began with the recent invention of certain new techniques and instruments, notably (1) the radio location method called "loran," (2) the instrument for rapidly measuring temperatures at various depths which is called the "bathythermograph," and (3) an instrument, invented by von Arx and named the "geomagnetic electrokinetograph," which determines the motion of ocean water by measuring the electric potentials induced in it because of its movement through the earth's magnetic field.

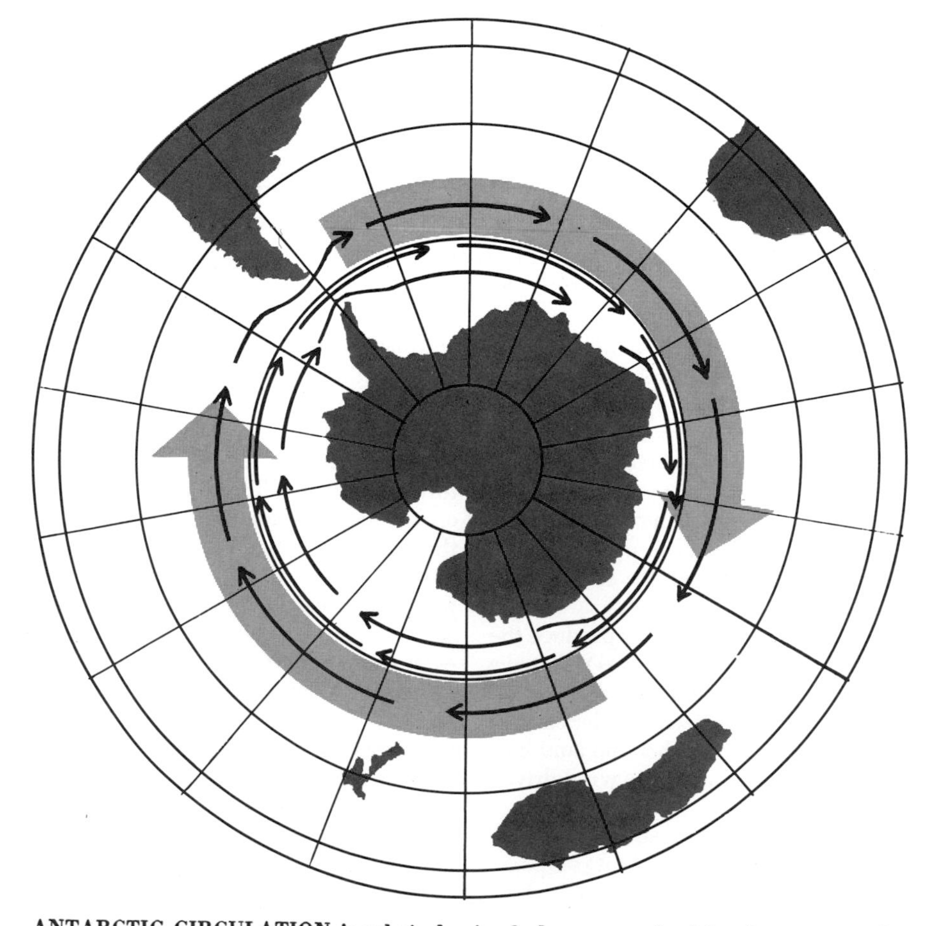

ANTARCTIC CIRCULATION is relatively simple because no land barriers prevent the waters from responding to the prevailing west winds. West-to-east current rings continent.

Resurveying the Gulf Stream with these techniques, Columbus O'Donnell Iselin and his collaborators at Woods Hole discovered that the Stream was narrower and much faster than had been thought. As their instruments and techniques improved, the current became even narrower and faster. They also found that the position and direction of the current varied from one cruise to the next. A five-ship expedition called Operation Cabot was organized by the U. S. Navy Hydrographic Office in 1950 to study the Gulf Stream more closely. This cruise detected a most important and dramatic phenomenon: the Gulf Stream meandered off the usual course to form a loop 250 miles long! Within two days the loop broke off and separated as an independent eddy [*see charts on page 67*]. The eddy then gradually weakened.

It is estimated that this single eddy injected some 10 million million tons of subarctic water from the North Atlantic into the subtropical Atlantic. Obviously such an immense transport of water, with its content of living organisms, must be of considerable importance to the biology of the sea. Possibly similar eddies of water from the south break off toward the north, injecting subtropical water into the colder part of the ocean.

Frederick Fuglister of Woods Hole, an artist who has been in oceanographic work since the war, later discovered some other unsuspected characteristics of the Gulf Stream. Plotting currents by means of temperature gradients measured with the bathythermograph, he found a pattern which suggested that the Gulf Stream consists of a number of long, narrow, separate ribbons, or filaments. They are not continuous over thousands of miles; as a rule one will peter out and another will start somewhere else. In other words, it appears that the concept of a single, continuous Gulf Stream all the way from Florida to Europe must be abandoned. Rather one must visualize the Stream as composed of high-speed filaments of current separated by countercurrents [*see chart on page 64*]. L. V. Worthington of Woods Hole, using all the modern tools, has substantially confirmed this picture with detailed cross-section studies. In one 30-mile cross section he found three separate major filaments, each flowing at better than three miles per hour. Gunther Wertheim, also of Woods Hole, further demonstrated the complexity and variability of the Gulf Stream by discovering that the transport of water by the Florida current section of the Stream doubled from one month to the next! He computed the movement of water from measurements of electric potential between Havana and Key West, made by attaching electrodes to the Western Union telegraph cable between those points.

Fuglister has satisfied himself that the Japanese current also can be interpreted as consisting of filaments; in fact almost everywhere we look the ocean weather seems extremely fickle. Henry Stommel, monitoring radio drift buoys near Bermuda, found the currents highly changeable; every sudden waxing or waning of the winds set up rotary currents.

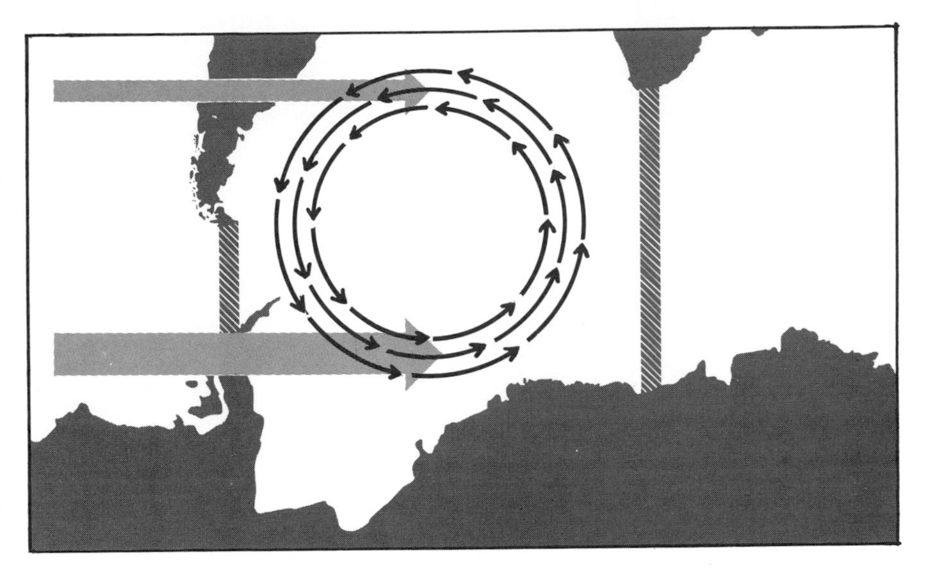

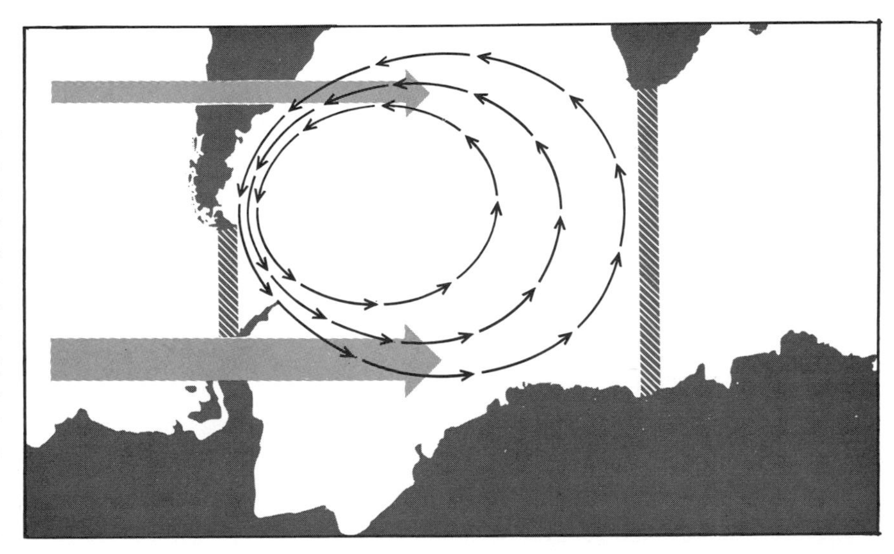

IMAGINARY SMALL OCEAN, made by enclosing part of the Antarctic waters with walls extending to South America and Africa, would circulate as shown at top if west winds in the south were stronger than those farther north. The effect of the earth's rotation would be to shift circulation pattern westward, as at bottom, compressing currents at the west wall.

My interpretation of the new look with regard to the ocean weather is something like this. The motion of water in the open sea is highly irregular and variable. If we release a drift buoy, we can expect the current to carry it something like half a mile in an hour, but the velocity and the direction will be quite different from one day to the next. This unsteady motion—the "noise" of the ocean circulation—represents in some way the response of the sea to the multiplicity of shocks it receives from the wind blowing on its surface. The response is not simple, and the underlying laws have not yet been recognized. The transient ocean weather, unlike the slow climatic circulation, apparently has no blow-by-blow counterpart in the circulation of the atmosphere.

The fine structure of the ocean currents can be tied in with the climatic circulation only in a general way. It evidently results from the fact that the broad circulation cannot dissipate all the energy received by the ocean from the wind, but just why the fine structure takes the forms it does is a problem awaiting further exploration.

THE CIRCULATION OF THE ABYSS

HENRY STOMMEL
July 1958

There is a good deal of talk these days about controlling the world's climate. Optimistic promoters of the earth sciences hold out visions of turning tropic deserts or arctic wastes into temperate and fertile plains. Of course we could not hope to do this by brute force. To deflect major wind systems or ocean currents, or to heat the outdoors, would call for engineering works on a scale that man cannot even dream of. But, some people suggest, perhaps if we knew enough about the mechanics of the atmosphere and the ocean circulations we might be able to find some critical time and place where a relatively small man-made disturbance could set off a snowballing reaction which would produce a major alteration in weather patterns. Actually this prospect is quite remote. It does add some spice, however, to a study such as oceanography.

The general circulation of the world's oceans is a matter of great interest, not only from various practical points of view—climate, fishing, dumping of ra-

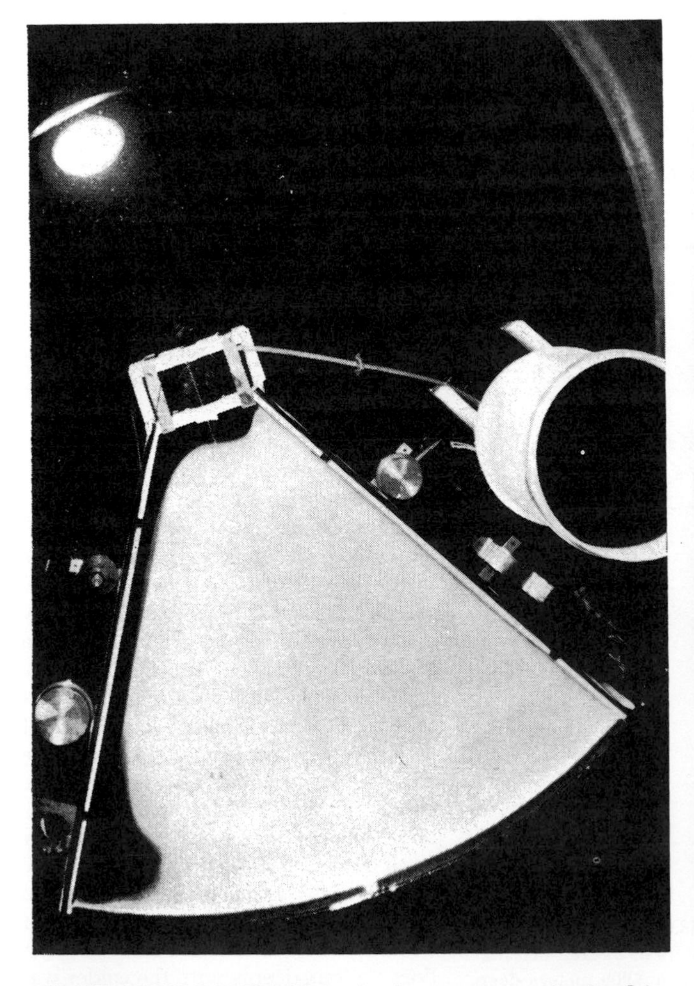

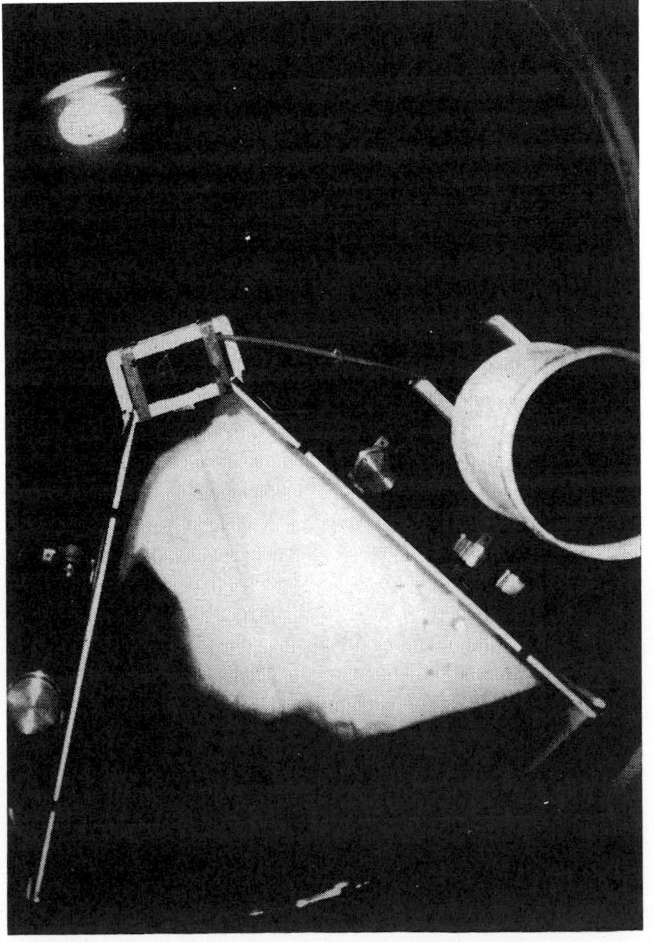

CIRCULATION PATTERNS in the deep ocean are simulated in a rotating, wedge-shaped basin. Ink poured in at the point of the wedge (the "pole") flows down along "western" edge (*left*), turns and flows over the rest of the bottom (*right*) toward the "pole".

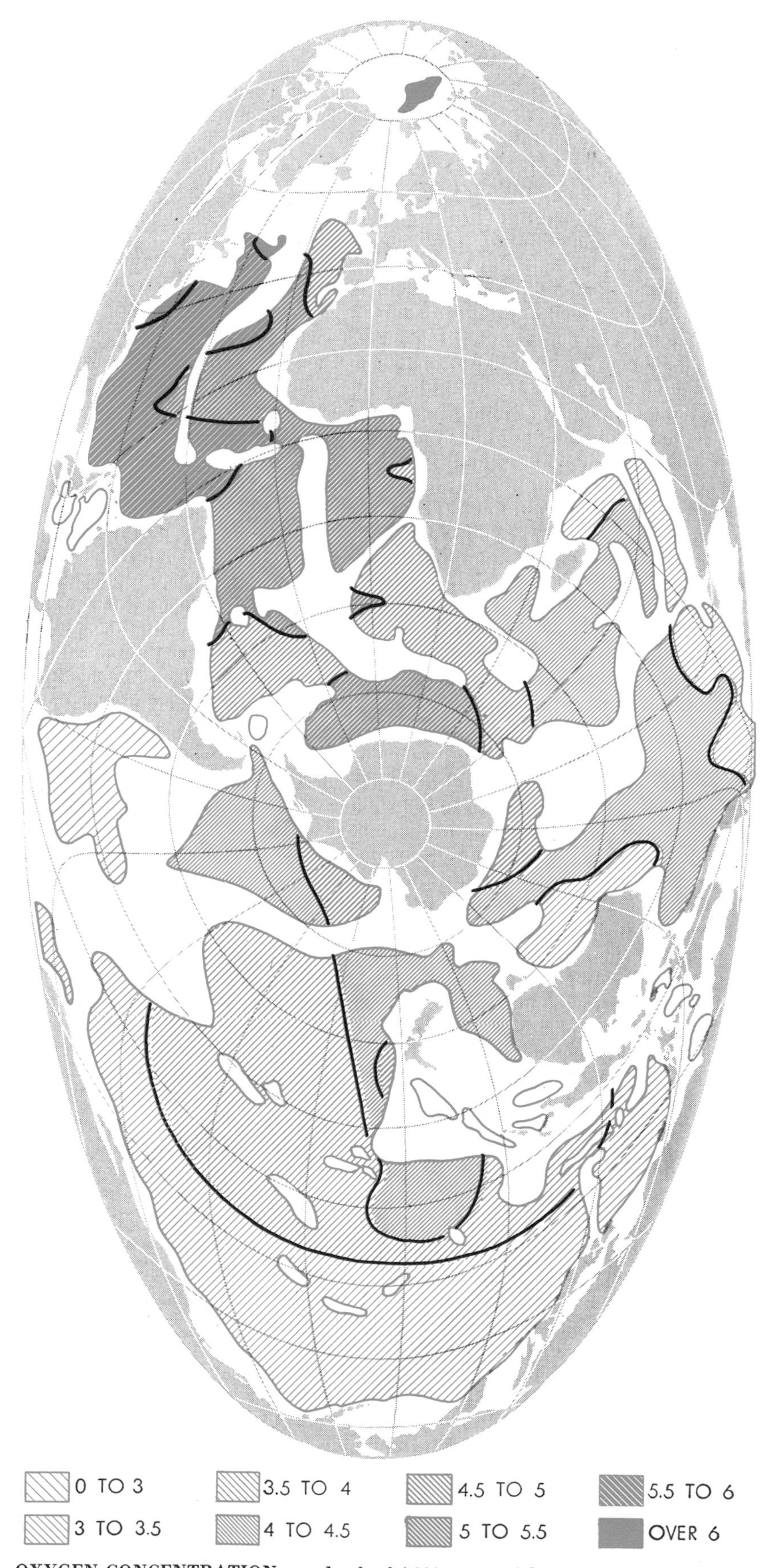

OXYGEN CONCENTRATION at a depth of 4,000 meters (about 13,000 feet) is indicated by colored areas on map. The numbers given in the key are milliliters of oxygen per liter of water. White areas are regions where the ocean bottom is less than 4,000 meters deep. This projection places the North Pole at top and the South Pole near the center of the map.

dioactive wastes and so forth—but primarily from the standpoint of understanding the dynamics and history of the planet on which we live. We know that the great surface currents (*e.g.*, the Gulf Stream) are important to us. Even more significant is the circulation of the oceanic water as a whole. And to get any sort of picture of this circulation we must find out about the movements of water in the ocean deeps. This is difficult to do: we have had no reliable means of getting direct, accurate measurements of the currents or the massive general flow at great depths. Our inferences have to be based mainly on indirect indications such as the comparative densities, salinities and pressures of the water at various places and various depths.

One item of evidence is the amount of dissolved oxygen in the waters of the abyss. Ocean water receives oxygen only at or near the surface—by direct contact with the atmosphere and by the action of photosynthesis in floating plants. After the water sinks to deeper levels it gradually loses some of its dissolved oxygen: the gas is consumed in chemical reactions with dead microorganisms. Therefore the amount of dissolved oxygen in a sample of deep water is a rough index to the "age" of the water since it sank from the surface.

Now when we measure the oxygen content of the deep water in various parts of the oceans, a distinctive pattern emerges [*see map at the left*]. The "youngest" deep waters (richest in oxygen) are found in the western North Atlantic and around the Antarctic Continent. The concentration in the other oceans diminishes with increasing distance from these sources: the water poorest in oxygen is that at the bottom of the vast and little explored basin off Peru. It thus appears that there are only two important regions where substantial amounts of water sink from the surface to the abyss: in the North Atlantic and around Antarctica. This water evidently spreads gradually into the Indian Ocean and finally northward into the Pacific.

The oxygen studies tell us, then, that there is a general sluggish circulation of water along the ocean bottoms from the North Atlantic and the Antarctic to the other oceans. We have another indication of this global flow. The deep water in all the oceans is very cold—only a few degrees above freezing. This is true even in the tropics; only the top 1,500 feet of water is warm. The warm surface water does not mix deeply with the colder water beneath it; this must mean that the

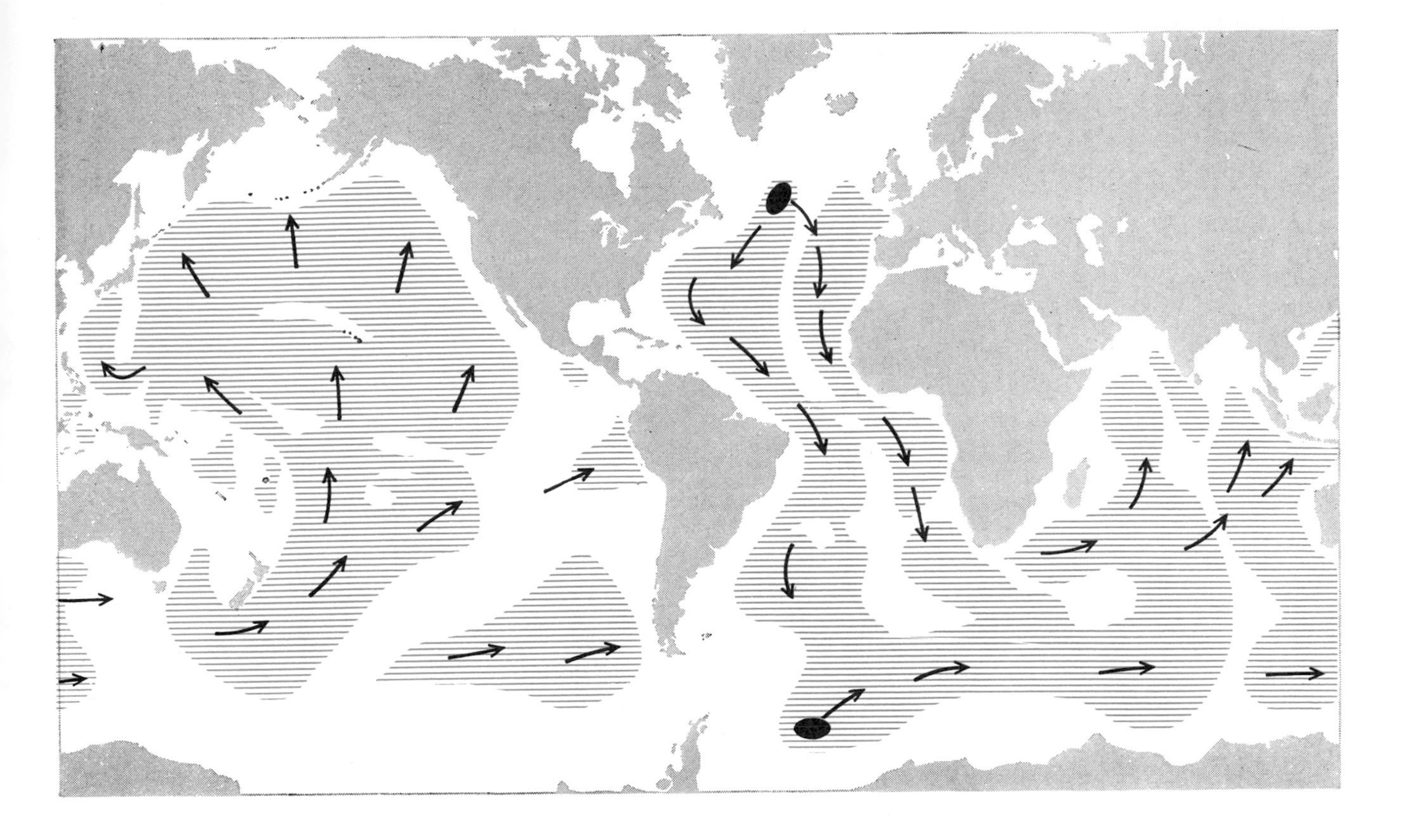

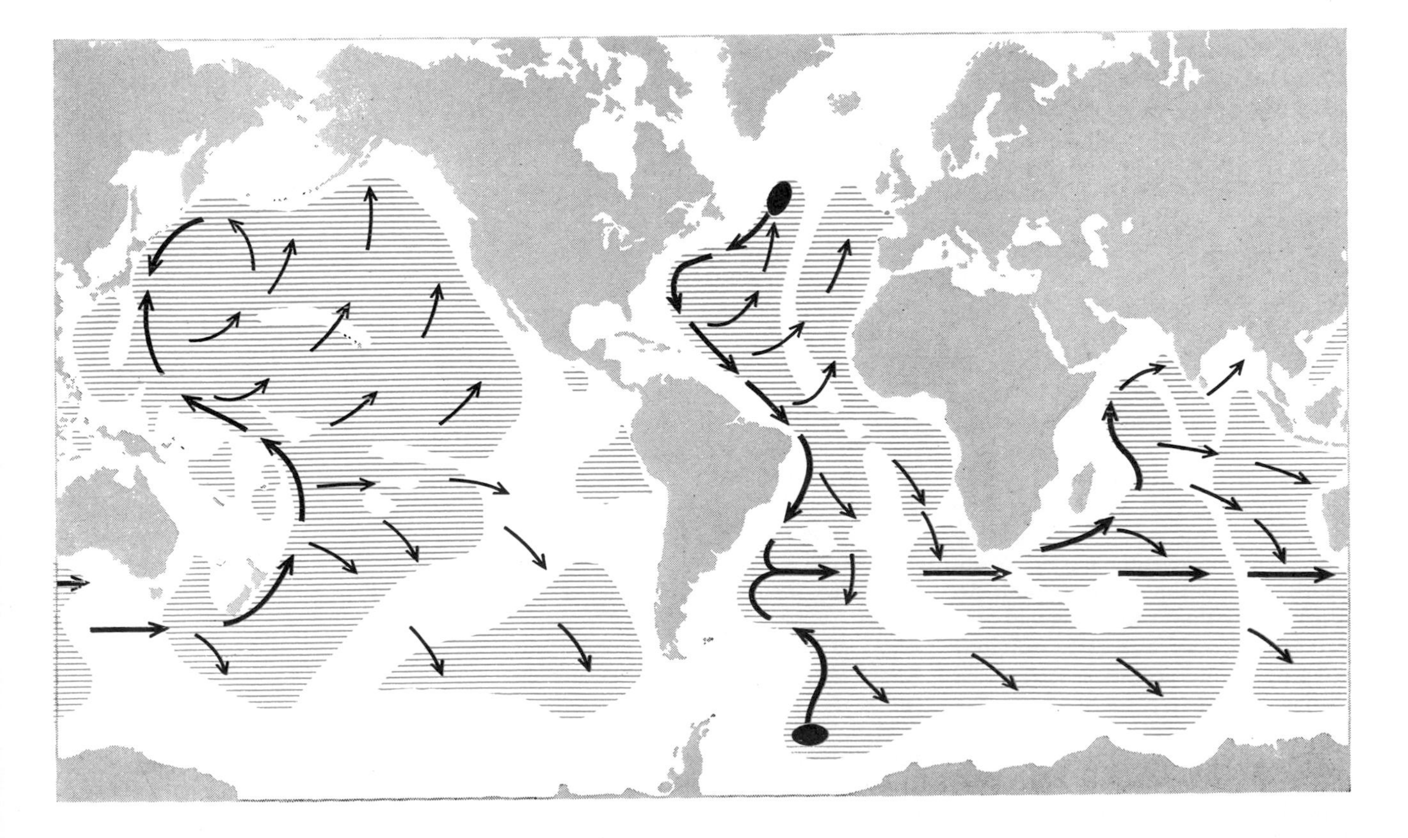

TWO THEORIES of the deep circulation are represented on these maps. On the older view (*top*) water flows in broad currents toward the Equator from its sources (*black ovals*) in the North Atlantic and the Antarctic, and then spreads through the Indian Ocean up into the Pacific. According to the author's theory (*bottom*) the flow away from the sources is confined to intense currents on the western edge of the oceans. These in turn feed a broad flow that carries water toward the poles in each of the ocean basins.

warm layer is held up by a continual upwelling of cold water from below. This slow rise of water (about an inch per day at mid-depths) represents a movement of water amounting to several thousand times the total daily discharge of the Mississippi River. Obviously all that water continually coming up from the depths must be replaced by an inflow into those depths—*i.e.*, it betokens a massive deep circulation.

The simplest circulation pattern that would fit the facts we have reviewed is a broad, sluggish, spreading current coming straight down the Atlantic and joining the current from the Antarctic to flow through the Indian Ocean and then up into the Pacific [*see upper map on the preceding page*]. Until recently most oceanographers accepted this general picture. But I believe that the rotation of the earth and considerations of fluid dynamics make another picture more likely. The general reasoning behind this new theory is as follows.

In the open ocean the flow of the bottom water must be different from that in a river. It should be more like the flow of air in the atmosphere. As every reader of weather maps knows, air does not travel directly from high-pressure to low-pressure areas. It circulates around the highs and lows, because of the so-called Coriolis force arising from the earth's rotation. That is, the air travels along the lines of equal pressure, or isobars. Now my own construction of the pattern of isobars in the deep ocean, based on dynamical considerations and on the requirements for the supply of upwelling water, shows a flow from the Equator toward the poles. The flow originating near the poles, it then appears, may be carried by strong currents along the western sides of the ocean basins, just as there are strong currents along these routes at the ocean surface [see the article "The Circulation of the Oceans," by Walter H. Munk, beginning on page 64]. We postulate a strong, deep current running down the western North Atlantic and up the western South Atlantic. When they come together, the currents merge and turn eastward, flowing up the eastern coasts of Africa and Asia [*see lower map on the preceding page*].

Two independent observations have partly confirmed this theoretical scheme. Last year a joint British-American oceanographic expedition found a strong deep current in the western Atlantic. It flows beneath the Gulf Stream but in the opposite direction, from north to south, just as the theory predicts. Detection of the current was made possible by a new device for investigating the motion of deep water, invented by the British oceanographer John C. Swallow. It is a float which is cleverly designed to seek a predetermined level under the surface and then to remain there, drifting with the current. The float has a small ultrasonic transmitter sending signals which locate its position. The expedition made the soundings off Charleston, S.C. They found that floats lowered to levels below 6,500 feet drifted southward at velocities between two and eight miles per day.

The other confirmation of the new circulation scheme was supplied by the German oceanographer Georg Wüst. Analyzing the pressure distributions in the South Atlantic, he was able to show that the flow of deep waters in that ocean is confined to relatively narrow streams along the continental slopes off Brazil and Argentina.

Thus we now can score two modest victories for the theory. I think it sporting to risk a few further predictions. We should find (1) a countercurrent underneath the Agulhas Current off the East Coast of Africa, (2) a narrow northward current in the western South Pacific, let us say along the slopes of the Tonga-Kermadec Trench, and (3) only a weak flow underneath the Japan current.

The theory has also been buttressed by some laboratory experiments at the Woods Hole Oceanographic Institution. Liquid in a flat-bottomed, rotating basin has the same type of circulation patterns as water on the spherical, rotating earth. Alan Faller at Woods Hole set up a wedge-shaped section in the basin to simulate a single ocean. The point of the wedge corresponds to a pole of the earth, the sides to the coastlines of the ocean and the outer perimeter to the Equator. When this sector was set into rotation and some dyed water was poured into the liquid at the "pole," the dye moved through the "ocean" just as the theory predicts. First it flowed in a narrow current along the "western" edge; at the "Equator" it turned and flowed slowly over the basin toward the center [*see photographs on page 71*].

With ingenuity you can reproduce this experiment yourself in a small tin dish on a phonograph turntable. The dish should be covered with a sheet of glass to prevent air currents from disturbing the water surface. Most turntables revolve in a direction opposite to that of the earth, so the boundary cur-

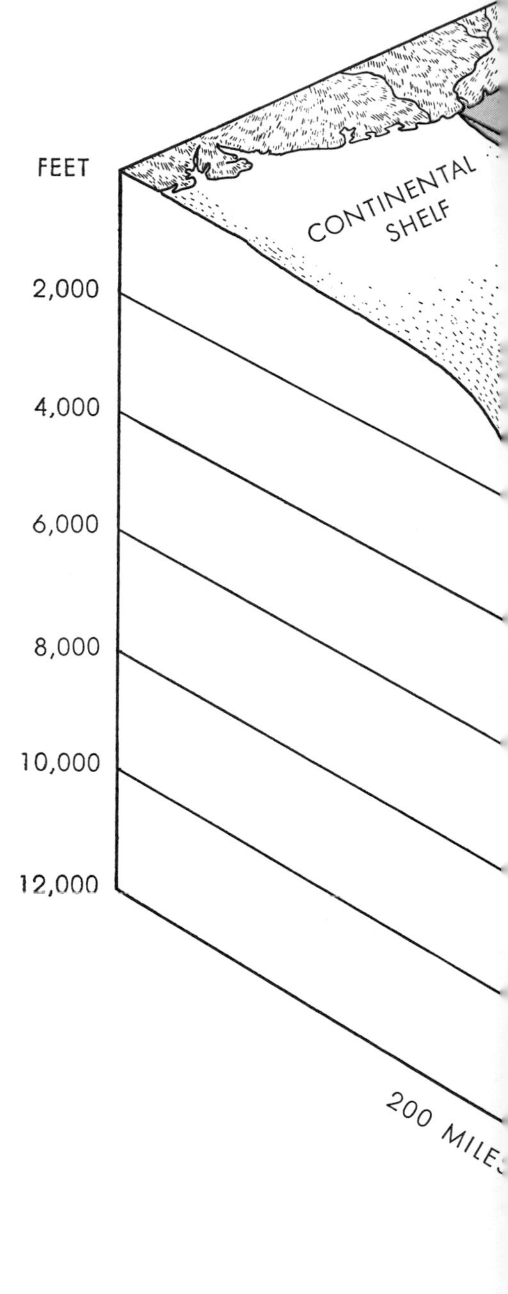

DEEP CURRENT flowing southward under the Gulf Stream off Charleston, S.C., was discovered last year by a joint British-

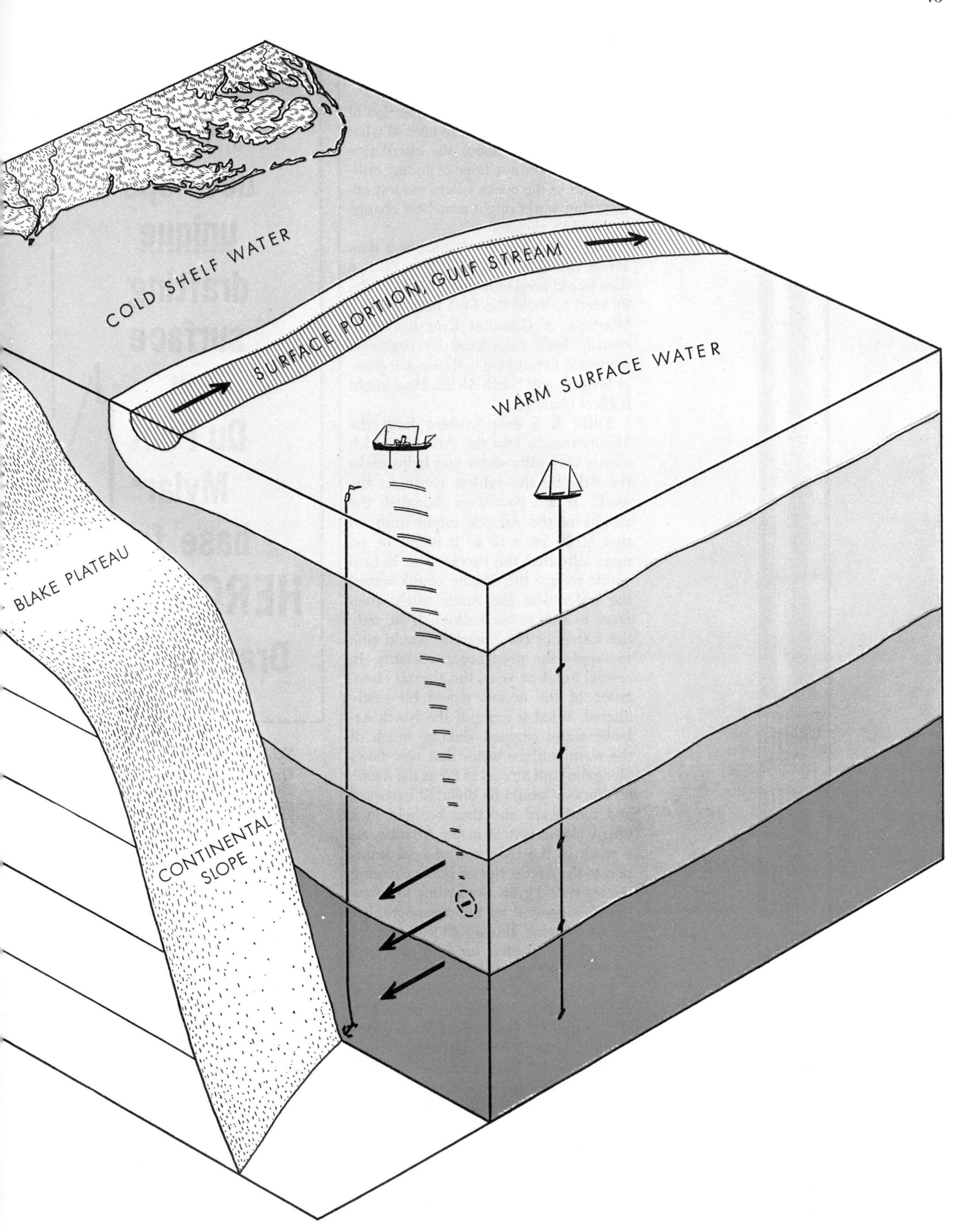

American oceanographic expedition. This cutaway view shows the British vessel *Discovery II* (*left*) receiving ultrasonic signals from a submerged float (*black bar*) which is drifting with the deep current. The American ship *Atlantis* (*right*) is recording other data such as temperature, salinity and density. The float is the first device with which undersea circulations can be measured directly.

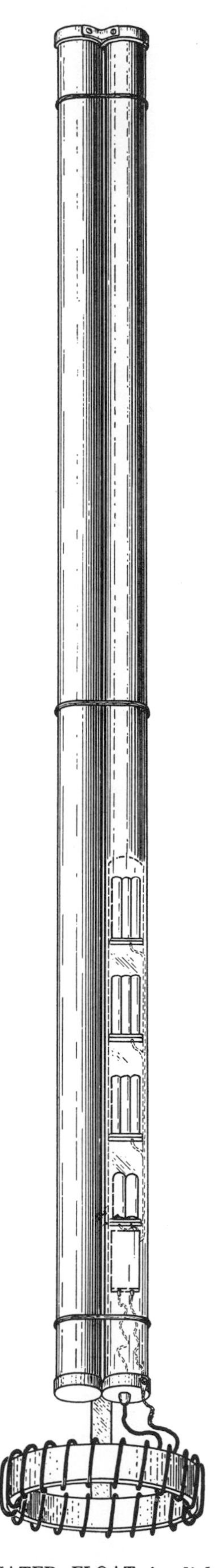

UNDERWATER FLOAT is slightly less compressible than sea water; thus it sinks until its density equals that of surrounding liquid. Oscillator in the tube at right sets up ultrasonic vibrations in the ring at bottom.

rent will probably show up on the "eastern" wall instead of the "western."

Now let us go back to the question of controlling climate. In view of what we have learned about the circulation pattern, is there any hope of finding critical areas in the ocean where modest engineering works might somehow change the world's climate?

The most attractive fantasy is a dam across the Straits of Gibraltar. Such a dam would need only about 10 times the fill used to build the Fort Peck Dam in Montana. A Gibraltar dam has occasionally been considered by engineers interested in bringing hydroelectric power to Spain and North Africa. How might it affect climate?

There is a deep current from the Mediterranean into the Atlantic which carries very salty water and helps make the Atlantic the saltiest ocean in the world. If this flow were dammed, the salinity of the Atlantic might drop, so that in 30 years or so it might be no more salty than the Pacific. This in turn would reduce the density of the water; the water near the Arctic might then cease to sink to the bottom. If so, only the waters of the Antarctic would sink to supply the deep ocean currents. In several hundred years the abyssal circulation of the oceans would be vastly altered. What is more, if the North Atlantic water stopped sinking, much of the warm surface water that now flows along the Gulf Stream as far as the Arctic off Norway would be diverted eastward and southward and thus be held in a nearly closed system in the Atlantic. As a result of the reduction of heat transport to the Arctic, the ice packs covering the sea would grow. According to a new theory suggested by the oceanographer Maurice Ewing, this would lead to a decline of the glaciers on land and to a general warming of the earth.

Common sense rebels against such an argument. It is hard to imagine so fantastic an effect from so small an intervention by man. And indeed the argument is loaded with unproved assumptions and tenuous speculations. We could construct an equally plausible argument that the same stratagem might cool rather than warm the earth. I cite this entertaining fantasy only to show that we need a great deal more information before we can begin to talk knowledgeably about altering the climate. All such speculations merely illustrate how little actual knowledge we have and how valuable it would be to develop a better quantitative understanding of the ocean circulation.

THE SARGASSO SEA

JOHN H. RYTHER
January 1956

For more than four centuries the Sargasso Sea has been wrapped in a legend which has frightened mariners and still fascinates armchair travelers. The seaweed to which the sea owes its name has always given it an aura of uniqueness and mystery. Early voyagers in it pictured the sea as a vast, impenetrable mass of floating vegetation in which hapless ships might become imprisoned with no hope of ever escaping. The legend still lives. As recently as 1952 Alain Bombard, the French physician who crossed the Atlantic alone in a life raft, carefully plotted his course to avoid the Sargasso Sea, because, as he said: "The whole area has always been a major navigational hazard, a terrible trap, where plant filaments and seaweed grip vessels in an unbreakable net."

Ship captains who regularly cruise to Bermuda, which lies in the middle of this sea, must smile at such accounts. Yet the Sargasso Sea remains an intensely interesting body of water, in many ways more interesting than the romantic but mistaken legends about it. Generations of scientists have sailed forth to study it, and the Sea has rewarded them with many unexpected discoveries. Most unexpected of all is the paradox that the Sargasso's masses of seaweed hide a biological wasteland. Contrary to what the floating vegetation might suggest, the Sargasso Sea is not a jungle teeming with life but one of the great oceanic deserts of the earth.

Christopher Columbus noticed the unusual plant life of these western waters on his first voyage across the Atlantic in 1492. He began to encounter floating seaweed not far west of the Azores, and by the time he reached mid-ocean there was "such an abundance of weeds that the ocean seemed to be covered with them." When the ship was becalmed for three days, his men grew alarmed, for they feared that the masses of vegetation covered coastal waters with submerged rocks and reefs—little realizing that the ocean bottom lay nearly three miles below them.

Columbus described the weeds in some detail in his log, and later explorers brought back further tales of these strange waters. Portuguese sailors gave the sea its name: air bladders on the floating seaweed reminded them of small grapes at home which they knew as "salgazo." Through the centuries the legend grew, as ship captains traveling between the Old World and the New reported encounters with the greatest accumulations of weeds they had ever seen. The legend became so firmly established that in 1897 the Sargasso Sea was described by the *Chambers' Journal for Popular Literature, Science and Arts* in these terms: "It seems doubtful whether a sailing vessel would be able to cut her way into the thick network of weeds even with a strong wind behind her. With regard to a steamer, no prudent skipper is ever likely to make the attempt, for it certainly will not be long before the tangling weeds would altogether choke up his screw and render it useless."

When William Beebe sailed on the much-heralded expedition of the *Arcturus* in 1925, the reading public was keyed up by lurid predictions of sea monsters that would be found in the great weed beds of the Sargasso. The expedition was an unbelievable disappointment. Not only was there a total absence of sea monsters, but in all the area of the Sargasso Sea over which he voyaged Beebe could find no patches of seaweed larger than a man's head!

Beebe was unlucky. The Sargasso Sea is rarely as barren as he found it along his route. Nonetheless scientists have known for a century that its reputation is greatly exaggerated. It is doubtful that the Sargasso's weed masses are ever dense enough to impede the progress of even the smallest vessel. And indeed the floating *Sargassum*, though intriguing enough in its own right, is no more than a surface outcrop of a great oceanic phenomenon.

What exactly is the Sargasso Sea? The scientific study of this huge sea without shores began with attempts to define its borders by charting the extent of the seaweed. In 1881 the German scientist O. Krümmel analyzed the reports of German sea captains, who for many years had been required to record their observations of drifting weeds in the Atlantic. He concluded that the Sargasso Sea covered an area of some 1,720,000 square miles—an area elliptical in shape and extending from the mid-Atlantic to near the North American coast.

In 1923 a Danish botanist, O. Winge, made a second attempt at the same problem. He had the advantage of information on regular collections of seaweed made with net tows by Danish ships plying the Atlantic, which gave more systematic data than the estimates of the German sea captains. Winge decided that the Sargasso Sea was considerably bigger than Krümmel had pictured it. He placed its eastern boundary near the Azores (at exactly the point where Columbus had located the first weed masses) and its southern boundary somewhere near the West Indies. The western and northern borders of the weed area, he found, shifted considerably from season to season; this he attributed to changing weather conditions.

In the 1930s and 1940s oceanographers made an altogether different approach to defining the Sargasso Sea. Columbus O'Donnell Iselin, then director of the Woods Hole Oceanographic Institution, pointed out that the circular system of currents in the North Atlantic Ocean would outline the boundaries of the Sargasso Sea more definitely than drifting seaweed could. These currents are the Gulf Stream and North Atlantic Current on the western and northern sides, the Canaries Current on the eastern side, and on the south the slow movement of water parallel to the equator which is known as the North Equatorial Drift. This ring of currents encloses a great eddy, some two million square miles in area, which rotates slowly clockwise under the influence of the earth's rotation. Detailed studies have now made clear that this eddy of surface water is the Sargasso Sea.

Because of variations in the currents, the borders of the Sea are not constant or sharply defined. The Gulf Stream shifts and meanders, hence it is small wonder that Winge found the weed boundaries shifting in this region. On the south the equatorial drift also changes position from season to season. And on the east the Canaries Current is so weak and diffuse that it can barely be detected, much less provide a barrier to the movement of water or seaweed.

We must therefore look below the surface to get a clearer picture of the Sargasso Sea. When we do, we find that temperature measurements mark out a distinct, clearly defined body of water. The Sargasso Sea is a huge lens of warm water, separated from the colder layers below by a zone of sharply changing temperature.

It is lens-shaped because the rotation of the eddy piles up water at the center (where the sea surface is about two feet higher than at the outer edges). At its deepest the layer that defines the Sargasso Sea goes down no more than about 3,000 feet. In other words, the Sea proper is a shallow body of fairly homogeneous water lying upon an ocean whose total depth is roughly five times as great as this layer. Tracing the borders of the lens by temperature measurements, oceanographers find it is bounded on the west and north by the Gulf Stream and North Atlantic Current, on the south by the equatorial drift and on the east roughly by a line which runs along the submerged mountain ridge in the middle Atlantic. The Sargasso's seaweed drifts almost 1,000 miles farther east, but from a hydrographic point of view the lens of water that defines the Sea ends here.

What makes the Sea's weeds collect in masses? Most commonly they lie in long parallel bands, sometimes stretching as far as the eye can see. Some of these formations undoubtedly are due to the major current systems, piling

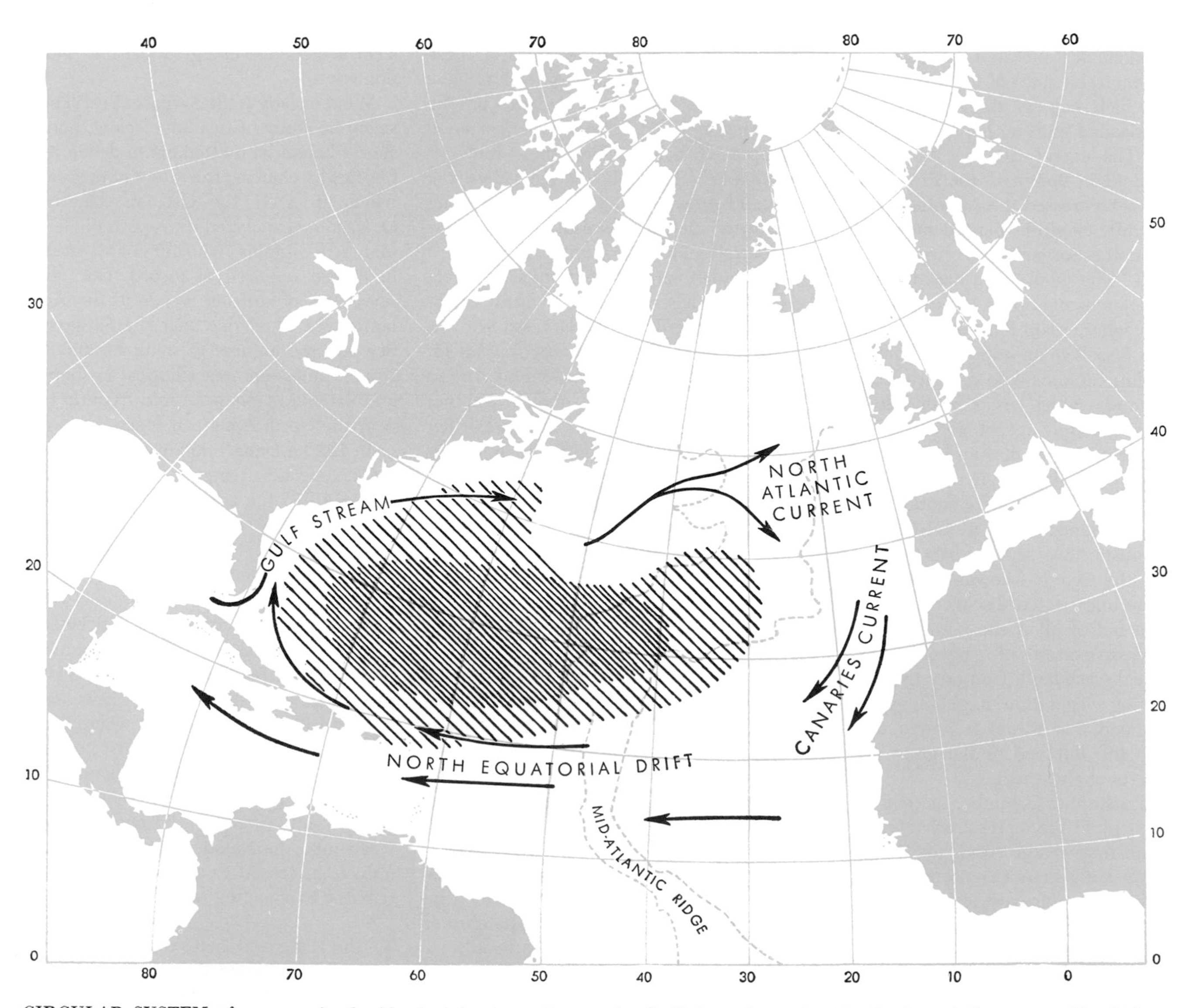

CIRCULAR SYSTEM of currents in the North Atlantic outline the Sargasso Sea. The darker shading represents the Sea as charted by O. Krümmel on the distribution of *Sargassum*. The lighter shading indicates a similar attempt made by the botanist O. Winge.

bands of weeds along the lines where water masses of different densities converge. But winds also can produce them. Winge reasoned that such bands might grow by accumulation as weeds sailed before the wind, picking up more and more weeds in their wake. However, the physicist Irving Langmuir, on a voyage across the Atlantic in 1927, noticed that when the wind veered about at right angles to its former direction, the seaweed bands re-formed in the new direction within 20 minutes. Since mere cohesion could not explain this rapid reorientation of weed streamers, Langmuir suggested that shifts in the flow of water, rather than the wind itself, must be responsible for the formation of the bands. He later demonstrated experimentally that the action of wind over open fetches of water produces counter-rotating eddies, and that between such eddies there are bands of sinking water where floating weeds would collect.

The Sargasso weeds themselves raise many interesting questions. The most intensive study of them was carried out between 1932 and 1935 by Albert E. Parr, then director of the Bingham

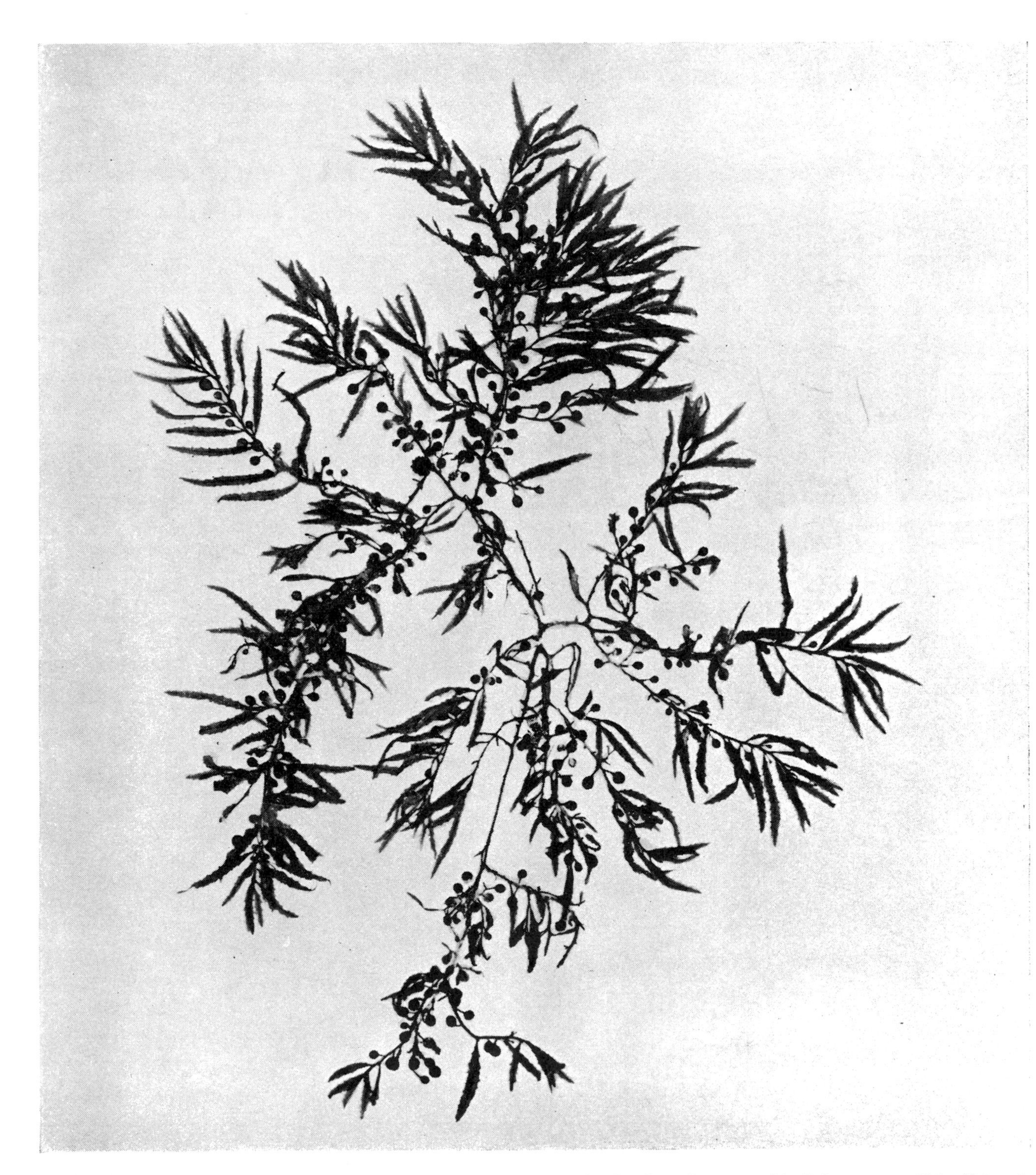

SARGASSUM FLUITANS, one of perhaps eight species of the weed, was collected in the Sargasso Sea by Albert E. Parr, then at the Bingham Oceanographic Laboratory at Yale University and now director of the American Museum of Natural History.

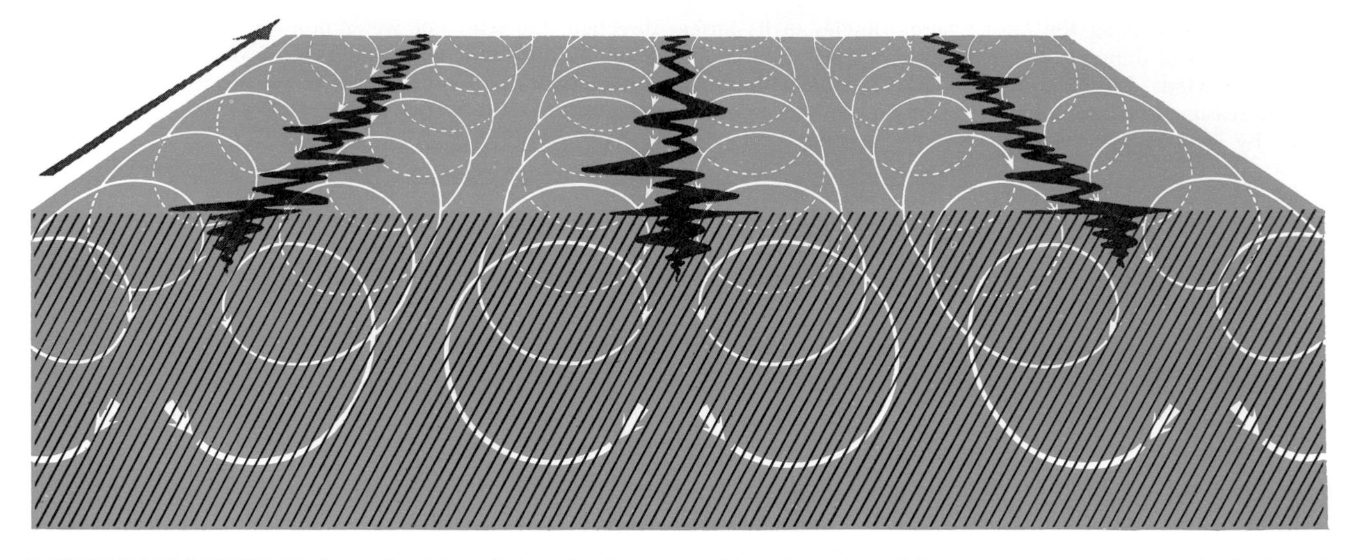

LANES OF SARGASSUM (*dark gray bands*) are believed to form as shown in this diagram. The drag of the wind (*arrow at upper left*) on the surface of the water sets up counter-rotating horizontal eddies (*white arrows*). The weed collects where the water sinks.

Oceanographic Laboratory at Yale University. In three cruises on the Woods Hole research ship *Atlantis*, covering 7,000 miles, he collected nearly 5,000 pounds of drifting Sargassum. More than 90 per cent of the weeds by bulk were of two floating species which are never found attached to the ocean bottom and lack organs for sexual reproduction. The question therefore arises: Where do the weeds come from, and how do they grow and reproduce?

Columbus theorized that the drifting weeds were torn loose from great submerged beds of plants near the Azores, and his theory was later shared by Alexander von Humboldt and other naturalists. But no such beds have ever been found, either near the Azores or Bermuda. Some botanists consequently have proposed that the weeds come from banks in the West Indies or the Gulf of Mexico. This theory too has been proved unlikely. From his sampling of the Sargasso Sea, Parr estimated that the Sea has an average standing crop of some seven million tons of weeds. No more than a small fraction of this crop could be supplied by all the available sites for beds along the entire Caribbean and Atlantic Coasts, even assuming that such beds exist. Moreover, it would probably take several years for weeds torn loose from the West Indies to drift far enough to span the whole Sargasso Sea, and weeds uprooted from their beds could not live more than a few months.

The floating Sargassum gives every evidence of growing, reproducing and

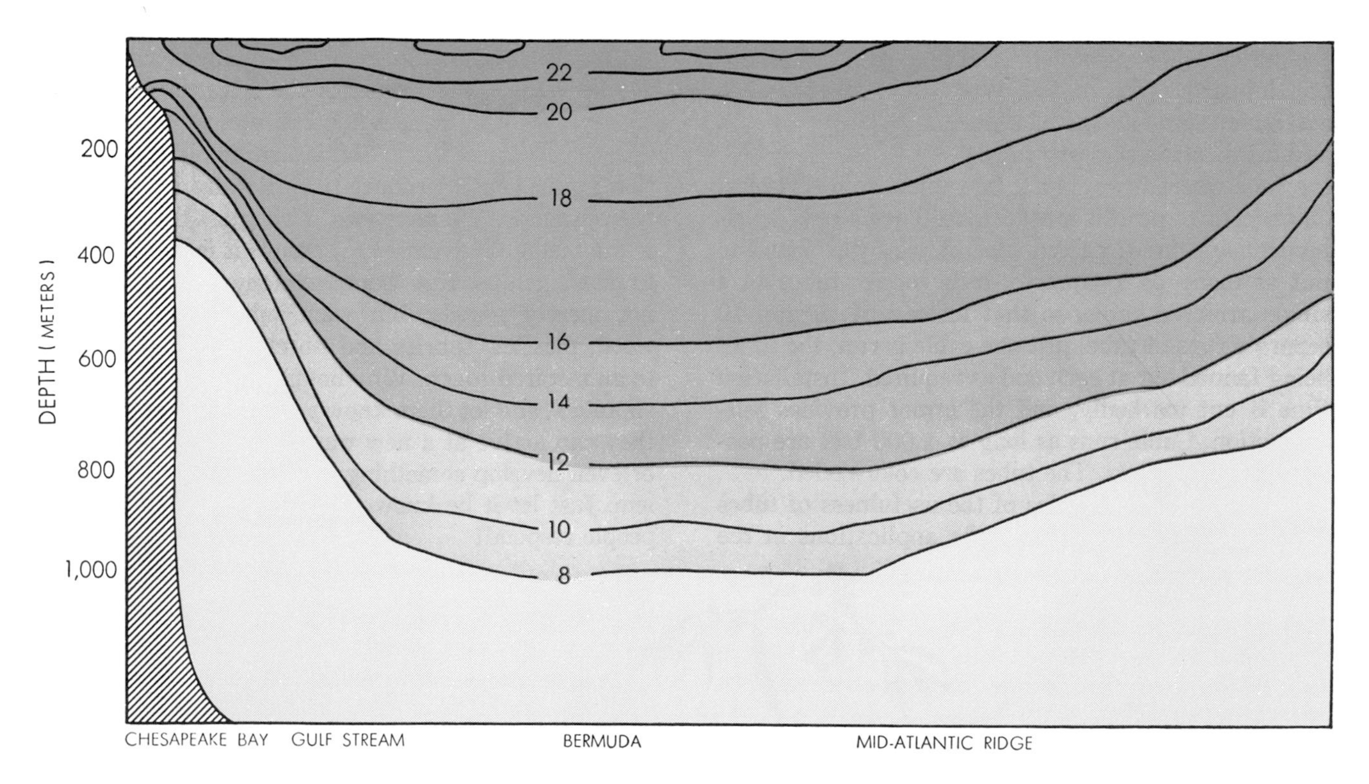

TEMPERATURE of the Atlantic is analyzed in this cross section along a line running due east from Chesapeake Bay. The contours, in degrees centigrade, indicate that the Sargasso Sea is a lens-shaped body of warm water. The vertical dimension is greatly exaggerated.

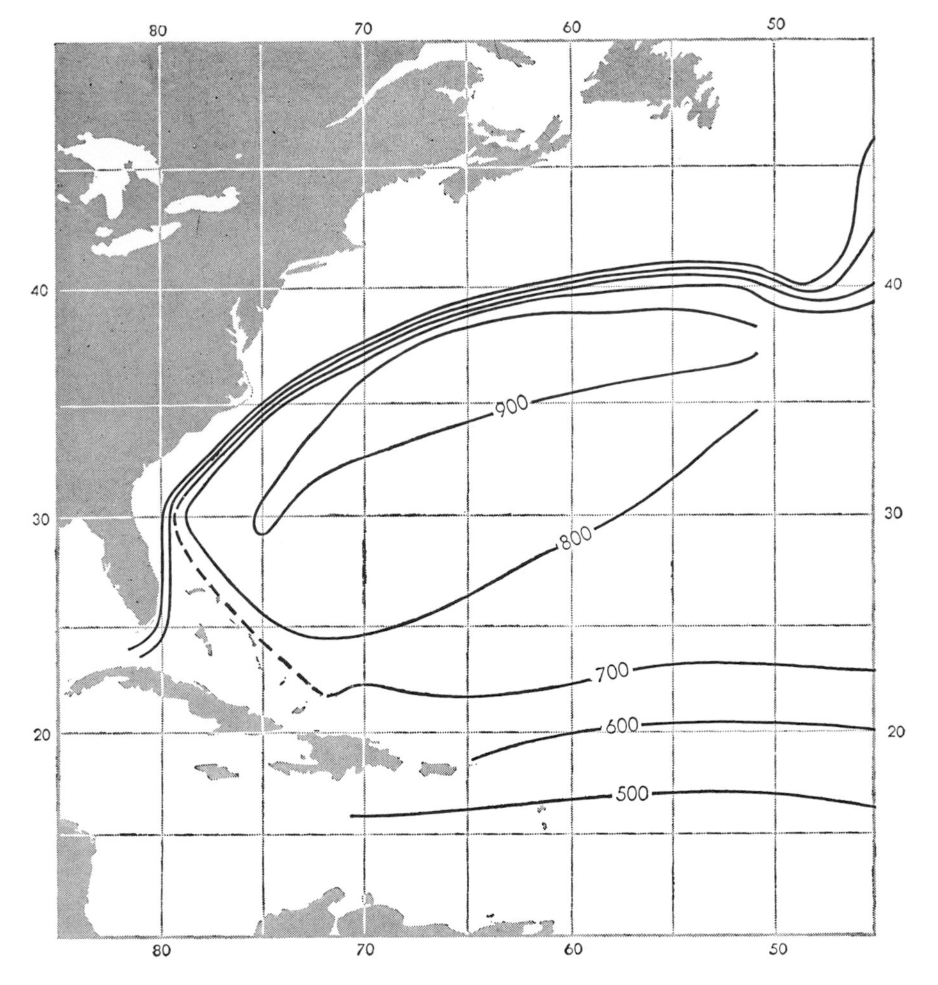

PLAN VIEW of the same Atlantic temperature structure again outlines the Sargasso Sea. Here the contours represent the depth in meters at which 10 degrees C. is encountered.

living an independent life in the Sea where it is found. It has a healthy color and shows new leaves and vigorous young shoots. For this reason and because of the absence of sexual fruiting bodies in the prevailing species, many oceanographers now favor the theory that the great bulk of the seaweed in the Sargasso Sea is native to the Sea itself. Its forebears may originally have come from beds on the bottom, but it has now evolved the ability to live a free, floating existence on the surface of the Sea. It can reproduce vegetatively—that is to say, by putting forth shoots which eventually break off as new plants.

Seaweed is not the only plant life in the Sargasso Sea. Like other seas it contains a subsurface floating population of the microscopic plants known as plankton. This material bulks much larger than the seaweed: estimates of the plankton production in the Sargasso range from 10 million to 100 million tons per day. Such a figure may seem huge, but it is small when one considers the vastness of the Sargasso. It amounts to something of the order of five hundredths of a gram of organic carbon per day for each square yard of the Sea's surface.

The Danish botanist Einer Steemann Nielsen, during a recent cruise of the research ship *Galathea*, made sample measurements of the plankton production in all the major oceans of the world. He found that the Sargasso Sea had the poorest production rate of all—only about one third of the average.

Here is the odd paradox. In spite of its show of life on the surface, the Sargasso Sea is in reality the most barren of waters. Alain Bombard was not far wrong when he wrote of the Sargasso as "a great dead expanse." The best evidence of its low biological content is the extreme clarity of its dark-blue waters.

There is a possibility that below the top 100 feet these clear waters, where sunlight can penetrate for an unusual distance, may contain more plant life than appears near the surface. But even allowing for this possibility, there can be no doubt that volume for volume the Sargasso Sea is the clearest, purest and biologically poorest ocean water ever studied.

THE ANTARCTIC OCEAN

V. G. KORT
September 1962

Ordinary maps usually show the Atlantic, Pacific and Indian oceans extending all the way to the frozen shores of Antarctica. To the oceanographer the water surrounding the Antarctic continent is not merely a confluence of three oceans but an ocean of itself, often known as the Antarctic, or Southern, Ocean. Although neither name has won international recognition, the latter is preferred by British and Soviet oceanographers. The Southern Ocean is unique in that it completely encircles the earth, unbroken by a continental land mass. In this great circumpolar expanse of water, driven eastward by the prevailing winds, the narrowest constriction is the 1,000 kilometers (620 miles) that separate South America from the Antarctic Peninsula. Elsewhere the distance between Antarctica and the nearest continent is more than 2,000 kilometers of open water.

The northern boundary of the Southern Ocean cannot be rigidly defined. Sometimes it is taken to be the region of the Antarctic Convergence, between 50 and 60 degrees south latitude, where surface waters flowing generally north converge with waters flowing generally south. The result is a marked change in temperature and salinity and an even sharper change in the character of the marine life [see "The Oceanic Life of the Antarctic," by Robert Cushman Murphy, Offprint 864]. According to another definition, which I prefer, the Southern Ocean can be regarded as extending northward to about 40 degrees south latitude, or approximately the southern coasts of Africa and Australia. Here there is another transition in flow,

SOVIET RESEARCH VESSEL *Ob* took part in the Soviet Marine Antarctic Expedition in the years 1956 to 1959. It and other vessels provided deep hydrologic observations from Antarctica to the coasts of South America, Africa, Tasmania and New Zealand.

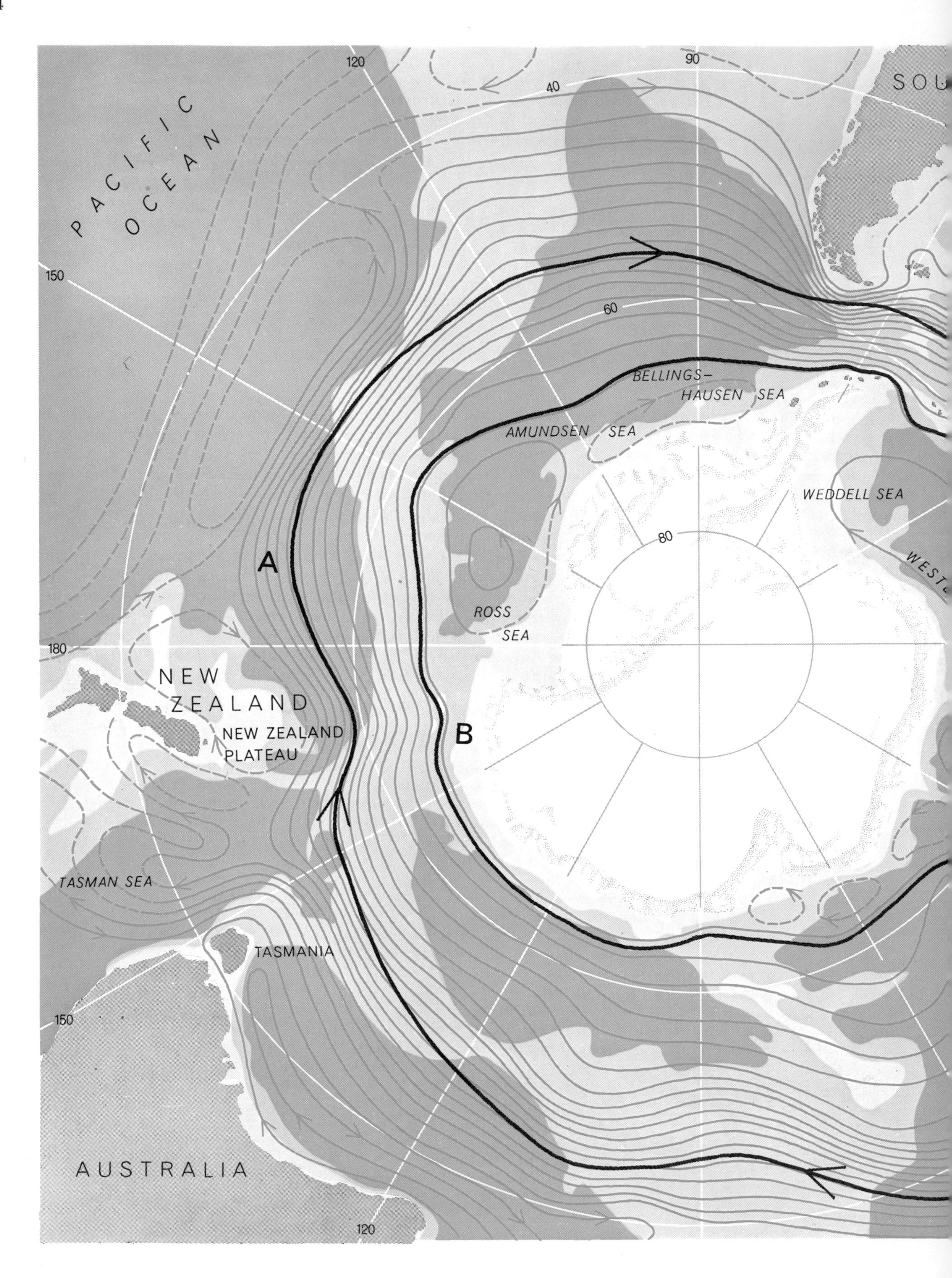

ANTARCTIC CURRENT FLOW has been computed from data collected during the International Geophysical Year by the Soviet Marine Antarctic Expedition. The volume of transport between adjacent flow lines is 10 million cubic meters per second. The Antarctic Convergence (*A*), where temperature and salinity change abruptly, coincides with the region of heaviest flow. The Antarctic Divergence (*B*) coincides with the region of minimum flow near the Antarctic coast. The Subtropical Convergence, which can be

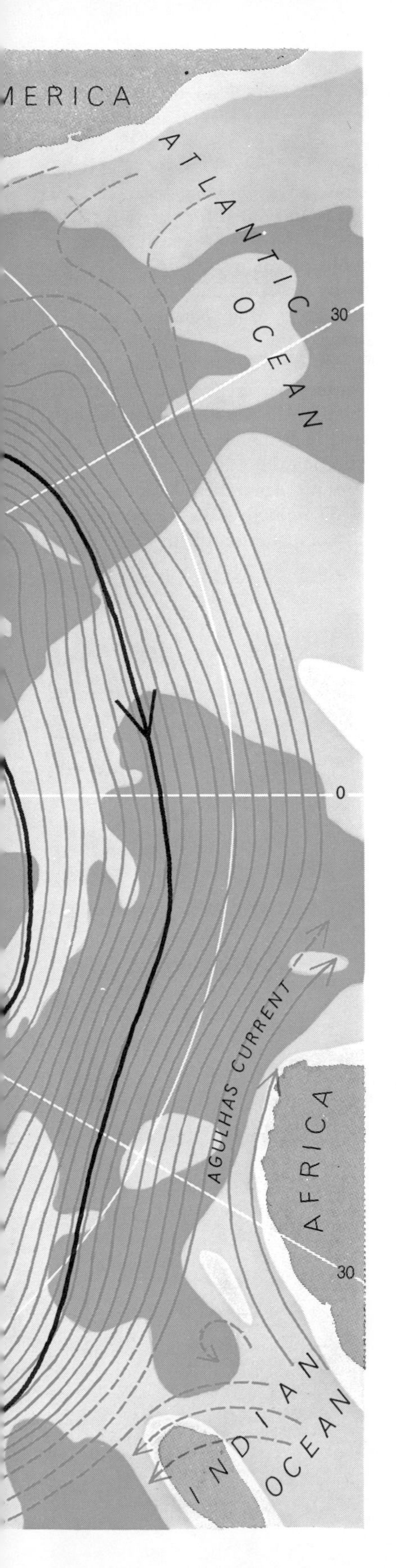

regarded as the northern boundary of the Southern Ocean, cannot be very accurately defined, but it generally follows the northern limit of the flow lines plotted here.

temperature and salinity designated the Subtropical Convergence. The Subtropical Convergence also marks the approximate limit of the northward drift of ice from the Antarctic. If 40 degrees south latitude is taken as the boundary, the Southern Ocean covers about 75 million square kilometers, nearly six times the area of the Antarctic continent. This is 22 per cent of the total area of all the oceans. The heat content of the Southern Ocean, however, is only 10 per cent of the oceanic total. It is apparent that this vast expanse of water, ranging in temperature from about −1.8 to 10 degrees centigrade (28.8 to 50 degrees Fahrenheit), exerts a great influence on the climate of the entire planet.

The interaction of the water masses of the Southern Ocean and the atmosphere over it was an important aspect of the program of the International Geophysical Year. Extensive oceanographic observations were conducted in Antarctic waters from 1956 to 1959 by the research ships of Argentina, Australia, France, Japan, New Zealand, the U.S. and the U.S.S.R.

One of the tasks of the Soviet Marine Antarctic Expedition was to obtain more precise values of water transport and current flow in the Southern Ocean. It had previously been established that the strong western winds set up a current that travels around the Antarctic continent. It was also known that frigid waters flow down from the shores of the continent and slowly travel northward along the bottom into the Atlantic, Pacific and Indian oceans. At the same time an equivalent volume of warmer water travels southward at intermediate depths to replace the water flowing northward below it [*see illustration on page 88*].

One of the first attempts at a quantitative estimate of Antarctic water transport was published in 1942 by Harald U. Sverdrup, then director of the Scripps Institution of Oceanography in La Jolla, Calif. His values were based on hydrologic observations carried out in cruises of the British research ship *Discovery* and the cruises of other research vessels. Although it is possible to make a rough estimate of surface currents by determining how much a ship is carried off its course, the oceanographer must resort to other methods for estimating the volume and velocity of currents below the surface. The standard method used by Sverdrup and others, including ourselves, involves measuring the temperature and salinity of the ocean at various depths. From these values one can calculate the relative fields of pressure, which must be converted to absolute fields. This is done by making assumptions as to the topography of the sea surface or the current velocity at some particular depth. The final step is to calculate the field of motion from that of pressure.

Sverdrup calculated the total west-to-east water transport in the Southern Ocean across three sections: between Antarctica and South Africa, between Antarctica and Tasmania (the island off the southern coast of Australia) and between Antarctica and South America (the Drake Passage). His values for total water transport in a layer from the surface to a depth of 3,000 meters across these sections were respectively 120 million cubic meters per second, 150 million cubic meters per second and 90 million cubic meters per second. The smallest of these values, that for the Drake Passage, is more than 400 times the volume of water carried by the Amazon, the world's largest river. It is evident that if 150 million cubic meters per second flows into the Pacific Ocean between Antarctica and Tasmania, some 60 million cubic meters must be diverted northward into the Pacific, since only 90 million cubic meters can be found moving into the Atlantic through the Drake Passage. One would also have to explain how this 60 million cubic meters per second is replenished.

To check Sverdrup's calculations, the Soviet Marine Antarctic Expedition used observations carried out primarily from the 12,000-ton diesel-electric research ship *Ob*. Deep hydrologic observations of temperature and salinity were made from the continental shelf of Antarctica to the coastal waters of South Africa, New Zealand (as well as Tasmania) and South America. The map at left shows in detail the path of the entire Antarctic Circumpolar Current as computed from our data. It can be seen that the maximum current flow is in the zone of the Antarctic Convergence.

In general our figures for transport across the three principal sections do not show such great variations as those computed by Sverdrup. Nevertheless, variations remain, and as I discuss them it may help the reader to follow the flow chart on the next page. For the three principal sections the average values in millions of cubic meters per second are:

Antarctica to Africa	190
Antarctica to Tasmania	180
Antarctica to Cape Horn	150

The flow of 190 million cubic meters per second represents water moving eastward from the Atlantic into the Indian Ocean, mainly between 38 and 56 degrees south latitude. To the north of this region, between 38 degrees and the African coast, about 25 million cubic meters of water per second is carried in the opposite direction—westward—from the Indian Ocean into the Atlantic by the Agulhas Current. South of 56 degrees south latitude an additional 10 million cubic meters per second is carried westward by the Antarctic Western Coastal Current.

If one now looks at the flow through the section between Antarctica and Tasmania one sees that the volume of water leaving the eastern end of the Indian Ocean is 180 million cubic meters per second, or 10 million less than that entering from the Atlantic. It appears, therefore, that 10 million cubic meters per second is diverted into the circulation of the Indian Ocean and partially makes up for the outflow carried west by the Agulhas Current. The balance of this current consists of 10 million cubic meters per second provided by the West Australia Coast Current and five million provided by river discharge and by Pacific water flowing through the Indonesian straits.

To the east of Tasmania the Antarctic Circumpolar Current branches and sends almost 40 million cubic meters per second northward into the Tasman Sea, which lies between Australia and New Zealand. Whereas 180 million cubic meters per second enters the Pacific Ocean (including the Tasman Sea) from the Indian Ocean, only 150 million cubic meters per second enters the Atlantic Ocean from the Pacific Ocean through the Drake Passage. The deficit of some 30 million cubic meters per second is accounted for partly by evaporation in the Pacific, partly by movement of water into the North Pacific and partly by transport back into the Indian Ocean across the northern coast of Australia.

Finally we observe that the 150 million cubic meters per second entering the South Atlantic through the Drake Passage is 40 million less than the volume leaving the Atlantic between South Africa and Antarctica. Of this deficit about 35 million cubic meters per second is compensated for by the Agulhas

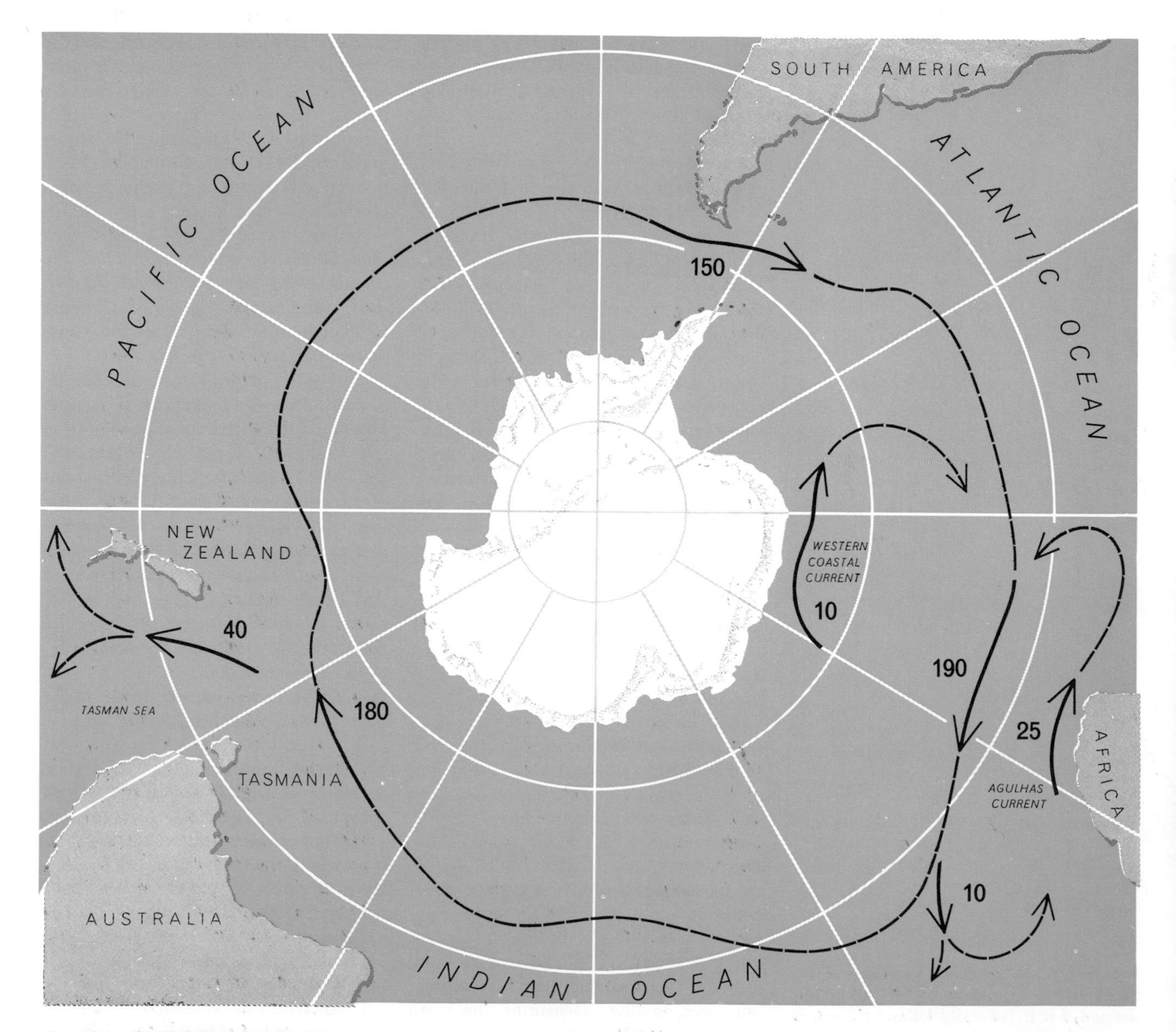

WATER TRANSPORT SUMMARY, in millions of cubic meters per second, shows how the Circumpolar Current varies in volume as it passes between Antarctica and other land masses. Largest diversion occurs where current meets the New Zealand Plateau.

Current and the Antarctic Western Coastal Current. The remaining five million or so is supplied by melted ice carried southward from the Arctic Ocean and by river discharge. In spite of the approximate nature of these transport figures, they provide a reasonable account of the water exchange between the great ocean systems.

A study of the transport pattern of the Antarctic Circumpolar Current reveals that all changes in its direction are closely correlated with changes in the topography of the ocean bottom. This is particularly clear in the region south of Tasmania. Under the influence of the western edge of the New Zealand Plateau a significant fraction (more than 20 per cent) of the Circumpolar Current is diverted northward into the Tasman Sea, and the remaining mass is deflected to the south. The result is a sharp southward displacement of the Antarctic Convergence.

Hydrologic data collected close to the Antarctic shelf indicate strong cyclonic (clockwise in the Southern Hemisphere) and anticyclonic circulation patterns in the vicinity of the Weddell, Ross, Amundsen and Bellingshausen seas. The largest of the anticyclonic patterns appears to lie in the region to the northeast of the Ross Sea. The most extensive cyclonic circulation is between Africa and Antarctica.

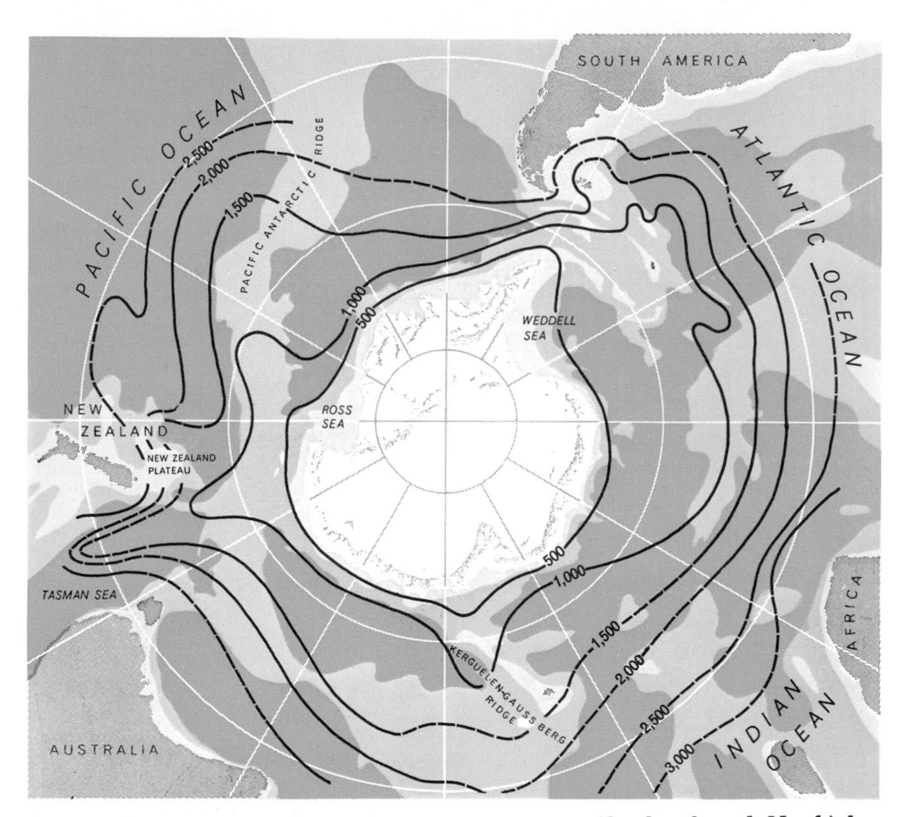

HEAT DISTRIBUTION in the Southern Ocean in summer (October through March) has been estimated by Soviet oceanographers. Figures represent the heat content, in thousands of calories, of a column of water a square centimeter in cross section extending from the surface to the ocean bottom. Heat content is proportional to number of degrees between the average water temperature and the freezing point of sea water, multiplied by the depth.

A second task of the Soviet Marine Antarctic Expedition was to estimate the meridional flow of deep water: toward and away from the Antarctic continent. More than 35 years ago Georg Wüst, who is now at the Lamont Geological Observatory of Columbia University, analyzed deep observations made in the Atlantic Ocean by the German research ship *Meteor* and concluded that Antarctic waters cross the Equator and penetrate as far as 40 degrees north. Soviet observations carried out in the Indian Ocean by the *Ob* indicate that Antarctic waters penetrate to the Arabian Sea and the Bay of Bengal, both of which lie some 10 to 20 degrees north of the Equator.

Because of the absence of a reliable theoretical model, the meridional transport is more difficult to compute than the circumpolar transport. Our semi-empirical method of computation depended on measurements of temperature and turbulent exchange of heat at various depths. When observations made on nine meridional sections were analyzed, the mean northward transport of bottom waters around the whole perimeter of the Antarctic continent turned out to be something more than 800 million cubic meters per second, or more than five times the flow rate of the Antarctic Circumpolar Current itself. To replace this huge outflow a mighty layer of deep warm water, equal in volume, crosses the perimeter of the Southern Ocean from the north.

Knowing the water exchange between the Southern Ocean and adjacent oceans and knowing water temperatures, it is possible to make a rough estimate of the amount of heat exchanged. First, however, it is necessary to determine the amount of solar radiation absorbed at the surface of the water and the amount of heat radiated from the surface back into the atmosphere. Data obtained in the third and fourth cruises of the *Ob* in summer (February) and winter (August), together with observations made at the Soviet Antarctic stations of Mirnyy and Lazarev and elsewhere, were used to estimate the radiation balance of the Southern Ocean to 40 degrees south latitude. It was found that the Southern Ocean gives off annually to the atmosphere nearly 34×10^{21} (34 followed by 21 zeros) gram calories and takes up from it only slightly more than 10×10^{21} gram calories. The heat given off serves to warm the frigid Antarctic air masses as they travel over the Southern Ocean on their way northward.

It is apparent that the main source of heat received by the Southern Ocean is the deep waters that carry it in from the north. Our estimates show that these waters annually bring in approximately 30×10^{21} gram calories. This value represents the heat released when some 800 million cubic meters per second of deep water at 1.2 degrees C. is chilled to the mean temperature of .2 degree prevailing near the coast of Antarctica. The equal volume of bottom water flowing northward carries with it about 5×10^{21} gram calories of heat.

These figures derived in different ways yield income and outgo energy values that are in reasonable, if not complete, balance. The income is 10×10^{21} gram calories of solar energy plus 30×10^{21} gram calories delivered by deep waters from the north, for a total of 40×10^{21} gram calories. The outgo includes roughly 34×10^{21} gram calories lost to the atmosphere plus 5×10^{21} gram calories carried north in bottom waters. In addition, some 4×10^{21} gram calories are required annually to melt the ice produced during the winter in the Southern Ocean and to warm Ant-

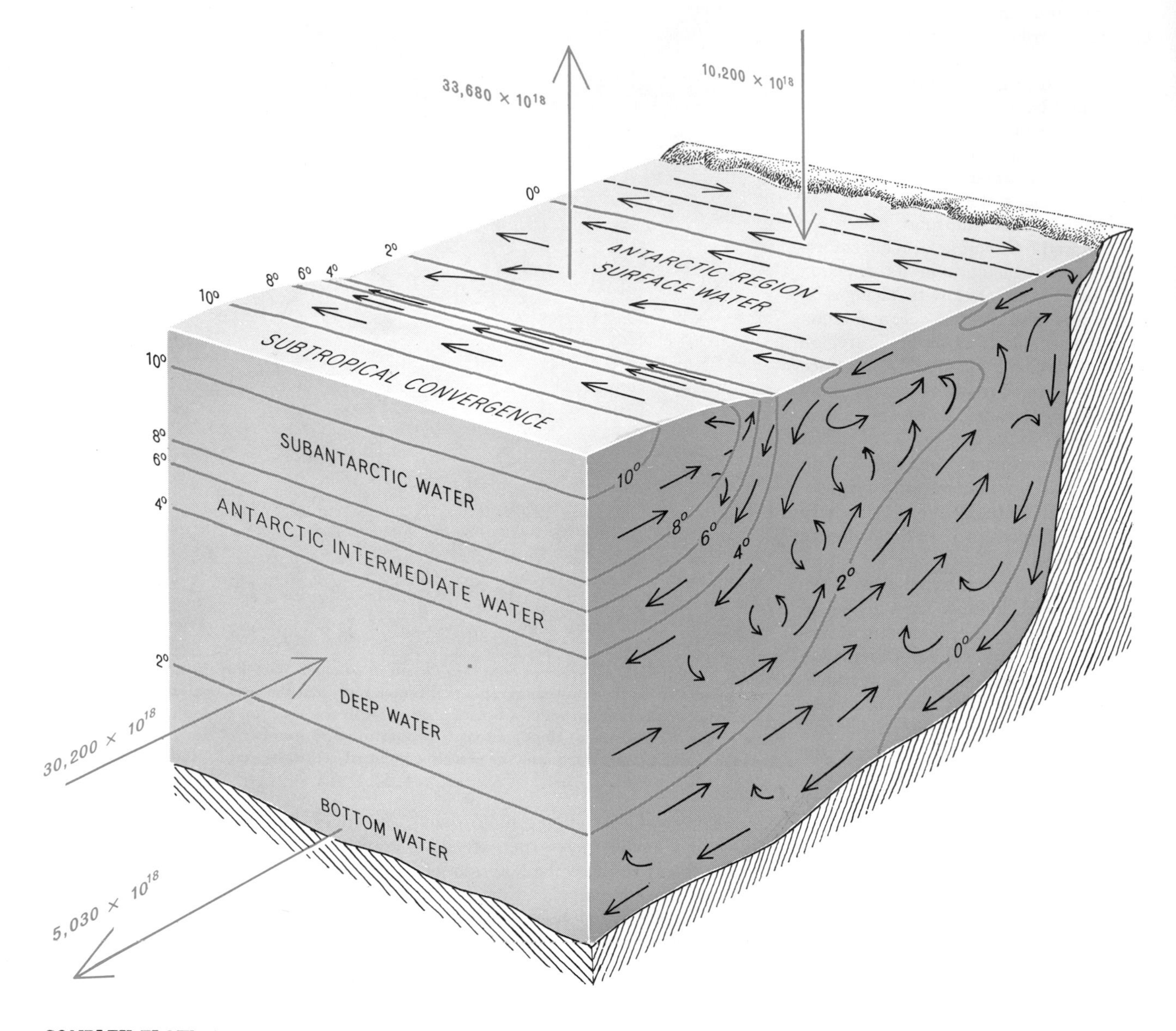

COMPLEX FLOW of currents in the Southern Ocean is depicted in this schematic view based on the work of Harald U. Sverdrup. Water cooled by the coastal ice masses of the Antarctic continent (*right*) sinks and flows northward along the bottom. An equal volume of warmer "Deep water" flows southward to replace it. The Antarctic Convergence is formed where south-flowing water of 8 to 10 degrees centigrade meets much colder surface water flowing away from the Antarctic. Figures at left in color, computed by the author and his associates, show the amount of heat, in calories, transported annually by the deep and bottom waters. Other figures in color show that the heat given off to the atmosphere by the Southern Ocean greatly exceeds the heat received from the sun.

arctic waters from − 1 degree C. in winter to + 1 degree C. in summer. The total outgo, therefore, is approximately 43×10^{21} gram calories.

One can see from the above analysis that the thermal influence of the Southern Ocean is two-sided. First, and most important, the Southern Ocean takes up from adjacent oceans a considerable amount of heat and expends it in warming up the cold Antarctic air masses, thereby exerting a decisive influence on the atmospheric circulation of the Southern Hemisphere. Second, the cold waters of the Southern Ocean penetrate into the adjacent oceans and noticeably cool their deep layers.

The interplay of water temperature and currents in the Southern Ocean can be brought out by charting the pattern of heat distribution in the water surrounding Antarctica. To do this one integrates two factors: the average water temperature from the surface to the bottom and the water depth. One can then calculate the heat content of a column of water a square centimeter in cross section extending from the surface to the bottom. In our calculations we regard the heat content of sea water at its freezing point (around −1.8 degrees C.) as being zero. Thus a cubic centimeter of water one degree above freezing is said to have a heat content of about one gram calorie. The pattern of heat distribution of the Southern Ocean in the Antarctic summer is shown in the map on page 87.

It can be seen that the coldest region of the Southern Ocean is the Weddell Sea, which lies within a huge cyclonic system. This system draws a vast quantity of water from the Antarctic Circumpolar Current and transports it southward to the great ice shelf of the Weddell

Sea, where the water is intensively cooled. As a result the region is literally a factory of cold waters.

Picked up again by the Antarctic Circumpolar Current, the refrigerated waters are carried eastward until, on reaching the New Zealand Plateau, some 40 million cubic meters per second is deflected into the Tasman Sea. The tongue of frigid water penetrating northward shows up clearly in the heat-distribution map. As the Circumpolar Current moves east into the South Pacific it is again deflected sharply northward under the combined influence of the Pacific-Antarctic Ridge and the anticyclonic circulation near the Ross Sea. The cold shelf waters of the Ross Sea also contribute to the outflow of cold water to the north. The net effect of these great diversions is to make the Pacific Ocean somewhat colder than the Indian Ocean and substantially colder than the Atlantic Ocean. According to the data of my colleague V. N. Stepanov, the annual mean heat content of the Pacific Ocean is 1,746,000 gram calories per square centimeter, compared with 1,783,000 for the Indian Ocean and 1,989,000 for the Atlantic Ocean. The mean heat content of the Southern Ocean is approximately half of these values, which accounts for the statement that it contains only about 10 per cent of the world's oceanic heat while representing 22 per cent of the world's oceanic area.

The exceptional intensity and stability of the Antarctic Circumpolar Current offer the investigator an attractive opportunity for testing theoretical models of the flow of a homogeneous fluid in a circular channel. It has been recognized for 50 years that the movement of the Antarctic waters from west to east is caused by the strong west winds that blow almost constantly in the belt between 40 and 60 degrees south latitude.

The first comprehensive description of the dynamics of the Southern Ocean was published 25 years ago by G. E. R. Deacon, now director of the National Institute of Oceanography in England. Subsequently Sverdrup developed the concept that the Antarctic Circumpolar Current is the sum of a pure wind-driven current in the surface layers of the ocean and of a gradient current, due to density variations, acting through the whole water mass. Sverdrup was also the first to call attention to the influence of bottom topography on the Circumpolar Current. Since then the leading students of the dynamics of the Southern Ocean have treated the Circumpolar Current as a current driven through a channel by a steady zonal wind.

The main objective of the theoretical studies, conducted primarily by U.S., French, German and Japanese investigators, has been to compute the volume of water transported by the Antarctic Circumpolar Current and to provide mathematical models describing how the current is influenced by bottom topography. Although such studies have usually led to unrealistically high values of water transport, they all confirmed that the transport is caused by wind.

The theoretical models developed prior to the IGY program left unanswered the question of whether the current is a continuous flow through the whole body of the ocean, from surface to bottom, or whether it has deep countercurrent or transverse circulation. The models also had difficulty explaining the mechanism of the formation of convergent and divergent zones of flow in the surface layers of the oceans; these zones are the Antarctic Convergence, the Antarctic Divergence and the Subtropical Convergence, which navigators in these regions have long recognized by changes in water and air temperature. Finally, the models could not assess quantitatively the influence of bottom topography on the Antarctic Circumpolar Current.

Data collected during the IGY program have made it possible to start solving some of these problems. For example, detailed deep observations of temperature and salinity, and of the distribution of oxygen and other elements, show that waters in the Antarctic Circumpolar Current move eastward through

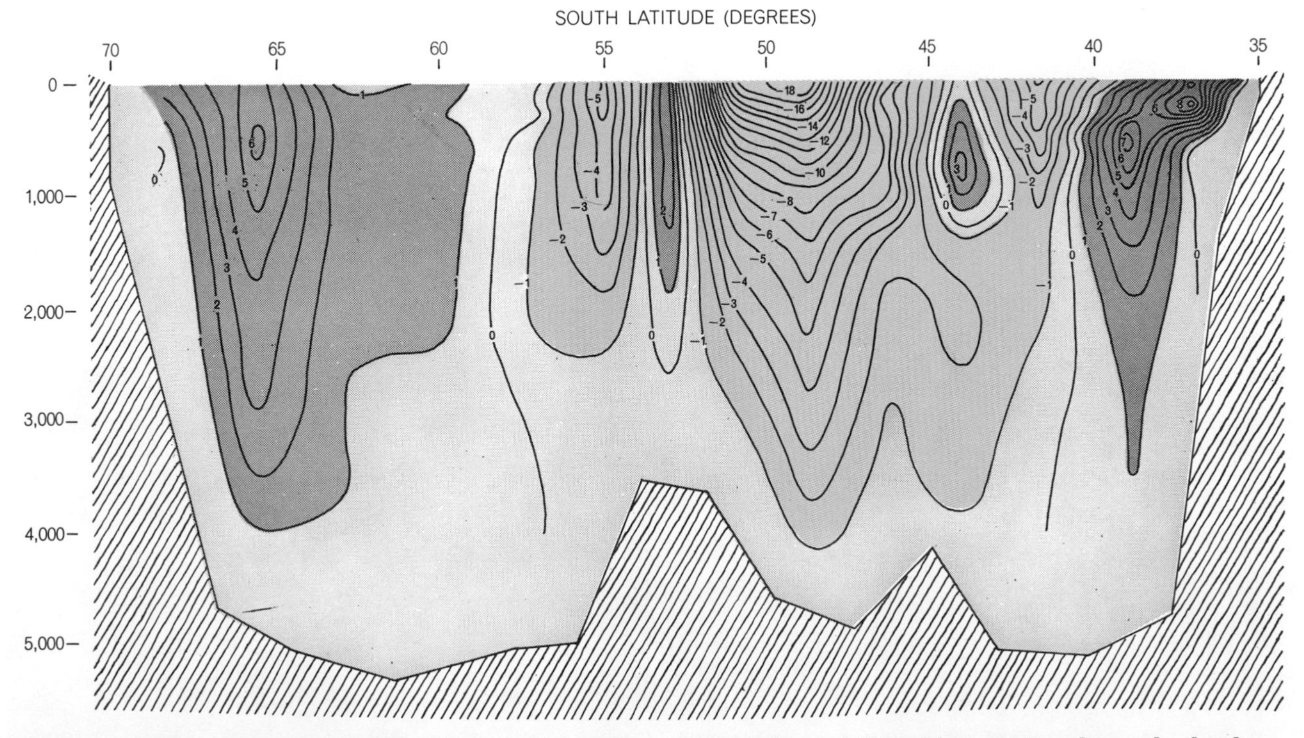

CURRENT VELOCITY DISTRIBUTION is shown for the oceanographic section between Antarctica (*left*) and South Africa. Figures give velocity in centimeters per second. The predominantly eastward flow is indicated by negative values and colored areas. The westward flow at right is produced by the Agulhas Current. Westward flow at left is the Antarctic Western Coastal Current.

the whole body of the Southern Ocean. The structure of this current, however, is complicated. The total flow breaks into separate streams with fast-moving cores. Elsewhere, in certain regions, countercurrents run westward. The illustration on page 89 shows a characteristic structure of the Circumpolar Current in the section between South Africa and Antarctica, along 20 degrees east longitude.

With the new evidence that the flow of Antarctic waters embraces the whole body of the ocean, it becomes easier to explain the role of large-scale submarine trenches and ridges. The bends in the stream lines, as charted on pages 84 and 85, are formed by the joint influence of bottom topography and the Coriolis force, which results in the deflection of the Antarctic Circumpolar Current to the north over a rising bottom and to the south over a falling bottom.

With these qualitative observations as a starting point, a new theoretical model of the Antarctic Circumpolar Current has been developed by V. M. Kamenkovitch, one of my colleagues at the Institute of Oceanology of the Soviet Academy of Sciences. The model yields values of water transport that agree reasonably well with the values actually observed in the Circumpolar Current.

Another worker at our institute, J. A. Ivanov, has been examining the formation of ocean "frontal" zones, such as

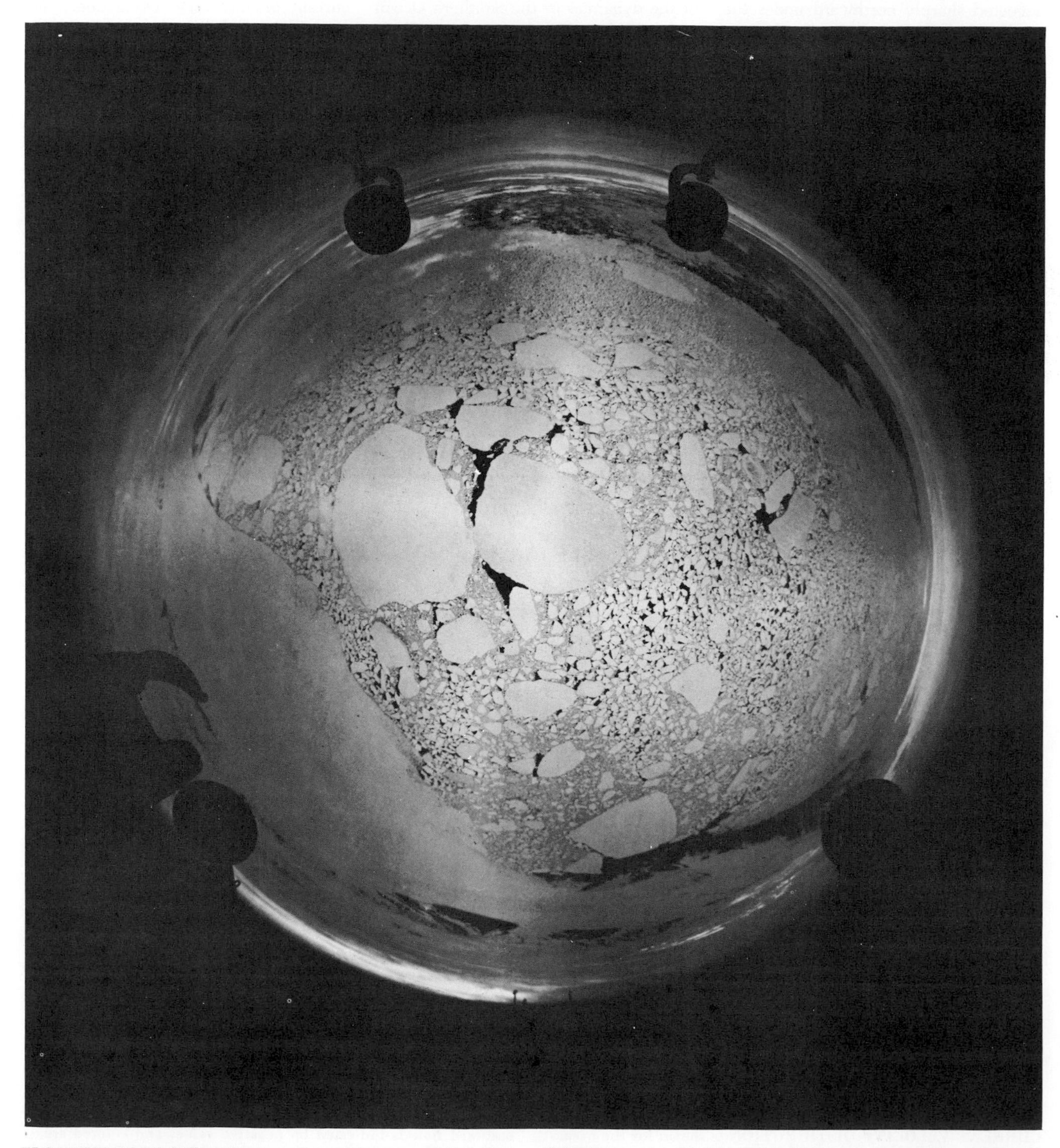

FLOATING PACK ICE at Moubray Bay, adjacent to Hallett Station on the Ross Sea, was photographed from 8,000 feet by Emil Schulthess, using his homemade "fish eye" camera. The round objects at the edge of the picture are the four wheels of the helicopter.

the zone of the Antarctic Convergence. He finds that seasonal changes in the geographical position of this zone correlate rather closely with seasonal changes in the meridional component of air pressure gradients over the Southern Ocean.

Harry Wexler, the late director of research for the U.S. Weather Bureau, concluded independently that the Antarctic Convergence is caused by meteorological factors. Wexler believed, moreover, that wind strength determines whether a frontal zone will be a convergence or a divergence. Klaus Wyrtki, a German investigator, has reached much the same conclusion. Thus, as a result of the combined efforts of the oceanographers of several countries, the dynamic processes in the Antarctic waters have received a more or less satisfactory physical explanation.

This does not mean, of course, that oceanographers have no more work to do in the Southern Ocean. I shall mention only one outstanding gap in our knowledge. The volume of water exchanged between the Southern Ocean and the oceans to its north varies considerably from season to season and from year to year. Such variability cannot help but influence the heat exchange of the Southern Ocean and therefore the atmospheric circulation over it. At present, however, we lack a quantitative estimate of this variability.

To fill this gap in knowledge we need a sharp increase in radiation measurements on research ships and at all island stations. There should be an international effort to make systematic deep observations over a period of many years in several sections across the Antarctic Circumpolar Current, repeating the measurements made most recently by the Soviet Marine Antarctic Expedition. In addition, periodic observations, at three-to-five-year intervals, should be made in large meridional sections extending from the coast of Antarctica to the northern latitudes in each of the adjoining oceans, the Atlantic, the Pacific and the Indian.

Such observations will provide the data necessary to reveal variations in the exchange of heat between the Southern Ocean and its neighbors to the north and to show how this heat exchange influences both atmospheric and oceanic circulations. A thorough understanding of these mechanisms will be of invaluable help in making long-range forecasts of weather and climate for the entire planet.

11

THE ARCTIC OCEAN

P. A. GORDIENKO
May 1961

Large regions of the U.S.S.R.—nearly a third of the country—lie within the Arctic Circle. South of this frigid zone lie the forests and prairies of the one-sixth of the earth's land that is encompassed by the borders of the Soviet Union. Great rivers—the Dvina, the Pechora, the Ob, the Yenisei, the Lena, the Yana, the Indigirka and the Kolyma—rise far to the south and run northward to the Arctic seas: the Barents, the Kara, the Laptev, the East Siberian and the Chukchi. The seas are arms of the Arctic Ocean, embayed by the northward-reaching peninsulas and archipelagoes of the Eurasian continent and hemmed in by the oceanic ice. During the few months of the year when the ice clears the straits and headlands, the rivers, seas and oceans provide the shortest over-water transportation route between the metropolitan centers of the country and the interior frontier.

The Northeast Passage from the Atlantic to the Pacific Ocean was one of the challenging objectives sought by the mariners of all the European powers during the age of discovery. For the economy of the Soviet Union the Northeast Passage constitutes a life line. Without it, the resources of the interior must be hauled at much greater cost overland, and shipping from Leningrad in the west to Kamchatka in the east must travel through the Panama or Suez canals—journeys of more than 14,000 miles to accomplish a distance only half as great. During the past 40 years Soviet explorers and scientists have filled in the last blank spots in the map of the Arctic and have opened the way to a full understanding of the relationship of this remote region to the geography and geophysics of the planet as a whole. Particularly during the past 15 years, large and small expeditions borne by icebreakers and oceanographic vessels have mapped the Arctic lands and waters; flying laboratories have made numerous landings on the ice pack; scientific settlements established on drifting ice islands have traced by their own wanderings the currents of the polar seas, and dozens of permanent observatories have been built on the outermost reaches of the land. As a result cargo vessels are proceeding east and west in increasing numbers each year between the European and Siberian centers of the Soviet economy. The centuries-old vision of the Northeast Passage has been realized, and man's knowledge of his planet has been considerably enriched.

Early Explorations

"If we compare Russia to a building," the 19th-century Arctic explorer Admiral Stepan O. Makarov wrote, "we see at once that it faces the Arctic Ocean." It was inevitable that Russia should play a prominent role in Arctic research. In this account of the most recent phases of that long effort it would be less than just not to begin with an excursion into the past and a brief account of the achievements of our forefathers and of the explorers from other nations who preceded us into the region.

Russians first appeared on the frozen shores of the Arctic in the 12th and 13th centuries. While the geographers of western Europe were writing imaginative treatises about fearsome monsters that inhabited the unexplored North, settlers from ancient Novgorod pushed through the wilderness and settled along the coast of the White Sea and later moved on to Murmansk. Russian hunters sailed across the icy Barents Sea in primitive boats, landed on Novaya Zemlya ("New Land") and Spitsbergen and followed the Arctic coast line to the East. Somewhat later eastward-migrating Cossacks crossed the Urals and penetrated into Siberia. Making their way across wild rivers, braving the polar night and the intense cold, these men charted lands that had never been visited by Europeans, established settlements and made the first contacts with the Chukchi, the aboriginal inhabitants of the Eurasian Arctic. Even the northernmost point of

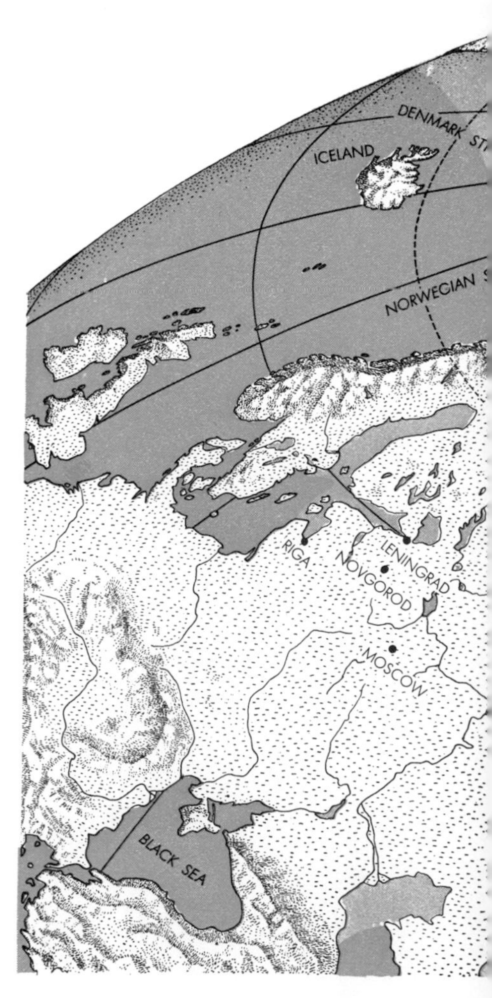

THE ARCTIC OCEAN, its islands and the lands adjoining it are shown from a

Asia, Cape Chelyuskin on the Taimyr Peninsula, was no barrier to the early Russian seafarers. They sailed around that cape some 350 years ago and left remains of their craft and dwellings on its eastern shore, where they were discovered only in 1940.

In the 16th century, with the riches of Cathay and India luring them on, the nations of Europe embarked on the age of discovery. Spain and Portugal soon pre-empted the warm-water routes to the East, and with the land routes all but closed by hostile Islam, the other nations of Europe looked to the Arctic for a northeast or a northwest passage to the Pacific. In the course of the next 300 years the search occupied the English, the Dutch and the French. They established colonies in the New World but they found no short routes to the East. The northern passage to Cathay remained a hope and a challenge.

It is little known that the scope of Russia's enterprises in her northern lands and in Siberia during the age of discovery was of the same magnitude. In 1648 a Cossack detachment under the leadership of Semen Dezhnev and the prospector Fedot Popov, looking for a way east from the mouth of the Kolyma River, rounded the Chukchi Peninsula and discovered the strait that separates Asia from America. This remarkable discovery proved that the two continents are separate and established the existence of the Northeast Passage. Under Peter the Great, in the early years of the 18th century, Russia expanded her domains in the East. The search for an alternative to the interminable and exhausting journey across the Siberian wilderness now assumed urgent priority. Peter planned and organized a tremendous geographical undertaking, the Great Northern Expedition of 1733 to 1743. Over the decade the various parties of the expedition succeeded in exploring and mapping the northern coast of Europe and Asia from the White Sea to Kamchatka. The names of Bering, Chirikov, Chelyuskin, the Laptev brothers, Pronchishchev and his wife Maria, Sterlegov, Ovtsin, Malygin and dozens of others live today in the maps of the north. Across land and sea these explorers proceeded to their objectives in the face of hardships and perils that can only be imagined by those who now venture into the Arctic with the protections and conveniences of modern technology. They made their way through the trackless wilderness on horseback and on foot; they set out to sea in ships they built themselves, and manned the oars when the storms carried away their sails. Yet they took soundings as they went and drew maps with a cartographic skill that is impressive even today. In 1741 the ships *St. Peter* and *St. Paul*, under the command of Vitus Bering and Alexei Chirikov, rounded the Kamchatka Peninsula and proceeded across the sea to Alaska.

Mikhail Lomonosov—scientist, histori-

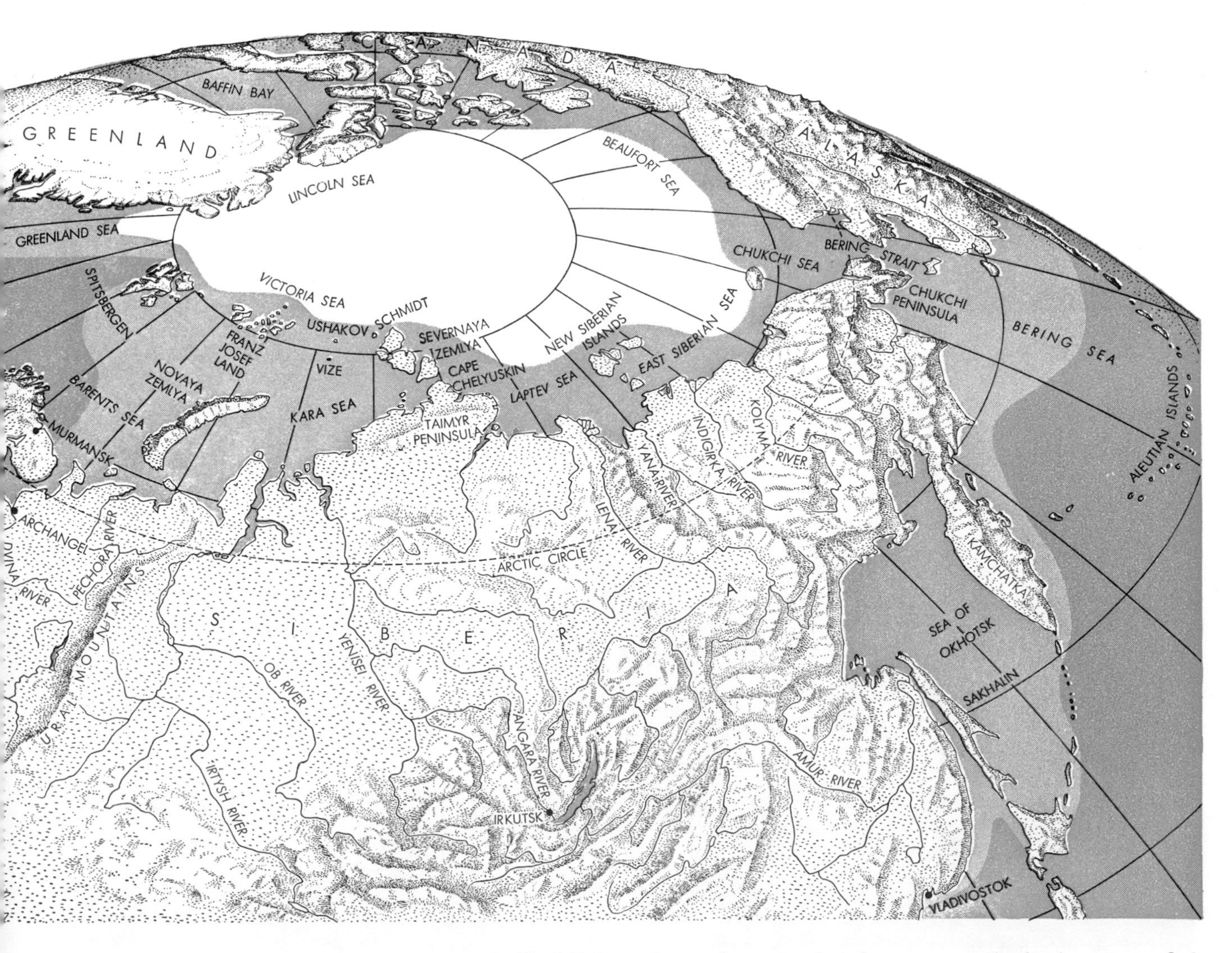

point high above central Asia. White area around the North Pole indicates the permanent ice pack; the light color represents waters that are frozen in winter but open to navigation in summer; dark color indicates seas that are open to navigation all year round.

ICE ISLAND NORTH POLE VI, a Soviet observatory established on the drifting ice, began a 700-mile journey in 1956 and ended it 27 months later. Hemispherical structures are laboratory tents; rectangular ones, living quarters. At right are radio antennas.

FISSURE IN ICE ISLAND, a constantly recurring threat, is dangerous because it may split campsite. Station buildings are therefore set on sled runners to facilitate relocating them. Action that breaks pieces off floe just as often fuses other pieces on.

an and reformer of the Russian language —set the capstone on this decade of adventure and achievement. In 1763 he brought the diverse findings of the Northern Expedition together in a comprehensive treatise under the name *Brief Account of Travels in the Northern Seas.* This work laid the foundations of the oceanography and meteorology of the polar region, the physics of Arctic ice and a theory of the aurora borealis.

In the first quarter of the 19th century Russians explored the Bering Sea and Bering Strait, discovered Kotzebue Sound (north of the Bering Strait off the western coast of Alaska), proved the existence of Wrangel Island in the Chukchi Sea and visited the New Siberian Islands and Novaya Zemlya. During this period a Russian expedition discovered the Antarctic continent. Then in 1878, with the financial backing of the merchant A. Sibiriakov (Peter's successors in power did not observe the bold precedents he set for the development of the Russian Empire), the Swedish explorer N. A. E. Nordenskjöld made the first passage across the northern waters, from the Barents Sea to the Pacific. Nordenskjöld's achievement, and the discovery of Franz Josef Land a few years earlier, inspired the organization of the first International Polar Year in 1882 and 1883. Russia participated by establishing a meteorological station at Novaya Zemlya and another in the delta of the Lena River.

In 1893 the Norwegian explorer Fridtjof Nansen took his ship the *Fram* on a daring voyage. Nansen was convinced that ocean currents carried ice from Siberia across the Arctic and through the Greenland Sea into the Atlantic. He advanced an ingenious scheme in which he proposed to use the forces of the Arctic rather than fight them. He would sail into the polar ice pack off the New Siberian Islands, anchor to a floe and allow the currents to carry his ship as they did the ice itself. In support of his hypothesis he pointed to the appearance on the Greenland coast of flotsam from Siberia—more particularly to the wreckage of the *Jeannette,* the ship of an ill-fated expedition commanded by George Washington DeLong of the U. S. Navy, which had been crushed in the ice off the New Siberian Islands in 1881. The drift of the *Fram* lasted three years and carried Nansen across the Arctic Ocean at high latitudes.

At the turn of the century the Arctic became an arena in which Englishmen, Americans, Italians, Norwegians, Russians, Austrians and Germans pitted their perseverance and courage against the ice and one another in an international contest to reach the Pole. Robert E. Peary won the prize for the U. S. in 1909.

The advent of the icebreaker, the airplane and the radio revolutionized Arctic exploration early in the 20th century. With the icebreaker the polar ice pack was no longer an impassable obstacle to navigation. The airplane opened previously inaccessible areas to reconnaissance; and radio not only alleviated the isolation of the Arctic explorer but also made possible the co-ordination of scientific work at opposite ends of the polar basin. But the knowledge was not yet dependable or comprehensive enough to equip man to live with safety in the Arctic region, to open up its resources and navigate its waters. It was not possible, for example, to forecast the weather and the related phenomena of the seasonal movement and drift of the ice. Roald Amundsen undertook in 1919 to repeat Nansen's drift on the *Fram,* embarking on the same course aboard the schooner *Maud.* But whereas it had taken Nansen only 40 days to drift from Novaya Zemlya to the New Siberian Islands, it took Amundsen more than a year to complete the same journey. Amundsen had assumed that the climatic and oceanic processes of the Arctic follow a regular cycle year after year. Actually they are subject to great and irregular fluctuations that yield a predictable pattern only in ultimate correlation with the cycles of solar activity. By chance Amundsen had chosen the worst possible time for his voyage. This, however, is hindsight that has been made possible by the most recent investigations in the Arctic.

Modern Exploration

The Soviet era of Russian Arctic exploration began soon after the October Revolution. Very much aware of the problems of the North, the Soviet Government, headed by V. I. Lenin, in one of its first acts in the year 1918 authorized hydrographic surveys to develop navigation in the Arctic seas. The immediate plan was to open the Kara Sea to shipping that would carry grain from the southerly steppes of Siberia, down the Ob and Yenisei rivers and westward through the Kara and the Barents seas to Archangel—a scheme analogous to the opening by the St. Lawrence Seaway of the plains of the American Middle West to ocean traffic. By 1921, 23 parties totaling 400 men had gone to work in the Arctic. Important maritime expeditions followed, and geophysical observatories were built in Novaya Zemlya, Franz Josef Land, the Severnaya Zemlya Archipelago and the New Siberian Islands.

By 1930 the original plan had been enlarged. No longer was the aim simply to open the Kara Sea to navigation; it was to open the entire Northeast Passage and to make the route feasible for scheduled traffic on a large scale. There were many authorities at home and abroad who categorically denied that the passage could be navigated in one season; previous expeditions, beginning with Nordenskjöld's, had required two years or more. The passage is icebound nine months of the year and is made hazardous by fog and heavy weather and by drifting ice even in the "open" months. The Kara Sea, while open three months of the year in favorable circumstances, holds its ice much longer than do other parts of the passage. But the Soviet Government backed the country's Arctic enthusiasts. It placed the program of polar exploration under the direction of a central agency that was ultimately to become the Arctic and Antarctic Research Institute. The institute assumed responsibility for all hydrometeorological, geophysical and physical-geographic investigations of the Arctic; schools to train engineers and scientific workers were organized; icebreakers were designed and ordered.

In 1932—aided by aerial reconnaissance, radio and the network of observatories that had been established—the icebreaker *Sibiriakov,* under the command of O. Schmidt and V. Voronin, sailed the entire Northeast Passage in a single season, demonstrating that there was now a practicable alternative to the route through the Panama Canal half a world away. A few years later the icebreaker *J. Stalin* underscored the feat by completing a trip from the Barents to the Bering Sea and back in a single season.

The voyage of the *Sibiriakov* (which had been appropriately named for Nordenskjöld's sponsor) was the prologue to even greater undertakings. Recognizing that the weather and ice conditions of the Arctic seas, across which the Northeast Passage lies, are directly influenced by the rest of the Arctic Ocean, the Arctic Institute now extended its research program to embrace the entire North Polar Basin.

A major vehicle for this undertaking is of course the airplane. Soviet polar pilots have mastered the art of locating places to land and making successful landings on the ice pack and thereby have vastly extended the range of scientific reconnaissance. In the spring of

HYDROLOGIST installs a winch on ice-island station preparatory to measuring submarine currents. Helicopter in the background supplies the station and is used for reconnaissance.

MARINE BIOLOGISTS lower specimen collector through floor of protective pit dug in ice island. Arctic waters are populated by crabs, medusas and plant and animal plankton.

1941 the Soviet airplane *N-169* made a historic sortie to the Pole of Inaccessibility—the point in the Arctic Ocean at 83 degrees 40 minutes North latitude and 170 degrees West longitude, which is the most distant from land in all directions—for the purpose of geophysical and oceanographic observations.

One of the most fruitful of the institute's innovations was the ice station North Pole I. In May of 1937 a hazardous airplane landing delivered a four-man team of scientists to a drifting ice floe in the heart of the Arctic Ocean. For the next nine months they lived on the floe, recording the sea and weather conditions that the floe encountered on its meandering course. When North Pole I, which was the forerunner of a series of ice stations, finally drifted to destruction in the Greenland Sea, an icebreaker picked up the team and its precious data. In the fall of the same year the icebreaker *G. Sedov* set out on an ice-locked drift that started in the Laptev Sea, traversed the Arctic Ocean along a line very close to the one followed by the *Fram* (but in higher latitudes) and ended after 812 days in the Greenland Sea. The oceanographic, geophysical and meteorological work of these expeditions into the heart of the Arctic Ocean, combined with the data issuing from the network of fixed stations ashore, began to yield an entirely new understanding of the weathermaking processes of the Arctic—and of the earth as a whole—even before the outbreak of World War II.

The Arctic and the Weather

The old theory regarded the Arctic as a kind of weather kitchen. The Arctic troposphere—the layer of the atmosphere nearest the earth—was believed to be the prime mover in the manufacture of temperate-zone weather. This uniformly cold, high-pressure mass of air, the so-called polar cap, seeped southward to interact with the warmer and wetter air there and generate the familiar cyclonic storms of the Temperate Zone. But it was supposed that the polar cap itself was only rarely penetrated by cyclones. Such cyclonic penetration as did take place must come mainly from the Atlantic side of the Arctic Basin rather than from the Pacific side.

This picture was reasonable but it was based upon incomplete evidence. We have found that the Arctic is frequently penetrated by cyclones. (The North Pole I station counted 78 days of deep cyclonic penetrations in one six-month period.) Secondly, cyclones enter from the Pacific side much more frequently than was thought. According to our present understanding, a high-pressure air mass does indeed blanket the Arctic. The cold, dense polar air mass builds up its identity particularly in the winter months. During this period the Arctic tropopause (the roof of the troposphere) descends to correspondingly low altitudes. But in the spring and summer the polar air mass warms, expands and rarefies, and the tropopause rises. Toward the end of April one year it rose to more than 34,000 feet from about half that height in only four days. As a result the warmer air from the south is able to invade the Arctic air mass and generate frontal storms within the Arctic region. At such times the tropopause of the Temperate Zone air is both higher and colder than the Arctic tropopause; the effect of this disparity is to set up vertical as well as horizontal movement, resulting in exchange of air between the two masses. The Arctic, far from generating the weather of the Northern Hemisphere, is itself subject to wide changes of weather that reflect fluctuations in the general circulation of the earth's atmosphere. In brief, the manufacture of weather is not merely an Arctic process but a global one. Such understanding of the Arctic weather now makes long-range weather forecasting a practical matter. When a high-latitude aeronautical expedition was being planned in 1954, for example, forecasters were able to take into account a large influx of warm air into the Arctic and correctly predict flying conditions.

World War II brought a temporary restriction in scientific investigation of the high-latitude Arctic. Nevertheless, regular aerial reconnaissance of ice conditions continued, with larger numbers of aircraft bringing larger areas of the Polar Basin under observation. The work of the Arctic Institute was resumed on an even larger scale as soon as the war ended. Shipping in the Northeast Passage now required three- to six-month forecasts of weather and ice conditions, and short-term (one- to 15-day) forecasts for each voyage and for each leg of the long journey. The expansion of polar aviation, not only for scientific purposes but also for the transportation

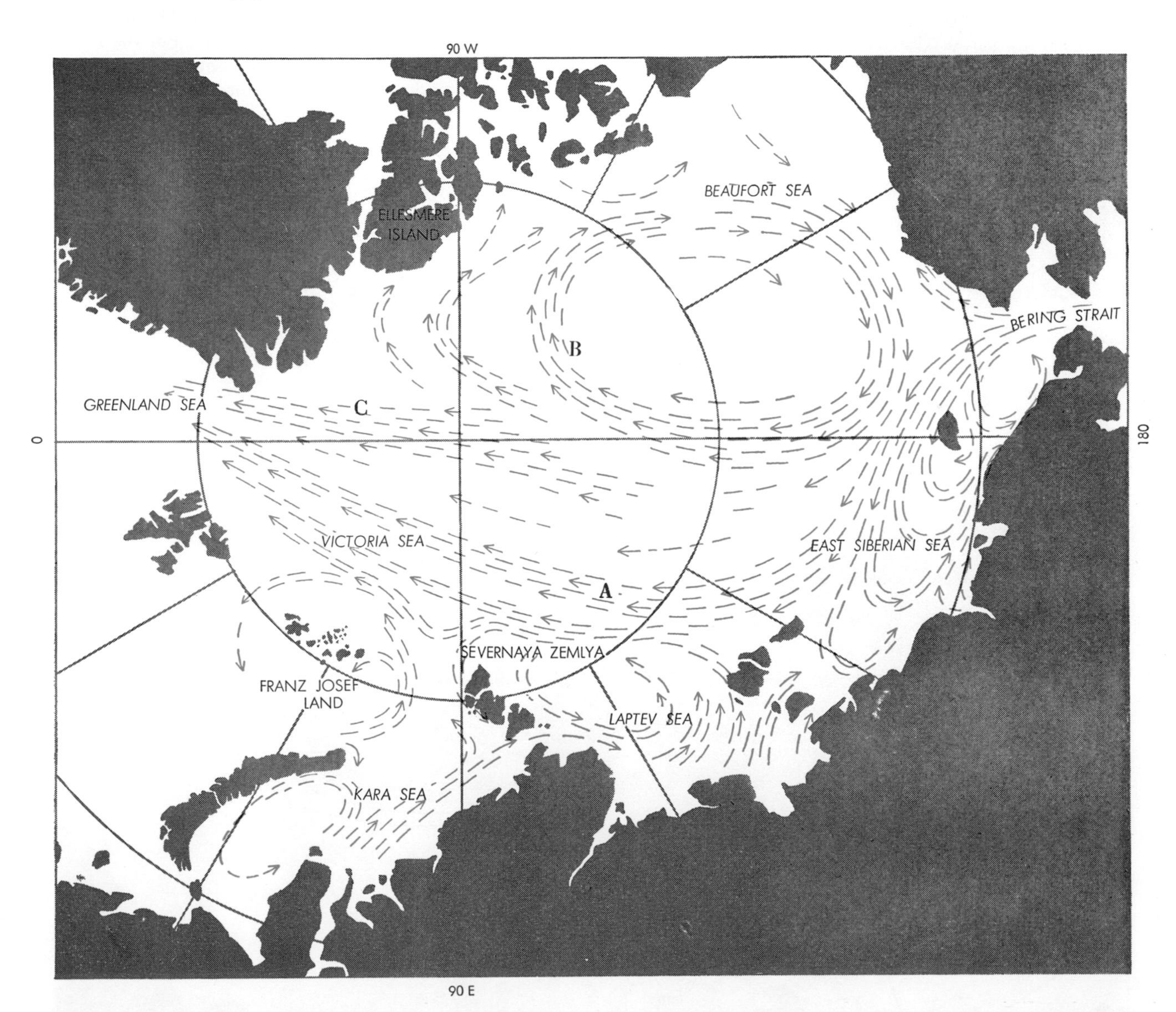

ARCTIC CURRENTS are traced by ice drifts, which originate chiefly in Siberian seas and follow three major systems. One (*A*) moves westward from East Siberian and Laptev seas; it has major branches east of Severnaya Zemlya and around Franz Josef Land. Young ice flows into both branches from the Kara Sea. So-called anticyclonic drift system (*B*), centering in Beaufort Sea, forms zone of stagnation in which ice moves slower, remains longer, grows thicker. A third system (*C*) originates in the Bering Strait and adjacent waters to the west and moves directly across the Pole. A branch carries ice around Ellesmere Island; main body converges with the first system and flows into the Greenland Sea. In passage across Arctic, ice may switch from one system to another.

of cargoes and passengers, called for a still finer-grained system of weather forecasting and improved knowledge of the earth's magnetic field and its strange high-latitude anomalies. Such practical objectives raised all kinds of fundamental questions. The scientific effort that the Arctic Institute has directed during the past 15 years has explored these questions on a broad front.

A network of some 100 fixed hydrometeorological stations and geophysical observatories on remote mainland points and on the islands of the Arctic makes continuous year-round observations. During the summer months the vessels of the ice patrol and the hydrological patrol extend the reach of this network to the edge of the retreating ice pack. These missions consist of groups of five to eight scientific workers, who spend three or four months crossing and recrossing the icy waters. Powerful new diesel-electric ships, carrying the larger parties of major expeditions, make it possible to penetrate to still more inaccessible areas. With the coming of spring, every year since 1948, the Arctic Institute has also launched a series of aerial expeditions. Each plane is a flying oceanographic laboratory manned by a staff of three to five scientists. Landing on the treacherous natural runways of carefully prospected ice floes, the men spend one to three days sounding the ocean depths, measuring the thickness of the ice and snow and the temperature of the waters below, taking samples from the ocean floor and making magnetic, astronomical and meteorological observations. The flights are necessarily suspended during the two or three dark months of the polar winter.

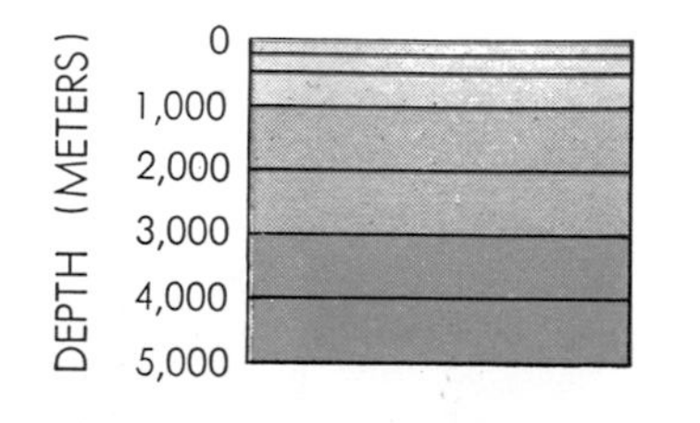

The Drift of the Ice

Much of the data about the drift of the ice, especially in the lower latitudes, is nowadays a by-product of the surveillance maintained on behalf of Arctic shipping. Shore stations keep the ice under constant observation. Technicians travel to specified ice floes by sea or air and set up radio markers to measure the speed and direction of drifts in the open seas and in the Arctic Ocean [*see illustration on page 101*]. For the six to 12 months of their useful lives these robot observatories intermittently broadcast

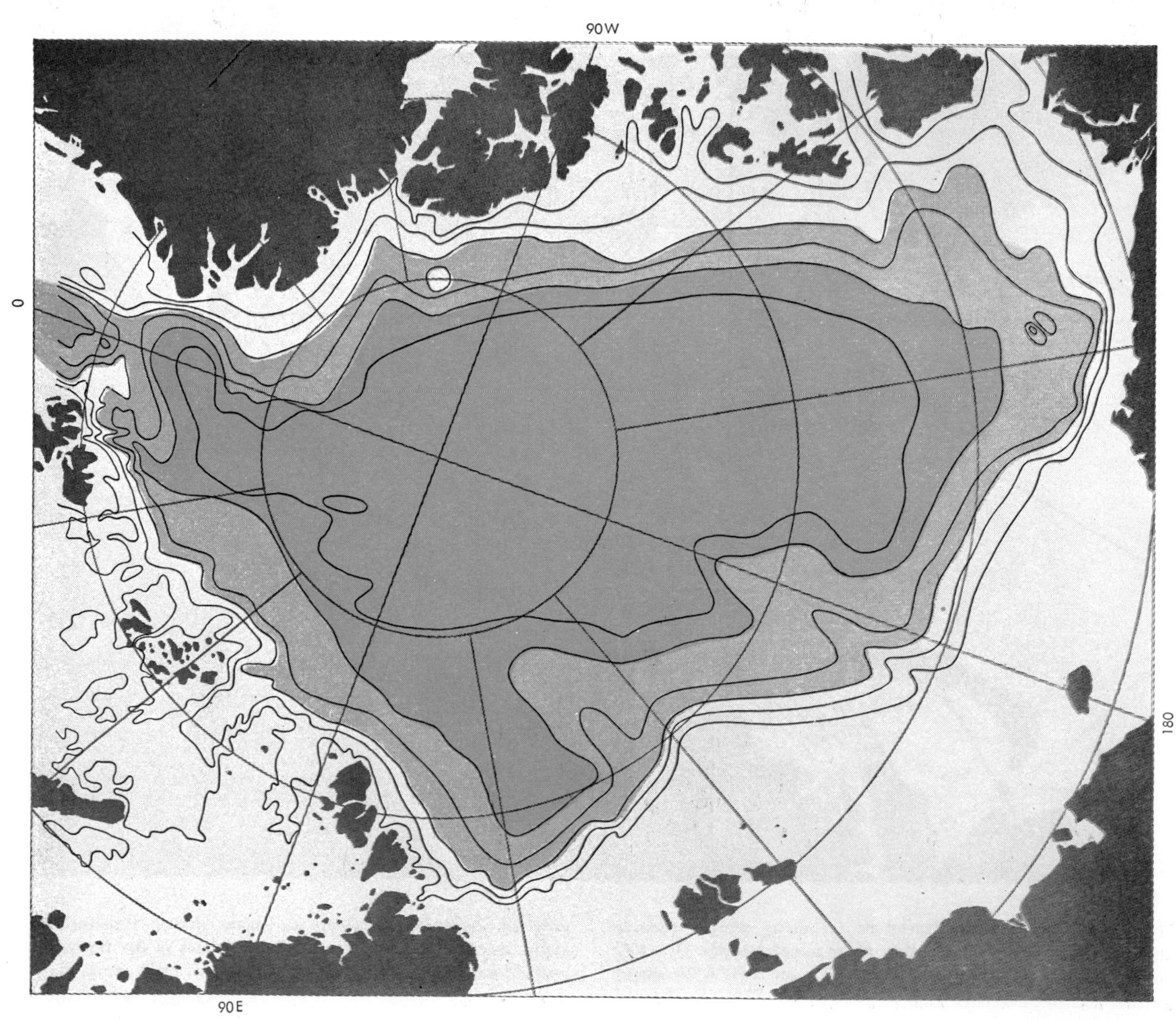

BATHYMETRIC CHART OF 1947 shows what was known of Arctic depths at that time. The area that lies within the contour nearest the Pole corresponds roughly to what had yet to be explored. View looks across the Pole from the Eurasian side of the region.

signals giving the velocity and direction of the wind and the temperature and pressure of the air. Directional fixes on the markers, recorded along with the information they transmit, chart the drift of the floe. By tracking 15 to 20 floes in various parts of the ocean at a time the shore stations keep a running account of drifts and climatic conditions.

The accumulation of this information over the years has now revised some long-standing misconceptions about the drifting of the Arctic ice. One notion was that the floes in the eastern Arctic seas rotated in closed and "eternal" trajectories. Another, based upon Nansen's experience, was that winds and currents carried the ice across the central Arctic toward the Greenland Sea in a steady stream. There is truth in both hypotheses, but our tracking stations have found that the drift pattern is far more complex.

There are three systems rather than one. The ice that enters all of them originates principally in the cold waters of the relatively shallow Siberian seas [*see illustration on page* 97]. One major drift sweeps westward from the East Siberian and Laptev seas, following a broad path between the 86th and 87th parallels, toward the Greenland Sea. This is the drift that carried the *Fram* on its voyage. Two loops branch off the mainstream. One, turning in the cyclonic or counterclockwise direction, is confined substantially to the Laptev Sea; the other swings, also cyclonically, around the islands in the region of Franz Josef Land, cutting through the Victoria and Barents seas north of Novaya Zemlya. Into both of these branches flows young ice from the Kara Sea.

A second system at the Pacific side of the Arctic—the so-called anticyclonic, or clockwise, drift system—describes a closed curve in the Beaufort Sea. This system forms a zone of stagnation in which the ice moves slower and remains for a longer time. As it rotates through a cycle of two or more years, the ice here grows thicker than elsewhere in the Arctic.

Both of these major systems may contribute some ice to a third, which originates in the East Siberian and Chukchi seas and in the Bering Strait. This system sweeps directly across the Pole and, converging with the first system, flows into the Greenland Sea. A branch of this third

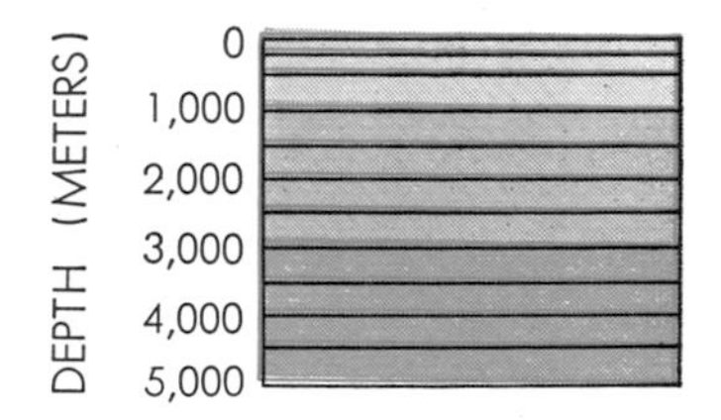

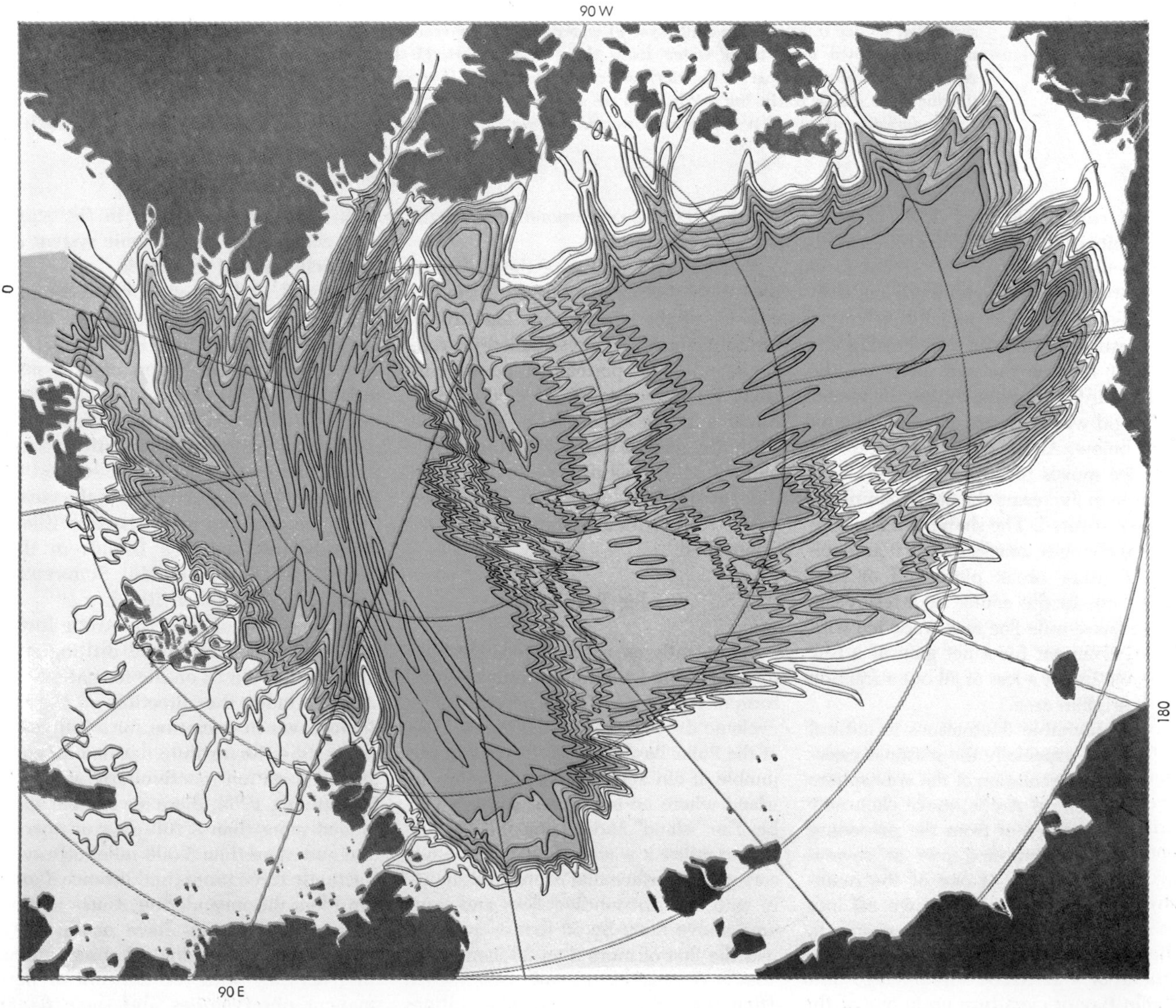

BATHYMETRIC CHART OF 1955 is far more detailed than earlier one on opposite page. Major discovery that emerged from additional data is existence of submarine Lomonosov Ridge, which, passing near Pole, extends from one continental shelf to the other.

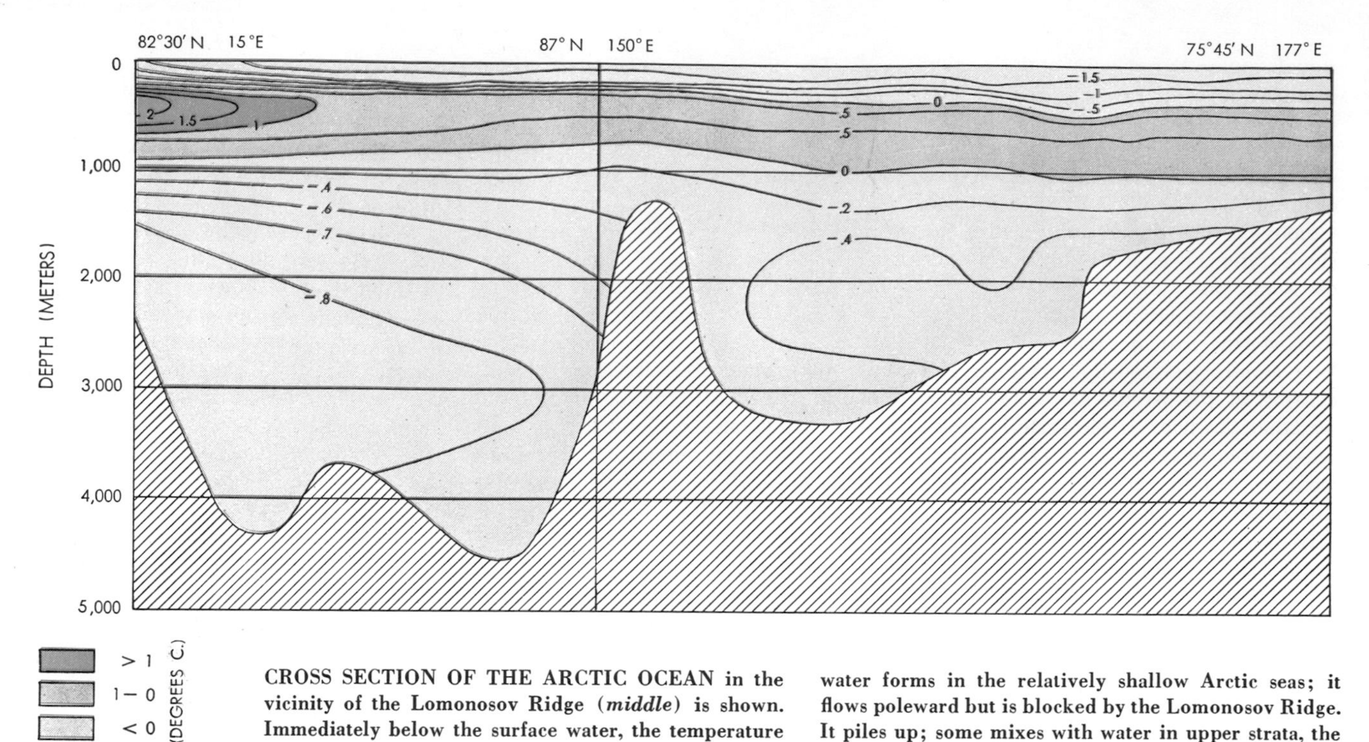

CROSS SECTION OF THE ARCTIC OCEAN in the vicinity of the Lomonosov Ridge (*middle*) is shown. Immediately below the surface water, the temperature of which is –1.8 to –1.5 degrees centigrade, lies the somewhat cooler intermediate cold layer of water. The next layer down consists of water from the North Atlantic; it is about 3 degrees C. (*darkest color*) when it enters the Polar Basin but cools to .6 or .8 degree (*next darkest color*) as it moves northward. Bottom water forms in the relatively shallow Arctic seas; it flows poleward but is blocked by the Lomonosov Ridge. It piles up; some mixes with water in upper strata, the rest spills over to other side of ridge. Newly discovered stratum of water from the Pacific enters basin through the Bering Strait and spreads beyond the Pole. It is somewhat colder than surface water and lies just below it. The Atlantic side of the basin is at left, the Pacific side at right. Temperatures are representative.

system carries ice floes down through the straits of the Canadian Arctic Archipelago to Baffin Bay.

On its course through the various drift systems an ice floe perpetually exchanges its substance with the sea. During the summer months the sun melts off the upper third of its mass; during the winter the frigid waters freeze on a like amount from below. As the cycle progresses the new ice moves toward the surface, and in three to five years the floe is completely reconstituted. The shape of a floe constantly changes; cracks divide it and collisions either break pieces off or fuse pieces on. In the course of a few years a 25-square-mile floe may trade ice with its environment for a net gain of a few square miles or a loss of all but a fraction of its original area.

The inevitable fluctuations in natural conditions, especially the seasonal variations in the circulation of the atmosphere over the central Arctic, make each new drift cycle different from the preceding one. After completing one or several cycles in an eddy of one of the mainstreams, a floe may be thrown off into the outward-flowing stream that goes to the Greenland Sea or pass through the straits of the Arctic Archipelago into Baffin Bay or even turn up in one of the outlying Arctic seas. The outflow into the North Atlantic is correspondingly irregular from year to year.

Our detailed studies of the drift pattern show that its main cause is the circulation of the atmosphere. The drift is also influenced by the so-called prevailing winds and permanent ocean currents. But these too are affected by fluctuations in the atmospheric circulation. With the general warming of the earth's climate the melting zone has been moving farther north. As the pack has retreated, its average thickness has diminished.

Ice Islands

As recently as 1946 Soviet investigators discovered a hitherto unrecognized form of Arctic ice. Flying over the anticyclonic drift system on the Pacific side of the Polar Basin, I. S. Kotov saw in the jumble of old and new sea ice below an island where no island was supposed to be. The "island" had an area of some 200 square miles; it was relatively flat, with a corrugated surface that contrasted sharply with the surrounding floes and towered above them by 30 feet or more. It was the first of more than 30 ice islands that have since been found. Observation from the air and by scientific stations established on the islands has shown that such formations originate in the stagnant zone of the anticyclonic system at the Pacific side of the Arctic.

When U. S. fliers rediscovered the first of these ice islands, they named them T-1, T-2 and T-3. In 1952 the U. S. established an ice station on T-3 and maintained it for a year, during which time the island drifted through a typical cycle of the anticyclonic drift system: from Ellesmere Island it traveled west to the Beaufort Sea, then north to the vicinity of the Pole and finally back to Ellesmere Island [see "Ice Islands in the Arctic," by Kaare Rodahl; SCIENTIFIC AMERICAN, December, 1954].

In the spring of 1950 the Arctic Institute established its second drifting station, North Pole II, on the ice pack. The station, under the direction of M. M. Somov, was in operation for a full year. Since 1954 the institute has had at least two such stations continuously at work. By January, 1959, these ice stations had totaled more than 3,400 days of operation and more than 5,000 miles of travel—actually three times that distance if one considers the meandering course of the islands. The stations have made more than 15,000 meteorological observations in synchrony with the network of permanent observatories and some 6,000 more independent observations. They have sent more than 8,000 radiosonde balloons into the upper atmosphere, measured the ocean depths at more than 3,000 points, made 40,000 readings of water temperatures and 100,000 readings of the speed and direction of various submarine currents. The synchronized observations are particularly important for understanding the general pattern of atmospheric circulation. Indeed, a synoptic weather map of the Northern Hemisphere would look downright peculiar nowadays in the absence of meteorological data gathered by the drifting ice stations.

The typical station consists of a cluster of sturdy houses not unlike small railroad cars. The houses are set up on sled runners so that they can be dragged by tractor or pushed by the personnel of the station should a crack threaten to split the campsite. Each foam-insulated cabin shelters three men in its 13 square meters; separate huts and hemispherical tents house the wardroom, kitchen and laboratories. A central diesel engine, supplemented by portable units, supplies the station's power and heat to the indoor living spaces; a small wind-driven generator serves the radio. Tractors, a helicopter, sledges and rubber boats provide transportation.

The 10 to 14 scientists who man each station are supported by a physician, a radio operator, a mechanic, a cook and a four-man helicopter crew. This permanent population, which is completely replaced each April, may occasionally be augmented by five- to 10-man groups that come for a month or two on special missions.

Much of the repetitive data is gathered automatically. A new vane-type meter, for example, measures the velocity and direction of deep submarine currents and records the information every hour or so for as long as two months between maintenance visits. Special integrating meters gauge surface currents and telegraph the information to instrument panels in the observers' huts. Other devices measure the melting and evaporation of the ice and monitor its thickness.

Since the station operates all day every day, and supply problems limit the size of the crew, a good deal of versatility is expected from each man. The magnetologist usually serves also as astronomer, the physician doubles as housekeeper, the meteorologist takes readings of the direct-heating power of the rays of the sun and the fliers assist the hydrologists and aerologists. All hands take turns at kitchen duty, which in addition to helping the cook in the traditional

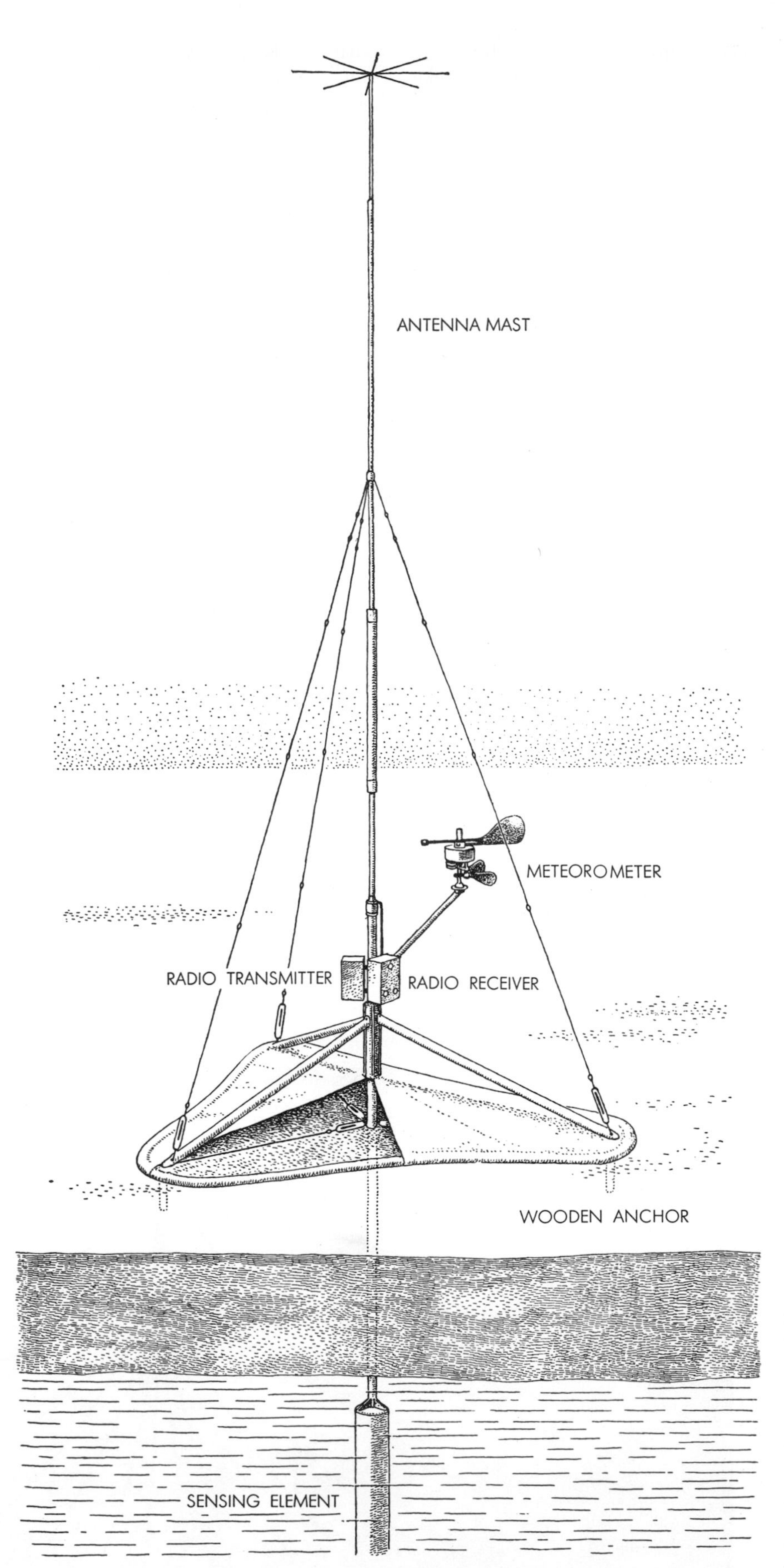

AUTOMATIC OBSERVATORY, of the type installed on selected ice floes, intermittently transmits data on ocean currents and temperatures, wind direction and velocity, air temperature and pressure. Tracking markers from land yields pattern of drift systems.

ways means supplying snow for the diesel and for such personal use as drinking and washing. And all lend a hand in hacking a runway out of the ice for the landing of airlifted fruits and vegetables, mail, reading matter, fuel, movies and specially designed clothing suited to the various seasons.

Whatever the season, life on a drifting ice station is severe. Frequent and lengthy snowstorms howl through the five-month polar night and the air temperature falls to 50 degrees below zero centigrade (90 degrees below freezing Fahrenheit). In summer persistent fogs throw a gray wall around one's head, and the otherwise welcome zero-degree "warmth" brings the nuisance of melting ice. At this period, with water standing in numerous deep (three feet or so) little lakes, the camp looks like a polar Venice. The thaw also brings the danger that the "land" on which one has lived for a whole year may suddenly split apart, that a black abyss a mile or more deep may open right under a hut. But man is wonderfully adaptable. One soon learns to live with the threat; the unpleasant expectancy subsides, although vigilance does not. As a member of the North Pole IV party in 1955 and 1956, I learned that the ice field will break apart on sunny days as well as during storms. Fortunately the crack does not always go through the camp!

The rigors of polar life are not altogether unrelieved. Those who take part in a drift through the Arctic night find a pleasant and reliable friend in the moon. The light suffusing the icy waters is so bright that in good weather we could confidently make excursions several kilometers away from camp and were always able to keep the camp in view. During the polar day visiting bears occasionally broke our isolation. We usually succeeded in chasing them off by firing rockets. Only once did we have to shoot one—when it began to wreck a laboratory tent.

Spring and summer bring a startling flare-up of life in the seemingly barren Arctic. Charming Arctic sparrows came to keep us company, and our hydrologists' jars were constantly catching tiny crabs, medusas, seaweed and curious representatives of the phytoplankton and zooplankton. Our biologists have found that the entire mass of water to depths of 10,000 or 15,000 feet, to the bottom itself, teems with microorganisms. The processes of life go on as actively in the Polar Basin as in the other oceans of the world. Toward the end of one polar night, at 86 degrees North, we found a small roe-filled cod in our Nansen net

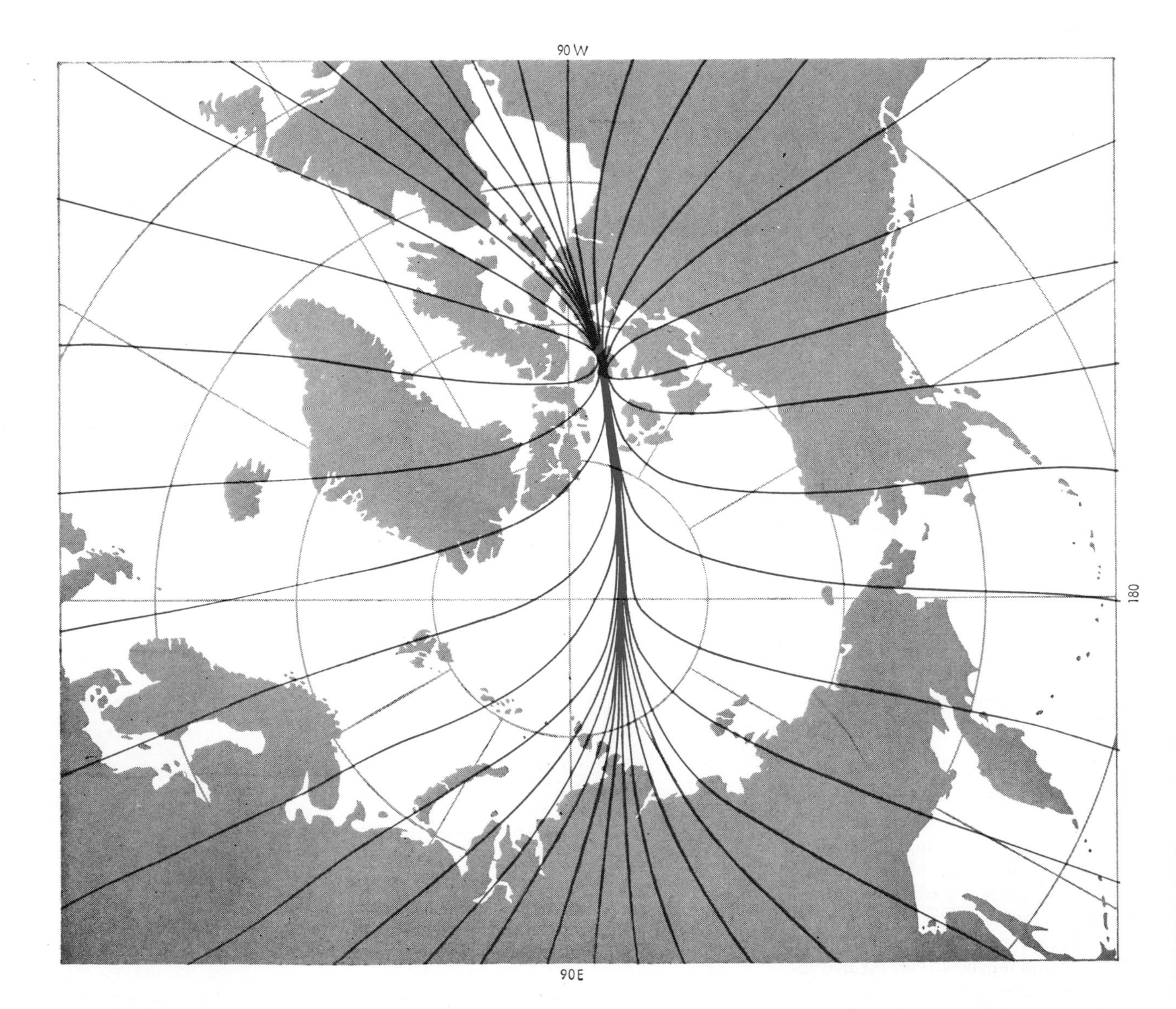

THE NORTH MAGNETIC POLE, once thought to be virtually a point in the Arctic Archipelago, has been shown by recent investigations to extend across the Polar Basin to the Taimyr Peninsula in Siberia. The colored lines represent magnetic meridians.

and thereby set some kind of record for catching a fish at such a latitude and in such a season.

The Waters of the Ocean

The oceanic fauna on the Atlantic side of the Arctic differ from the fauna on the Pacific side, and those found in one layer of ocean are not found in another. Among the factors that account for these differences are the variations in the temperature and circulation of the layers of ocean. Of the major waters of the Arctic Ocean the surface water is ubiquitous; it is identified by its relatively low temperature (about −1.8 degrees C. in winter and −1.5 degrees in summer) and low salinity (about .3 per cent). Between 150 and 600 feet below the surface lies the so-called intermediate layer. Its temperature, always a little lower than that of the surface layer, falls from about −1.6 degrees in summer to −1.9 degrees in winter, with water from the surface layer becoming heavier and sinking downward. The same exchange with the surface waters increases the salinity of the intermediate layer; the salt is not bound in the formation of the ice at the surface and is carried downward to the intermediate layer.

On the Pacific side of the basin a newly discovered layer of water separates the surface waters from the intermediate cold layer. Hydrochemical and hydrobiological research has shown that these somewhat warmer waters enter the Polar Basin through the Bering Strait and occupy a stratum between 250 and 350 feet below the surface.

The great layer of Atlantic waters, which lies below the intermediate cold water, enters the Polar Basin in a powerful submarine current that travels mainly between Greenland and Spitsbergen. This current replaces much of the water that the Arctic Ocean loses through the surface currents and ice floes that sweep out of the Polar Basin. Upon entering the basin, part of the Atlantic current moves eastward, hugging the Eurasian continental shelf, and part spreads north. The eastern branch flows into the shallow Arctic seas, where it cools and sinks and flows along depressions in the sea bottom into the deep ocean. When the Atlantic current enters the polar region, it has a high salinity and a temperature of three to four degrees C. and is almost 2,000 feet thick. But as it rolls farther into the basin through the shallow seas and mingles with the colder Arctic waters it cools and becomes a thinner stratum of water. By the time it reaches the Pole its temperature is not above .8 degree, and as it continues east into the Beaufort Sea its temperature falls to between .6 and .8 degree.

Beneath the Atlantic waters is the great mass of cold (−.8 to −1 degree) bottom water that fills the submarine depressions. According to earlier accounts the bottom water was formed mainly in the Greenland Sea as a result of winter cooling and the sinking of colder strata. We now know that much of the water in this layer forms, largely from Atlantic water, at the edge of the shelf of the Eurasian continent and in the shallow Arctic seas, whence it flows into the depressions of the Polar Basin. In its poleward movement the bottom water encounters the great submarine barrier of the Lomonosov Ridge, discovered and investigated by Soviet polar explorers in 1948 and 1949. Welling up on the Atlantic side of this undersea mountain range, the bottom water mingles with the warmer Atlantic waters above and, spilling over the barrier, fills

NUCLEAR-POWERED "LENIN" is late addition to Soviet fleet of icebreakers. Ships are used for scientific research and for clearing paths through ice for commercial vessels.

the depressions in the ocean floor on the Pacific side.

The Lomonosov Ridge

The Lomonosov Ridge reaches like a gigantic bridge about 900 miles from the Asian continental shelf, some 250 miles north of the New Siberian Islands, past the vicinity of the Pole to the edge of the continental shelf of North America near Ellesmere Island. Its peaks rise to within 3,200 feet of the surface, and it divides the abyss of the Arctic Ocean into two parts. On the Atlantic side the ridge slopes steeply to a bottom that is more than 13,000 feet down. On the other side of the barrier the slopes descend to somewhat more than 11,000 feet. Thus the ridge towers from 6,000 to 11,000 feet above the basins.

The discovery of the Lomonosov Ridge is of unparalleled importance to an understanding of every aspect of the Arctic region. The ridge plays a decisive role in determining the circulation and exchange of water among the different parts of the ocean, in setting the pattern of the ice drift (especially the anticyclonic drift on the Pacific side) and in establishing the major provinces of life in the Arctic waters. In the light of this discovery, geologists have also had to revise their view of the history of the earth's crust in the northern regions. The topography and specimen cores taken from the slopes of the ridge have persuaded Y. Gakkel to the conclusion that the Lomonosov Ridge has been the site of volcanic activity in the past and that renewed volcanic activity is within the realm of possibility in the future. V. N. Saks has advanced the hypothesis that the Arctic was once dry land that sank not later than the Devonian period, about 325 million years ago.

In recent years the building of sturdy and powerful diesel-electric icebreakers has made it possible for Soviet scientists to undertake extended oceanographic expeditions in hitherto inaccessible regions of the Arctic, notably the northern part of the Greenland Sea and the adjacent reaches of the Arctic Ocean. In this vicinity the icebreaker *Litke* in 1955 sounded the greatest known depth in the Arctic Ocean, a depth of 17,880 feet, and discovered a rift running northward at depths of 9,190 to 12,800 feet.

Geomagnetism

Navigators in the high latitudes have always been troubled by the odd behavior of their magnetic compasses caused by apparent irregularities and asymmetries in the magnetic field of the earth. Early magnetic maps had been drawn on the assumption, based upon hopeful guesses, that the North Magnetic Pole is virtually a point. Accordingly it was expected that the compass needle, which dips more steeply as it approaches the Magnetic Pole, would point straight down, or very nearly so, at the Magnetic Pole itself. But data from the North Pole I, the *G. Sedov* and other expeditions showed that the compass needle points straight down for a very long distance across the Arctic Ocean, from a point northwest of the Taimyr Peninsula to another point in the Arctic Archipelago. This discovery first inspired the hypothesis that there is a second North Magnetic Pole, tentatively located at 86 degrees North latitude and 182 degrees East longitude. More refined observation has disposed of this idea. The map of the magnetic field now shows the magnetic meridians running close together in a thick bunch of lines from the North Magnetic Pole in the Arctic Archipelago to Siberia.

Our program of geomagnetic studies, in co-ordination with that of the International Geophysical Year, has been extended to high altitudes. This work has clarified the interaction of the magnetic field with the charged particles emitted by the sun, especially with respect to the formation of the belts of radiation that surround the earth, the generation of magnetic storms and the production of auroral displays. Today our stations in the Arctic conduct regular simultaneous observations with our stations in the Antarctic. Along with its great theoretical interest, the program has given improved accuracy to navigation charts of the Arctic regions.

New Lands

In completion of the task begun in the age of discovery, the Arctic Institute has also made some significant revisions in the geography of the Arctic. The map shows the newly discovered Schmidt, Vize and Ushakov islands in the Kara Sea, and new bays, straits and islands in the Severnaya Zemlya Archipelago. Missing from the map are some places that have been "undiscovered." Makarov Land, Gillis Land and Petermann Land, all in the vicinity of Franz Josef Land, and Sannikov Island and Andreev Land in the Laptev and East Siberian seas have had to be erased from the map. In each of these cases either accumulations of ice or thick fogs over open water had congealed in the heads of imaginative cartographers as land masses.

With its broadly conceived and rigorously prosecuted program of exploration and research, the Arctic Institute has shown that an enormous region of the earth's surface and correspondingly large realms of the unknown may be brought within the compass of human understanding in a very few years. Moreover, the often perilous undertakings have been carried out with a high regard for the safety of the personnel; despite gales, snowstorms, fogs and treacherous ice floes, not a single life has been lost during the entire postwar program. The data thus far amassed by the institute's expeditions and ice stations fill more than 120 volumes; the list of books, monographs and articles that is emerging from that data already exceeds 600 titles. And while the main purpose of these expeditions was to delve into the mysteries of nature, they also served to train young scientists in nearly every branch of polar science. For the most part, these men are in love with the Arctic and see great prospects for its development. Many of them plunged into Arctic exploration immediately upon leaving school and have already devoted 20 and 30 years of their lives to this work. They have shown themselves to be worthy of their predecessors.

Full-scale Arctic navigation is now a reality. Guided by aerial reconnaissance, the way through the ice having been cleared by icebreakers, and with ice movements and weather changes monitored by the network of observatories, hundreds of ships are carrying many thousands of tons of cargo through the Northeast Passage on a regular and dependable basis through a season that has been extended to a full four months.

As recently as 30 years ago more than half of the total area of the Polar Basin was unexplored, and 16 per cent was still terra incognita only 15 years ago. Today, disappointing as this may be to young geographers, the area of the blank spots on the map of the Polar Basin has shrunk to almost nothing. At the same time, to the regret of the older explorers and the understandable pleasure of the younger ones, there are still blank spots elsewhere in the Arctic. The ocean, the air and the ionosphere still hold many mysteries. Responding to that challenge, the Soviet ice stations North Pole VIII and North Pole IX are even now adrift in the Arctic Ocean. And the nuclear-powered icebreaker *Lenin* has given Soviet scientists a new capability in the task of subjugating the Arctic and exploiting it for the needs of the community.

12

SALT AND RAIN

A. H. WOODCOCK

October 1957

When you stand at the rail of a ship on a bright sunny day with a fresh wind blowing and watch the sparkling whitecaps dancing on the sea, rain is likely to be one of the farthest subjects from your mind. But those whitecaps have begun to interest meteorologists intensely. There is reason to suspect that they are responsible, in an unexpected way, for much of the rainfall that nourishes life on the earth. We have arrived at this suspicion through laboratory studies of the bubbles that form their spray.

The chain of reasoning starts from one of the key questions of meteorology: What makes clouds turn into rain? More specifically, the question is: How do the tiny droplets of a cloud coalesce into water drops big enough to fall as rain?

If cloud droplets themselves could fall to the ground, our planet would have fewer water problems. As everyone knows, a cloud at ground level (*i.e.*, fog)

CLOUD DROPLETS AND RAINDROP are enlarged some 100 diameters in this schematic comparison. The cloud droplets (*dots*) are about .001 inch in diameter; the raindrop (*arc*), about .1 inch. One raindrop consists of a million to eight million cloud droplets.

deposits moisture on leaves, twigs and other surfaces. The grass grows greener and longer under the drip zone of a tree or bush, and even under a telephone line. (This observation has led one cloud physicist to propose that water might be trapped for reservoirs by stringing wires on mountains frequently shrouded in cloud.) But over most of the earth clouds usually float high in the air. Their droplets fall so slowly that they evaporate long before they might reach the ground. To come down as rain the droplets in a cloud must grow to drops that fall at least 20 times faster. The problem that has occupied cloud physicists is to learn just how and under what conditions droplets grow into raindrops.

Let us see what gives rise to the cloud droplets themselves. The beam of a searchlight pointed upward at night shows that even apparently clear air is actually a "soup" of particles. The air may contain anywhere from 10,000 to 100,000 particles per cubic inch. When the relative humidity is high, water vapor condenses on many particles and begins to form droplets; this accounts for the haziness of the air on a muggy day and for the poor visibility you may have noticed while flying in an airplane below a cloud. Actual cloud materializes when the humidity reaches a certain critical value which turns most of the dust particles into water droplets.

Under the right conditions the cloud droplets will combine rapidly into raindrops; a concentration of 10,000 cloud droplets per cubic inch yields one raindrop per 10 cubic inches. There are two general theories about how this takes place.

One is the ice-crystal theory, originally developed by Tor Bergeron of Sweden and Walter Findeisen of Germany. In the cold upper regions of a high cloud the droplets are supercooled. If ice crystals are present, they evaporate the droplets and then absorb the vapor, much as crystals of calcium chloride and other drying agents absorb moisture. The ice crystals, feeding on the cloud droplets, may grow to large size and fall as snow or melt to rain.

This process is the basis of artificial rainmaking by means of dry ice and other crystals. The exciting experiments and discoveries of Irving Langmuir, Vincent J. Schaefer and Bernard Vonnegut in the laboratories of the General Electric Company led to the first clear-cut information on how nature makes rain [see "Cloud Seeding," by Bernard Vonnegut; SCIENTIFIC AMERICAN, January,

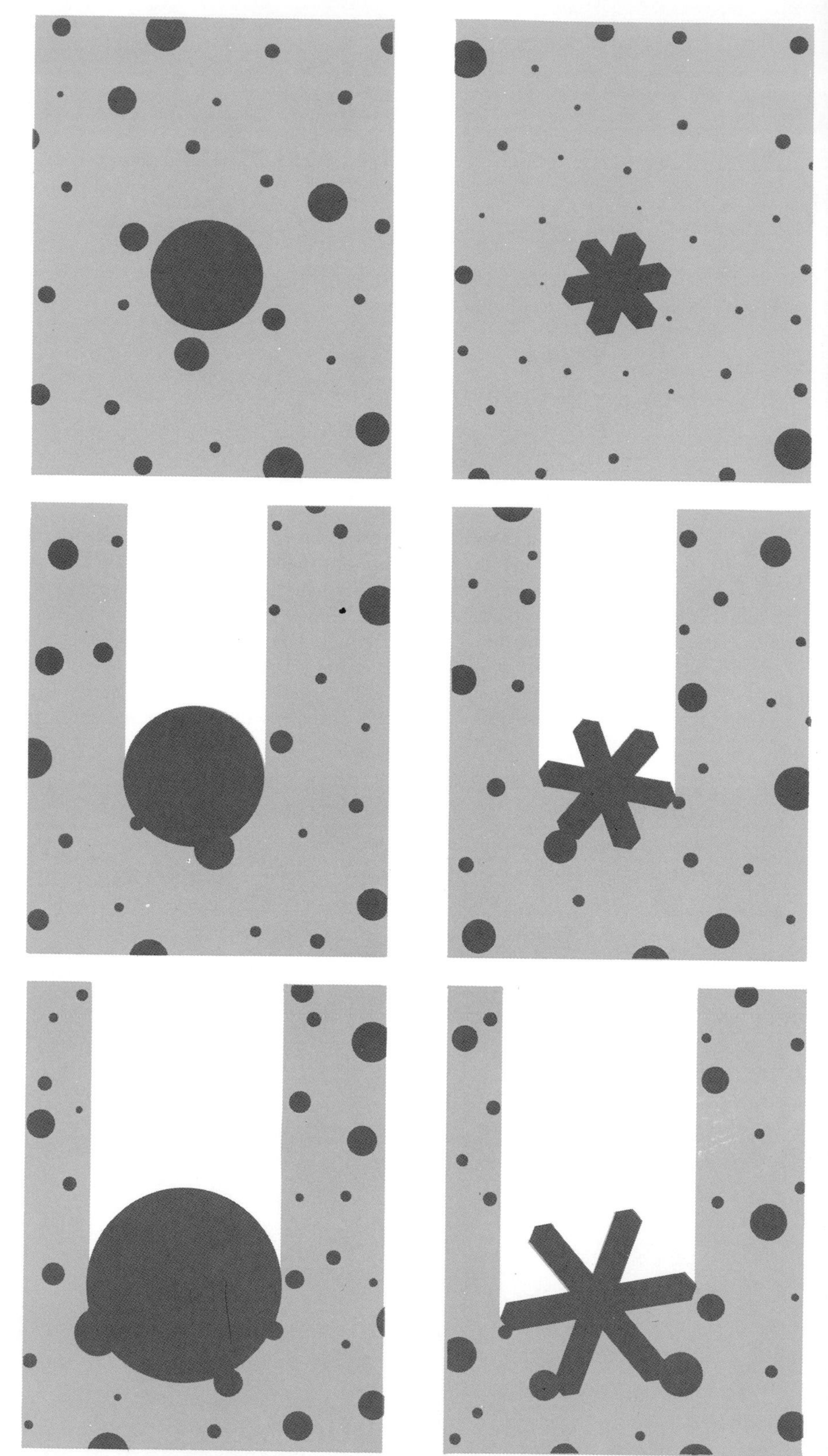

TWO PROCESSES by which cloud droplets grow into raindrops or snow are outlined. In the first picture of the sequence at left droplets of various sizes have formed by condensing on nuclei such as smoke or salt particles. The largest nuclei form the largest droplets. In the second and third pictures a large droplet grows by collision with small droplets until it is heavy enough to fall to the ground. In the first picture of the sequence at right a snow crystal forms when a droplet freezes or when a freezing nucleus is present. The droplets in the vicinity of the crystal tend to evaporate, providing water vapor for the crystal to grow. In the second and third pictures the crystal has grown large enough to fall and to grow by coalescence with the droplets. It may melt into a raindrop or reach the ground as snow.

1952]. But it soon became evident that the ice-crystal process is not the only way. Rain can fall from warm clouds as well as cold. How is it generated in clouds that lack ice crystals and supercooled droplets?

We must find some other mechanism that can combine droplets into big drops, and this brings us to the second theory. It suggests that large particles in clouds grow into raindrops by sweeping up the smaller droplets. The particles of dust or dustlike material in our atmosphere have a great range of sizes: if we magnified them so that the smallest were the size of a BB pellet, the largest would be as big as a bathtub. The big particles form comparatively large cloud droplets, and these of course fall faster than the small ones. As they move through the cloud, they pick up the smaller droplets in their path, just as a rolling drop of mercury gathers up any mercury drops it encounters. Thus the larger dust nuclei in a cloud can grow to the size of a raindrop. Each big droplet has plenty of smaller ones to feed on, because BB-sized nuclei outnumber the bathtub-sized ones about 100,000 to 1. A cloud will produce rain, according to this theory, when it contains sufficient moisture and a suitable number of giant nuclei.

It may be that some rains are triggered by ice crystals, some by giant nuclei and some by both combined. Very probably both processes play a part in many rains. We are therefore led to the interesting question: What are the giant nuclei and where do they come from? This question gave rise to the researches I mentioned at the beginning of the article. My associates and I at the Woods Hole Oceanographic Institution, and other groups in the U. S., Sweden and Australia, have been investigating the possibility that the rain-generating giant nuclei are salt particles from the sea.

The problem is being explored along several lines, with the objects of determining how much salt is taken up from the sea by winds, how much is carried inland over the continents, what the sizes of the salt particles are, whether they serve as nuclei for raindrops, how salty inland rains are, and so on.

First, there can be no doubt that winds blowing over the oceans pick up a substantial load of salt particles. Ordinary sea winds carry from 10 to 100 pounds of sea salt per cubic mile of air; storm winds may bear as much as 1,000 pounds or more per cubic mile. Secondly, it is equally plain that the winds transport a great deal of salt from the sea over land. The corrosion of steel towers testifies to the saltiness of our inland atmosphere. A gentle, invisible rain of salt falls constantly on the land. We can taste the salt on pine needles, and in the early morning light we see the salt particles glittering like jewels on spider webs. Systematic surveys have verified that salt particles, large and small, are spread through the atmosphere, from the ground up to high altitudes. Cloud-physics groups at the University of Chicago, at the University of Stockholm and in Australia have found them over much of the interior of the U. S., Sweden and Australia.

Next, there is statistical evidence of a relationship between the amount of salt carried inland from the sea and the amount of salt in our rainfall. This is based on measurements of the average salt content of sea air, the amount of sea air blown in over our coastlines in a year, the country's total yearly rainfall and the average saltiness of the rain (as sampled by C. Junge at 60 stations distributed throughout the U. S.). It turns out that the actual "salt-fall" in rain (3.8 pounds per acre per year) is remarkably close to the theoretical total deposit of salt (4.8 pounds per acre) that would be expected from the calculation of the salt load brought in over our shores by the winds.

Let us look at the salt particles in some detail. We have collected them in the air on glass slides held out of an airplane: the slides soon become fogged with salt crystals or droplets. Although the particles are tiny (some amounting to as little as a millionth of a millionth of an ounce), we can weigh them by an indirect method which entails exposing them to moisture. Salt greedily takes up water from the air, as anyone who has dealt with a salt shaker on a humid morning is well aware. A salt crystal kept in damp air collects enough water to dissolve completely into a droplet. The amount of water it will collect depends

SALT PARTICLE SIZE (TIMES 100)		RAINDROP SIZE (TIMES 10)		APPROXIMATE NUMBER PER CUBIC YARD OF AIR
	DIAMETER (MILLIMETERS)		DIAMETER (MILLIMETERS)	
•	.44	●	5.8	3,000
•	.66	●	8.7	1,000
●	.96	●	12.4	400
●	1.41	●	18.7	100
●	2.06	●	26.8	1

SIZE OF RAINDROPS is directly related to the size of salt particles on which they condense. This relationship was observed in photographs such as those on pages 108 and 109.

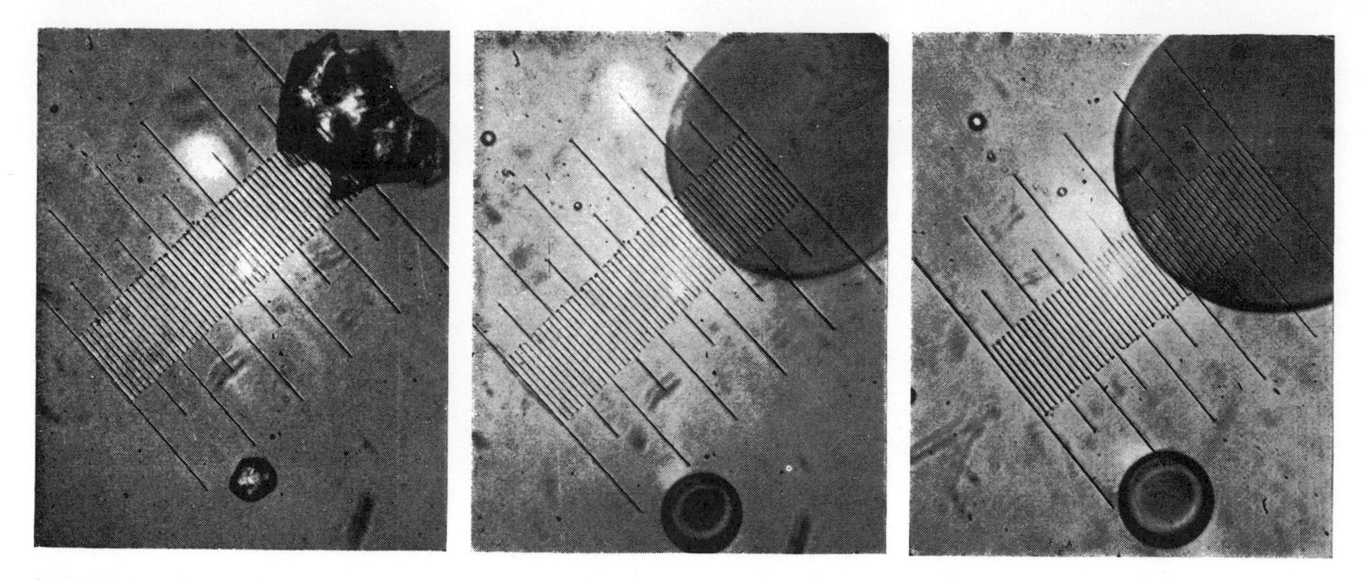

SALT PARTICLES grow into droplets in these two sequences of three photomicrographs each. The photomicrographs were made by placing slides bearing salt particles in a chamber in which the humidity could be controlled. The salt particles in the sequence

on its weight [*see illustrations on these two pages*]. Thus we can calculate the weight of a salt particle from the size of the droplet it forms in air of a given relative humidity.

By "reading" the weights of sea-salt particles in this way we have been able not only to measure the salt load of the air at various locations and altitudes but also to relate the particles to raindrops. We have learned, for one thing, that there is a direct parallel between the sizes of salt particles in the air and the sizes of raindrops [*see chart on page 107*]. Rain usually has a salt concentration of about one part per million by weight. Assuming that this ratio holds in the individual drops, we can compute the size of raindrop that a salt particle of a given size should form. The larger the particle, the larger the raindrop, of course. Analysis of ocean rain with the help of a raindrop "spectrograph," an instrument developed by Australian workers to sort out the drops according to size, has shown that the larger drops do indeed contain more salt than the small ones.

All of this certainly seems to indicate that salt particles act as nuclei to produce raindrops and precipitation. The idea gains further support from a finding that the number of drops per unit volume in rain over the sea is about the same as the number of salt particles in ocean air.

We come now to the question of how salt particles get into the air from sea water. A natural place to look for the mechanism is in whitecaps or surf. Their bursting bubbles are known to shoot small droplets into the air. Members of our laboratory have examined the process by high-speed photography and other studies and learned some important facts.

When a bubble in the froth of a whitecap bursts, it shoots into the air a jet which breaks into many droplets. The starting speed of these droplets ranges from 10 to 80 miles per hour. Because of the air resistance, the droplets rise no higher than eight inches from the surface. But this gives time enough for the water to evaporate and leave tiny salt particles floating in the air, where they can be swept up by the wind. The bubbles range from microscopic size up to about one tenth of an inch in diameter. Small bubbles vastly outnumber the larger ones, which accounts for the fact that small salt particles are so much more numerous than large ones in the air.

The discovery that salt particles fired into the air by bubbles on the sea surface may be largely responsible for our

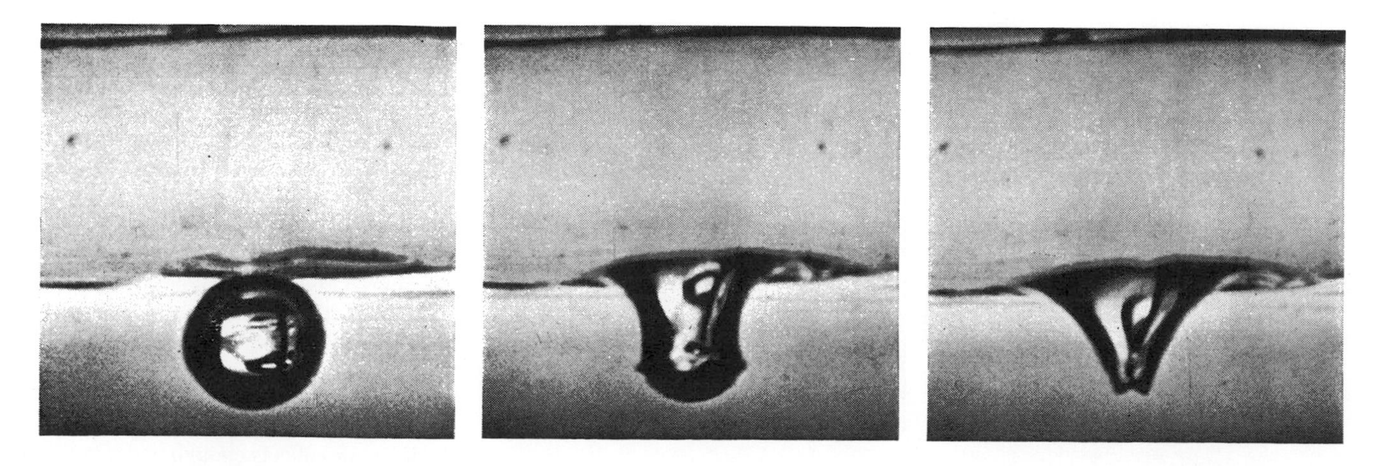

BURSTING BUBBLE fires droplets of water into the air in this sequence of six high-speed photographs made at the Woods Hole Oceanographic Institution. This is the main mechanism by which salt particles rise from the surface of the sea. In the first photo-

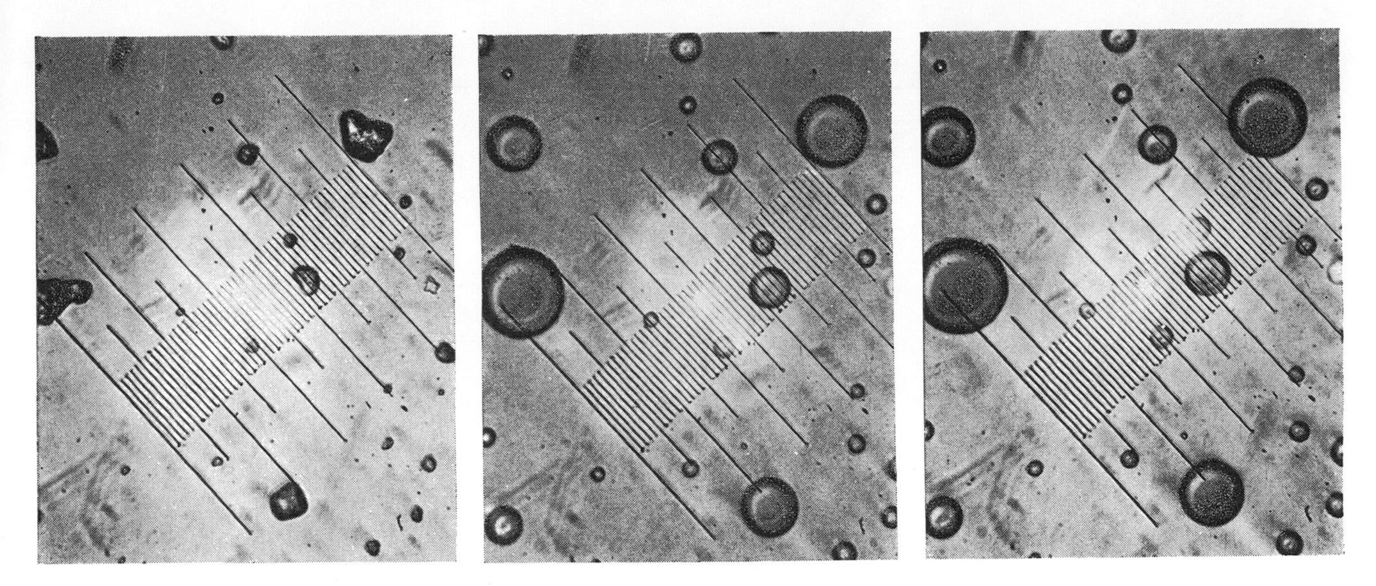

at left were obtained from concentrated sea water; the particles in the sequence at right, from droplets in the air at an altitude of 2,000 feet. In the first photograph of each sequence the relative humidity is 32 per cent; in the second, 84 per cent; in the third, 96 per cent.

rain presents a number of exciting questions for study. Among other things, we would like to know a good deal more about what factors control the formation of bubbles on the surface of the sea. Besides the winds that kick up whitecaps, other important factors seem to be the fall of raindrops and the melting of snowflakes, both of which processes produce bubbles. Since these three phenomena (wind, rain and snow) are so variable, air masses arriving over land from the sea must vary a great deal in salt load. Does saltier air produce more rain? The answers to such questions may bring us closer to being able to control rainfall than we have been up to now.

Basic research often yields by-products which are just as interesting as the main question being pursued. It was so in our studies. Studying the droplets shot from the sea surface, we discovered that they were electrified. In some manner not yet understood, the bubble-bursting mechanism separates charges, so that most of the ejected droplets are positively charged. Duncan C. Blanchard of our laboratory measured their charges by suspending them in an electric field (the technique originated by the Nobel prize physicist Robert A. Millikan). On the basis of his measurements he calculated that the charged droplets ejected from the oceans may account for a major part of the electrification of the atmosphere.

Another unexpected and important discovery, made in our laboratory by Charles H. Keith, was that droplets ejected from a water surface covered with organic matter come out coated with this material. Since the seas in many parts of the world have organic films floating on their surface, this finding suggests that the oceans may supply a substantial part of the food of plants. It has been known that airborne droplets sometimes contain small sea organisms and parts of plankton. The notorious "red tide" along our Gulf seacoast, an occasional phenomenon produced by great swarms of reddish plankton, sends up airborne droplets containing a substance very irritating to mucous membranes, as shore dwellers and fishermen have discovered to their distress.

Looking at those innocent whitecaps on the sea, one would hardly imagine that they are important causes of our planet's rain, the electrification of our atmosphere and the distribution of plant nutrients from the sea to the land. Whether they turn out in the end to be more or less important than we now think them, they have led us into an eye-opening new field of research.

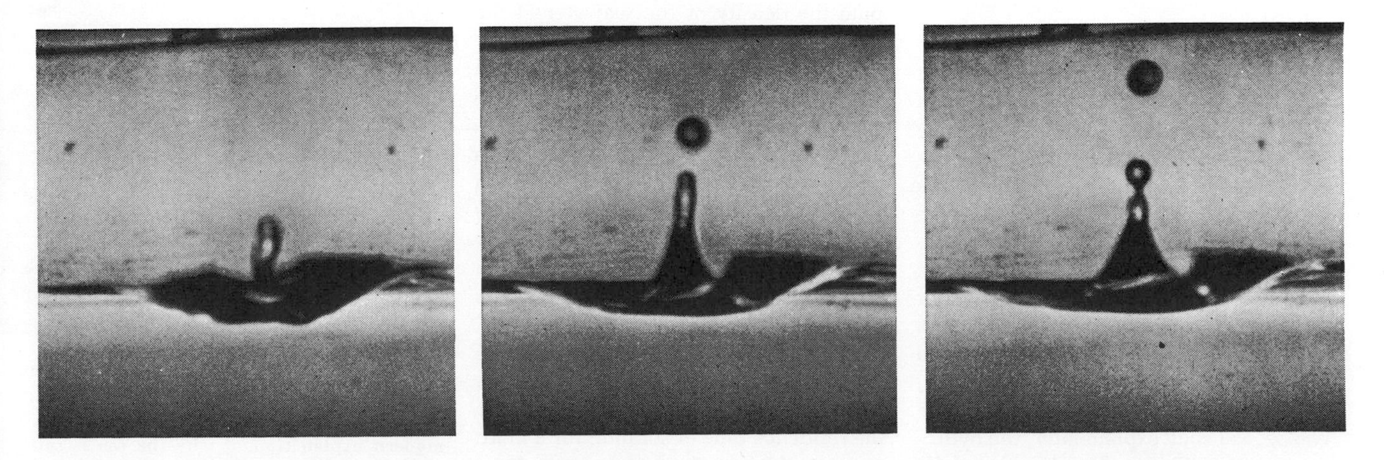

graph the bubble is intact. In the second the bubble breaks through the surface. In the third it becomes almost conical. In the fourth a tiny jet of water begins to form. In the fifth a droplet breaks away from the jet. In the sixth another droplet begins to form.

13

WHY THE SEA IS SALT

FERREN MACINTYRE
November 1970

According to an old Norse folktale the sea is salt because somewhere at the bottom of the ocean a magic salt mill is steadily grinding away. The tale is perfectly true. Only the details need to be worked out. The "mill," as it is visualized in current geophysical theory, is the "mid-ocean" rift that meanders for 40,000 miles through all the major ocean basins. Fresh basalt flows up into the rift from the earth's plastic mantle in regions where the sea floor is spreading apart at the rate of several centimeters per year. Accompanying this mantle rock is "juvenile" water—water never before in the liquid phase—containing in solution many of the components of seawater, including chlorine, bromine, iodine, carbon, boron, nitrogen and various trace elements. Additional juvenile water, equally salty but of somewhat different composition, is released by volcanoes that rim certain continental margins, such as those bordering the Pacific, where the sea floor seems to be disappearing into deep trenches [*see illustration on these two pages*].

The elements most abundant in juvenile water are precisely those that cannot be accounted for if the solids dissolved in the sea were simply those provided by the weathering of rocks on the earth's surface. The "missing" elements, such as chlorine, bromine and iodine, were once called "excess volatiles" and were attributed solely to volcanic emanations. It is now recognized that juvenile water may have nearly the same chlorinity as seawater but is much more acid due to the presence of one hydrogen ion (H^+) for every chloride ion (Cl^-). In due course, as I shall explain later, the hydrogen ions are removed and replaced by sodium ions (Na^+), yielding the concentration of ordinary salt (NaCl) that constitutes 90-odd percent of all the "salt" in the sea.

The chemistry of the sea is largely the chemistry of obscure reactions at extreme dilution in a strong salt solution, where all the classical chemist's "distilled water" theories and procedures break down. The father of oceanographic chemistry was Robert Boyle, who demonstrated in the 1670's that fresh waters on the way to the sea carry small amounts of salt with them. He also made the first attempt to quantify saltiness by drying seawater and weighing the residue, but his results were erratic because some of the constituents of sea salt are volatile. Boyle found that a better method was simply to measure the specific gravity of seawater and from this estimate the amount of salt present. Since the distribution of density in the sea is important to oceanographers, the same calculation is routinely performed today in reverse: the salinity is deduced by measuring the electrical conductivity of a sample of seawater, and from this and the original temperature of the sample one can compute the density of the seawater at the point the sample was taken.

In 1715 Edmund Halley suggested that the age of the ocean and thus of the world might be estimated from the rate of salt transport by rivers. When this proposal was finally acted on by John Joly in 1899, it gave an age of some 90 million years. The quantity that Joly measured (total amount of x in ocean divided by annual river input of x) is now recognized as the "residence time" of the constituent x, which is an index of an element's relative chemical activity in the ocean. Joly's value is about right for the residence time of sodium; for a more reactive element (in the ocean environment) such as aluminum the residence time is as brief as 100 years.

Not quite 200 years ago Antoine Laurent Lavoisier conducted the first analysis of seawater by evaporating it slowly and obtaining a series of compounds by fractional crystallization. The first compound to settle out is calcium carbonate ($CaCO_3$), followed by gypsum ($CaSO_4 \cdot 2H_2O$), common salt (NaCl), Glauber's salt ($Na_2SO_4 \cdot 10H_2O$), Epsom salts ($MgSO_4 \cdot 7H_2O$) and finally the chlorides of calcium ($CaCl_2$) and mag-

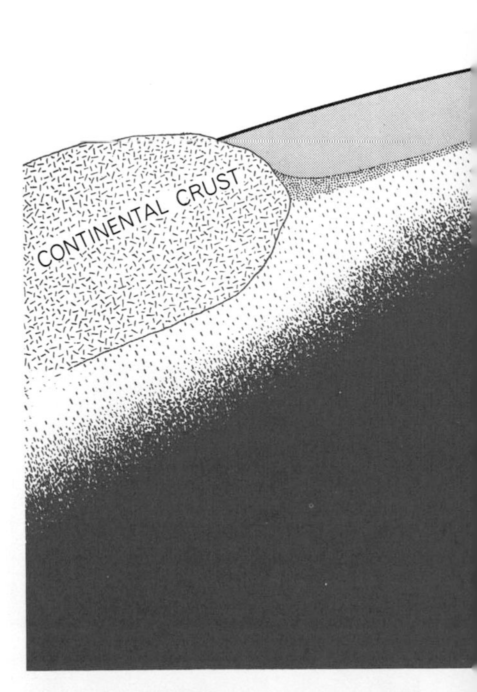

MAGIC SALT MILL at the bottom of the sea, imagined in the old Norse folktale, turns out to be not so fanciful after all. The modern explanation of why the sea is salt invokes the concept of the "mid-ocean" rift and sea-floor spreading, as depicted here in cross section. The rift is a weak point be-

nesium ($MgCl_2$). Lavoisier noted that slight changes in experimental conditions gave rise to large shifts in the relative amounts of the various salts crystallized. (In fact, some 54 salts, double salts and hydrated salts can be obtained by evaporating seawater.) To get reproducible results for even the total weight of salt one must remove all organic matter, convert bromides and iodides to chlorides, and carbonates to oxides, before evaporating. The resulting weight, in grams of salt per kilogram of seawater, is the salinity, S‰. (The symbol ‰ is read "per mil.")

In actual practice the total weight of salt in seawater is nowadays never determined. Instead the amount of chloride ion is carefully measured and a total for all other ions is computed by applying the "constancy of relative proportions." This concept dates back to the middle of the 19th century, when John Murray eliminated confusion about the multiplicity of salts by observing that individual ions are the important thing to talk about when analyzing seawater. Independently A. M. Marcet concluded from many measurements that various ions in the world ocean were present in nearly constant proportions, and that only the absolute amount of salt was variable. This constancy of relative proportions was confirmed by Johann Forchhammer and again more thoroughly by Wilhelm Dittmar's analysis of 77 samples of seawater collected by H.M.S. *Challenger* on the first worldwide oceanographic cruise. These 77 samples are probably the last ever analyzed for all the major constituents. Their average salinity was close to 35‰, with a normal variation of only ±2‰.

In the 86 years since Dittmar reported eight elements, 65 more elements have been detected in seawater. It was recognized more than a century ago that elements present in minute amounts in seawater might be concentrated by sea organisms and thereby raised to the threshold of detectability. Iodine, for example, was discovered in algae 14 years before it was found in seawater. Subsequently barium, cobalt, copper, lead, nickel, silver and zinc were all detected first in sea organisms. More recently the isotope silicon 32, apparently produced by the cosmic ray bombardment of argon, has been discovered in marine sponges.

There are also inorganic processes in the ocean that concentrate trace elements. Manganese nodules (of which more below) are able to concentrate elements such as thallium and platinum to detectable levels. The cosmic ray isotope beryllium 10 was recently discovered in a marine clay that concentrates beryllium. In all, 73 elements (including 13 of the rare-earth group) apart from hydrogen and oxygen have now been detected directly in seawater [*see illustration on page 113*].

It is only in the past 40 years that geochemists have become interested in the chemical processes of the sea for what they can tell us about the history of the earth. Conversely, only as geophysicists have pieced together a comprehensive picture of the earth's history has it been possible to bring order into marine chemistry.

The earth's present atmosphere and ocean are not primordial but have been liberated from chemical and mechanical entrapment in solid rock. Perhaps four billion years ago, or a little less, there was (according to many geophysicists) a "grand catastrophe" in which the earth's core, mantle, crust, ocean and atmo-

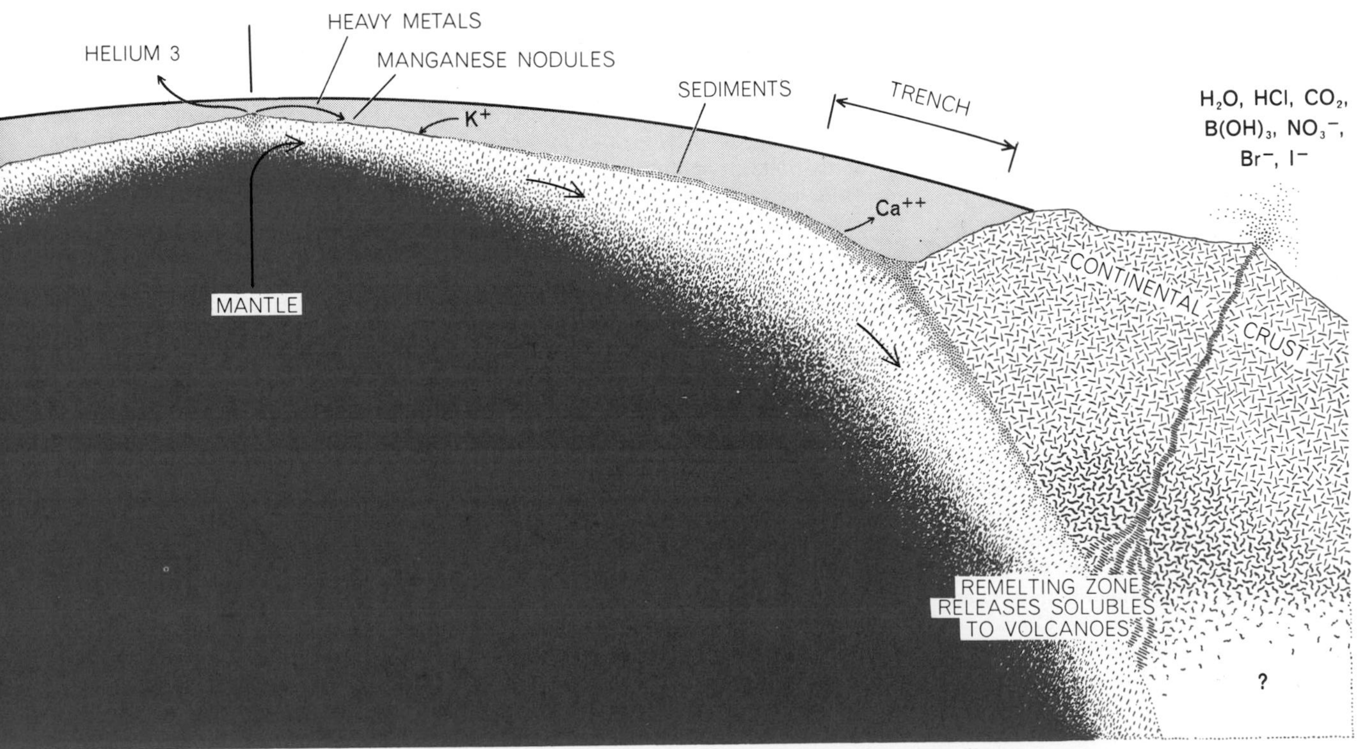

tween rigid plates, or segments, in the earth's crust. Although the driving mechanism is not yet understood, the plates move apart a few centimeters a year as fresh basalt from the plastic mantle flows up between them. The new basalt releases "juvenile" water (water never before in liquid form) and a variety of elements, including heavy metals that become incorporated in manganese nodules and the rare isotope helium 3, which escapes finally into space. At the continental margin (*right*) the lithospheric plate is subducted, forming a trench and carrying accumulated sediments with it. (The plate apparently thickens en route as plastic basalt "freezes" to its underside.) As it descends the plate remelts and releases soluble elements and ions that are ejected into the atmosphere by volcanoes. They maintain the saltiness of the sea and together with weathered crustal rock, such as granite, provide the stuff of sediments.

sphere were differentiated from an original homogeneous accumulation of material. Estimates of water released during the catastrophe range from a third to 90 percent of the present volume of the ocean. The catastrophe is not finished even yet, since differentiation of the mantle continues in regions of volcanic activity. Most of the exhalations of volcanoes and hot springs are simply recycled ground water, but if only half of 1 percent of the water released is juvenile, the present production rate is sufficient to have filled the entire ocean in four billion years.

There is evidence that the salinity of the ocean has not changed greatly since the ocean was formed; in any event the salinity has been nearly constant for the past 200 million years (5 percent of geologic time). The composition of ancient sediments suggests that the ratio of sodium to potassium in seawater has risen from about 1 : 1 to its present value of about 28 : 1. Over the same period the ratio of magnesium to calcium has risen from roughly 1 : 1 to 3 : 1 as organisms removed calcium by building shells of calcium carbonate. It is significant, however, that the total amount of each pair of ions varied much less than the relative amounts.

If we look at rain as it reaches the sea in rivers, we find a distinctly nonmarine mix to its ions. If we catch it even earlier as it tumbles down young mountains, the differences are even more pronounced. This continual input of water of nonmarine composition would eventually overwhelm the original composition of the ocean unless there were corrective reactions at work.

The overall geochemical cycle that keeps the marine ions closely in balance involves a complex interchange of material over decades, centuries and millenniums among the atmosphere, the ocean, the rivers, the crustal rocks, the oceanic sediments and ultimately the mantle [*see "a" in illustration on page 114*]. Because this overall picture is too general to be of much use, we abstract bits from it and call them thalassochemical models (*thalassa* is the Greek word for "sea"). One model involves simply the cyclic exchange of sea salt between the rivers and the sea; the cycle includes the transport of salt from the sea surface into the atmosphere, where salt particles act as condensation nuclei on which raindrops grow [*see "b" in illustration on page 114*]. This process accounts for more than 90 percent of the chloride and about 50 percent of the sodium carried to the sea by rivers.

Another useful abstraction is the "steady state" thalassochemical model. If the ocean composition does not change with time, it must be rigorously true that whatever is added by the rivers must be precipitated in marine sediments [*see "c" in illustration on page 114*]. Oceanic residence times computed from sedimentation rates, particularly for reactive trace metals, agree well with the input rates from rivers. Unfortunately residence times do not reveal the mechanism by which an element is removed from seawater. For residence times greater than a million years it is often helpful to invoke the "equilibrium" model, which deals only with the rate of exchange between the ocean and its sediments [*see "d" in illustration on page 114*].

To understand how the earth maintains its geochemical poise over a billion-year time scale we must return to the circle of arrows—the weathering and "unweathering" processes—of the geochemical cycle. This circle starts with primordial igneous rock, squeezed from the mantle. Ignoring relatively minor heavy metals such as iron, we can assume that the rock consists of aluminum, silicon and oxygen combined with the alkali metals: potassium, sodium and calcium. The resulting minerals are feldspars (for example $KAlSi_3O_8$). Rainwater picks up carbon dioxide from the air and falls on the feldspar. The reaction of water, carbon dioxide and feldspar typically yields a solution of alkali ions and bicarbonate ions (HCO_3^-) in which is suspended hydrated silica (SiO_2). The residual detrital aluminosilicate can be approximated by the clay kaolinite: $Al_2Si_2O_5(OH)_4$ [*see Step 1 in illustration on page 115*]. A mountain stream carries off the ions and the silica. The kaolinite fraction lags behind, first as a friable surface on weathering rock, then as soil material and finally as alluvial deposits in river valleys. If the stream evaporates in a closed basin, such as one finds in the western U.S., the result is a "soda lake" containing high concentrations of carbonates and amorphous silica.

In mature river systems the kaolinite fraction reaches the sea as suspended sediment. Encountering an ion-rich environment for the first time, the aluminosilicate must reorganize itself into new minerals. One such mineral, which seems to be forming in the ocean today, is the potassium-containing clay illite [*see Step 2 in illustration on page 115*]. These "clay cation" reactions may take decades or centuries. They are poorly understood because graduate students who study them invariably leave before the reactions are complete. The net effect of such reactions is to tie up and remove some of the potassium and bicarbonate ions, along with aluminum, silicon and oxygen.

A biologically important reaction, usually confined to shallow water, allows marine organisms to build shells of calcium carbonate, which precipitates when calcium (Ca^{++}) and bicarbonate ions react. If dilute hydrochloric acid is present (it is released by volcanoes), it reacts even more rapidly with bicarbonate, forming water and carbon dioxide and leaving free the chloride ion. When marine organisms die and sink to about 4,000 meters, they cross the "lysocline," below which calcium carbonate redissolves because of the high pressure. We have now traced the three metallic ions removed from igneous rock to three separate niches in the ocean. Sodium remains dissolved, potassium precipitates in clays on the deep-sea floor and calcium precipitates in shallow water as biogenic limestone: coral reefs and calcareous oozes.

Ages pass and the geochemical cycle rolls on, converting ocean-bottom clay into hard rock such as granite. When old sea floor finally reaches a region of high pressure and temperature under a continental block, it still contains some free ions that can react with the clay to reconstitute hard rock. A score of reaction

CURRENTLY RECOVERED FROM SEAWATER
ELEMENTS IN SHORT SUPPLY
RANGE OF BIOLOGICALLY CAUSED CHANGE
RANGE OF ANALYSES
METALS CONCENTRATED IN MANGANESE NODULES

COMPOSITION OF SEAWATER has been a challenge to chemists since Antoine Laurent Lavoisier made the first analyses. The logarithmic chart on the opposite page shows in moles per kilogram the concentration of 40 of the 73 elements that have been identified in seawater. A mole is equivalent to the element's atomic weight in grams; thus a mole of chlorine is 35 grams, a mole of uranium 238 grams. Only four elements are now recovered from the sea commercially: chlorine, sodium, magnesium and bromine. Recovery of other scarce elements is not promising unless biological concentrating techniques can be developed. Manganese nodules are a potential source of scarce metals but gathering them from the deep-sea floor may not be profitable in this century.

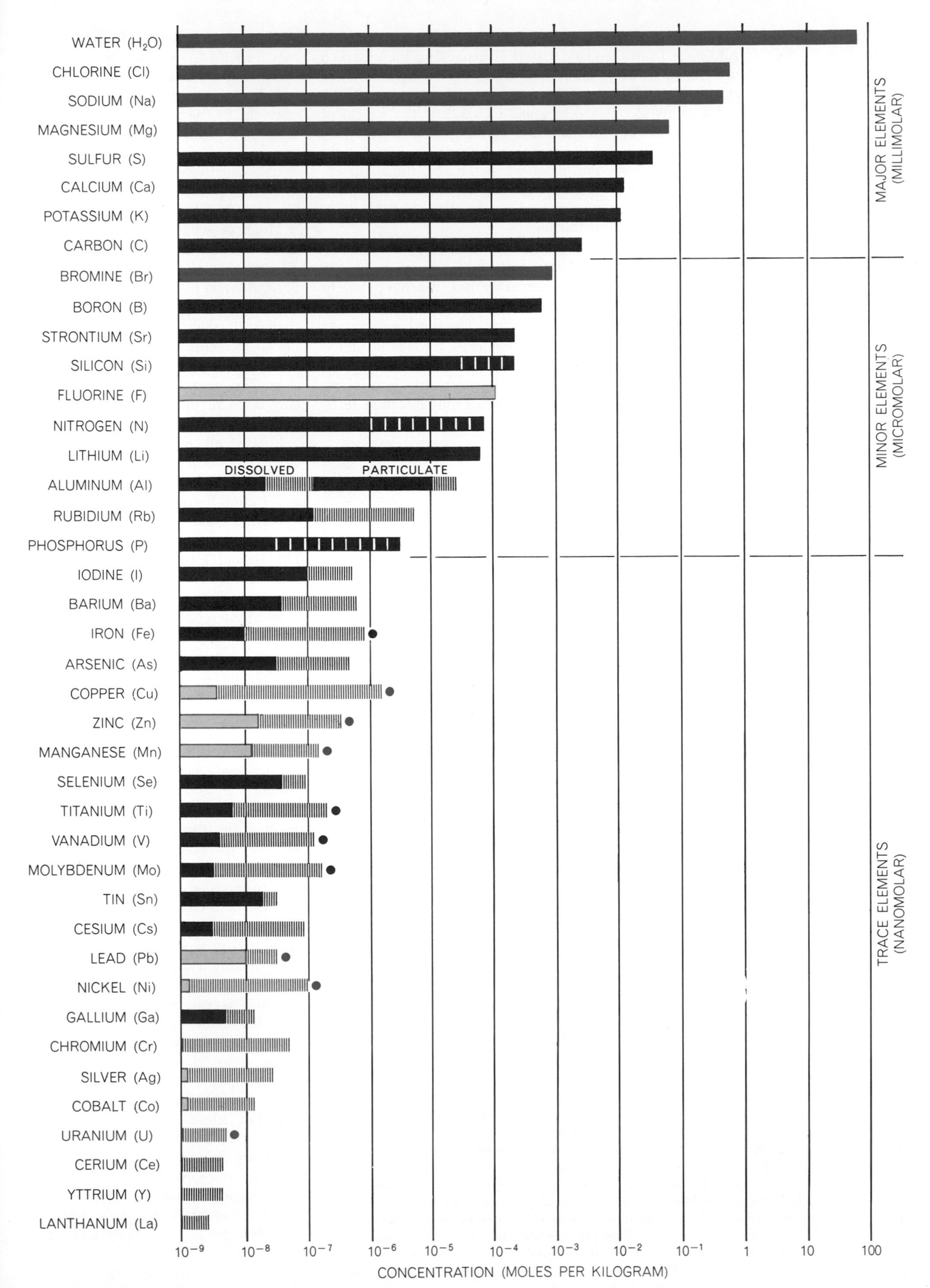
WATER (H_2O)
CHLORINE (Cl)
SODIUM (Na)
MAGNESIUM (Mg)
SULFUR (S)
CALCIUM (Ca)
POTASSIUM (K)
CARBON (C)
BROMINE (Br)
BORON (B)
STRONTIUM (Sr)
SILICON (Si)
FLUORINE (F)
NITROGEN (N)
LITHIUM (Li)
ALUMINUM (Al)
RUBIDIUM (Rb)
PHOSPHORUS (P)
IODINE (I)
BARIUM (Ba)
IRON (Fe)
ARSENIC (As)
COPPER (Cu)
ZINC (Zn)
MANGANESE (Mn)
SELENIUM (Se)
TITANIUM (Ti)
VANADIUM (V)
MOLYBDENUM (Mo)
TIN (Sn)
CESIUM (Cs)
LEAD (Pb)
NICKEL (Ni)
GALLIUM (Ga)
CHROMIUM (Cr)
SILVER (Ag)
COBALT (Co)
URANIUM (U)
CERIUM (Ce)
YTTRIUM (Y)
LANTHANUM (La)
DISSOLVED
PARTICULATE
MAJOR ELEMENTS (MILLIMOLAR)
MINOR ELEMENTS (MICROMOLAR)
TRACE ELEMENTS (NANOMOLAR)
10^{-9}
10^{-8}
10^{-7}
10^{-6}
10^{-5}
10^{-4}
10^{-3}
10^{-2}
10^{-1}
1
10
100
CONCENTRATION (MOLES PER KILOGRAM)

schemes are possible. In Step 3 in the illustration on the opposite page I have chosen to build a "granite" from equal parts of potassium feldspar, sodium feldspar, potassium mica and quartz. (Notice that calcium is missing because it has dissolved from the sediments during their descent into the deep-ocean trenches that carry the sediments under the continental blocks.) The reaction written in Step 3 uses up all the silica formed in Step 1.

The goal of this geochemical exercise has now been reached. First, we have shown that of all the substances that enter the ocean, only sodium and chlorine remain abundantly in solution. Of the other elements, the amount remaining in solution is less than a hundredth of the amount delivered to the ocean and precipitated from it. Second, we have made a start at explaining the observed sodium-potassium ratios: in basalt this ratio is about 1 : 1, in seawater 28 : 1 and in granite 1 : 1.2. If the weight of sodium tied up in granite were about 140 times as great as the weight of sodium dissolved in the sea, the slight excess of potassium over sodium in granite would explain the sea's deficiency in potassium.

We now have working models for thinking about the circulation of the major elements, but we have barely scratched the true complexity and subtlety of seawater. The sources and sinks of the minor elements are now being explored. In many cases we can only guess at what the natural marine form of an element is because our detection techniques either convert all forms to a common form for analysis or miss some forms completely. Moreover, certain ions seem to behave capriciously in the ocean. For example, at the *p*H (hydrogen-ion concentration) of seawater, vanadium should appear as $VO_2(OH)_3^{--}$, an ion with a double negative charge; instead it seems to exist in positively charged form, perhaps as VO_2^+.

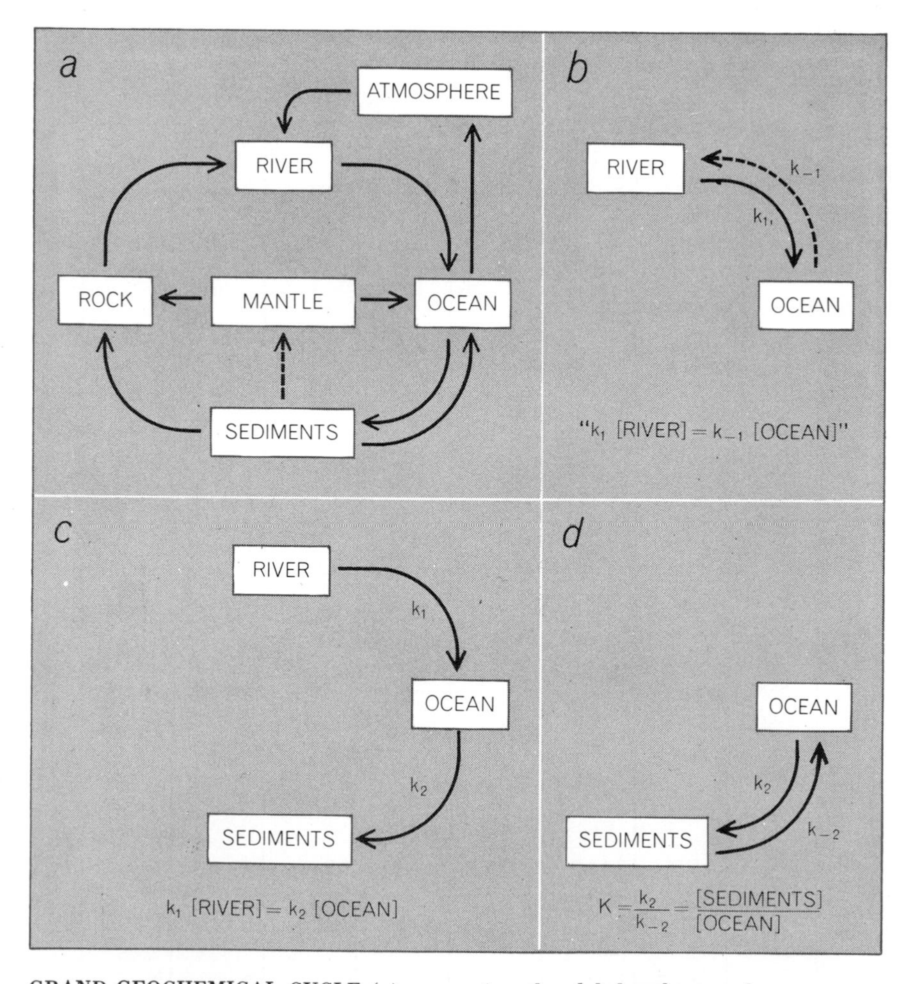

GRAND GEOCHEMICAL CYCLE (*a*) **summarizes the global pathways taken sooner or later by the three-score elements that pass through the ocean and maintain its saltiness. The three "thalassochemical" models** (*b, c, d*) **abstracted from it are more helpful when trying to understand the rate laws governing the transport of specific elements. The rate constants,** *k*, **are expressed as a fraction: one over some number of years. The brackets enclose concentrations of the element being studied, specified according to its environment. The "cyclic" model** (*b*) **accounts for 90 percent of the chloride in river water. Its rate law is in quotation marks because extra factors, such as the area of the ocean, must be incorporated. The "steady state" model** (*c*) **works well for reactive trace metals; the reciprocal of** k_2 **is simply the residence time in the ocean. The "equilibrium" model** (*d*) **seems the most appropriate for the hydrogen ion** (H^+) **and the ions of the major metals, such as sodium.**

Much of what is known about elements in the sea can be summarized in an oceanographer's periodic table [*see illustration on page 116*]. The usefulness of the usual kind of periodic table to the chemist is that it arranges chemically similar elements in vertical columns and presents behavioral trends in horizontal rows. The oceanographer's table shows how these regularities are disrupted in the ocean environment.

First of all, more than a dozen elements have never been detected in seawater, although two of them (palladium and iridium) exist in parts per billion in marine sediments and another (platinum) is present in manganese nodules. The second interesting feature of the oceanographer's table is the tendency for the "upper" and "outer" elements, those in the raised wings, so to speak, to be the most plentiful in the sea. The "upper" tendency simply reflects the greater cosmic abundance of light elements. (Lithium, beryllium and boron, however, are fairly scarce even cosmically.)

The "outer" trend can be explained in quantum-mechanical terms by the presence or absence of electrons in *d* orbitals, the electron shells principally involved in forming complexes. Elements in the first three columns at the left have no *d* orbitals; those in the last four columns at the right have full *d* orbitals. Both characteristics favor weak chemical bonds, with the result that these two groups of elements tend to ionize readily and remain in solution, either by themselves or in simple combination with oxygen and hydrogen. In contrast, the elements in the center of the table with partially filled *d* orbitals form strong chemical bonds and compounds that precipitate readily; thus they can exist only at low concentration in solution. For silver and the surrounding group of metals the most stable complexes are formed with the most abundant seawater ion: chloride. Most of the other elements that are hungry for *d* electrons form their complexes with oxygen, or oxygen plus some protons (hydrogen nuclei).

Ordinarily the oxidation state of metals

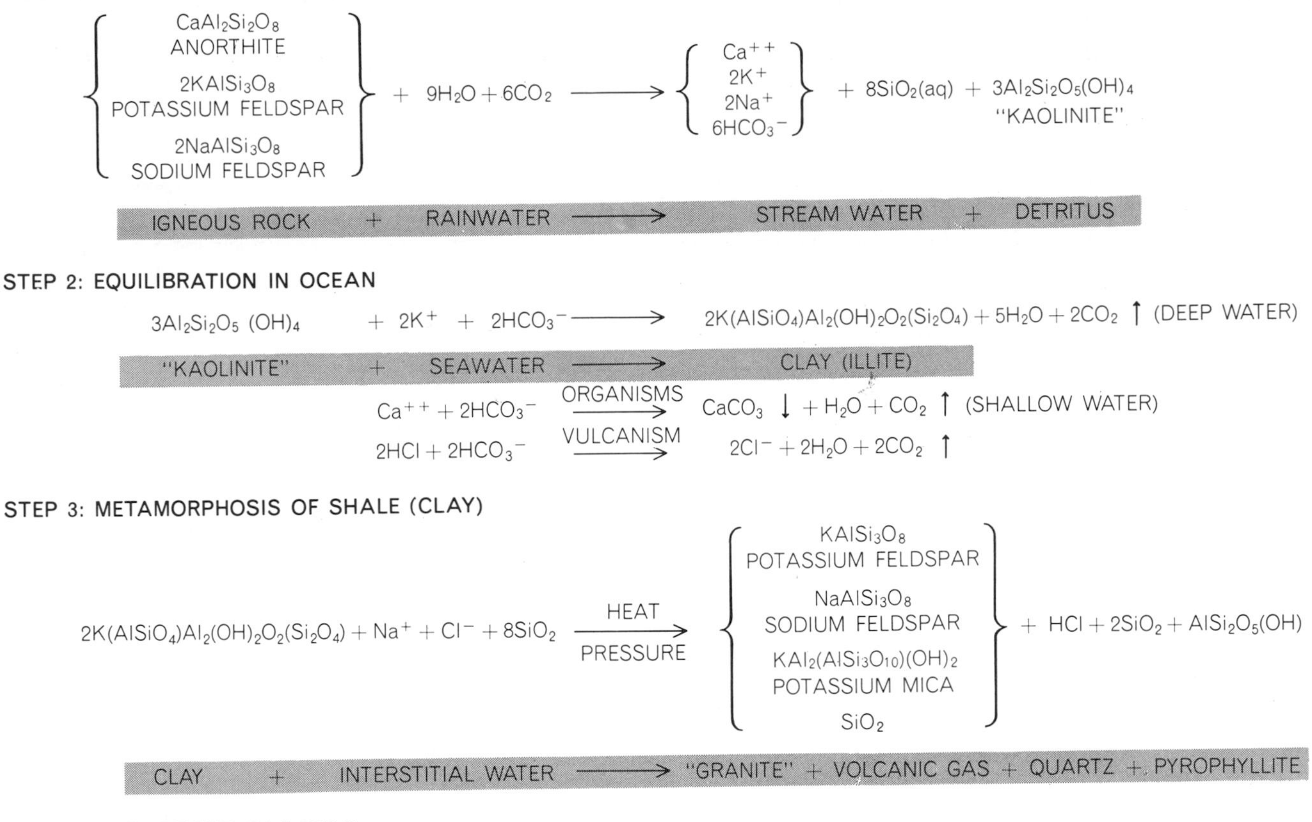

ONLY SALT REMAINS after the ocean "laboratory" has finished processing the complex of chemicals removed from igneous rock by rainwater containing dissolved carbon dioxide. Step 1 yields a solution of alkali ions and bicarbonate (HCO_3^-) ions in which hydrated silica (SiO_2) and aluminosilicate detritus are suspended. In crystalline form the aluminosilicate would be kaolinite. In the ocean (*Step 2*) the "kaolinite" is complexed with potassium ions (K^+) to form illite clay. Marine organisms use the calcium ion (Ca^{++}) to make calcium carbonate shells, which form sediments in shallow water. Hydrochloric acid (HCl), injected by undersea volcanoes, reacts with bicarbonate ions, returning some carbon dioxide to the atmosphere. In Step 3 clay is metamorphosed into "granite." Sodium chloride (*Step 4*) remains. Although some of this sequence is hypothetical, something very similar seems to take place.

avid for *d* electrons would be determined by the oxidation potential of seawater, which is a measure of its ability to extract electrons from a substance just as its *p*H is a measure of its ability to extract protons. The oxidation potential of seawater has the high value of .75 volt, enabling it to extract the maximum possible number of electrons from nearly all elements except the noble metals (platinum group) and the halogens (fluorine family).

Surprisingly, however, the oxidation potential of seawater does not seem to control the oxidation states of many metals that have partially filled *d* shells. One reason is that most reactions proceed by a mechanism in which only a single electron is transferred at a time. Such transfers occur most readily when the reactants are adsorbed on surfaces where atomic geometry and electric-charge distribution are able to expedite the redistribution of electrons (hence the utility of catalysts, which provide such surfaces). But surfaces of any kind are few and far between in the ocean, and (with the exception of manganese nodules) those that do exist are poor catalysts. A second reason for the failure of the sea's oxidation potential to control valence states is that organisms sometimes excrete electron-rich substances, which then remain in that reduced state in spite of seawater's apparent capacity to oxidize them.

Manganese nodules are porous chunks of metallic oxides up to several centimeters in diameter, widely distributed over the ocean floor. They evidently exist because they are autocatalytic for the reaction that produces them. Because of their porous structure, nodules have a surface area of as much as 100 square meters per gram. The autocatalytic property seems to extend to an entire suite of metals that coprecipitate with manganese: iron, cobalt, nickel, copper, zinc, chromium, vanadium, tungsten and lead. Nodules found on the flanks of oceanic ridges contain significant concentrations of metals, such as nickel, that are scarce in seawater itself. This suggests that the nodules are collecting juvenile metals as the metals leak from the mantle at the fissure of the ridge. One would like to know why the nodule metals are present in oxide form rather than, as one would expect, in carbonate form.

The level of the discussion so far might best be called thalassopoetry. The discussion can be made more serious in two ways. One approach—the "geochemical balance"—has employed a computer to follow in detail as many as 60 elements as they move through the geochemical cycle, from igneous rock back

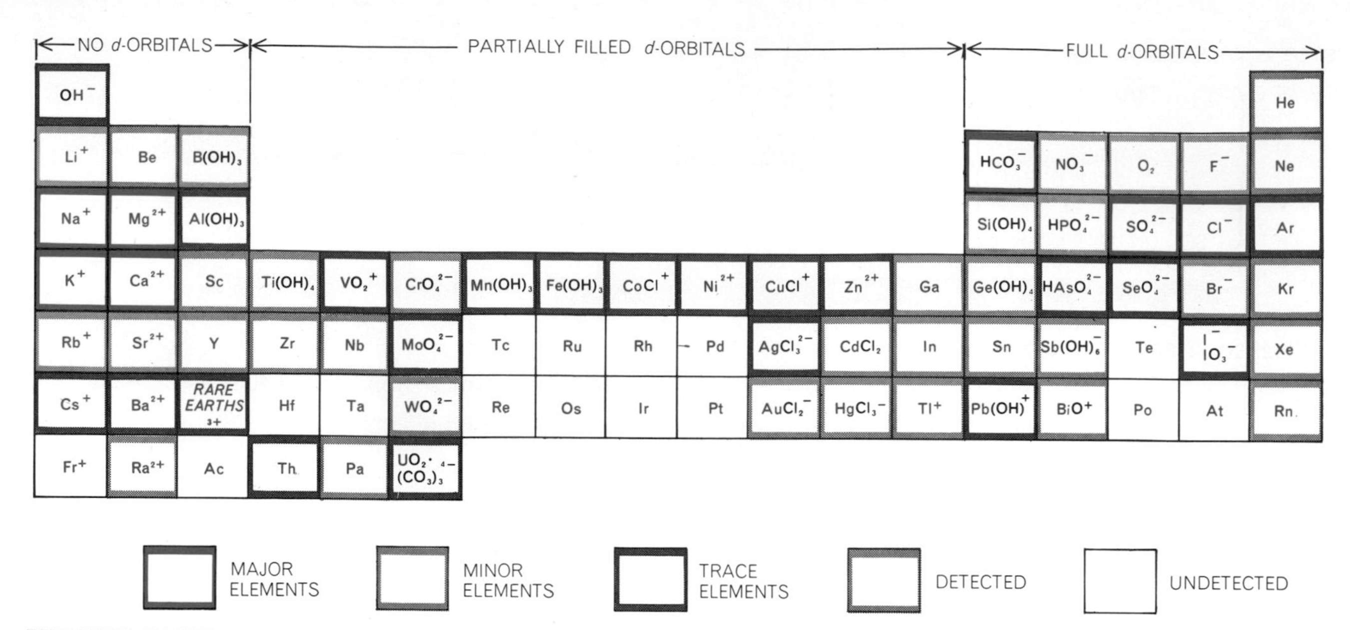

PERIODIC TABLE, as prepared by the "thalassochemist," shows the form in which the detectable elements appear in seawater. In each box the element normally found in that place in the usual periodic table is shown in color; the elements associated with it are in black. Thus carbon appears predominantly as HCO_3^-, arsenic as $HAsO_4^{2-}$ and so on. The superscripts show the number of positive or negative charges carried by each ion. Iodine's two forms, I^- and IO_3^-, are about equally common. Except for the noble gases (*last column at right*), all the elements dissolved in the sea must be present as ions. When an element (other than a noble gas) is shown by itself, without a plus or minus charge, it means that its preferred ionic form in seawater is not yet established.

to metamorphosed sediments. In the second approach the actual chemistry of each element is followed by applying the thermodynamic methods of Josiah Willard Gibbs to systems regarded as being near equilibrium. This effort was launched by Lars Gunnar Sillén of Sweden and has been pursued by Robert M. Garrels of Northwestern University and by Heinrich D. Holland of Princeton University.

Of course no chemist in his right mind would talk seriously about equilibria in a system of variable temperature, pressure and composition that was poorly stirred, had variable inputs and contained living creatures. On the other hand, the observed uniformity of the ocean and the long periods available for reacting suggest that at least the major components are sufficiently close to equilibrium to make an investigation worthwhile. (We *know* the minor constituents are not in equilibrium.)

The equilibrium approach is based on Gibbs's phase rule, which states that the number of phases (P) possible in a system of C components at equilibrium is given by the equation $P = C + 2 - F$, where F is the number of "degrees of freedom," or quantities that may be independently varied without changing the number of phases or their composition (although F may change their relative proportions). The 2 enters the equation because only two variables, temperature and pressure, are important in most chemical reactions.

One of Sillén's most comprehensive ocean models has nine components: water, hydrochloric acid, silica, three hydroxides (aluminum, sodium and potassium), carbon dioxide and the oxides of magnesium and calcium. Observation of sea-floor sediments, aided by laboratory studies, suggests that a nine-phase ocean will result [*see illustrations on opposite page*]. If C and P both equal nine, the phase rule states that the number of degrees of freedom (F) must equal two. Logically these are temperature (which can vary over the oceanic range from −2 degrees Celsius to 30 degrees) and the chloride ion concentration (which can shift over the normal oceanic range without changing the composition of the stable phases).

A diagrammatic view of how the nine components sort themselves into phases is shown in the bottom illustration on the opposite page. Note that the liquid phase contains ions not listed either as components or phases (for example H^+ and OH^-). Thermodynamics need not consider them explicitly because they do not vary independently; their concentrations are fixed by the equilibrium constants that connect the observed phases. Thus $H_2O = H^+ + OH^-$. Moreover, one knows that the product of H^+ and OH^- is a thermodynamic constant, which equals 10^{-14} mole per liter. Similar relations tie the entire system into a comprehensible whole, so that when all the calculations are performed one has discovered the equilibrium concentrations of five cations (H^+, Na^+, K^+, Mg^{++} and Ca^{++}) and four anions (Cl^-, OH^-, HCO_3^- and CO_3^{--}).

It may seem peculiar to discuss an "atmosphere" containing only water vapor and carbon dioxide. One could easily add oxygen and nitrogen to the list of components. Since they would add no new phases, they would raise the number of degrees of freedom from two to four ($9 = 11 + 2 - 4$). The two new F's would be the total atmospheric pressure and the ratio of oxygen to nitrogen. In the study of the ocean, however, the partial pressure due to carbon dioxide is more significant than the total pressure of the atmosphere. Moreover, the presence of gaseous oxygen and nitrogen has little importance for the inorganic environment of the ocean, so that it is simpler to omit them and just as "real."

Suppose now we perturb the equilibrium of the model ocean by assuming that a submerged volcano has suddenly released enough hydrochloric acid (HCl) to double the amount of chloride ion (Cl^-). The dissociation of hydrochloric acid releases enough H^+ ions to raise the total number of hydrogen ions in the ocean from the former equilibrium value of 10^{-8} mole per liter to $10^{+.3}$. This excess of hydrogen ions almost immediately pushes all the available carbonate ions (CO_3^{--}) to bicarbonate ions (HCO_3^-) and the latter to carbonic acid (H_2CO_3). These shifts, however, only slightly depress the pH, which remains high until the slow circulation of the ocean brings the hydrogen ions in direct contact with the clay sediments on the sea floor.

The structure of clay is such that oxygen atoms at the free corners of polyhedrons carry unsatisfied negative charges, which attract positive ions [*see top illustration on next page*]. Because the ocean is so rich in sodium ions (Na^+), they occupy most of the corners of clay polyhedrons. When the excess hydrogen ions come in contact with the clay, they quickly replace the sodium ions and set them adrift. This fast reaction is limited in scope because the surface and interlayer ion-exchange capacity of clay is not very great. Much more capacity is provided when the structure of the clay is rearranged; for example, the conversion of montmorillonite to kaolinite also consumes hydrogen atoms and releases sodium. Given sufficient time—centuries—such rearrangements inexorably take place, and the *p*H of the ocean slowly drifts back to its equilibrium value. The charge on the excess chloride introduced by the volcano will then be balanced not by H^+ but by Na^+. This slow equilibration mechanism can be regarded as the ocean's "*p*H-stat" (in analogy with "thermostat"). This clay-cation model suggests that the *p*H of the ocean has been constant over the span of geologic time and that hence the carbon dioxide content of the atmosphere has been held within narrow limits.

If only the *p*H-stat were available for leveling surges in *p*H, the ocean might be subjected to violent local fluctuations. For fast response *p*H control is taken over by a carbonate buffer system [*see bottom illustration on next page*]. In fact, until recently oceanographers neglected the clay-cation reactions and assumed that the carbonate-buffer system almost completely determined the *p*H of the ocean.

One might think that if the carbon dioxide content of the atmosphere were to decrease, carbon dioxide would flow from the sea into the atmosphere, leading to a general depletion of all carbonate species in the ocean and eventually to the dissolution of some carbonate sediments. In actuality something quite different happens because the carbonate system is its own source of hydrogen ions. Removal of carbon dioxide from water reduces the concentration of carbonic acid (H_2CO_3), the hydrated form of carbon dioxide. Replacement of this acid from bicarbonate ions requires a hydrogen ion, which can only be obtained by converting another bicar-

COMPONENTS (C)	PHASES (P)	VARIABLES (F)
H_2O	1 GAS	TEMPERATURE
HCl	2 LIQUID	Cl^-
SiO_2	3 QUARTZ (SiO_2)	
$Al(OH)_3$	4 KAOLINITE (t-o CLAY)	
NaOH	5 MONTMORILLONITE (Na-t-o-t CLAY)	
KOH	6 ILLITE (K-t-o-t CLAY)	
MgO	7 CHLORITE (Mg-t-o-t CLAY)	
CO_2	8 CALCITE ($CaCO_3$)	
CaO	9 PHILLIPSITE (Na-K FELDSPAR)	

NINE MAJOR COMPONENTS IN SEA can, to a first approximation, be combined into nine distinctive phases to satisfy the "phase rule" that governs systems in equilibrium. The rule, formulated in the 19th century by Josiah Willard Gibbs, prescribes the number of phases *P*, components *C* and degrees of freedom *F* in such a system: $P = C + 2 - F$. When the number of phases and components are equal, the number of degrees of freedom, *F*, must be two, which allows both the temperature and the chloride-ion concentration to vary without altering the number of phases. In the clay-containing phases (*4, 5, 6, 7*) the letter "t" stands for a tetrahedral crystal structure; the letter "o" stands for an octahedral structure.

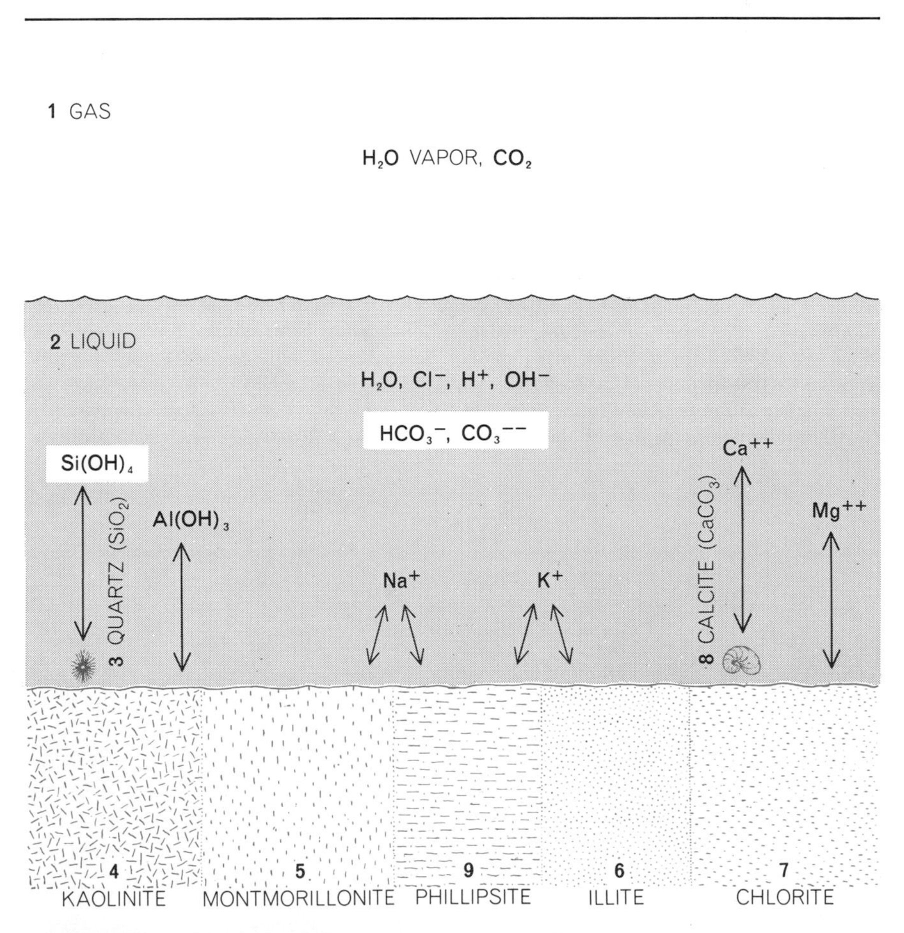

EQUILIBRIUM OCEAN MODEL, consisting of nine phases and nine components, shows how the principal constituents of the ocean distribute themselves among the atmosphere, the ocean and the sediments. Three of the constituents (HCO_3^-, CO_3^{--} and $Si(OH)_4$) are not included among nine listed components but appear as equilibrium products of those that are listed, as do seven ions (H^+, K^+, Na^+, Ca^{++}, Mg^{++}, Cl^-, OH^-). Two of the solids are shown as biological "precipitates": "quartz" (*3*) in the form of silicate structures built by radiolarians and "calcite" (*8*) in the form of calcium carbonate chambers built by foraminifera. The method of precipitation is unimportant as long as the product is stable. The equilibrium model goes far to explain why the ocean has the composition it does.

bonate ion to carbonate. The overall reaction is $2HCO_3^- \rightarrow H_2CO_3 + CO_3^{--}$. Thus instead of dissolving existing sediments, removing carbon dioxide from the sea may actually precipitate carbonate. This reaction can be seen in the "whitings" of the sea over the Bahama Banks, where cold deep water, rich in dissolved carbon dioxide and calcium, is forced to the surface and warmed. As carbon dioxide escapes into the air, the *p*H drops and aragonite ($CaCO_3$) precipitates, turning large areas of the ocean white with a myriad of small crystals.

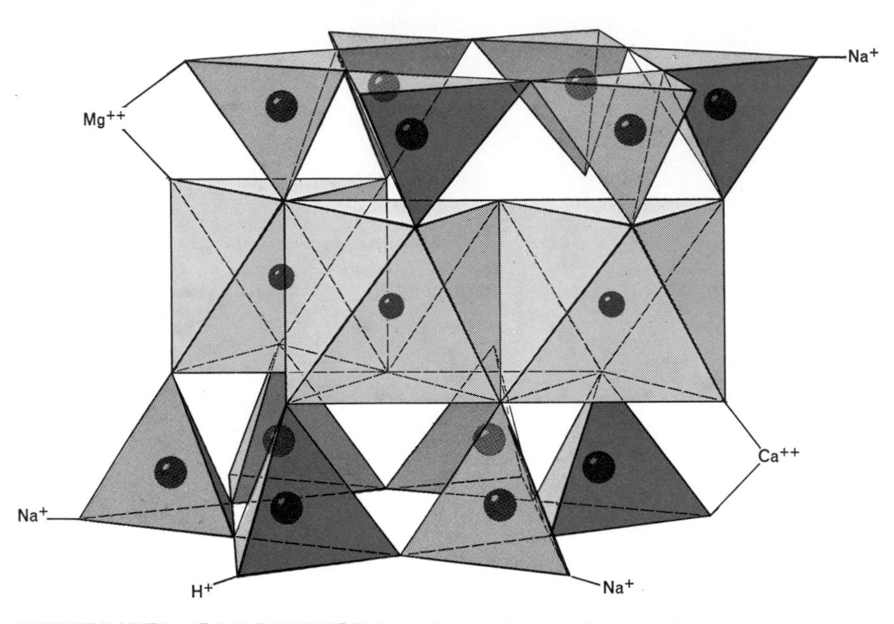

THREE-LAYER CLAY PARTICLE has a layer of octahedrons sandwiched between two layers of tetrahedrons. Each octahedron consists of an atom of aluminum surrounded by six closely packed atoms of oxygen. Each tetrahedron consists of a silicon atom surrounded by four atoms of oxygen. The polyhedrons are tied into layers at shared corners where a single oxygen atom is bonded to a silicon atom on one side and to an aluminum atom on the other. At the free corners the oxygen atoms bear unsatisfied negative charges that attract cations such as sodium ($Na+$) and potassium ($K+$). If the hydrogen-ion concentration should rise in the vicinity of clay, free hydrogen ions tend to be exchanged for sodium ions, which are released. In addition, many doubly charged metal ions can replace Si^{4+} at the centers of tetrahedrons and Si^{4+} can replace Al^{3+} in the octahedrons. Whenever this occurs, another cation is bound to the structure to conserve charge. Such reactions apparently exert considerable control over the ocean's composition and hydrogen-ion concentration.

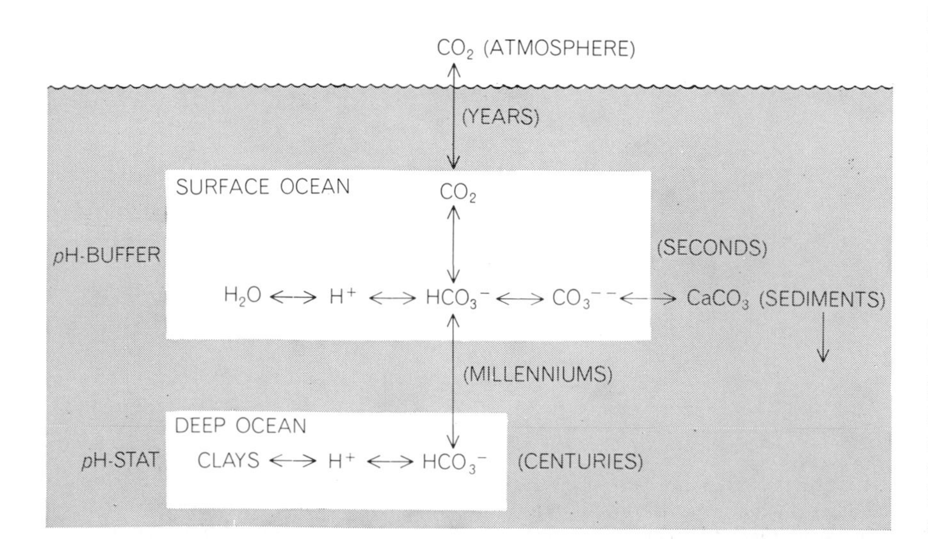

HYDROGEN-ION CONCENTRATION, or *p*H, of the ocean is controlled by two mechanisms, one that responds swiftly and one that takes centuries. The first, the "*p*H-buffer," operates near the surface and maintains equilibrium among carbon dioxide, bicarbonate ion (HCO_3^-), carbonate ion (CO_3^{--}) and sediments. The slower mechanism, the "*p*H-stat," seems to exert ultimate control over *p*H; it involves the interaction of bicarbonate ions and protons (H^+) with clays. Clay will accept protons in exchange for sodium ions (primarily).

The reaction above conserves charge, which means that the "alkalinity"—the traditional name for the concentration of sodium ion ("alkali") needed to balance this negative charge—is also conserved. The "carbonate alkalinity," defined as the bicarbonate concentration plus twice the carbonate concentration, is useful because it remains fixed even when the relative amounts of the two species vary.

The system can be visualized with the help of the illustration on the opposite page, which is the "Bjerrum plot" for carbonic acid at constant alkalinity. It takes its name from Niels Bjerrum, who introduced such plots in 1914; it shows the interrelations between the various compounds in the world carbonate system as a function of *p*H. Although the diagram ignores variations of pressure, temperature and salinity, it displays the essential features of the system.

The Bjerrum plot facilitates a semiquantitative discussion of the relation of atmospheric carbon dioxide to oceanic carbon dioxide. Over the next 20 years we shall burn enough fossil fuel to double the amount of carbon dioxide in the atmosphere from 320 parts per million to 640. On the plot this is indicated by shifting the line *A*, corresponding to 320 parts per million, to position *B*, 640 parts per million.

To produce this shift some 2.5×10^{18} grams of carbon dioxide must be added to the atmosphere. If the altered atmosphere were to come to equilibrium with the ocean, the *p*H of the ocean would drop from its present value of 8.15 to 7.89—still well within the range tolerated by marine organisms. This cannot happen, however, because the total mass of carbon dioxide in the ocean (Σ in the Bjerrum plot) plus the carbon dioxide in the atmosphere would have to increase from its present value, 128.9×10^{18} grams, to 138.3×10^{18} grams. The difference, 9.4×10^{18} grams, is nearly four times the amount added to the atmosphere.

The long-term equilibration process for such an atmospheric doubling can be broken down into two steps. First the *p*H-buffer system operates: 2.5×10^{18} grams, or 2 percent of the total mass, is added to the world system at constant alkalinity. The result of this step is the line *C* in the diagram, corresponding to a total mass of 131.4×10^{18} grams, an atmospheric carbon dioxide content of 390 parts per million and an oceanic *p*H of 8.08. Next, if the ocean has time to equilibrate with its sediments, the *p*H-stat will operate, returning the system to *p*H 8.15 at a constant total mass. The re-

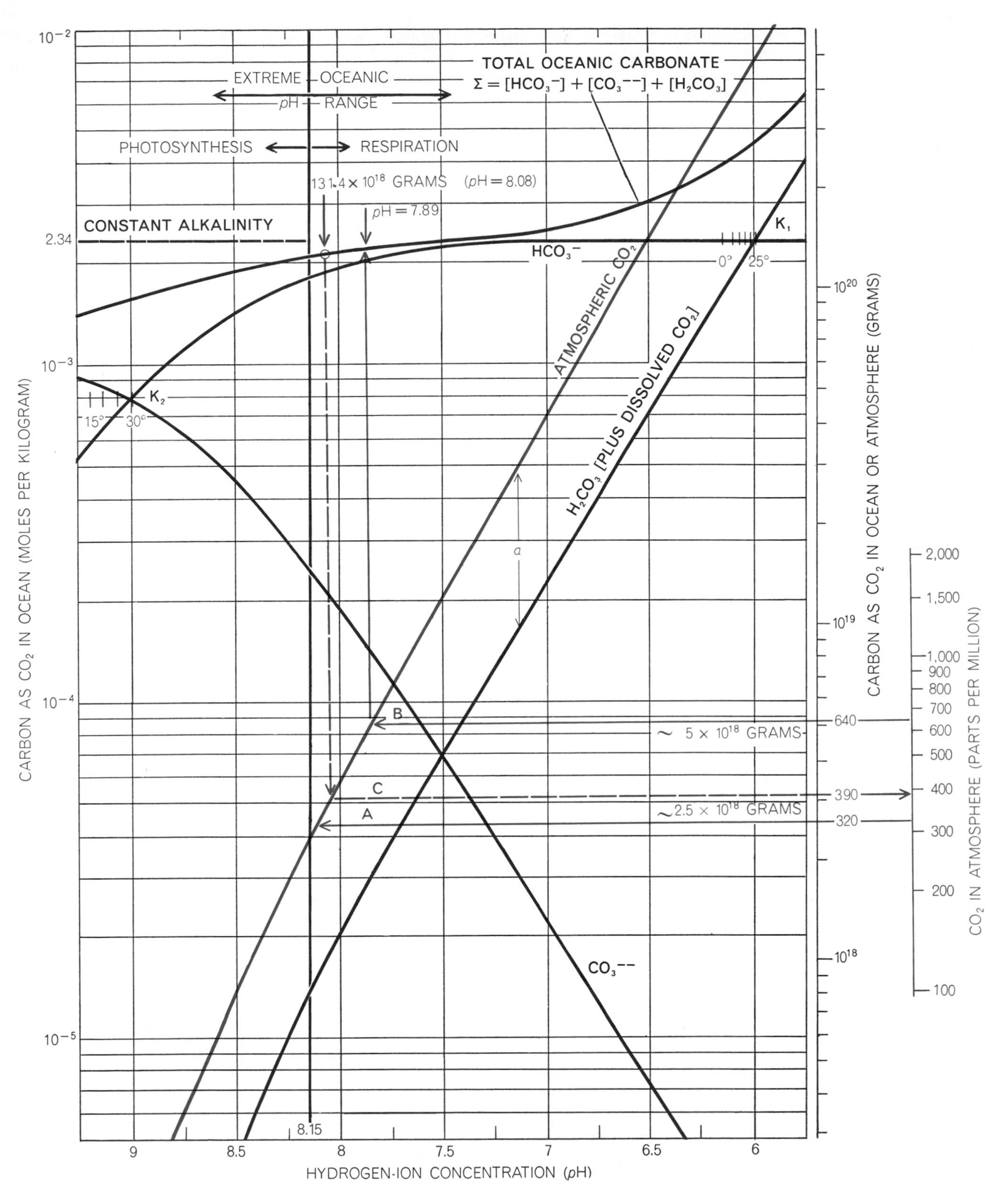

OCEANIC CARBONATE SYSTEM can be represented by a "Bjerrum diagram" that shows how carbonate in its several forms varies with the ocean's pH, or hydrogen-ion concentration. The diagram is plotted for a constant "carbonate alkalinity" of 2.34×10^{-3} moles of carbonate per kilogram of seawater (*scale at left*). "System point" K_1 shows where the concentrations of bicarbonate ion (HCO_3^-) and carbonic acid (H_2CO_3) are equal. At K_2 the concentrations of bicarbonate and carbonate (CO_3^{--}) are equal. The exact locations of K_1 and K_2 are shown for a range of temperatures (in degrees Celsius) at constant conditions of salinity and pressure. The top curve, Σ, is the sum of oceanic carbonate in all its forms. The normal pH of the ocean is 8.15. The two short arrows at top mark the normal biological limits: at 7.95 the available oxygen has been consumed by respiration; at 8.35 photosynthesis has removed so much carbon dioxide that absorption from the atmosphere rises sharply. The limits of oceanic pH lie between 7.45 and 8.6. The amount of carbon dioxide in the atmosphere (*colored curve and scale at far right*) is related to the amount of carbon dioxide dissolved in the ocean by alpha (α), the average worldwide solubility of carbon dioxide in seawater. The consequences of doubling the carbon dioxide in the atmosphere from 320 parts per million (*A*) to 640 parts (*B*) are discussed in the text of the article, as is line *C*.

sult of this step is that the alkalinity rises by 2 percent, which in terms of the Bjerrum plot means that the system will return to normal except that all the numbers on the concentration axes will be multiplied by 1.02. The long-range effect of a sudden doubling of the atmosphere's carbon dioxide, therefore, is to increase the ultimate value 2 percent, from 320 parts per million to 326, and some of that increase will ultimately find its way into vegetation and humus.

It is obvious that rates are crucial in the global distribution of carbon dioxide. The wind-stirred surface layer of the sea exchanges carbon dioxide rapidly with the atmosphere, requiring less than a decade for equilibration. Because this layer is only about 100 meters deep it contains only a tiny fraction of the ocean's total volume. Large-scale disposal of atmospheric carbon dioxide therefore requires that the gas be dissolved and transported to deep water.

Such vertical transport takes place almost exclusively in the Weddell Sea off the coast of Antarctica. Every winter, when the Weddell ice shelf freezes, the salt excluded from the newly formed ice increases the salinity and hence the density of the water below. This ice-cold water, capable of containing more dissolved gas than an equal volume of tropical water, cascades gently down the slope of Antarctica to begin a 5,000-year journey northward across the bottom of the ocean. The carbon dioxide in this "antarctic bottom water" has plenty of time to come to equilibrium with clay sediments.

Enough fossil fuel has been burned in the past century to have raised the carbon dioxide content of the atmosphere from about 290 parts per million to 350 parts. Since the actual level is now 320 parts per million, about half of the carbon dioxide put into the air has been removed. Although proof is lacking, a principal removal agent is undoubtedly antarctic bottom water. The process is so slow, however, that the carbon dioxide content of the atmosphere may reach 480 parts per million before the end of the century. By then it should be clear

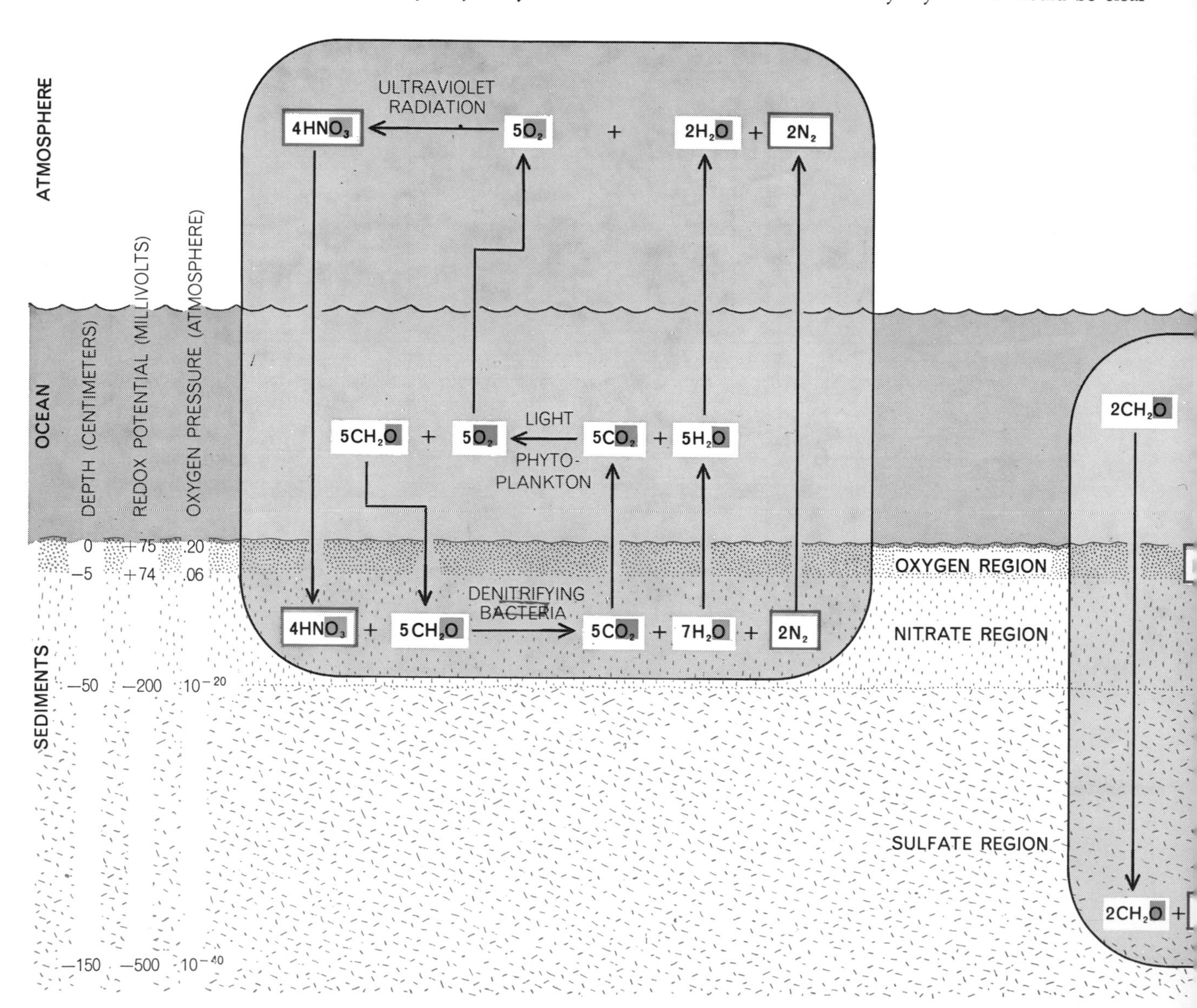

BACTERIA IN MARINE SEDIMENTS, although scarce by terrestrial soil standards, play a major role in replenishing the oxygen of the atmosphere and in limiting the accumulation of organic sediments. The bacteria concerned are buried in fine-grained sediments from several centimeters to several tens of centimeters below the ocean floor, with limited access to free oxygen for respiration. Thus deprived, they use the oxygen in nitrates and sulfates to oxidize organic compounds, represented by CH_2O. The actual reactions are far more complex than indicated here. The net result, however, is that denitrifying bacteria (*left*) release free nitrogen and convert carbon to a form (carbon dioxide) that can be reutilized by phytoplankton. These organisms, in turn, release free oxy-

if man's inadvertent global experiment (altering the atmosphere's carbon dioxide content) will have the predicted effect of changing the earth's climate. In principle an increase in atmospheric carbon dioxide should reduce the amount of long-wavelength radiation sent back into space by the earth and thus produce a greenhouse effect, slightly raising the average world temperature.

Having described an equilibrium model of the ocean that neglected the atmosphere's content of nitrogen and oxygen, I should not leave the reader with the impression that the continued presence of these two gases in the atmosphere is independent of the ocean. If the ocean were truly in equilibrium with the atmosphere, it would long since have captured all the atmospheric oxygen in the form of nitrates, both in solution and in sediments. This catastrophe has apparently been averted by the intervention of certain marine bacteria that have the happy faculty of releasing nitrogen gas from nitrate compounds and of converting the oxygen to a form that can later be liberated by phytoplankton.

The story is this. A variety of high-energy processes in the atmosphere continuously break the triple chemical bond that holds two nitrogen atoms together in a nitrogen molecule (N_2). The bonds can be broken by ultraviolet photons, by cosmic rays, by lightning and by the explosions in internal-combustion engines. Once dissociated, nitrogen atoms can react with oxygen to form various oxides, which are then carried to the ground by rainfall. In the soil these oxides are useful as fertilizer. Ultimately large amounts of them reach the sea. They do not, however, accumulate there and no one is really sure why.

The best guess is that denitrifying bacteria in oceanic sediments use the oxygen of nitrate to oxidize organic molecules when they run out of free oxygen [*see left half of illustration on these two pages*]. The nitrogen is released directly as a gas, which goes into solution but is available for return to the atmosphere. The oxygen emerges in molecules of water and carbon dioxide. The carbon dioxide is assimilated by phytoplankton, which build the carbon into organic compounds and release the oxygen as dissolved gas, also available for return to the atmosphere. Without these coupled biological processes the atmospheric fixation of nitrogen would probably exhaust the world's oxygen supply in less than 10 million years. Nevertheless, the amount of nitrogen returned to the atmosphere from the sediments is so small that we may never be able to measure it directly: the yearly return is less than one two-thousandth of the total nitrogen dissolved in the sea.

Another little-known epicycle in the global oxygen cycle probably has the effect of limiting the net accumulation of carbon in the form of oil-bearing shale, tar sands and petroleum. After denitrifying bacteria have consumed the nitrate in young sediments, sulfate bacteria begin oxidizing organic matter with the oxygen contained in sulfates [*see right half of illustration on these two pages*]. The product, in addition to water and carbon dioxide, is hydrogen sulfide, the foul-smelling compound that characterizes environments deficient in oxygen. In undisturbed mud the hydrogen sulfide never reaches the surface because it is inorganically reoxidized to sulfate as soon as it comes in contact with free oxygen. It seems likely that the bacterial turnover of oxygen in sulfate is so rapid that half of the world's oxygen passes through this epicycle in about 50,000 years.

The global activities of man have now reached such a scale that they are beginning to have a profound effect on marine chemistry and biology. We are learning that even the ocean is not large enough to absorb all the waste products of industrial society. The experiment involving the release of carbon dioxide is now in progress. DDT, only 25 years on the scene, is now found in the tissues of animals from pole to pole and has pushed several species of birds close to extinction. The concentration of lead in plants, animals and man has increased tenfold since tetraethyl lead was first used as an antiknock agent in motor fuels. And high levels of mercury in fish have forced the abandonment of some commercial fisheries. (Lead and mercury are systemic enzyme poisons.) Of the total petroleum production some 2 percent gets slopped into the sea in half a dozen major accidents each year. (At least six of the rare gray whales died last year after migrating through the oil slick off Santa Barbara caused by the blowout of a well casing belonging to the Union Oil Company.) Conceivably a persistent oil film could change the surface reflectivity of the ocean enough to alter the world's energy balance. The rapid increase in the use of nitrogen fertilizers leaves a nitrate excess that runs into rivers, lakes and ultimately reaches the sea. The sea can probably tolerate the runoff indefinitely but along the way the nitrogen creates algal "blooms" that are hastening the dystrophication of lakes and estuaries.

It is fashionable today to view the ocean as the last global frontier, waiting only technological "development." Thermodynamically it is easier to extract fresh water from sewage than from seawater. Ecologically it is wiser to keep our concentrated nutrients on land than to dilute them beyond recall in the ocean. Sociologically, and probably economically, it makes more sense to process our junkyards for usable metals than to mine the deep-sea floor. The task is to persuade our engineers and business companies that working with sewage and junk is just as challenging as oceanography and thalassochemistry.

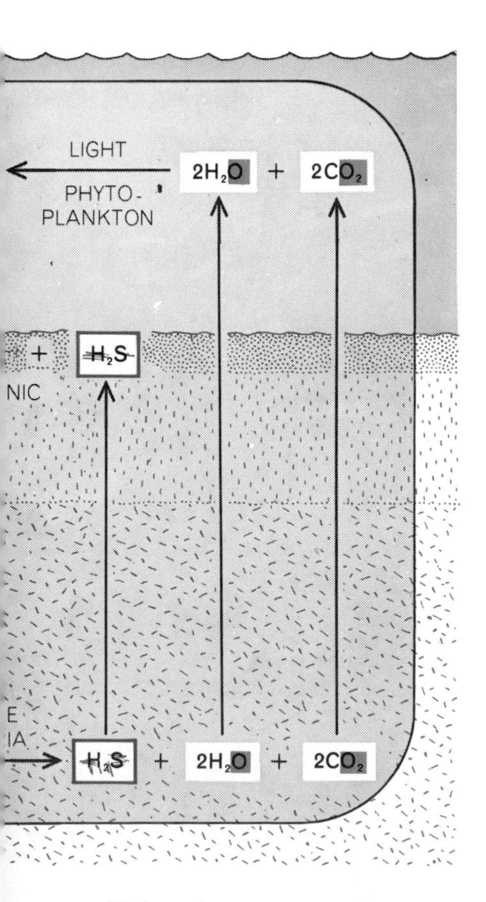

gen. Without the cooperative effort of these two groups of organisms the oxygen in the atmosphere might all be fixed by high-energy processes within some 10 million years. The sulfate bacteria (*right*) play a role in the recycling of sulfur and oxygen.

III

THE FLOOR OF THE OCEAN

III

THE FLOOR OF THE OCEAN

INTRODUCTION

On a recent trip along the coast of southern California, my wife and I drove for many miles behind a colorfully painted van carrying four young surfers and their surfboards. Because the van had an out-of-state license plate, we did not think it odd that it stopped or slowed down on arriving at each new beach along the coastal highway. We did, however, become curious at the dispatch of a young man from the van, who, on two occasions, ran down almost to the swash zone and critically viewed the form of the beach and inspected a handful of sand. Twice our intrepid surveyor returned to the van, and the young people drove on. At the third stop, the young man was again dispatched to reconnoiter, but this time he waved to his friends in the van, signalling them to pull off the road and park. Apparently, his experience by sight and touch had told him that this was *the* beach for surfing. Being somewhat curious at this point, we stopped and asked him how he had come to select this particular beach, inasmuch as no real surfing waves were breaking on any of the beaches, and, as I learned later, none were expected until that afternoon.

His explanation was: "The steeper the beach and the rougher the sand, the better the surfing." As for the predicted afternoon arrival of surf waves along the coast, he had picked up that information from a gas station attendant who, I suspect, had learned of it from the Coast Guard.

The point of relating this brief encounter is that beaches are characterized by all who use, enjoy, or study them—surfers, sunbathers, and sedimentologists alike—in terms of their quality and usefulness. It is interesting that this young man had, through his cumulative but nonscientific experience, reached the same conclusions regarding beach processes that more learned investigators have reached using sophisticated research methods: namely, that a steeply sloping beach and coarse sand (if such is naturally available) are indicative of a high-energy environment—in short, good surfing.

All of us are concerned, whether we realize it or not, with the floor of the ocean, and the several articles in this part of the book serve to show how, why, and where this concern manifests itself. Beaches make a good starting point: they are at our doorstep, as we look seaward; they are familiar to many and known to all.

During the coming decade, this nation can anticipate two developments that will place greater human demands on our presently crowded beaches: more people and more leisure time. In conflict with the expanding use of beaches for recreation is the increasing demand for coastal areas by industry and by real-estate developers. Power companies, for example, prefer to locate their nuclear power plants along the shores of oceans and large lakes in order to obtain a large, readily available supply of cooling water. Although the water being returned to the ocean or large lake may or may not be chemically altered, it is most assuredly warmer coming out than when it went in. This introduction of warm water into the beach zone creates changes in the ecological regime, some of which are deleterious to water quality. The swiftly flowing water coming from the plant also interferes with the natural coastwise transport of beach sand, so much so that some beaches may become "starved" for sand and turn into rocky stretches of coast, and other beaches may fall victim to overly rapid deposition of sand, causing the beach front to move far seaward from previously built homes, bathhouses, piers, and even the ubiquitous hot-dog stands.

In some of the most densely populated parts of the United States, we will find it necessary to construct new beaches and to widen old ones. To accomplish this, massive funding will be necessary. In constructing new beaches, artificial wave traps, in the form of offshore breakwaters, will need to be built. Perhaps, old automobile bodies could be used to form such offshore breakwaters, thus providing for the utilitarian disposition of unsightly accumulations of junked cars.

Many others are concerned with the nature of beaches: the military commander of an amphibious landing force, the sand and gravel supplier, the highway engineer, the fisherman, the amusement park operator, the lawyer, and the construction engineer. Although much research is presently being done on the technical aspects of the beach environment by universities, the U.S. Army Corps of Engineers, the Navy, and numerous state agencies, much research remains to be done on beach use, planning, development, and relating community and social needs to this thin strip of mobile sand touching the sea. An understanding of the processes controlling deposition and erosion is basic to planning effective utilization by all sectors of the public.

Beyond the beach zone (which is rarely more than a few hundred yards wide), and extending seaward to a depth of between 400 and 600 feet, lies the continental shelf. Although the widths of continental shelves vary from as much as a hundred miles (as off a part of the coast of Texas) to hardly more than a narrow step in the seafloor (as off the west coast of South America), the aggregate area of the submerged continental shelves of the world is equal to an area about the size of Africa. This vast piece of shallow, underwater "real estate" has, in recent years, become increasingly important for national defense, as a source of mineral resources (including oil, gas, and sulfur), as a site to lay power cables and oil pipelines, as a dumping ground for wastes, and, of course, as the site of major fisheries.

Although we still know very little about the deep geological structure of continental shelves, and virtually nothing at all about some shelves off Africa and Asia, the recent boom in offshore oil exploration has provided marine geologists with considerable knowledge of the structures of the continental shelf off the Gulf and Atlantic coasts of the United States, off western Europe, and off some parts of southeast Asia. This is largely because of extensive surveys conducted by oil and, to a lesser extent, hard-mineral companies, although some oceanographic institutions have also done considerable shelf exploration in American waters.

During the 1930s and 1940s, exploration of the continental shelf off the East Coast of North America, conducted by the late Henry C. Stetson of Harvard, established the fact that sedimentary strata are exposed on the walls of submarine canyons at the edge of the continental shelf. Stetson acquired samples of the exposed submarine rock by lowering large dredges on steel cables and then pulling the dredges up the sloping faces of the canyon walls, breaking off large chunks of rock in the process. Study of the fossils in the samples established that *these* shelf rocks were deposited during the Tertiary and Cretaceous geological periods, and that they resemble sandstones of similar age that crop out on land farther south in the coastal plain between New Jersey and Virginia. Although Stetson's discovery was of much interest to marine geologists, it was of even greater interest to petroleum geologists. The latter immediately saw the economic potential of continental-shelf rocks as petroleum reservoirs. While Stetson was busy dredging the submarine canyon walls, Maurice Ewing, a marine geophysicist, made the first of many geophysical profiles of the Atlantic shelf, and by the beginning of World War II, these two academic researchers had provided a general picture of the dimensions and composition of the Atlantic continental shelf of the United States. Encouraged by Stetson's and Ewing's findings, the Standard Oil Company of New Jersey, early in the 1940s, drilled a deep well on the outer barrier island forming Cape Hatteras. Because offshore drilling had not been perfected at that time, it was believed that the best way to confirm the stratigraphy, or sequence of beds, was to drill as far seaward from the mainland as possible—thus, the site at Cape Hatteras. The well, named Hatteras Light No. 1, is of historical interest, for although it failed to produce oil, it did establish the nature and geologic age of the underlying strata by penetrating slightly more than 10,000 feet of ancient sediments, most of which were similar to those that Stetson had found farther north in submarine canyons. When the drilling was terminated, the well had penetrated a few feet into hard crystalline rock—basement rock—much like some of those crystalline rocks cropping out in the eastern United States. By combining this information with the data obtained by study of dredge samples and geophysical records, the sedimentary origin of the shelf was confirmed. In the years since then, many additional surveys have been made, and we now have a reasonably detailed knowledge of the material composing the shelf and, thus, its probable origin.

Kenneth Emery has reviewed the several modes of origin of continental shelves in his instructive article "The Continental Shelves," the second article in this section of the book. He has also shown the value of carbon-14 dating of certain shells and plant material that once lived at or near sea level. By plotting the age of such organic remains against the depth below present sea level at which they are found, marine scientists are now able to provide an historical record of the major changes in sea level during the late ice age and, thus, to speculate with some reasonable certainty about the times of man and animal migrations across the formerly dry continental shelves.

During the present decade, we can anticipate further scientific investigation and industrial exploitation on the continental shelf; we can also expect to encounter some difficult legal and political problems. The continental shelf may, one day, become of more concern to the lawyer and the politician than to the marine scientist.

Let us turn our attention briefly to the enigmatic submarine canyons that are found on many continental shelves of the world. We speak of them as enigmatic simply because oceanographers have not yet conclusively proved the precise process, or processes, by which they have been formed.

Few New Yorkers sunning themselves on their favorite Long Island beach are aware that, only a few dozen miles seaward, there is a canyon on the floor of the ocean that rivals the Grand Canyon of the Colorado River. This canyon, known as the Hudson Submarine Canyon, cuts deeply into the continental shelf off New York. It is not the only one of its kind, however: numerous submarine canyons have been charted along the outer edge of the continental shelf from George's Bank off Massachusetts all the way south to Florida. What processes have led to the formation of these canyons? Several hypotheses have been put forward, as Bruce Heezen so ably reports in "The Origin of Submarine Canyons." While the scientific controversies over their mode of origin are interesting in themselves, submarine canyons are of timely interest for several other reasons.

In this age of nuclear submarines, which can be used as mobile underwater missile-launching platforms, the defense of any nation is complicated by the existence of underwater pathways to launching sites near its shores. Although sophisticated countermeasures have been perfected for locating submarines that hide behind coral reefs, behind acoustically opaque water barriers, and over the edges of submarine canyons, the newest types of nuclear submarines are capable of diving to such depths as are found in the upper reaches of many submarine canyons; in doing so, they may be difficult, if not impossible, to detect. Indeed, it is conceivable that, by charting a secretive path up the Hudson Submarine Canyon, a deep-diving submarine could approach undetected almost to the doorstep of the great port of New York.

Submarine canyons are also coming to the attention of municipalities and federal agencies who have refuse of one sort or another that they wish to dump at sea. Casting trash, outdated munitions, and other assorted debris into submarine canyons is safer by far than dumping it on shelves and slopes. Occasional mud slides, and the currents of turbid, silty water that are known to be active in many canyons, would readily serve to blanket the disposed trash with layers of muddy sediments. F. P. Shepard, of the Scripps Institution of Oceanography, has shown that, in some submarine canyons along the Pacific Coast of the United States, a steady flow of sand constantly moves down the canyon floor to be deposited, finally, on the deep ocean bottom. Surely, it makes sense to cast our accumulated trash where it will be further removed and, thus, be less likely to interfere with food resources on the shelf or to be returned to shore by shelf currents. Fortunately, continental shelves in many parts of the world are cut by submarine canyons; although some of these canyons are not where we would most like to see them present, they are common enough and widely enough distributed to be used as dumping grounds.

Canyons that cut through continental shelves are not the only ones present beneath the sea. For example, a large and well-developed submarine canyon originates in midocean in the vicinity of Greenland and extends southward in the North Atlantic into abyssal deeps exceeding 12,000 feet. This submarine canyon possesses a dendritic configuration and resembles, in many ways, the physiographic erosional pattern of continental streams and rivers. At this time, we can only speculate about the way in which such major deep-sea canyons were formed; much research remains to be done before we will be able to demonstrate their origins. It is most likely, however, that deep-sea canyons serve as pathways for the flow of colder and denser water. Moreover, midocean canyons with only a slight gradient may serve as channel ways for the dispersal of fine-grained sediments and turbidity currents. Perhaps in the decade ahead we shall learn some of the answers to our many questions about these strange undersea features and how they relate to the oceanic milieu.

In his article "The Deep-Ocean Floor," Henry W. Menard provides a synthesis of the status of our observations about the deep floor of the ocean. His article introduces the answer to the riddle of continental drift.

For at least a hundred years, geologists have speculated about the possibility that continents are floating "islands" of lighter rock adrift on a "sea" of denser crustal rocks. First led to the belief that the continents were once in contact by noting the matching outlines of Africa and South America, geologists and oceanographers have debated about whether or not the continents have actually drifted and, if so, by what grand mechanism. Menard's lucid explanation of spreading centers, tectonic plates, and the role of magnetic evidence reveals that the ocean floor does indeed spread and the continents do indeed drift. The reader will note that no single observation, or series of observations, and no single mechanistic *a priori* concept, led to this important discovery. The concept of global plate tectonics and ocean-floor spreading was developed through the contributions of many scientists. Moreover, ocean-floor spreading was accepted by the sceptics only after studies of long cores obtained by the research vessel *Glomar Challenger* from critical parts of the deep-sea floor were shown to verify this theory. In the century since the circumnavigating cruise of the *Challenger*, the single most important oceanographic discovery has been that of ocean-floor spreading.

In addition to explaining sea-floor spreading, global tectonics has provided us with explanations of the origin of deep trenches and mid-ocean ridges. Consider the observations of Robert L. Fisher and Roger Revelle in their article "The Trenches of the Pacific": written only fifteen years ago, their descriptive commentary on the deep trenches of the Pacific—particularly the Tonga-Kermadec Trench, which is located in the South Pacific near Samoa—provided, even then, hints of sea-floor spreading. Indeed, the varying heights of seamounts, bathymetric profiles showing trench shapes, and deep geophysical anomalies all suggested evidence that was subsequently shown to match the requirements for tectonic spreading. Their paper, as a descriptive exposition of the morphology and distribution of trenches, remains instructive reading. Perhaps the message to be gained in comparing and contrasting "The Deep-Ocean Floor" with "The Trenches of the Pacific" is that scientific discovery results only through an understanding of the fundamental processes, and not through the mere acquisition of data, no matter how detailed those data may be.

The acceptance of ocean-floor spreading has now led some geologists to recognize that some geologic terrain presently exposed on land was once deep beneath the sea. In his paper "The Afar Triangle" (an area in Ethiopia around the south end of the Red Sea), Haroun Tazieff reports that many of the geologic features of that area are true deep-sea features—an example being Mount Asmara, a large, flat-topped volcano that originally formed under water as a guyot. It is not surprising, then, that geologists are beginning to search on land for other expressions of former deep-sea features. Although marine sediments now exposed in many land areas have long been recognized as such, they are essentially deposits of shallow-water environments. In the light of Tazieff's observations, we may expect many marine geologists to turn their attention to such exotic areas as the Afar Triangle in the decade ahead.

The ocean records its history through the layer-by-layer deposition of organic and inorganic sediments upon its floor. To the skilled marine geologist and micropaleontologist engaged in the analysis and study of cored samples of oceanic deposits, changes in grain size and mineral components, changes in the species of microscopic plant and animal fossils, and changes in the composition of interstitial water together provide evidence of the ocean's history and, to a lesser extent, the history of the land as well. Inasmuch as pelagic sediments are more or less continually deposited on the floor of the sea, these changes may reflect the shift from a warm to a cold climate, the uplift and erosion of major mountain ranges, and the eruption of volcanoes. Even faunal and floral evolution is recorded in the fossils of certain animals and plants that left their remains on the sea floor. An adage well known to freshman geology students, that "the present is the key to the past," is well applied in its use by marine micropaleontologists who study the small fossils found in core sediments. Although the field of micropaleontology is too broad to be reviewed here, I would like to comment briefly on the role of the micropaleontologist who devotes his efforts to the study of ocean-floor cores. Of the several kinds of microfauna commonly found in deep-sea sediments, none is more useful than the Foraminifera, an order of microscopic shell-bearing protozoans. These tiny creatures range from pole to pole, and they are readily related to major bodies of water on the basis of their temperature tolerance. Those of most interest to oceanographers in their study of earth history are the planktonic species. Such forms are common in the upper waters of the world ocean, and it is only after death that their tests, or shells, sink to the floor of the ocean. Thus, if we measure the temperature of the body of water in which a given species is presently living, and we find fossil forms of the same species buried in sediments below that body of water, we may assume, cautiously, that the temperature of the surface waters, during the time those sediments were deposited, was probably much the same as it is today. Of course, there are exceptions to this rule: foraminiferal tests may be transported by turbidity currents, or a given species of foraminiferan may have adapted to slow changes in water temperature with the passage of time. Nevertheless, the study of modern foraminiferans, both planktonic and benthic, has provided the biostratigrapher, in his analysis of ancient sediments, and the marine micropaleontologist, in his analysis of very young sediments, with a key to deciphering the ocean environment at the time of deposition. In recent years, the application of the mass spectrograph to determine the oxygen-16 and oxygen-18 content of sediments, and to compute the ratio of one isotope to the other, has provided an additional clue to past thermal history of the ocean, because the tiny foraminiferan combines these two isotopes in its calcium carbonate shell at a ratio that correlates to the temperature of the surrounding water. Through application of isotopic data, marine micropaleontologists are able to determine, within surprisingly narrow limits, the water temperatures of the past. Furthermore, additional isotopic studies employing carbon-14 dating —carbon is also an element in the shell—have made it possible for oceanographers to date the events of the past 50,000 years or so. One of the most critical findings of such research by marine micropaleontologists is that the last ice age ended about 11,000 years before the present, rather than about 20,000 years ago, as was previously thought. Obviously, this new finding has required that scientists reconsider their conclusions about the recent migrations of man.

In turning our attention from microscopic studies to marine research of larger dimensions, we might consider the way in which oceanographers determine the thickness of the sedimentary carpet on the floor of the ocean. Relying on small explosions to generate a sound source, the marine geophysicist can determine the refraction and reflection of the sound waves as they are so influenced by the sediments and the denser rock beneath them. On first consideration, the use of geophysics may appear to be solely in the province of the academician. This is not the case. Marine geophysical surveys are routinely conducted for the purpose of locating salt domes and folded strata at depth beneath the continental shelf, such geologic structures favoring the accumulation of petroleum, natural gas, and sulfur. Geophysical surveys are also employed to locate solid rock beneath a veneer of sediments, so that bridges may be built upon solid foundations. One of the most interesting examples of applied geophysics was the recent survey conducted by Robert P. Meyer at the University of Wisconsin, wherein the thickness and distribution of polluted muds flooring a major portion of Green Bay, Wisconsin, were determined. His research provided clear evidence of the source and quantity of unchecked pollution.

In the search for new mineral resources to feed the smelters and refineries of industrial America, shallow bays and coastal areas must be surveyed for new lode and placer deposits of strategic and industrial metals. The geophysical technique provides a simple method by which industrial marine scientists and engineers may readily assess the potential for mineral accumulation. Obviously, the presence of petroleum and mineral deposits on the ocean floor, which may be suggested by geophysical evidence, can only be positively ascertained by drilling. Nevertheless, the use of marine geophysics in oceanic exploration will continue to expand during the next decade, and trained geophysicists will be needed to fill the demand for such exploration.

"The Origin of the Oceans," by Sir Edward Bullard, is the concluding article for this section because it masterfully brings together the several pieces of oceanography's grand puzzle—and makes them fit.

BEACHES

WILLARD BASCOM
August 1960

Beaches are natural playgrounds partly under the sun and partly under the sea where people can swim and surfboard, sun themselves and study other people. This human activity tends to obscure the fact that the beach itself is constantly in motion, quietly changing its configuration and restlessly shifting its position, grain by grain, until huge masses of sand have been moved. On a small scale and in a matter of hours the sand castles disappear and the footprints are erased; on a large scale, after days and months, the height of the sand around the rocks changes, the waves break in new places and the beach becomes broader or narrower. Indeed, over a period of years large quantities of sand may arrive or depart, posing complex problems of conservation for the people who want to enjoy the beach. This dynamic quality was incorporated by the late Columbia University physiographer Douglas W. Johnson into a definition: "A beach is a deposit of material which is in transit either alongshore or off-and-on shore." Thus three elements make a beach: a quantity of rocky material, a shore-line area in which it moves and a supply of energy to move it.

Most of the beach material along the coasts of the U. S. consists of light-colored sand—the product of the weathering of granite rock into its two main constituents, quartz and feldspar [see "Sand," by Ph. H. Kuenen; SCIENTIFIC AMERICAN Offprint 803]. Americans therefore tend to think of beaches as stretches of white sand. But white sand is no more required to make a beach than is the sun or sun bathers. Many Tahitians and Hawaiians think that a proper beach is made of black sand—the result of the disintegration of dark volcanic rocks. Along much of the coast of England and much of the French Riviera the beach is composed of small flat stones called shingle. The word beach may originally have meant shore lines made of this shingle. In parts of Labrador, Alaska and Argentina the beaches consist of large cobbles from four to 12 inches in diameter. The beach of the Pacific Coast of Lower California is made of two materials: a flat sandy portion that is exposed only at low tide, while above and behind the sand great cobble ramparts rise in steep steps to a height of 30 feet or more. Nor are stones and sand the only beach materials. At Fort Bragg, Calif., a small pocket beach consists entirely of old tin cans washed in from the city's nearby oceanic dump and arranged by the sea in the usual beach forms. It seems that a beach can be made of almost any material of reasonable size and density that is present in quantity. Because the principles involved in the motion of beaches are much the same regardless of the material, the word sand will be used in this article for all beach materials.

The Work of the Waves

A casual observer thinks of a beach as the sandy surface above water. The student of beaches takes a broader view and includes all of the area in which the sand moves. For him the beach extends from a depth of 30 feet below the water level at the lowest tide to the edge of the permanent coast. The latter may consist of a cliff, sand dunes or man-made structures, none of them really permanent, but all more enduring than the beach as seen at any one time. The offshore boundary of 30 feet below the low-tide water level is the depth beyond which ordinary water motion does not have sufficient energy to move the sand. A beach also has limits in the alongshore direction. A point of land or a stream may make such a boundary.

The waves and currents of the water provide the third element that must be present to make a beach: the energy to keep the sand in transit. At some beaches the wind also moves quite a lot of sand, but its direct effects will be ignored here. Of course it is the wind blowing on the surface of the sea that creates the waves, and so the wind ultimately causes all the movement of beaches. The waves rolling in on the beach may have originated far out at sea. The faster the wind, the longer it blows and the greater the distance over which it blows, the larger the waves it raises. If the storm is near the shore, the waves will be steep and may very quickly change the configuration of the beach. Normally, however, the waves that shape the beach have moved out from under the winds that generated them and are longer, lower and more regular than the wind waves. Such waves are called swell, and they travel away from the storm in all directions with very little loss of energy. Since a storm is almost always taking place somewhere at sea, the swell is constantly molding all the beaches around an ocean.

Swell is described by its height (the vertical distance between the trough and the crest of a wave), by its length (the horizontal distance between crests) and by its period (the time in seconds between crests observed at a given point). As swell moves into shallow water a remarkable change occurs. When the depth equals half a wavelength, the waves are said to "feel bottom." The velocity and length decrease, and the height increases; only the period remains the same. Rolling farther inshore, the waves rise higher and finally topple over and break. The result is surf—a turbulent mass of water [see the article "Ocean Waves," by Willard Bascom, beginning on page 45].

It is the action of waves in shallow

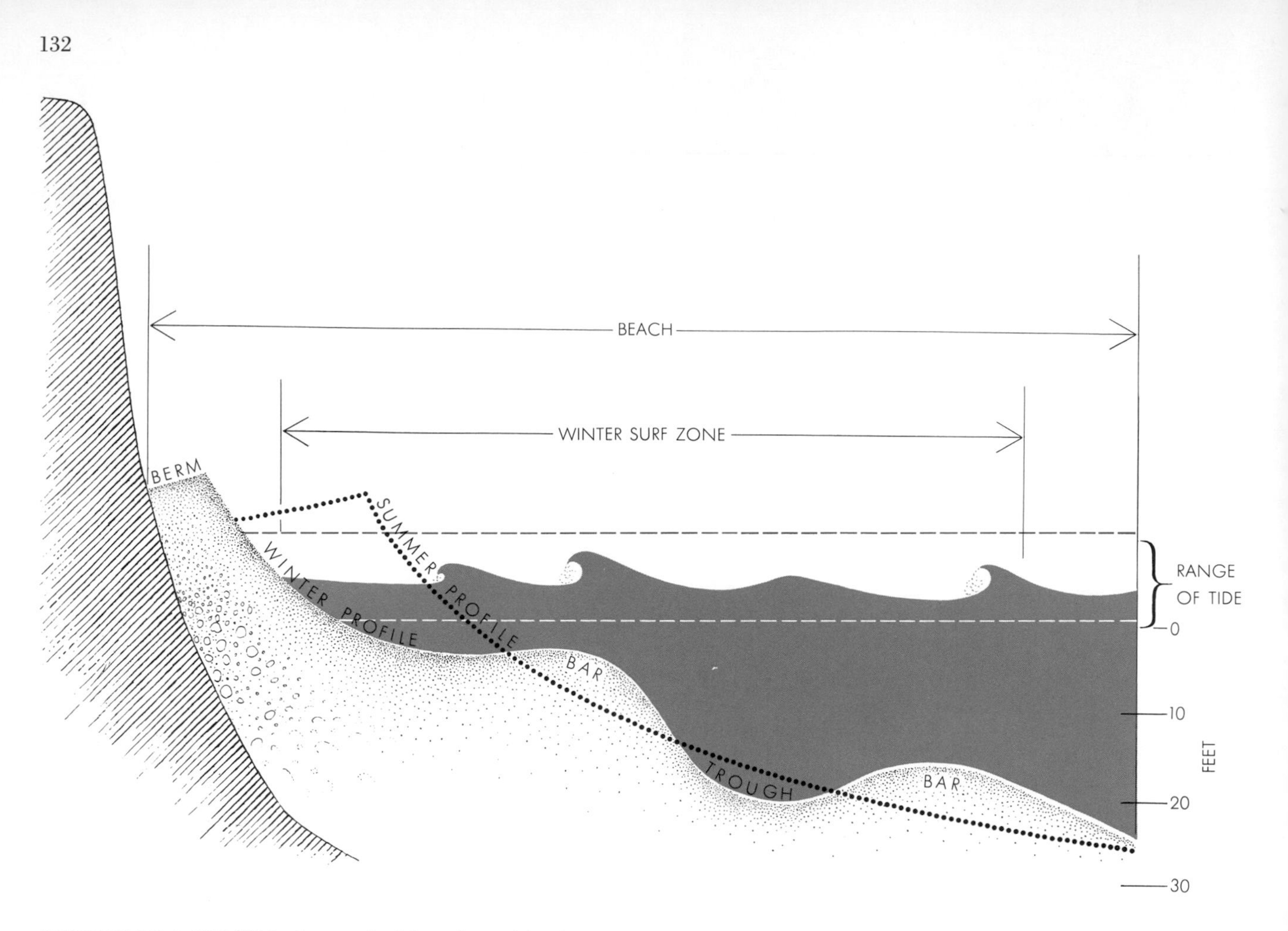

PROFILE OF A BEACH is characterized by a berm (the deposit of material at the top of the beach) and bars. In winter heavy surf removes sand from the berm and deposits it on the bars; in summer, light surf builds the berm. Vertical scale is exaggerated 25 times.

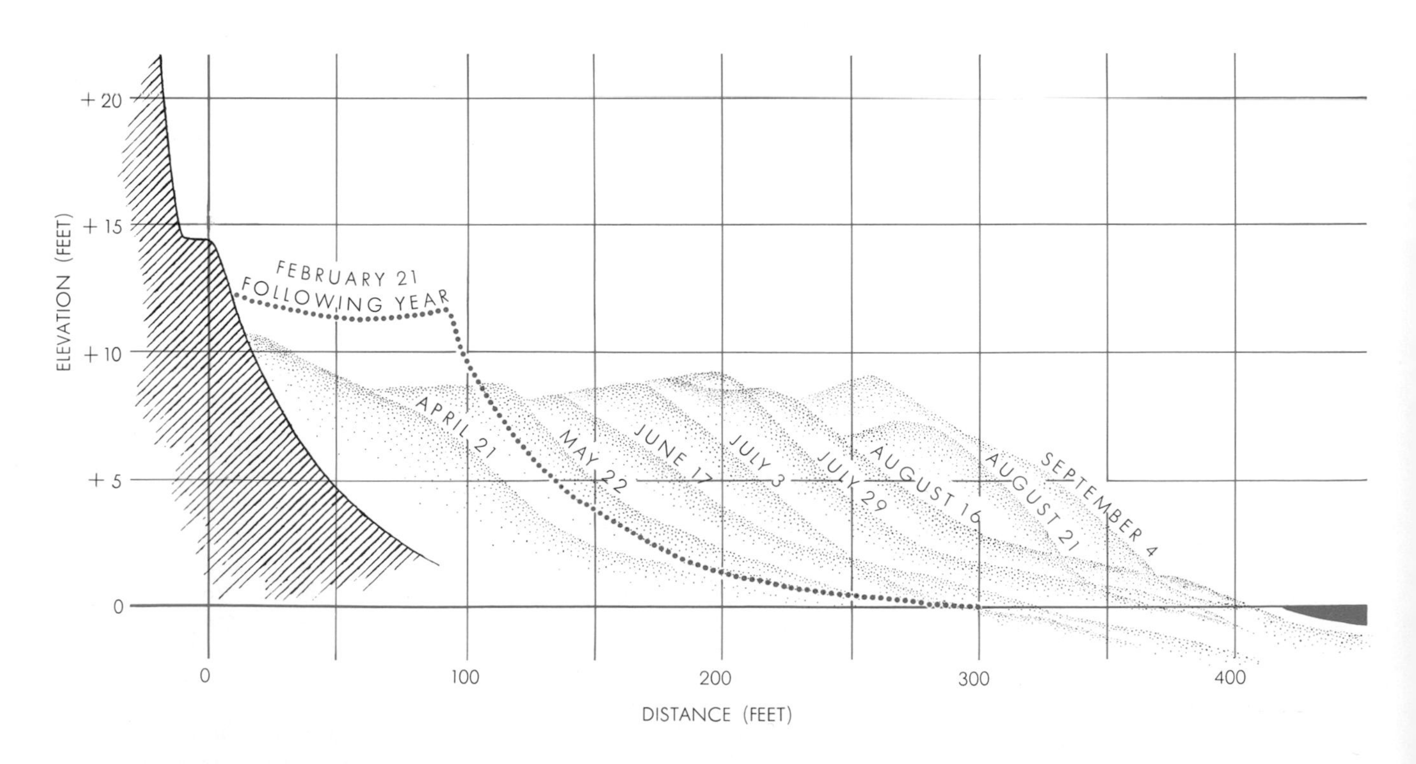

GROWTH OF THE BERM at Carmel, Calif., during the spring and summer is indicated by this series of dated slopes, based on actual measurements. Vertical dimension is exaggerated 10 times. The dotted line shows how berm was cut back during following winter.

water that changes the beach. The basic mechanism is simply the lifting of the sand grain by the turbulence that accompanies the passage of a wave and the free fall of the grain to the bottom as the wave loses its lifting force. Since a sand grain is lighter under water than in air by an amount equal to the weight of water it displaces, the water does not need great energy to lift it. Moreover, the grain settles back rather slowly because of the viscosity and turbulence of the water. While in suspension, a sand grain tends to move with the water, and currents of very low velocity will displace it. Each time a grain is lifted it lands in a slightly different location. Because uncounted millions of sand grains are continually being picked up and relocated, the beach shifts its position.

The Measurement of Beaches

During the years 1945 to 1950 John D. Isaacs, now of Scripps Institution of Oceanography, and I were employed on the "Waves" project of the University of California to study the beaches of the U. S. Pacific Coast. Under the direction of Morrough P. O'Brien, dean of engineering and a member of the Beach Erosion Board of the Army Corps of Engineers, we spent virtually all our time, winter and summer, observing the interaction of waves and beaches. To measure the height and period of the waves, we installed a dozen or so ocean-wave meters offshore, connected by armored submarine cable to recorders on the beach. With radio-controlled cameras we made photographs of the surf simultaneously from the beach, from nearby cliffs and from an aircraft directly above. We threw dye into the surf to determine the nature of its currents.

We spent most of our time, however, making repeated "profiles" of various beaches. This involved going out in an amphibious vehicle, the DUKW, or "duck," of World War II, to beyond the 30-foot low-tide depth and making numerous soundings as we came in to the beach face. Moving at three knots, we would keep the duck lined up with two marker poles set at right angles to the shore line. At short intervals I would heave the sounding lead, read the depth and call the results into a radio transmitter. Isaacs would be listening on shore, about 1,000 feet down the beach from the poles and watching the duck's progress through a surveyor's transit. The poles, the duck and the transit made a right triangle, with Isaacs sighting along the hypotenuse. When he saw the

AMPHIBIOUS TRUCK, the Army "duck" of World War II, rides a 12-foot breaker during a University of California survey of beaches. By piloting the duck straight toward the shore the workers aboard it were able to sound the profile of the beach. This beach is at Carmel.

RIP CHANNELS are marked by the dark lanes in the surf in this aerial photograph of a beach near Monterey, Calif. The channels are formed when a series of waves raises the level of the water inside a bar, and the outrushing water cuts a series of notches in the bar. The water in such channels can flow as fast as four knots.

BERM

BEACH FACE

PATHS OF SAND PARTICLES ON BEACH FACE

LITTORAL CURRENT

PATH OF UNDERWATER PARTICLES

DIRECTION OF SWELL IN DEEP WATER

LITTORAL CURRENT, a current running parallel to the beach, is set up when waves move toward the beach at an oblique angle. Under such conditions sand grains lifted by the surf, normally moved at right angles to the beach, are transported with the current.

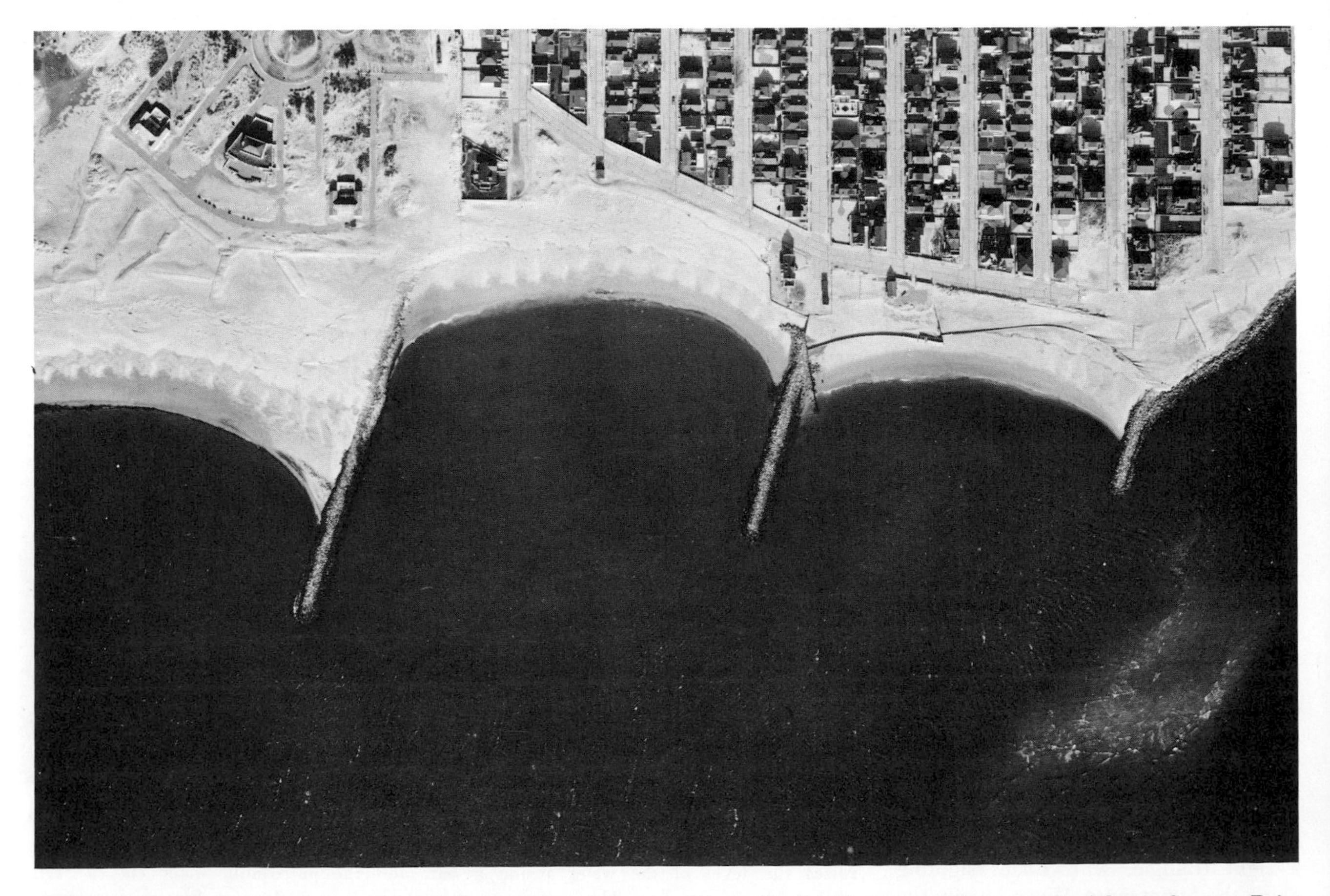

GROINS, dams of stone or wood jutting out from the beach, are widely used to retard the erosion of beaches by littoral currents. The groins in this photograph are on the Atlantic Coast at Point Lookout on Long Island. The current moves from left to right.

lead weight splash, he would read the angle on the transit and call it to an assistant who recorded depths and angles together. By plotting the depth at each distance from the shore line, we could draw a profile of the beach.

Somewhat surprisingly we obtained good profiles even in rough surf, partly because the duck is a fine beach-and-surf craft. It is 32 feet long, has six wheels and was originally designed for moving cargo from ship to shore and inland during amphibious operations. In breakers the front wheels tend to "hook" the crest of the wave, hanging down in front of the shoreward-tumbling water so that the vehicle is carried in like a surfboard. In the zone where it is only partly afloat the wheels and the propeller may drive it at the same time. The air pressure of its big tires is controlled from the cab, and when the vehicle reaches the steep beach-face, the pressure can be adjusted to achieve enough traction for the vehicle to grind its way upward.

On the northern California beaches in winter we often surveyed beaches where the breakers were 12 feet high, and occasionally, having misjudged the waves before starting out, we found ourselves amid breakers half again as high—a remarkable experience in a relatively small craft. Using the reliable duck we recorded changes in beach profiles from winter to summer on more than 30 West Coast beaches. We also kept a record of the ease with which the duck could move about on the part of beach above water. On the hard beaches north of the Columbia River it could travel as fast as it could on a highway—if the beach was being eroded. A day later, however, when the waves had altered the delicate balance of sand transport and deposited a new layer of soft sand, the duck could not exceed 10 miles per hour. We found that the beach face in the zone between high and low tide seemed to be the only place that retained its hardness and its degree of slope over relatively long periods.

Later we correlated our slope and wave measurements with the results of elaborate samplings of sand-grain size. This showed, as even a casual observer will note, that steeper beaches are usually composed of larger sand grains. Our studies also showed that factors not quite so apparent enter into the picture. For example, beaches that are partly protected from the swell will be steeper than beaches composed of sand of the same size that are exposed to it.

The underwater slopes are quite different from those of the beach face, and are usually described in a different way

BREAKWATER AT SANTA BARBARA, CALIF., causes sand transported by a littoral current running from west to east (*i.e.* from left to right) to deposit in the sheltered water. A dredge in center of harbor moves sand beyond the large pier at right.

BREAKWATER AT SANTA MONICA, CALIF., runs parallel to the shore (*bottom right*). Originally it was thought that the littoral current running inside the breakwater would carry sand past it, but instead the sand was deposited in the quiet water.

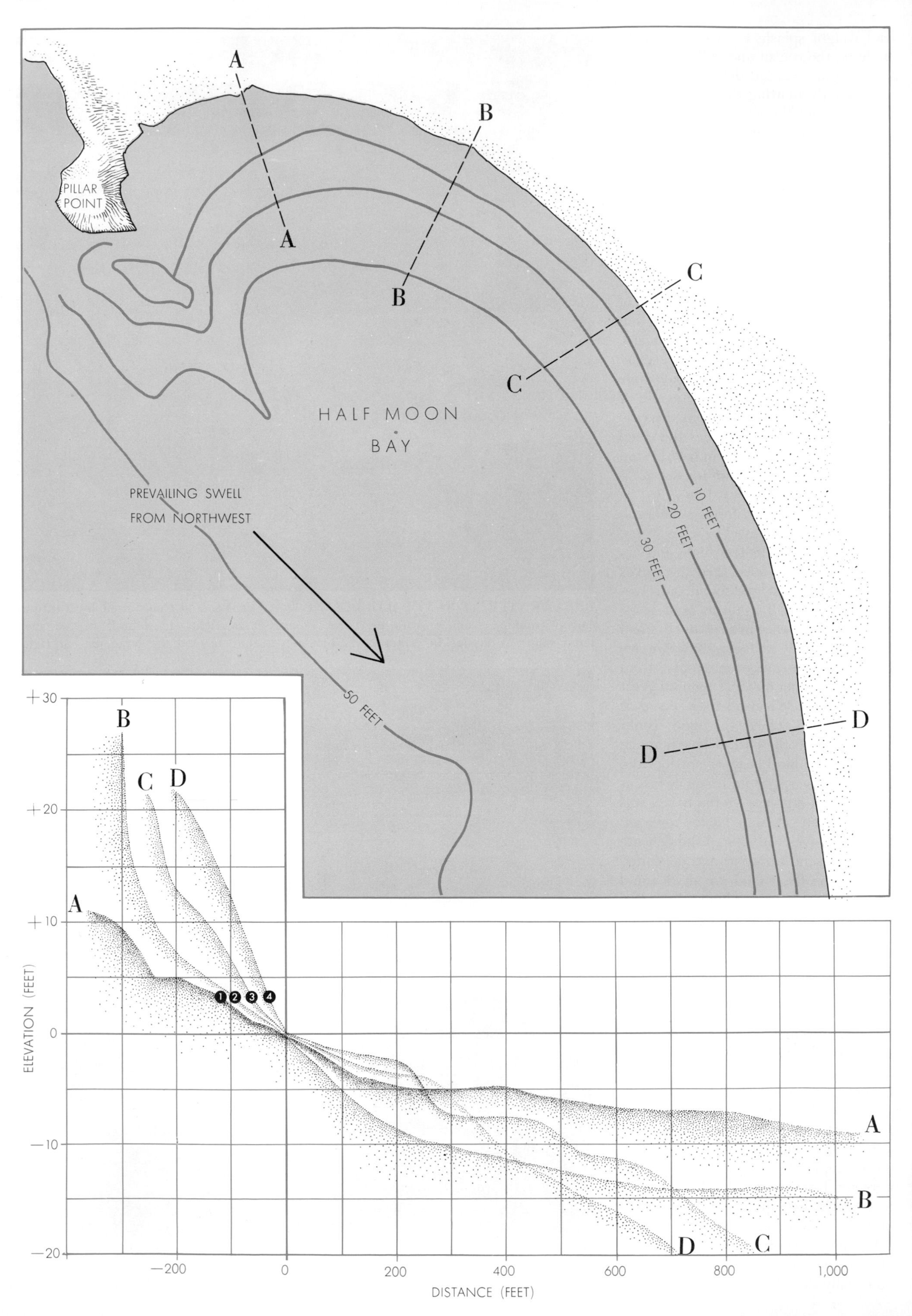

A
B
PILLAR POINT
A
C
B
C
HALF MOON BAY
PREVAILING SWELL FROM NORTHWEST
10 FEET
20 FEET
30 FEET
50 FEET
D
D
+30
+20
+10
0
−10
−20
B
C
D
A
A
B
D
C
ELEVATION (FEET)
−200
0
200
400
600
800
1,000
DISTANCE (FEET)

to take account of the substantial irregularities between the waterline and the seaward boundary. In this scheme of classification a flat beach has an average underwater slope with a vertical rise of less than one foot in 75 feet of horizontal distance; a steep beach has a gradient steeper than one foot in 50 feet.

A large part of the movement of beach material consists in an exchange between offshore underwater ridges, or bars, and the berm, the nearly horizontal deposit of material at the top of the beach onshore [*see top illustration on page 132*]. Bars may be considered as products of erosion, since they appear when violent wave action cuts back the berm and deposits the beach material in neat ridges offshore. Because they are associated with storm conditions, and since more storms occur in winter than in summer, bars are regarded as a normal feature of the beach profile in winter. All beaches exposed to the ocean swell (as well as beaches on such large bodies of water as the Great Lakes) have them, and beaches with a slope of less than one foot in 50 frequently have two or more. Essentially continuous bars 10 to 20 miles long are commonplace on the Pacific Coast north of Cape Mendocino.

Beach investigators do not know exactly how bars are formed. They have noted that the creation of bars is somehow related to wave steepness. Using an experimental wave-channel, J. W. Johnson of the University of California found that bars always formed on model beaches when the ratio of wave height to wavelength was steeper than .03, and that bars never appeared when the ratio was less than .03. In nature the numbers seem to be different, but the principle is undoubtedly the same. Since the tiny forces that cause differential motion of individual sand grains are hard to detect amid the general turbulence, the exact manner in which variations in steepness cause the sand to move landward or seaward remains elusive.

The bars in their turn have a decided effect upon waves. The outer slope of a bar is relatively steep, and this abrupt rise of the bottom causes the larger waves to break. The waves often re-form in the trough between bars and proceed toward shore as smaller waves, breaking on the shallower inner bars or on the beach face. The smaller waves in a train of waves of irregular heights will not break on the outer bar. Thus a bar tends to act as a wave filter, breaking and reducing the higher waves and passing waves that are below a certain height. On Pacific Coast beaches that are exposed to the full force of the waves, the top of the innermost bar is usually about a foot below the low-tide water level, the top of the second bar is at a depth of seven feet and the third bar is 13 feet deep. Large swell from a nearby storm will produce violent breakers as high as 30 feet on the outermost bar. On a beach with a gentle slope and a series of bars the waves will re-form and break again and again as they move in, creating a surf zone as much as a mile wide.

After the storm season the steepness of the waves decreases and they begin to move the sand toward the shore. The material from the outer bars fills in the troughs, and soon the beach profile shows no bars. The material from the inner bar migrates to the berm, building it seaward. Except on very flat beaches the berm usually has a well-defined edge, or crest, and its method of growth can be readily observed. As each wave reaches the beach face, it uses up its remaining energy in a thin swash of water that runs up the beach face carrying sand with it. Depending on how permeable and how saturated the sand is, a certain amount of the water sinks into the beach and does not return to the sea as backrush. Thus the transport capacity of the returning water is less than that of the uprushing water, and sand is added to the berm. When conditions are precisely right, a berm may grow as much as six inches an hour, or 10 feet a day. The berm of the beach at Carmel, Calif., which we studied for several years, is about 300 feet wider in September than it is in April, the months which respectively mark the end of the calm season and the storm season.

Large waves build a higher berm than small ones do. At Monterey Bay, Calif., the crest of the berm on the exposed beach at Fort Ord has reached 16 feet above the low-tide water level, whereas the berm of the beach a few miles away, which is protected by a headland from large waves, is six feet lower. Paradoxically the storm seas that remove the summer berm often leave a higher berm of their own at the back of the beach. This berm may remain clearly visible throughout the summer.

The material that stormy seas remove from the berm ordinarily returns during calm seasons. However, an occasional very large storm or a tsunami (a "tidal" wave) may strip a beach face of sand and carry the material to depths so great that the normal waves cannot reach it and return it. At Long Branch, N.J., and Santa Barbara, Calif., the Beach Erosion Board of the Army Corps of Engineers dumped mounds of sand at depths of 38 and 18 feet in the hope that waves would move the sand onto the berm. Unfortunately the sand, like that removed by a great storm, was too deep for normal waves to pick up.

When higher-than-average waves break in quick succession and raise the water level inside a bar, the water rushes back so energetically to sea that it sometimes breaches the bar at a narrow place, producing a so-called rip channel. From then on much of the excess water hurled over the bar by the breakers moves along the beach until it reaches the channel, where it flows out as a rip current [*see bottom illustration on page 133*]. The current can be dangerous, since it flows directly out to sea with a velocity as high as four knots, considerably faster than a man can swim. Fortunately rip currents are confined to relatively narrow channels, and the bather can get out of them simply by swimming a short distance parallel to the shore. On some beaches the lifeguards mark rip currents (they flow on the surface and can be seen from shore by a trained observer), often moving the warning signs several times a day to keep up with the migration of the currents along the beach.

The rip current appears in popular mythology as the fearsome "undertow," which is otherwise a pure invention of the imagination. The undertow is said to flow outward beneath the surf and so pull swimmers out to sea. Experiments with dye markers at beaches marked with undertow warning signs have repeatedly shown that no such current exists. The water does, of course, move in and out along the bottom with every wave, but anyone being pulled seaward would be carried shoreward after six sec-

EFFECT OF A HEADLAND on a beach eroded by a littoral current is depicted in this chart made by the University of California "Waves" project. The headland is Pillar Point and the sheltered area is Half Moon Bay. The profiles of the beach at the locations marked on the map at the top of the opposite page are traced in the graph below the map. The average size of the sand grains at each location and the slope of the bottom are listed in the table below.

REFERENCE POINTS	SAND SIZE (MILLIMETERS)	SLOPE
❶	.17	1:41
❷	.20	1:30
❸	.39	1:13.5
❹	.65	1:8

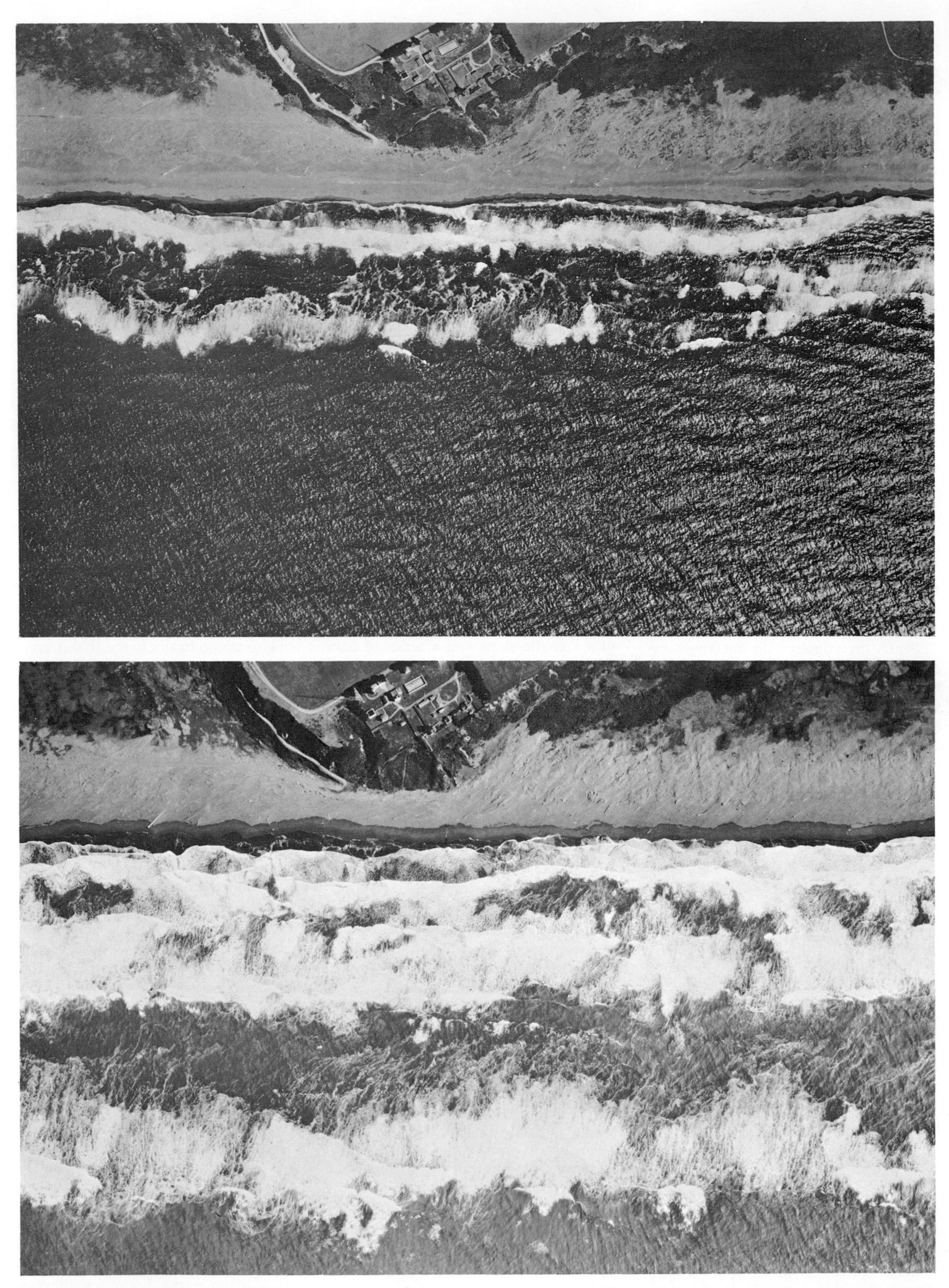

TWO VERTICAL AERIAL PHOTOGRAPHS show the beach at Table Bluff, Calif., during a light surf (*top*) and during a heavy surf (*bottom*). The first light band below the beach in the top photograph is waves breaking on the beach; the second light band is waves breaking on a sand bar. The light band at the bottom of bottom photograph is larger waves breaking on a deeper bar farther out.

onds or so by the other half of the cycle. Unlike rip currents, undertow can be dismissed as a danger to swimmers.

Cusps and Ripple Marks

A beach feature that offers intriguing problems for the investigator is the cusp. This is a crescent-shaped depression that occasionally forms in regularly spaced series along the beach face. The triangular apex (the horn) at which two crescents join points seaward. A "bay" of sand, which may be deep and narrow or broad and shallow, lies between the horns. Cusps vary in length from a few feet to hundreds of feet and their relief may exceed six feet or be so shallow that they are barely discernible. Although investigators have studied these curious beach forms for years and have constructed many hypotheses about them, no one has produced a generally accepted explanation of how they begin, why they are so regular and why they have the dimensions they have. These factors are almost certainly related to the character of the waves that form them. Perhaps cusps develop when the waves have a "balanced" steepness, so that neither erosion nor deposition of the sand occurs.

Ripple marks, or sand ripples, are roughly parallel arrays of wave-shaped ridges and troughs that are formed in the sand on the sea bottom by the action of the water. Ordinarily their "wavelength" is three or four inches, and they are about an inch deep, but giant ripples something like desert sand dunes and with wavelengths of more than 10 feet have been observed under the surf. Since ripple crests are usually parallel to the wave crests, paleographers have used fossil ripples in sandstone to establish the orientation of ancient beaches.

Ripple marks may begin to form around a pebble or any other small prominence on the bottom. The oscillating wave-motion creates horizontal vortices first on one side of the pebble, then on the other, and a small ridge of sand begins to accumulate. At a certain distance on either side of the original ridge the vortex currents diminish and deposit sand so that another ridge forms and the ripple pattern grows. Water motion due entirely to waves produces symmetrical ripples, but if a current is superimposed on the wave motion, the sides of the ripples are steeper on the lee side. Ripple marks come and go with the changing wave conditions, and they migrate with the currents. Although they, too, have been extensively studied, the relationship between current velocity, sand-grain size and the dimensions of the ripple marks remains unexplained.

The Preservation of Beaches

The major problems of beach conservation are created not by the seasonal movement of sand onshore and offshore, but by the motion of sand parallel to the shore. On some coasts, alongshore or littoral currents, which arise when waves strike the shore at an angle, annually transport millions of tons of sand, eroding one beach and building up another. The largest waves create the strongest littoral currents because they contain the most energy. They "feel" bottom well out from shore, and if they approach

CUSPS are a series of crescent-shaped depressions along a beach. The cusps in this aerial photograph are at El Segundo, Calif.; they are 60 feet across. There is no satisfactory explanation of how cusps are formed, though they are undoubtedly related to wave action.

shore at an angle, they tend to be refracted or bent by the underwater contours so that the wave front becomes parallel to the shore line. Waves are often incompletely refracted, especially if the angle between the deep-water wave fronts and the shore line is great, or if the water becomes shallow abruptly. Consequently waves often strike the shore at an angle.

Along coasts where the prevailing waves arrive in this way, a littoral current flows constantly. These currents usually flow too slowly to move the sand grains by themselves. Turbulence in the surf zone, however, keeps the sand in suspension, and even a low-velocity alongshore current is able to transport large quantities of material. On the beach face the sand particles carried by the uprush describe an arc in the direction of the alongshore current so that each wave moves them along the beach a little way. In shallow water, where the waves can lift sand and move it back and forth, the littoral current gives the sand grains a saw-toothed motion [*see top illustration on page 134*]. With every oscillation the sand moves sideways, and there is no force to return it to its original position. As a result the sand travels downstream as on a conveyor belt, the belt having the width of the surf zone and the velocity of the littoral current.

Many seaside areas afflicted with littoral currents have serious problems of beach erosion. The California coast north of Point Conception faces west, directly into the Pacific Ocean winds and swell, and there is no appreciable alongshore current. But south of this point the shore line turns abruptly to the east, so that these same winds and waves strike the shore at an angle. As a result an almost continuous current moves sand to the east. Any structure that interrupts the flow acts like a dam, and the beaches immediately to the west grow while those to the east are stripped of sand by the waves and currents. Of the several structures that produce this effect, the most interesting is the breakwater at Santa Barbara, for there the amount of sand carried along the coast can be measured. Sand moving from the west past the end of the breakwater abruptly encounters the deeper, quiet water that the breakwater was built to create. There the cessation of turbulence causes all the suspended particles to be deposited just inside the end of the breakwater [*see top illustration on page 135*].

Frequent surveys of the changes in the volume of sand in the spit have revealed the rate of deposition, which equals the rate of transport of sand along the shore. On an average day about 800 cubic yards of sand are dumped in the harbor and under storm conditions four times that much will arrive. To keep the sand from filling the harbor and to prevent damaging erosion on the shore beyond, a dredge pumps the sand from the spit across the quiet water to the downstream beach. Once again it is exposed to wave action, and the littoral current carries it along the coast until eventually it reaches Santa Monica.

That city also needed quiet water for a yacht harbor. Because of the difficulties associated with the damming of the sand by a conventional breakwater, the city built a wave barrier consisting of a straight line of rocks parallel to the shore

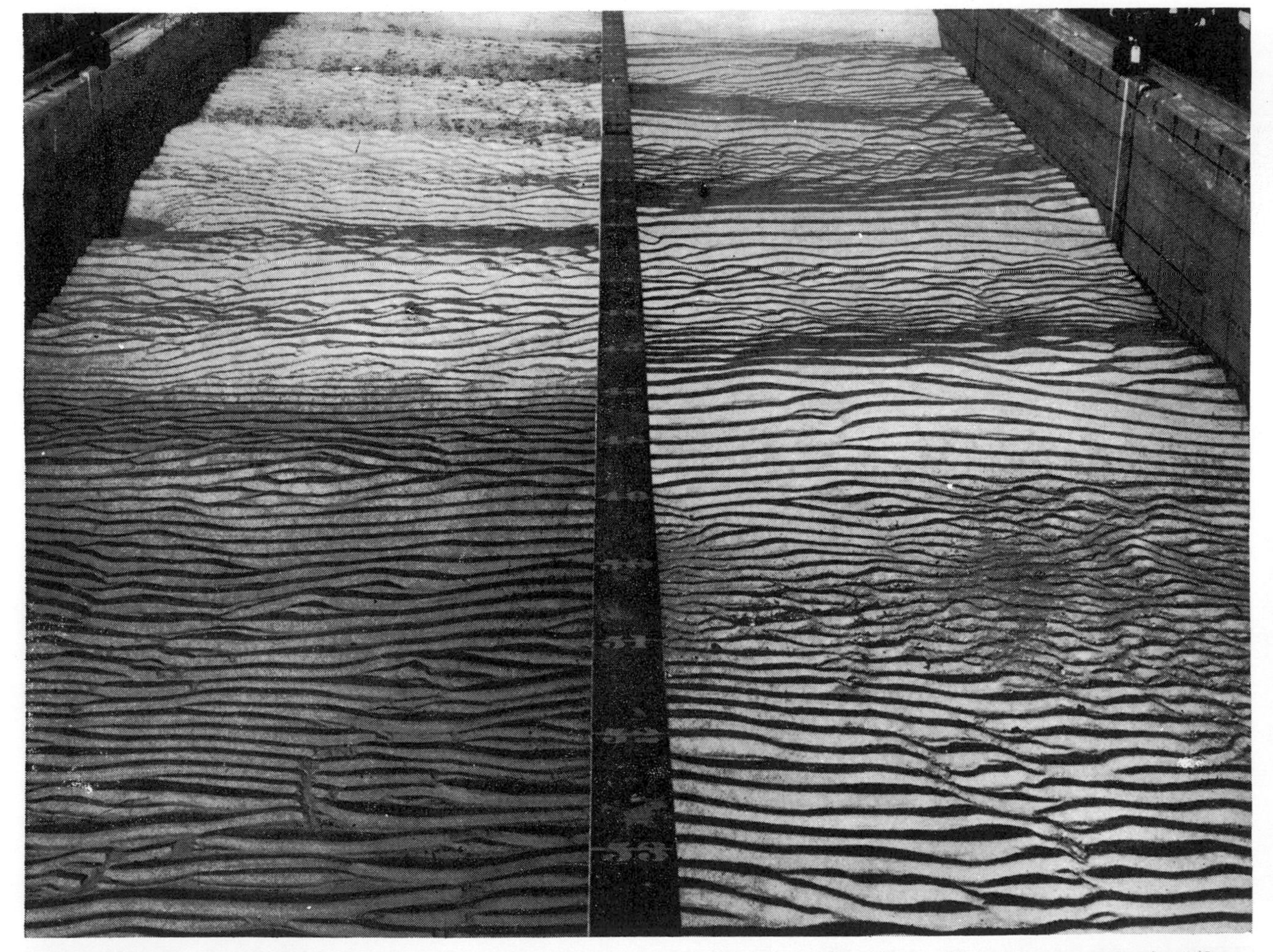

RIPPLE MARKS appear in the sand at the bottom of two experimental wave-tanks of the Beach Erosion Board in Washington, D.C. The tank at left contains fine sand; the tank at right, coarse sand. After the sand had been subjected to wave action for 60 hours, the tanks were drained and the configuration of the sand was photographed. Three sand bars have formed in the tank at right.

and several hundred yards out [*see bottom illustration on page 135*]. The sand was expected to flow past in the wide space between the breakwater and the shore. It did not. The sand simply stopped moving in the quiet water, and the beach started to build outward toward the rock wall. Downstream from the structure the beaches retreated. Now Santa Monica must also employ a dredge to put the sand back into circulation.

A similar littoral sand-transport system operates along the southern shore of Long Island on the East Coast. Prevailing winds and the unrefracted waves from the North Atlantic sweep the sand along the shore from Montauk Point at the island's eastern tip to the Rockaway spit at the entrance of New York Harbor. Montauk is rapidly eroding, and Rockaway spit is (or was for a considerable period before the present shore-line structures were built) building at the rate of 200 feet a year.

If no action is taken on erosion problems, everyone shares the erosion. But as soon as one part of the shore is protected, the remainder of the shore must supply the sand. Nevertheless for many years the customary way to stop the retreat of a beach was to build groins, that is, dams made of rocks or wooden piling a few feet high and a few hundred feet long, jutting out from the beach face to stop the passing sand.

Along some coasts groins have been constructed at regular intervals for many miles, each supporting a curving beach that spills over its end, giving the beach a cuspate appearance. The sand still flows, but it is retained temporarily on each little segment of beach. Groins can hold sand if they are properly engineered, but their effect is local and temporary. They are no longer the preferred means of maintaining a beach, for it has been found that they are usually less effective and more expensive than a "beach-nourishment" program. In such programs new sand is supplied to the system from inland dunes or from the bottoms of nearby lagoons. For example, the famous Waikiki Beach in the Hawaiian Islands was recently rebuilt with sand that was trucked in from dunes 14 miles away.

This change of opinion about the best way to maintain beaches is illustrated by the problem now facing the state of New Jersey. The configuration of the coast is such that refracted Atlantic Ocean swell strikes heavily on the New Jersey coast's most prominent point, near Barnegat Inlet. Littoral currents move the sand away from the point in both directions, and the point is eroding rapidly. In the past 50 years nearly $50 million has been spent on shore works in an attempt to stabilize the shore line. The present annual rate of expenditure is more than $2 million, and the results are not entirely satisfactory. Some parts of the shore have long since been stripped of sand; others are still retreating.

The Beach Erosion Board has studied the New Jersey problem and has proposed a project to develop adequate recreational beaches and to prevent further erosion. This project would nourish all the beaches along the coast by supplying new sand to the beaches in the vicinity of Barnegat Inlet. The sand would come partly by truck from inland locations and partly by pumping from Barnegat Bay; wave action and littoral currents would be relied upon to distribute it along the coast. The initial investment would be $28 million, but the program would require less than $1 million per year to maintain the beaches thereafter.

Sixty-six other shore-line construction projects, costing a total of over $100 million, have been planned for the shores of the U. S., and about half are completed or are well under way. The preservation of valuable coastal land, the maintenance of usable harbors and the development of recreational activities require an understanding of the ways of moving sand. In these enterprises the knowledge gained by the scientific study of beaches will play a central role.

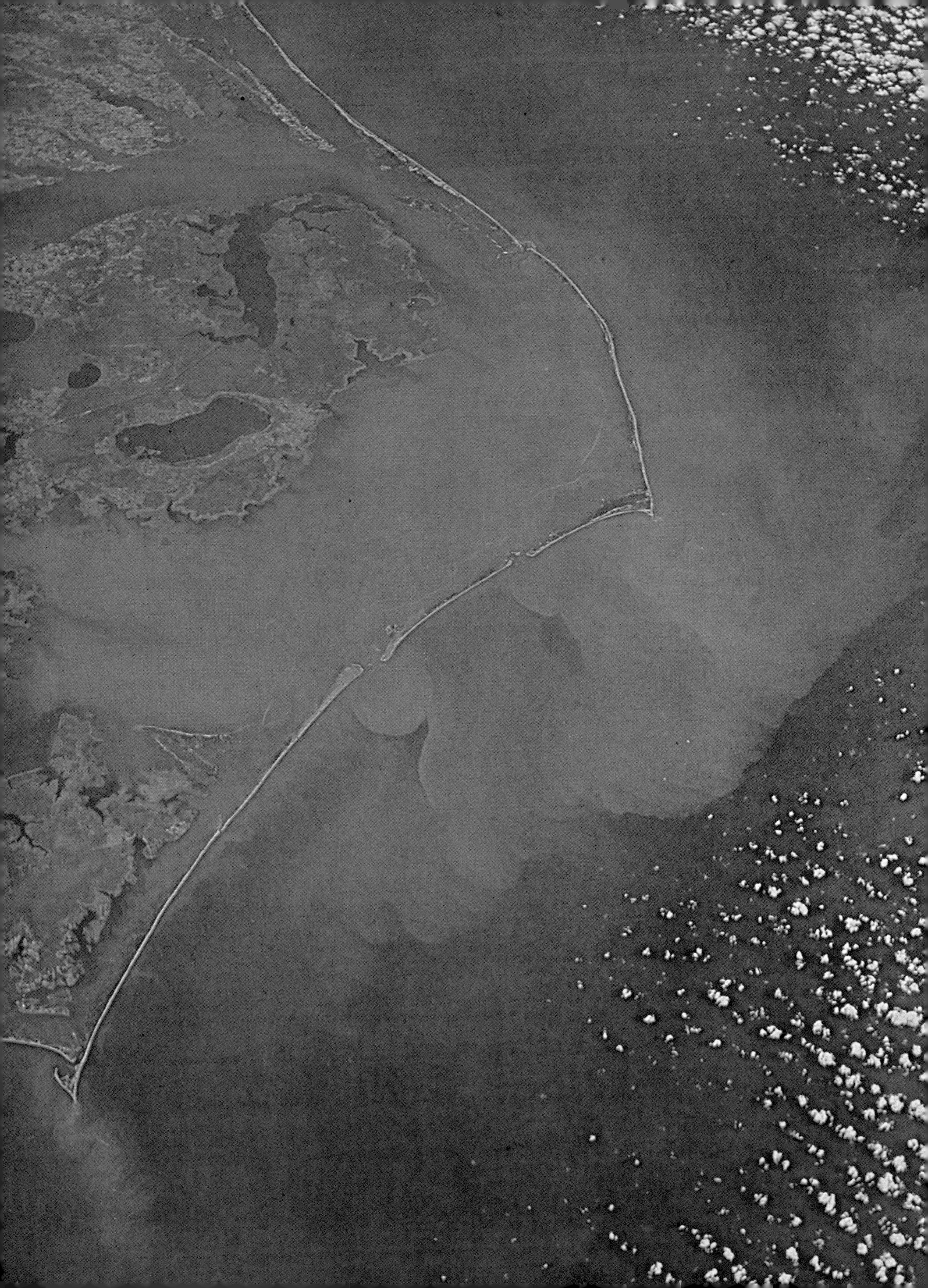

THE CONTINENTAL SHELVES

K. O. EMERY
September 1969

The continental shelves were the first part of the sea floor that was studied by man, chiefly as an aid to navigation and fishing. Perhaps the earliest recorded observation was one made by Herodotus about 450 B.C. "The nature of the land of Egypt is such," he wrote, "that when a ship is approaching it and is yet one day's sail from the shore, if a man try the sounding, he will bring up mud even at a depth of 11 fathoms." A more recent example is found in the diary of a 19th-century seaman: "An old captain once told me to take a cast of the lead at 4 a.m. We were bound to Hull from the Baltic. He came on deck before breakfast and on showing him the arming of the lead, which consisted of sand and small pebbles, I was surprised to see him take a small pebble and put it in his mouth. He tried to break it with his teeth. I was very curious and asked him why he did so. He told me that the small pebbles were called Yorkshire beans, and if you could break them you were toward the westward of the Dogger Bank; if you could not, you were toward the eastward."

Fishing success often depends on knowledge of the kind of bottom frequented by particular fish and on the avoidance of rocky areas that can catch and tear nets. As a result governmental agencies routinely chart bottom topography and materials to aid the fishing industry, but the successful fisherman generally keeps much additional information to himself. Similarly, the production of oil and gas from the continental shelves during the past two decades has led governmental and international agencies to make broad geological surveys, which help to guide the oil industry to the areas of greatest economic promise. The oil companies make studies that are much more detailed and so expensive that the results are considered proprietary, at least until after exploitation rights are secured.

During World War II submarines took a large toll of the ships that crossed the continental shelves, mainly at the approaches of ports. The effectiveness of acoustical detection equipment on submarine hunting ships was much increased by a knowledge of bottom materials and their effects: long ranges over sand, short ones over mud, confusing echoes over rock or coral. Accordingly charts of bottom sediments were compiled by the American and German navies for many areas of the world. This problem had not arisen earlier because both submarines and the search gear of surface ships were too primitive during World War I, and it may not be important in the future owing to the greatly increased sophistication of submarines and to the different role they may play in any future war.

The conflict between disseminating information and keeping it secret is about what one would expect in an environment that is both economically and militarily important. The recent political interest in the sovereignty of the ocean is also to be expected, considering the way the economic potential of the shelves has often been exaggerated in recent years. Thus it is not surprising that there have been a number of proposals to redefine the continental shelf so as to extend it seaward to whatever distance and to whatever depth are necessary to give a nation access to the resources presumably lying or hidden there. In 1953, before the world developed its present large appetite for seafood and minerals, an international commission defined the continental shelf, shelf edge and continental borderland as: "The zone around the continent extending from the low-water line to the depth at which there is a marked increase of slope to greater depth. Where this increase occurs the term shelf edge is appropriate. Conventionally the edge is taken at 100 fathoms (or 200 meters), but instances are known where the increase of slope occurs at more than 200 or less than 65 fathoms. When the zone below the low-water line is highly irregular and includes depths well in excess of those typical of continental shelves, the term continental borderland is appropriate."

Somewhat similar, but shorter, definitions are presented by most textbooks of geology. Where a depth limit is given it is 100 fathoms, an inheritance from the time when navigational charts had only three depth contours: 10, 100 and 1,000 fathoms. On a global basis the edge of the continental shelf ranges in depth from 20 to 550 meters, with an average of 133 meters; the shelf ranges in width from zero to 1,500 kilometers, with an average of 78 kilometers [*see illustration on next two pages*].

CONTINENTAL SHELF in the Atlantic Ocean off Cape Hatteras is delineated by puffy clouds that show where cold surface water on the eastern edge of the shelf meets warmer surface water. The boundary is near the edge of the shelf, which at this point averages about 120 meters in depth. The picture also shows turbid water moving from Pamlico Sound into the ocean, where it is carried northward by a fringe of the Gulf Stream that lies atop the shelf. The photograph was taken from *Apollo 9* on March 12, 1969, at an altitude of 134 statute miles; the view extends 175 miles in the north-south direction. The astronauts on this mission were James A. McDivitt, David R. Scott and Russell L. Schweikart.

The continental shelves underlie only 7.5 percent of the total area of the oceans, but they are equal to 18 percent of the earth's total land area. A geological understanding of this huge region requires a knowledge of its topography, sediments, rocks and geologic structure. For nearly all shelves there is some information about topography; for perhaps a fourth of them something can be said about the surface sediments, but the rocks and geologic structure are known for less than 10 percent. Detailed knowledge is far less available. The best-known large areas are the ones off the U.S., eastern Canada, western Europe and Japan—in short, the shelves next to countries where scientific knowledge is well developed and freely disseminated. Smaller areas of knowledge are found where oil companies have worked (such as parts of northern South America, parts of Australia and the Persian Gulf) and where oceanographic institutions have conducted repeated operations (northwestern Alaska, the Gulf of California, northwestern Africa, the shelf off Argentina, the Red Sea and the Yellow Sea). Recently some developing countries have effectively closed their shelves to foreign scientific studies; these areas are fated to remain unknown and unexploited for the foreseeable future.

The information that is most costly and most difficult to obtain concerns the underlying rocks and the structural geology of the continental shelves. This information is essential for understanding the origin and most of the history of the shelves. Data about the surface topography and sediments, which are readily accessible, tell only the late history.

Samples of bedrock have been dredged from the surface of many shelves, mainly from the top of small projecting hills and the sides and heads of submarine canyons that incise the shelves. Additional rock samples have come from the top of the adjacent continental slopes. Care must be taken in deciding whether the rock samples are from outcrops, whether they are loose pieces that were deposited by ancient streams or glaciers, or whether they were rafted to their present location by ice, kelp, marine animals or man. The decision is usually based on the size of the piece, on the presence or absence of fresh fractures, on the similarity of lithologic types within a given dredging area or between adjacent dredging areas, and on the amount of tension of the dredge cable. It is helpful to have submarine photographs, which may reveal rock outcrops in the dredging area. Rock also can be sampled by coring: by dropping or forcing a heavily weighted pipe into the bottom. This method can show the dip of the strata, and it can sample rock that is covered by sediments if the sediment is thin or if it is first removed by hydraulic jetting. A better but more expensive method of rock sampling is provided by well-drilling methods. Many holes have been drilled for geological information in shelf areas of structural interest; they provide good information on the sequence and depth of strata, the date of original deposition and geologic structure. In addition several thousand oil wells have been sunk into shelves, but most have been drilled in abnormal geologic structures such as salt domes and folds.

Geophysical methods provide excellent, although indirect, data from which geological cross sections can be constructed. These methods include seismic reflection and refraction, as well as measurements of geomagnetism and gravity. Each method has its advantages, but the most generally successful one is seismic reflection. In practice a ship traverses the shelf and produces a loud acoustical signal in the water at intervals of a few seconds. The chief source of sound energy a few years ago was a chemical explosive, usually dynamite; other sources are now preferred because they are

CONTINENTAL SHELVES underlie about 7.5 percent of the total ocean; all together they occupy an area roughly equal to that of Europe and South America combined, or some 10 million square miles. The shelf is defined as the zone around a continent extending from

cheaper, easier to trigger accurately and greatly reduce the danger both to the operators on the ship and to the fish in the sea. These newer energy sources include electric spark, compressed air and propane gas. Although part of the sound energy is reflected from the sea floor, much of it enters the bottom to be reflected upward from various layers of rock under the bottom. The reflected energy is received by hydrophones trailed behind the ship; the signal is amplified, filtered and recorded on continuously moving paper tapes. This method, termed continuous seismic-reflection profiling, is rapid and can yield information from depths of several kilometers under the sea floor, making it possible to construct geological cross sections. When the interpretations are supplemented by dredging or drilling, they provide the best information now available about the structure of continental shelves.

When existing geological and geophysical information is assembled on a worldwide basis, it shows that continental shelves can be classified into two main types by composition: those that are underlain by sedimentary strata and those underlain by igneous and metamorphic rocks. A large majority of the world's continental shelves mark the top surface of long, thick prisms of sedimentary strata [*see illustration on next two pages*]. Many of the prisms are held in position against the continents by long, narrow fault blocks. Such is true of almost the entire perimeter of the Pacific Ocean, where tectonic activity has also produced deep trenches that are parallel to the base of the continental slope. In some areas, such as the West Coast of the U.S., a single geologic dam is known to have extended for thousands of kilometers along the coast. Locally part of the dam rises above sea level to form the granitic Farallon Islands that lie immediately off San Francisco. These rocks are some 100 million years old, but they were thrust up to form the dam only about 25 million years ago. Elsewhere, as in the Yellow Sea of Asia, half a dozen such fault dams or fold dams

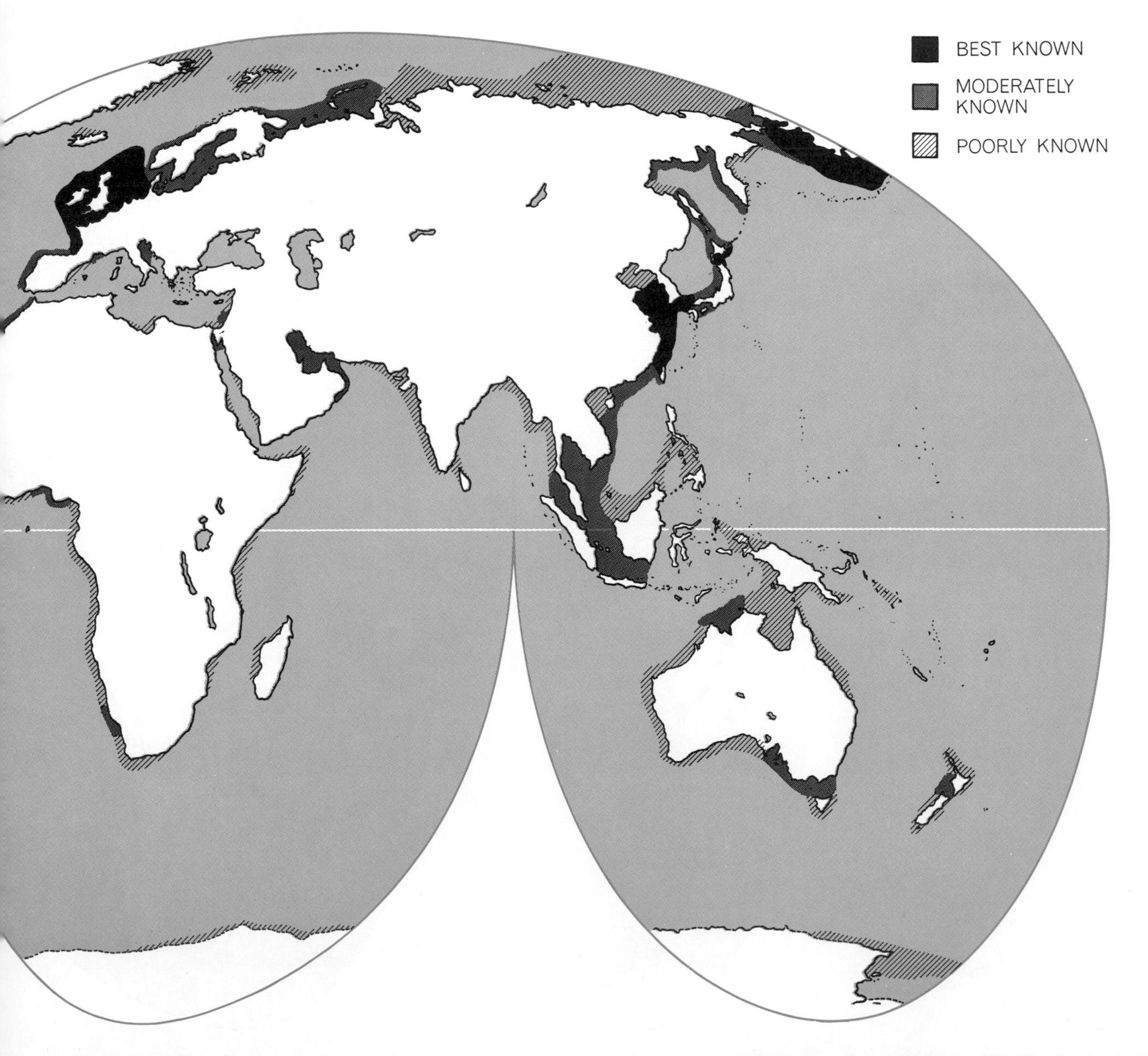

the low-water line to the depth at which the ocean bottom slopes markedly downward. Conventionally the edge of the shelf is taken to lie at 100 fathoms, or 200 meters, but a more accurate average value for all continents is about 130 meters. Worldwide the shelf has an average width of 78 kilometers. The illustration indicates the present state of knowledge of the continental shelves of the world.

have risen in the last 500 million years or so, each dam in turn causing the ponding of sediments from the land. There was a similar dam off the entire length of the East Coast of the U.S. between 270 and 60 million years ago; in time the trench on its landward side was filled with sediments that subsequently spilled over the former dam to build a continental slope that is held in place only by the angle of rest of the sediments. That this angle is unstable is shown by numerous landslides and erosion features recorded in seismic profiles across the continental slope and rise.

The continental shelf off the western part of the Gulf Coast of the U.S. is held in place by a diapir dam: a dam formed by the upward movement of salt from a bed that is buried several kilometers deep and is about 150 million years old. Seismic profiles and dredgings show the presence of still another kind of dam in the eastern Gulf of Mexico and off the southeastern coast of the U.S. This is an algal reef that dates from 130 million years ago and was succeeded by a coral reef off Florida at some time before 25 million years ago; even today the Florida keys are bordered by a living coral reef. Similar tectonic and biogenic dams elsewhere in the world have trapped huge quantities of sediments in the geological past.

Shelf areas underlain by igneous and metamorphic rocks are found on top of the tectonic dams. Off Maine, however, glacial erosion has removed the sedimentary rocks that once covered such a dam [*see illustration on pages 148 and 149*]. Other shelves underlain by igneous and metamorphic rocks are known, but most of them appear to be at high latitudes where glacial erosion has been effective. Nevertheless, even at high altitudes probably most of the shelves are underlain by sedimentary rock. In a sense we can consider the shelves whose shape is chiefly due to glacial or wave erosion as youthful ones (or rejuvenated ones); the shelves that are mainly depositional, with a thick prism of sediments on top of igneous and metamorphic rocks, can be regarded as mature ones. The sediments have built the shelves upward during the concurrent sinking of the edges of the continents. Perhaps more important from the viewpoint of real estate, these sediments have increased the size of the continents, widening them as much as 800 kilometers in areas where rivers have brought much sediment from the land and where tectonic or other dams effectively prevent the escape of the sediment to the ocean basins. This dammed sediment, as well as the sediment held only by its angle of rest, has an estimated average thickness of about two kilometers, yielding a total

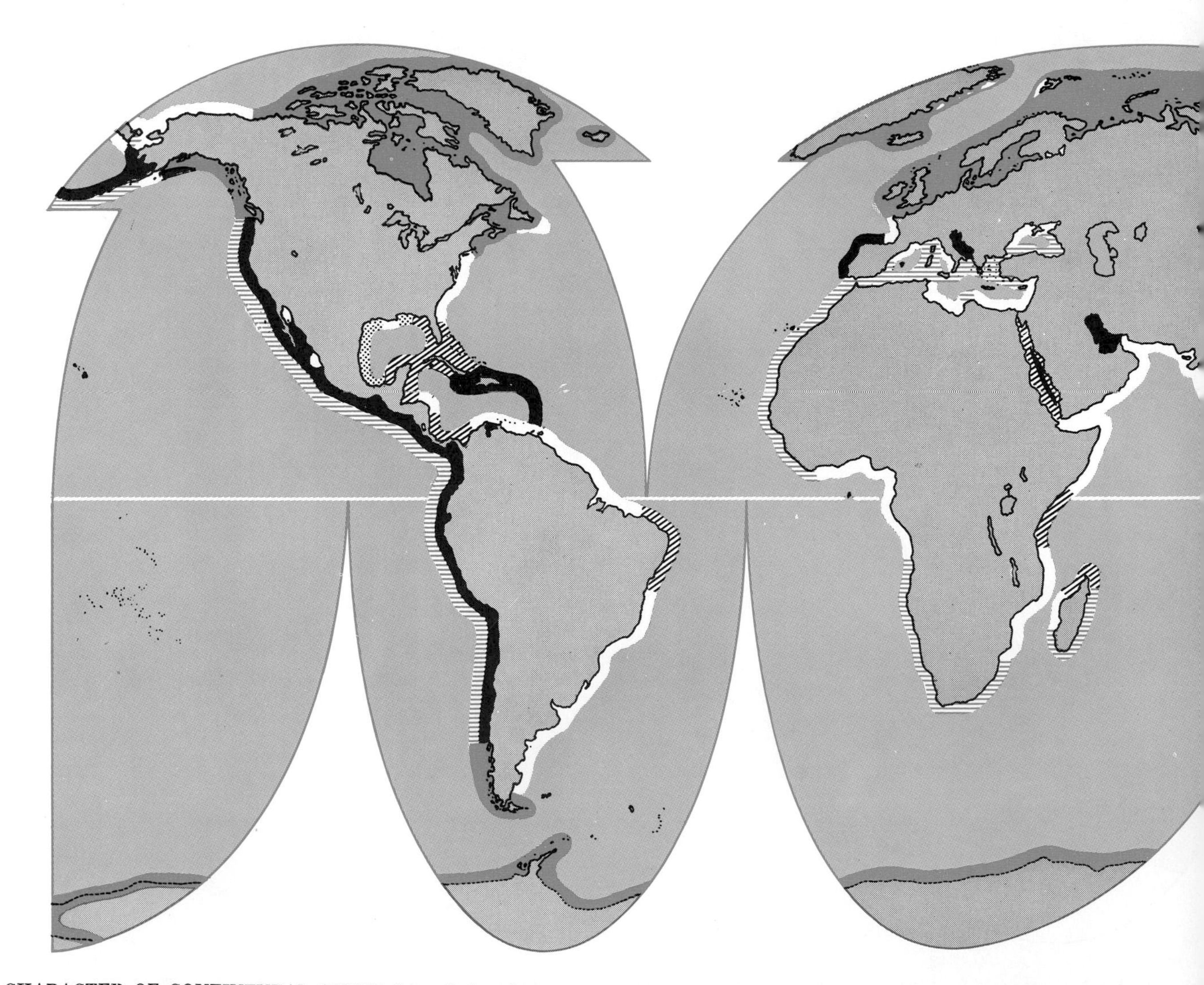

CHARACTER OF CONTINENTAL SHELF depends largely on how the shelf was formed. Six types of shelf, classified by origin, are indicated in this worldwide map. Many shelves are deposited behind three kinds of dams: tectonic dams, formed by geological uplift or upwelling of lava; reef dams, created by marine organisms, and diapir dams, which are pushed upward by salt domes.

volume of sedimentary strata under continental shelves of about 50 million cubic kilometers.

Perhaps the most dramatic period in the history of the continental shelves was the million-year passage of the Pleistocene epoch when the sea level changed in response to the waxing and waning of the continental glaciers. At their maximum the glaciers appear to have been so extensive as to have stored in the form of ice enough water to lower the surface of the ocean nearly 150 meters below the present level. Four major lowerings of the sea level were produced by the four main glaciations, with minor lowerings caused by secondary fluctuations of climate and ice volume. Limited investigations with special seismic equipment off the East Coast of the U.S. show four or five somewhat irregular acoustical reflecting surfaces near the top of the shelf sediments. These reflecting surfaces probably can be explained by erosion and sand deposition at stages of low sea level. Cores from these beds probably would provide much interesting information about Pleistocene climates and Pleistocene chronology. For the present, however, our data for glacial effects on the continental shelf are restricted largely to surface sediments and topography.

About 50 years ago most textbooks of geology led the reader to believe that sediments became progressively finer in texture with distance from shore: gravel and sand at the shore, coarse sand grading to fine sand across the shelf, and finally silt and clay (the "mud line") at the shelf edge. Bottom-sediment charts compiled during World War II, however, showed that this simple pattern is very rare, and that the size of sediment grains is unrelated to the distance from shore. The examination of actual samples showed that most of the shelves are floored with coarse sands that commonly are stained by iron and contain the empty shells of mollusks that live only close to shore in shallow depths. Broken shells or shell sand are particularly abundant at the outer edge of the shelf and on small submerged hills that are relatively inaccessible to detrital sediments. Some of these same areas contain glauconite and phosphorite, minerals that are precipitated from seawater, but so slowly that they are obscured or diluted beyond recognition where detrital minerals are present. The only areas that exhibit a consistent seaward decrease of grain size are those between the shore and depths of 10 or 20 meters—in short, whatever areas are shallow enough to be ruled by the waves. At greater depths the sediments are too deep to be reached by new supplies of sand. These sections

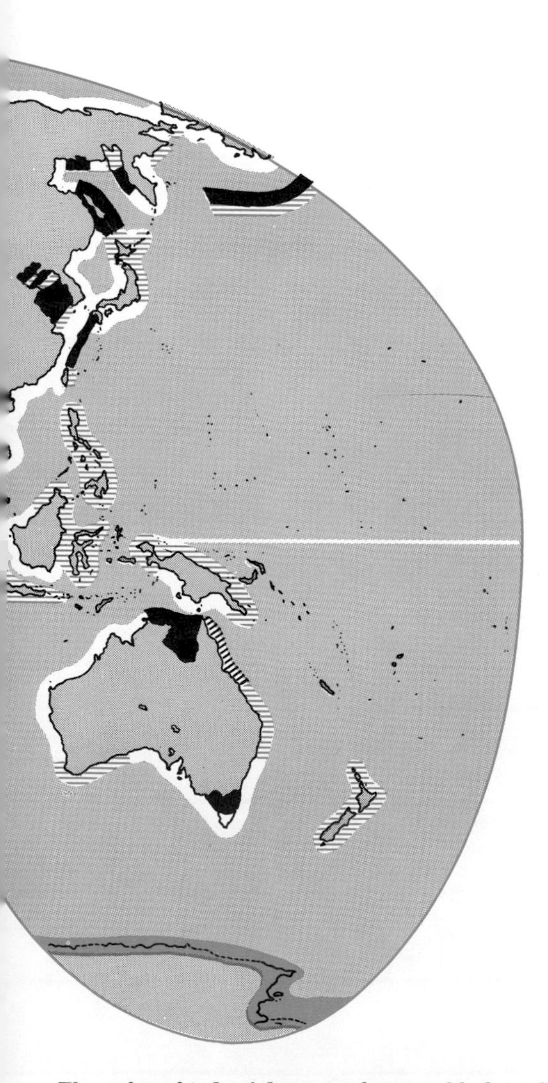

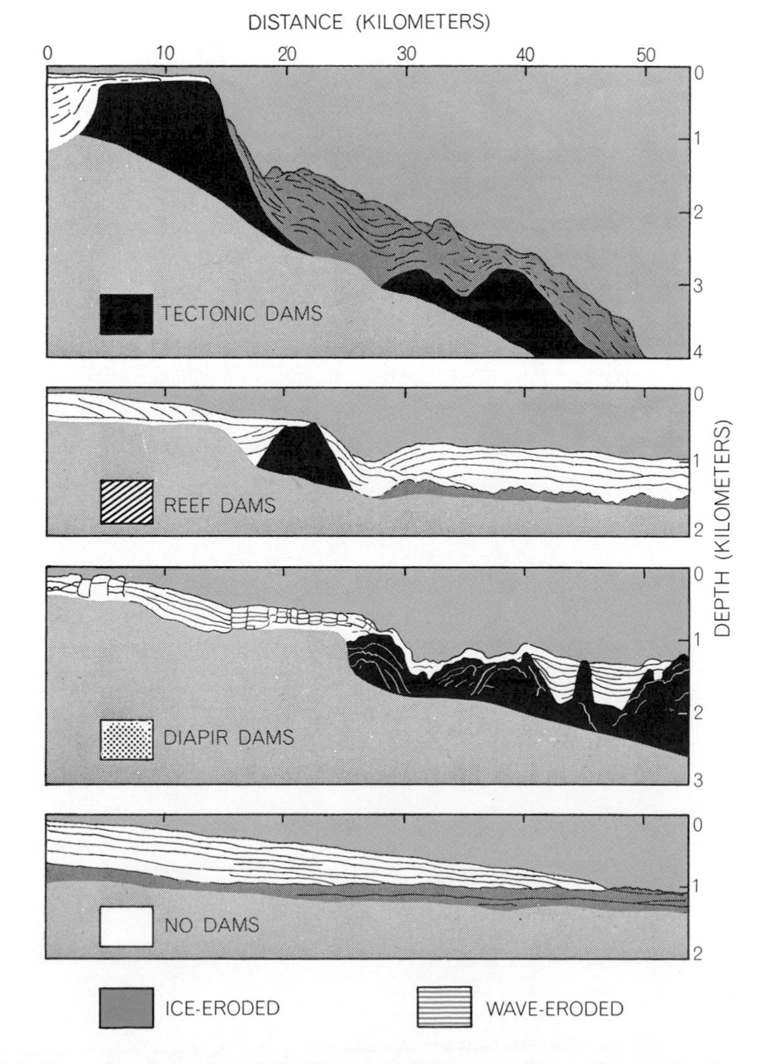

These three kinds of dams are shown in black in the typical shelf cross sections at the right; a simple damless shelf is also depicted. The vertical scale is exaggerated six times. Sediment deposited before formation of the dam is shaded gray; sediment deposited subsequently is unshaded. All four kinds of shelf structure may be eroded by waves (*hatched color on map*) or by ice (*solid color*).

of the shelves are also bypassed by contemporary silts and clays that remain in suspension en route to deeper or quieter waters.

The sediments on about 70 percent of the world's continental-shelf area have been laid down in the past 15,000 years, since the last glacial lowering of the sea level. The rest of the shelf is floored by silts at the mouths of large rivers, in quiet waters behind barriers and in shelf basins, by recent shell debris and by chemically deposited minerals. This means that when the sea level was low, the entire shelf was exposed and the rivers deposited sands on the then broader coastal plain and transported their silts and clays to the ocean. At that time ocean waves, with no shelf to reduce their height, were probably higher at the shore than they are today, with the result that shore sediments were probably coarser. The broad expanse of lowland favored the development of ponds and marshes, which were partly filled with debris from the forests and meadows that extended unbroken from the inland areas across what is now the sea floor. Freshwater peat now submerged in the ocean has been sampled at 10 sites off the eastern U.S. and at many other sites on the shore; similar peats have been found off Europe, Japan and elsewhere.

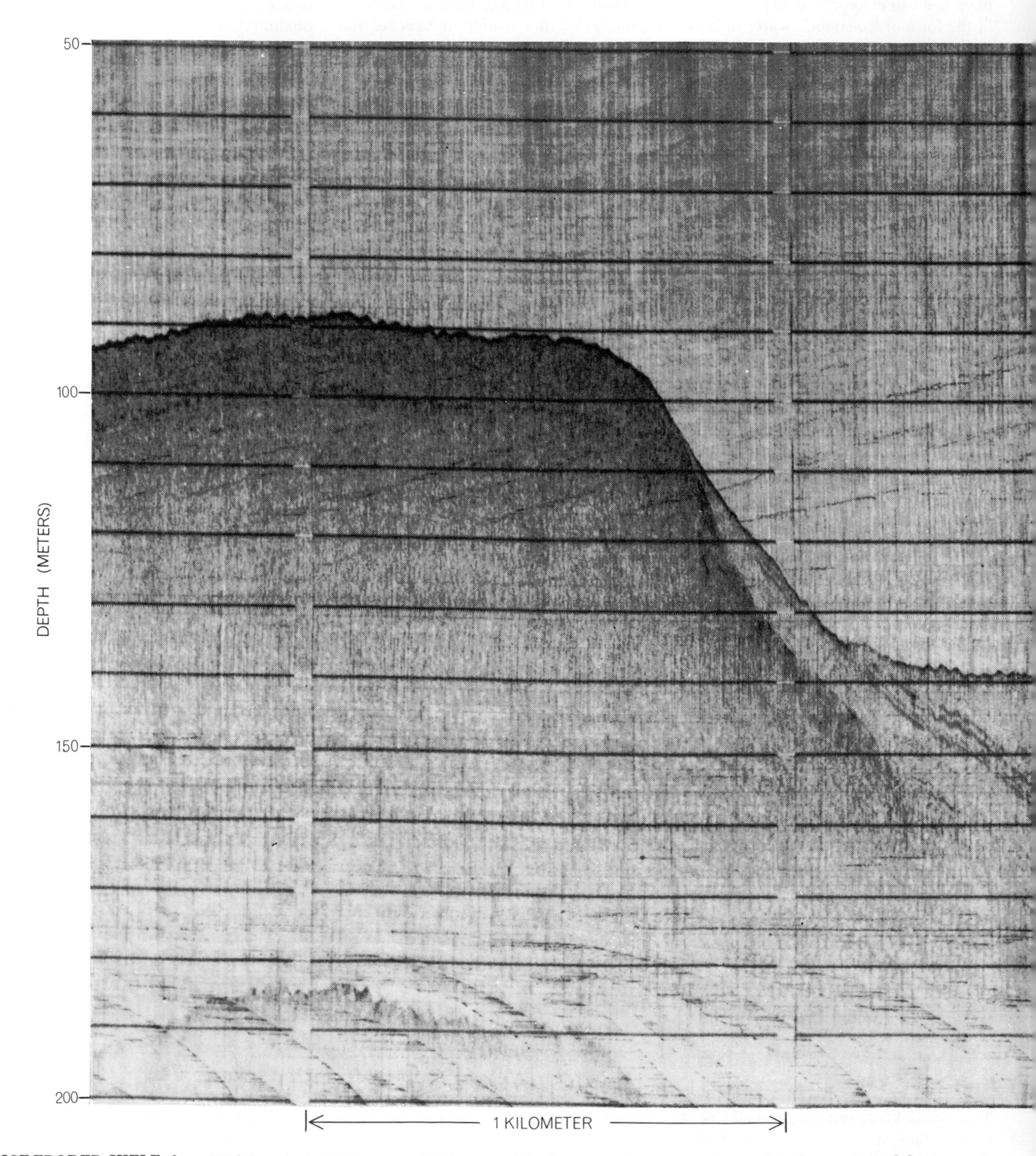

ICE-ERODED SHELF about 100 kilometers off the coast of Maine, landward of Georges Bank, is shown in this seismic-reflection record. The deep trough was gouged out of the basement rock by ice some 15,000 years ago during the last glaciation. Subsequently the trough

Pollen analysis shows a succession from tundra to boreal spruce and pine some 12,000 years ago, followed by oak and other Temperate Zone deciduous trees about 8,500 years ago; the deciduous trees flourished until the site was submerged. Birds once flew among the trees in many areas where fish now swim.

The vegetation attracted many animals, but only their heavier bones are preserved or are readily detected by dredging. Nearly 50 teeth of mammoths and mastodons have been collected off the East Coast of the U.S., along with the bones of the musk ox, giant moose, horse, tapir and giant ground sloth. Similar finds have been reported off Europe and Japan.

Carbon-14 dates have been obtained for more than 50 samples of shallow-water material from the shelf off the East Coast of the U.S. The materials include salt-marsh peat, oölites (concentrically banded calcium carbonate pellets that typically form only in warm, shallow, agitated seawater) and the shells of oysters and other mollusks (which live in only a few meters of water but whose empty shells are found as deep as 130 meters). The dates and depths make it possible to draw a curve showing the changes of sea level in an-

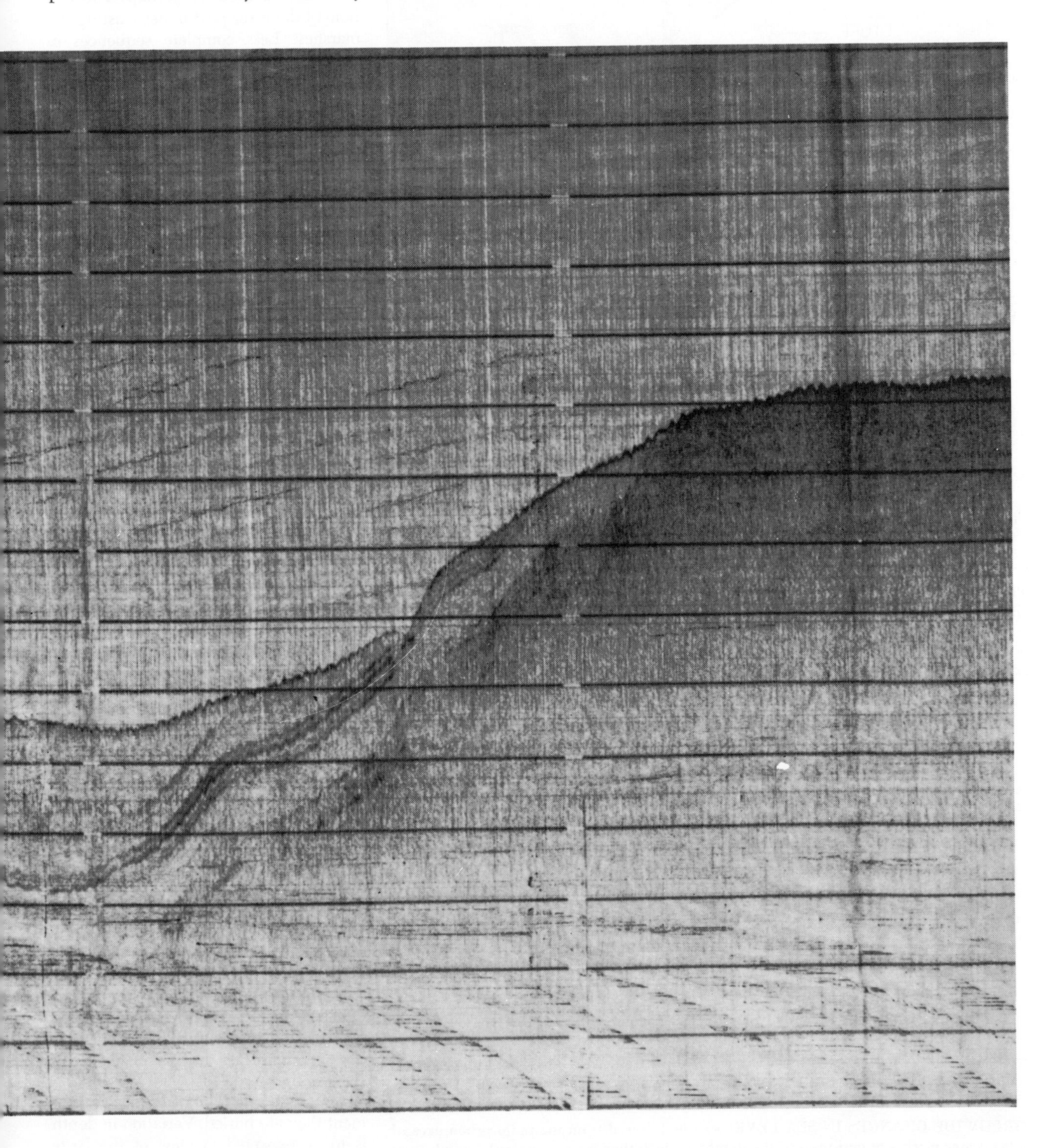

became filled with sediments about 30 meters thick. The recording was made this past July with high-frequency seismic equipment aboard the *Dolphin*, a vessel operated by the U.S. Geological Survey in cooperation with the Woods Hole Oceanographic Institution.

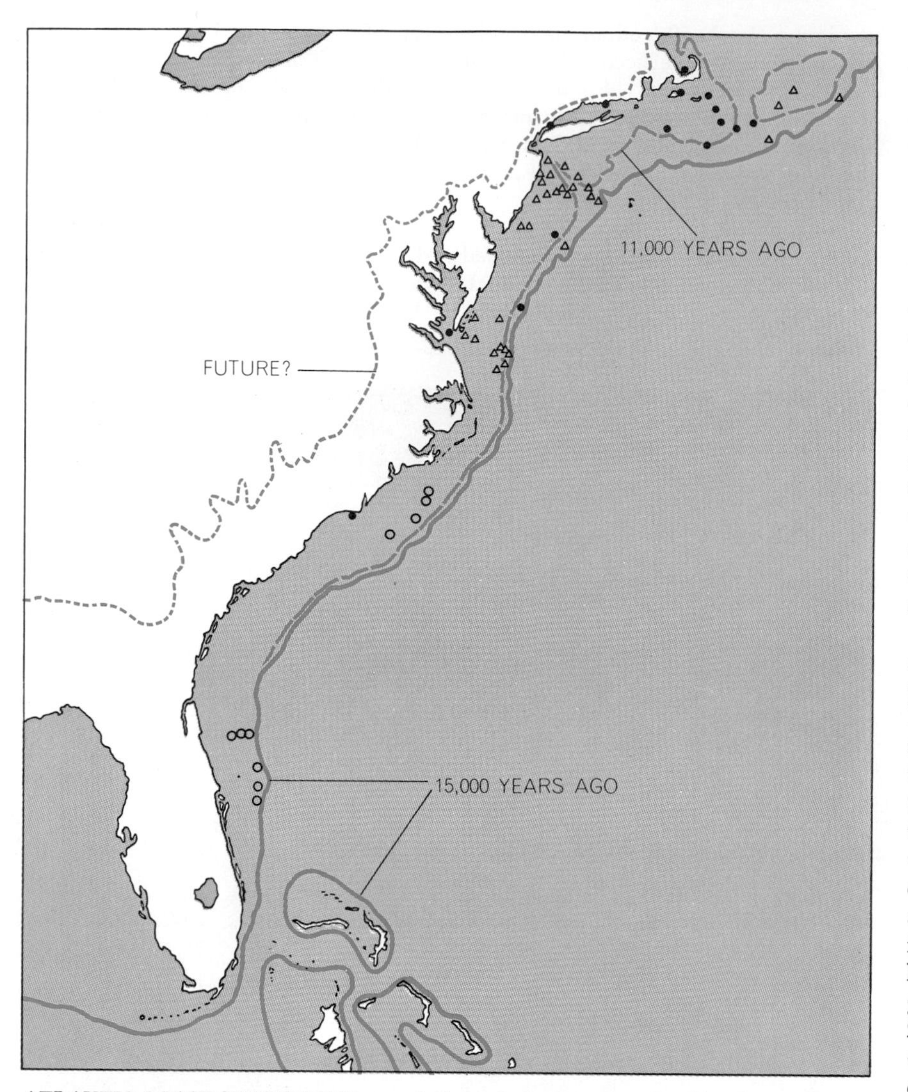

ATLANTIC COAST SHORELINE has varied greatly in the past and will undoubtedly continue to in the future. This illustration compares the shoreline of 15,000 and 11,000 years ago with the probable shoreline if all the ice at the poles were to melt. Confirmation that the continental shelf was once laid bare is found in discoveries of elephant teeth (*triangles*), freshwater peat (*dots*) and the shallow-water formations called oölites (*circles*).

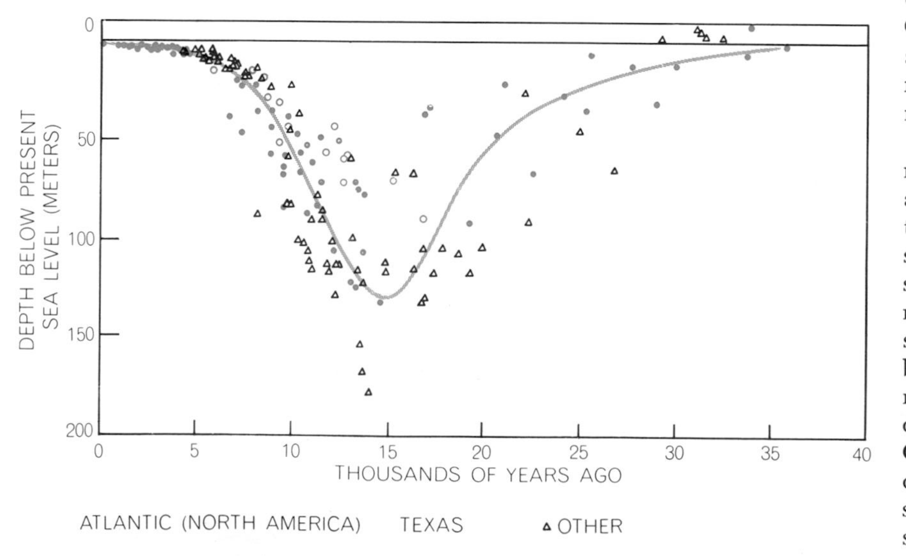

WORLDWIDE CHANGES IN SEA LEVEL can be inferred from the radiocarbon ages of shallow-water marine organisms and the depth at which they were recovered. Samples are from the Atlantic shelf of North America, the Texas shelf and other parts of the world. The depth inconsistency of the Texas samples implies that the shelf there has been uplifted.

cient times [*see bottom illustration at left*].

Apparently the sea was near its present level about 35,000 years ago and began to recede about 30,000 years ago. The level dropped by 130 meters or more 15,000 years ago; then it rose rather rapidly to within about five meters of the present level 5,000 years ago. The slow rise during the past 5,000 years has been documented by perhaps 100 carbon-14 dates for peat under existing salt marshes. Less complete sequences of dates for similar samples from elsewhere in the world show a sea-level curve resembling the one for the East Coast. Only the samples from the shelf in the western Gulf of Mexico provide a different curve, which suggests that this part of the gulf shelf was uplifted about 40 meters during the past 10,000 years.

Early men of the Clovis culture (characterized by fluted stone projectile points) appeared in North America some 12,000 years ago, when the sea level was still very low. What is more reasonable than to suppose such men ranged over the forested lowland that is now continental shelf? Game, fish and oysters were abundant. How were they to know or care that in a few thousand years the area was to be drowned by the advancing sea, any more than New Yorkers know or care that when the remaining glaciers melt, the ocean will rise to the 20th story of tall buildings? [*see top illustration at left*]. The search for traces of early man far out on the shelf began with the discovery of what may be the remains of an oyster dinner on a former beach off Chesapeake Bay, a site that is now 43 meters below sea level. This discovery was made from the Woods Hole Oceanographic Institution's research submarine *Alvin;* many similar discoveries will probably be made during the next decade.

Submerged barrier beaches are common on the continental shelf, but they are easily confused with the sand waves that are formed by strong currents. More spectacular and of certain origin are the submerged sea cliffs and terraces that mark the temporary stillstands of the sea level. Most of the shelves that have been studied have four to six such terraces, but the recognition of the terraces depends on their width and sharpness. On gently sloping shelves the terraces are almost imperceptible; on steep shelves they are narrow or absent; on shelves receiving a large supply of sediment they are buried. Variation in depth is to be expected in view of the large variation in depth of the most prominent terrace of all—the edge of the shelf. Passing through the terraces are channels cut by streams that flowed across the shelf when the sea level was low. Most of these channels have been filled by sediment; they can be recognized only by seismic profiling and by drilling on the shore at the mouths of stream valleys. Probably hundreds of channels cross the continental shelves of the U.S., but only a dozen are known. One channel, the one cut by the Hudson River off New York, is so large it is not yet filled with sediment.

At the seaward end of the channels, near the edge of the shelf, the channels are replaced by the heads of submarine canyons that continue down the continental slope to depths of several kilometers. The continuation of the submarine canyons to depths far below the maximum probable lowering of the sea level means that the canyons must have been formed by some process that operates under the ocean surface. Although the matter is still the subject of debate, most of the evidence favors the view that the canyons were excavated by turbidity currents: currents that arise when sediment slips down a slope and becomes mixed with overlying water, thereby increasing its density so that it continues down the slope, often at high speed. Today the shelf off the East Coast of the U.S. is only slightly modified by submarine canyons; only the heads of the canyons indent the shelf edge. When the sea level was at its lowest, the canyons were probably important factors in sedimentation. The shelf off the West Coast of the U.S. is so narrow that the heads of many canyons reach almost to the shore. In those areas the canyons serve to trap and divert sand that is moved along in the shore zone under the influence of wind-driven waves and their associated currents. As a result the sand that is brought to the shore by streams and cliff erosion is only temporarily added to the beaches; eventually it moves seaward through the canyons in the form of slow sand glaciers or rapid turbidity currents.

The water above the continental shelves is complex in composition and movement because it is shallow and close to the land. Large rivers contribute so much fresh water that they dilute the ocean, but they also increase the local concentration of calcium, phosphate, silica and nitrate—precisely those elements and compounds that elsewhere in the ocean have been reduced to low concentration by incorporation into marine plants and animals. Continental-shelf waters that are distant from river mouths are sometimes saltier than the open sea because their rate of evaporation is high. Local variations in salinity (and therefore density) control the direction of currents on the shelf. For example, the low salinity at the mouth of a river means a higher sea level near the shore than farther out on the shelf, leading to a flow toward the right (when one is facing the ocean in the Northern Hemisphere).

Just at the shore, however, the longshore currents are mainly controlled by the angle at which waves intersect the beach, which in turn is a function of the wind direction. As a result the cur-

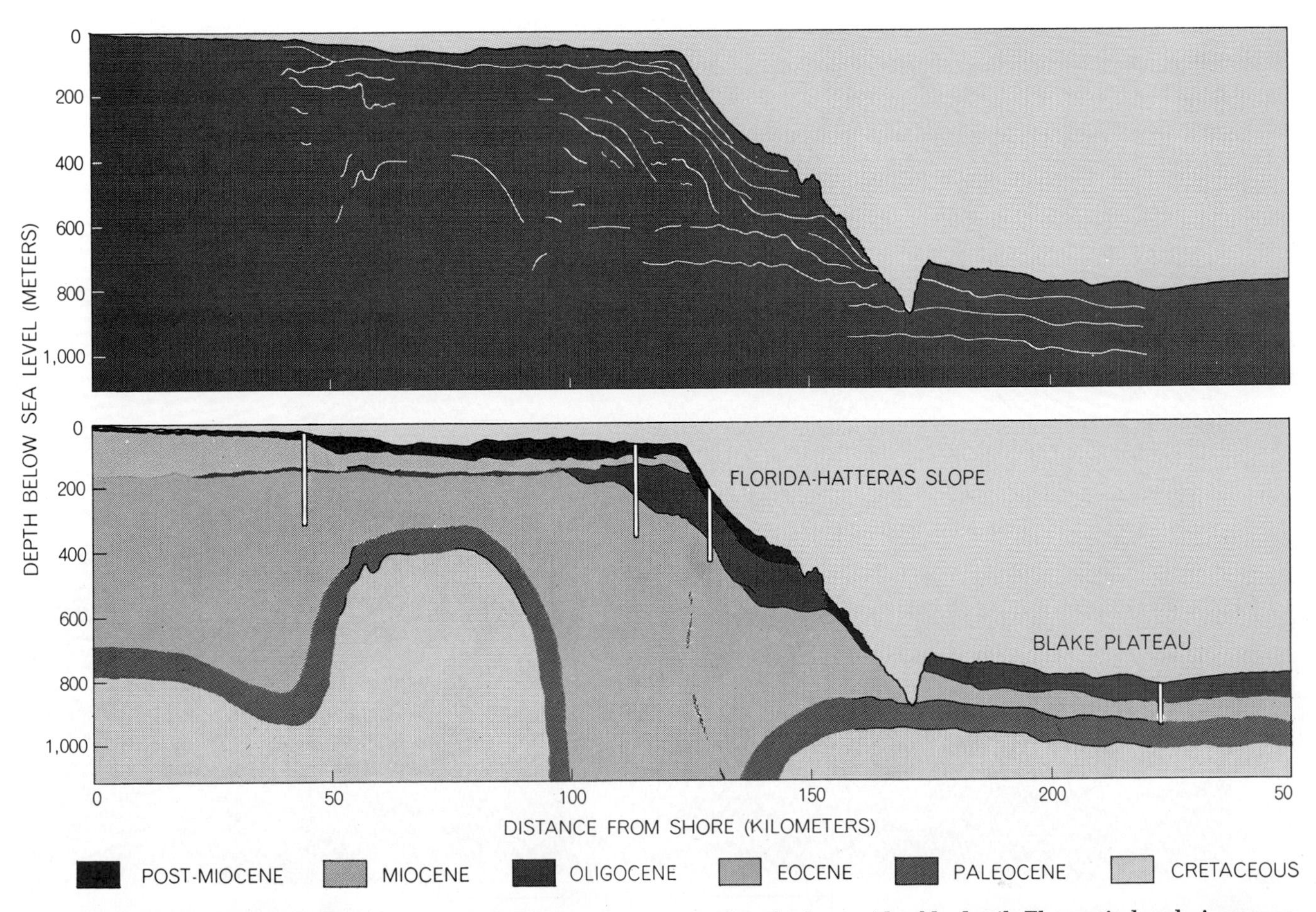

SHELF OFF JACKSONVILLE, FLA., has been studied by two geophysical methods: seismic reflection (*top*) and drilling (*bottom*). Seismic studies can show only the general nature of the stratigraphy. Cores obtained by the JOIDES project, a drilling study conducted by a consortium of institutions, made it possible to map the stratigraphy in considerable detail. The vertical scale is exaggerated 67 times. The approximate termination dates for the various geologic periods are as follows: Miocene, 10 million years ago; Oligocene, 25 million years ago; Eocene, 40 million years ago; Paleocene, 55 million years ago; Cretaceous, 65 million years ago.

rent in the wave zone may be northward, the current on the inner half of the shelf may be southward and the current on the outer shelf may be northward again (for example where an oceanic current such as the Gulf Stream runs along the edge of the shelf). Where rivers bring water to the ocean there must be a general current component toward the ocean at the surface; this induces a return flow toward the land at the bottom [*see top illustration on page 153*]. Thus the sediment on the sea floor may be moved landward often working its way into the mouths of estuaries. This means estuaries are truly ephemeral features, receiving sediments from both rivers and the open shelf.

Temperature zones on land are mainly a function of latitude, with secondary modifications resulting from winds whose direction may change seasonally or may be controlled by topography. Similarly, ocean water is cooled at high

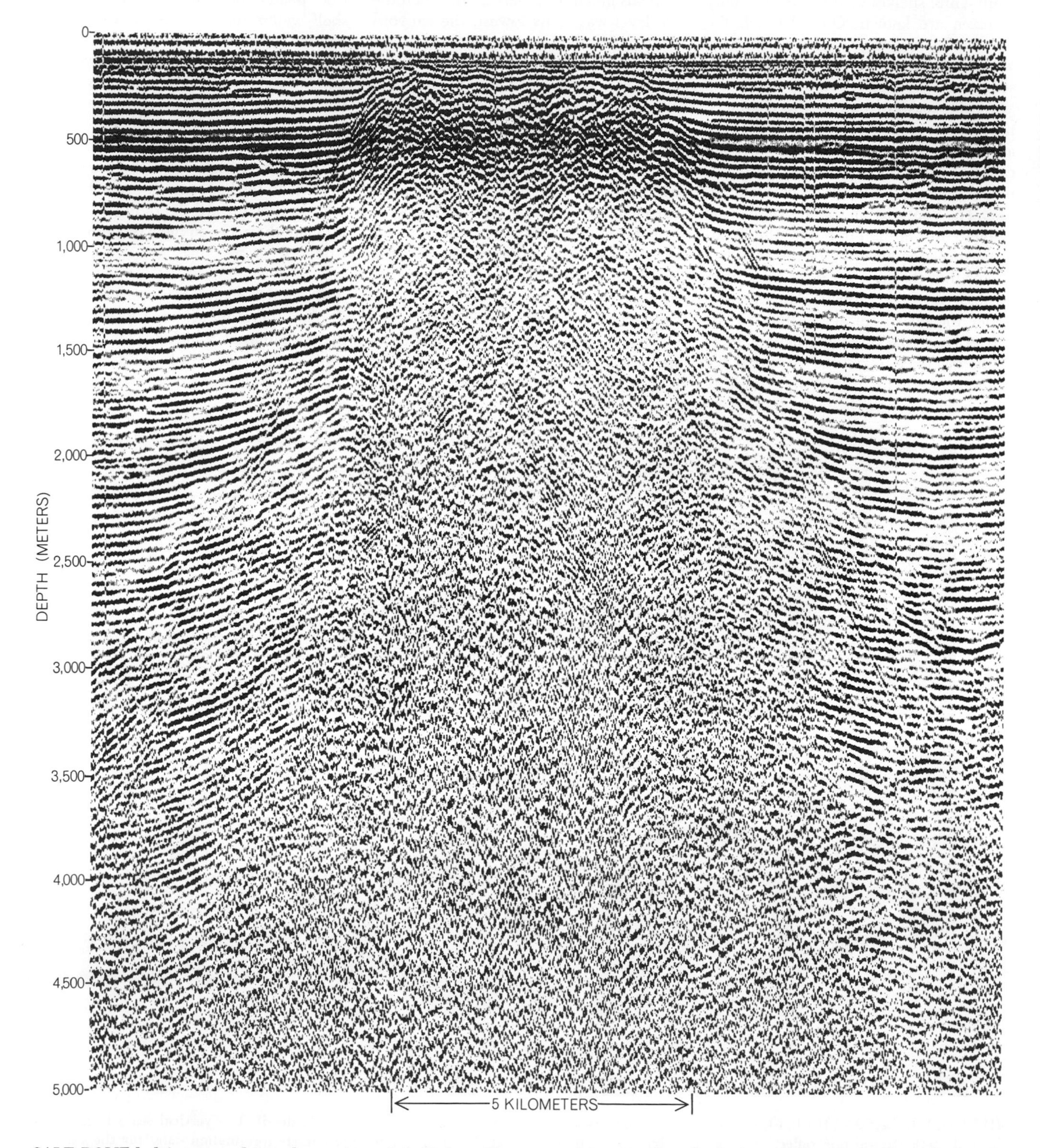

SALT DOME bulging upward into the continental shelf about 10 miles south of Galveston in the Gulf of Mexico is shown in this seismic record. The water is so shallow (between 10 and 20 meters) that the reflection from the surface of the shelf is virtually at the top of the recording. Geologists can discern significant features in such a record down to a depth of about 3,000 meters. The record was made by the Teledyne Exploration Company. The salt dome was subsequently drilled and was found to contain hydrocarbons.

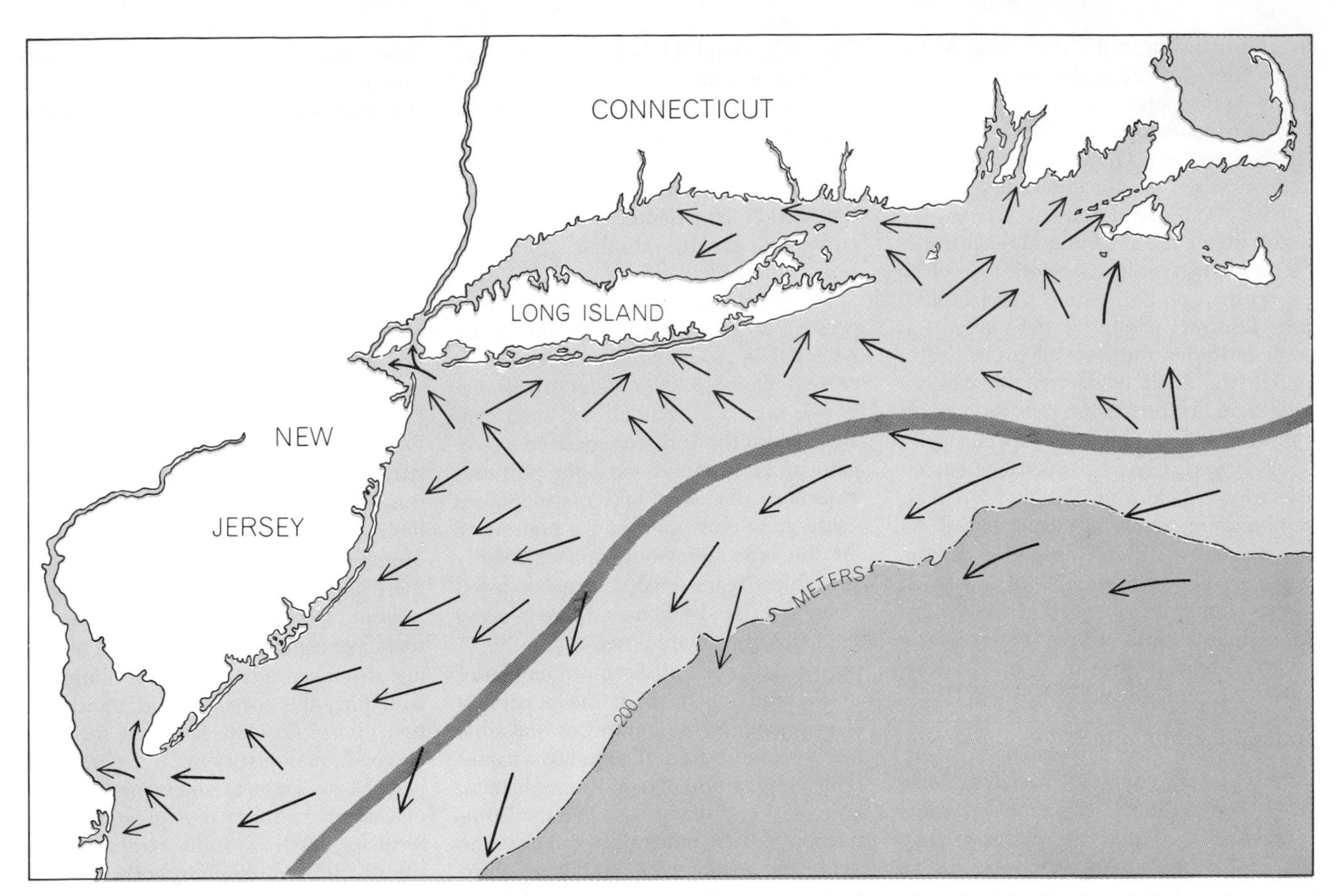

BOTTOM CURRENTS, indicated by arrows, can be traced with the help of simple plastic devices called bottom-drifters. The gray band marks the boundary between landward flow and seaward flow. The broken line represents the edge of the continental shelf.

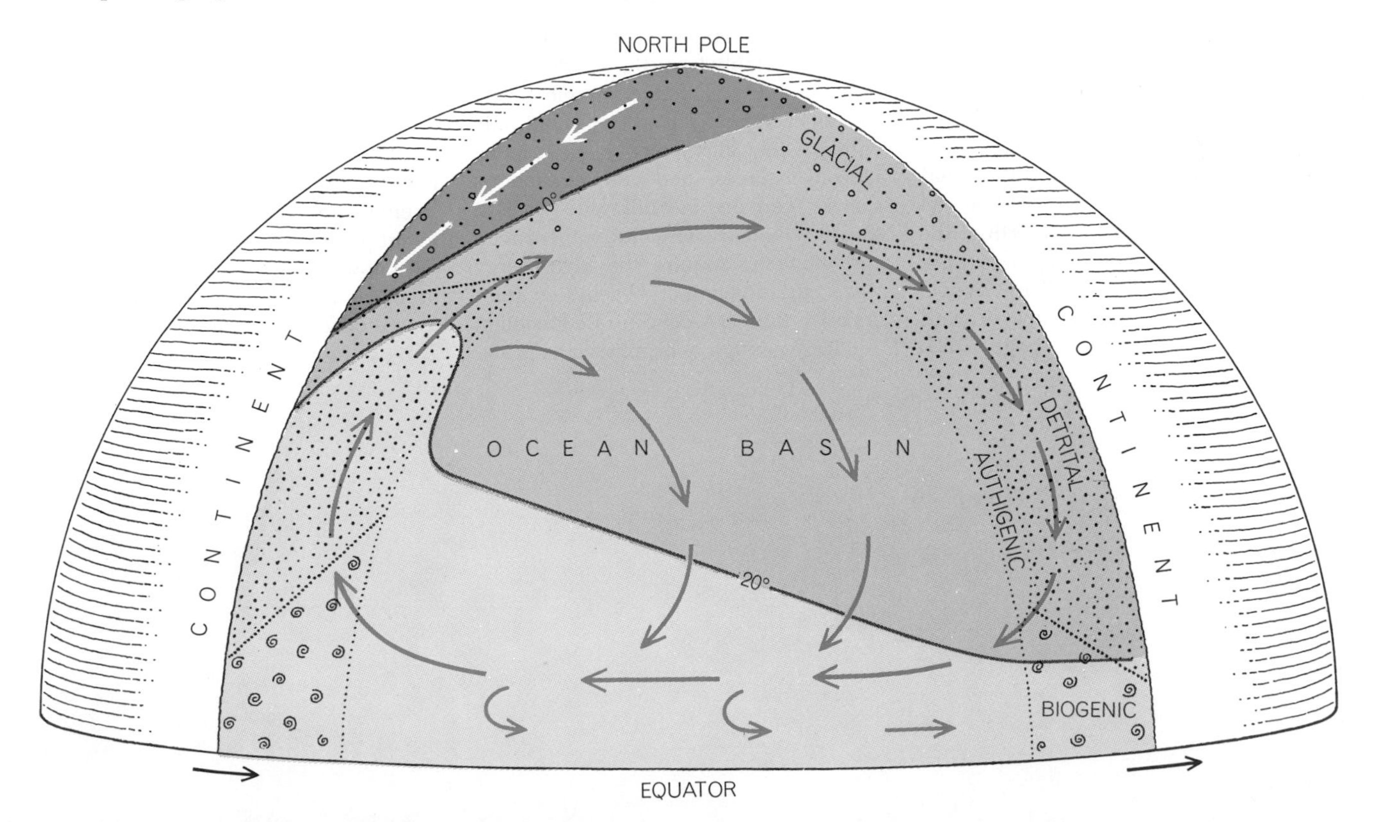

CHARACTER OF SHELF SEDIMENTS around an ocean basin, shown here schematically, is heavily influenced by oceanic currents. The rotation of the earth produces a clockwise flow in the Northern Hemisphere, so that the western edge of the basin, up to a certain latitude, is warmer than the eastern edge. Thus biogenic sediments extend farther north on the west than on the east. The effect would be greater except for a counterflow of arctic water on the western side of the basin. Detrital sediments are the typical outwash of continents. Authigenic sediments are minerals that come out of solution under suitable conditions and fall to the ocean floor.

latitudes and heated at low ones. At the same time, however, the pattern of currents in the open ocean displaces the climatic zones in a clockwise direction in the Northern Hemisphere and counterclockwise in the Southern Hemisphere. The displacement causes the water above the shelf at middle latitudes to be warmer on the western side of an ocean than on the eastern side [*see bottom illustration on preceding page*]. At high latitudes the flow of arctic water makes the shelf colder on the western side than on the eastern one. As a result the correlation of animal species with shelf latitude shows a displacement on the opposite sides of oceans. Moreover, temperature zones are compressed on the western side and expanded on the eastern. The movements of currents, waves and tides above the shelves are so complex that they have received little study compared with those of the deep ocean. Much fieldwork is needed.

The present great interest in exploring the world's continental shelves flows from their potential economic exploitation. About 90 percent of the world's marine food resources, now extracted at the rate of $8 billion per year, comes from the shelves and adjacent bays [see the article "The Food Resources of the Ocean," by S. J. Holt, beginning on page 356]. Most of this is fish for human and animal consumption; the remainder is largely used for fertilizer.

Second in economic importance is petroleum and natural gas from the shelf; their present annual value is about $4 billion, representing nearly a fifth of the total world production of these substances [see the article "The Physical Resources of the Ocean," by Edward Wenk, Jr., beginning on page 347]. Currently about $1 billion worth of oil and gas a year is extracted from the shelves off the U.S., and much of the rest was developed by American companies with interests abroad. It is safe to predict that the future production from the world's continental shelves will increase at a greater rate than production from wells drilled on land.

The third marine resource in terms of present annual production and future potential is lowly sand and gravel. At present about $200 million worth per year is mined for landfill and road construction in the U.S., for concrete aggregate in Britain and for both purposes elsewhere. As cities and megalopolises continue to grow and show a preference for the coastal regions, and as readily available stream deposits are exhausted or are overlain by houses, there is every sand and gravel will increase greatly. prospect that the offshore production of

We read much about the possibility of economic exploitation of valuable heavy minerals from the sea floor, namely ilmenite, rutile, zircon, tin, monazite, iron, gold and diamonds. The total production of these minerals from below the sea is now less than $50 million per year. Production may increase, particularly in the case of tin, but it is decreasing for iron. Prospects for gold are not very hopeful, and diamonds have never been mined profitably from the sea floor. The basic problem is that economic placer deposits of tin and gold are found only within a few kilometers of the original igneous sources, and few continental shelves contain metalliferous igneous sources. Similarly, ilmenite, rutile, zircon and monazite require the high-energy wave environment of beaches to form deposits that are concentrated enough and large enough to be mined at a profit. When ancient beach deposits are submerged, even if they are not buried under worthless sediment or mixed with it, the cost of mining increases substantially. They will probably be mined in the future but not until they are economically competitive with shoreline deposits. This could come about either through a rise in prices, resulting from a diminution of known deposits on land, or when more efficient mining and separation methods are devised for the marine environment.

Phosphorite is present in large quantities on shelves off southern California, Peru, southeastern Africa, northeastern Africa and Florida. It can be mined off the U.S., but it has to compete with high-grade land deposits in Florida, Montana, Idaho and Wyoming (where there is about a 1,000-year supply at present rates of mining). Most investigators have concluded that the cost of mining at sea exceeds the cost of mining on land plus the costs of land transportation. Some deposits far from the U.S., however, may justify mining, particularly because some of them may be near places where there is a great need for fertilizer, such as India. Unfortunately the distribution of phosphorite in these areas is poorly known, and little or no effort is currently being expended on their investigation.

The would-be exploiter of the ocean will do well to remember the words of the old Newfoundland skipper, "We don't be *takin'* nothin' from the sea. We has to sneak up on what we wants and wiggle it away." Nevertheless, the continental shelves, when they are properly investigated, promise to greatly increase our knowledge of the earth's history and to become a steadily more important source of food and raw materials.

16

THE ORIGIN OF SUBMARINE CANYONS

BRUCE C. HEEZEN
August 1956

During the last 30 years or so the ocean bottom has become the main frontier of exploration on the surface of our planet. The vast area hidden in the darkness of the deep sea (nearly three quarters of the earth's surface) is slowly being explored and mapped for the first time, lately with the cogent help of the echo sounder. It has been an adventure full of surprises. We have "seen" enough to realize that if the oceans were somehow drained of water, they would expose a landscape as varied as the continents—mountains taller than Everest, plains as vast as the Russian steppes, gorges rivaling the Grand Canyon.

All this has brought a crowd of exciting questions. We know that the topography of the continents has been shaped in large part by river and wind erosion. What forces could have molded features so similar in the deep sea? Of all the enigmas, none has been more spiritedly debated than the mystery of the undersea canyons [see "Submarine Canyons," by Francis P. Shepard; SCIENTIFIC AMERICAN, April, 1949]. Off the shores of all the continents the underwater slopes are cut by great channels which look exactly like river canyons on land. They course through the sea bottom for hundreds of miles and invariably debouch onto a broad plain, like a river mouth opening to a delta. What process cut these gorges in the ocean? Many theories were proposed: all had a touch of the incredible. Now it seems that the question can at last be answered.

Two hypotheses have received foremost consideration. One suggested that the canyons were cut by actual rivers during ice ages when the oceans were lower than now and the continental shelves stood above sea level. This theory has always faced the formidable objection that it demanded the removal of a seemingly incredible amount of water from the oceans (by way of glacial ice piled on the continents). The hypothesis now seems to have been squelched by certain recent discoveries: namely, that some submarine canyons, notably the one extending from the mouth of the Hudson River, run out to deep ocean basins which could never have stood above water.

The other hypothesis is that the canyons were eroded by submarine flows, or avalanches, of silt-laden water, called "turbidity currents." This is an old theory with a checkered career. But the idea has been confirmed and built into a convincing picture by a series of rather dramatic discoveries and tests in the past decade.

More than three quarters of a century ago a Swiss engineer suggested that a canyon in the bed of Lake Léman might have been cut by heavy, silt-laden water flowing in along the bottom of the River Rhône. This suggestion received scant attention, but 20 years ago the Harvard University geologist Reginald A. Daly revived the idea as an explanation of the canyons in the oceans. His prestige encouraged several investigators to make experiments with models to test the possibility of turbidity currents. They obtained some evidence, though not conclusive, that such flows could occur. Moreover, engineers working on Boulder Dam observed some natural turbid underflows in the newly created Lake Mead behind the dam. Submarine geologists remained unconvinced, however, that this process could account for the huge canyons in the sea. It was difficult to believe that currents of muddy water along the ocean bottom could be powerful enough to erode these great gorges.

Then the matter was given an entirely new slant by discoveries from another quarter. In the summer of 1947 the research vessel *Atlantis* surveyed the deep-sea bottom between Bermuda and the Azores with two new instruments developed during the war: a continuously recording echo sounder which traces the profile of the ocean floor, and a corer which takes samples of the sea bed 40 to 60 feet thick. The *Atlantis* discovered a vast plain more than 200 miles wide in the deepest part of the western Atlantic. Cores extracted from this plain yielded a major surprise: whereas the deep-sea bed elsewhere consisted of thick deposits of fine oozes and clays, here the bed was mainly composed, down to the depth of the 15-foot core, of sand and silt very like the material on the beaches of New England!

The significance of these facts was not realized at first. But it became clearer after another exploration by the *Atlantis* in the summer of 1949. This voyage traced the Hudson submarine canyon for some 200 miles from the edge of the North American continental shelf to the western Atlantic plain mentioned above. Cores taken along this route showed that the canyon had been eroded through beds of clay many millions of years old, and that the floor of the canyon contained not only sand and gravel but also shells of clams which grow only in shallow waters near shore. The canyon's floor and walls, like the plain on which it debouched, were overlaid with only a thin layer of recent ooze.

All this strongly suggested that the canyon and the plain had been formed by a submarine river or currents of some kind. One could imagine that during the Ice Age streams of water from melting glaciers washed vast quantities of sand and gravel from the interior of

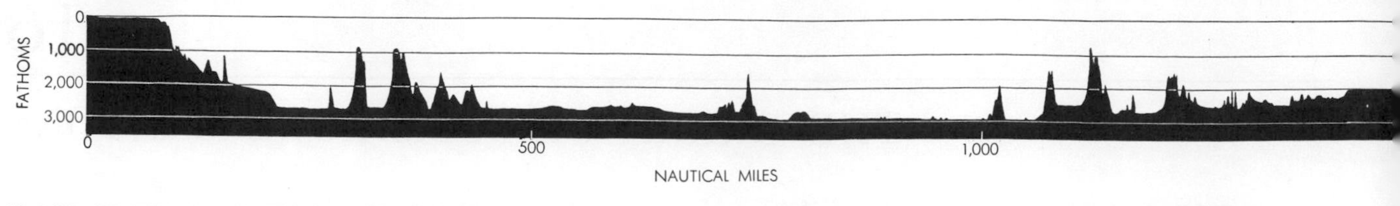

FLOOR OF THE ATLANTIC from Martha's Vineyard (*left*) to Gibraltar (*right*) is traced in profile. The abyssal plain, which extends from about 750 miles to about 1,000 miles off the U. S. East Coast, is remarkable for its flatness and for the sand and silt deposits

North America to the Atlantic coast, that masses of this material in the water flowed by gravity as currents down the slope at the edge of the continental shelf, that in so doing they eroded canyons in the slope, and that eventually the submarine streams deposited their sand and gravel in the deep-sea basins, building plains as a river builds a delta.

Such currents need not necessarily have been confined to glacial ages. On a smaller scale, one may suppose, occasional flows of turbid water may still take place on the sea bottom today. Sand and gravel accumulating on the steep continental slope, for example, might slide down the slope as avalanches from time to time. How could one find evidence of such a flow? It occurred to Maurice Ewing and the writer that the telegraph cables lying across the floor of the Atlantic might give a clue. Many of the transatlantic cables have been there for nearly a century. If strong submarine flows of the kind imagined do in fact occur, some of them may have broken the cables. An inspection of the record of cable breaks was obviously called for.

We did not have to look far. On November 18, 1929, a severe earthquake shook the Grand Banks in the Atlantic south of Newfoundland—an area which has a greater concentration of submarine cables than any other in the world. The cables broke on a wholesale scale, and the breaks were naturally attributed at the time to the quake itself. But a study of the timetable of the breaks discloses a remarkable fact. While the cables lying within 60 miles of the epicenter of the quake broke instantly, farther away the breaks came in a delayed sequence. For more than 13 hours after the earthquake, cables farther and farther to the south of the epicenter went on breaking one by one in regular succession. Each break was downslope from the one before, and the last took place in the deep ocean basin 300 miles from the epicenter.

It seems quite clear that this series of events must indicate a submarine flow: the quake set in motion a gigantic avalanche of sediment on the steep continental slope which broke the cables one after another as it rushed downslope and flowed onto the abyssal plain. Fortunately the automatic machines monitoring telegraph transmission recorded the precise time of the break in each cable, so that we can calculate the speed of the turbidity current that flowed down the slope. It ranged from about 50 miles per hour on the steep slope to about 15 miles per hour on the plain.

The deduction that such a current flowed over the cables is supported by reports of engineers on the cable repair ships, who had much difficulty finding the broken cables (presumably because they were deeply buried) and had to replace lost sections as much as 200 miles long. But for conclusive proof of our deduction, we decided to examine the scene itself. If our interpretation was correct, there should be a layer of silt of a certain thickness on the sea floor where the deepest cable break had occurred; Philip H. Kuenen of the Netherlands indeed calculated just how thick this layer should be. For several years hurricanes and instrument failures thwarted our efforts to get cores at this site, but we were finally able to obtain

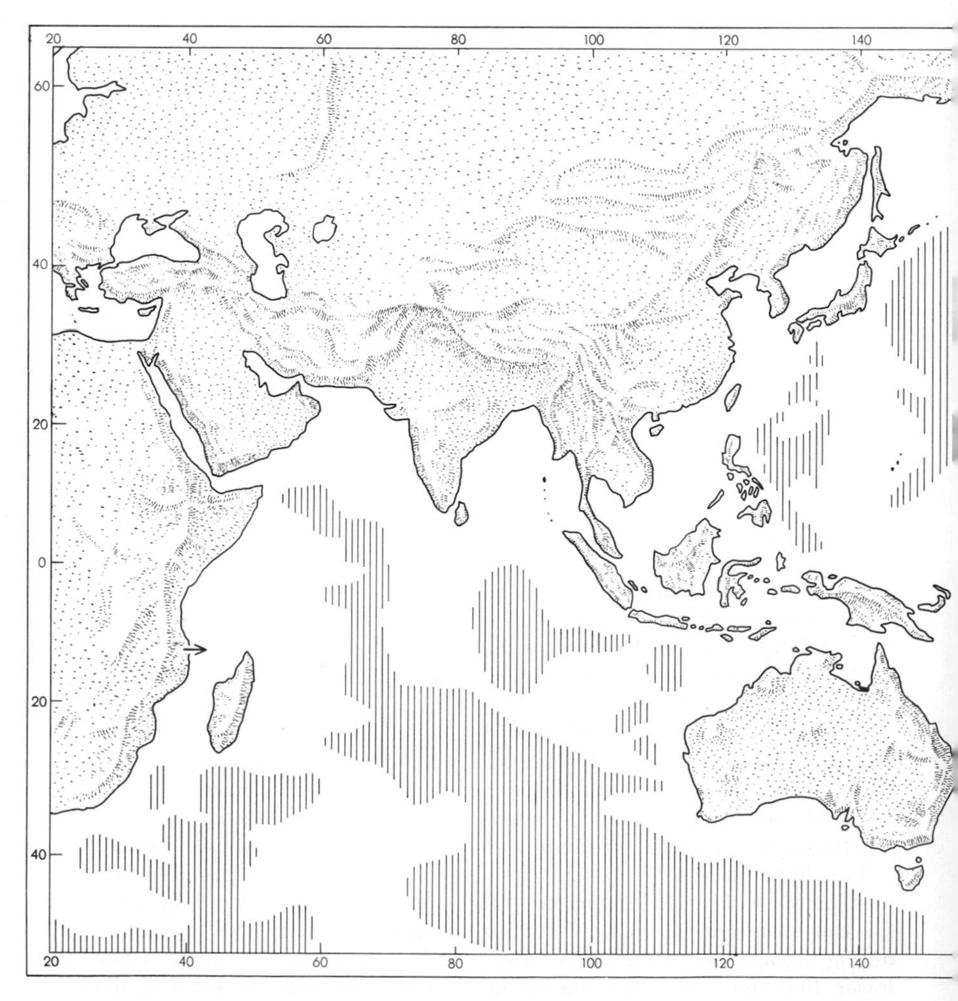

HISTORICAL TURBIDITY CURRENTS appear as arrows on this map of the oceans of the world. The shaded areas represent regions which are inaccessible to turbidity currents

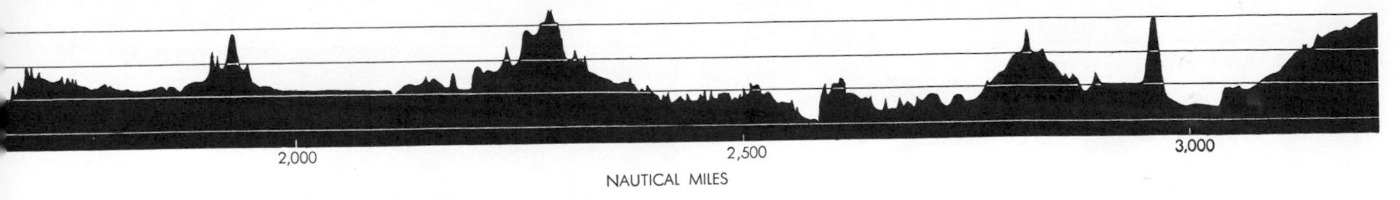

typical of shallow-water areas. These deposits were presumably washed out onto the plain by turbidity currents. The only one of the numerous seamounts to reach above the ocean surface is represented by the Azores. To the left of the Azores is Atlantis Seamount.

some in 1952, and much to our delight the thickness of the silt proved to be exactly as predicted.

In 1954 another earthquake in the Mediterranean yielded evidence almost as impressive as the case at Grand Banks. On September 9 of that year the disastrous quake centered at Orléansville in Algeria caused the breakage of five cables on the floor of the Mediterranean, lying at distances from 40 to 70 miles from the North African coast. The timing of the breaks showed that these, too, broke in succession according to their distance downslope in the sea; the presumed turbidity current traveled at from 40 to 5 miles per hour. Near the deepest break the repair crew found sand—apparently washed down from the coast.

In the Caribbean off the mouth of the Magdalena River of Colombia there is a network of submarine canyons. A submarine cable which passes across these canyons has been broken 17 times since it was laid in 1930. On two of these occasions jetties flanking the river mouth slid into the sea. It can be concluded that the earth slides that carried away the jetties were transformed into turbidity currents which flowed down the canyon and deposited their load of sediments in the sea basin. Recent explorations have shown that this basin has an extremely smooth floor and contains sands, shallow-water shells and debris of land plants.

Evidence of the existence of turbidity currents has accumulated from many other sources. Several years ago the *Atlantis* took two cores from the bottom of the Puerto Rican Trench—the deepest place in the Atlantic Ocean. The deposit here included limy shells of shallow-water organisms, and, most important, it contained fragments of limy marine plants which require sunlight for life. This material, transport-

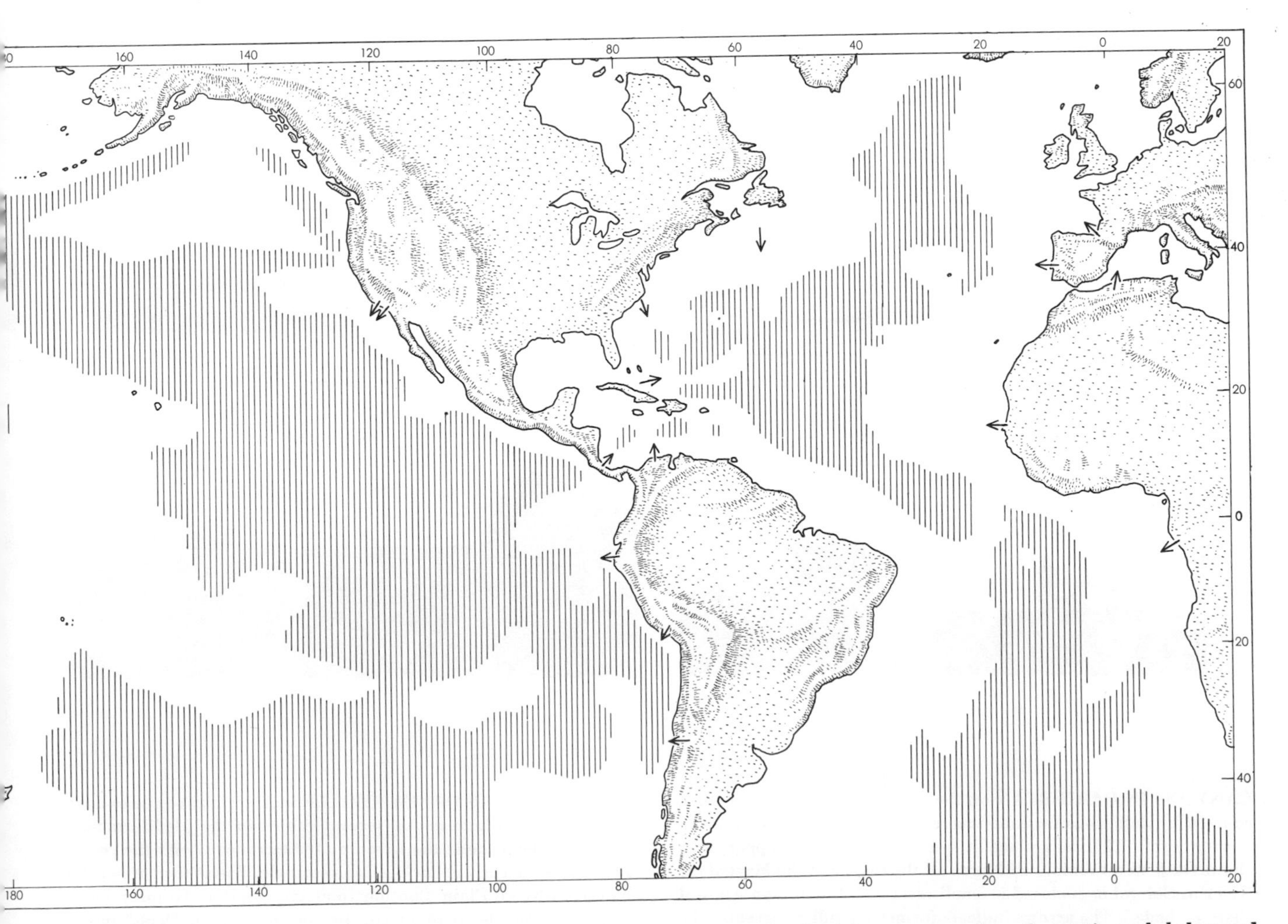

coming from shallow water. The height of these regions and the surrounding topography cut them off from the flows which have taken sand and marine organisms from the continental shelves and transported them to many of the deepest parts of the ocean floor.

CANYONS IN CARIBBEAN off the coast of Colombia have been coursed by numerous turbidity currents which have broken a submarine cable 17 times since it was laid in 1930. Upper map shows the general region, with area off the mouth of the Magdalena River in Colombia enclosed in small rectangle. Colombian Abyssal Plain is shaded. The arrows indicate former turbidity currents; the parallel lines, the path of profile at bottom. Lower map gives detail in the rectangle. The paths of turbidity currents are marked by broken lines. The breaks in the cable are indicated by the solid lines.

EARTHQUAKE IN ORLEANSVILLE in Algeria gave rise to a turbidity current which broke several cables in the Mediterranean. The upper map shows the western Mediterranean, with the region affected by the current enclosed in rectangle. The shaded area is the Balearic Abyssal Plain. Contour lines give depth in fathoms. Parallel lines show the path of the profile at bottom. Detail in the marked area is shown in the lower map. Lines represent cables, with breaks marked by open dots. Arrows show the probable path of the current. Epicenter of the quake is indicated by dot and circles.

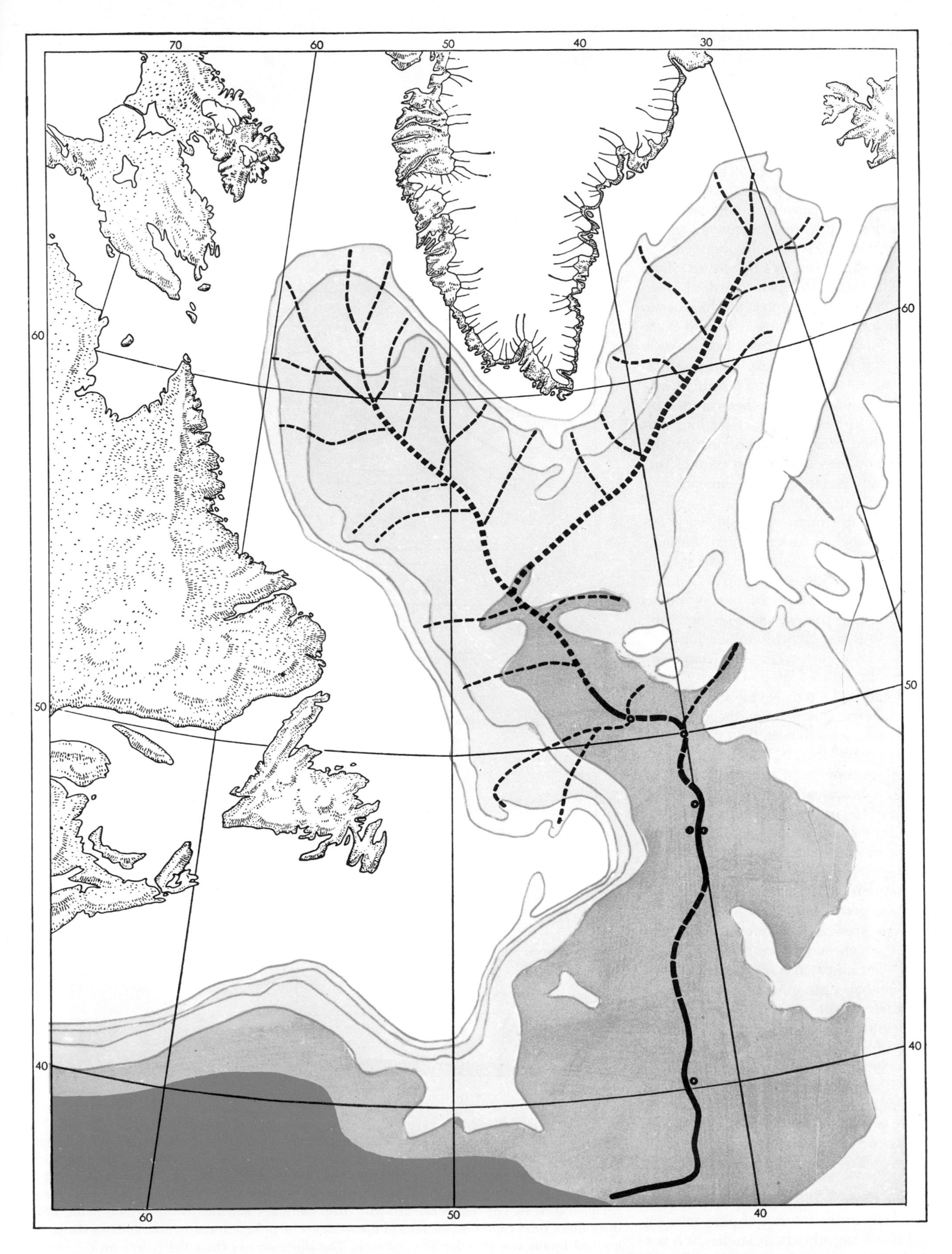

MID-OCEAN CANYON begins off Greenland and extends south to about the 38th parallel. The solid line traces sounded portions; the heavy broken line, unsounded connections between known sections. The heavy dotted lines represent the probable extension of the canyon; the light broken lines, possible tributaries. The small circles mark sites of breaks in transatlantic cables. The depth at various points is indicated by blue shading. The darkest tone represents a depth of about 2,500 fathoms; the lightest tone, about 500.

ed from the shore to the deepest part of the ocean, was striking proof that currents must flow over the sea floor. The flat floor of the Puerto Rican Trench, suggesting a plain formed by flow down its slopes, had led us to expect that material from shallow waters would be found there, but this expectation in no way diminished our joy in the discovery.

Not all the canyons in the sea stem from the mouths of land rivers. Indeed one of the longest is a mid-ocean canyon that runs from the area between Iceland and Greenland to a point south of Newfoundland. This canyon in the deep ocean basin, two to four miles wide and 25 to 100 fathoms deep, was discovered in 1949 and extensively explored in 1952. Its floor contains sand and possibly gravel. The canyon may have been eroded by periodic turbidity currents set into motion when freezing conditions in the Greenland area produced masses of turbid, heavy water—full of sediment, very salty and extremely cold.

There are many possible ways in which turbidity currents might be generated—freezing conditions, triggering by earthquakes, the piling up of deposits on steep ocean slopes by rivers, and so on. Granting the existence of such currents, it is not difficult to see how repeated flows, each originating from nearly the same spot atop a slope, could gradually erode a canyon in the ocean.

Aside from explaining the origin of the submarine canyons, the evidence for the existence of turbidity currents is interesting on several other counts. For one thing, it indicates a way in which organic material is transported from the continental coasts to feed fish and other organisms in the deep sea. For another, it suggests how organic sediments may accumulate in sea basins and eventually become petroleum. Much of the petroleum in the world is found in ancient sands which once were part of the sea floor. Turbidity currents therefore offer a clue concerning where to look for oil; the concept has in fact been applied successfully in exploration of the oil fields of southern California. Still another interesting aspect of these currents has to do with developments in the atomic age: for example, it is possible that an offshore atomic blast might trigger a turbidity current which would spread radioactive debris throughout an entire ocean basin.

The study of the sea floor is carried on primarily in the interest of pure science, but like many other basic studies, it is not without commercial, economic, military, social and political implications.

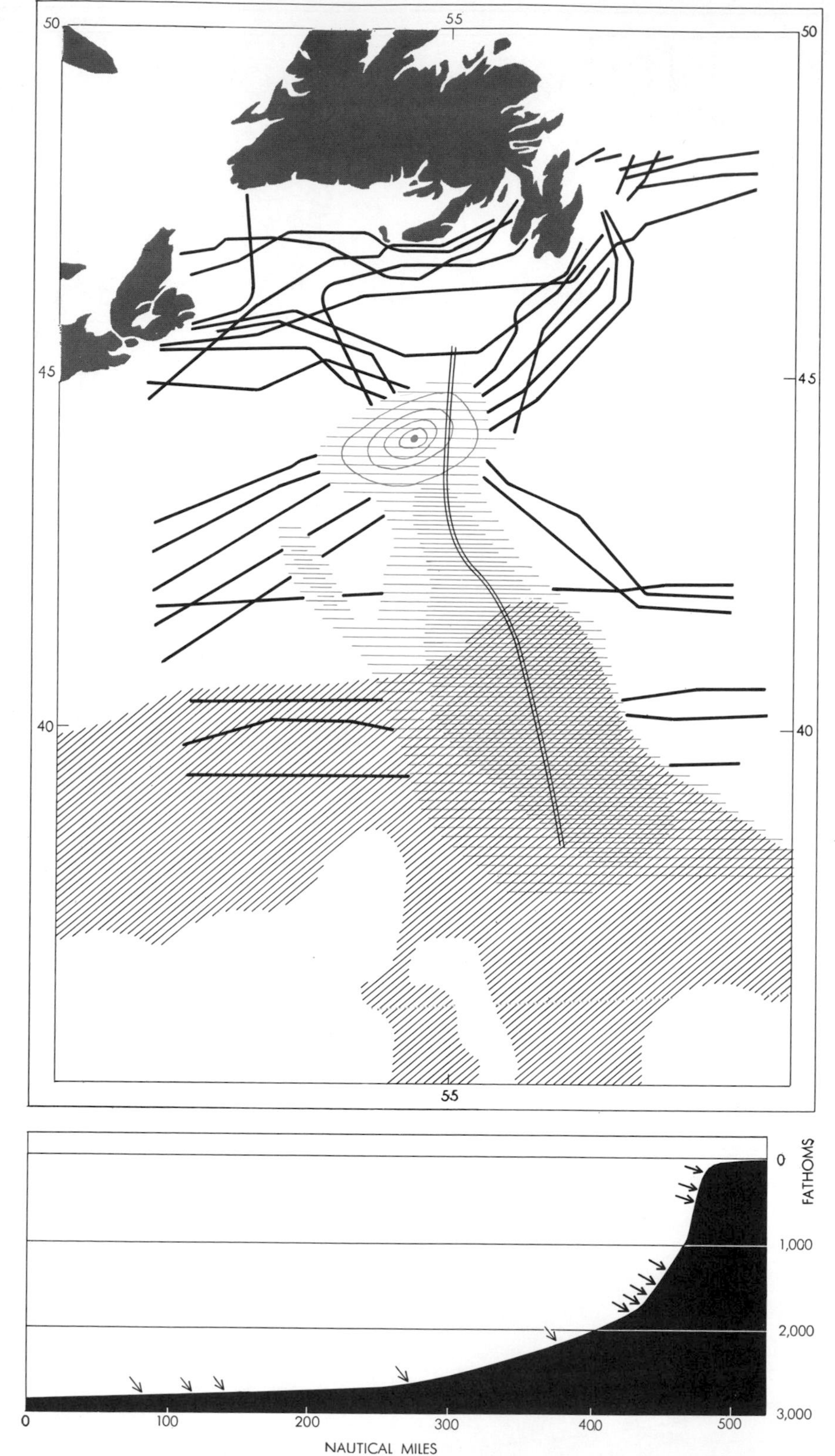

GRAND BANKS EARTHQUAKE, which occurred in 1929, caused a turbidity current that broke cables as far as 300 miles away from the epicenter of the disturbance. The epicenter is indicated on the map at the top of this page by the colored dot surrounded by circles. The turbidity current was strongest in the dark-colored portion and weaker in the lighter-colored peripheral areas. The region marked with diagonal black hatching is the abyssal plain. The solid black lines represent sections of the cables. The profile at the bottom follows the path marked by the two parallel lines on map. The short arrows show the points on the slope where the cables broke. On the profile the epicentral area is between the three arrows at upper right and the five arrows below them. The current reached a speed of 55 knots per hour.

THE DEEP-OCEAN FLOOR

H. W. MENARD
September 1969

Oceanic geology is in the midst of a revolution. All the data gathered over the past 30 years—the soundings of the deep ocean, the samples and photographs of the bottom, the measurements of heat flow and magnetism—are being reinterpreted according to the concept of continental drift and two new concepts: sea-floor spreading and plate tectonics (the notion that the earth's crust consists of plates that are created at one edge and destroyed at the other). Discoveries are made and interpretations developed so often that the scientific literature cannot keep up with them; they are reported by preprint and wandering minstrel. At such a time any broad synthesis is likely to be short-lived, yet so many diverse observations can now be fitted into a coherent picture that it seems worthwhile to present it.

Before continental drift, sea-floor spreading and plate tectonics captured the imagination of geologists, most of them conceived the earth's crust as being a fairly stable layer enveloping the earth's fluid mantle and core. The only kind of motion normally perceived in this picture was isostasy: the tendency of crustal blocks to float on a plastic mantle. The horizontal displacement of any geologic feature by as much as 100 kilometers was considered startling. This view is no longer consistent with the geological evidence. Instead each new discovery seems to favor sea-floor spreading, continental drift and plate tectonics. These concepts are described elsewhere in this book [see the article "The Origin of the Oceans," by Sir Edward Bullard, beginning on page 196]. Here I shall recapitulate them briefly to show how they are related to the actual features of the deep-ocean floor.

According to plate tectonics the earth's crust is divided into huge segments afloat on the mantle. When such a plate is in motion on the sphere of the earth, it describes a circle around a point termed the pole of rotation (not to be confused with the entire earth's pole of rotation). This motion has profound geological effects. When two plates move apart, a fissure called a spreading center opens between them. Through this fissure rises the hot, plastic material of the mantle, which solidifies and joins the trailing edge of each plate. Meanwhile the edge of the plate farthest from the spreading center—the leading edge—pushes against another plate. Where that happens, the leading edge may be deflected downward so that it sinks into a region of soft material called the asthenosphere, 100 kilometers or more below the surface. This process destroys the plate material at the same rate at which it is being created at its trailing edge. Many of the fissures where plate material is being created are in the middle of the ocean floor, which therefore spreads continuously from a median line. Where the plates float apart, the continents, which are embedded in them, also drift away from one another.

The most obvious consequence of this process on the ocean floor is the symmetrical seascape on each side of a spreading center. As two crustal plates move apart (at a rate of one to 10 centimeters per year), the basaltic material that wells up through the spreading center between them splits down the middle. The upwelling in some way produces a ridge, flanked on each side by deep ocean basins and capped by long hills and mountains that run parallel to the crest. The flow of heat from the earth's interior is generally high along the crest because dikes of molten rock have been injected at the spreading center. A spreading center may also open under a continent. If it does, it produces a linear deep such as the Red Sea or the Gulf of California. If it continues to spread, or if the spreading center opens in an existing ocean basin, the same symmetrical seascape is ultimately formed.

This symmetry extends to less tangible features of the ocean floor such as the magnetic patterns in the basalt of the slopes on either side of the mid-ocean ridge. As the plastic material reaches the surface and hardens, it "freezes" into it the direction of the earth's magnetic field. The earth's magnetic field reverses from time to time, and as each band of new material moves outward across the ocean floor it retains a magnetic pattern shared by a corresponding band on the other side of the ridge. The result is a matching set of parallel bands on both sides. These patterns provide evidence of symmetrical flows and make it possible to date them, since they correspond to similar patterns on land that have been reliably dated by other means.

The steepness of the mid-ocean ridge is determined by a balance between the rate at which material moves outward from the spreading center and the rate at which it sinks as it ages after solidifying. The rate of sinking remains fairly constant throughout the ocean basin, and it seems to depend on the age of the

PILLOW LAVA (as pictured on the following page) assumes its rounded shape because it cools rapidly in ocean water. This flow lies on the western slope of the mid-ocean ridge in the South Atlantic at a depth of 2,650 meters. Flows like this erupt from the many volcanic vents and fissures that are created as the ocean floor spreads out from the mid-ocean ridges in the form of vast crustal plates. The photograph was made under the direction of Maurice Ewing of the Lamont-Doherty Geological Observatory.

crust. It can be calculated if the age of the oceanic crust (as indicated by the magnetic patterns) is divided into the depth at which a particular section lies. Such calculations show that the crust sinks about nine centimeters per 1,000 years for the first 10 million years after it forms, 3.3 centimeters per 1,000 years for the next 30 million years, and two centimeters per 1,000 years thereafter. Not all the crust sinks: on the southern Mid-Atlantic Ridge the sea floor has remained at the same level for as long as 20 million years.

The rate at which the sea floor spreads varies from one to 10 centimeters per year. Therefore fast spreading builds broad elevations and gentle slopes such as those of the East Pacific Rise. The steep, concave flanks of the Mid-Atlantic Ridge, on the other hand, were formed by slow spreading.

Whether the slopes are steep or gentle, the trailing edge of the plate at the spreading center is about three kilometers higher than the leading edge on the other side of the plate. The reason for this difference in elevation is not known. Heating causes some elevation and cooling some sinking, but the total relief appears much too great to be attributed to thermal expansion. Cooling might account for the relatively rapid sinking observed during the first 10 million years, but continued sinking remains a puzzle.

A decade ago scanty information suggested that the mid-ocean ridge in both the Atlantic and Pacific was continuous, with a few branches. More complete surveys have revealed that crustal plates have ragged edges. Instead of extending unbroken for thousands of kilometers, a mid-ocean ridge at the trailing edges of two crustal plates forms a zigzag line consisting of many short segments connected by fracture zones to other ridges, trenches, young mountain ranges or crustal sinks. The fracture zones connecting the ridge segments are associated with what are called "transform" faults. They provide important clues to the history of a plate. Because they form some of the edges of the plate, they delineate the circle around its pole of rotation, thereby indicating the direction in which it has been moving.

From what has been said so far it might appear that the spreading centers are fixed and stationary. The constantly repeated splitting of the new crust at the spreading center produces symmetrical continental margins, symmetrical magnetic patterns on the ocean floor, symmetrical ridge flanks and even

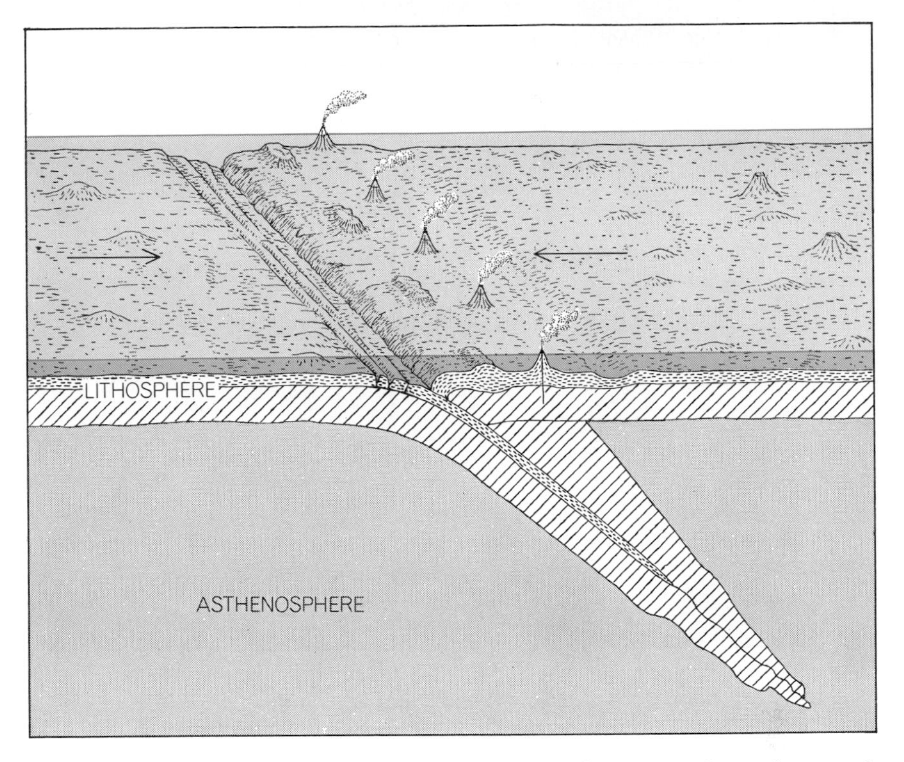

TRENCH IS CREATED where the leading edge of a plate that emerges from a fast spreading center collides with another plate. Because the combined speed of the two is more than six centimeters per year neither can absorb the impact by buckling. Instead one crustal plate (*in lithosphere*) plunges under the other to be destroyed in the asthenosphere, a hot, weak layer below. The impact produces volcanoes, islands and a deep, such as the Tonga Trench. Beside a trench are cracks that are produced by bending of the crust.

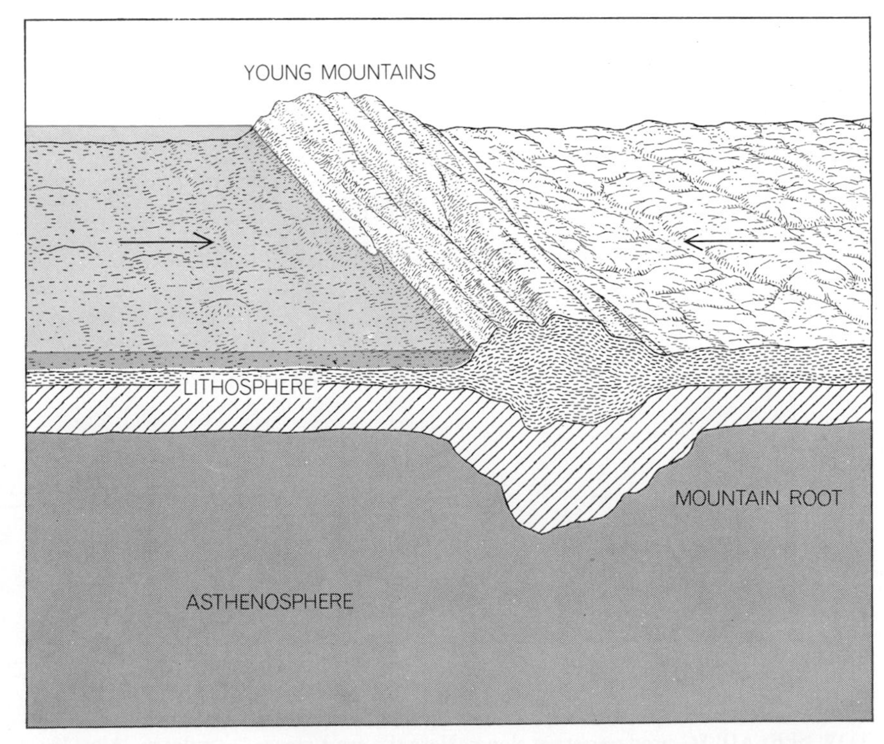

MOUNTAIN RANGE IS FORMED when the leading edges of two plates come together at less than six centimeters per year. Instead of colliding catastrophically, so that one plate slides under the other, both plates buckle, raising a young mountain range between them. The range consists of crustal material that folds upward under the compression exerted by the two plates (and also downward, forming the root of the mountain). Such ranges can be identified because they contain cherts and other material typical of the ocean bottom.

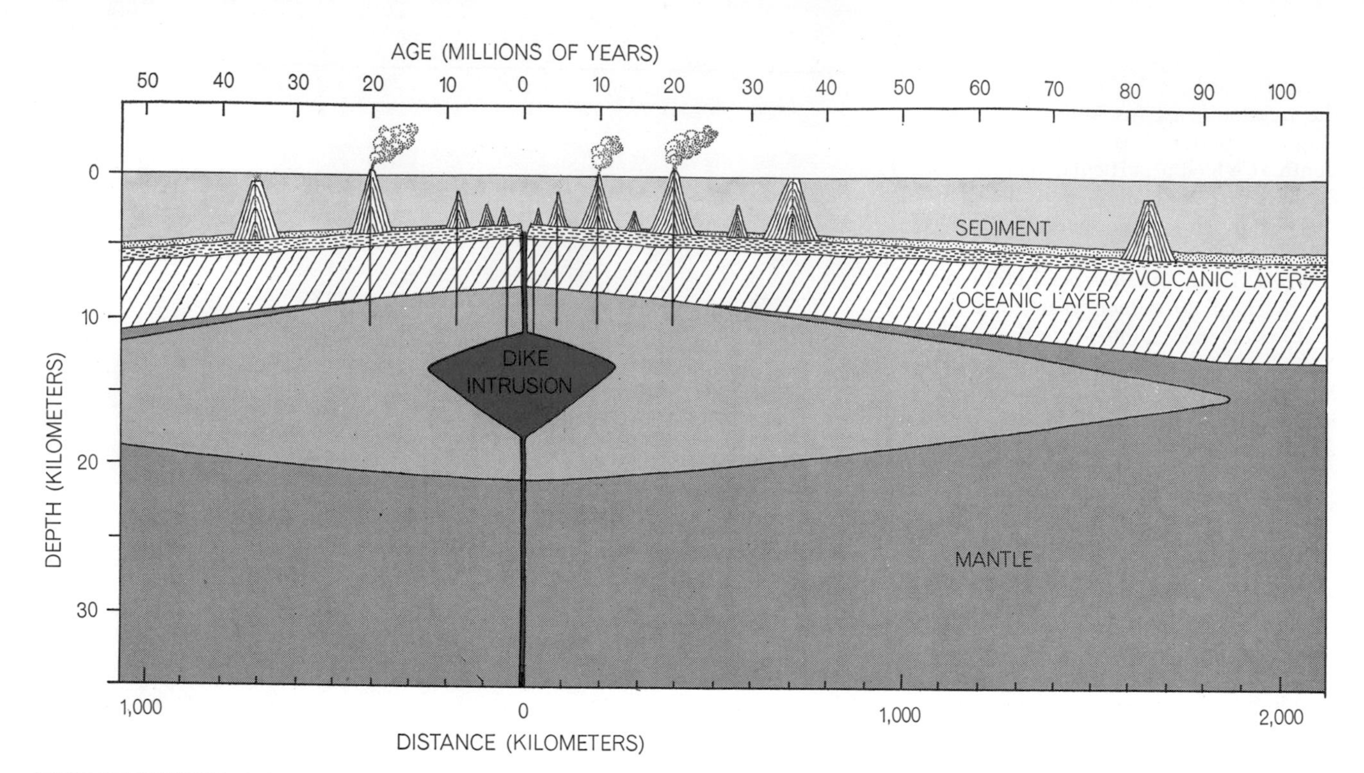

FAST SPREADING of the sea floor is revealed by gentle slopes. The sea floor is created at a spreading center that leaks molten rock from several dikes intruding from a pool in the low-density mantle (*light shading*). As the molten rock emerges it cools and adheres to the crust sliding away on each side of the fissure. If the crust moves at more than three centimeters per year, the slopes are gradual because spreading, which is horizontal, is rapid compared with the sinking of the crust. The balance between the two determines the steepness of the slope. Fast spreading also produces a thin volcanic layer because material moves so quickly from the fissure that it cannot accumulate. Islands built by eruptions are distant from the center because they grow on rapidly moving crust.

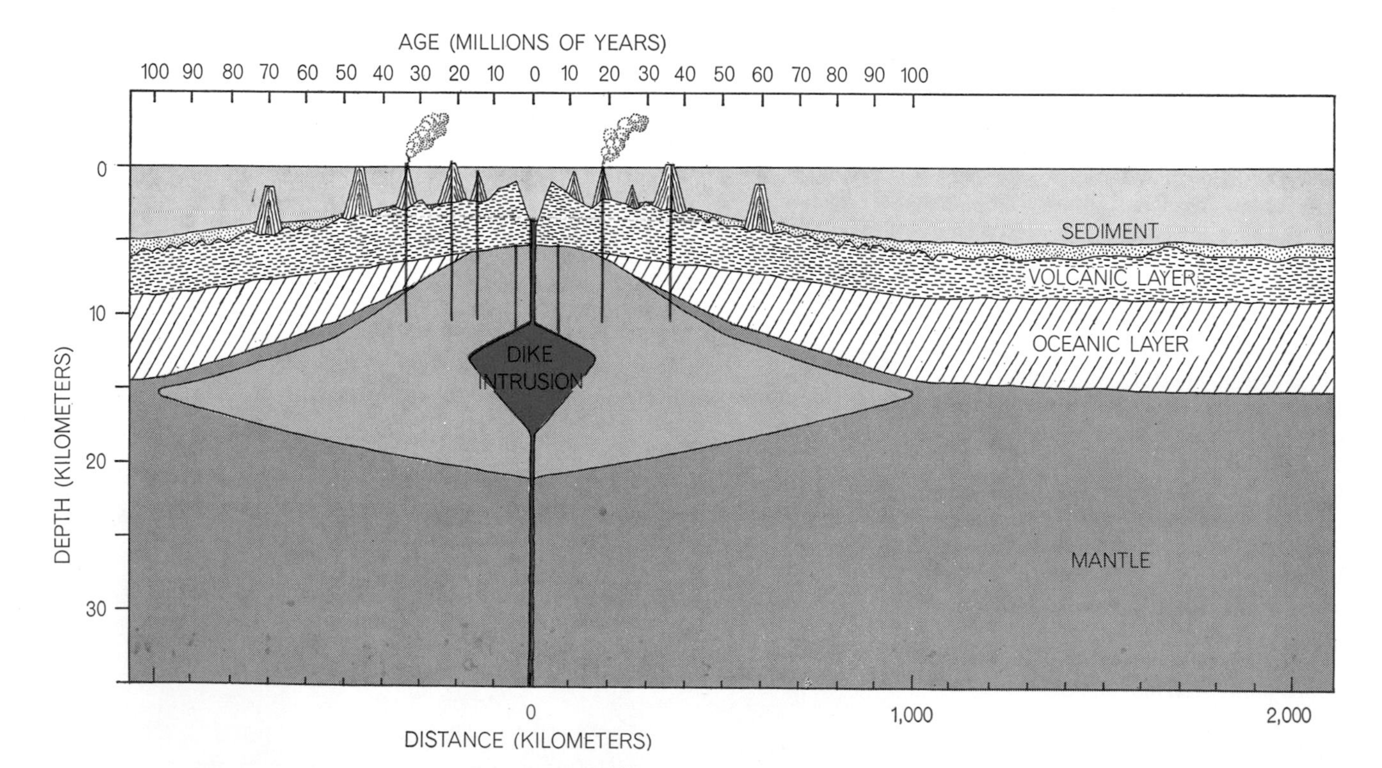

SLOW SPREADING produces steep slopes. Here the crust moves less than three centimeters per year; consequently sinking dominates the slope-forming process and produces steep escarpments. The volcanic layer is thicker at a slow center because material has time to accumulate. Mountains and volcanoes are high near the spreading center because the crust moves so slowly that the lava piles up. After 100 million years crust produced from a slow spreading center has strong similarity to crust from a fast center. Both kinds of crust have sunk to a depth of five kilometers. The oceanic layer is about five kilometers thick. Both fast- and slow-spreading crust are covered by the same kind of sediment. Slow spreading occurs mainly in the Atlantic, fast spreading in the Pacific.

symmetrical mountain ranges. More often than not, however, it has been found that the spreading center itself moves. Oddly enough such movement gives rise to the same symmetrical geology. All that is required in order to maintain the symmetry is that the spreading center move at exactly half the rate at which the plates are separating. If it moved faster or slower, the symmetry of the magnetic patterns would be destroyed.

Imagine, for instance, that the plate to the east of a spreading center remains stationary as the plate to the west moves. Since the material welling up through the fissure splits down the middle, half of it adheres to the stationary plate and the other half adheres to the moving plate. The next flow of material to well up through the split thus appears half the width of the spreading center away from the stationary plate. The flow after that appears a whole width of the spreading center away from the stationary plate, and so on. In effect the spreading center is migrating away from the stationary plate and following the moving one. If the speed of the spreading center exceeded half the speed of the migrating plate, however, a kind of geological Doppler effect would set in: the bands of the magnetic pattern would be condensed in the direction of the moving plate, and they would be stretched out in the direction of the stationary plate [*see illustration below*].

It might seem unlikely that the spreading center would maintain its even rate of speed and remain exactly between the two plates. W. Jason Morgan of Princeton University observes, however, that there is no impediment to such motion, provided only that the crust splits where it is weakest (which is where it split before, at the point where the hot dike was originally injected). As a result the spreading center is always exactly between two crustal plates whether it moves or not.

Moving spreading centers account for some of the major features of the ocean floor. The Chile Rise off the coast of South America and the East Pacific Rise are adjacent spreading centers. Since there is no crustal sink between them, and new plate is constantly being added on the inside edge of each rise, at least one of the centers must be moving, otherwise the basin between them might fold and thrust upward into a mountain range or downward into a trench. Similarly, the Carlsberg Ridge in the Indian Ocean and the Mid-Atlantic Ridge are not separated by a crustal sink and hence one of them must be moving.

A moving center may have created the ancient Darwin Rise on the western edge of the Pacific basin and also the modern East Pacific Rise. As in the case of the Carlsberg Ridge and the Mid-Atlantic Ridge, the existence of two vast spreading centers on opposite sides of the ocean with no intervening crustal sink has puzzled geologists. If such centers can move, however, it is possible that the spreading center in the western Pacific merely migrated all the way across the basin, leaving behind the ridges of the Darwin Rise. In this way one rise could simply have become the other. Many other examples exist, and Manik Talwani of the Lamont-Doherty Geological Observatory proposes that all spreading centers move.

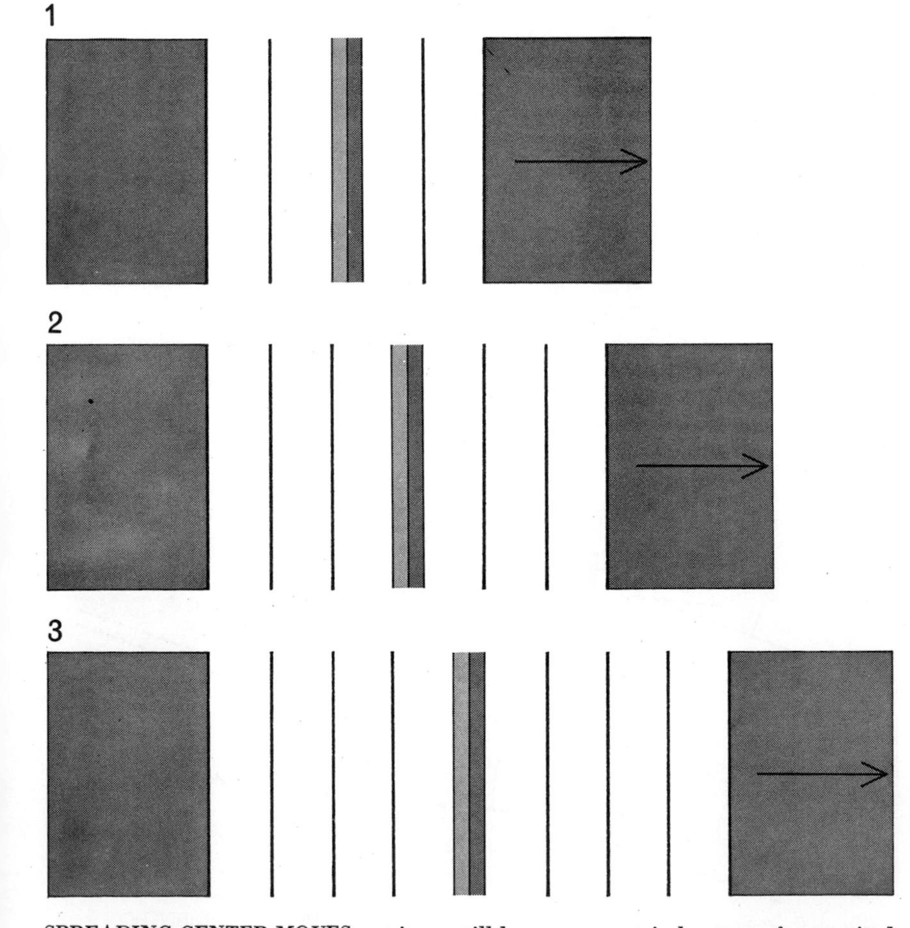

SPREADING CENTER MOVES, yet it can still leave a symmetrical pattern of magnetized rock. The molten material emerging from a spreading center becomes magnetized because as it cools it captures the prevailing direction of the earth's periodically changing magnetic field. In the instance illustrated here the right-hand plate moves out to the right while the left-hand plate remains stationary. In *1* hot material from a dike arrives at the surface, cools and splits down the middle. In *2* the next injection of material arrives in the crevice between the two halves of the preceding mass of rock. The new mass is therefore half the width of the preceding mass farther from the stationary plate than the preceding mass of material itself was. In *3* the new material has cooled and split in its turn and another mass has appeared that is a whole width farther from the left-hand plate. As long as the center moves at half the speed at which the right-hand plate moves away the magnetic bands remain symmetrical. If plate moved faster or more slowly, they would be jumbled.

As a plate forms at a spreading center it consists of two layers of material, an upper "volcanic" layer and a lower "oceanic" one. Lava and feeder dikes from the mantle form the volcanic layer; its rocks are oceanic tholeiite (or a metamorphosed equivalent), which is rich in aluminum and poor in potassium. The oceanic layer is also some form of mantle material, but its precise composition, density and condition are not known. Farther down the slope of the ridge the plate acquires a third layer consisting of sediment.

The sediment comes from the continents and sifts down on all parts of the basin, accumulating to a considerable depth. It is mixed with a residue of the hard parts of microorganisms that is called calcareous ooze. Below a certain depth (which varies among regions) this material dissolves, and only red clay and other resistant components remain.

For reasons only partly known the sediment is not uniformly distributed. At the spreading center the newly created crust is of course bare of sediment, and within 100 kilometers of such a center the calcareous ooze is rarely thick enough to measure. The ooze accumulates at an average rate of 10 meters for each million years, during which time the plate moves horizontally from 10,000 to 100,000 meters and sinks 100 meters. Where the red clay appears, it accumulates at a rate of less than one meter per million years.

The puzzle deepens when one considers that sediment on oceanic crust older than 20 million years stops increasing in thickness after it sinks to the depth where the calcareous ooze dissolves. Indeed, in many places the age of the oldest sediment is about the same as the volcanic layer on which it lies. It would therefore seem that almost all the deep-ocean sediment accumulates in narrow zones on the flanks of the mid-ocean ridges. If this is correct, it has yet to be explained.

The volcanic layer forms mainly at the spreading center. Volcanoes and vents on the slopes of the mid-ocean ridge contribute a certain amount of oceanic tholeiite to it. It can be said in general that the thickness of the volcanic layer decreases as the spreading rate increases. If the crust spreads slowly, the material has time to accumulate. Fast spreading reduces this time and therefore the accumulation. The conclusion can be drawn that the rate at which the volcanic-layer material is discharged is nearly constant. These relations are based on only 10 observations, but they apply to spreading rates from 1.4 centimeters to 12 centimeters per year, and to thicknesses from .8 kilometer to 3.8 kilometers.

The total flow of volcanic material from all active spreading centers is about four cubic kilometers per year—four times the flow on the continents. Not the slightest sign of this volcanism on the ocean bottom can be detected at the surface of the ocean, with one possible exception: late in the 19th century a ship reported seeing smoke rising from the waters above the equatorial Mid-Atlantic Ridge. The British oceanographer Sir John Murray remarked that he hoped the smoke signified the emergence of an island, since the Royal Navy needed a coaling station at that point.

Like the volcanic layer, the oceanic

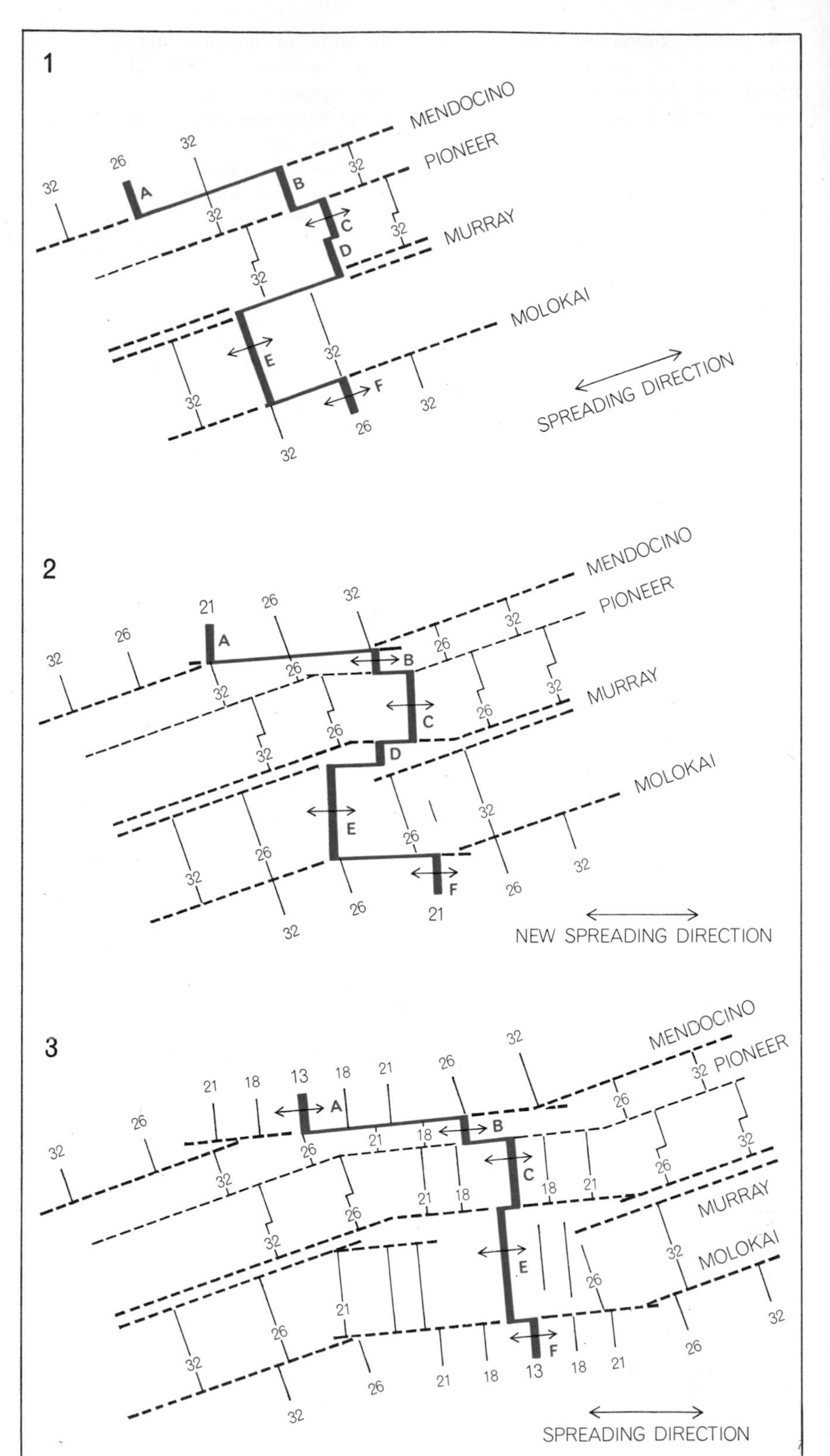

HOW A PLATE MOVES is revealed by the patterns of magnetic bands (time of formation is indicated by numbers) and by the relation between ridges and faults in the northeastern Pacific. At the time illustrated in *1* material from the Murray fault and other faults connected segments of the ridge (*indicated by letters*) offset from one another by plate motion. At the time shown in *2* the spreading direction changed. The readjustment of the plates has shortened the Pioneer-Mendocino ridge segment (*B*) while lengthening and reorienting the Pioneer, Mendocino and Molokai transform faults. In *3* the Mendocino fault remains the same length, but between most of the other ridges faults have been shortened or have almost disappeared as ridge segments tended to rejoin one another. Between Murray and Molokai faults ridge has jumped eastward, and one segment (*D*) has vanished.

layer forms at the spreading center. Acoustical measurements of the thickness of this layer at spreading centers, on the flanks of mid-ocean ridges and on the deep-ocean floor show, however, that at least part of the oceanic layer evolves slowly from the mantle rather than solidifying quickly and completely at the spreading center. At the spreading center the thickness of the layer depends on how fast the ocean floor moves. In regions such as the South Atlantic, where the floor spreads at a rate of two centimeters per year, no oceanic layer forms within a few hundred kilometers of the spreading center. Farther away from the spreading center the oceanic layer accumulates rapidly, reaching a normal thickness of four to five kilometers on the flank of the mid-ocean ridge. A spreading rate of three centimeters per year is associated with an oceanic layer roughly two kilometers deep at the center that thickens by one kilometer in 13 million years. A plate with a spreading rate of eight centimeters per year is three kilometers deep at the center and thickens by one kilometer in 20 million years. Thus the thinner the initial crust is, the faster the thickness increases as the crust spreads.

As a plate flows continuously from the spreading center, faulting, volcanic eruptions and lava flows along the length of the mid-ocean ridge build its mountains and escarpments. This process can be most easily observed in Iceland, a part of the Mid-Atlantic Ridge that has grown so rapidly it has emerged from the ocean. A central rift, 45 kilometers wide at its northern end, cuts the island parallel to the ridge. The sides of the rift consist of active, steplike faults. There are other step faults on the rift floor, which is otherwise dominated by a large number of longitudinal fissures. Some of these fissures are open and filled with dikes. Fluid lava wells up from the fissures and either buries the surrounding mountains, valleys and faults or forms long, low "shield" volcanoes. Two hundred such young volcanoes, which have been erupting about once every five years over the past 1,000 years, dot the floor of the rift. Thirty of them are currently active.

Just as a balance between spreading and sinking shapes the slopes of the mid-ocean ridges, so does a balance between lava discharge and spreading build undersea mountains, hills and valleys. High mountains normally form at slow centers where spreading proceeds at two to 3.5 centimeters per year. In contrast, a spreading center that opens at a rate of five to 12 centimeters per year produces long hills less than 500 meters high. This relationship is a natural consequence of the long-term constancy of the lava discharge. Over a short period of time, however, the lava discharge may fluctuate or pulsate, a picture suggested by the fact that the thick volcanic layer associated with slow spreading can consist of volcanic mountains (which represent copious flow) separated by valleys covered by a thinner volcanic layer.

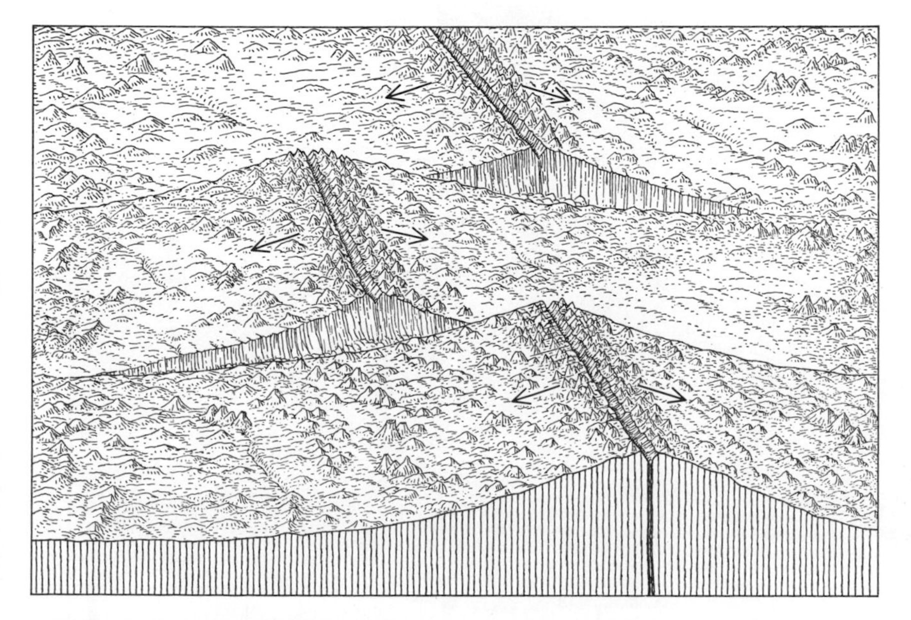

RIDGE-RIDGE TRANSFORM FAULT appears between two segments of ridge that are displaced from each other. Mountains are built, earthquakes shake the plate edges and volcanoes erupt in such an area because of the forces generated as plates, formed at the spreading centers under the ridges, slide past each other in opposite directions. On outer slopes of the mid-ocean ridges, however, this intense seismic activity appears to subside.

The volcanic activity and faulting that first appear near the spreading centers decrease rapidly as the plate ages and material moves toward its center, but volcanic activity in some form is never entirely absent. Small conical volcanoes are found on crust only a few hundred thousand years old near spreading centers, and active circular volcanoes such as those of Hawaii exist even in the middle of a plate. It would appear that the great cracks that serve as conduits for dikes and lava flows are soon sealed as a plate ages and spreads. Volcanic activity is then concentrated in a few central vents, created at different times and places.

Many of these vents remain open for tens of millions of years, judging by the size and distribution of the different classes of marine volcanoes. First, the biggest volcanoes are increasingly big at greater distances from a spreading center, which means they must continue to erupt and grow as the crust ages and sinks, even when the age of the crust exceeds 10 million years. In most places, in fact, a volcano needs at least 10 million years in order to grow large enough to become an island. Volcanoes that discharge lava at a rate lower than 100 cubic kilometers per million years never become islands because the sea floor sinks too fast for them to reach the sea surface.

Other volcanoes drifting with a spreading ocean floor may remain active or become active on crust that is 100 million years old (as the volcanoes of the Canary Islands have). Normally, however, volcanoes become inactive by the time the crust is 20 to 30 million years old. This is demonstrated by the existence of guyots, drowned ancient island volcanoes that were submerged by the gradual sinking of the aging crust. Guyots are found almost entirely on crust that is more than 30 million years old, such as the floor of the western equatorial Pacific.

Traditionally it has been thought that marine volcanoes spew lava from a magma chamber located deep in the mantle. Some volcanoes have a top composed of alkali basalt, slopes with transitional basalt outcrops and a base of oceanic tholeiite, and it was therefore assumed that the lava in the magma chamber became differentiated into components that, rather like a pousse-café, separated into several layers of different kinds of material, each of which followed the layer above it up the spout. The emergence of plate tectonics and continental drift as respectable concepts have now brought this view of volcanic action into question.

It remains perfectly possible for a volcano to drift for tens of millions of years over hundreds of kilometers while tapping a single magma chamber embedded deep in the mantle. The motion of the plates, however, suggests another hypothesis. According to this view, the volcano and its conduit drift along with the crust as the conduit continually taps different parts of a relatively stationary magma that is ready to yield various kinds of lava whenever a conduit appears. In actuality the composition of the lava usually changes only slightly after the first 10 to 20 kilometers of drifting. Although the older hypothesis is still reasonable, the newer one must also be considered because it explains the facts equally well.

In addition to their characteristic volcanoes and mountains, spreading centers are marked by median valleys, which in places such as the North Atlantic or the northwestern Indian Ocean are deeper than the surrounding region. These rifts are commonly found in centers opening at a rate of two to five centimeters per year. The deepest rifts, which may go as deep as 1,000 to 1,300 meters below the surrounding floor, are associated with spreading at three to four centimeters per year. Only one valley is known to be associated with spreading at five to 12 centimeters per year. Although rifts are not found in all spreading centers, they usually do appear in conjunction with a slow center. Both of these features are also associated with volcanic activity.

The mid-ocean ridges, as we have noted, seldom run unbroken for more than a few hundred kilometers. They are interrupted by fracture zones, and the segments are shifted out of line with respect to one another. These fracture zones run at right angles to the ridge and connect the segments. Where they lie between the segments they are termed ridge-ridge transform faults, which are the site of intense geological activity. As the two edges of the fault slide past each other they rub and produce earthquakes. The slope of a transform fault drops steeply from the crest of one ridge segment to a point halfway between it and the adjacent segment and then climbs to the top of the adjacent segment, reflecting the fact that the crust is elevated at the spreading center and subsides at some distance from it [*see illustration on page 167*].

Like spreading centers, fracture zones have their own complex geology. In these

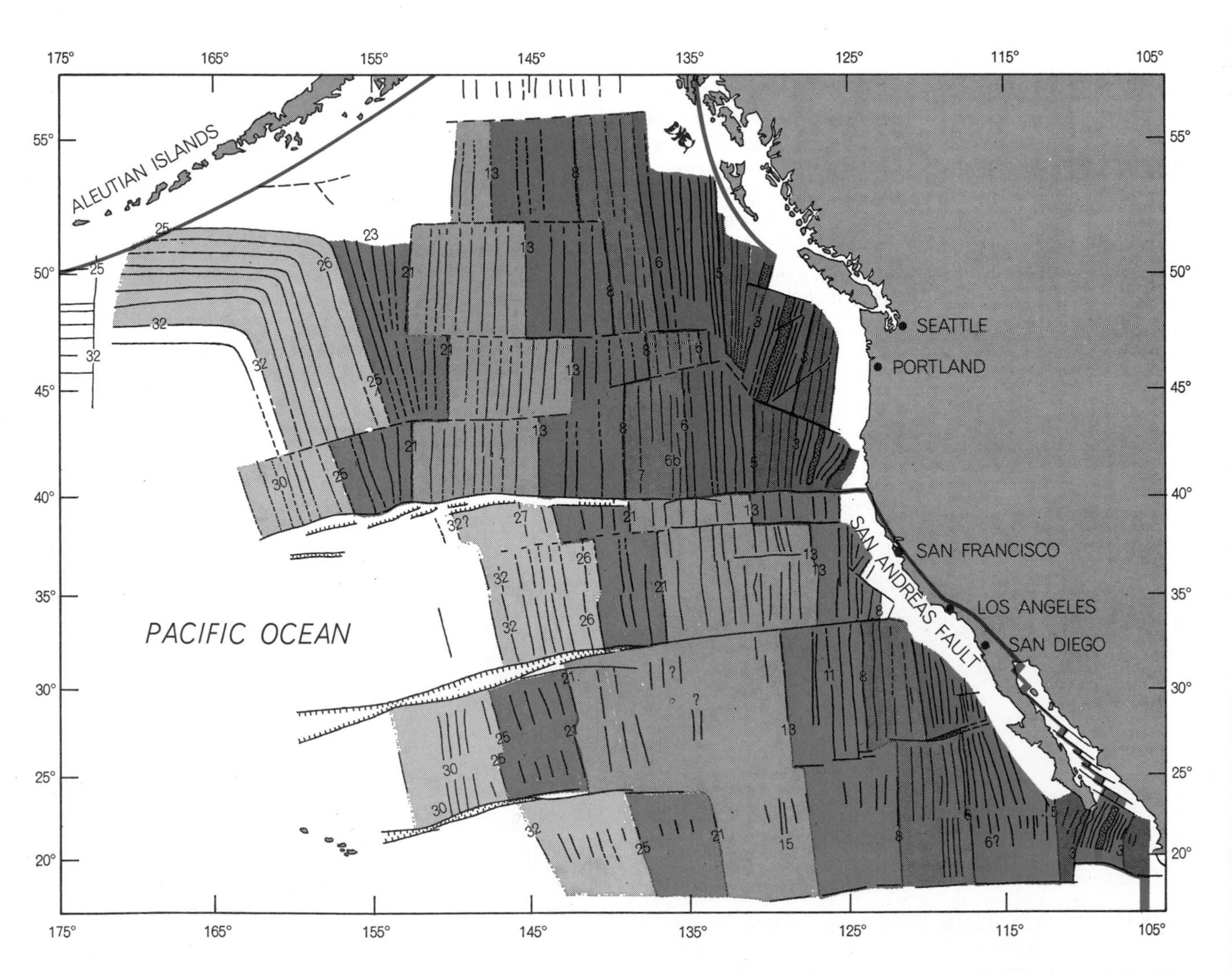

MAGNETIC PATTERNS reveal how the plate forming the floor of the northeastern Pacific has moved. Its active eastern edge now stretches from Alaska through California (where it forms the San Andreas fault) to the Gulf of California. In the gulf spreading centers break into short segments joined by active faults. Plate motion is opening the gulf and moving coastal California in the direction of the Aleutians. To the south lies the Great Magnetic Bight, formed by three plates that spread away from one another.

areas the ridges stand as much as several kilometers high, and the troughs are equally deep. It appears that the same volcanic forces that shape the main ridges produce the mountains and valleys of the faults. As fracture zones open they slowly leak lava from hot dikes. At the same time the crust sinks away from the fault line, and this balance produces high mountains.

Beyond the spreading centers the fracture zones become the inactive remains of earlier faulting. The different rates at which these outer flanks of the mid-ocean ridge sink do produce some vertical motion as the scarps of the fracture zone decay. This may account for the few earthquakes in these areas. I should emphasize that it is not known if horizontal motion is also absent from such dead fracture zones. It is not necessary, however, to postulate such motion in order to explain existing observations.

A fracture zone can become active again at any time, but if it does so, it becomes the side of a smaller new plate rather than part of the trailing edge of an old one. If the flank fracture zones are as quiescent as they appear to be, then the plates forming the earth's crust are large and long-lived. If these fracture zones were active, on the other hand, it could only be concluded that each one marked the flank of a small, elongated plate.

The direction the plate is moving can be deduced from the magnetic pattern that runs at right angles to the fractures in the fracture zone. When the direction of plate motion changes, the direction in which the fracture zone moves also changes. This change in direction can be most clearly seen in the northeastern Pacific, where our knowledge is most detailed. On this part of the ocean floor the changes of direction have taken place at the same time in many zones, indicating that the entire North Pacific plate has changed direction as a unit [*see illustration on page 168*].

On the bottom of the Pacific and the North Atlantic the magnetic patterns are sometimes garbled. Old transform faults may have vanished if short segments of spreading center have been united by reorientation. By the same token new transform faults may have formed if the change in plate motion has been too rapid to be accommodated by existing motions. Thus fracture zones may be discontinuous. They may start and stop abruptly, and the offset of the magnetic patterns may change from place to place along them without indicating any activity except at the former edges of plates.

Some patterns are even harder to interpret. Douglas J. Elvers and his colleagues in the U.S. Coast and Geodetic Survey discovered an abrupt boomerang-shaped bend in the magnetic pattern south of the Aleutians. The arms of this

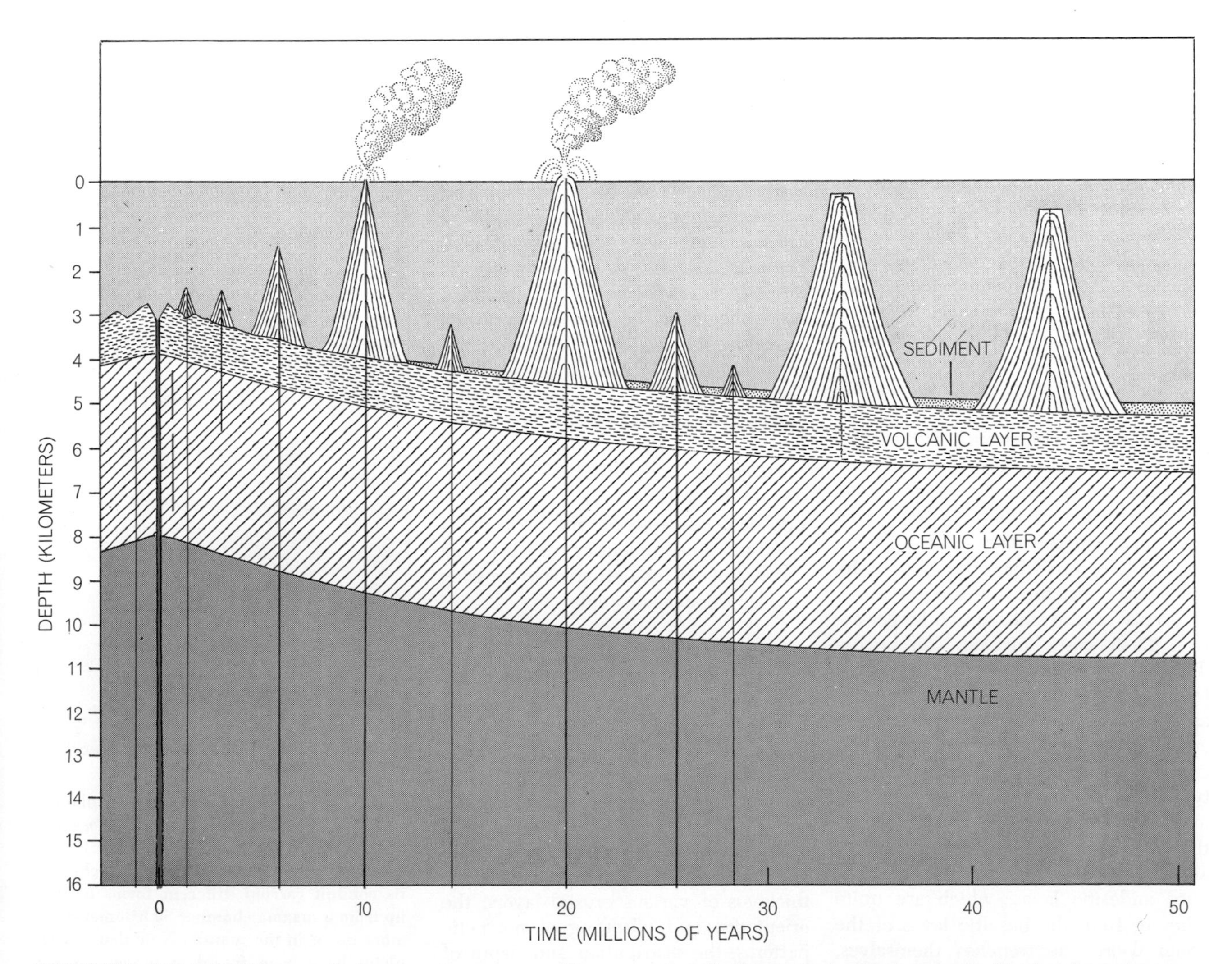

UNDERSEA VOLCANOES normally begin to rise near spreading centers. Then they ride along on the moving plate as they grow. If a volcano rises fast enough to surmount the original depth of the water and the sinking of the ocean floor, it emerges as an island such as St. Helena in the South Atlantic. To rise above the water an undersea volcano must grow to a height of about four kilometers in 10 million years. Island volcanoes sink after 20 to 30 million years and become the sediment-capped seamounts called guyots.

configuration, which Elvers calls the Great Magnetic Bight, are offset by fracture zones at right angles to the magnetic pattern. The Great Magnetic Bight seems to have required impossible forms of sea-floor spreading and plate movement. However, Walter C. Pitman III and Dennis E. Hayes of the Lamont-Doherty Geological Observatory, among others, have been able to show that the configuration is fully understandable if it is assumed that the trailing edges of three plates met and formed a Y. Similar complex patterns have now been found in the Atlantic and the Pacific. Indeed, if the transform faults are perpendicular to the spreading centers and two spreading rates are known, both the orientation and the spreading rate of the third spreading center can be calculated even before it is mapped. One can also calculate the orientation and spreading rate for a third center that has already vanished in a trench.

As a crustal plate grows, its leading edge is destroyed at an equal rate. Sometimes this edge slides under the oncoming edge of another plate and returns to the asthenosphere. When this happens, a deep trench such as the Mariana Trench and the Tonga Trench in the Pacific is formed. In other areas the movement of the crust creates young mountain ranges. Xavier Le Pichon of the Lamont-Doherty Geological Observatory concludes that the occurrence of one event or the other is a function of the rate at which plates are moving together. If the rate is less than five to six centimeters per year, the crust can absorb the compression and buckles up into large mountain ranges such as the Himalayas. In these ranges folding and overthrusting deform and shorten the crust. If the rate is higher, the plate breaks free and sinks into the mantle, creating an oceanic trench in which the topography and surface structure indicate tension.

Several crustal sinks are no longer active. Their past, however, can be deduced from their geology. Large-scale folding and thrust-faulting can be taken as evidence of the former presence of a crustal sink, although such deformation can also arise in other ways. Certain types of rock may also indicate the formation of a trench. The arcs of islands that lie parallel to trenches, for instance, are characterized by volcanoes that produce andesitic lavas, which are quite different from the basaltic lavas of the ocean floor. The trenches themselves, and the deep-sea floor in general, are featured by deposits of graywackes and cherts. These rocks are commonly found exposed on land in the thick prisms of sediment that lie in geosynclines: large depressed regions created by horizontal forces resembling those generated by a drifting crustal plate. Thus the presence of some or all of these types of rock may indicate the former existence of a crustal sink. This linking of marine geology at spreading centers with land geology at crustal sinks is becoming one of the most fruitful aspects of plate tectonics. Still, crustal sinks are by no means as informative about the history of the ocean floor as spreading centers are, because in such sinks much of the evidence of past events is destroyed. Even if the leading edge of a plate was once the side of a plate (or vice versa), there would be no way to tell them apart.

At the boundary between the land and the sea a puzzle presents itself. The sides of an oceanic trench move together at more than five centimeters per year, and it would seem that the sediment sliding into the bottom of the trench should be folded into pronounced ridges and valleys. Yet virtually undeformed sediments have been mapped in trenches by David William Scholl and his colleagues at the U.S. Naval Electronics Laboratory Center. Furthermore, the enormous quantity of deep-ocean sediment that has presumably been swept up to the margins of trenches cannot be detected on sub-bottom profiling records. There are many ingenious (but unpublished) explanations of the phenomenon in terms of plate tectonics. One of them may conceivably be correct. According to that hypothesis, the sediments are intricately folded in such a way that the slopes and walls of trenches cannot be detected by normal survey techniques, which look at the sediments from the ocean surface and along profiles perpendicular to the slopes. This kind of folding could be detected only by trawling a recording instrument across the trench much closer to the bottom or by crossing the slope at an acute angle.

The concepts of sea-floor spreading and plate tectonics allow a quantitative evaluation of the interaction of many important variables in marine geology. By combining empirical observation with theory it is possible not only to explain but also to predict the thickness and age of sediments in a given locality, the scale and orientation of topographic relief, the thickness of various crustal layers, the orientation and offsetting of magnetic patterns, the distribution and depth of drowned ancient islands, the occurrence of trenches and young mountain ranges, the characteristics of earthquakes, and many other previously unrelated and unpredictable phenomena. This revolution in marine geology may take some years to run its course. Ideas are changing, and new puzzles present themselves even as the old ones are solved. The only certainty is that the subject will never be the same again.

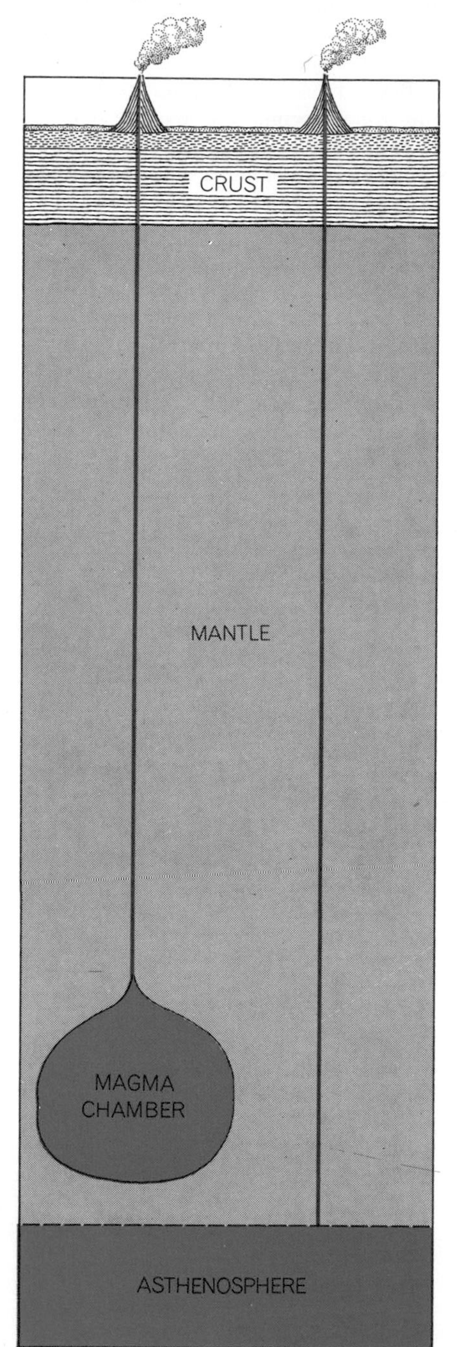

UNDERSEA ERUPTIONS can be explained in two ways, both consistent with observed facts. Since a volcano consists of different kinds of rock, it was originally thought that its conduit carried different forms of lava up from a magma chamber 50 kilometers or more down in the mantle. Now that crustal plates have been found to move, another theory must be considered. According to this idea, the conduit reaches through the mantle and taps several different kinds of magma at different places in the asthenosphere.

THE TRENCHES OF THE PACIFIC

ROBERT L. FISHER AND ROGER REVELLE
November 1955

On April 28, 1789, Lieutenant William Bligh, commanding H.M.S. *Bounty*, had a memorable quarrel in the Pacific Ocean with his senior warrant officer, one Fletcher Christian, as a result of which they parted company and sailed off in opposite directions—Christian in the *Bounty* and Bligh in the ship's longboat. This historic mutiny took place near the great volcano of Tofua in the Friendly Islands, better known today as the Tonga Islands. Bligh and Christian were well acquainted with the fact that the oceanic topography around these islands was somewhat unusual—full of treacherous shoals and narrow interisland passages. But they could not know, for methods of deep-sea sounding had not yet been invented, how unusual it really was, nor that this place would one day yield one of the most remarkable discoveries in the history of ocean-going exploration.

Beneath the placid sea east of the Tonga Islands yawns a monstrous chasm nearly seven miles deep. A hundred years after the *Bounty* episode another British vessel first plumbed its depths. Surveying the ocean bottom around the islands, Pelham Aldrich, commanding H.M.S. *Egeria*, was surprised to find that on two occasions his sounding lead did not touch bottom until 24,000 feet of wire had been paid out. Aldrich's discovery prompted other nations to send out expeditions to explore the Tonga undersea abyss. Eventually they traced out a great trench running from the Tonga Islands south to the Kermadec Islands [*see map on page 175*]. The deepest sounding made recently by the research vessel *Horizon* of the Scripps Institution of Oceanography, is some 35,000 feet. The immense chasm plunges about 6,000 feet farther below sea level than Mount Everest rises above it!

The Tonga-Kermadec Trench is now known to be but one member in a vast chain of deep, narrow trenches which lie like moats around the central basin of the Pacific [*see map*]. All of them parallel island archipelagoes or mountain ranges on the coasts of continents. Along the coast of South America, the drop from the top of the Andes to the bottom of the offshore trench is more than 40,000 feet. And the length of the undersea troughs is no less remarkable than their depth: some are 2,000 miles long.

These great gashes in the sea floor are so unlike anything on land that they are difficult for us as land animals to visualize. It is hard to grasp the reality of a chasm so deep that seven Grand Canyons could be piled on one another in it, and so long that it would extend from New York to Kansas City. Yet these are the dimensions of the Tonga-Kermadec Trench.

The size and peculiar shape of the Pacific trenches stir our sense of wonder. What implacable forces could have caused such large-scale distortions of the sea floor? Why are they so narrow, so long and so deep? What has become of the displaced material? Are they young or old, and what is the significance of the fact that they lie along the Pacific "ring of fire"—the zone of active volcanoes and violent earthquakes that encircles the vast ocean?

Although the trenches are still only sketchily explored, some tentative answers to these questions can be gleaned from the information already obtained. We can take the Tonga-Kermadec Trench as a typical example.

The Trench lies on a long, nearly straight, north-south line east of the Tonga and Kermadec archipelagoes. At its northern end it has a slight hook. It begins there as a gentle, spoon-shaped depression, runs southeasterly between Tonga and Samoa, then turns and deepens, strikes south for 1,200 miles and finally shoals and disappears at a point north of New Zealand. In its deepest central portion the Trench is very narrow—no more than five miles wide. The chasm is V-shaped, but the arm of the V on the island side is considerably steeper than on the seaward side: on the landward western wall the slopes average from 16 to 30 per cent—*i.e.*, in places they are steeper than the 24-per-cent-average slope of the sides of the Grand Canyon at Bright Angel. In longitudinal section the Trench consists of deep depressions separated by saddles; it looks like beads on a string, or peaks and saddles on an upside-down mountain range.

The islands on the western lip of the Trench appear to be part of the same crustal structure. They lie in two lines on a thousand-mile-long ridge atop the Trench's western slope. The islands of the Polynesian kingdom of Tonga are capped with limestones, laid down in shallow water during the last era of geologic time. These islands rest on broad shelves of drowned coral, 180 to 360 feet deep, and they rise in a series of terraces to a few hundred feet above sea level. West of the limestone islands, separated from them by a shallow trough, is a chain of submarine volcanoes and high volcanic islands. The volcanoes are explosive, rather than of the quiet Hawaiian variety. They have contributed great quantities of ash and cinders to the surrounding sea floor. Five of the island volcanoes have erupted during the last hundred years, and the danger of further explosions has forced the government of Tonga to evacuate the inhabitants. There are also active volcanoes below the sea surface. One of them, Falcon Bank, rises several hundred feet above the sea during an eruption; indeed, this bank is commonly called Falcon Island. After each eruption waves quickly erode the erupted lava, and within a few years the volcano is submerged again.

The floor of the Tonga-Kermadec Trench is rocky and seems to be nearly bare of sediments. During the Scripps Institution *Capricorn* expedition of 1952-1953, a core barrel with a heavy lead weight, which because of difficulties with the winch was dragged along the sea floor for several hours before it could be raised, came up badly battered by the bottom rocks. The heavy steel bail holding the instrument had been bent, and the lead weight looked as if it had been beaten with a hammer and chisel. Small fragments of black volcanic rock were embedded in the lead.

On the seaward slope of the Trench a single volcanic cone rises smoothly 27,000 feet, to within 1,200 feet of the sea surface. Just below its summit is a broad flat bench, tilted to the westward. Further study of this great cone, one of the highest mountains on earth, might tell us much about the history of the trench. Almost certainly the flat bench was cut by waves when the topmost part of the peak was above sea level. If shallow-water fossils could be recovered from the summit, we could fix the time when submergence occurred, and perhaps when the bench began to be tilted. This in turn might give us information about the rate of downward bending of the trench floor.

The Tonga Trench, as we have said, is typical of the great trenches in the Pacific. Some of the other giant furrows are the Aleutian, Kurile, Japan, Marianas, Philippine and Java Trenches on the northern and western sides and the Acapulco and Peru-Chile Trenches on the eastern side of the ocean. It is a remarkable and probably significant fact that the deepest trenches all have about the same maximum depth. The record sounding so far was one estimated to be somewhere between 35,290 and 35,640 feet, made in the trench southeast of the Mariana Islands. Appropriately enough this depth was measured by H.M.S. *Challenger*, the modern namesake of the famous ship whose voyage around the world in the 1870s marks the beginning of modern oceanography [see the article "The Voyage of the *Challenger*," by Herbert S. Bailey, Jr., beginning on page 20]. The original *Challenger* actually discovered the Marianas depression, and for many years it was known as the Challenger Deep.

All the deep trenches seem to be generally V-shaped in cross section, although some are slightly flattened at the very bottom; in the Japan and Philippine trenches this flat portion is two to 10 miles wide. Some shallower trenches and trenchlike depressions are U-shaped, with extremely flat bottoms over broad areas, as if they had been partly filled with sediments. If the V-shaped trenches contain any sediments, the layer cannot be more than a few hundred feet thick.

Direct exploration of the trenches is most difficult. Their great depth and extreme narrowness present formidable obstacles. To lower a dredge or other heavy sampling apparatus to the bottom of the deeper trenches, the ship needs a tapered wire rope of the strongest steel and a powerful, specially designed winch. Only three such winches exist today. One was built for the Swedish *Albatross* Expedition of 1948-49 and was later used on the Danish *Galathea* Expedition of 1950-52; another is installed on the Scripps Institution's research vessel *Spencer F. Baird*; the third is on the U.S.S.R. research ship *Vitiaz*. The winch drum on the *Baird* carries 40,000 feet of wire rope. When this wire was paid out in the Tonga Trench with a heavy core barrel on the end, the strain at deck level was 12 tons.

A single lowering of a dredge or core barrel takes many hours. It is complicated by the problem of keeping a small ship in position in a rolling sea, often under the influence of strong and unpredictable currents and shifting winds. The hazards of fouling the wire or of machinery breakdown under the heavy strain are always present, with the possible loss of the precious cable. Such a loss would be crippling, and much of the investment in time and effort required to send a scientific ship to a remote part of the world would also be lost.

If sounding and sampling of the bottom are difficult, drilling to find what lies beneath the bottom of the trenches is quite impossible, with present techniques. For such explorations we must depend on indirect methods—studies of earthquake waves, measurements of gravity anomalies, the flow of heat through the crust and the magnetic properties of the buried rocks.

The zone of trenches is the scene of our planet's most intense earthquake activity. Nearly all the major earthquakes, especially those originating at great depths, occur in this zone. The deepest-focus earthquakes are associated with the deepest and steepest trenches. This strongly suggests that the seat of the trench-producing forces lies far below the earth's surface.

The earthquakes may, indeed, be responsible for the fact that a line of explosive volcanoes parallels the trenches. Some investigators have proposed that the heat produced near the focus of an earthquake melts the surrounding rocks, and that the melted material rises and is eventually ejected by the volcanoes.

Seismic refraction studies give us another clue to the nature of the crust under the trenches. These investigations have shown that beneath the trenches (Tonga and others) the outer crust is less than one third as thick as under the continents. We therefore arrive at the important conclusion that the crustal structure under the trenches is oceanic and not continental.

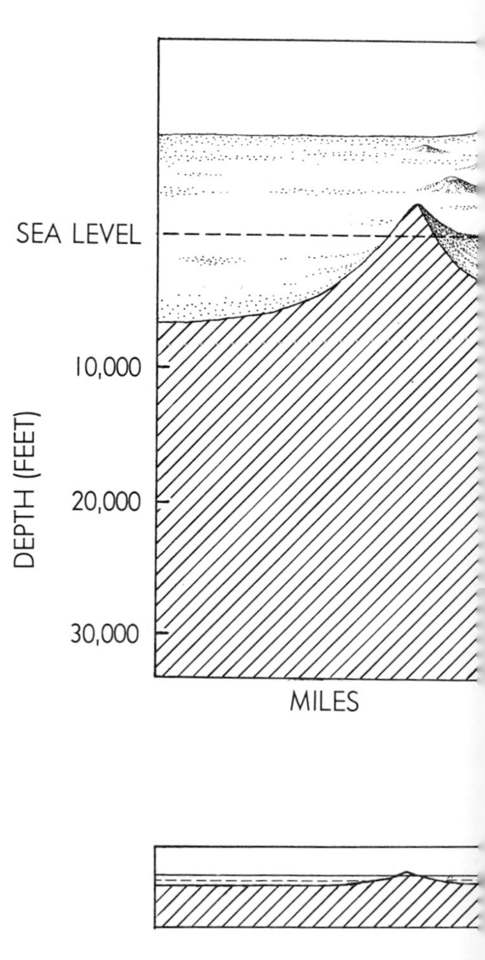

TONGA TRENCH would appear as in the upper drawing if viewed northward from a point in the central Tonga Islands, and if vertical distances were exaggerated by a

The most striking phenomenon associated with the trenches is a deficiency in gravity. The force of gravity depends on the mass of matter between the surface and some great depth in the earth. In general this force at any given latitude is about the same in ocean basins as in the continents, despite the fact that the volume of rock under a continental area is greater than under an equal area of the ocean. Evidently the continents "float" high above the deep sea floor, like rafts of light material in a heavier medium. Within the continents themselves, there is usually little difference in gravity between high mountains and low plains, and it is commonly supposed that the mountains are underlain by a larger thickness of light material than the plains. This state of the earth's crust is called isostatic equilibrium.

Measurements of gravity near trenches show pronounced departures from the expected values. These gravity anomalies are among the largest found on earth. It is clear that isostatic equilibrium does not exist near the trenches. The trench-producing forces must be acting against the force of gravity to pull the crust under the trenches downward!

What may these forces be? Here studies of heat flow in the crust suggest a possible answer. It has long been known that there is a small, steady flow of heat from the earth's depths outward toward the surface. Most of this heat is generated by the disintegration of radioactive elements in the crust and the mantle just beneath the crust. Near the surface of the earth the heat is transported outward principally by conduction, but at greater depths there may be a slow upward movement of the hot rock itself, carrying heat toward the surface. If rock at these depths moves upward in some regions of the earth, there must be other regions where cold rock moves downward. This movement would reduce the outward flow of heat. Now measurements near the floor of the Acapulco Trench show that the flow of heat there is less than half the average for the earth's surface (the average being about 250 calories per year per square inch of surface). So it may be that relatively cool rocks are slowly moving downward under the trench. Such a downward flow would tend to drag the crust down with it and may well account for the formation of the trench. If this process is occurring, the earth's mantle should be cooler under the trench than elsewhere. Magnetic measurements suggest that this is in fact the case, but they are too few so far to be conclusive.

Speculating from what we know, we may imagine that a trench has the following life history. Forces deep within the earth cause a foundering of the sea floor, forming a V-shaped trench. The depth stabilizes at about 35,000 feet, but crustal material, including sediments, may continue to be dragged downward into the earth. This is suggested by the

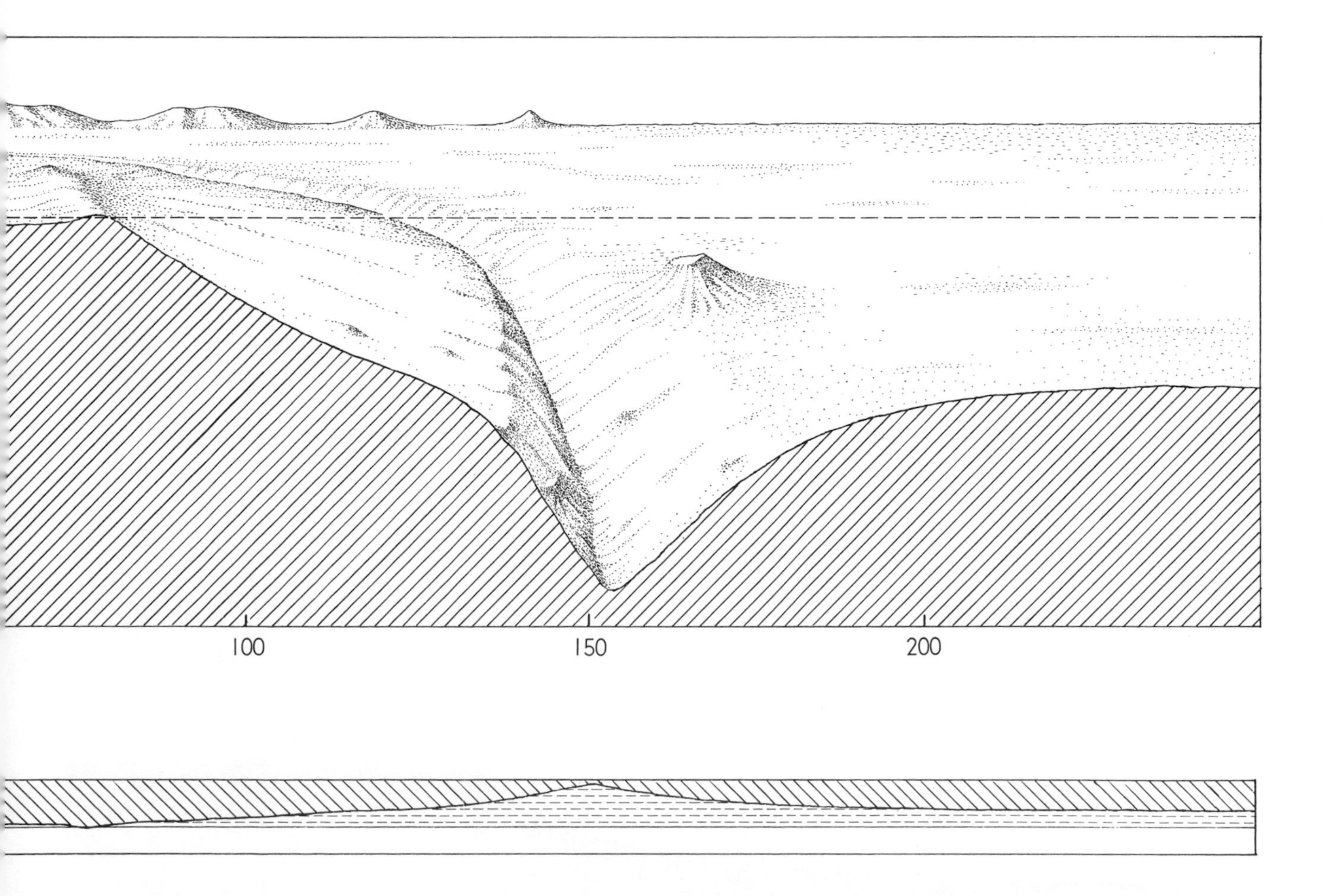

factor of 10 in comparison with horizontal. (In the lower drawing the cross section is shown without vertical exaggeration.) The exposed land mass in the left foreground is the island of Kao, a dormant volcano. In the distance are the Samoan Islands. The flat-topped seamount on the eastern slope of the trench is one of the highest mountains in the world. Its summit, which is tilted down toward the west about one degree, was probably worn flat by wave action when trench was shallower and mountain above water.

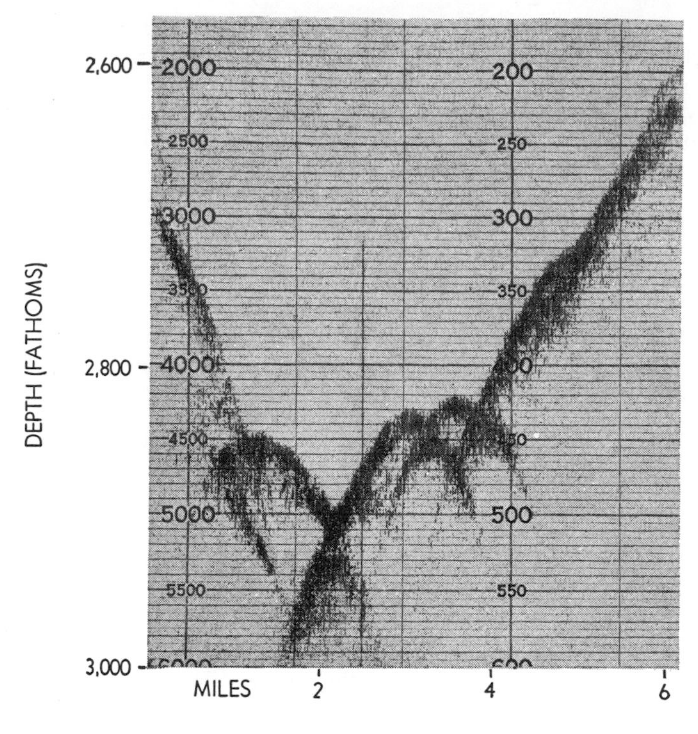

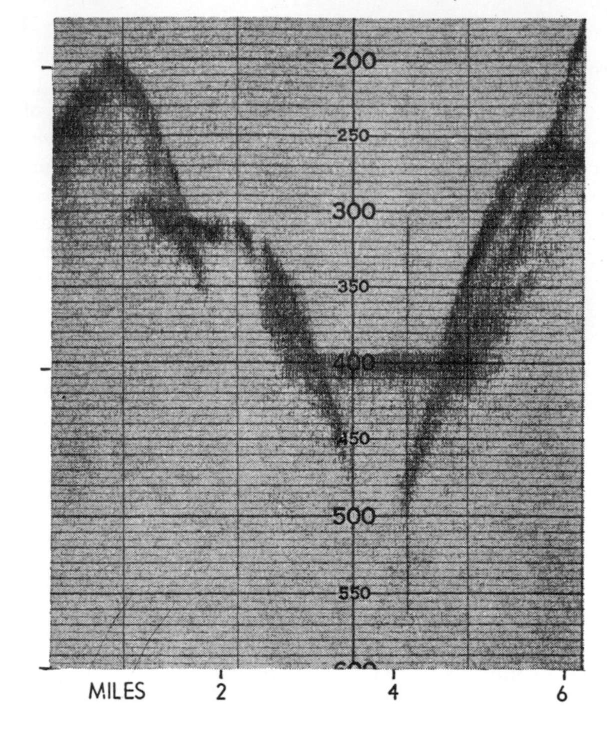

ACAPULCO TRENCH is revealed in cross section by echo soundings. Near Acapulco (*left*), the bottom is V-shaped, with little sediment, and 2,930 fathoms deep. Near Manzanillo (*right*), it is flat at 2,795 fathoms. Numbers on the records do not represent fathoms.

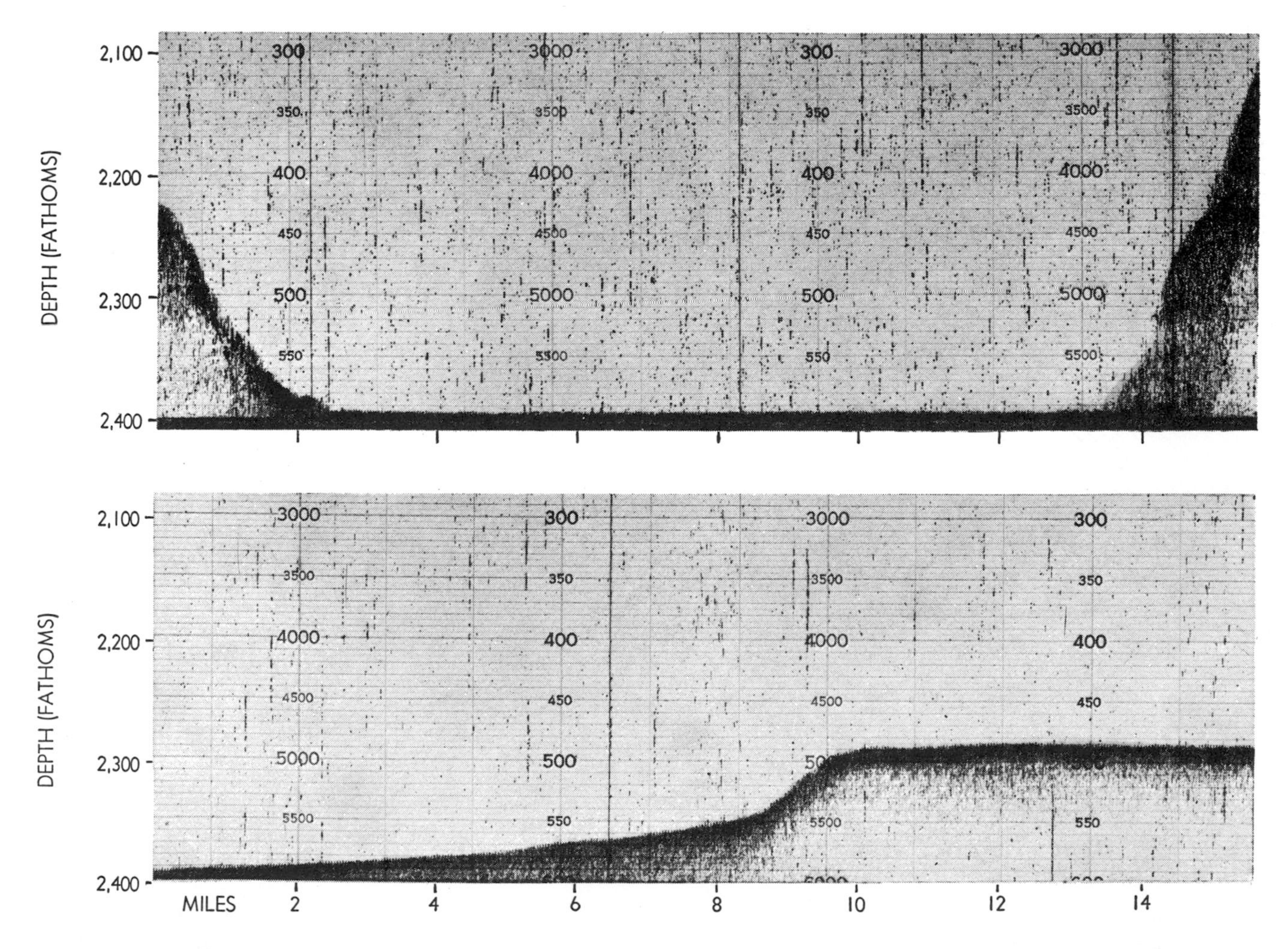

CEDROS TROUGH, a short trench off the coast of Lower California, is traced in cross section by the upper echo-sounding and longitudinally by the lower. The bottom is flat at 2,395 fathoms and measures 11 miles across at the point where measurement was made.

fact that the deepest trenches contain virtually no sediments, although they are natural sediment traps. During this stage in the trench's history there is violent volcanic and earthquake activity.

In a later stage the internal forces pulling or squeezing the crust downward under the trench become less active, and the trench begins to fill up with sediments. It acquires a flat bottom and a U shape as the accumulating sediments cover the topographic irregularities. The sediments may eventually pile up so high that the top of the pile rises above the sea, forming islands, when isostatic equilibrium is finally restored. The topmost sediments will be rock of the kind that is deposited in shallow water, like the limestones of Tonga and the Marianas.

Another process also may come into play if a thick layer of sediments accumulates. Because of their lower heat conductivity, the sediments would form an insulating blanket along the former trench. This would block the heat flow from the interior and cause a temperature rise which would partly melt the deep rocks. The melted material might then move upward and transform the heavy existing rock and the lower part of the sedimentary layer into light, granite-like rock. The thickness of the crust therefore should increase.

Some geophysicists have suggested that it was by such a sequence of events, occurring repeatedly during the geologic past, that the continents grew, at the expense of the ocean basins. The question then arises: Where on the continents are the ancient, filled-in trenches?

One naturally thinks at once of the long, narrow structures, called geosynclines, where sediments piled up and mountain ranges developed by compression and folding. Were some geosynclines originally deep trenches such as now exist on the sea floor? It has usually been thought that this is not so, because most of the sediments in geosynclines appear to have been laid down in shallow water rather than in deep trenches. However, this appearance may in some cases be an illusion. Sediment samples collected from even the deepest trenches resemble in many ways deposits laid down in shallow water.

It is true that the sedimentary rocks in geosynclines contain no recognizable fossils of deep-sea animals. But trenches have little life that could leave a distinctive record. The depths of a trench are completely dark, except for the fee-

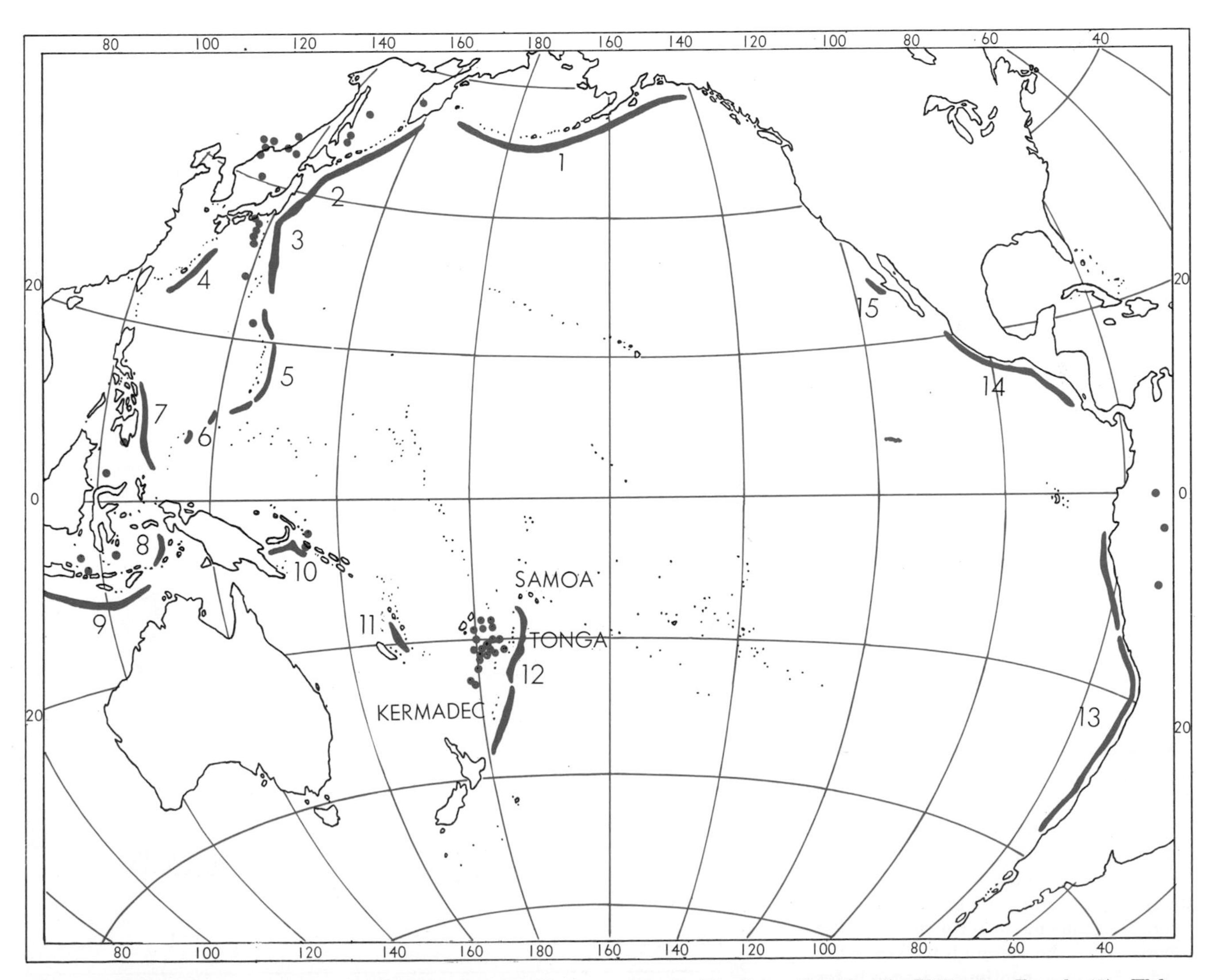

RING OF TRENCHES around the central basin of the Pacific is shown in color. The colored dots represent deep centers of earthquake activity. Numbered serially are: Aleutian Trench (1), Kurile Trench (2), Japan Trench (3), Nansei Shoto Trench (4), Marianas Trench (5), Palau Trench (6), Philippine Trench (7), Weber Trough (8), Java Trough (9), New Britain Trench (10), New Hebrides Trench (11), Tonga-Kermadec Trench (12), Peru-Chile Trench (13), Acapulco-Guatemala Trench (14), Cedros Trough (15).

ble and flickering light produced by luminous organisms, and no plants can live there. The animals and bacteria of the abyss must gain their sparse food supply from plant and animal remains settling slowly from the upper layers of the sea. The waters are very cold: about 36.5 degrees Fahrenheit now, though they may have been some 20 degrees warmer in the geologic past. The pressure at the bottom of a trench is, of course, enormous—more than eight tons per square inch.

The Danish *Galathea* Expedition several years ago dredged up a few animals from the floors of trenches more than 30,000 feet deep. The principal animals recovered were sea cucumbers and a type of sea anemone, neither of which would be likely to leave distinctive fossils. Some worms, clams and crustaceans also were obtained, together with beautiful glass sponges.

Materials which are usually supposed to be deposited only in shallow water have actually been found on the floor of some of the deep trenches. The *Galathea* recovered fine gray sand, pebbles, cobbles and land-plant debris from the floor of the Philippine Trench. The Lamont Geological Observatory of Columbia University found, in cores from the Puerto Rico Trough, the skeletons of plants and animals that live only at shallow depths. In the flat-bottomed northern part of the Acapulco Trench one core contained soft black mud, high in organic debris and stinking of hydrogen sulfide, while in other cores layers of gray, green and brownish sand and silt were interbedded with charred woody fragments and fine green mud.

However, it is clear that some geosynclines, notably those along what are now the Appalachian Mountains, could not have been deep sea trenches, for they contain deposits from marshes and flood plains interbedded with marine sediments, and therefore the deposits must have been laid down in shallow water.

The question remains: Where are the trenches of yesteryear? Are we living in an exceptional geologic era; are the apparently young trenches of the present day unusual formations that have had no counterparts during most of geologic time? Such a speculation would be repugnant to many geologists, because it would be difficult to reconcile with the doctrine that the present is the key to the past. We must continue to search for ancient trenches—on the deep-sea floor, in the marginal shallow water areas and on the continents themselves.

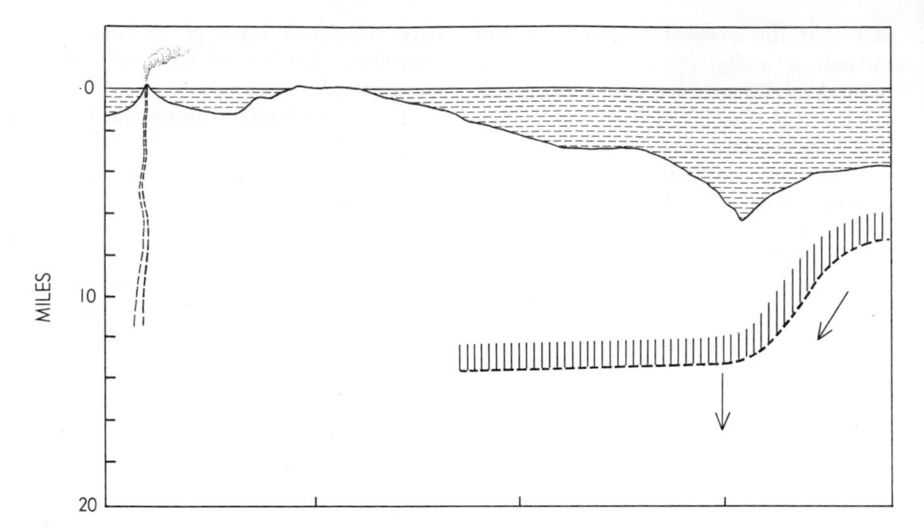

POSSIBLE LIFE HISTORY of a trench is outlined in the three diagrams on this page. Here a force deep within the earth pulls down the floor of the ocean to form a V-shaped trench.

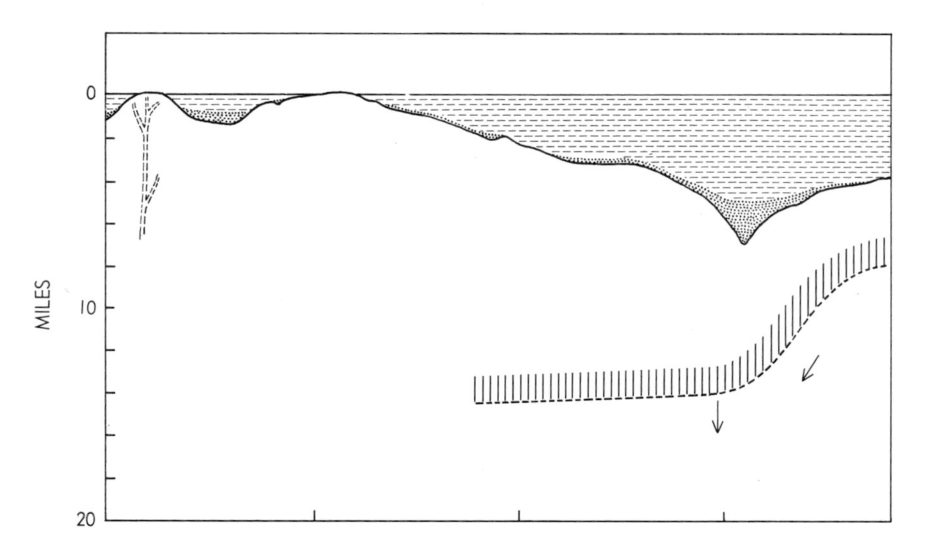

SEDIMENT COLLECTS in the bottom of the trench during the second stage, when the deep force weakens and relaxes its downward pull. Bottom is now flat and V changes to U.

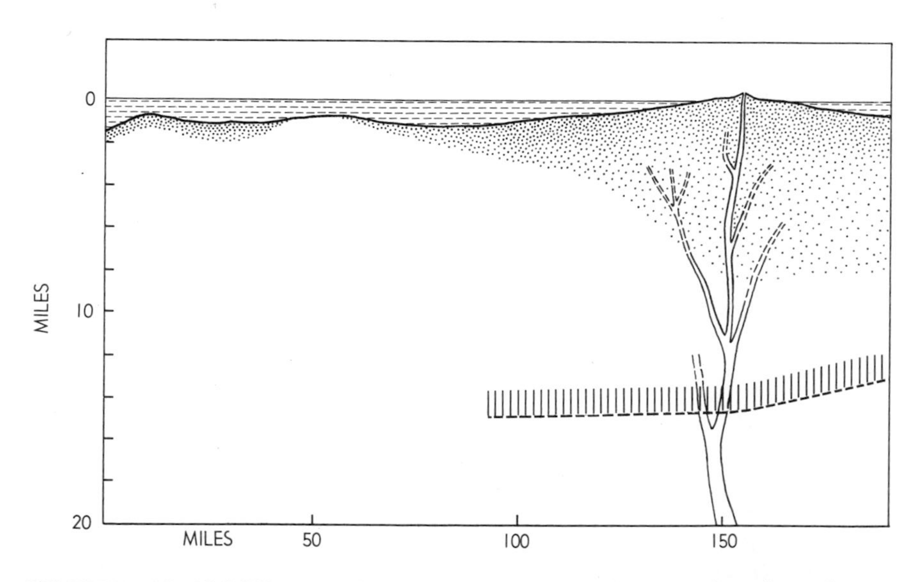

SEDIMENT BUILDS UP and, with isostatic adjustment of the crust, rises above the surface of the sea. Deep molten rock rises through the sediment and is released by volcanoes.

THE AFAR TRIANGLE

HAROUN TAZIEFF
February 1970

In the northeastern part of Ethiopia, at the juncture of the Red Sea and the Gulf of Aden, lies a region known as the Afar triangle. It is a wild and rugged country, featured by below-sea-level deserts, towering escarpments, fissures, volcanoes and craters. Few men have explored the region, and until recently it was terra incognita as far as its geology and even its exact geography were concerned. Now, however, the discovery of detailed evidence for the drift of continents and the growth of oceans has focused considerable interest on the Afar triangle [see the article "The Origin of the Oceans," by Sir Edward Bullard, beginning on page 196]. The triangle seems to be a focal point for new oceans in the making. What is more, whereas elsewhere the process that is producing continental separation is hidden in the depths of the ocean, here we can see it taking place in direct view on dry land.

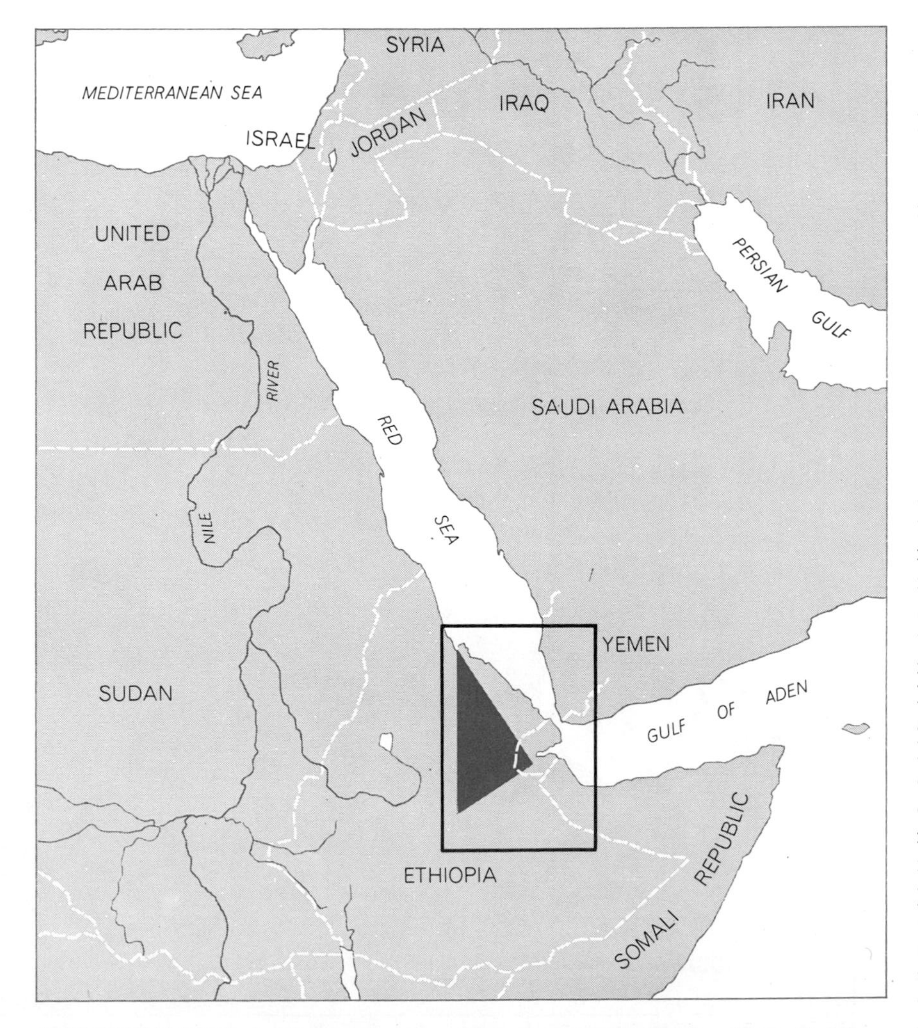

AFAR TRIANGLE (*color*) is in northeastern Ethiopia where the Red Sea rift, the Carlsberg Ridge of the Indian Ocean and the Rift Valley system of East Africa meet. Its northern section was once submerged, and part is still below sea level (*see pages 178 and 179*).

The theory that the earth's present continents were once united in a great land mass and have gradually drifted apart is now generally accepted. Evidence collected over the past 10 years, mainly by exploration of the worldwide system of ridges running along the middle of the oceans, has outlined a convincing picture of how the continents were separated [see "The Confirmation of Continental Drift," by Patrick M. Hurley; Scientific American Offprint 874]. In brief, the process seems to be as follows. The material of the continents is a layer of comparatively light sialic rock, resting on a denser basaltic magma underneath. Stresses in the earth's crust may crack the sialic layer, producing faults and fissures that can be as much as 20 meters wide. Then molten magma wells up into the fissure, sometimes spilling out over the surface. The magma hardens into solid rock, thus holding apart the separated sialic blocks. Over long periods of time the same stresses create new fissures parallel to the old ones; these fissures too are filled with magma. Examination of the oceanic ridge has shown that it is composed of parallel strips of hardened magma, indicating that the crust was repeatedly fissured along the axis of the ridge and that the continental blocks thus moved farther and farther apart. The upwelled strips of basalt are distinguishable from one another by differences in the direction of their magnetic polarity, which reflect changes in the earth's magnetic field with time.

The overall conclusion from studies of the mid-oceanic ridges is that the continents have been moving apart at an average rate of a few centimeters per year. In the South Atlantic, for instance, it appears that the ocean has been widening at the rate of 2.5 centimeters per year, which indicates that the South American continent began to separate from Africa approximately 180 million to 200 million years ago.

Recent worldwide explorations of the ocean bottoms have shown that a rise running along the middle of the Indian Ocean, known as the Carlsberg Ridge, has a branch extending into the Gulf of Aden. The Red Sea bottom similarly has an axial ridge with physical properties like those of the oceanic ridges. The Gulf of Aden and Red Sea rifts, which are perpendicular to each other, meet in the Afar triangle. And the same region also lies at the northern end of the system of rift valleys that runs down the eastern side of the African continent. Once these facts were recognized, it was suddenly realized that the largely uncharted Afar triangle might offer an extraordinary opportunity for investigating the origin of new oceans.

The Afar triangle is one of the world's most forbidding regions. In addition to the fact that its terrain is all but impassable, the area is extremely hot; we were to find that the temperature rises to as high as 134 degrees Fahrenheit in the shade in summer and 123 degrees in winter. The region is inhabited only by nomadic tribes of fierce repute; the young warriors are said to mutilate male victims to offer trophies to their women, and they have been known to massacre armed parties for their weapons. Several exploring parties in the 19th century were slaughtered by the tribesmen.

Of the expeditions that had explored the Afar triangle before we began to survey it in 1967, the best-known was one carried out in the spring of 1928 by two Italians, Tullio Pastori and G. Rosina, a British mining engineer, L. M. Nesbitt, and half a dozen Ethiopians. The expedition's leader was Pastori, a hardy ore prospector who earlier had explored various parts of the region (and who, at the age of 60, in 1943 escaped from a British prisoner-of-war camp in Kenya and made his way on foot all the way to Alexandria on the Mediterranean coast). Nesbitt, who wrote a report on the 1928 expedition, vividly described the difficulties and dangers the expedition had encountered. It was apparently the first journey along the entire length of the Afar triangle, and Nesbitt's information provided a basis for the only detailed maps of the region that were available before our expeditions. The 1928 party was not equipped, however, to undertake a geological survey.

In 1967 we organized a team of specialists to study the geology of the Afar triangle with all possible thoroughness. Our group includes investigators from several universities in Italy, France and the U.S. It consists of three petrologists

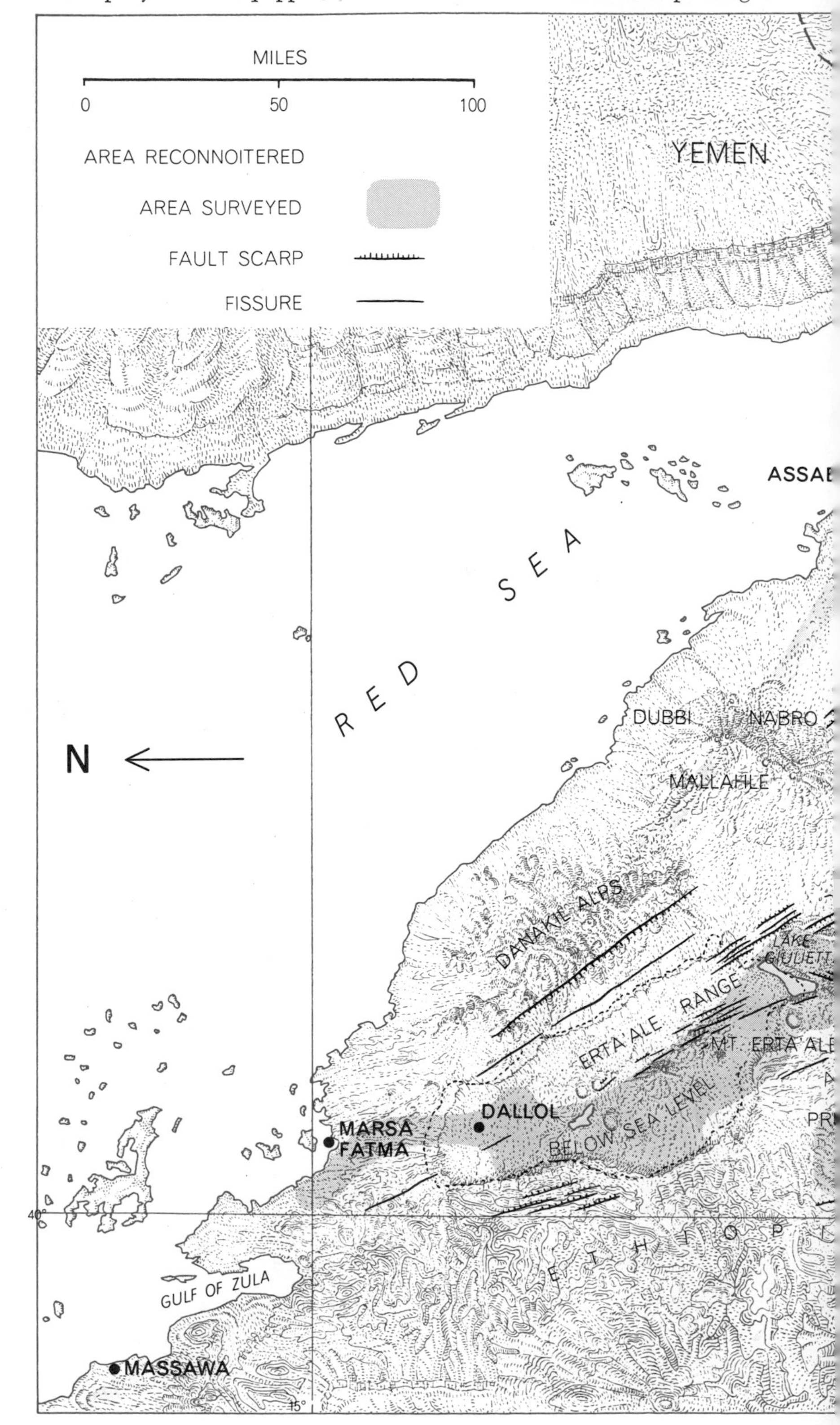

GEOLOGICAL SURVEY of the Afar triangle was conducted by the author and his associates beginning in 1967. In this map north is at the left and east at the top. The most inten-

(Giorgio Marinelli, Franco Barberi and Jacques Varet), four geochemists (Giorgio Ferrara, Sergio Borsi, J. L. Cheminée and Marino Martini), one tectonic geologist (Gaetano Giglia), two students of the geology of recent times (Hugues Faure and Colette Roubet), a geophysicist (Guy Bonnet), an oceanographer (Enrico Bonatti of the University of Miami) and a volcanologist (myself).

We have now completed three expeditions to the Afar triangle, all during the comparatively cool winter season (in 1967–1968, 1968–1969 and 1969–1970). Our party was unarmed and was not molested by the local people. For our explorations we have had the invaluable help of a helicopter, without which it would simply have been impossible to do serious, comprehensive fieldwork in that rugged country. So far we have

sive work was done in and near the area below sea level, between the Danakil Alps and the Ethiopian escarpment. In addition the group scouted widely by helicopter, landing more than 1,000 times to gather rock samples for later analysis and to study formations.

MASSIVE FAULT BLOCKS characterize the terrain of the Afar triangle. The aerial photograph above shows a region near 12 degrees north latitude, where faulting has caused huge segments of surface to subside below the level of the surrounding plateau. The depressed structures are whitened in places by salt or gypsum.

STAIRSTEP LANDSCAPE of the eastern shore of Lake Giulietti (*below*) is further evidence of the extensive downfaulting that has shaped much of the Afar triangle. Such long, depressed blocks, bordered by fault zones, are called graben structures by geologists; the graben structures here are among the world's most spectacular.

spent a total of 13 weeks in the field and have carefully mapped the geological structures and petrology of some 12,000 square miles in the northern half of the triangle. We have crisscrossed this entire area with the helicopter and have landed at more than 1,000 places to sample rocks and examine the tectonic structures on the ground.

On its northeast side the Afar triangle is separated from the Red Sea by a series of heights produced by deformations of the earth crust at faults and by volcanic action. Reading from north to south, these include a tectonic horst (upraised block) called the Danakil Alps, a mountain range formed by sedimentary and intrusive rocks, a volcanic massif composed of an active volcano and three volcanic piles crowned with calderas (craters of collapsed volcanoes) and, at the southern end, another mountainous horst. On the south side the triangle is bounded by the tall Somali scarp, from 4,900 to 6,500 feet high. On the west, the third side of the triangle, stands the huge Ethiopian escarpment, towering in some places to more than 13,000 feet. Here clifftops stand higher above the valley floor below them than anywhere else in the world. As for the floor of the triangle itself, it rises gently from about 400 feet below sea level near its northern end to more than 3,300 feet above sea level at the Somali end some 300 miles south.

What is one to make of this strange and spectacular landscape? Until very recently most geologists believed the Afar triangle was a funnel-shaped widening, produced in some unexplained manner, of the Great Rift Valley of East Africa. Our studies of the region's geology have led us to a completely different interpretation. The facts, as we have observed them on the scene, indicate that the floor of the triangle is actually a part of the Red Sea! In fact, the triangle from its northern apex down to the Ghubbet al Kharab at the western end of the Gulf of Aden and the Gulf of Tadjoura is a southwestern continuation of the central trough of the Red Sea that fades out close to 15 degrees south latitude. The tectonic trends of the Red Sea are evident and are geologically active throughout the area. There are none of the trans-rift structures or other formations that were once believed to account for the Afar triangle. It appears that as recently as some tens of thousands of years ago at least the northern half of this area was covered with seawater, with only the Danakil Alps and the high volcanoes standing above water as islands, and that most of the land has since been raised above sea level by tectonic uplifts through earthquakes, volcanic action and the rise of basaltic magma that filled the fissures.

The evidence that the Afar triangle is a part of the Red Sea floor, not some bizarre widening of the Ethiopian rift, can be seen on every hand. To begin with, we found that the observable facts contradicted other explanations of the region's topography. It had been suggested, for example, that the lofty escarpment on the west side of the triangle was produced by a downfolding of the high plateau, followed by erosion of the resulting hillside. Our observations turned up three important objections to this idea: (1) the blocks of faulted crust in the lower part of the escarpment are tilted westward—in the direction opposite to what would be expected in a downfold; (2) the supposed erosion should have deposited a vast amount of sediment in the triangle's closed basin, but almost none was found there; (3) the basin is filled with more than 3,000 feet of evaporites (salt formations deposited by the evaporation of seawater), and the basin's wall plunged down below this material at a steep angle, again indicating that the western boundary of the triangle was formed by slippage of the crust along a fault rather than by downwarping.

It had also been suggested that south of the Danakil Alps a belt of big calderas, apparently running north and south, was a continuation of the main Ethiopian rift and was a major active feature of the entire region. Field investigations show that no north-south belt exists, and that all the big calderas are located on a graben (a depressed section of crust) running north-northeast and south-southwest. This observation is an important one, because it again demonstrates that the Afar triangle is a part of the Red Sea and not an extension of the Ethiopian rift.

We found innumerable signs that the topography of the Afar triangle has been created by violent events that have occurred in very recent times, geologically speaking, and are still in progress. The entire northern half of the triangle shows clear evidence of extensive faulting of the crust and active crustal movement along the faults. North of latitude 13 degrees 10 minutes north all these faults are aligned along the axis of the triangle in the north-northwest to south-southeast direction [*see illustration on pages 178 and 179*]. South of 13 degrees 10 minutes north down to the Ghubbet al Kharab (11 degrees 30 minutes north) the same direction prevails, although there are also many faults and fissures running northwest-southeast and east-west. Along much of this part of the triangle the evidences of crustal movement are in plain view as wide-open fissures, horsts and grabens that form a classic graben structure of steps down the sides of a major depression. Even where the graben structure is now hidden under deposits of volcanic material and evaporites (but is still detectable by geomagnetic mapping), the fault axis is shown visibly by potash salt domes, explosion craters, boiling springs and other signs of volcanic activity below the surface. These eruptions follow the same line in the north-northwest to south-southeast direction.

This direction is precisely that of the Red Sea, so that the whole of the northern part of the Afar triangle can be regarded as part of the sea. All the evidence suggests that the waters of the Red Sea extended south-southeastward as far as the Somali scarp in the geologically recent past and that the present absence of water in the Afar triangle is only a temporary phase in the development of the ocean.

That the fissuring and displacements of the crust are going on actively at the present time is shown by several signs we were able to observe. Here and there we found fresh faults cutting through very young structures, such as alluvial fans and cones of active volcanoes. The volcano called Erta'Ale, the most active in all East Africa, has its main cone sliced by tectonic fissures that are parallel to the Red Sea axis. This is an exceptional phenomenon; volcanic fissures are usually not parallel but radial. We ourselves actually witnessed a significant event during our study: an earthquake in the middle of the depression on March 26, 1969, produced an appreciable slippage of the crust down a fault in the north-northwest to south-southeast direction.

Further evidence that the Afar triangle is actually a scene of oceans in the making came from the examination of the rocks themselves. Our samples gave every indication that the triangle's central trough contains no sialic (that is, continental) rock. The rocks, all very young, of the Erta'Ale volcanic range, which runs parallel to the Red Sea central trough, are preponderantly basaltic and typical of the rocks of oceanic ridges. From analysis of more than 100

FLAT-TOPPED VOLCANO, Mount Asmara is composed of shards of volcanic glass such as are formed during underwater volcanic explosions. It resembles the numerous guyots, or submerged oceanic mountains, whose level summits are usually attributed to wave erosion. Because Mount Asmara was formed under water, it may be that a flat top is instead a feature common to all such volcanoes.

CINDER CONE near Lake Giulietti was built up at a point that overlies one of the innumerable fault lines found in the Afar triangle. A subsequent horizontal shift of one fault block has moved the far half of the cinder cone 100 meters ahead of the near half.

specimens from this range we estimate that its composition is more than 90 percent basalts, the rest consisting of varieties of volcanic rock: dark trachytes (8 percent) and rhyolites (.5 percent). Chemically these rocks all show an evolutionary relationship, forming a practically continuous series from olivine basalts to the dark trachytes and rhyolites. The relationship has been confirmed by analysis of their strontium; the ratio of the rare isotope strontium 87 to the common one, strontium 88, in all these rocks is uniform and about the same as that in oceanic basalts. Another index also shows clearly that they were derived from oceanic magmas. The differentiation into the rock varieties in the series apparently came about mainly through gravity separation of components as the original magmas evolved in a series of steps, marked by a distinct iron enrichment in the middle stages. The end products are glassy trachytes and rhyolites, giving evidence that they were produced in a highly fluid state.

We found that a parallel volcanic range in the triangle, the Alayta range, has the same character, structurally and petrographically, as Erta'Ale. To sum up, the petrology of both ranges seems to indicate that they were born as upwellings through fissures in the crust, that their parent material was basaltic magma such as is characteristic of ocean floors, and hence that there is no sialic crust immediately under these ranges. All of this suggests that in the northern part of the Afar triangle continental blocks have already been severed from each other. Evidently this area is a part of the Red Sea ocean-forming system that is separating Arabia from the African continent. Similarly, our survey of the region indicates that the Danakil Alps and two smaller horsts we have detected south of latitude 13 degrees (one close to the Ethiopian escarpment and the other in the middle of the Afar triangle) are continental structures in the process of being split off and separated from the Ethiopian plateau.

Many signs that the northern region of the Afar triangle was covered with seawater in quite recent times turned up in our on-the-ground explorations. On a terrace at the foot of the Ethiopian scarp Faure found a stone axe that was encrusted with seashells, indicating that the sea had covered it after it was abandoned. This axe is from the Acheulean period, not more than 200,000 years ago. We found coral reefs of the Quaternary period (geologically the most recent) in the lava fields of the region. Scattered about on the floor of the triangle are ash rings, resembling the one at Diamond Head in Hawaii, such as are known to be formed under water; these consist entirely of the shards of volcanic glass called hyaloclastites, which are typical of underwater basaltic eruptions. We even discovered a flat-topped cone, now standing on the dry beds of the triangle, that bears every resemblance to the famous guyots, or flat-topped seamounts, of ocean floors.

The ash rings were particularly interesting to me because I had previously witnessed the formation of two such structures from submarine eruptions, one in the Azores in 1957 and another south of Iceland in 1965. I found that the ash ring is formed as the result of secondary steam explosions from a volcanic eruption under water. The primary eruption tears the molten magma into pieces and hurls them into the water. Because of their large surface-to-volume ratio (a ratio much larger than it would be for a quiet lava flow) the lumps of hot lava transfer enough heat to the water immediately around them to turn it into superheated steam. This steam generates secondary explosions that shatter the lumps into tiny pieces of glass (hyaloclastites) and throw them high into the air—from half a kilometer to more than a kilometer above sea level (which is a great deal higher than material is tossed by the usual volcanic explosions of the Hawaiian or Strombolian type on land). The tiny fragments falling back around the volcano's vent form a large rim that is frequently horizontal and highly regular, because the fragments are deposited under water. We found that the Afar ash rings looked very fresh, and the most eroded one (presumably the oldest) was overlain with corals and shells of marine animals of the Pleistocene period, all of which adds to the evidence that the region was covered by the sea not long ago.

Our examination of the apparent guyot we found in the depression seems to cast a new light on the origin of seamounts. Hundreds of these flat-topped submarine mountains, with their tops in

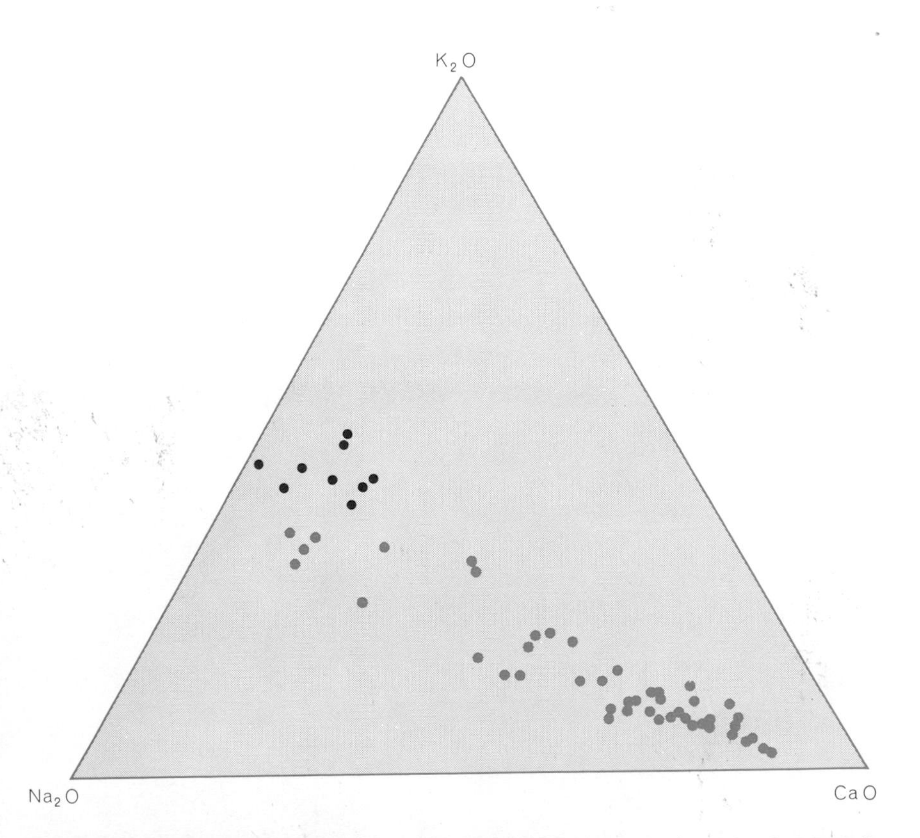

DIFFERENTIATION DIAGRAM shows that the chemical composition of volcanic rock from Mount Erta'Ale (*colored dots*), midway between the Ethiopian escarpment and the Danakil Alps, is different from the composition of the rock from the Pierre Pruvost volcano (*black dots*), which lies close to the escarpment. The Pruvost lavas result either from a "contamination" of deep magmas by the sialic material in the earth's crust or from a melting of the crust itself. The lack of such contamination in the Erta'Ale lavas suggests that, as the Alps and the escarpment have drifted apart, they have split open the earth's crust along a vast rift zone, with the result that no crust remains below the Erta'Ale range.

BLACK RIBBONS of fresh basalt in the Erta'Ale mountains mark zones where molten rock has poured out of fissures in the floor of the Afar triangle. The basalt is chemically similar to the magmas that have welled up from the rifts in the earth's mid-oceanic ridges.

LOW VOLCANIC CONE that marks the northern end of the Erta'-Ale mountain range is further evidence that much of the area was submerged in the recent geologic past. The shards of glass that make up the cone are formed during an underwater eruption as lumps of hot lava turn the water about them into superheated steam. The steam explodes the lumps into hundreds of fragments.

some cases as much as 1,000 fathoms or more below the sea surface, have been discovered in the world ocean. To explain their flat tops, it is generally supposed that the peaks formerly stood above the surface of the water and were eroded flat by the waves, and that the tops are now submerged because the sea bottom sank or the ocean level rose. We found that the truncated cone resembling a seamount in the Afar triangle was built in a way that would account for its flat top simply by the method of its construction. This pile, called Mount Asmara, is about 1,200 feet high and tapers from a width of a mile and a quarter at the base to two-thirds of a mile at the top. It was apparently formed by a buildup of layer on layer of beautiful golden hyaloclastites, which consist of palagonitized olivine basalt and are clearly of submarine origin. Such a process, produced by successive eruptions from a volcanic vent, can account for Mount Asmara's flat top. I am wondering if many of the seamounts in the oceans may not have been built in the same way under water. Perhaps their tops never have in fact been above water but may someday rise out of the ocean as the building process goes on. At all events, it appears that Mount Asmara in the not distant past was totally submerged, with its base more than 1,600 feet below sea level.

If we suppose that the Afar triangle is a part of the Red Sea, the coast of Arabia matches very well the contour of the "coastline" of the part of the African continent from which it is assumed to be separating; the match is at least as good as that between Africa and South America on opposite sides of the Atlantic.

We still have to explain why the axis of the Afar rift is displaced somewhat westward from that of the central ridge in the Red Sea and why the two troughs are now separated. That question will have to await our further explorations in the southern half of the depression. It seems possible that the Gulf of Aden ridge and rift system, thrusting into the depression at right angles to the Red Sea axis, may be exerting a powerful influence that could account for the displacement.

Meanwhile the information obtained so far about the Afar triangle raises an interesting economic question. Because of the absence of sialic crust below the axes of active volcanic ranges of basaltic composition and the probable closeness of the hot mantle to the surface in the northern part of the triangle, a great deal of heat flows into the underground rock strata there. These strata are highly porous and absorb a vast amount of fresh water that drains into the floor of the triangle from the surrounding highlands in the rainy season. Consequently it seems likely that subterranean fields of superheated water and steam underlie parts of this desert region where impermeable strata prevent them from escaping into the atmosphere. If they could be tapped, they might supply millions of kilowatt-hours of cheap electricity per year. This reservoir could supply power to the nearby seaports (Assab, Massawa, Djibouti) to support large new industries (aluminum and other metallurgies, petrochemistry, fertilizers, canneries) in which electricity is the main cost factor. With ores and other raw materials shipped to these ports at low cost, the price of finished products would also be low, and the area could be expected to have a tremendous economic growth. This almost desert region might be an industrial megalopolis in the future—a future far less remote than the geologic one, when the Gulf of Aden and the Red Sea will have expanded into new oceans.

SALT PLAIN lies near the northern apex of the Afar triangle. Each year, when rainfall in the highlands to the east and west drains into this low area, the plain is covered by pools of brine. These are too shallow, however, to hinder the passage of salt caravans.

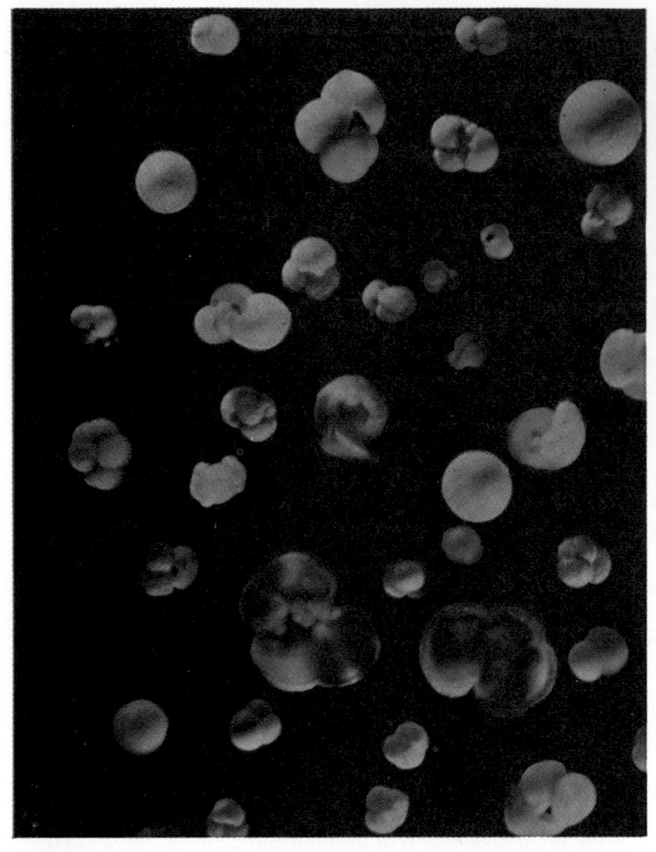

FORAMINIFERA are the most important microfossils in the study of earth history. Various planktonic species produced the fossil shells shown here at a magnification of about 20 diameters.

GLOBIGERINA PACHYDERMA, a foraminifer, lives in the Arctic Ocean. In this photomicrograph light from below shows details of chambers in the shells. Magnification here is some 60 diameters.

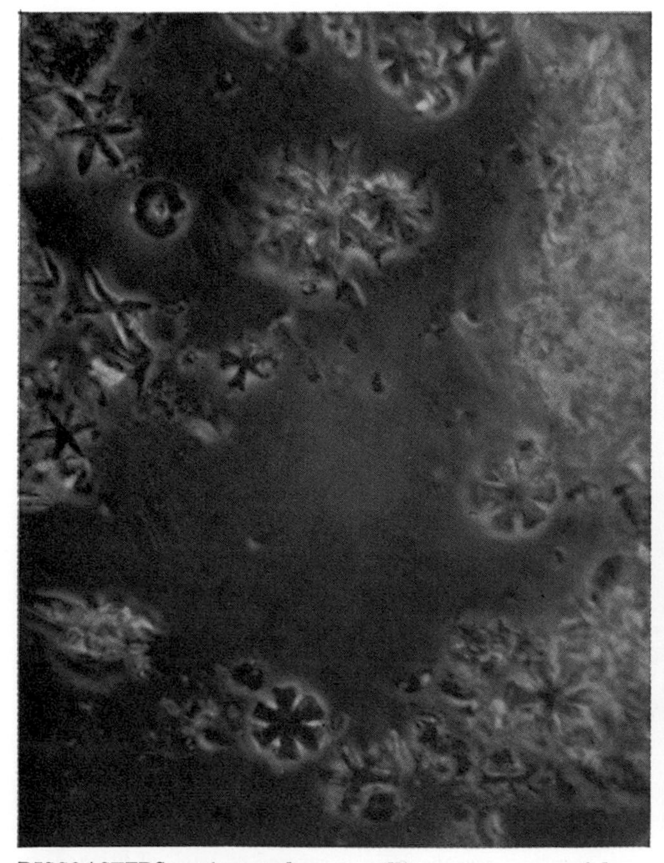

DISCOASTERS, extinct perhaps a million years, are useful as guide fossils. Forms with six rays shown here lived about 40 million years ago. They are magnified approximately 1,000 diameters.

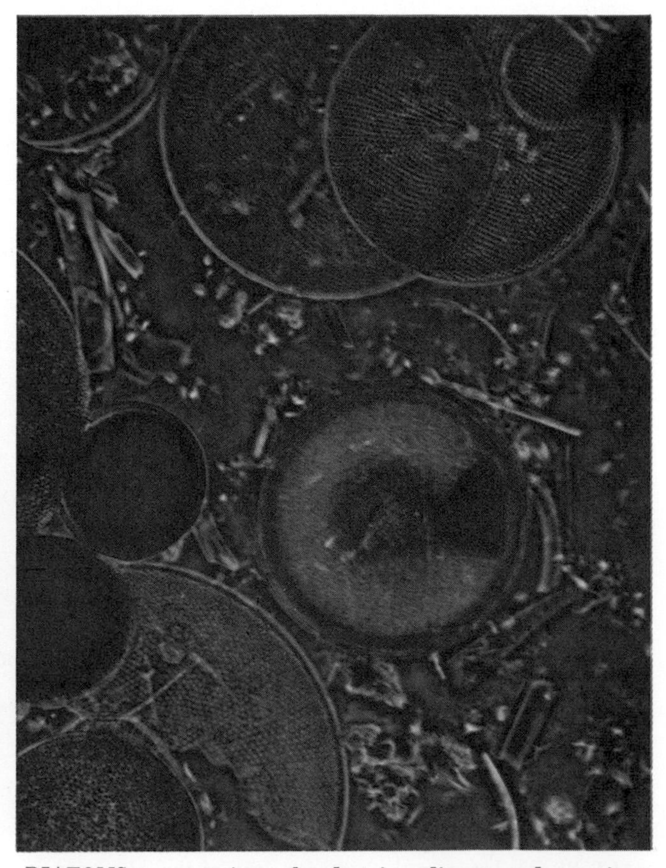

DIATOMS are sometimes abundant in sediments and sometimes missing. These were illuminated by polarized light. Magnification is 200 diameters. Photomicrographs were made by Roman Vishniac.

MICROPALEONTOLOGY

DAVID B. ERICSON AND GOESTA WOLLIN

July 1962

The words "fossil" and "paleontology" usually evoke pictures of dinosaur bones or other good-sized pieces of vertebrate skeleton. This article is concerned with micropaleontology, which deals with fossils of entirely different magnitude. They are the shells, or, more properly, the skeletons, of minute aquatic animals. None of these microfossils can be recognized without the help of a strong magnifying glass; some must actually be examined in the electron microscope. Their minuteness makes them especially useful for geological research. They can be brought up unbroken—and in enormous numbers—by a narrow coring pipe or even by an oil-well drill. Found both in the ocean floor and in dry-land formations that were once covered by water, microfossils have long served oil prospectors as stratigraphic markers. More recently they have begun to furnish important information about processes of change in the structure of the earth's surface and in the earth's climate.

To meet the needs of the micropaleontologist an organism must have characteristics other than small size. First, of course, the organism must build a skeleton, or some hard part durable enough to fossilize under normal conditions of sedimentation. The fossils in a group should be distinguished as to genus and preferably species. This implies some complexity of organization. Species that flourished during the shortest periods of time, and over the greatest geographical area, are the best geological indicators. They make it possible to differentiate sharply among layers of sedimentary rock of different age and to match strata in widely separated parts of the earth.

In order to obtain evidence of past climate and other environmental conditions the investigator must think of fossils as once living animals with specialized adaptations to their particular surroundings. He then tries to reconstruct those surroundings by analogy with the ecological requirements of near relatives of the ancient organisms that are alive today. As might be expected, the method becomes more difficult as the evolutionary distance between fossils and living organisms increases. A paleoecologist must be a good detective and find meaning in all sorts of apparently trivial and irrelevant observations. Certain potentially useful microfossils are just beginning to attract serious study. Previously they were ignored because they are so very small. The coccoliths, disk-shaped plates of calcium carbonate, are one example. They are planktonic, which means that the ocean currents in which they float have carried them to many quarters of the globe and that they have been settling to the bottom for some 500 million years. One cubic centimeter of sediment can contain 800 million of them; they must be enlarged 800 diameters to be seen at all. Because they were found in the topmost layer of sediment on the ocean floor, it was realized that the organism depositing them must be extant. For a long time what the organism was remained a mystery; the finest collecting nets could not catch it. Eventually it was located not in a man-made trap but in the filtering apparatus of the common marine animal called *Salpa*. Recently the electron microscope has revealed that coccoliths possess an astonishingly complex structure. When they have been more thoroughly described and classified, they should be helpful in correlating sedimentary strata from continent to continent as well as in the oceans.

Another group of tiny organisms, now extinct, were the star-shaped discoasters. A little larger than coccoliths, their world-wide distribution in deeper sediments suggests that they too were planktonic. Although known for almost 100 years, they have not been applied in stratigraphy because of the mistaken notion that all known species lived continuously from 60 million years ago to the time of their disappearance. In reality they evolved fairly rapidly, and some species make excellent guide fossils. The earliest discoaster stars had as many as 24 rays, or points. By 20 million years ago the number of rays had declined to six, and the last forms had only five slender rays. The exact time of their extinction is unknown, but if, as is suspected, it was just before the Pleistocene (the geological period of the last ice age), five-rayed discoasters will occupy an important position as reliable indicators for this important boundary in geologic time.

In their day-to-day work micropaleontologists deal chiefly with diatoms, radiolaria, conodonts, ostracodes and foraminifera. All of these except the diatoms are large enough to be studied in a low-power microscope at a magnification of between 30 and 100 diameters—an important consideration when hundreds of samples must be examined each day, as is the practice in some oil-company laboratories.

Among the microfossils the diatoms and the radiolaria compete for the first place in beauty. Both secrete shells of opal (silica combined with some water). In the lacy skeletons of radiolaria the opal looks like clear spun glass; in diatom shells it takes on a jewel-like quality, often displaying a many-colored fire that must be seen to be appreciated. Many species of diatoms live exclusively in fresh water, whereas all radiolaria live in salt water. In consequence diatom fossils are somewhat more informative; from the species found in a sedimentary layer it is usually possible to tell whether the sediment was formed in a lake or in the sea. There are a great many species of diatoms and radiolaria; diatoms can occur locally in such fantastic numbers that they produce fairly thick layers of sediment, known as diatomite, consisting almost entirely of their fossils. The opal skeletons of both diatoms and radiolaria, however, tend to dissolve in water, and they cannot be counted on to show up in a particular region. Often they are missing from just those places where an index fossil is most needed.

Conodonts are small tooth-shaped or platelike objects with one or more points. Like vertebrate teeth, they are made of calcium phosphate. Although many "genera" and "species" have been described since they were discovered more than a century ago, no one knows what kind of animal produced them. Whatever it was, it became extinct some 240 million years ago during the Triassic period. The variation in the form of conodonts from level to level in older sediments makes them particularly use-

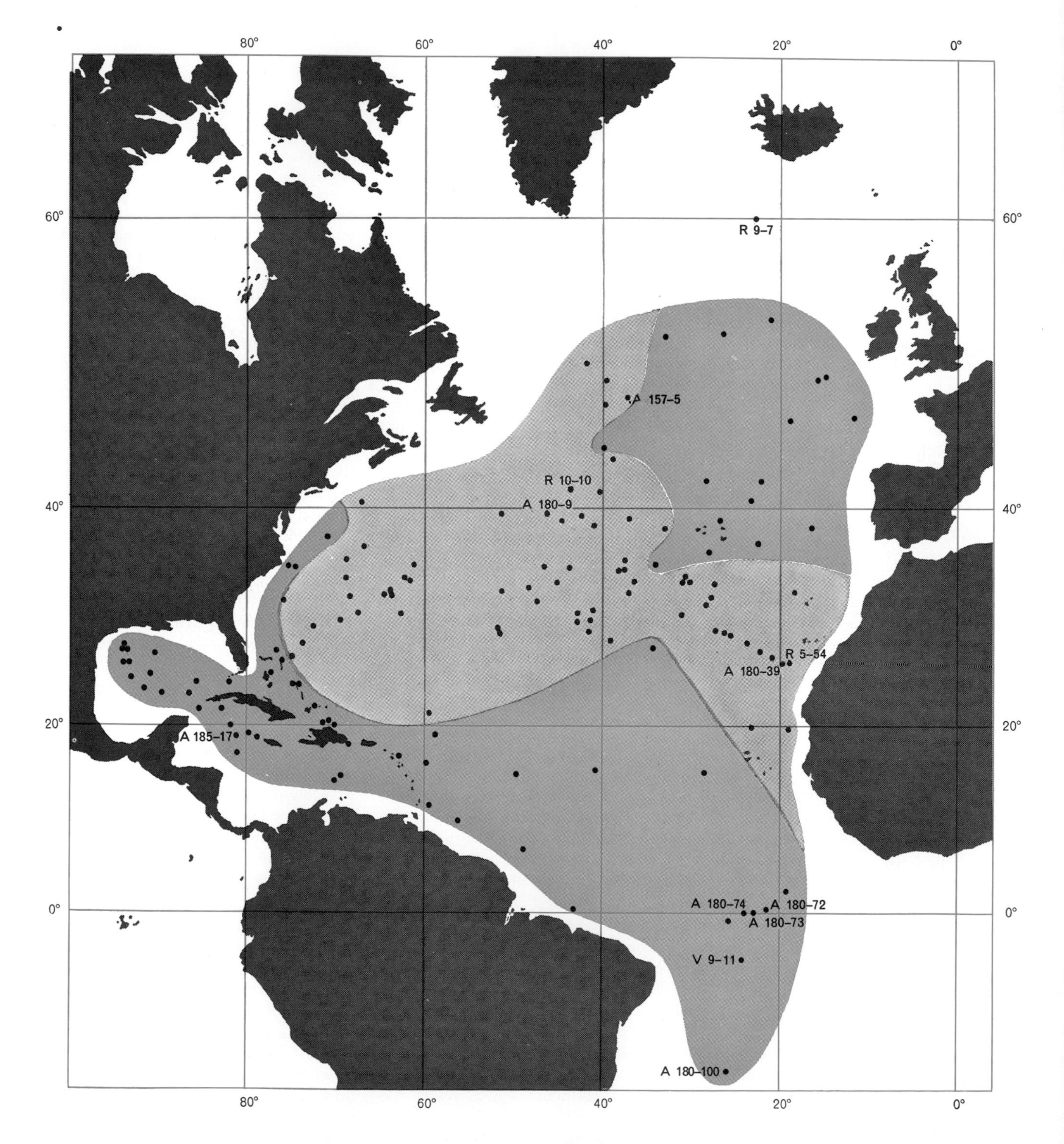

COILING DIRECTIONS of living *Globorotalia truncatulinoides*, a planktonic foraminifer, define three provinces in the Atlantic Ocean. Most of the spiral-shaped shells of this species found on the sea floor coil to the left in the gray region and to the right in the colored areas. The dots mark sites where cores have been taken. The letters and numbers identify the cores diagramed on the following pages. Core V 9–11, brought up just south of the Equator, carries the Pleistocene record back at least 600,000 years.

ful to the petroleum geologist. Because of their small size they often come up undamaged in rock cuttings from oil wells or exploratory holes.

The ostracodes present no mysteries. These odd little relatives of crabs and lobsters are very much alive today and flourish wherever there is enough water, whether it is fresh, brackish or salty. They are the only crustaceans that have two valves, or shells, which makes them look like tiny clams. The valves vary in length from half a millimeter (a fiftieth of an inch) to three or four millimeters. They first appear in the geological column in sediments deposited at least 450 million years ago in the Cambrian period of the early Paleozoic era. In the course of evolution the shells have varied greatly in shape and ornamentation, and many species have lived only for short intervals of geologic time. Knowledge of the present distribution of living genera in open salt water, sounds, estuaries, lagoons and lakes helps in analyzing past conditions when similar genera show up in ancient sediments.

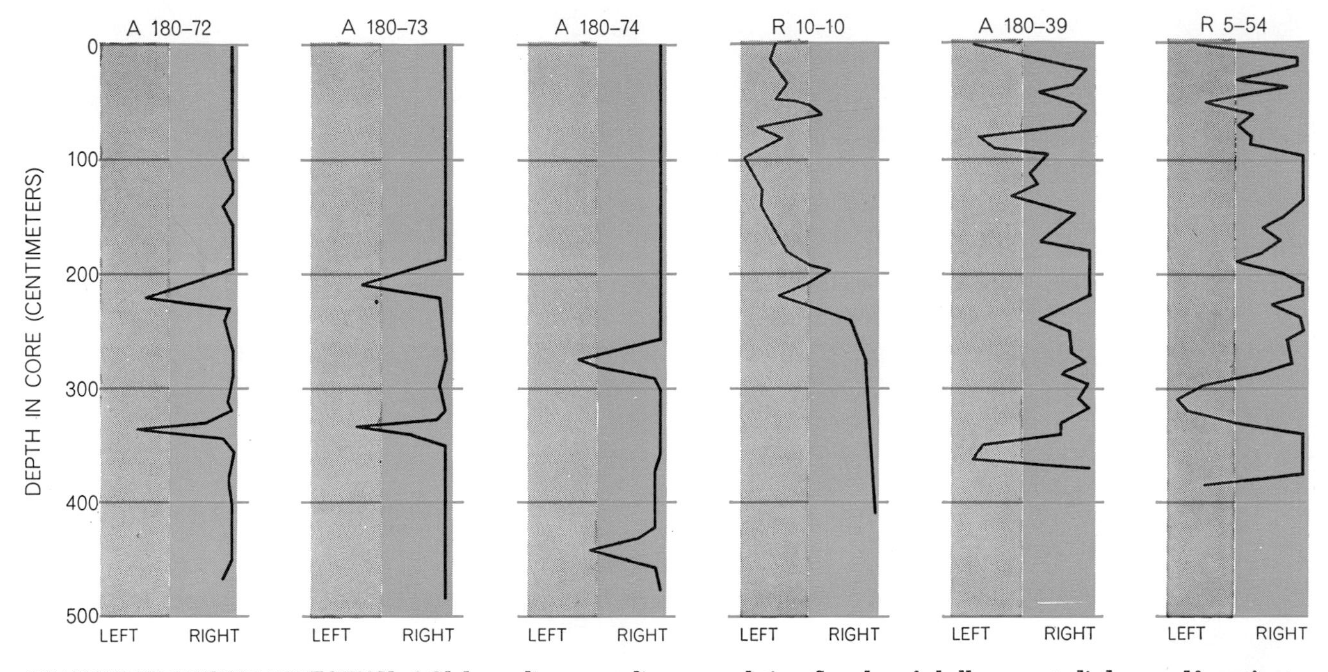

CHANGES IN COILING DIRECTION of ***Globorotalia truncatulinoides*** **with depth in cores make it possible to correlate cores from different locations. The correlation is readily apparent in the three cores A 180–72, –73 and –74. Cores A 180–39 and R 5–54 also show correlation. Samples of shells were studied every 10 centimeters or so from top of cores to bottom. Variation in each diagram is from 100 per cent left-coiling at left to equal ratio in the middle to 100 per cent right-coiling at right. The older shells are deeper.**

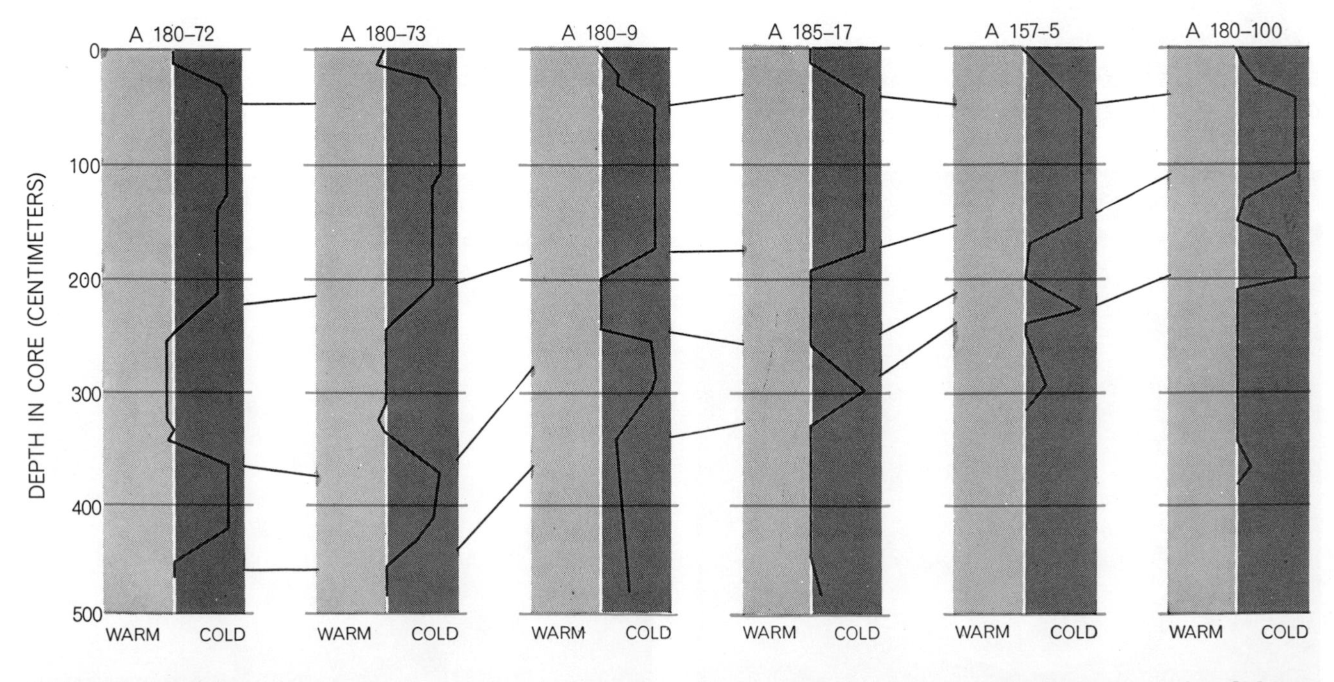

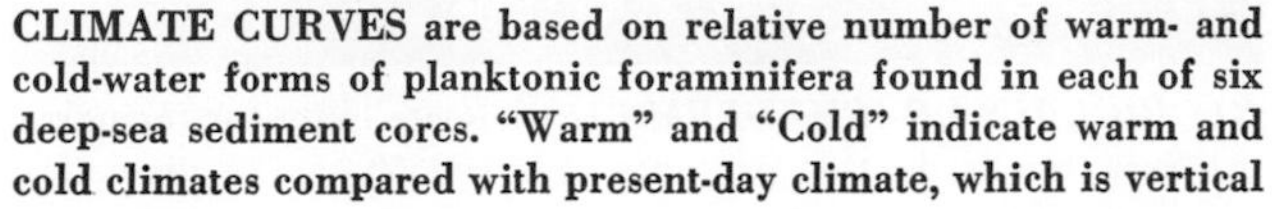
CLIMATE CURVES are based on relative number of warm- and cold-water forms of planktonic foraminifera found in each of six deep-sea sediment cores. "Warm" and "Cold" indicate warm and cold climates compared with present-day climate, which is vertical line in center of each diagram. Thin lines connect faunal changes believed to have occurred at the same time in the various locations. Obviously rates of sedimentation have differed widely. Such curves provide the data for establishing a chronology of the ice ages.

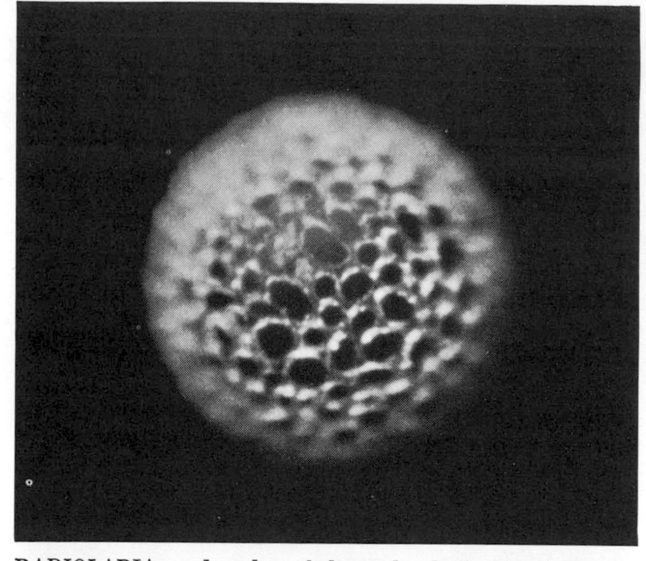

RADIOLARIA produce lacy skeleton that looks like clear spun glass. It is made of opal, which tends to dissolve in water.

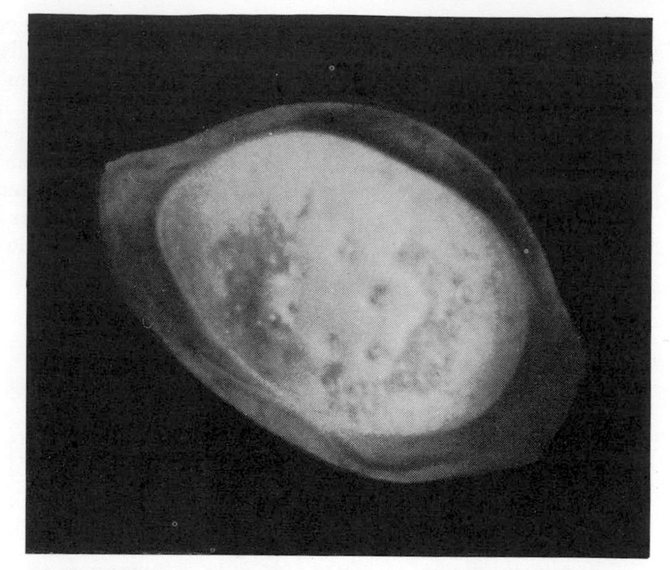

OSTRACODES are related to crabs and lobsters but have two valves, or shells, that closely resemble shells of tiny clams.

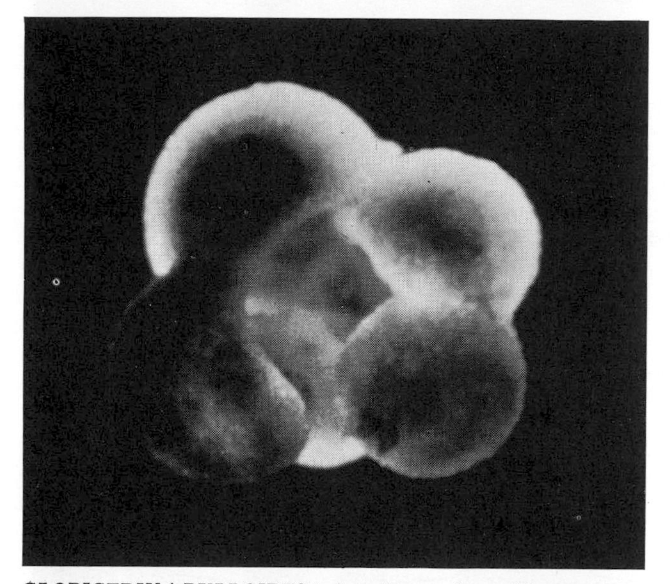

GLOBIGERINA BULLOIDES is found in cool to cold water. This and forms that follow are all species of planktonic foraminifera.

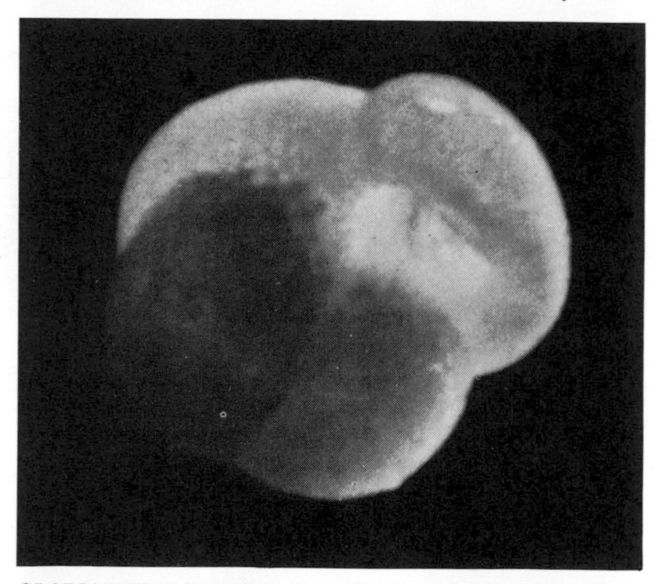

GLOBIGERINA INFLATA is a cool-water form found in middle latitudes. It occurs only in sediments of the Pleistocene epoch.

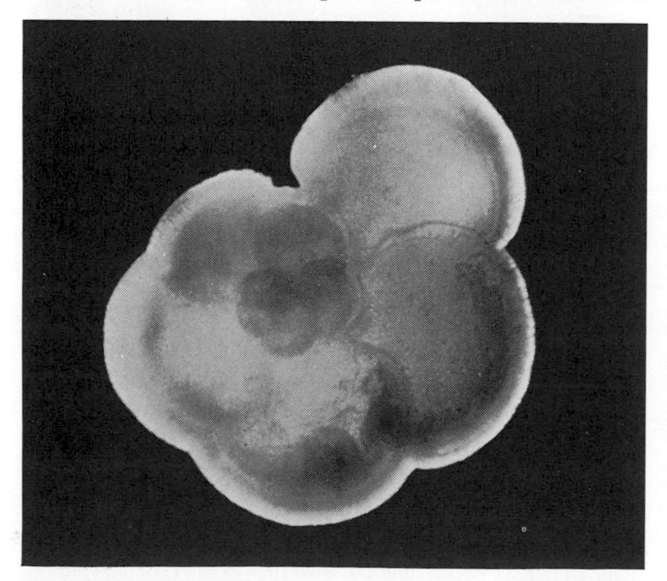

GLOBOROTALIA MENARDII is found in waters of mid-latitudes and tropics. All Pleistocene forms of this species coiled to left.

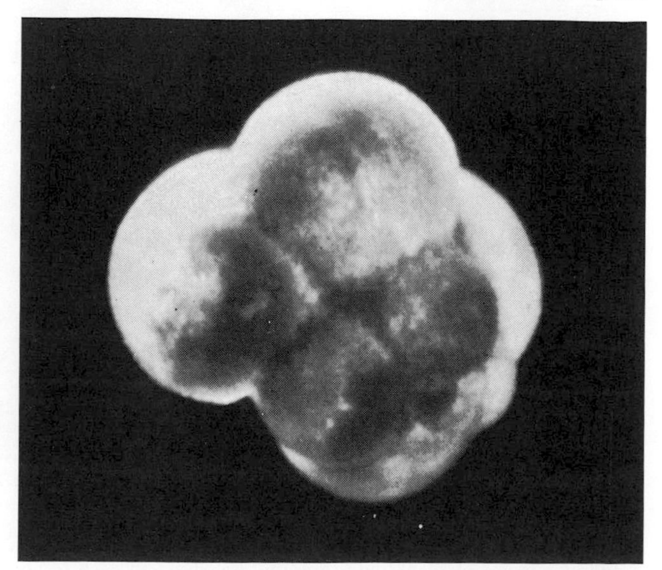

GLOBIGERINA PACHYDERMA is a climate indicator for northern seas. Shells on this page are enlarged 60 to 120 diameters.

The ostracodes have only one drawback: they are not nearly so numerous as the universal favorites among microfossils, the foraminifera.

The very popularity of foraminifera greatly enhances their usefulness. Over the years a mountain of information of all sorts has accumulated, and most of it is readily available. Descriptions of genera and species have been brought together in an enormous catalogue published by the American Museum of Natural History, which now includes 69 volumes and is growing constantly as descriptions of new species are added.

The foraminifera are protozoa, or single-celled animals, that build tests (shells) of various materials. On the basis of the mode of construction the animals are divided into two large groups. Calcareous species construct tests of calcium carbonate precipitated directly from sea water; arenaceous species build their shells of sand grains, flakes of mica, sponge spicules or even small discarded calcareous tests of other species, any of which they can cement together with secretions of calcium carbonate or iron salts. The size of the shell varies widely with the species. Some long-extinct giant foraminifera exceeded 15 centimeters (six inches) in diameter. The majority of fossil foraminifera range from .2 millimeter to two millimeters.

The architectural unit of the test is the chamber. A few species have only one chamber, but most build anywhere from two to several hundred. On this basic principle of structure the foraminifera improvise endlessly. To duplicate all the strangely shaped chambers and their intricate arrangements would tax the ingenuity of a topologist. Geometric versatility has made possible all the thousands of different species that have come and gone during the past 500 million years.

Almost all the foraminifera species live on the ocean bottom. Although some attach themselves permanently to rocks, most of them potter along at a few millimeters per hour, pushing themselves by means of pseudopodia extruded through minute pores in the shells. Clearly this way of life does not make for wide distribution. Beginning some 100 million years ago in the Upper Cretaceous period, however, a few types became planktonic. Although they constitute only about 1 per cent of the known species, the enormous volume of living space open to them has permitted great proliferation: individuals of the planktonic forms make up almost 99 per cent of the fossils found in ocean sediments. In some places accumulation of the shells has produced thick deposits of chalk. The white cliffs of Dover and Normandy are such deposits, now uplifted and partly eroded. Today large areas of the bottom of all the oceans are receiving a slow but constant rain of discarded tests of planktonic foraminifera, which make up on the average 30 to 50 per cent of all bottom sediments.

Shells of the important planktonic species have fairly simple forms. The dominant theme is a series of chambers arranged in a spiral. As the animal grows it adds chambers of steadily increasing size. In most species the general form resembles that of a small shell. Like snail shells, some tests coil to the left and others coil to the right, the two kinds being mirror images of each other.

Because of the rapid succession of distinctive species throughout successive geological ages, foraminifera are ideally suited to the needs of the petroleum geologist, who must deal with many kinds of thick sedimentary rocks, some of them heavily folded and faulted. If one were asked to invent a class of ideal fossils for identifying and matching strata, it would be hard to improve on the foraminifera. It is small wonder that most of the hundreds of paleontologists working for oil companies devote full time to the study of these microfossils.

Foraminifera are a good deal more than tags for sedimentary rocks. To workers in pure geology, and particularly to those in the hybrid branch known as marine geology, they furnish invaluable keys to the remote past. So many samples of ocean sediment have been examined by now and their microfossils classified that it is possible to chart the approximate distribution of the most common species of planktonic foraminifera. The charts show that some species live only in low latitudes, others are most abundant in middle latitudes and

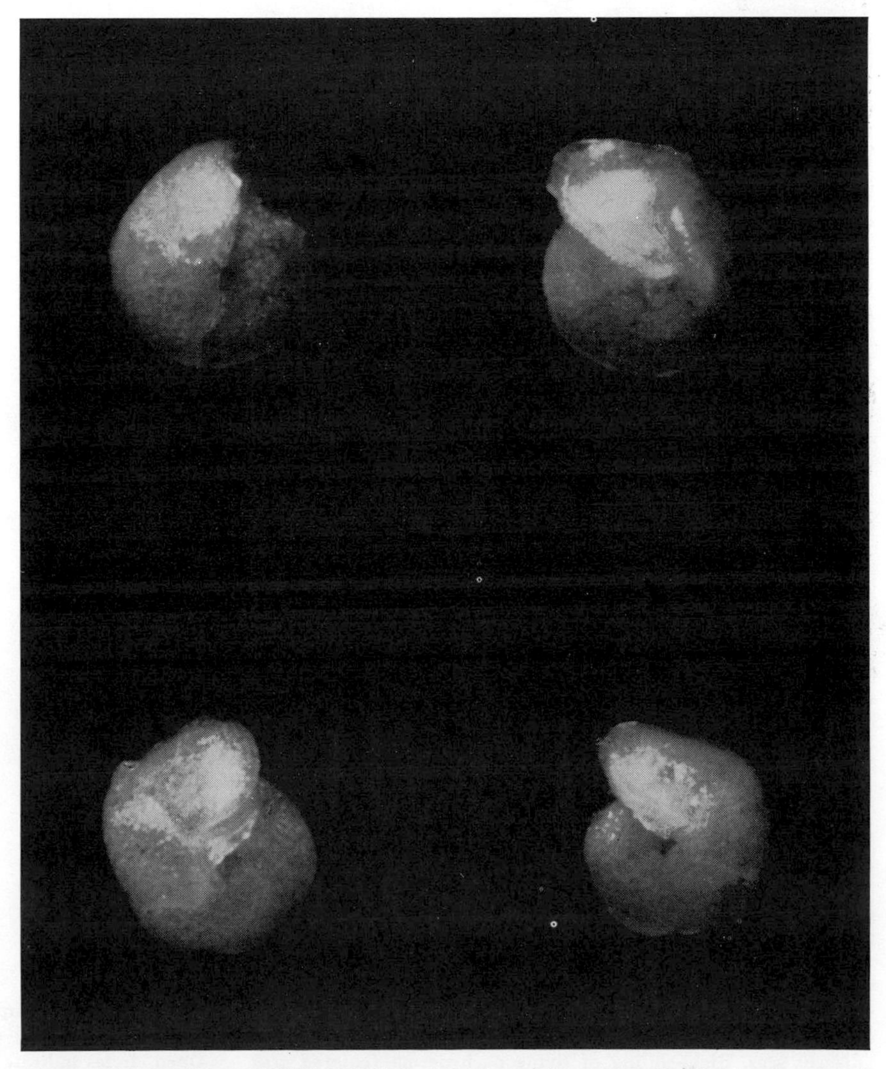

LEFT- AND RIGHT-COILING SHELLS of *Globorotalia truncatulinoides* appear at left and right respectively. Coiling direction is apparently associated with climatic conditions.

still others live in high latitudes. (One species, *Globigerina pachyderma*, ranges up to the North Pole.) Evidently water temperature plays an important part in the distribution of the various species.

If so, the animals must live near the surface; only there does the temperature vary significantly with latitude. The fact that living foraminifera are caught in plankton nets towed through the photic zone (the sunlit top 100 meters of the ocean) seems to bear this out. But the surface samples long presented something of a problem. Their shells are thin-walled and transparent, whereas shells of the same species taken from the bottom are almost all heavily encrusted with calcium carbonate or calcite.

Until recently it was supposed that the material precipitates on the empty tests after they settle to the bottom. This hypothesis had fatal flaws, however. For one thing, oceanographers generally agree that at depths of several thousand meters in the ocean calcium carbonate dissolves far more rapidly than it precipitates. Our laboratory at the Lamont Geological Observatory of Columbia University finally undertook a close examination of the distribution of calcite on individual shells, which furnished the clue to the correct explanation. We discovered that calcite is thickest on the earliest formed chambers and diminishes steadily in the chambers farther out on the growth spiral. This can only mean that the living foraminifera precipitate the calcite and that the thin-walled specimens in the photic zone are immature.

Recently Allan W. H. Bé of Lamont has caught heavily encrusted tests containing living foraminifera by towing plankton nets at depths of more than 500 meters. They provide the final proof that at least some species mature and reproduce well below the photic zone. Evidently the embryos rise to the photic zone, fatten on diatoms and other photosynthetic organisms there and then sink to a lower level, where they complete their life cycle. The pattern does not conflict with the idea that temperature variations in the upper layers of water determine the geographical distribution of various species; each individual passes a critical period of its life in the photic zone. On the other

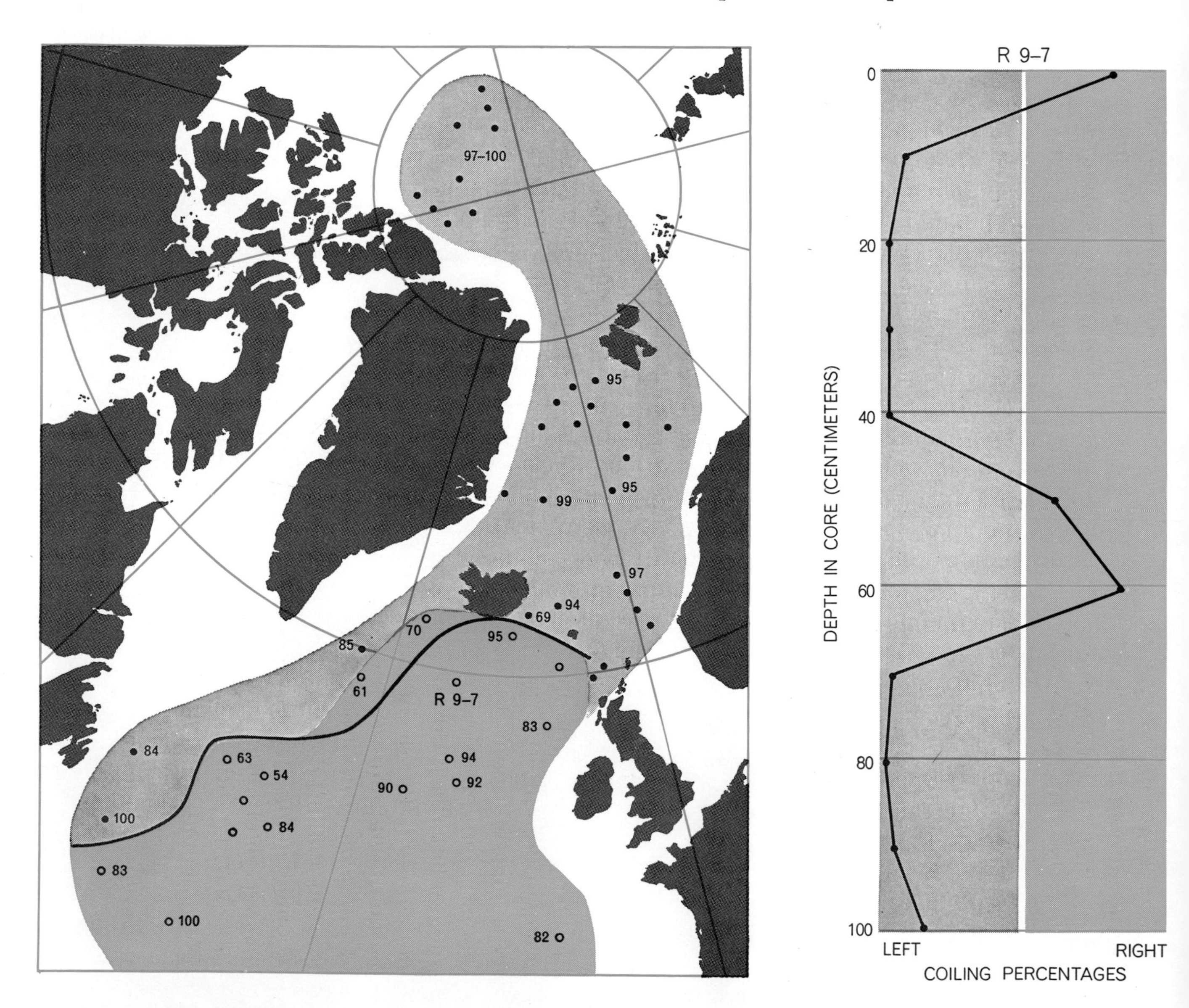

CLIMATE AND COILING DIRECTION of *Globigerina pachyderma* seem to be closely correlated in northernmost Atlantic and adjacent seas. Right-coiling of living pachyderma (*colored area on map*) is associated with warmer water, left-coiling (*gray area on map*) with coldest water. Position of the isotherm marking an average surface temperature of 7.2 degrees centigrade in April (*black line on map*) closely follows border of right-left provinces. Open circles indicate dominance of right-coiling at tops of cores; closed circles mean left-coiling. Percentages of shells coiling in dominant direction are shown for some cores. Curves at right

hand, it suggests a possible explanation for a rather puzzling distribution of the species *Globorotalia menardii.* Fossils of this group are extremely abundant in sediments south and southwest of the Canary Islands, whereas they are entirely absent from the region to the north and northeast. Yet the Canaries current flows southwestward through the area and would sweep all *G. menardii* out of the region south of the islands if they spent their entire lives in the photic zone. We believe that the animals sink below the Canaries current as they approach maturity and enter a deep countercurrent that returns them and the new generation of embryos to the northeast. As yet no one has attempted to detect the countercurrent directly, but mathematical analysis of deep circulation in the Atlantic suggests it should be there.

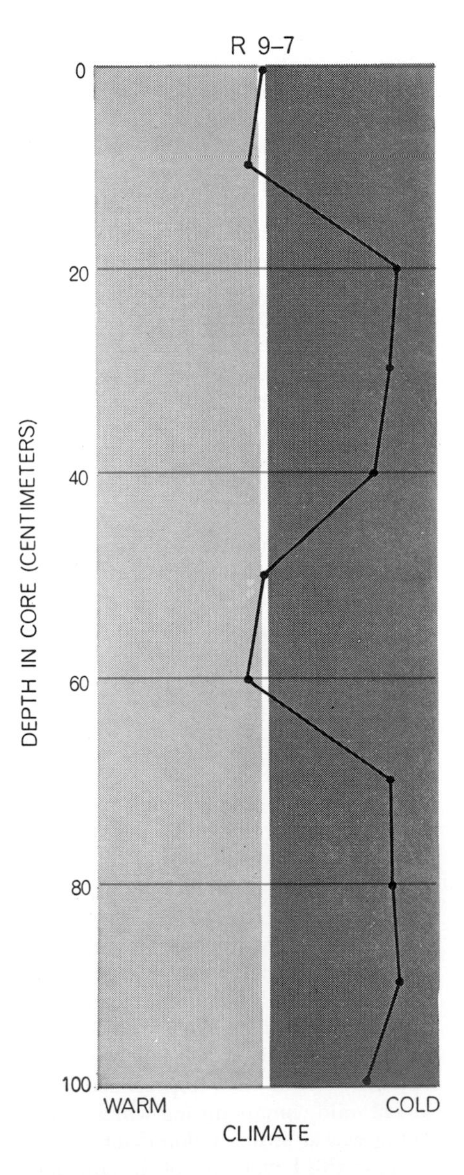

show (*left curve*) relative percentages of left- and right-coiling *G. pachyderma* at 10-centimeter intervals in core R 9–7, and abundance of cold and warm forms of all other planktonic foraminifera in core (*right*).

The discovery that foraminifera live part of their lives deeper than 500 meters has an important bearing on recent attempts to estimate ancient water temperatures from measurements of oxygen isotopes in fossil shells. The technique is based on the fact that in warmer waters there is a slight preponderance of heavier isotopes in the shells and in cooler waters a slight preponderance of lighter isotopes. If the relative abundance of the isotopes is to be significant, the depth at which the shells incorporated their oxygen must be known. Deep waters at every latitude are quite cold.

Since warm-water and cold-water species have not changed greatly in the past million years or so, the micropaleontologist can use the fossils to obtain a quite objective picture of changing climatic conditions during the period. Cores of sediment from ocean bottoms furnish a continuous record of geological and climatic events in contrast with the garbled record available from the distorted layers of rock on dry land. From our study of planktonic foraminifera in more than 1,000 cores in our laboratory, we feel that we have been able to arrive at the first accurate set of dates for the most recent glacial and interglacial periods. Carbon-14 analysis of the fossils shows that the last ice age ended 11,000 years ago instead of 20,000 years ago, as had previously been thought. The new date, now generally accepted, may change some ideas about the rate of human evolution. Judging from the average rate of sedimentation, the last part of the last glaciation began approximately 60,000 years ago, after an interstadial period (an interglacial of short duration) of 30,000 years [*see illustration on next page*]. The preceding glacial period lasted only 20,000 years, whereas the interglacial before it appears to have gone on for 110,000 years. This is as far back as reliable core data go, although a single core from the equatorial Atlantic appears to carry the Pleistocene record back at least 600,000 years.

Since the last glaciation ended 11,000 years ago, and since the minimum interglacial period seems to have been 30,000 years, it would seem that man can look forward to at least 20,000 more years of the present mild climate, if not to even warmer weather. If a warmer climate should melt the glaciers that remain today, the sea level would rise, but probably no more than 10 meters or so. This would be a considerable nuisance—it would put much of New York City under

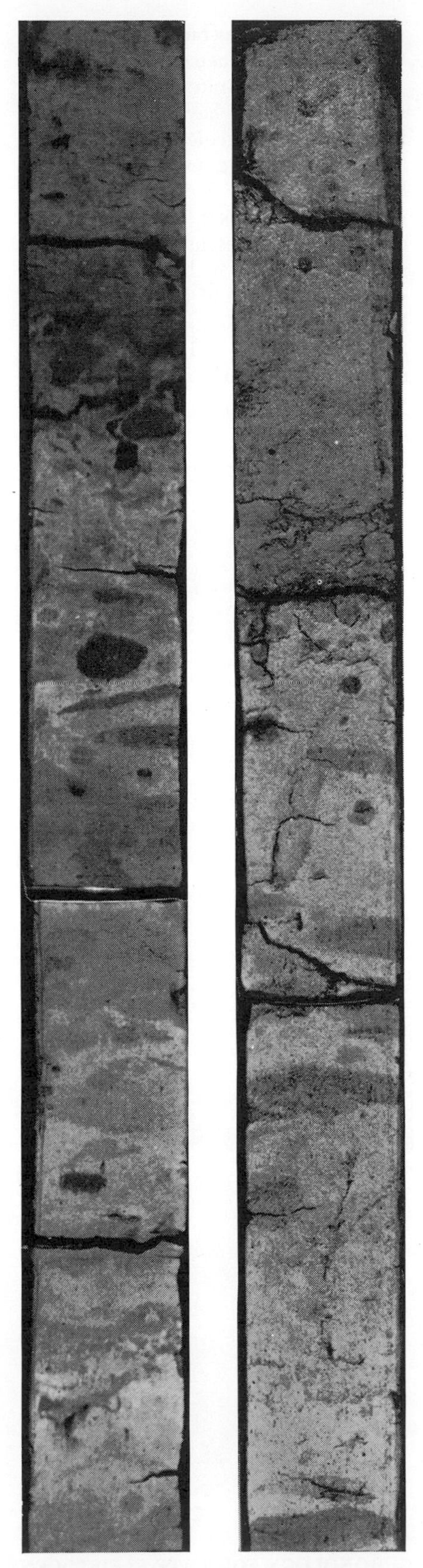

DEEP-SEA CORES represent a vertical cross section of ocean-bottom sediments. These are small parts of cores A 180–72 (*left*) and A 180–73 (*right*). Sometimes layers are quite apparent in cores; at other times no layers can be seen with the naked eye, but the microscope reveals sharp changes in the fossil contents of the undersea sediments.

water, for example—but it would hardly threaten the existence of mankind.

Arriving at accurate dates for the late Pleistocene would have been far easier if it were true, as geologists formerly believed, that marine sediments accumulate everywhere at the rate of one centimeter per 1,000 years. Then each sample core would represent the same time scale. Our studies of hundreds of cores show that the rate of accumulation can be that show, but in many places it is as much as 50, 100 and even 250 centimeters in 1,000 years, depending largely on the underwater topography. "Turbidity" currents, consisting of silt-laden water denser than surrounding water, frequently run down even gentle ocean-bottom slopes, depositing several meters of mud in a few hours in some places; in other places they may scour away sediments accumulated over thousands of years [see the article "The Origin of Submarine Canyons," by Bruce C. Heezen, beginning on page 155]. Another phenomenon responsible for large variations is slumping, in which the sediments simply slide away down a steep slope. The removal of upper layers of sediments by these processes is not all bad, however; it has brought 100-million-year-old fossils near enough to the sea floor to be reached by our coring tubes. So far we have not brought up a core in which the entire upper part of the Pleistocene record has been removed, leaving the older ice-age sediments intact. Several cores of this type, or longer cores from places already sampled, would carry our time scale back to the beginnings of human evolution.

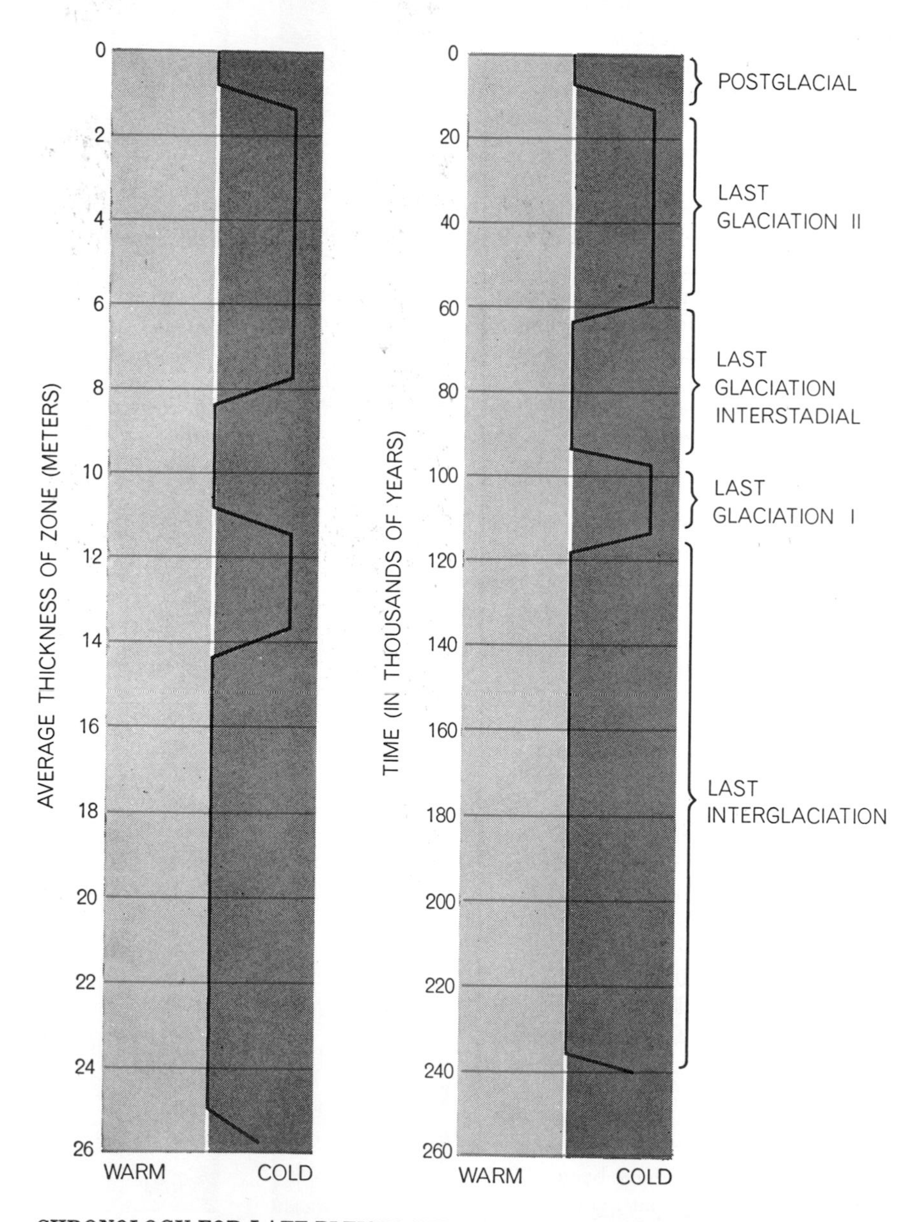

CHRONOLOGY FOR LATE PLEISTOCENE (*right*) is drawn from micropaleontological studies of planktonic foraminifera in 108 deep-sea sediment cores and on extrapolations of the rate of sedimentation from 37 carbon-14 dates in 11 of the cores. The authors consider this the most accurate timetable available for the past 240,000 years. Present-day climate is at center line of diagrams. The average-thickness curve (*left*) shows greater rate of sedimentation during glaciations than during interglacial or interstadial periods. This is caused by lowering of seas, which exposes continental shelf. Rivers thus carry material from continents out to edge of shelf and dump land sediments into deep sea, whereas in warmer times sea is higher and continental sediments are deposited on shelf.

The coiling direction of shells of *Globorotalia truncatulinoides* is proving extremely useful in matching sediments from various locations. In our laboratory we have discovered that the ratio between right-coiling and left-coiling shells varies from place to place in such a way as to define three distinct geographical provinces in the North Atlantic [*see top illustration on page 189*]. Carbon-14 dating of samples from cores shows that the present pattern of distribution has persisted for about 10,000 years. Evidently some environmental factor has maintained the pattern in spite of the mixing effect of general ocean circulation. Going back in time by determining coiling ratios in fossil samples taken every 10 centimeters in cores, we find that the pattern of distribution changed rather suddenly from time to time during the late Pleistocene, presumably in response to changing currents or shifting water masses. Although we cannot yet say just what the changes were, we can match layers in different cores by means of the coiling-direction ratios. Petroleum geologists in Europe and India have applied this same method to other species to match strata penetrated by oil wells.

In the case of *Globigerina pachyderma*, the dominant direction of coiling follows closely the temperature of surface water in the northernmost Atlantic. Again we find evidence of shifts in the boundary between the left-coiling (cold water) and right-coiling (warm water) populations. Here we believe the shifts were determined directly by temperature changes in the late Pleistocene [*see illustration on pages 192 and 193*]. The coiling changes in a typical core show a period of cold climate preceded by a time of mild climate during which right-coiling was as strongly dominant as it is today. In this lower zone of the core and at the top of the core we find abundant fossils of various other warm-water species; these are absent in the intervening zone of left-coiling. (A direct causal relation between coiling direction and

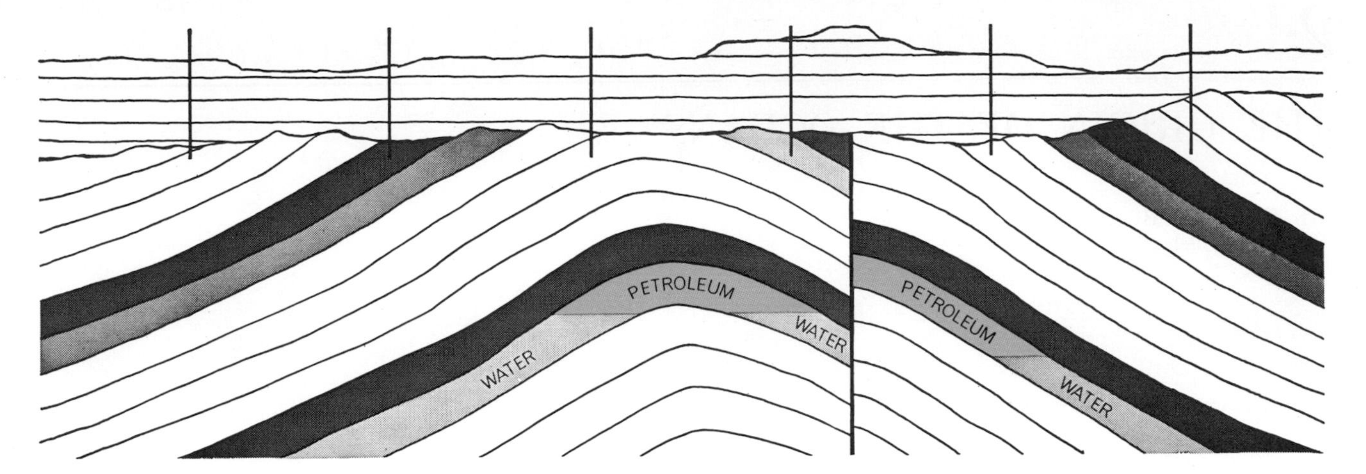

CROSS SECTION OF OIL POOL shows "slim holes," or exploratory bore holes (*vertical lines*), passing through a series of nearly horizontal sediments and into folded, faulted and partly eroded beds of an earlier era. Two beds of same dark gray tone are both shale, which is likely to lie over oil. Microfossils from bottoms of slim holes show relative ages of beds and thereby reveal existence of fold and fault. The deeper shale bed is of course much older. Although slim holes have not struck oil, they have shown presence of a dome that in this case contains oil, as well as the fault that has moved part of the oil-bearing stratum upward.

temperature is hardly conceivable. If both the direction of coiling and the temperature tolerance are determined genetically, the two characteristics may be associated because their genes are linked.)

Such changes in coiling direction of *Globigerina pachyderma* at lower levels in sediment cores provide insight into oceanographic conditions in the North Atlantic during the late Pleistocene. For example, left-coiling dominates from top to bottom of all cores taken in the area of present-day left-coiling. This indicates that the inflow of relatively warm Atlantic water into the Norwegian Sea was never greater in the late Pleistocene than it is today. Again, the change to left-coiling directly below the tops of cores in the area of present-day right-coiling implies that the inflow of warm Atlantic water decreased during the last glaciation. As yet we cannot tell whether this resulted from a decrease in energy of the general circulation of the North Atlantic or from the lower sea level that accompanied the glaciations. The latter would make the submarine ridge between Iceland and Scotland a more effective barrier to influx. We hope that further study of the foraminifera in sediment cores will answer this and similar questions. Since ocean circulation, particularly that of the North Atlantic, must have had a powerful influence on Pleistocene climates, a better understanding of this circulation may yet provide the basis for a satisfactory theory of the cause of ice ages.

Finally, our microfossil studies have thrown new light on the origin of the Atlantic Ocean. No cores from any ocean have yielded fossils older than the late Cretaceous period, about 100 million years ago. It is particularly suggestive that the rather thorough sampling of the bottom of the Atlantic during the past 15 years has yielded no older fossils. Can we conclude from this that the Atlantic basin came into existence in its present form at some time during the Cretaceous period? To admit this possibility is to question the widely held belief in the permanence of continents and ocean basins. For a definitive answer we shall probably have to wait until samples have been raised from one or more deep borings in the Atlantic. In the meantime the accumulating evidence suggests that a drastic reorganization of that part of the earth's surface now occupied by the ocean basins took place about 100 million years ago.

21

THE ORIGIN OF THE OCEANS

SIR EDWARD BULLARD
September 1969

The earth is uniquely favored among the planets: it has rain, rivers and seas. The large planets (Jupiter, Saturn, Uranus and Neptune) have only a small solid core, presumably overlain by gases liquefied by pressure; they are also surrounded by enormous atmospheres. The inner planets are more like the earth. Mercury, however, has practically no atmosphere and the side of the planet facing the sun is hot enough to melt lead. Venus has a thick atmosphere containing little water and a surface that, according to recent measurements, may be even hotter than the surface of Mercury. Mars and the moon appear to show us their primeval surfaces, affected only by craters formed by the impact of meteorites, and perhaps by volcanoes. Only on the earth has the repetition of erosion and sedimentation —"the colossal hour glass of rock destruction and rock formation"—run its course cycle after cycle and produced the diverse surface that we see. The mountains are raised and then worn away by falling and running water; the debris is carried onto the lowlands and then out to the ocean. Geologically speaking, the process is rapid. The great plateau of Africa is reduced by a foot in a few thousand years, and in a few million years it will be near sea level, like the Precambrian rocks of Canada and Finland. All trace of the original surface of the earth has been removed, but as far back as one can see there is evidence in rounded, water-worn pebbles for the existence of running water and therefore, presumably, of an ocean and of dry land.

The obvious things that no one comments on are often the most remarkable; one of them is the constancy of the total volume of water through the ages. The level of the sea, of course, has varied from time to time. During the ice ages, when much water was locked up in ice sheets on the continents, the level of the sea was lower than it is at present, and the continental shelves of Europe and North America were laid bare. Often the sea has advanced over the coastal plains, but never has it covered all the land or even most of it. The mechanism of this equilibrium is unknown; it might have been expected that water would be expelled gradually from the interior of the earth and that the seas would grow steadily larger, or that water would be dissociated into hydrogen and oxygen in the upper atmosphere and that the hydrogen would escape, leading to a gradual drying up of the seas. These things either do not happen or they balance each other.

The mystery is deepened by the almost complete loss of neon from the earth; in the sun and the stars neon is only a little rarer than oxygen. The neon was presumably lost when the earth was built up from dust and solid grains because neon normally does not form compounds, but if that is so, why was the water not lost too? Water has a molecular weight of 18, which is less than the atomic weight of neon, and thus should escape more easily. It looks as if the water must have been tied up in compounds, perhaps hydrated silicates, until the earth had formed and the neon had escaped. Water must then have been released as a liquid sometime during the first billion years of the earth's history, for which we have no geological record. The planet Mercury and the moon would have been too small to retain water after it was released. Mars seems to have been able to retain a trace, not enough to make oceans but enough to be detectable by spectroscopy.

These speculations about the early history of the earth are open to many doubts. The evidence is almost nonexistent, and all one can say is, "It might have been that. . . ." The great increase in understanding of the present state and recent history of the ocean basins that we have gained in the past 20 years is something quite different. For the first time the geology of the oceans has been studied with energy and resources commensurate with the tremendous task. It turns out that the main processes of geology can be understood only when the oceans have been studied; no amount of effort on land could have told us what we now know. The study of marine geology has unlocked the history of the oceans, and it seems likely to make intelligible the history of the continents as well. We are in the middle of a rejuvenating process in geology comparable to the one that physics experienced in the 1890's and to the one that is now in progress in molecular biology.

The critical step was the realization that the oceans are quite different from the continents. The mountains of the oceans are nothing like the Alps or the Rockies, which are largely built from folded sediments. There is a world-encircling mountain range—the mid-ocean ridge—on the sea bottom, but it is built entirely of igneous rocks, of basalts that have emerged from the interior of the earth. Although the undersea mountains have a covering of sediments in

RED SEA and the Gulf of Aden represent two of the newest seaways created by the worldwide spreading of the ocean floor. In this photograph, taken at an altitude of 390 miles from the spacecraft *Gemini 11* in September, 1966, the Red Sea separates Ethiopia (*at left*) from the Arabian peninsula (*at right*). The Gulf of Aden lies between the southern shore of Arabia and Somalia. The excellent fit between drifting land masses is depicted in the illustrations on page 199.

many places, they are not made of sediments, they are not folded and they have not been compressed.

A cracklike valley runs along the crest of the mid-ocean ridge for most of its length, and it is here that new ocean floor is being formed today [*see illustration on next two pages*]. From a study of the numerous earthquakes along this crack it is clear that the two sides are moving apart and that the crack would continually widen if it were not being filled with material from below. As the rocks on the two sides move away and new rock solidifies in the crack, the events are recorded by a kind of geological tape recorder: the newly solidified rock is magnetized in the direction of the earth's magnetic field. For at least the past 10,000 years, and possibly for as long as 700,000 years, the north magnetic pole has been close to its present location, so that the magnetic field is to the north and downward in the Northern Hemisphere, and to the north and upward in the Southern Hemisphere. As the cracking and the spreading of the ocean floor go on, a strip of magnetized rock is produced. Then one day, or rather in the course of several thousand years, the earth's field reverses, the next effusion of lava is magnetized in the reverse direction and a strip of reversely magnetized rocks is built up between the two split halves of the earlier strip. The reversals succeed one another at widely varying intervals; sometimes the change comes after 50,000 years, often there is no change for a million years and occasionally, as during the Permian period, there is no reversal for 20 million years. The sequence of reversals and the progress of spreading is recorded in all the oceans by the magnetization of the rocks of the ocean floor. The message can be read by a magnetometer towed behind a ship.

We now have enough examples of these magnetic messages to leave no doubt about what is happening. It is a truly remarkable fact that the results of magnetic surveys in the South Pacific can be explained—indeed predicted—from the sequence of reversals of the direction of the earth's magnetic field known from magnetic and age measurements, made quite independently on lavas in California, Africa and elsewhere. The only adjustable factor in the calculation is the rate of spreading. Such worldwide theoretical ideas and such detailed agreement between calculation and theory are rare in geology, where theories are usually qualitative, local and of little predictive value.

The speed of spreading on each side of a mid-ocean ridge varies from less than a centimeter per year to as much as eight centimeters. The fastest rate is the one from the East Pacific Rise and the slowest rates are those from the Mid-Atlantic Ridge and from the Carlsberg Ridge of the northwest Indian Ocean. The rate of production of new terrestrial crust at the central valley of a ridge is the sum of the rates of spreading on the two sides. Since the rates on the two sides are commonly almost equal, this sum is twice the rate on each side and may be as much as 16 centimeters (six inches) per year. Such rates are, geologically speaking, fast. At 16 centimeters per year the entire floor of the Pacific Ocean, which is about 15,000 kilometers (10,000 miles) wide, could be produced in 100 million years.

When the mid-ocean ridges are examined in more detail, they are found not to be continuous but to be cut into sections by "fracture zones" [*see top illustration on page 202*]. A study of the earthquakes on these fracture zones shows that the separate pieces of ridge crest on the two sides of a fracture zone are not moving apart, as might seem likely on first consideration. The two pieces of ridge remain fixed with respect to each other while on each side a plate of the crust moves away as a rigid body; such a fracture is called a transform fault. The earthquakes occur only on the piece of the fracture zone between the two ridge crests; there is no relative motion along the parts outside this section.

If two rigid plates on a sphere are spreading out on each side of a ridge that is crossed by fracture zones, the relative motion of the two plates must be a rotation around some point, termed the pole of spreading. The "axis of spreading," around which the rotation takes place, passes through this pole and the center of the earth. The existence of a pole of spreading and an axis of spreading is geometrically necessary, as was shown by Leonhard Euler in the

PROBABLE ARRANGEMENT of continents before the formation of the Atlantic Ocean was determined by the author with the aid of a computer. The fit was made not at the present coastlines but at the true edge of each continent, the line where the continental shelf (*dark brown*) slopes down steeply to the sea floor. Overlapping land and shelf areas are reddish orange; gaps where the continental edges do not quite meet are dark blue. At present the entire western Atlantic is moving as one great plate carrying both North America and South America with it. At an earlier period the two continents must have moved independently.

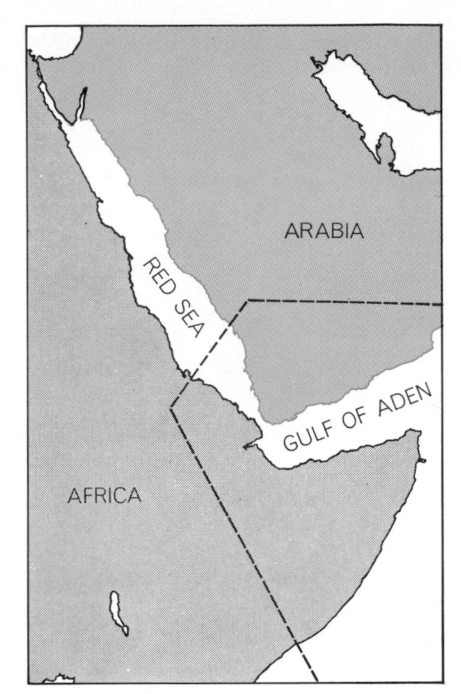

RUPTURE OF MIDDLE EAST is being caused by the widening of the Red Sea and the Gulf of Aden. Some 20 million years ago the Arabian peninsula was joined to Africa, as evidenced by the remarkable fit between shorelines (*see illustration below*). The area within the *Gemini 11* photograph on page 197 is shown by the broken lines.

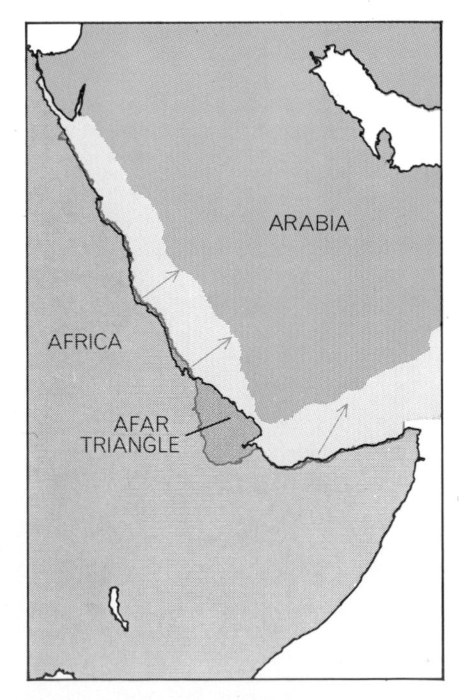

FIT OF SHORELINES of Arabia and Africa works out most successfully if the African coast (*black*) is left intact and if the Arabian coast (*color*) is superposed in two separate sections. In the reconstruction a corner of Arabia overlaps the "Afar triangle" in northern Ethiopia, an area that now has some of the characteristics of an ocean floor.

18th century. If the only motion on the fracture zones is the sliding of the two plates past each other, then the fracture zones must lie along circles of latitude with respect to the pole of spreading, and the rates of spreading at any point on the ridge must be proportional to the perpendicular distance from the point to the axis of spreading [*see bottom illustration on page 202*].

All of this is well verified for the spreading that is going on today. The rates of spreading can be obtained from the magnetic patterns and the dates of the reversals. The poles of spreading can be found from the directions of the fracture zones and checked by the direction of earthquake motions. It turns out that the ridge axes and the magnetic pattern are usually almost at right angles to the fracture zones. This is not a geometrical

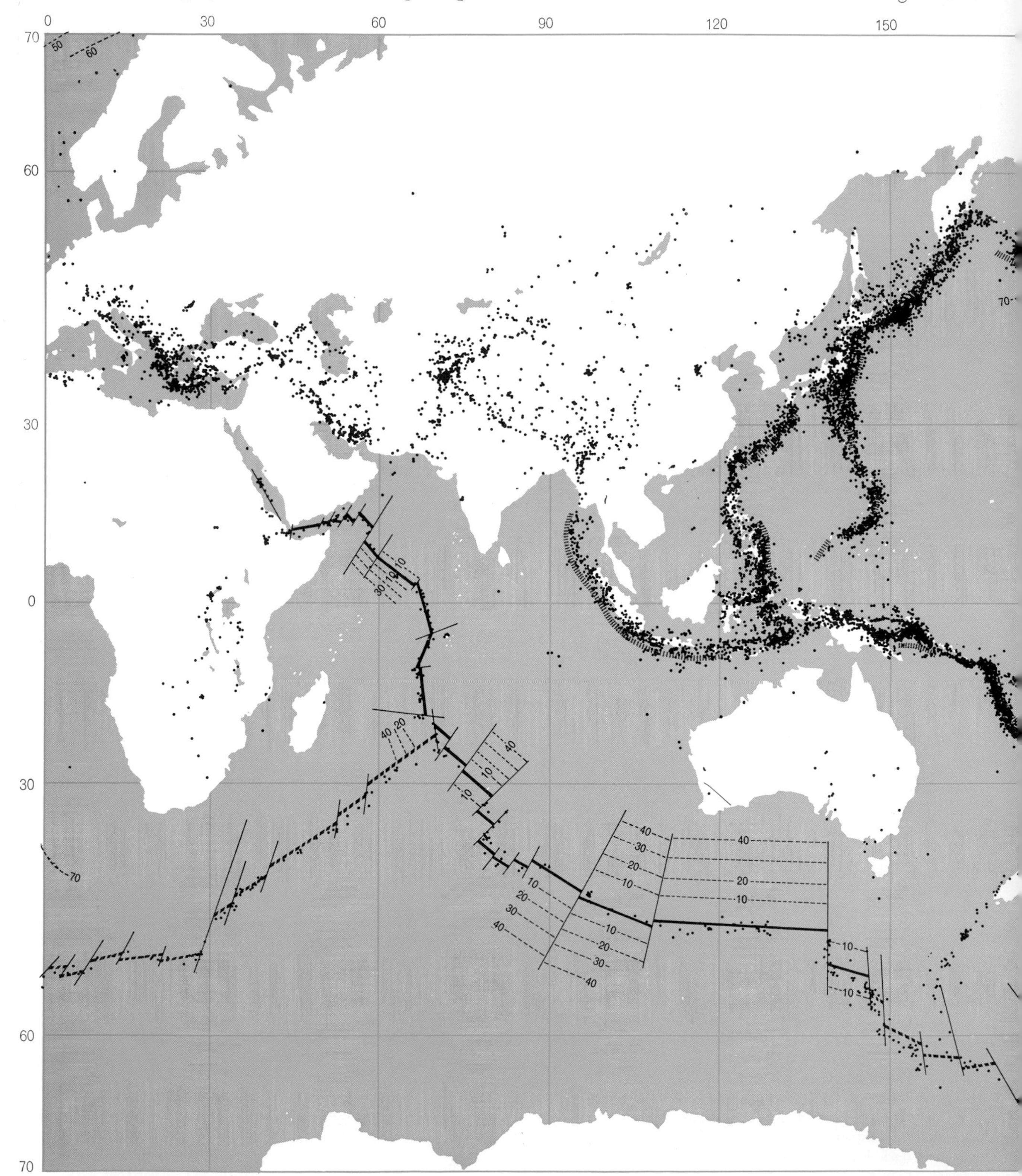

OCEANIC GEOLOGY has turned out to be much simpler than the geology of the continents. New ocean bottom is continuously being extruded along the crest of a worldwide system of ridges (*thick black lines*). The present position of material extruded at intervals of 10 million years, as determined by magnetic studies, appears as broken lines parallel to the ridge system, which is offset by fracture

necessity, but when it does happen it means that the lines of the ridge axes and of the magnetic pattern must, if they are extrapolated, go through the pole of spreading. If the ridge consists of a number of offset sections at right angles to the fracture zones, the axes of these sections will converge on the pole of spreading. It is one of the surprises of the work at sea that this rather simple geometry embraces so large a part of the facts. It seems that marine geology is truly simpler than continental geology and that this is not merely an illusion based on our lesser knowledge of the oceans.

The regularity of the magnetic pattern suggests that the ocean floor can move as a rigid plate over areas several thousand kilometers across. The thickness of the rigid moving plate is quite uncertain,

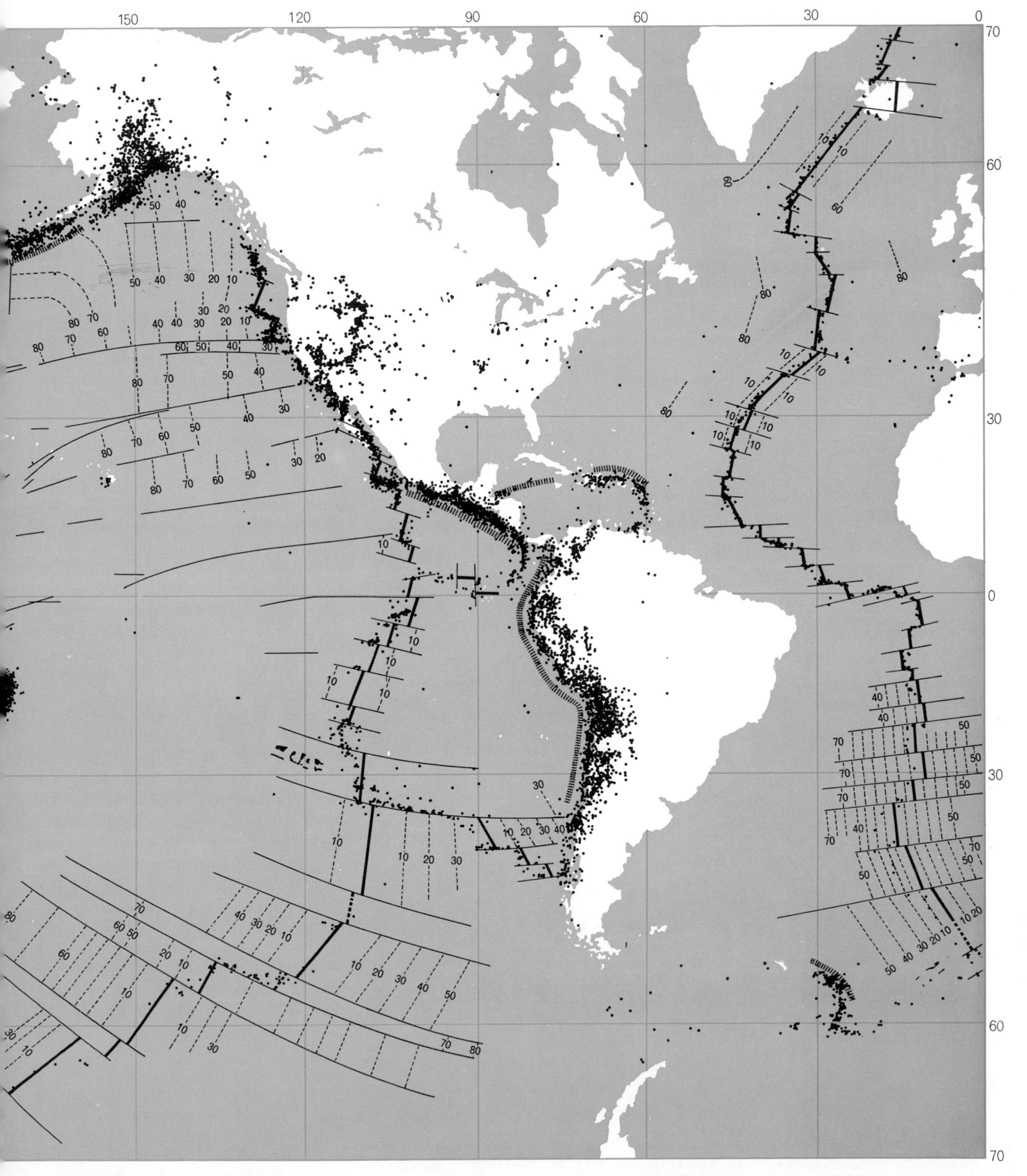

zones (*thin black lines*). Earthquakes (*black dots*) occur along the crests of ridges, on parts of the fracture zone and along deep trenches. These trenches, where the ocean floor dips steeply, are represented by hatched bands. At the maximum estimated rate of sea-floor spreading, about 16 centimeters a year, the entire floor of the Pacific Ocean could be created in perhaps 100 million years.

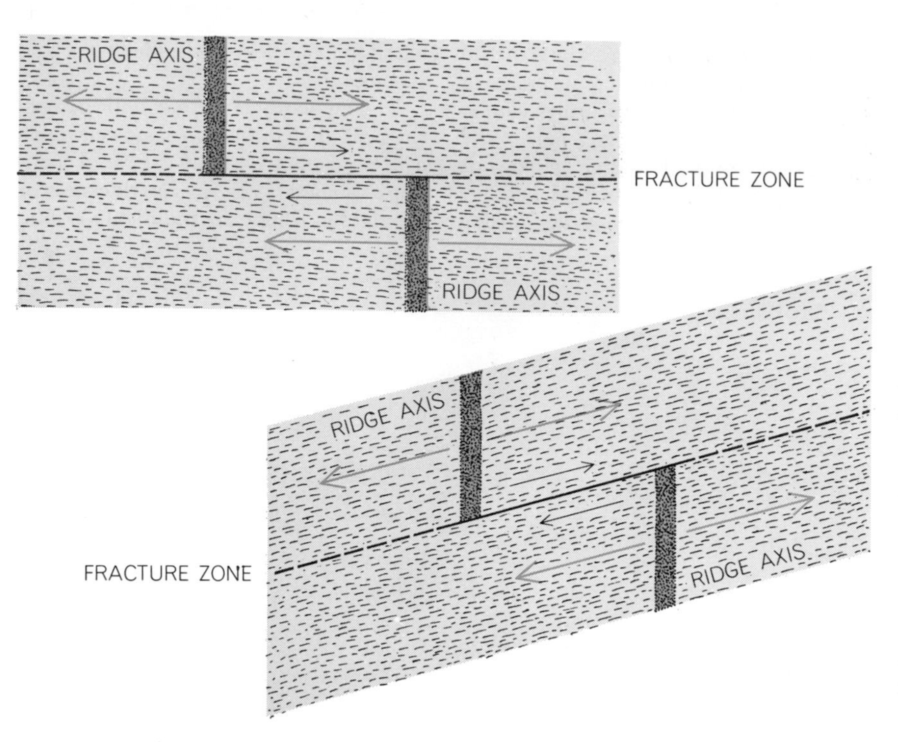

MOTION AT AXIS OF RIDGE consists of an opening of an axial crack (*vertical bands*) where two plates separate (*arrows*). Often the ridge is offset by a fracture zone, making a transform fault where one plate slips past another. The motion must be parallel to the fracture zone. It is usually at right angles to the ridge (*upper left*) but need not be (*lower right*).

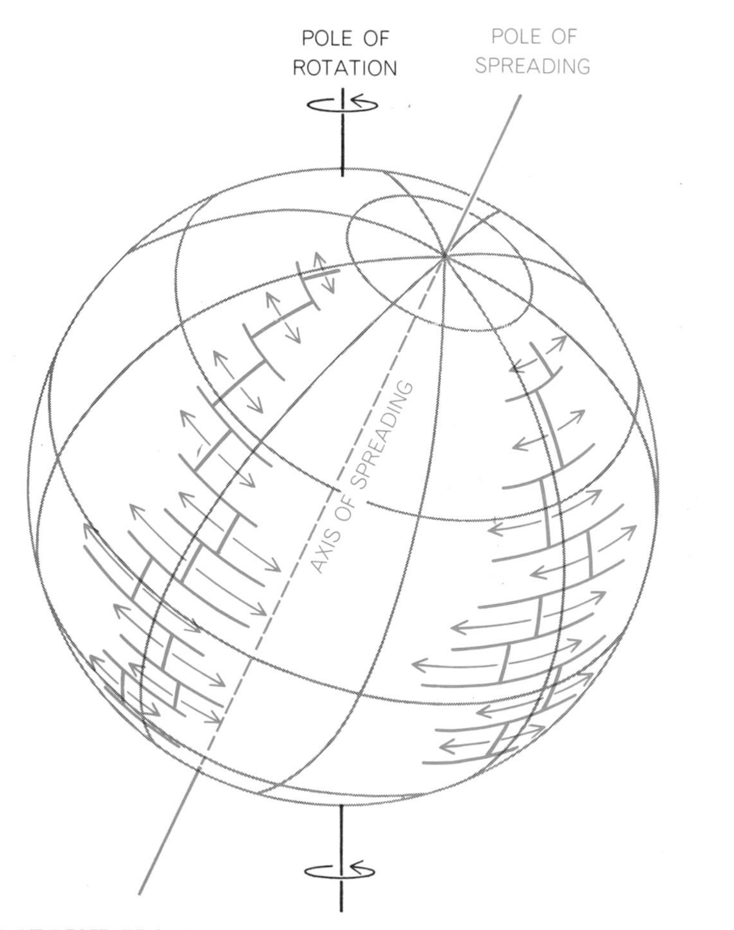

MOTION OF RIGID PLATES on a sphere requires that the plates rotate around a "pole of spreading" through which passes an "axis of spreading." Plates always move parallel to the fracture zones and along circles of latitude perpendicular to the axis of spreading. The rate of spreading is slowest near the pole of spreading and fastest 90 degrees away from it. The spreading pole can be quite remote from the sphere's pole of rotation.

but a value between 70 and 100 kilometers seems likely. If this is so, the greater part of the plate will be made of the same material as the upper part of the earth's mantle—probably of peridotite, a rock largely composed of olivine, a silicate of magnesium and iron, $(Mg, Fe)_2SiO_4$. The basaltic rocks of the oceanic crust will form the upper five kilometers or so of the plate, with a veneer of sediments on top.

What happens at the boundary of an ocean and a continent? Sometimes, as in the South Atlantic, nothing happens; there are no earthquakes, no distortion, nothing to indicate relative motion between the sea floor and the bordering continent. The continent can then be regarded as part of the same plate as the adjacent ocean floor; the rocks of the continental crust evidently ride on top of the plate and move with it. In other places there is another kind of coast, what Eduard Suess called "a Pacific coast." Such a coast is typified by the Pacific coast of South America. Here the oceanic plate dives under the continent and goes down at an angle of about 45 degrees. On the upper surface of this sloping plate there are numerous earthquakes—quite shallow ones near the coast and others as deep as 700 kilometers inland, under the continent. The evidence for the sinking plate has been beautifully confirmed by the discovery that seismic waves from shallow earthquakes and explosions, occurring near the place where the plate starts its dive, travel faster down the plate than they do in other directions. This is expected because the plate is relatively cold, whereas the upper mantle, into which the plate is sinking, is made of similar material but is hot.

Little is known of the detailed behavior of the plate; further study is vital for an understanding of the phenomena along the edges of continents. Near the point where the plate turns down there is an ocean deep, whose mode of formation is not precisely understood, but if a plate goes down, it is not difficult to imagine ways in which it could leave a depression in the sea floor. It is probable that, as the plate goes down, some of the sediment on its surface is scraped off and piled up in a jumbled mass on the landward side of the ocean deep. This sediment may later be incorporated in the mountain range that usually appears on the edge of the continent. The mountain range bordering the continent commonly has a row of volcanoes, as in the Andes. The lavas from the volcanoes are frequently composed of andesites, which are different from the lavas of the mid-ocean ridges in that they contain more silica. It may reasonably be supposed that they are formed by the partial melting of the descending plate at a depth of about 150 kilometers. The first material to melt will contain more silica than the remaining material; it is also possible that the melted material is contaminated by granite as it rises to the surface through the continental rocks.

In many places the sinking plate goes down under a chain of islands and not under the continent itself. This happens in the Aleutians, to the south of Indonesia, off the islands of the Tonga group, in the Caribbean and in many other places. The volcanoes are then on the islands and the deep earthquakes occur under the almost enclosed sea behind the chain or arc of islands, as they do in the Sea of Japan, the Sea of Okhotsk and the Java Sea.

The destruction of oceanic crust explains one of the great paradoxes of geology. There have always been oceans, but the present oceans contain no sediment more than 150 million years old and very little sediment older than 80 million years. The explanation is that the older sediments have been carried away with the plates and are either piled up at the edge of a continent or are carried down with a sinking plate and lost in the mantle.

The picture is simple: the greater part of the earth's surface is divided into six plates [*see illustration on next two pages*]. These plates move as rigid bodies, new material for them being produced from the upper mantle by lava emerging from the crack along the crest of a mid-ocean ridge. Plates are destroyed at the oceanic trenches by plunging into the mantle, where ultimately they are mixed again with the material whence they came. The scheme is not yet established in all its details. Perhaps the greatest uncertainty is in the section of the ridge running south of South Africa; it is not clear how much of this is truly a ridge and a source of new crust and how much is a series of transform faults with only tangential motion. It is also uncertain whether the American and Eurasian plates meet in Alaska or in Siberia. It appears certain, however, that they do not meet along the Bering Strait.

A close look at the system of ridges, fracture zones, trenches and earthquakes reveals many other features of great interest, which can only be mentioned here. The Red Sea and the Gulf of Aden appear to be embryo oceans [*see illustration on page 197*]. Their floors are truly oceanic, with no continental rocks; along their axes one can find offset lengths of crack joined by fracture zones, and magnetic surveys show the worldwide magnetic pattern but only the most recent parts of it. These seas are being formed by the movement of Africa and Arabia away from each other. A detailed study of the geology, the topography and the present motion suggests that the separation started 20 million years ago in the Miocene period and that it is still continuing. If this is so, there must have been a sliding movement along the Jordan rift valley, with the area to the east having moved about 100 kilometers northward with respect to the western portion. There must also have been an opening of the East African rift valley by 65 kilometers or so.

The first of these displacements is well established by geological comparisons between the two sides of the valley, and it should be possible to verify the second. The reassembly of the pieces requires that the southwest corner of Arabia overlap the "Afar triangle" in northern Ethiopia [*see bottom illustration on page 199*]. This area should therefore be part of the embryo ocean. The fact that it is dry land presented a substantial puzzle, but recently it has been shown that the oceanic magnetic pattern extends over the area; it is the only land area in the world where this is known to happen. It seems likely that the Afar triangle is in some sense oceanic. The results of gravity surveys, seismic measurements and drilling will be awaited with interest. On this picture Arabia and the area to the north comprise a small plate separate from the African and Asian plates. The northern boundary of this small plate may be in the mountains of Iran and Turkey, where motion is proceeding today.

A number of other small plates are known. There is one between the Pacific coast of Canada and the ridge off Vancouver Island; it is probable that this is being crumpled at the coast rather than diving under the continent. Farther south the plate and the ridge from which it spread may have been overrun by the westward motion of North America. The ridge appears again in the Gulf of California, which is similar in many ways to the Red Sea and the Gulf of Aden. From the mouth of the Gulf of California the ridge runs southward and is joined by an east-west ridge running through the Galápagos Islands. The sea floor bounded by the two ridges and the trench off Central America seems to constitute a separate small plate.

For the past four million years we can date the lavas on land with enough

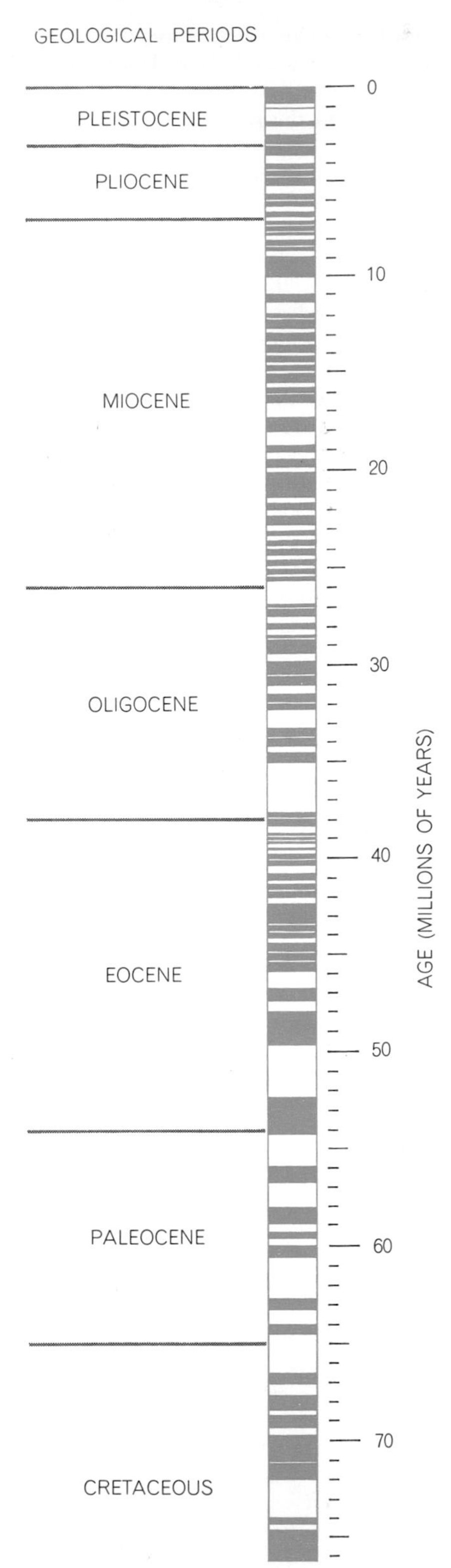

REVERSALS of the earth's magnetic field can be traced back more than 70 million years using magnetic patterns observed on the sea floor. The timetable of reversals for the most recent four million years was obtained by dating reversals in lava flows on land. Extrapolations beyond that assume that the sea floor spread at a constant rate. Colored bars show periods when the direction of the magnetic field was as it is now.

accuracy to give a timetable of magnetic reversals that can be correlated with the magnetic pattern on the sea bottom. For this period the rates of spreading from the ridges have remained constant. Further back we have a long series of reversals recorded in the ocean floor, but we cannot date them by comparison with lavas on land because the accuracy of the dates is insufficient to put the lavas in order. A rough guess can be made of the time since the oldest part of the magnetic pattern was formed by assuming that the rates have always been what they are today. This yields about 70 million years in the eastern Pacific and the South Atlantic. In fact the spacings of the older magnetic lineations are not in a constant proportion in the different oceans. The rates of spreading must therefore vary with time when long periods are considered. Directions of motion have also changed during this period, as can be seen from the departure of the older parts of some of the Pacific fracture zones from circles of latitude around the present pole of spreading. A change of direction is also shown by the accurate geometrical and geochronological fit that can be made between South America and Africa [*see illustration on page 198*]. A rotation around the present pole of spreading will not bring the continents together; it is therefore likely that in the early stages of the separation motion was around a point farther to the south.

The ideas of the development of the earth's surface by plate formation, plate motion and plate destruction can be checked with some rigor by drilling. If they are correct, drilling at any point should show sediments of all ages from the present to the time at which this part of the plate was in the central valley of the ridge. Under these sediments there should be lavas of about the same age as the lowest sediments. From preliminary reports of the drilling by the JOIDES project (a joint enterprise of five American universities) it seems that this expectation has been brilliantly verified and that the rate of spreading has been roughly constant for 70 million years in the South Atlantic. Such studies are of great importance because they will give firm dates for the entire magnetic pattern and provide a detailed chronology for all parts of the ocean floor.

The process of consumption of oceanic crust at the edge of a continent may proceed for tens of millions of years, but if the plate that is being consumed carries a continental fragment, then the consumption must stop when the fragment reaches the trench and collides with the continent beyond it. Because

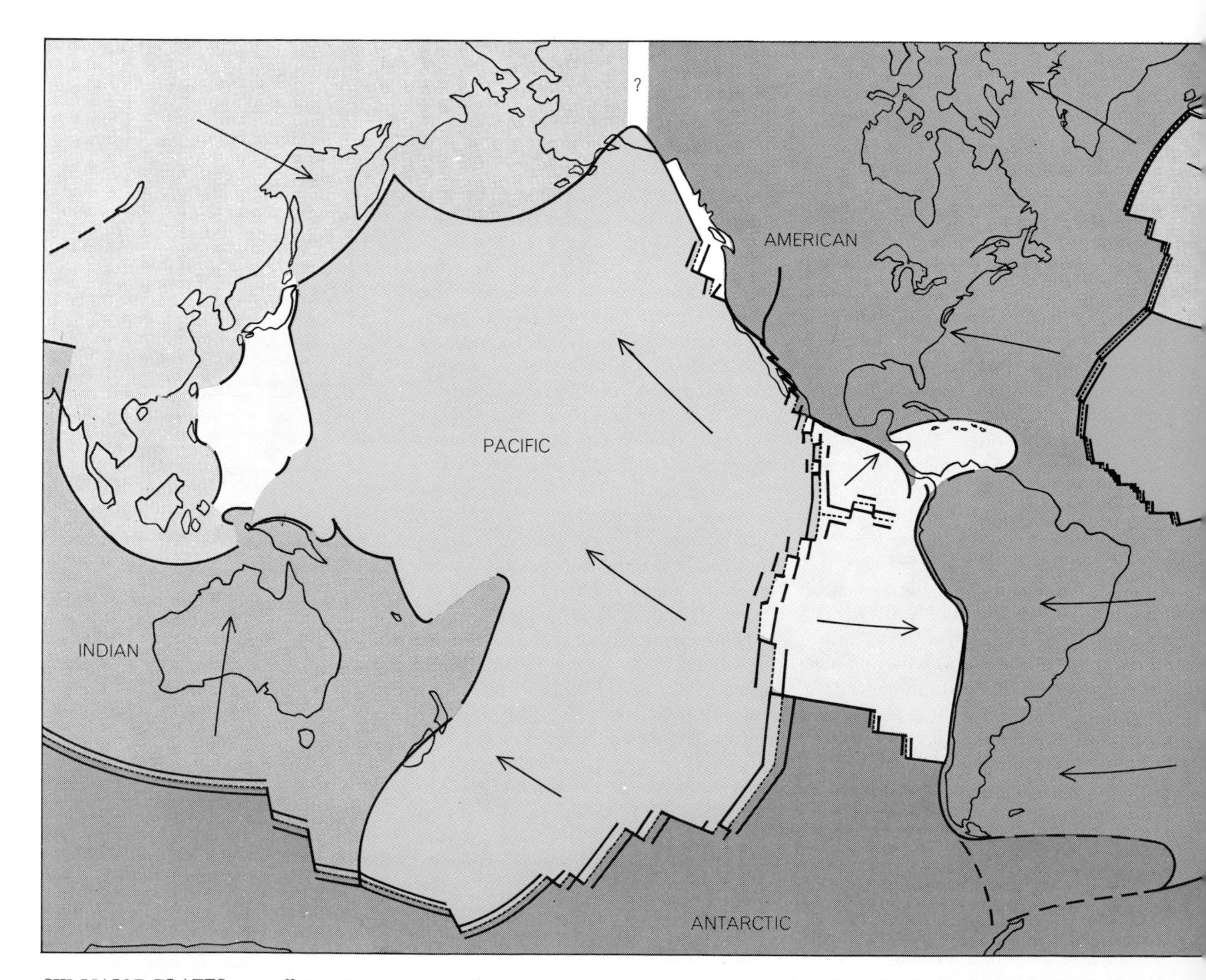

SIX MAJOR PLATES are sufficient to account for the pattern of continental drift inferred to be taking place today. In this model the African plate is assumed to be stationary. Arrows show the direction of motion of the five other large plates, which are generally bounded by ridges or trenches. Several smaller plates, unnamed, also appear. In certain areas, particularly at the junction of

the fragment consists of relatively light rocks it cannot be forced under a continent. The clearest example is the collision of India with what was once the southern margin of Asia. Paleomagnetic work shows that India has been moving northward for the past 100 million years. If it is attached to the plate that is spreading northward and eastward from the Carlsberg Ridge (which runs down the Indian Ocean halfway between Africa and India), then the motion is continuing today. This motion may be the cause of the earthquakes of the Himalayas, and it may also be connected with the formation of the mountains and of the deep sediment-filled trough to the south of them. The exact place where the joint occurs is far from clear and needs study by those with a detailed geological knowledge of northern India.

It seems unlikely that all the continents were collected in a single block for 4,000 million years and then broke apart and started their wanderings during the past 100 million years. It is more likely that the processes we see today have always been in action and that all through geologic time there have been moving plates carrying continents. We must expect continents to have split many times and formed new oceans and sometimes to have collided and been welded together. We are only at the beginning of the study of pre-Tertiary events; anything that can be said is speculation and is to be taken only as an indication of where to look.

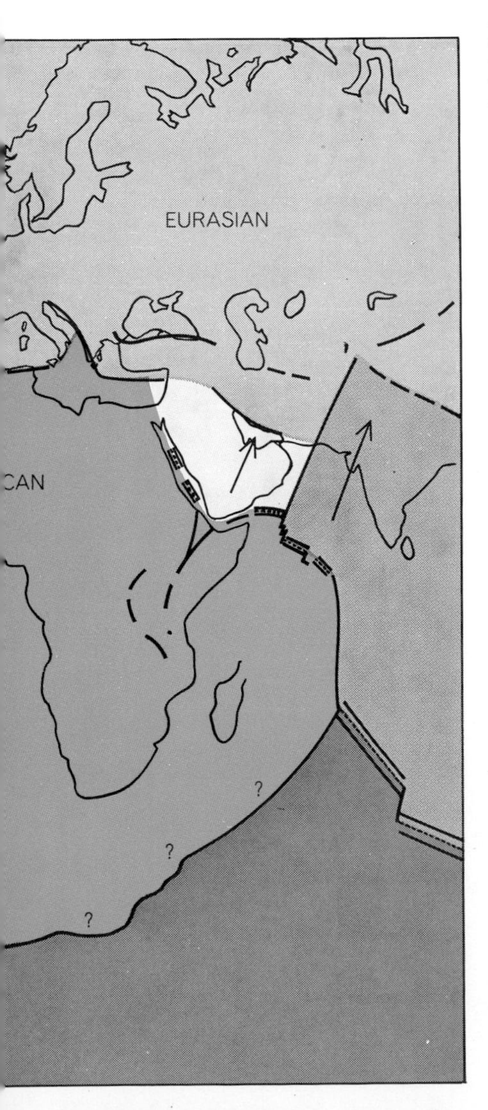

the American and Eurasian plates and in the region south of Africa, it is hard to say just where the boundaries of the plates lie.

It is virtually certain that the Atlantic did not exist 150 million years ago. Long before that, in the Lower Paleozoic, 650 to 400 million years ago, there was an older ocean in which the sediments now in the Caledonian-Hercynian-Appalachian mountains of Europe and North America were laid down. Perhaps this ocean was closed long before the present Atlantic opened and separated the Appalachian Mountains of eastern North America from their continuation in northwestern Europe.

The Urals, if they are not unique among mountain ranges, are at least exceptional in being situated in the middle of a continent. There is some paleomagnetic evidence that Siberia is a mosaic of fragments that were not originally contiguous; perhaps the Urals were once near the borders of an ocean that divided Siberia from western Russia. Similarly, it is desirable to ask where the ocean was when the Rockies were being formed. A large part of California is moving rapidly northward, and the entire continent has overrun an ocean ridge; clearly the early Tertiary geography must have been very different from that of the present. Such questions are for the future and require that the ideas of moving plates be applied by those with a detailed knowledge of the various areas.

A history of the oceans does not necessarily require an account of the mechanism behind the observed phenomena. Indeed, no very satisfactory account can be given. The traditional view, put forward by Arthur Holmes and Felix A. Vening-Meinesz, supposes that the upper mantle behaves as a liquid when it is subjected to small forces for long periods and that differences in temperature under oceans and continents are sufficient to produce convection cells in the mantle—with rising currents under the mid-ocean ridges and sinking ones under the continents. These hypothetical cells would carry the plates along as on a conveyor belt and would provide the forces needed to produce the split along the ridge. This view may be correct; it has the advantage that the currents are driven by temperature differences that themselves depend on the position of the continents. Such a back-coupling can produce complicated and varying motions.

On the other hand, the theory is implausible in that convection does not normally happen along lines. It certainly does not happen along lines broken by frequent offsets, as the ridge is. Also it is difficult to see how the theory applies to the plate between the Mid-Atlantic Ridge and the ridge in the Indian Ocean. This plate is growing on both sides, and since there is no intermediate trench the two ridges must be moving apart. It would be odd if the rising convection currents kept exact pace with them. An alternative theory is that the sinking part of the plate, which is denser than the hotter surrounding mantle, pulls the rest of the plate after it. Again it is difficult to see how this applies to the ridge in the South Atlantic, where neither the African nor the American plate has a sinking part.

Another possibility is that the sinking plate cools the neighboring mantle and produces convection currents that move the plates. This last theory is attractive because it gives some hope of explaining the almost enclosed seas, such as the Sea of Japan. These seas have a typical oceanic floor except that the floor is overlain by several kilometers of sediment. Their floors have probably been sinking for long periods. It seems possible that a sinking current of cooled mantle material on the upper side of the plate might be the cause of such deep basins. The enclosed seas are an important feature of the earth's surface and urgently require explanation; in addition to the seas that are developing at present behind island arcs there are a number of older ones of possibly similar origin, such as the Gulf of Mexico, the Black Sea and perhaps the North Sea.

The ideas set out in this attempt at a history of the ocean have developed in the past 10 years. What we have is a sketch of the outlines of a history; a mass of detail needs to be filled in and many major features are quite uncertain. Nonetheless, there is a stage in the development of a theory when it is most attractive to study and easiest to explain, that is while it is still simple and successful and before too many details and difficulties have been uncovered. This is the interesting stage at which plate theory now stands.

IV

OCEANIC LIFE

IV

OCEANIC LIFE

INTRODUCTION

At the beginning of the semester, a colleague of mine—an imminent marine biologist—was assigned to advise those students seeking additional information on the several courses being offered in biology and to answer their questions on biological oceanography as a career.

The afternoon counseling session passed quietly. My friend had almost completed his period of duty when a young freshman approached and inquired about two of the courses. The young man, obviously pleased at having located the fountainhead of information on marine biology, inquired further of my colleague about what it was like to be a marine biologist. For the next half hour, my fellow professor held forth on everything from the importance of proper laboratory techniques to the origin of life. Finally he turned to the young student—sure that he had covered the subject—and asked if there was anything else the student wished to know. The young man stared at him a moment and said, "But what does a marine biologist *do*?"

This incident has reminded me many times since that, although we may specify in detail *what* a scientist is, and *how* scientists relate to their science, we are guilty of not reporting what they *do*.

Preferring to let each speak for himself, as it were, I have selected thirteen articles on marine biology, each written by a specialist in marine life or by an extraordinarily capable reporter. The reader will readily observe that the first article, John D. Isaacs's "The Nature of Oceanic Life," is the only one that presents an overview. Isaacs has synthesized much basic information, and he has provided a single theme: that the food cycle in the sea is such that it relates all marine life. Although perhaps not obvious at first glance, Isaacs's article is a good example of what all marine biologists must eventually *do*: organize, interpret, and relate observations into a meaningful, coherent scheme. This suggests that marine biologists—like all scientists—must be thinkers first and technicians second.

Each of the remaining articles in this part of the book concentrates on one particular research topic. Moreover, each study is shown to have an objective or purpose, a technique of investigation, and a conclusion. The last and most important part, the conclusion, may be firm or only tentative, depending on the completeness of study or the research philosophy of the investigator. Taken together, these articles tell us what marine biologists *do*; more importantly, they tell us what these scientists think.

Sir James Gray, in "How Fishes Swim," suggests that, although basic research on the simple mechanics of swimming is interesting to the academic scientist, such research also leads to applications that are of importance to commercial fishermen, dam-construction engineers, conservationists, and even housewives. For example, the economically important salmon must return to its spawning stream in order to lay its eggs and keep the fishery producing. With the introduction of hydroelectric and flood-control dams on rivers in which salmon are known to spawn, the wildlife conservationists have required that fish ladders be constructed to provide migrating salmon with a way of bypassing the dams on their upstream migrations. The engineer, in designing such fish ladders, must know what the swimming strength of the salmon is—basically, how many horsepower are generated by an individual salmon. As Gray points out, because there are few data from field observations, he found it necessary to build a rotating device in his laboratory at Cambridge in order to observe how fish swim and to measure their motions and their rates of movement. Such research, far removed from the natural stream or ocean habitat of a fish, provides, nevertheless, basic information on a life process. Such laboratory research is becoming quite commonplace, and many marine biologists now spend a considerable part of their time in carefully controlled laboratory experimentation. At the time of the *Challenger,* marine biologists were solely concerned with collecting and identifying marine life. A century later, the research objectives are largely to develop a thorough understanding of physiological phenomena, ecological responses, and basic behavioral processes.

John Magnuson, a fish specialist at the University of Wisconsin, has recently begun a major research investigation on tuna. Although much of his research is carried out at sea—he is studying tuna in the waters around Hawaii and their responses to variations in water temperature there—the germinal idea behind the research was originally conceived through measurements on aquarium-held fish in the Laboratory of Limnology at Madison, some 5000 miles from Hawaii! Again, in Magnuson's investigation, we see application of very basic research. Helping salmon swim upstream to spawn and locating tuna by thermal surveys are two approaches for helping solve the world food crisis—they are also good examples of what marine biologists *do.*

With few exceptions, fishermen still employ fishing techniques that were used in Biblical times. The greatest effort in conducting commercial fishing operations is expended in the search for schools of fish that are large enough to warrant exploitation. Although the size of fish schools is not as important to the commercial fisheries of the Iron Curtain countries because they are not based on the profit motive, such pragmatic commercialism is an important consideration of the rest of the world's major fishing nations. For East and West alike, however, the need for additional protein is basic.

It is evident, then, that man should learn why and how fishes school; for if he can achieve such understanding, and in some way induce scattered fish to join in a school at his command, the harvest would be bountiful indeed.

Evelyn Shaw, in "The Schooling of Fishes," presents a timely appraisal of our knowledge of fish schooling. Shaw points out in her observations of a school of fish that common speed, synchronous response, and constancy of spacing suggest some sort of central control—a master switch, as it were—for the entire school. But since there is apparently no mechanism for central control, to what is the schooling response related? In seeking an answer to this question, she too has turned to controlled and carefully monitored experiments in the laboratory. Although no firm conclusions are at hand regarding the entire schooling system, some very promising hints are available. For example, behavior of young fish suggests that they may be susceptible to "instruction," or at least to preplanned experience, which could provide man with a means of inducing schooling. This research is exciting, not only as a contribution to science, but also as a contribution in mankind's race to feed the hungry millions.

Likewise, Arthur Hasler has shown, in "The Homing Salmon," that salmon, in their search for the stream of their birth and for the site of spawning, rely upon their sense of smell to detect the minute quantities of one or more trace elements in the water and, thus, to guide them home. Hasler has confirmed the olfactory response in his laboratory studies using salmon in special tanks with runways. Using the chemical morpholine in only one of the runways, fingerling response was proven.

Subsequent to these experiments, Hasler and his associates developed small sound transmitters about the size of lipstick containers, which were forced into captured fish. By following the fish using a sonic hydrophone, the migration path was charted in detail. Such research improves man's knowledge of fish behavior and helps the commercial fisherman to locate his quarry more easily. Such research also suggests that marine biologists must also become familiar with acoustical physics and electronics.

Supported by the U.S. Air Force, Brian B. Boycott ("Learning in the Octopus") has studied the behavior of the octopus and has concluded that this interesting marine mollusc can be taught to attack crabs in the presence of selected objects and to leave them alone in the presence of others. Although research on the octopus has shown that the level of learning can be affected—even controlled—by surgically removing lobes or parts of lobes of the brain, the most interesting aspect of these investigations has been in providing clues to short- and long-term memory. Such research as Boycott describes is truly at the edge of the frontier of research. Again, it is noteworthy that such research is also laboratory-oriented.

Evidence of the superiority of marine animals over man in matters of navigation is found in Archie Carr's article "The Navigation of the Green Turtle." For several years, I have used the navigational feats of the green turtle—and particularly the "voyage" that this cumbersome animal makes in swimming from the coast of Brazil to Ascension Island in the middle of the South Atlantic in order to lay eggs—as an introduction to lectures on navigation. Clearly, the green turtle must be the champion navigator of all marine animals, for he not only locates the tiny island, but he also returns to a specific part of a certain beach!

Is the green turtle only a scientific curiosity—and a gourmet's delight? Hardly, for even the British Government has recently been forced to acknowledge concern for the turtle's preservation. This development came about several years ago when Her Majesty's Government made plans to bulldoze much of Aldabra Island, a tiny speck of an atoll in the Indian Ocean, in order to build a large airfield. Because this island is one of the few remaining nesting sites for the green turtle, marine biologists and conservationists voiced alarming objection to such plans, and in the end, the airfield project was abandoned. Even so, the green turtle is threatened with extinction in our lifetime unless strong measures to ensure its preservation are enforced.

Continuing the theme of animal behavior, Perry Gilbert, in "The Behavior of Sharks," presents new evidence of the nature of shark attacks on man and on oceanic prey. This brief article does much to eliminate the myths that have come to be associated with sharks; but more importantly, it shows the value of controlled experimentation on the habits of sharks as animals of prey. The young biologist will note that the study of shark behavior first requires a fundamental knowledge of the peculiar optical system of the shark, the complex *lateralis* nerve system, and the olfactory nerves.

Shark research is of considerable interest to the public, now that scuba diving has become such a popular sport. Aside from such sporting activity, we need to learn more about shark behavior in order to provide foolproof protective measures for aviators downed at sea, as well as for ocean and cable engineers who must work in shark-infested waters in order to perform their duties. Although much chemical and biological research has been done over the past twenty-five years to provide a reliable shark repellent, none of those yet devised has proven completely successful. Unfortunately, less than 100% success in this development is not acceptable—certainly not to the diver who must rely upon it.

Jacques Millot ("The Coelacanth") has brought together the interests of the paleontologist and those of the marine zoologist in his story of the discovery, in 1938, of a "living fossil," the coelacanth, a primitive fish long thought to be extinct. Although once distributed widely about the world (fossil coelacanths are found in Spitzbergen, England, and Madagascar), the only known present-day locality is in the waters off South Africa, chiefly those in the area of the Mozambique Channel. One might argue that the discoveries to date only reflect the area in which an intensive search has been made, and that, perhaps, the coelacanth will be found elsewhere in the world ocean. Surely, the coelacanth is not the only example of an ancient marine animal, long thought to be extinct, that may yet to be living in remote waters. The challenge is exciting.

In considering the Antarctic Ocean as one of the frontiers for the sea-going marine biologist, we have chosen to include Robert C. Murphy's article "The Oceanic Life of the Antarctic" and Johan T. Rudd's "The Ice Fish." Murphy presents a broad overview of the entire life spectrum of the Antarctic and provides additional proof of the tremendous food resources available there. Rudd, on the other hand, brings our attention to one single small group of species, the very strange ice fish, whose blood is colorless.

When we consider the tremendous food resources available to man in the Antarctic, we may question why they have not been tapped already. In the several years that I have asked this question of my zoologist colleagues, I have never received an answer that I consider plausible. Why is it that, in this world of starvation for millions, we have made no attempt to exploit the bountiful food of the Antarctic—save for the whales—to satisfy the protein needs of hungry peoples? Let us hope that, in the decade ahead, this question will be asked loud enough to be heard and often enough to receive positive attention.

The food potential of the ocean is vast. One may readily visualize the great protein resource suggested by Robert S. Dietz in "The Sea's Deep Scattering Layers." The deep scattering layers, first noticed on acoustic depth recorders, rise at night and fall during the day. Careful study of the records and selective deep trawling have clearly shown the layers to be composed of nocturnal marine organisms, primarily small lantern fish and small shrimplike crustaceans.

Research has been initiated to survey the density of these organisms off the coast of California by means of bathyscaph dives. The results so far indicate that an entirely new food resource exists in the deep scattering layers, and that, although the depths (1700 to 2300 feet below the surface) preclude conventional trawling, they are not inaccessible to modern fishing trawlers, which could be rigged to net these abundant small fish. Dietz's article presents an excellent example of how research on acoustic sounding, much of it sponsored by the U.S. Navy, can provide critical spin-off in such an important area as the location of potential food resources. As is true of the biologist engaged in salmon research, the marine biologist studying deep-sea life would do well to be prepared in electronics and acoustical physics.

This part of the book includes two articles that draw our attention to a major tragedy: the slaughter of the great whales. Willis Pequegnat's article, "Whales, Plankton and Man" describes a food pyramid that begins with plankton on the bottom and ends with man on top: the plankton provide food for the shrimplike krill, which in turn provide food for whales, which are eaten by man. Placing man at the top of the pyramid is, according to Pequegnat, totally unnecessary. His novel suggestion that man harvest the krill directly, bypassing the whale, seems both compassionate and highly desirable when we consider the tragedy of the whale as presented by Scott McVay in "The Last of the Great Whales." Here, we are brought face to face with the sobering fact that man has, through his own stupidity, brought the great whales to the point of extinction. Although some of the blame must lie with the United States because of our desire to avoid upsetting an uneasy peace with those most guilty, the USSR and Japan, all of the maritime nations of the world are guilty, and have succumbed to apathy and to permissiveness on an international scale. Pequegnat is not alone in suggesting a realistic alternative to the slaughter so well documented by McVay; many marine scientists have made similar cries. Unfortunately, those cries have not been heeded.

In concluding these introductory comments on life in the sea, we should reconsider the question posed by the young student mentioned earlier: "But what do marine biologists *do*?" From a cursory review of these articles on marine life, it appears that, among other things, they measure the swimming speeds of fishes; they design special laboratory facilities that simulate conditions in nature; they travel all over the world to conduct specific research at specific locales; they spend hours simply watching and recording the motions of fish swimming about in a tank; they go beneath the surface of the sea as scuba-diving scientists to observe, record, and study marine life; they follow and track fish and other animals using acoustic devices; they teach the octopus to respond to artificial stimuli; they perform brain surgery on animals; they follow green turtles on long migrations using sophisticated radio transmitters; they predict the movement and concentration of commercial fish; they become better-than-average photographers, carpenters, deck hands, and shipboard and laboratory artisans; they study man-eating sharks at close range from the relative safety of protective cages; they experiment with chemical clues that fish use to find their spawning sites; they become specialists in the optical systems of sharks; they search for "living fossils"; they perform public relations duties in working with nonscientists; they read and (it is hoped) write in at least two foreign languages; they have a good working knowledge of electronics, navigation, and netting; they alert their colleagues and the public to the plight of the whale; they study microscopic organisms so small that their identification requires the application of electron microscopy; they study penguins, seals, petrels, krill, terns, and squid; they advise national and international conservation and regulatory agencies; they speculate about why the fish of one singularly strange group of species have no red blood cells and no hemoglobin like all other vertebrates; they descend in diving chambers and submersibles to count and identify strange fish in the dark depths of the sea; they cut their shoes to ribbons sampling remote reefs; and they search for new ways to apply their results in serving mankind. That—for a starter—is what they *do*.

THE NATURE OF OCEANIC LIFE

JOHN D. ISAACS
September 1969

I plan to take the reader on a brief tour of marine life from the surface layers of the open sea, down through the intermediate layers to the deep-sea floor, and from there to the living communities on continental shelves and coral reefs. Like Dante, I shall be able to record only a scattered sampling of the races and inhabitants of each region and to point out only the general dominant factors that typify each domain; in particular I shall review some of the conditions, principles and interactions that appear to have molded the forms of life in the sea and to have established their range and compass.

The organisms of the sea are born, live, breathe, feed, excrete, move, grow, mate, reproduce and die within a single interconnected medium. Thus interactions among the marine organisms and interactions of the organisms with the chemical and physical processes of the sea range across the entire spectrum from simple, adamant constraints to complex effects of many subtle interactions.

Far more, of course, is known about the life of the sea than I shall be able even to suggest, and there are yet to be achieved great steps in our knowledge of the living entities of the sea. I shall mention some of these possibilities in my concluding remarks.

A general discussion of a living system should consider the ways in which plants elaborate basic organic material from inorganic substances and the successive and often highly intricate steps by which organisms then return this material to the inorganic reservoir. The discussion should also show the forms of life by which such processes are conducted. I shall briefly trace these processes through the regions I have indicated, returning later to a more detailed discussion of the living forms and their constraints.

Some organic material is carried to the sea by rivers, and some is manufactured in shallow water by attached plants. More than 90 percent of the basic organic material that fuels and builds the life in the sea, however, is synthesized within the lighted surface layers of open water by the many varieties of phytoplankton. These sunny pastures of plant cells are grazed by the herbivorous zooplankton (small planktonic animals) and by some small fishes. These in turn are prey to various carnivorous creatures, large and small, who have their predators also.

The debris from the activities in the surface layers settles into the dimly lighted and unlighted midlayers of the sea, the twilight mesopelagic zone and the midnight bathypelagic zone, to serve as one source of food for their strange inhabitants. This process depletes the surface layers of some food and particularly of the vital plant nutrients, or fertilizers, that become trapped below the surface layers, where they are unavailable to the plants. Food and nutrients are also actively carried downward from the surface by vertically migrating animals.

The depleted remnants of this constant "rain" of detritus continue to the sea floor and support those animals that live just above the bottom (epibenthic animals), on the bottom (benthic animals) and burrowed into the bottom. Here filter-feeding and burrowing (deposit-feeding) animals and bacteria rework the remaining refractory particles. The more active animals also find repast in mid-water creatures and in the occasional falls of carcasses and other larger debris. Except in unusual small areas there is an abundance of oxygen in the deep water, and the solid bottom presents advantages that allow the support of a denser population of larger creatures than can exist in deep mid-water.

In shallower water such as banks, atolls, continental shelves and shallow seas conditions associated with a solid bottom and other regional modifications of the general regime enable rich populations to develop. Such areas constitute about 7 percent of the total area of the ocean. In some of these regions added food results from the growth of larger fixed plants and from land drainage.

With the above bare recitation for general orientation, I shall now discuss these matters in more detail.

The cycle of life in the sea, like that on land, is fueled by the sun's visible light acting on green plants. Of every million photons of sunlight reaching the earth's surface, some 90 enter into the net production of basic food. Perhaps 50

NEW EVIDENCE that an abundance of large active fishes inhabit the deep-sea floor was obtained recently by the author and his colleagues in the form of photographs such as the one on the opposite page. The photograph was made by a camera hovering over a five-gallon bait can at a depth of 1,400 meters off Lower California. The diagonal of the bait can measures a foot. The larger fish are mostly rat-tailed grenadiers and sablefish. The fact that large numbers of such fish are attracted almost immediately to the bait suggests that two rather independent branches of the marine food web coexist in support of the deep-bottom creatures by dead material: the rain of fine detritus, which supports a variety of attached filter-feeding and burrowing organisms, and rare, widely separated falls of large food fragments, which support active creatures adapted to the discovery and utilization of such food.

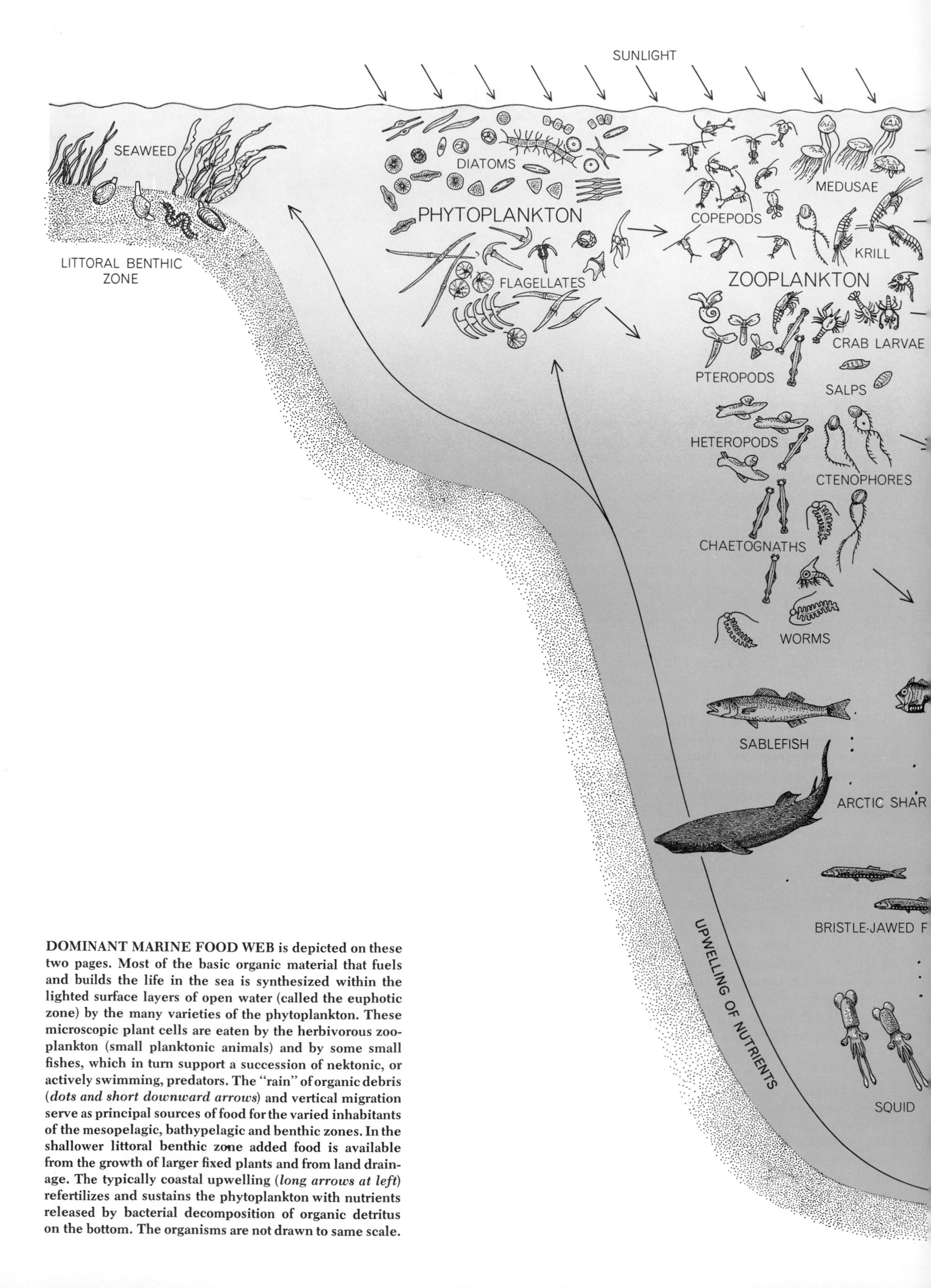

DOMINANT MARINE FOOD WEB is depicted on these two pages. Most of the basic organic material that fuels and builds the life in the sea is synthesized within the lighted surface layers of open water (called the euphotic zone) by the many varieties of the phytoplankton. These microscopic plant cells are eaten by the herbivorous zooplankton (small planktonic animals) and by some small fishes, which in turn support a succession of nektonic, or actively swimming, predators. The "rain" of organic debris (*dots and short downward arrows*) and vertical migration serve as principal sources of food for the varied inhabitants of the mesopelagic, bathypelagic and benthic zones. In the shallower littoral benthic zone added food is available from the growth of larger fixed plants and from land drainage. The typically coastal upwelling (*long arrows at left*) refertilizes and sustains the phytoplankton with nutrients released by bacterial decomposition of organic detritus on the bottom. The organisms are not drawn to same scale.

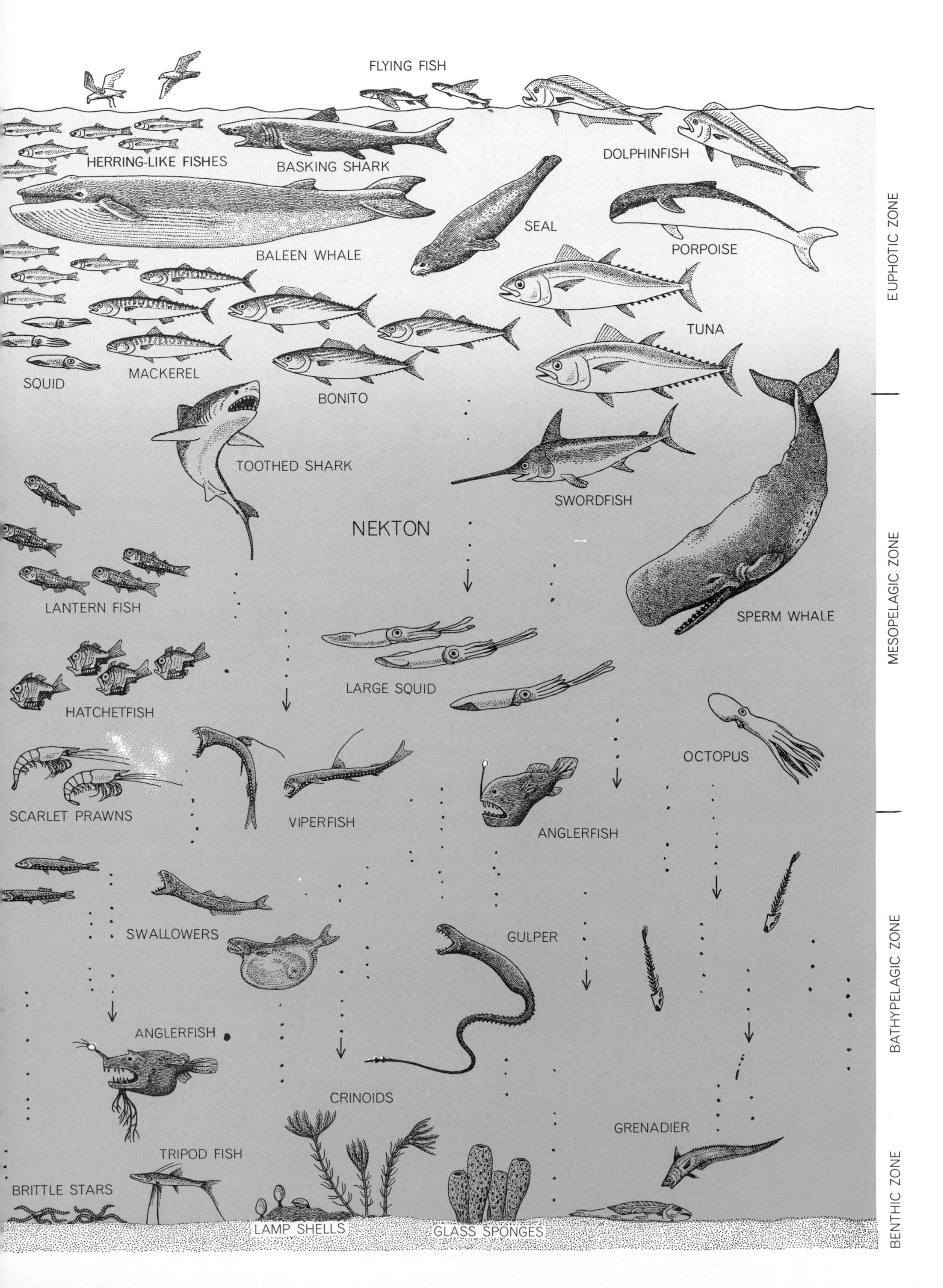
FLYING FISH
HERRING-LIKE FISHES
BASKING SHARK
DOLPHINFISH
SEAL
BALEEN WHALE
PORPOISE
TUNA
SQUID
MACKEREL
BONITO
TOOTHED SHARK
SWORDFISH
NEKTON
LANTERN FISH
SPERM WHALE
LARGE SQUID
HATCHETFISH
OCTOPUS
SCARLET PRAWNS
VIPERFISH
ANGLERFISH
SWALLOWERS
GULPER
ANGLERFISH
CRINOIDS
GRENADIER
TRIPOD FISH
BRITTLE STARS
LAMP SHELLS
GLASS SPONGES
EUPHOTIC ZONE
MESOPELAGIC ZONE
BATHYPELAGIC ZONE
BENTHIC ZONE

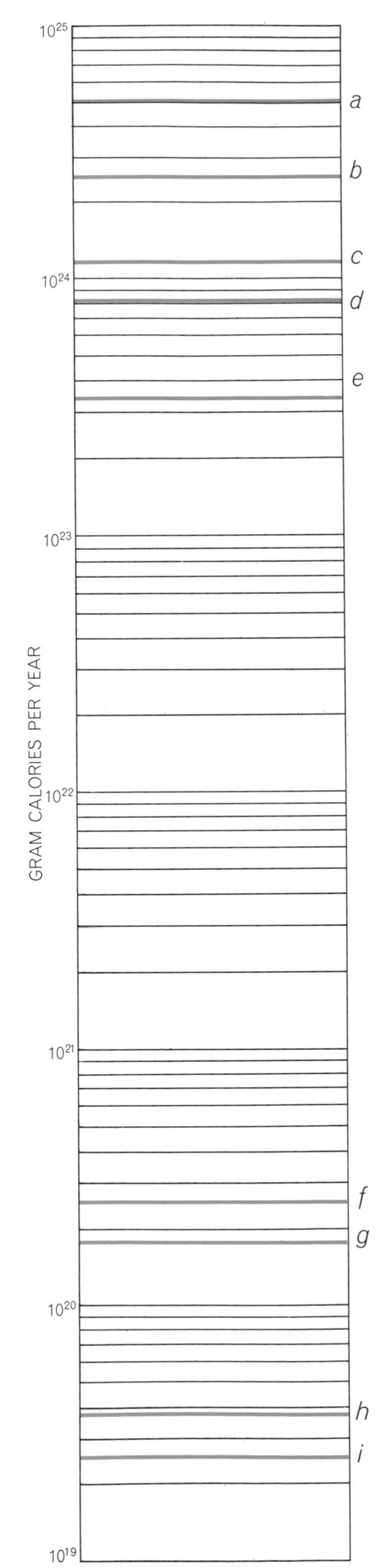

of the 90 contribute to the growth of land plants and about 40 to the growth of the single-celled green plants of the sea, the phytoplankton [*see illustration at left*]. It is this minute fraction of the sun's radiant energy that supplies the living organisms of this planet not only with their food but also with a breathable atmosphere.

The terrestrial and marine plants and animals arose from the same sources, through similar evolutionary sequences and by the action of the same natural laws. Yet these two living systems differ greatly at the stage in which we now view them. Were we to imagine a terrestrial food web that had developed in a form limited to that of the open sea, we would envision the land populated predominantly by short-lived simple plant cells grazed by small insects, worms and snails, which in turn would support a sparse predaceous population of larger insects, birds, frogs and lizards. The population of still larger carnivores would be a small fraction of the populations of large creatures that the existing land food web can nurture, because organisms in each of these steps pass on not more than 15 percent of the organic substance.

In some important respects this imaginary condition is not unlike that of the dominant food web of the sea, where almost all marine life is sustained by microscopic plants and near-microscopic herbivores and carnivores, which pass on only a greatly diminished supply of food to sustain the larger, more active and more complex creatures. In other respects the analogy is substantially inaccurate, because the primary marine food production is carried out by cells dispersed widely in a dense fluid medium.

This fact of an initial dispersal imposes a set of profound general conditions on all forms of life in the sea. For comparison, the concentration of plant food in a moderately rich grassland is of the order of a thousandth of the volume of the gross space it occupies and of the order of half of the mass of the air in which it is immersed. In moderately rich areas of the sea, on the other hand, food is hundreds of times more dilute in volume and hundreds of thousands of times more dilute in relative mass. To crop this meager broth a blind herbivore or a simple pore in a filtering structure would need to process a weight of water hundreds of thousands of times the weight of the cell it eventually captures. In even the densest concentrations the factor exceeds several thousands, and with each further step in the food web dilution increases. Thus from the beginnings of the marine food web we see many adaptations accommodating to this dilution: eyes in microscopic herbivorous animals, filters of exquisite design, mechanisms and behavior for discovering local concentrations, complex search gear and, on the bottom, attachments to elicit the aid of moving water in carrying out the task of filtration. All these adaptations stem from the conditions that limit plant life in the open sea to microscopic dimensions.

It is in the sunlit near-surface of the open sea that the unique nature of the dominant system of marine life is irrevocably molded. The near-surface, or mixed, layer of the sea varies in thickness from tens of feet to hundreds depending on the nature of the general circulation, mixing by winds and heating [see the article "The Atmosphere and the Ocean," by R. W. Stewart, beginning on page 35]. Here the basic food production of the sea is accomplished by single-celled plants. One common group of small phytoplankton are the coccolithophores, with calcareous plates, a swimming ability and often an oil droplet for food storage and buoyancy. The larger microscopic phytoplankton are composed of many species belonging to several groups: naked algal cells, diatoms with complex shells of silica and actively swimming and rotating flagellates. Very small forms of many groups are also abundant and collectively are called nannoplankton.

The species composition of the phytoplankton is everywhere complex and varies from place to place, season to season and year to year. The various regions of the ocean are typified, however, by

PRODUCTIVITY of the land and the sea are compared in terms of the net amount of energy that is converted from sunlight to organic matter by the green cells of land and sea plants. Colored lines denote total energy reaching the earth's upper atmosphere (*a*), total energy reaching earth's surface (*b*), total energy usable for photosynthesis (*c*), total energy usable for photosynthesis at sea (*d*), total energy usable for photosynthesis on land (*e*), net energy used for photosynthesis on land (*f*), net energy used for photosynthesis at sea (*g*), net energy used by land herbivores (*h*) and net energy used by sea herbivores (*i*). Although more sunlight falls on the sea than on the land (by virtue of the sea's larger surface area), the total land area is estimated to outproduce the total sea area by 25 to 50 percent. This is primarily due to low nutrient concentrations in the euphotic zone and high metabolism in marine plants. The data are from Walter R. Schmitt of Scripps Institution of Oceanography.

dominant major groups and particular species. Seasonal effects are often strong, with dense blooms of phytoplankton occurring when high levels of plant nutrients suddenly become usable or available, such as in high latitudes in spring or along coasts at the onset of upwelling. The concentration of phytoplankton varies on all dimensional scales, even down to small patches.

It is not immediately obvious why the dominant primary production of organic matter in the sea is carried out by microscopic single-celled plants instead of free-floating higher plants or other intermediate plant forms. The question arises: Why are there no pelagic "trees" in the ocean? One can easily compute the advantages such a tree would enjoy, with its canopy near the surface in the lighted levels and its trunk and roots extending down to the nutrient-rich waters under the mixed layer. The answer to this fundamental question probably has several parts. The evolution of plants in the pelagic realm favored smallness rather than expansion because the mixed layer in which these plants live is quite homogeneous; hence small incremental extensions from a plant cell cannot aid it in bridging to richer sources in order to satisfy its several needs.

On land, light is immediately above the soil and nutrients are immediately below; thus any extension is of immediate benefit, and the development of single cells into higher erect plants is able to follow in a stepwise evolutionary sequence. At sea the same richer sources exist but are so far apart that only a very large ready-made plant could act as a bridge between them. Although such plants could develop in some other environment and then adapt to the pelagic conditions, this has not come about. It is difficult to see how such a plant would propagate anyway; certainly it could not propagate in the open sea, because the young plants there would be at a severe disadvantage. In the sea small-scale differential motions of water are rapidly damped out, and any free-floating plant must often depend on molecular diffusion in the water for the uptake of nutrients and excretion of wastes. Smallness and self-motion are then advantageous, and a gross structure of cells cannot exchange nutrients or wastes as well as the same cells can separately or in open aggregations.

In addition the large-scale circulation of the ocean continuously sweeps the pelagic plants out of the region to which they are best adapted. It is essential that some individuals be returned to renew the populations. More mechanisms for this essential return exist for single-celled plants than exist for large plants, or even for any conventional spores, seeds or juveniles. Any of these can be carried by oceanic gyres or diffused by large-scale motions of surface eddies and periodic counterflow, but single-celled plants can also ride submerged countercurrents while temporarily feeding on food particles or perhaps on dissolved organic material. Other mechanisms of distribution undoubtedly are also occasionally important. For example, living marine plant cells are carried by stormborne spray, in bird feathers and by well-fed fish and birds in their undigested food.

No large plant has solved the many problems of development, dispersal and reproduction. There *are* no pelagic trees, and these several factors in concert therefore restrict the open sea in a profound way. They confine it to an initial food web composed of microscopic forms, whereas larger plants live attached only to shallow bottoms (which comprise some 2 percent of the ocean area). Attached plants, unlike free-floating plants, are not subject to the aforementioned limitations. For attached plants all degrees of water motion enhance the exchange of nutrients and wastes. Moreover, their normal population does not drift, much of their reproduction is by budding, and their spores are adapted for rapid development and settlement. Larger plants too are sometimes found in nonreproducing terminal accumulations of drifting shore plants in a few special convergent deep-sea areas such as the Sargasso Sea.

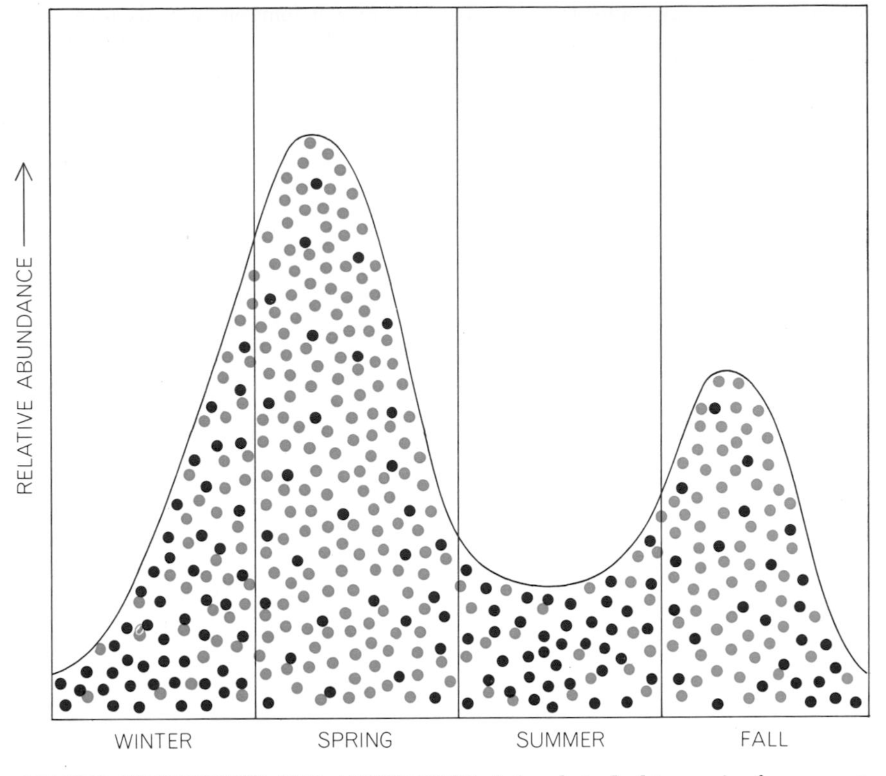

SPECIES COMPOSITION AND ABUNDANCE of the phytoplankton varies from season to season, particularly at high latitudes. During the winter the turbulence caused by storms replenishes the supply of nutrients in the surface layers. During this period flagellates (*black dots*) tend to dominate. In early spring the increase in the amount of sunlight reaching the surface stimulates plant growth, and diatoms (*colored dots*) are stimulated to grow. Later in spring grazing by zooplankton and a decrease in the supply of nutrients caused by calmer weather result in a general reduction in the phytoplankton population, which reaches a secondary minimum in midsummer, during which time flagellates again dominate. The increased mixing caused by early autumn storms causes a rise in the supply of nutrients and a corresponding minor surge in the population of diatoms. The decreasing sunlight of late fall and grazing by zooplankton again reduce the general level of the plant population.

Although species of phytoplankton will populate only regions with conditions to which they are adapted, factors other than temperature, nutrients and light levels undoubtedly are important in determining the species composition of phytoplankton populations. Little is understood of the mechanisms that give rise to an abundance of particular species under certain conditions. Grazing herbivores may consume only a part of the size range of cells, allowing certain sizes and types to dominate temporarily. Little is understood of the mechanisms that give rise to an abundance of particular species under certain conditions. Chemical by-products of certain species probably exclude certain other species. Often details of individual cell behavior are probably also important in the introduction and success of a species in a particular area. In some cases we can glimpse what these mechanisms are.

For example, both the larger diatoms and the larger flagellates can move at appreciable velocities through the water. The diatoms commonly sink downward, whereas the flagellates actively swim upward toward light. These are probably patterns of behavior primarily for increasing exchange, but the interaction of such unidirectional motions with random turbulence or systematic convective motion is not simple, as it is with an inactive particle. Rather, we would expect diatoms to be statistically abundant in upward-moving water and to sink out of the near-surface layers when turbulence or upward convection is low.

Conversely, flagellates should be statistically more abundant in downwelling water and should concentrate near the surface in low turbulence and slow downward water motions. These effects seem to exist. Off some continental coasts in summer flagellates may eventually collect in high concentrations. As they begin to shade one another from the light, each individual struggles closer to the lighted surface, producing such a high density that large areas of the water are turned red or brown by their pigments. The concentration of flagellates in these "red tides" sometimes becomes too great for their own survival. Several species of flagellates also become highly toxic as they grow older. Thus they sometimes both produce and participate in a mass death of fish and invertebrates that has been known to give rise to such a high yield of hydrogen sulfide as to blacken the white houses of coastal cities.

Large diatom cells, on the other hand, spend a disproportionately greater time in upward-moving regions of the water and an unlimited time in any region where the upward motion about equals their own downward motion. (The support of unidirectionally moving objects by contrary environmental motion is observed in other phenomena, such as the production of rain and hail.) Diatom cells are thus statistically abundant in upwelling water, and the distribution of diatoms probably is often a reflection of the turbulent-convective regime of the water. Sinking and the dependence of the larger diatoms on upward convection and turbulence for support aids them in reaching upwelling regions, where nutrients are high; it helps to explain their dominance in such regions and such other features of their distribution as their high proportion in rich ocean regions and their frequent inverse occurrence with flagellates. Differences in adaptations to the physical and chemical conditions, and the release of chemical products, probably reinforce such relations.

In some areas, such as parts of the equatorial current system and shallow seas, where lateral and vertical circulation is rapid, the species composition of phytoplankton is perhaps more simply a result of the inherent ability of the species to grow, survive and reproduce under the local conditions of temperature, light, nutrients, competitors and herbivores. Elsewhere second-order effects of the detailed cell behavior often dominate. Those details of behavior that give rise to concentrations on any dimensional scale are particularly important to all subsequent steps in the food chain.

All phytoplankton cells eventually settle from the surface layers. The depletion of nutrients and food from the surface layers takes place continuously through the loss of organic material, plant cells, molts, bodies of animals, fecal pellets and so forth, which release their content of chemical nutrients at

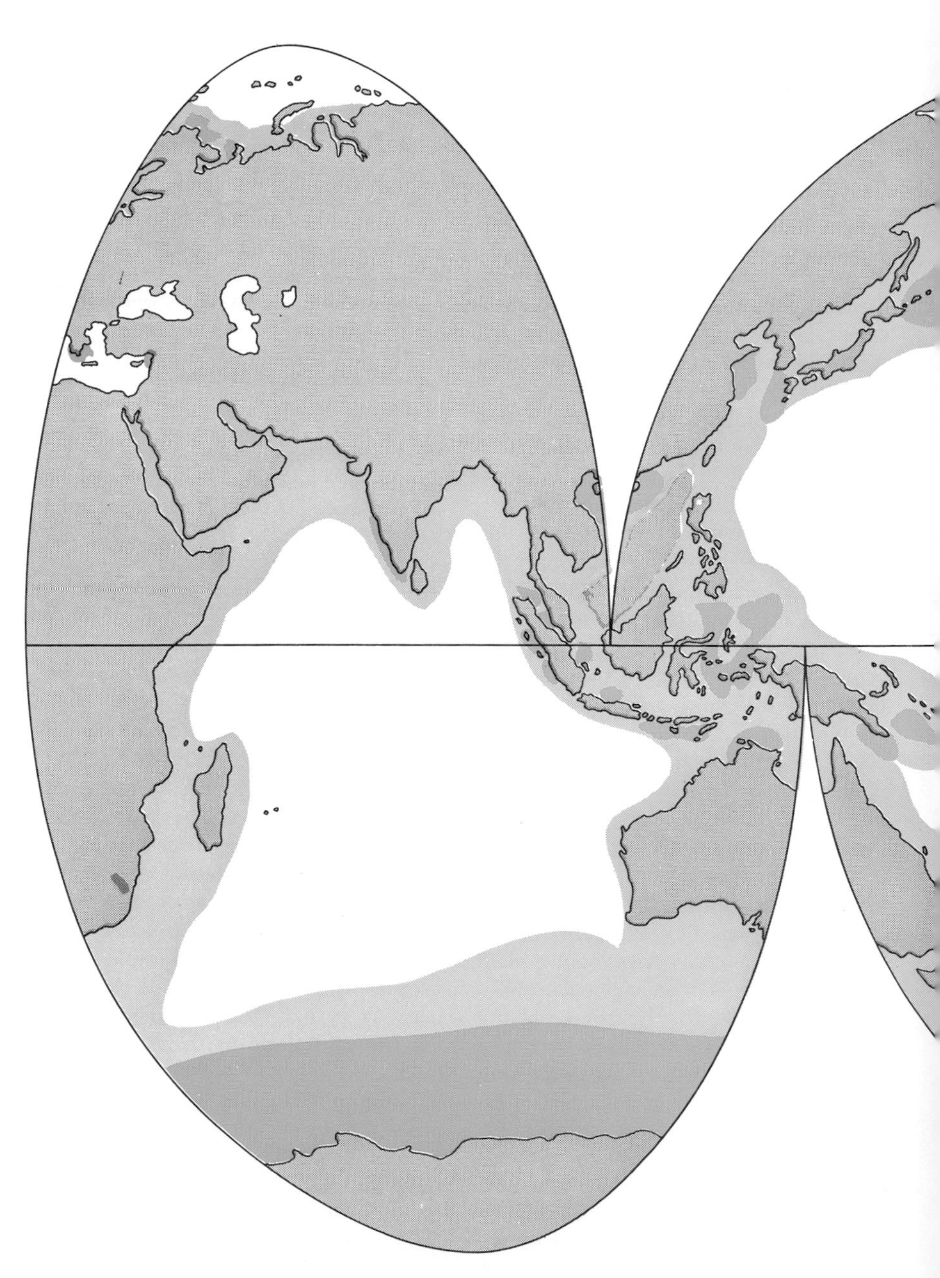

FAVORABLE CONDITIONS for the growth of phytoplankton occur wherever upwelling or mixing tends to bring subsurface nutrients up to the euphotic layer of the ocean. This map,

various depths through the action of bacteria and other organisms. The periodic downward migration of zooplankton further contributes to this loss.

These nutrients are "trapped" below the level of light adequate to sustain photosynthesis, and therefore the water in which plants must grow generally contains very low concentrations of such vital substances. It is this condition that is principally responsible for the comparatively low total net productivity of the sea compared with that of the land. The regions where trapping is broken down or does not exist—where there is upwelling of nutrient-rich water along coasts, in parts of the equatorial regions, in the wakes of islands and banks and in high latitudes, and where there is rapid recirculation of nutrients over shallow shelves and seas—locally bear the sea's richest fund of life.

The initial factors discussed so far have placed an inescapable stamp on the form of all life in the open sea, as irrevocably no doubt as the properties and distribution of hydrogen have dictated the form of the universe. These factors have limited the dominant form of life in the sea to an initial microscopic sequence that is relatively unproductive, is stimulated by upwelling and mixing and is otherwise altered in species composition and distribution by physical, chemical and biological processes on all dimensional scales. The same factors also limit the populations of higher animals and have led to unexpectedly simple adaptations, such as the sinking of the larger diatoms as a tactic to solve the manifold problems of enhancing nutrient and waste exchange, finding nutrients, remaining in the surface waters and repopulating.

The grazing of the phytoplankton is principally conducted by the herbivorous members of the zooplankton, a heterogeneous group of small animals that carry out several steps in the food web as herbivores, carnivores and detrital (debris-eating) feeders. Among the important members of the zooplankton are the arthropods, animals with external

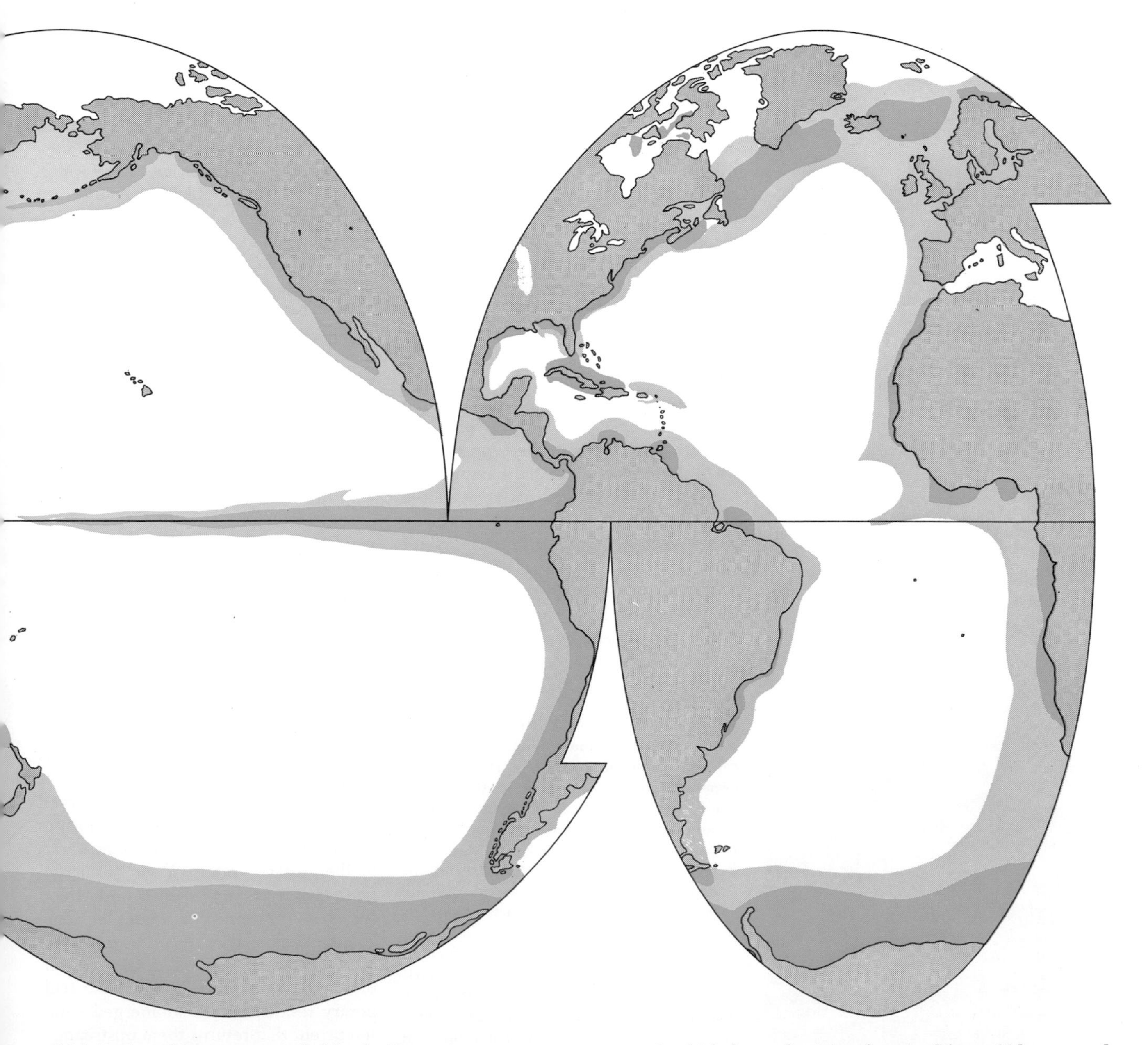

which is adapted from one compiled by the Norwegian oceanographer Harald U. Sverdrup, shows the global distribution of such waters, in which the productivity of marine life would be expected to be very high (*dark color*) and moderately high (*light color*).

skeletons that belong to the same broad group as insects, crabs and shrimps. The planktonic arthropods include the abundant copepods, which are in a sense the marine equivalent of insects. Copepods are represented in the sea by some 10,000 or more species that act not only as herbivores, carnivores or detrital feeders but also as external or even internal parasites! Two or three thousand of these species live in the open sea. Other important arthropods are the shrimplike euphausiids, the strongest vertical migrators of the zooplankton. They compose the vast shoals of krill that occur in high latitudes and that constitute one of the principal foods of the baleen whales. The zooplankton also include the strange bristle-jawed chaetognaths, or arrowworms, carnivores of mysterious origin and affinities known only in the marine environment. Widely distributed and abundant, the chaetognaths are represented by a surprisingly small number of species, perhaps fewer than 50. Larvae of many types, worms, medusae (jellyfish), ctenophores (comb jellies), gastropods (snails), pteropods and heteropods (other pelagic mollusks), salps, unpigmented flagellates and many others are also important components of this milieu, each with its own remarkably complex and often bizarre life history, behavior and form.

The larger zooplankton are mainly carnivores, and those of herbivorous habit are restricted to feeding on the larger plant cells. Much of the food supply, however, exists in the form of very small particles such as the nannoplankton, and these appear to be available almost solely to microscopic creatures. The immense distances between plant cells, many thousands of times their diameter, place a great premium on the development of feeding mechanisms that avoid the simple filtering of water through fine pores. The power necessary to maintain a certain rate of flow through pores or nets increases inversely at an exponential rate with respect to the pore or mesh diameter, and the small planktonic herbivores, detrital feeders and carnivores show many adaptations to avoid this energy loss. Eyesight has developed in many minute animals to make possible selective capture. A variety of webs, bristles, rakes, combs, cilia and other structures are found, and they are often sticky. Stickiness allows the capture of food that is finer than the interspaces in the filtering structures, and it greatly reduces the expenditure of energy.

A few groups have developed extremely fine and apparently quite effective nets. One group that has accomplished this is the Larvacea. A larvacian produces and inhabits a complex external "house," much larger than its owner, that contains a system of very finely constructed nets through which the creature maintains a gentle flow [*see illustration on page 224*]. The Larvacea have apparently solved the problem of energy loss in filtering by having proportionately large nets, fine strong threads and a low rate of flow.

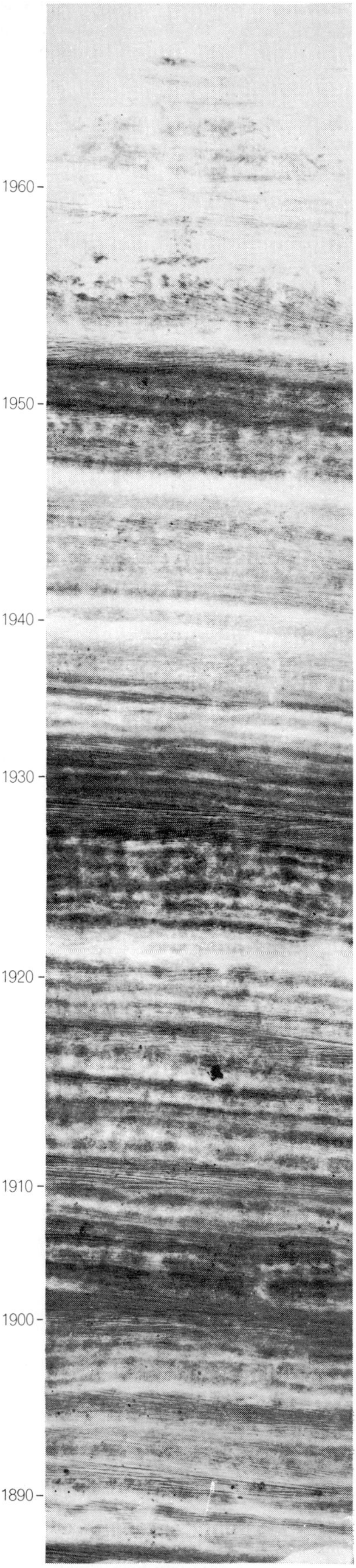

RARE SEDIMENTARY RECORD of the recent annual oceanographic, meteorological and biological history of part of a major oceanic system is revealed in this radiograph of a section of an ocean-bottom core obtained by Andrew Soutar of the Scripps Institution of Oceanography in the Santa Barbara Basin off the California coast. In some near-shore basins such as this one the absence of oxygen causes refractory parts of the organic debris to be left undecomposed and the sediment to remain undisturbed in the annual layers called varves. The dark layers are the densest and represent winter sedimentation. The lighter and less dense layers are composed mostly of diatoms and represent spring and summer sedimentation.

The composition of the zooplankton differs from place to place, day to night, season to season and year to year, yet most species are limited in distribution, and the members of the planktonic communities commonly show a rather stable representation of the modes of life.

The zooplankton are, of course, faced with the necessity of maintaining breeding assemblages and, like the phytoplankton, with the necessity of establishing a reinoculation of parent waters. In addition, their behavior must lead to a correspondence with their food and to the pattern of large-scale and small-scale spottiness already imposed on the marine realm by the phytoplankton. The swimming powers of the larger zooplankton are quite adequate for finding local small-scale patches of food. That this task is accomplished on a large scale is indirectly demonstrated by the observed correspondence between the quantities of zooplankton and the plant nutrients in the surface waters. How this large-scale task is accomplished is understood for some groups. For example, some zooplankton species have been shown to descend near the end of suitable conditions at the surface and to take temporary residence in a submerged countercurrent that returns them upstream.

There are many large and small puz-

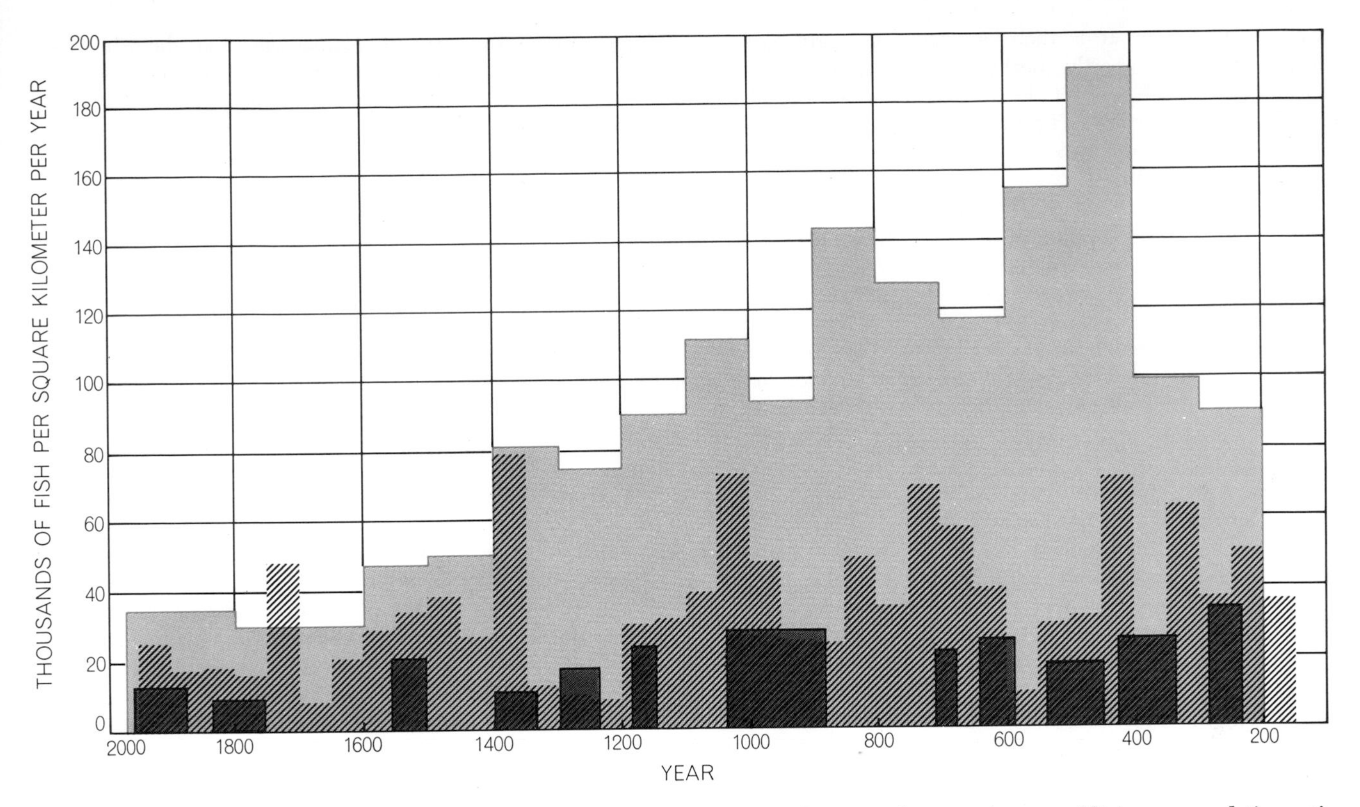

ESTIMATED FISH POPULATIONS in the Santa Barbara Basin over the past 1,800 years were obtained for three species by counting the average number of scales of each species in the varves of the core shown on the opposite page. Minimum population estimates for fish one year old and older are given for Pacific sardines (*gray*), northern sardines (*colored areas*) and Pacific hake (*hatched*).

zles in the distribution of zooplankton. As an example, dense concentrations of phytoplankton are often associated with low populations of zooplankton. These are probably rapidly growing blooms that zooplankton have not yet invaded and grazed on, but it is not completely clear that this is so. Chemical repulsion may be involved.

The concentration of larger zooplankton and small fish in the surface layers is much greater at night than during the day, because of a group of strongly swimming members that share their time between the surface and the mesopelagic region. This behavior is probably primarily a tactic to enjoy the best of two worlds: to crop the richer food developing in the surface layers and to minimize mortality from predation by remaining always in the dark, like timid rabbits emerging from the thicket to graze the nighttime fields, although still in the presence of foxes and ferrets. Many small zooplankton organisms also make a daily migration of some vertical extent.

In addition to its primary purpose daily vertical migration undoubtedly serves the migrating organisms in a number of other ways. It enables the creatures to adjust their mean temperature, so that by spending the days in cooler water the amount of food used during rest is reduced. Perhaps such processes as the rate of egg development are also controlled by these tactics. Many land animals employ hiding behavior for similar kinds of adjustment. Convincing arguments have also been presented to show that vertical migration serves to maintain a wide range of tolerance in the migrating species, so that they will be more successful under many more conditions than if they lived solely in the surface layers. This migration must also play an important part in the distribution of many species. Interaction of the daily migrants with the water motion produced by daily land-sea breeze alternation can hold the migrants offshore by a kind of "rectification" of the oscillating water motion. More generally, descent into the lower layers increases the influence of submerged countercurrents, thereby enhancing the opportunity to return upstream, to enter upwelling regions and hence to find high nutrient levels and associated high phytoplankton productivity.

Even minor details of behavior may strongly contribute to success. Migrants spend the day at a depth corresponding to relatively constant low light levels, where the movement of the water commonly is different from that at the surface. Most of the members rise somewhat even at the passage of a cloud shadow. Should they be carried under the shadow of an area rich in phytoplankton, they migrate to shallower depths, thereby often decreasing or even halting their drift with respect to this rich region to which they will ascend at night. Conversely, when the surface waters are clear and lean, they will migrate deeper and most often drift relatively faster.

We might simplistically view the distribution of zooplankton, and phytoplankton for that matter, as the consequence of a broad inoculation of the oceans with a spectrum of species, each with a certain adaptive range of tolerances and a certain variable range of feeding, reproducing and migrating behavior. At some places and at some times the behavior of a species, interacting even in detailed secondary ways with the variable conditions of the ocean and its other inhabitants, results in temporary, seasonal or persistent success.

There are a few exceptions to the microscopic dimensions of the herbivores in the pelagic food web. Among these the herrings and herring-like fishes are able to consume phytoplankton as a substantial component of their diet. Such an adaptation gives these fishes access to many times the food supply of the more carnivorous groups. It is therefore no surprise that the partly herbivorous fishes comprise the bulk of the world's fisheries [see the article "The Food Resources of the Ocean," by S. J. Holt, beginning on page 356].

The principal food supplies of the pelagic populations are passed on in incremental steps and rapidly depleted quantity to the larger carnivorous zooplankton, then to small fishes and squids, and ultimately to the wide range of larger carnivores of the pelagic realm. In this region without refuge, either powerful static defenses, such as the stinging cells of the medusae and men-o'-war, or increasing size, acuity, alertness, speed and strength are the requirements for survival at each step. Streamlining of form here reaches a high point of development, and in tropical waters it is conspicuous even in small fishes, since the lower viscosity of the warmer waters will enable a highly streamlined small prey to escape a poorly streamlined predator, an effect that exists only for fishes of twice the length in cold, viscous, arctic or deep waters.

The pelagic region contains some of the largest and most superbly designed creatures ever to inhabit this earth: the exquisitely constructed pelagic tunas; the multicolored dolphinfishes, capturers of flying fishes; the conversational porpoises; the shallow- and deep-feeding swordfishes and toothed whales, and the greatest carnivores of all, the baleen whales and some plankton-eating sharks, whose prey are entire schools of krill or small fishes. Seals and sea lions feed far into the pelagic realm. In concert with these great predators, large carnivorous sharks await injured prey. Marine birds, some adapted to almost continuous pelagic life, consume surprising quantities of ocean food, diving, plunging, skimming and gulping in pursuit. Creatures of this region have developed such faculties as advanced sonar, unexplained senses of orientation and homing, and extreme olfactory sensitivity.

These larger creatures of the sea commonly move in schools, shoals and herds. In addition to meeting the needs of mating such grouping is advantageous in both defensive and predatory strategy, much like the cargo-ship convoy and submarine "wolf pack" of World War II. Both defensive and predatory assemblages are often complex. Small fishes of several species commonly school together. Diverse predators also form loosely cooperative groups, and many species of marine birds depend almost wholly on prey driven to the surface by submerged predators.

At night, schools of prey and predators are almost always spectacularly illuminated by bioluminescence produced by the microscopic and larger plankton. The reason for the ubiquitous production of light by the microorganisms of the sea remains obscure, and suggested explanations are controversial. It has been suggested that light is a kind of inadvertent by-product of life in transparent organisms. It has also been hypothesized that the emission of light on disturbance is advantageous to the plankton in making the predators of the plankton conspicuous to *their* predators! Unquestionably it does act this way. Indeed, some fisheries base the detection of their prey on the bioluminescence that the fish excite. It is difficult, however, to defend the thesis that this effect was the direct factor in the original development of bioluminescence, since the effect was of no advantage to the individual microorganism that first developed it. Perhaps the luminescence of a microorganism also discourages attack by the light-avoiding zooplankton and is of initial survival benefit to the individual. As it then became general in the population, the effect of revealing plankton predators to their predators would also become important.

The fallout of organic material into the deep, dimly lighted mid-water supports a sparse population of fishes and invertebrates. Within the mesopelagic and bathypelagic zones are found some of the most curious and bizarre creatures of this earth. These range from the highly developed and powerfully predaceous intruders, toothed whales and swordfishes, at the climax of the food chain, to the remarkable squids, octopuses, euphausiids, lantern fishes, gulpers and anglerfishes that inhabit the bathypelagic region.

In the mesopelagic region, where some sunlight penetrates, fishes are often countershaded, that is, they are darker above and lighter below, as are surface fishes. Many of the creatures of this dimly lighted region participate in the daily migration, swimming to the upper layers at evening like bats emerging from their caves. At greater depths, over a half-mile or so, the common inhabitants are often darkly pigmented, weak-bodied and frequently adapted to unusual feeding techniques. Attraction of prey by luminescent lures or by mimicry of small prey, greatly extensible jaws and expansible abdomens are common. It is, however, a region of Lilliputian monsters, usually not more than six inches in length, with most larger fishes greatly reduced in musculature and weakly constructed.

There are some much larger, stronger and more active fishes and squids in this region, although they are not taken in trawls or seen from submersibles. Knowledge of their existence comes mainly from specimens found in the stomach of sperm whales and swordfish. They must be rare, however, since the slow, conservative creatures that are taken in trawls could hardly coexist with large numbers of active predators. Nevertheless, populations must be sufficiently large to attract the sperm whales and swordfish. There is evidence that the sperm whales possess highly developed long-range hunting sonar. They may lo-

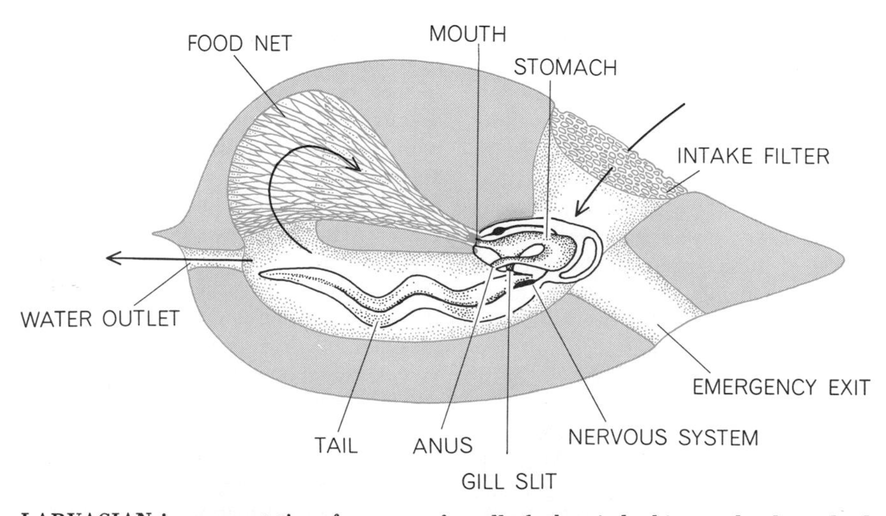

LARVACIAN is representative of a group of small planktonic herbivores that has solved the problem of energy loss in filtering, apparently without utilizing "stickiness," by having proportionately large nets, strong fine threads and a low rate of water flow. The larvacian (*black*) produces and inhabits a complex external "house" (*color*), much larger than its owner, which contains a system of nets through which the organism maintains a gentle flow. In almost all other groups simple filters are employed only to exclude large particles.

cate their prey over relatively great distances, perhaps miles, from just such an extremely sparse population of active bathypelagic animals.

Although many near-surface organisms are luminescent, it is in the bathypelagic region that bioluminescence has reached a surprising level of development, with at least two-thirds of the species producing light. Were we truly marine-oriented, we would perhaps be more surprised by the almost complete absence of biological light in the land environment, with its few rare cases of fireflies, glowworms and luminous bacteria. Clearly bioluminescence can be valuable to higher organisms, and the creatures of the bathypelagic realm have developed light-producing organs and structures to a high degree. In many cases the organs have obvious functions. Some fishes, squids and euphausiids possess searchlights with reflector, lens and iris almost as complex as the eye. Others have complex patterns of small lights that may serve the functions of recognition, schooling control and even mimicry of a small group of luminous plankton. Strong flashes may confuse predators by "target alteration" effects, or by producing residual images in the predators' vision. Some squids and shrimps are more direct and discharge luminous clouds to cover their escape. The luminous organs are arranged on some fishes so that they can be used to countershade their silhouettes against faint light coming from the surface. Luminous baits are well developed. Lights may also be used for locating a mate, a problem of this vast, sparsely populated domain that has been solved by some anglerfishes by the development of tiny males that live parasitically attached to their relatively huge mates.

It has been shown that the vertebrate eye has been adapted to detect objects in the lowest light level on the earth's surface—a moonless, overcast night under a dense forest canopy—but not lower. Light levels in the bathypelagic region can be much lower. This is most probably the primary difference that accounts for the absence of bioluminescence in higher land animals and the richness of its development in the ocean forms.

The densest populations of bathypelagic creatures lie below the most productive surface regions, except at high latitudes, where the dearth of winter food probably would exhaust the meager reserves of these creatures. All the bathypelagic populations are sparse, and in this region living creatures are less than one hundred-millionth of the water volume. Nevertheless, the zone is of immense dimensions and the total populations may be large. Some genera, such as the feeble, tiny bristle-jawed fishes, are probably the most numerous fishes in the world and constitute a gigantic total biomass. There are some 2,000 species of fishes and as many species of the larger invertebrates known to inhabit the bathypelagic zone, but only a few of these species appear to be widespread. The barriers to distribution in this widely interconnected mid-water region are not obvious.

The floor of the deep sea constitutes an environment quite unlike the mid-water and surface environments. Here are sites for the attachment of the larger

CHAMPION FILTER FEEDER of the world ocean in terms of volume is the blue whale, a mature specimen of which lies freshly butchered on the deck of a whaling vessel in this photograph. The whale's stomach has been cut open with a flensing knife to reveal its last meal: an immense quantity of euphausiids, or krill, each measuring about three inches in length. The baleen whales are not plankton-filterers in the ordinary sense but rather are great carnivores that seek out and engulf entire schools of small fish or invertebrates. The photograph was made by Robert Clarke of the National Institute of Oceanography in Wormley, England.

invertebrates that filter detritus from the water. Among these animals are representatives of some of the earliest multicelled creatures to exist on the earth, glass sponges, sea lilies (crinoids)—once thought to have been long extinct—and lamp shells (brachiopods).

At one time it was also thought that the abyssal floor was sparsely inhabited and that the populations of the deep-ocean floor were supplied with food only by the slow, meager rain of terminal detrital food material that has passed through the surface and bathypelagic populations. Such refractory material requires further passage into filter feeders or through slow bacterial action in the sediment, followed by consumption by larger burrowing organisms, before it becomes available to active free-living animals. This remnant portion of the food web could support only a very small active population.

Recent exploration of the abyssal realm with a baited camera throws doubt on the view that this is the exclusive mechanism of food transfer to the deep bottom. Large numbers of active fishes and other creatures are attracted to the bait almost immediately [*see illustration on page 214*]. It is probably true that several rather independent branches of the food web coexist in support of the deep-bottom creatures: one the familiar rain of fine detritus, and the other the rare, widely separated falls of large food particles that are in excess of the local feeding capacity of the broadly diffuse bathypelagic population. Such falls would include dead whales, large sharks or other large fishes and fragments of these, the multitude of remnants that are left when predators attack a school of surface fish and now, undoubtedly, garbage from ships and kills from underwater explosions. These sources result in an influx of high-grade food to the sea floor, and we would expect to find a population of active creatures adapted to its prompt discovery and utilization. The baited cameras have demonstrated that this is so.

Other sources of food materials are braided into these two extremes of the abyssal food web. There is the rather subtle downward diffusion of living and dead food that results initially from the daily vertical migration of small fish and zooplankton near the surface. This migration appears to impress a sympathetic daily migration on the mid-water populations down to great depths, far below the levels that light penetrates. Not only may such vertical migration bring feeble bathypelagic creatures near the bottom but also it accelerates in itself the flux of dead food material to the bottom of the deep sea.

There must also be some unassignable flux of food to the abyssal population resulting from the return of juveniles to their habitat. The larvae and young of many abyssal creatures develop at much shallower levels. To the extent that the biomass of juveniles returning to the deep regions exceeds the biomass of spawn released from it, this process, which might be called "Faginism," constitutes an input of food.

Benthic animals are much more abundant in the shallower waters off continents, particularly offshore from large rivers. Here there is often not only a richer near-surface production and a less hazardous journey of food to the sea floor but also a considerable input of food conveyed by rivers to the bottom. The deep slopes of river sediment wedges are typified by a comparatively rich population of burrowing and filtering animals that utilize this fine organic material. All the great rivers of the world save one, the Congo, have built sedimentary wedges along broad reaches of their coast, and in many instances these wedges extend into deep water. The shallow regions of such wedges are highly productive of active and often valuable marine organisms. At all depths the wedges bear larger populations than are common at similar depths elsewhere. Thus one wonders what inhabits the fan of the Congo. That great river, because of a strange invasion of a submarine canyon into its month, has built no wedge but rather is depositing a vast alluvial fan in the two-mile depths of the Angola Basin. This great deep region of the sea floor may harbor an unexplored population that is wholly unique.

In itself the pressure of the water at great depths appears to constitute no insurmountable barrier to water-breathing animal life. The depth limitations of many creatures are the associated conditions of low temperature, darkness, sparse food and so on. It should perhaps come as no surprise, therefore, that some of the fishes of high latitudes, which are of course adapted to cold dark waters, extend far into the deep cold waters in much more southern latitudes. Off the coast of Lower California, in water 1,200 to 6,000 feet deep, baited cameras have found an abundance of several species of fishes that are known at the near surface only far to the north. These include giant arctic sharks, sablefish and others. It appears that some of the fishes that have been called arctic species are actually fishes of the dark cold waters of the seas, which only "outcrop" in the Arctic, where cold water is at the surface.

I have discussed several of the benthic and epibenthic environments without pointing out some of the unique features the presence of a solid interface entails. The bottom is much more variable than the mid-water zone is. There are as a result more environmental niches for an organism to occupy, and hence we see organisms that are of a wider range of form and habit. Adaptations develop for hiding and ambuscade, for mimicry and controlled patterns. Nests and burrows can be built, lairs occupied and defended and booby traps set.

Aside from the wide range of form and function the benthic environment elicits from its inhabitants, there are more fundamental conditions that influence the nature and form of life there. For example, the dispersed food material settling from the upper layers becomes much concentrated against the sea floor. Indeed, it may become further concentrated by lateral currents moving it into depressions or the troughs of ripples.

In the mid-water environment most creatures must move by their own energies to seek food, using their own food stores for this motion. On the bottom, however, substantial water currents are present at all depths, and creatures can await the passage of their food. Although this saving only amounts to an added effectiveness for predators, it is of critical importance to those organisms that filter water for the fine food material it contains, and it is against the bottom interface that a major bypass to the microscopic steps of the dominant food web is achieved. Here large organisms can grow by consuming microscopic or even submicroscopic food particles. Clams, scallops, mussels, tube worms, barnacles and a host of other creatures that inhabit this zone have developed a wide range of extremely effective filtering mechanisms. In one step, aided by their attachment, the constant currents and the concentration of detritus against the interface, they perform the feat, most unusual in the sea, of growing large organisms directly from microscopic food.

Although the benthic environment enables the creatures of the sea to develop a major branch of the food web that is emancipated from successive microscopic steps, this makes little difference to the food economy of the sea. The sea is quite content with a large population of tiny organisms. From man's standpoint, however, the shallow benthic environment is an unusually effective producer of larger creatures for his food, and he widely utilizes these resources.

Man may not have created an ideal environment for himself, but of all the environments of the sea it is difficult to conceive of one better for its inhabitants than the one marine creatures have created almost exclusively for themselves: the coral islands and coral reefs. In these exquisite, immense and well-nigh unbelievable structures form and adaptation reach a zenith.

An adequate description of the coral reef and coral atoll structure, environments and living communities is beyond the scope of this article. The general history and structure of atolls is well known, not only because of an inherent fascination with the magic and beauty of coral islands but also because of the wide admiration and publicity given to the prescient deductions on the origin of atolls by Charles Darwin, who foresaw much of what modern exploration has affirmed.

From their slowly sinking foundations of ancient volcanic mountains, the creatures of the coral shoals have erected the greatest organic structures that exist. Even the smallest atoll far surpasses any of man's greatest building feats, and a large atoll structure in actual mass approaches the total of all man's building that now exists.

These are living monuments to the success of an extremely intricate but balanced society of fish, invertebrates and plants, capitalizing on the basic advantages of benthic populations already discussed. Here, however, each of the reef structures acts almost like a single great isolated and complex benthic organism that has extended itself from the deep poor waters to the sunlit richer surface. The trapping of the advected food from the surface currents enriches the entire community. Attached plants further add to the economy, and there is considerable direct consumption of plant life by large invertebrates and fish. Some of the creatures and relationships that have developed in this environment are among the most highly adapted found on the earth. For example, a number of the important reef-building animals, the corals, the great tridacna clams and others not only feed but also harbor within their tissues dense populations of single-celled green plants. These plants photosynthesize food that is then directly available within the bodies of the animals; the plants in turn depend on the animal waste products within the body fluids, with which they are bathed, to derive their basic nutrients. Thus within the small environment of these plant-animal composites both the entire laborious nutrient cycle and the microscopic food web of the sea appear to be substantially bypassed.

There is much unknown and much to be discovered in the structure and ecology of coral atolls. Besides the task of unraveling the complex relationships of its inhabitants there are many questions such as: Why have many potential atolls never initiated effective growth and remained submerged almost a mile below the surface? Why have others lost the race with submergence in recent times and now become shallowly submerged, dying banks? Can the nature of the circulation of the ancient ocean be deduced from the distribution of successful and unsuccessful atolls? Is there circulation within the coral limestone structure that adds to the nutrient supply, and is this related to the curious development of coral knolls, or coral heads, within the lagoons? Finally, what is the potential of cultivation within these vast, shallow-water bodies of the deep open sea?

There is, of course, much to learn about all marine life: the basic processes of the food web, productivity, populations, distributions and the mechanisms of reinoculation, and the effects of intervention into these processes, such as pollution, artificial upwelling, transplantation, cultivation and fisheries. To learn of these processes and effects we must understand the nature not only of strong simple actions but also of weak complex interactions, since the forms of life or the success of a species may be determined by extremely small second- and third-order effects. In natural affairs, unlike human codes, *de minimis curat lex*—the law *is* concerned with trivia!

Little is understood of the manner in which speciation (that is, the evolution of new species) occurs in the broadly intercommunicating pelagic environment with so few obvious barriers. Important yet unexpected environmental niches may exist in which temporary isolation may enable a new pelagic species to evolve. For example, the top few millimeters of the open sea have recently been shown to constitute a demanding environment with unique inhabitants. Further knowledge of such microcosms may well yield insight into speciation.

As it has in the past, further exploration of the abyssal realm will undoubtedly reveal undescribed creatures including members of groups thought long extinct, as well as commercially valuable populations. As we learn more of the conditions that control the distribution of species of pelagic organisms, we shall become increasingly competent to read the pages of the earth's marine-biological, oceanographic and meteorological history that are recorded in the sediments by organic remains. We shall know more of primordial history, the early production of a breathable atmosphere and petroleum production. Some of these deposits of sediment cover even the period of man's recorded history with a fine time resolution. From such great records we should eventually be able to increase greatly our understanding of the range and interrelations of weather, ocean conditions and biology for sophisticated and enlightened guidance of a broad spectrum of man's activities extending from meteorology and hydrology to oceanography and fisheries.

Learning and guidance of a more specific nature can also be of great practical importance. The diving physiology of marine mammals throws much light on the same physiological processes in land animals in oxygen stress (during birth, for example). The higher flowering plants that inhabit the marine salt marshes are able to tolerate salt at high concentration, desalinating seawater with the sun's energy. Perhaps the tiny molecule of DNA that commands this process is the most precious of marine-life resources for man's uses. Bred into existing crop plants, it may bring salt-water agriculture to reality and nullify the creeping scourge of salinization of agricultural soils.

Routine upstream reinoculation of preferred species of phytoplankton and zooplankton might stabilize some pelagic marine populations at high effectiveness. Transplanted marine plants and animals may also animate the dead saline lakes of continental interiors, as they have the Salton Sea of California.

The possible benefits of broad marine-biological understanding are endless. Man's aesthetic, adventurous, recreational and practical proclivities can be richly served. Most important, undoubtedly, is the intellectual promise: to learn how to approach and understand a complex system of strongly interacting biological, physical and chemical entities that is vastly more than the sum of its parts, and thus how better to understand complex man and his interactions with his complex planet, and to explore with intelligence and open eyes a huge portion of this earth, which continuously teaches that when understanding and insight are sought for their own sake, the rewards are more substantial and enduring than when they are sought for more limited goals.

23

HOW FISHES SWIM

SIR JAMES GRAY
August 1957

The submarine and the airplane obviously owe their existence in part to the inspiration of Nature's smaller but not less attractive prototypes—the fish and the bird. It cannot be said that study of the living models has contributed much to the actual design of the machines; indeed, the boot is on the other foot, for it is rather the machines that have helped us to understand how birds fly and fish swim [see "Bird Aerodynamics," by John H. Storer; SCIENTIFIC AMERICAN Offprint 1115]. But engineers may nevertheless have something to learn from intensive study of the locomotion of these animals. Some of their performances are spectacular almost beyond belief, and raise remarkably interesting questions for both the biologist and the engineer. In this article we shall consider the swimming achievements of fishes and whales.

Looking at the propulsive mechanism of a fish, or any other animal, we must note at once a basic difference in mechanical principle between animals and inanimate machines. Nearly all machines apply power by means of wheels or shafts rotating about a fixed axis, normally at a constant speed of rotation. This plan is ruled out for animals because all parts of the body must be connected by blood vessels and nerves: there is no part which can rotate freely about a fixed axis. Debarred from the use of the wheel and axle, animals must employ levers, whipping back and forth, to produce motion. The levers are the bones of its skeleton, hinged together by smooth joints, and the source of power is the muscles, which pull and push the levers by contraction.

The chain of levers comprising a vertebrate's propulsive machine appears in its simplest form in aquatic animals. Each vertebra (lever) is so hinged to its neighbors that it can turn in a single plane. In fishes the backbone whips from side to side (like a snake slithering along the ground), in whales the backbone undulates up and down. A swimming fish drives itself forward by sweeping its tail sidewise; as the tail and caudal fin are bent by the resistance of the water, the forward component of the resultant force propels the fish [*see drawing on page 230*]. As the tail sweeps in one direction, the front end of the body must tend to swing in the opposite direction, since it is on the opposite side of the hinge, but this movement is usually small—partly because of the high moment of inertia of the front end of the body and partly on account of the resistance which the body offers to the flow of water at right angles to its surface. Thus the head end of the fish acts as a fulcrum for the tail, operating as a flexible lever.

At the moment when the tail fin sweeps across the axis of propulsion, it is traveling rapidly but at a constant speed. During other phases of its motion the speed changes, accelerating as the tail approaches the axis and decelerating after it passes the axis. The whole cycle can be regarded as comparable to that of a variable propeller blade which periodically reverses its direction of rotation and changes pitch as it does so.

How efficient is this propulsion system? Can the oscillating tail of a fish approach in efficiency the steadily running screw propellers that drive a submarine, in terms of the ratio of speed to applied power? To attempt an answer to this question we must first know how fast a fish can swim. Here the biologist finds himself in an embarrassing position, for our information on the subject is far from precise.

As in the case of the flight of birds, the speed of fish is a good deal slower than most people think. When a stationary trout is startled, it appears to move off at an extremely high speed. But the human eye is a very unreliable instrument for judging the rate of this sudden movement. There are, in fact, very few reliable observations concerning the maximum speed of fish of known size and weight. Almost all the data we have are derived from studies of fish under laboratory conditions. These are only small fish, and in addition there is always some question whether the animals are in as good athletic condition as fish in their native environment.

With the assistance of a camera, a number of such measurements have been made by Richard Bainbridge and others in our zoological laboratory at the University of Cambridge. They indicate that ordinarily the maximum speed of a small fresh-water fish is about 10 times the length of the fish's body per second; these speeds are attained only briefly at moments of great stress, when the fish is frightened by a sudden stimulus. A trout eight inches long had a maximum speed of about four miles per hour. The larger the fish, the greater the speed: Bainbridge found, for instance, that a trout one foot long maintained a speed of 6.5 miles per hour for a considerable period [*see table on page 232*].

It is by no means easy to establish a fair basis of comparison between fish of different species or between different-sized members of the same species. Individual fish—like individual human beings—probably vary in their degree of athletic fitness. Only very extensive observations could distinguish between average and "record-breaking" performances. On general grounds one would expect the speed of a fish to increase

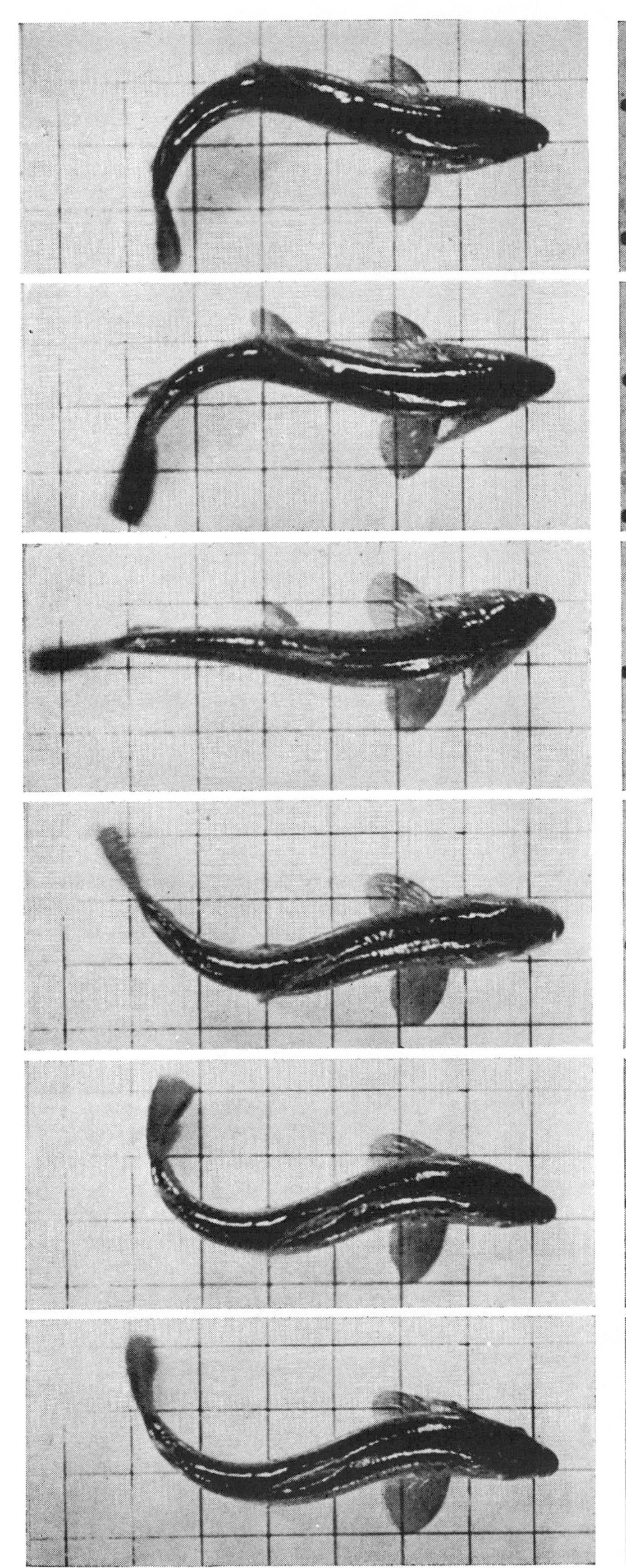

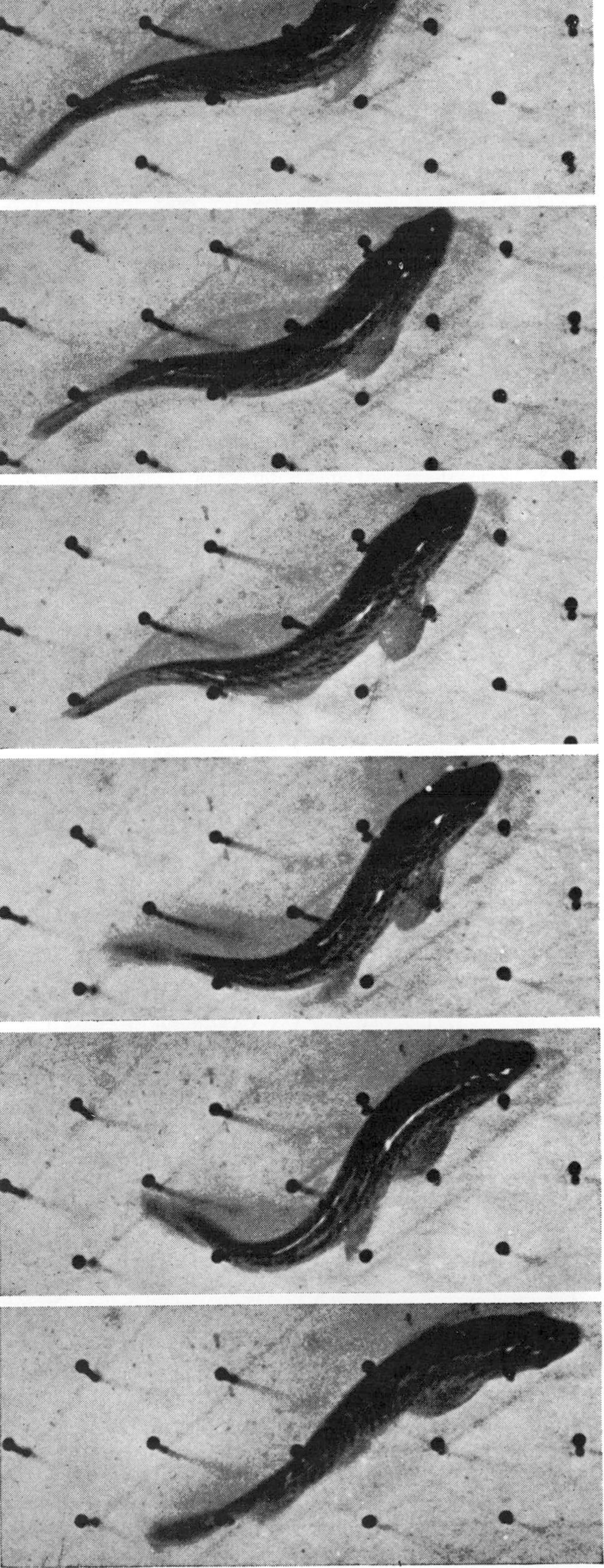

HOW FISH EXERTS FORCE against its medium is illustrated by these two sequences of photographs showing trout out of the water. In the sequence at left the fish has been placed on a board marked with squares; it wriggles but makes no forward progress. In the sequence at right the fish has been placed on a board covered with pegs; its tail pushes against the pegs and moves it across the board.

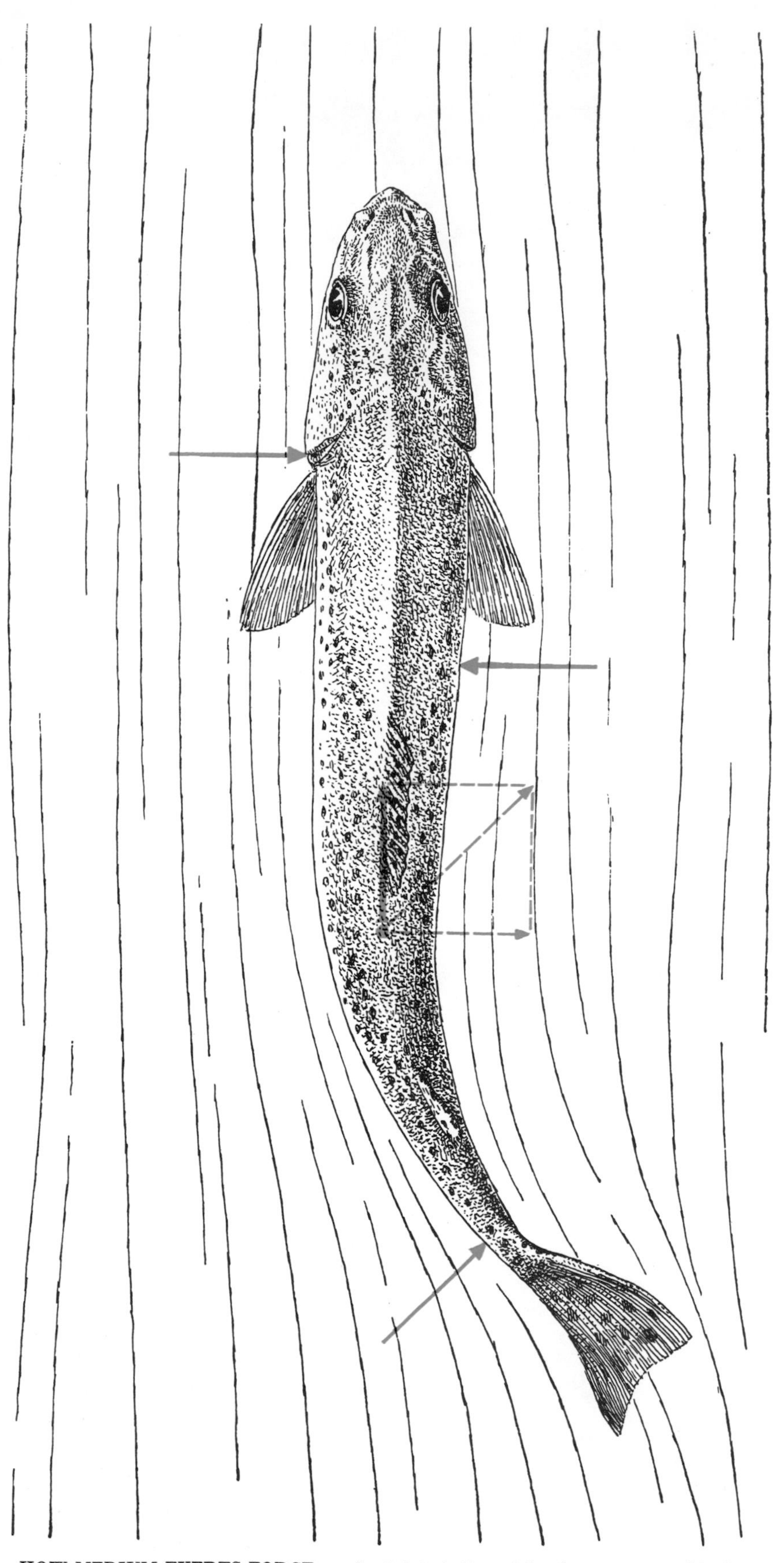

HOW MEDIUM EXERTS FORCE on the fish is indicated by the arrows on this drawing of a trout. As the tail of the fish moves from right to left the water exerts a force upon it (*two diagonal arrows*). The forward component of this force (*heavy vertical arrow*) drives the fish forward. The lateral component (*broken horizontal arrow*) tends to turn the fish to the side. This motion is opposed by the force exerted by the water on sides of the fish.

with size and with the rapidity of the tail beat. Bainbridge's data suggest that there may be a fairly simple relationship between these values: the speed of various sizes of fish belonging to the same species seems to be directly proportional to the length of the body and to the frequency of beat of the tail—so long as the frequency of beat is not too low. In all the species examined the maximum frequency at which the fish can move its tail decreases with increase of length of the body. In the trout the maximum observed frequencies were 24 per second for a 1.5-inch fish and 16 per second for an 11-inch fish—giving maximum speeds of 1.5 and 6.5 miles per hour respectively.

The data collected in Cambridge indicate a very striking feature of fish movement. Evidently the power to execute a sudden spurt is more important to a fish (for escape or for capturing prey) than the maintenance of high speed. Some of these small fish reached their maximum speed within one twentieth of a second from a "standing" start. To accomplish this they must have developed an initial thrust of about four times their own weight.

This brings us to the question of the muscle power a fish must put forth to reach or maintain a given speed. We can calculate the power from the resistance the fish has to overcome as it moves through the water at the speed in question, and the resistance in turn can be estimated by observing how rapidly the fish slows down when it stops its thrust and coasts passively through the water. It was found that for a trout weighing 84 grams the resistance at three miles per hour was approximately 24 grams—roughly one quarter of the weight of the fish. From these figures it was calculated that the fish put out a maximum of about .002 horsepower per pound of body weight in swimming at three miles per hour. This agrees with estimates of the muscle power of fishes which were arrived at in other ways. It seems reasonable to conclude that a small fish can maintain an effective thrust of about one half to one quarter of its body weight for a short time.

As we have noted, in a sudden start the fish may exert a thrust several times greater than this—some four times its body weight. The power required for its "take-off" may be as much as .014 horsepower per pound of total body weight, or .03 per pound of muscle. The fish achieves this extra force by a much more violent maneuver than in ordinary swimming. It turns its head end sharply to one side and with its markedly flexed tail executes a wide and powerful sweep against the water—in short, the fish takes off by arching its back.

This sort of study of fishes' swimming performances may seem at first sight to be of little more than academic interest. But in fact it has considerable practical importance. The problem of the salmon industry is a case in point. The seagoing salmon will lay eggs only if it can get back to its native stream. To reach its spawning bed it must journey upriver in the face of swift currents and sometimes hydroelectric dams. In designing fish-passes to get them past these obstacles it is important to know precisely what the salmon's swimming capacities are.

Contrary to popular belief, there is little evidence that salmon generally surmount falls by leaping over them. Most of the fish almost certainly climb the falls by swimming up a continuous sheet of water. Very likely the objective of their

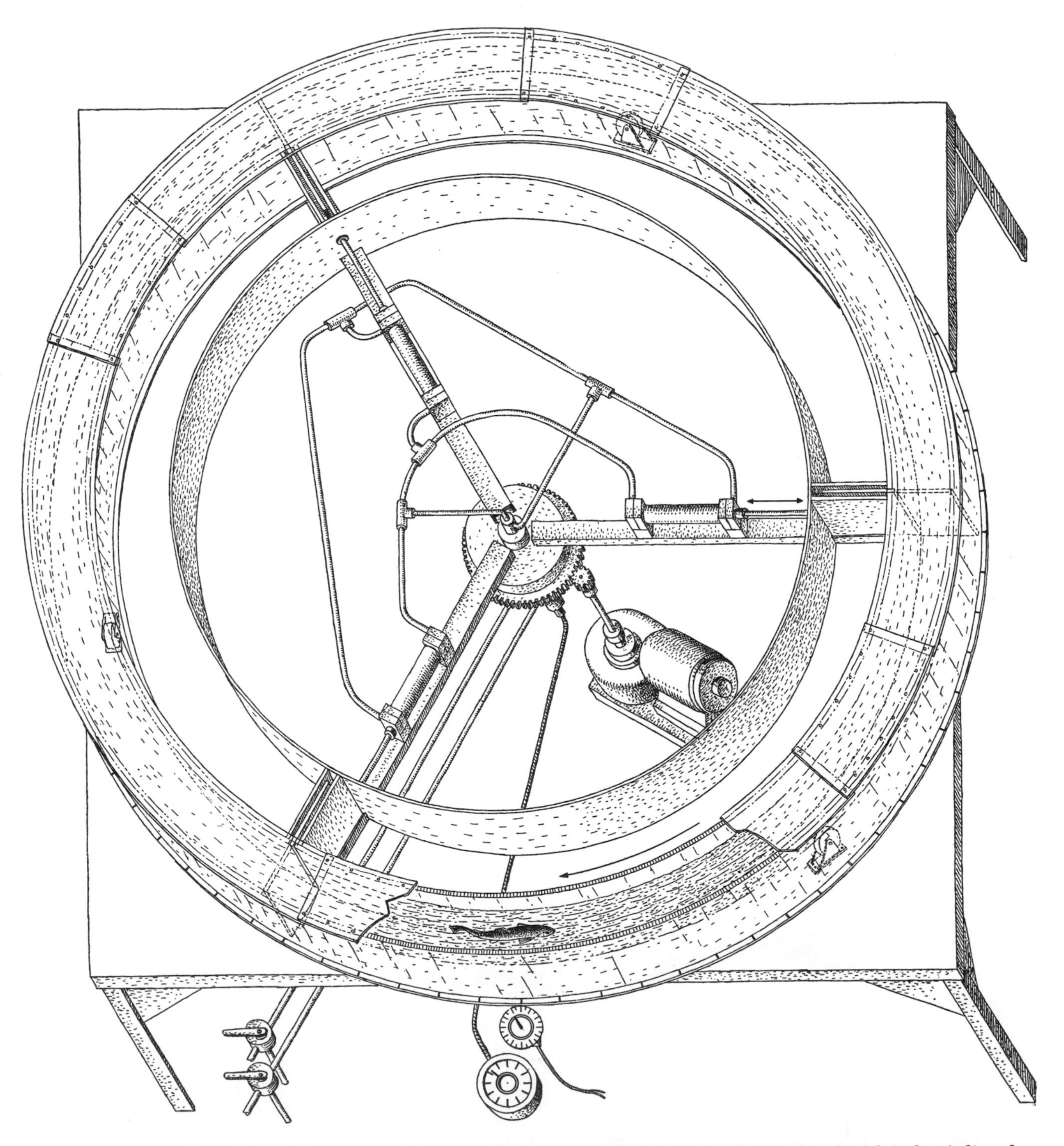

SPEED OF FISHES WAS MEASURED in this apparatus at the University of Cambridge. The fish swims in a circular trough which is rotated by the motor at right center. The speed of the trough is adjusted so that the swimming fish is stationary with respect to the observer. The speed of the fish is then indicated on the speedometer at bottom. Above the speedometer is a clock. When the apparatus is started up, the water is made to move at the same speed as the trough by doors which open to let the fish pass.

SPECIES	LENGTH (FEET)	MAXIMUM OBSERVED SPEED (FEET PER SECOND)	MAXIMUM OBSERVED SPEED (MILES PER HOUR)	RATIO OF MAXIMUM SPEED TO LENGTH
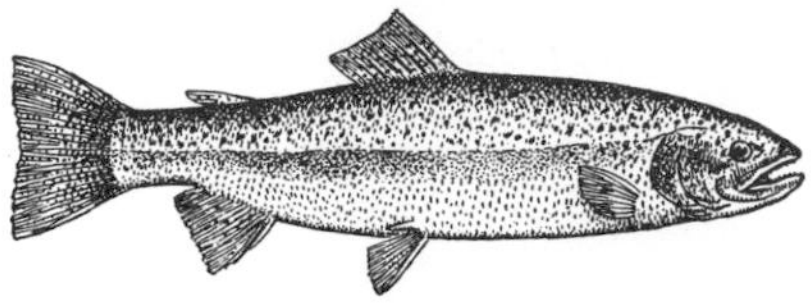 TROUT	.656 .957	5.552 10.427	3.8 6.5	8.5 11
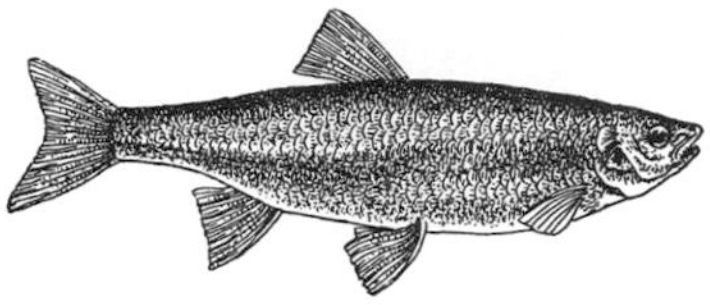 DACE	.301 .594 .656	5.229 5.552 8.812	3.6 3.8 5.5	17.8 9 13.5
 PIKE	.529 .656	6.850 4.896	4.7 3.3	13 7.5
 GOLDFISH	.229 .427	2.301 5.552	1.5 3.8	10.3 13
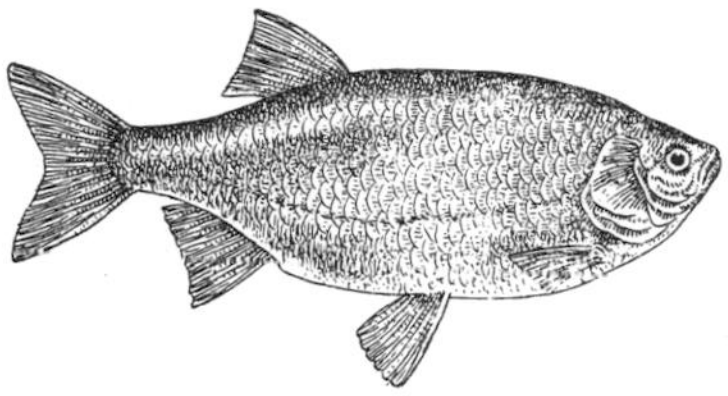 RUDD	.730	4.240	2.9	6
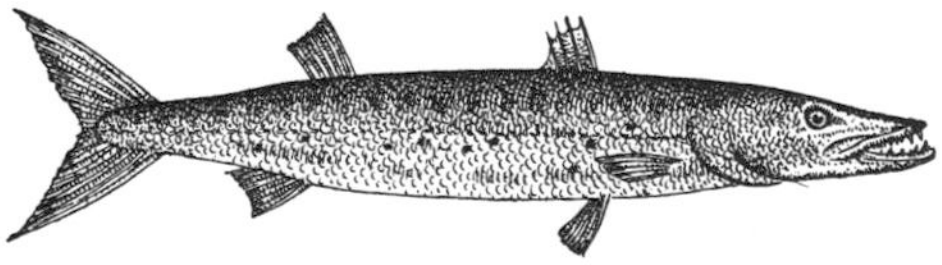 BARRACUDA	3.937	39.125	27.3	10
 DOLPHIN	6.529	32.604	22.4	5
 WHALE	90	33	20	.33

SPEED OF FISHES IS LISTED in this table. The speed of the first five fishes from the top was measured in the laboratory; that of the barracuda, dolphin and whale in nature. The barracuda is the fastest known swimmer. Whale in the table is the blue whale.

leap at the bottom of the fall is to pass through the fast-running water on the surface of the torrent and reach a region of the fall where the velocity of flow can be negotiated without undue difficulty. The brave and prodigious leaps into the air at which spectators marvel may well be badly aimed attempts of the salmon to get into the "solid" water!

A salmon is capable of leaping about six feet up and 12 feet forward in the air; to accomplish this it must leave the water with a velocity of about 14 miles per hour. The swimming speed it can maintain for any appreciable time is probably no more than about eight miles per hour. Accurate measurements of the swimming behavior of salmon in the neighborhood of falls are badly needed—and should be possible to obtain with electronic equipment.

At this point it may be useful to summarize the three main conclusions that have been reached from our study of the small fish. Firstly, a typical fish can exert a very powerful initial thrust when starting from rest, producing an acceleration about four times greater than gravity. Secondly, at times of stress it can exert for a limited period a sustained propulsive thrust equal to about one quarter or one half the weight of its body. Thirdly, the resistance exerted by the water against the surface of the moving fish (*i.e.*, the drag) appears to be of the same order as that exerted upon a flat, rigid plate of similar area and speed. Fourthly, the maximum effective power of a fish's muscles is equivalent to about .002 horsepower per pound of body weight.

Such is the picture drawn from studies of small fishes in tanks. It has its points of interest, and its possible applications to the design of fish-passes, but it poses no particularly intriguing or baffling hydrodynamic problems. Recently, however, the whole matter of the swimming performance of fishes was given a fresh slant by a discovery which led to some very puzzling questions indeed. D. R. Gero, a U. S. aircraft engineer, announced some startling figures for the speed of the barracuda. He found that a four-foot, 20-pound barracuda was capable of a maximum speed of 27 miles per hour! This figure not only established the barracuda's claim to be the world's fastest swimmer but also prompted a new look into the horsepower of aquatic animals.

A more convenient subject for such an examination is the dolphin, whose attributes are somewhat better known than those of the barracuda. (The only essential difference between the propulsive machinery of a fish and that of a dolphin, small relative of the whale, is that the dolphin's tail flaps up and down instead of from side to side.) The dolphin is, of course, a proverbially fast swimmer. More than 20 years ago a dolphin swimming close to the side of a ship was timed at better than 22 miles per hour, and this speed has been confirmed in more recent observations. Now assuming that the drag of the animal's body in the water is comparable to that of a flat plate of comparable area and speed, a six-foot dolphin traveling at 22 miles per hour would require 2.6 horsepower, and its work output would be equivalent to a man—of the same weight as the dolphin—climbing 28,600 feet in one hour! This conclusion is so clearly fantastic that we are forced to look for some error in our assumptions.

Bearing in mind the limitations of animal muscle, it is difficult to endow the dolphin with much more than three tenths of one horsepower of effective output. If this figure is correct, there must be something wrong with the assumption about the drag of the animal's surface in the water: it cannot be more than about one tenth of the assumed value. Yet the resistance could have this low value only if the flow of water were laminar (smooth) over practically the whole of the animal's surface—which an aerodynamic or hydrodynamic engineer must consider altogether unlikely.

The situation is further complicated when we consider the dolphin's larger relatives. The blue whale, largest of all the whales, may weigh some 100 tons. If we suppose that the muscles of a whale are similar to those of a dolphin, a 100-ton whale would develop 448 horsepower. This increase in power over the dolphin is far greater than the increase in surface area (*i.e.*, drag). We should therefore expect the whale to be much faster than the dolphin, yet its top speed appears to be no more than that of the dolphin—about 22 miles per hour. There is another reason to doubt that the whale can put forth anything like 448 horsepower. Physiologists estimate that an exertion beyond about 60 or 70 horsepower would put an intolerable strain on the whale's heart. Now 60 horsepower would not suffice to drive a whale through the water at 20 miles per hour if the flow over its body were turbulent, but it would be sufficient if the flow were laminar.

Thus we reach an impasse. Biologists are extremely unwilling to believe that

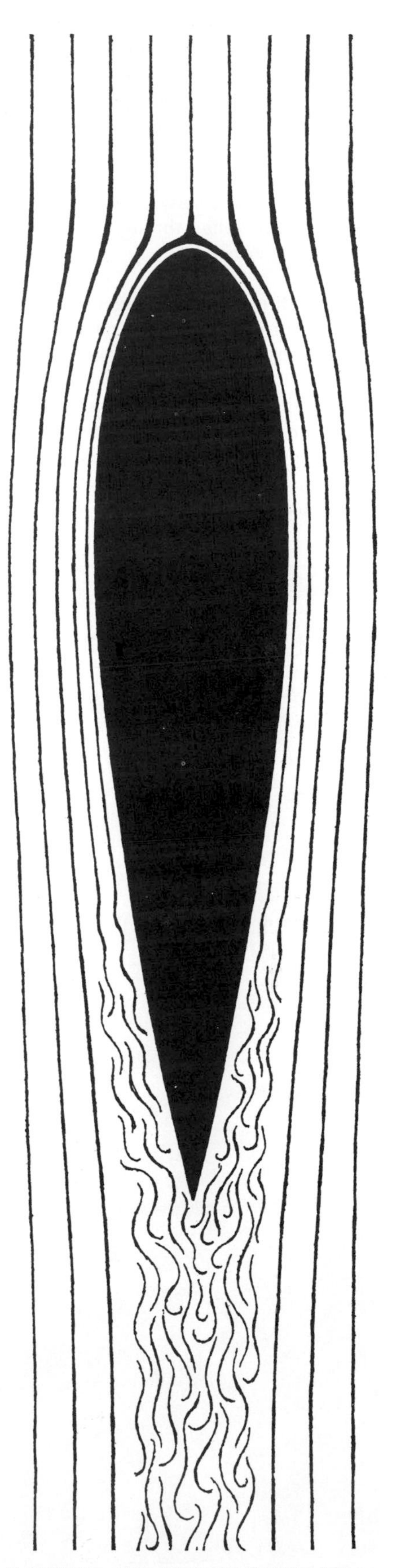

LAMINAR AND TURBULENT FLOW are depicted in this diagram of a streamlined body passing through the water. The smooth lines passing around the body indicate laminar flow; the wavy lines, turbulent flow.

fishes or whales can exert enough power to drive themselves through the water at the recorded speeds against the resistance that would be produced by turbulent flow over their bodies, while engineers are probably equally loath to believe that laminar flow can be maintained over a huge body, even a streamlined body, traveling through the water at 20 miles per hour.

Lacking direct data on these questions, we can only speculate on possible explanations which might resolve the contradiction. One point that seems well worth re-examining is our assumption about the hydrodynamic form of the swimming animal. We assumed that the resistance which the animal (say a dolphin) has to overcome is the same as that of a rigid body of the same size and shape moving forward under a steady propulsive force. But the fact of the matter is that the swimming dolphin is not a rigid body: its tail and flukes are continually moving and bending during each propulsive stroke. It seems reasonable to assume, therefore, that the flow of water over the hind end of the dolphin is not the same as it would be over a rigid structure. In the case of a rigid model towed through the water, much of the resistance is due to slowing down of the water as it flows past the rear end of the model. But the oscillating movement of a swimming animal's tail accelerates water in contact with the tail; this may well reduce or prevent turbulence of flow. There is also another possibility which might be worth investigation. When a rigid body starts from rest, it takes a little time for turbulence to develop. It is conceivable that in the case of a swimming animal the turbulence never materializes, because the flukes reverse their direction of motion before it has an opportunity to do so.

It would be foolish to urge these speculative suggestions as serious contributions to the problem: they can only be justified insofar as they stimulate engineers to examine the hydrodynamic properties of oscillating bodies. Few, if any, biologists have either the knowledge or the facilities for handling such problems. The questions need to be studied by biologists and engineers working together. Such a cooperative effort could not fail to produce facts of great intrinsic, and possibly of great applied, interest.

DOLPHINS (called porpoises by seamen) were photographed by Jan Hahn as they swam beside the bow of the *Atlantis*, research vessel of the Woods Hole Oceanographic Institution, in the Gulf of Mexico. The speed of these dolphins was about 11 miles per hour.

THE SCHOOLING OF FISHES

EVELYN SHAW
June 1962

For sea gulls, fishermen and other predators the propensity of certain species of fish to assemble in large schools is a great convenience. A school of fish is something more, however, than a crowd of fish; it is a social organization to which the fish are bound by rigorously stereotyped behavior and even by anatomical specialization. Schooling fishes do not merely live in close proximity to their kind, as many other fishes do; they maintain, during most of their activities, a remarkably constant geometric orientation to their fellows, heading in the same direction, their bodies parallel and with virtually equal spacing from fish to fish. Swimming together, approaching, turning and fleeing together, all doing the same thing at the same time, they create the illusion of a huge single animal moving in a sinuous path through the water.

This peculiar social organization has no leaders. The fish traveling at the leading edge of a school frequently trade places with those behind. When the school turns abruptly to the right or left, the fish on that flank become the "leaders," and what was the leading edge becomes a flank. Except in the execution of such a turn and during feeding—when the school formation may break up completely—the fish swim parallel to one another. The distances between fish may vary as individuals swim along at different and changing speeds, particularly in a slower moving, loose school. When a school is startled, for example, by a predator or an observer, it closes ranks immediately and the fish-to-fish spacing becomes equal and fixed as the entire school takes flight.

Even in schools of as many as a million fish, all members are of a similar size. Speed increases with size and the fish of a species therefore tend to sort themselves out by size and by generation in the sea. Schools can take many shapes and usually have a third dimension, being a few fish or many fish deep. From above they may appear rectangular or elliptical or amorphous and changeable. Some species form schools of characteristic shape. The Atlantic menhaden, for example, can be easily recognized from the air because their schools move through the water like a giant amoeboid shadow, often changing course but never breaking apart.

The speed and synchronization of response, the parallel orientation and the constancy of spacing among members of a school inevitably suggest that their behavior is integrated by some central control system that makes each "think" of changing course at exactly the same moment. Of course, there is no such central control system. Nor is it possible to explain the simultaneity of the members' actions as response to external stimuli from the environment. From time to time the fish do respond, as other animals do, to such stimuli as food and change of light intensity. Environmental conditions, however, do not explain the high degree of synchronized parallel movement that the members of a school display moment after moment, day after day. In fact, the great stability of schools, persisting through the most varied environmental conditions, suggests that the school organization must be dominated by internal factors.

Schooling is easily enough explained as an instinct. The term implies a causal factor—saying, in effect, that fishes school because they have an instinct to school. This tautology does not explain much, even when it is amplified by the more sophisticated statement that the behavior is inborn, unlearned and characteristic of the species. Many animals exhibit clear-cut, species-specific patterns of behavior, and it is useful to seek these out and compare them as they appear in related species. Such inquiry leaves equally interesting questions unanswered. In the present instance it does not explain what brings about the concerted action of the fish in a school. This requires, above all, study of the behavior as it unfolds in the developing organism. With growth and particularly with the maturation of the sensory system, the relation between the organism and its environment changes. The life history of the individual, however typical of its species, has a profound role in the molding of the behavior of the mature animal and holds the principal clues to the mechanism that governs its interaction with its social and physical environment. So far this approach to the schooling of fishes has only made the mystery more intriguing.

With progress on the question of how fishes school, one can also hope for some light on why fishes school. No other line of study has disclosed what function this highly organized social behavior serves in the perpetuation of the species that have adopted it.

In my own work at the Marine Biological Laboratory at Woods Hole, Mass., at the Woods Hole Oceanographic Institution, at the Bermuda Biological Station and at the Lerner Marine Laboratory on Bimini in the Bahamas, I have attempted to overcome the difficulty of study in the field by bringing fishes into the laboratory for observation and experiment. Life begins for most species of schooling fish in the plankton, where the eggs drift untended and abandoned by the school that laid and fertilized them in its passage. The eggs develop into embryos and the embryos into larvae, or "fry," which are capable of some feeble swimming movement. They grow, they ma-

SCHOOL OF HERRING was photographed by Ron Church near San Diego, Calif. The majority of herring caught in the Pacific Ocean are used to make fish oil and fish meal. This school, originally headed straight for the camera, has begun to turn to its right.

SCHOOL OF MULLET, which are common in the waters off Florida, was photographed there by Jerry Greenberg. A member of the order Mugiliformes, the mullet is an oceanic fish, and its distribution is primarily on both sides of the temperate South Atlantic.

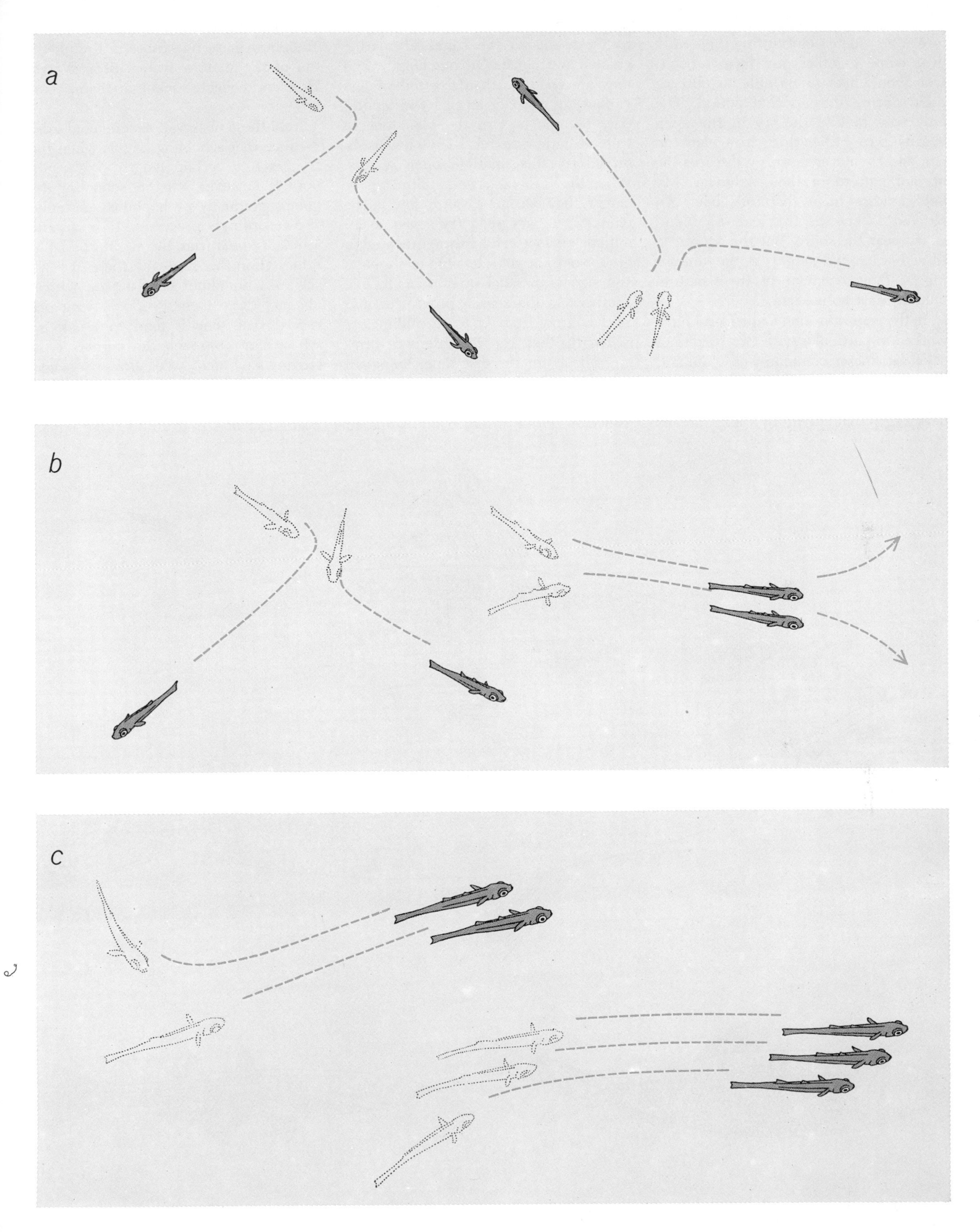

SCHOOLING ACTIVITY OF JUVENILE FISHES, or fry, develops as they grow. When newly hatched fry five to seven millimeters in length (*top*) approach the head, tail or side of other fry to within five millimeters, they dart away. At eight to nine millimeters (*middle*) two fry school momentarily if one has approached the tail of the other, but a side approach or one to the head still makes them dart away. As the fry grow from a length of about nine millimeters to 10.5 (*bottom*), the head-to-tail approach becomes predominant and two fry will school for five to 10 seconds; they later begin to school for short periods in threes and fours.

ture and at some point during their early lives come together and form schools. One would like to be able to observe them during this epochal period. The only way to find the fry in the open oceans is to gather them in a plankton net, and the net necessarily disrupts the normal pattern of their behavior. My field studies have therefore been restricted to species that can be found as fry near the shore. But the fry are so tiny that crucial stages in the unfolding of their behavior in their natural habitat must go unseen.

In the waters around Cape Cod I have worked in particular with two species of *Menidia,* known commonly as whitebait, spearing or silversides. During late spring and early summer they spawn heavy eggs that adhere by sticky threads to rocks and to the stems of marine grasses and algae. On hatching, when they are no more than five millimeters (about a quarter of an inch) in length, they become part of the plankton. In spite of patient search I have never observed fry this small in open waters. When they grow to seven millimeters or longer, they become easier to find in the plankton. I have seen fry seven to 10 millimeters long randomly aggregated in groups but not yet schooling or showing any sign of parallel orientation to one another. As the season progresses and as they grow from 11 to 12 millimeters in length, they can be observed forming schools for the first time, lining up in parallel, with 30 to 50 fry to the school. During the summer of 1960 my associates and I observed an estimated 10,000 of these tiny fishes in the plankton of the shallow waters near Woods Hole and collected many of them.

From these observations one could deduce that schooling begins when the fry reach a certain length. We could not tell, however, whether schooling develops gradually or happens suddenly. We accordingly proceeded to rear some 1,000 *Menidia* from the egg in the laboratory. For the study of these fry we set up a doughnut-shaped tank with a channel three inches wide, having observed that schools tend to break up when they approach the corners of a rectangular tank. We took care also to observe them in constant light and through a one-way mirror. We were reassured to find that under these condi-

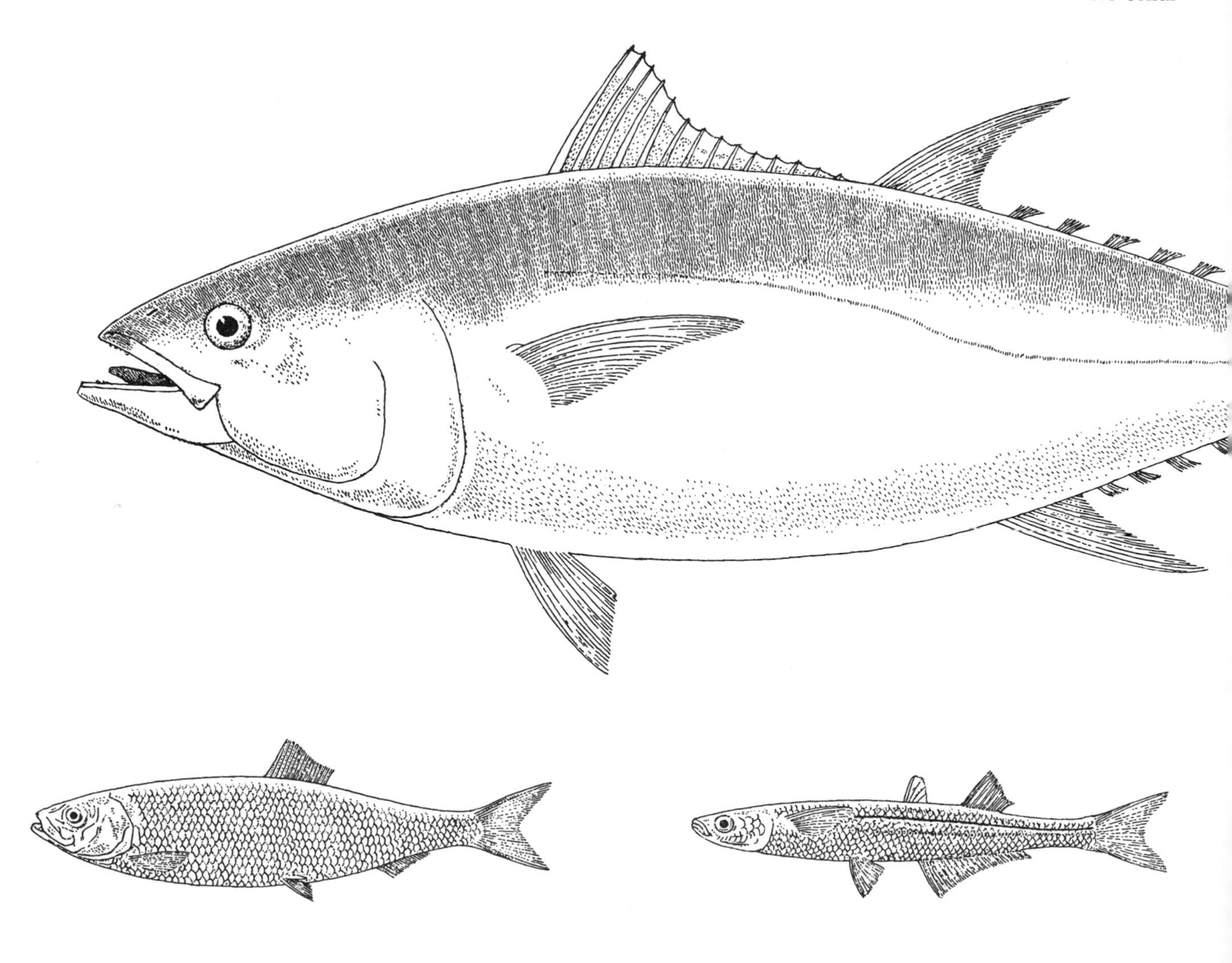

REPRESENTATIVE SCHOOLING FISH shown on these two pages are a tuna (*Thunnus thynnus*), at top left; a herring (*Clupea harengus*), at bottom left; a silverside (*Menidia menidia*), second from bottom left; a mackerel (*Scomber scombrus*), at top right; and a

tions schooling appeared in laboratory-reared fry when they grew to the same size as the smallest schooling fry observed in the sea.

The close-up and constant surveillance in the laboratory showed that schooling unfolds gradually in characteristic patterns of fish-to-fish approach and orientation. Newly hatched fry, five to seven millimeters in length, would approach the head, the tail or the side of other fry to within five millimeters and then dart away. At eight to nine millimeters in length, a fry would approach the tail of another fry and, when the two fry were one to three centimeters apart, they would swim on a parallel course for a second or two. If either fry approached the other head on at an angle, however, each would dart off rapidly in the opposite direction. At about nine millimeters in length the head-to-tail approach became predominant, and the fry would now swim on parallel courses for five or 10 seconds. When they reached a length of 10 to 10.5 millimeters, one fry would approach the tail of another and both fry would briefly vibrate their entire bodies. This curious behavior would terminate with the two fry swimming off in tandem, or in parallel, for 30 to 60 seconds, occasionally joined by three or four other fry in the formation of a recognizable little school. The number that would engage in this behavior increased to 10 or so when the fry reached a length of 11 to 12 millimeters. With the distances from fish to fish ranging from 10 to 35 millimeters, the school was a ragged one. By the time the fry had grown to 14 millimeters the fish-to-fish spacing became less variable, ranging from 10 to 15 millimeters, and there was less shifting about in the school.

Schooling behavior can therefore be described as developing initially from the interaction of two tiny fry. As they grow older and larger, the head-on approach gives way to the head-to-tail approach; the two fry tend to swim forward in parallel instead of fleeing from one another, and they are joined by increasing numbers of individuals in the formation of the incipient school.

At this point some speculation is in order, particularly if it suggests specific hypotheses for exploration by observation and experiment. During the head-on approach, one may suppose, each fry sees a changing visual pattern:

jack (*Caranx hippos*). The fish are not drawn to scale. Tuna have been known to reach a length of 14 feet. The mackerel averages 14 to 18 inches, and the jack about two feet. The silverside grows to six inches; the herring may reach a length of 12 inches.

an oval mass (the head) and bright black spots (the eyes) coming closer and closer. The stimulus becomes too intense and each fry veers off in flight. The tail-on approach, in contrast, presents a quite different, although changing, pattern. This time it is a small silvery stripe and a transparent tail, swishing rhythmically and steadily moving away. The approaching fry follows. The leading fry may see, out of the rear edge of its eye, only a vague image of the follower. In each case the visual stimulus is moderate to weak in intensity, and the two fry swim forward together.

T. C. Schneirla of the American Museum of Natural History has postulated that, in general, mild stimuli attract and strong stimuli repel, and that most animals tend to approach the source of a mild stimulus and withdraw from the source of a strong one, even if they have had no prior experience with these conditions. Our fry had had considerable time to accumulate experiences of mutual encounter. We could not be certain, however, about the nature and impact of such experiences. A natural question therefore arose: Is such experience essential to the nature of schooling behavior? Or, to let the question suggest an experiment: Will fishes show schooling if they are taken away from their species-mates and raised in isolation? One must be cautious, however, in interpreting the results of such an experiment. On finding that a given behavioral trait appears in an animal that has been reared in isolation, some students of animal behavior are ready to conclude that the trait must be innate or instinctive and to close the book on further investigation at that point. Perhaps the pitfall lies in the word "isolation." No animal can grow up in a total vacuum of experience. In the case of the fry we proceeded to rear away from their species-mates, it was clear that each one had experience of itself (although we coated the bowls with paraffin so that the fry could not see their own reflection), of the water in its bowl, of the *Artemeia* shrimp on which it dined and of such stimuli as reached it from the world outside its bowl.

The mortality among the fry we isolated in this fashion proved to be extremely high. Only four out of 400 survived to schooling size in the first season and only nine out of 87 in the second. Apparently the fry need one another in the earliest larval stage, but we do not yet know why. The one noticeable difference between those reared in the community of their fellows and those reared alone seems to show up in the initiation of their feeding behavior. The fry in our laboratory communities began to feed two or three days after hatching, while they still carried a large yolky sac on their abdomen, whereas their siblings in isolation evidently starved to death. When we placed fry in isolation a week after hatching and after they had begun feeding, we secured a somewhat higher survival rate and, it turned out, a different and still enigmatic result when it came to observing the emergence of their schooling behavior.

As soon as the first four fry reared in isolation reached schooling size, we placed them in the company of schooling fish in community tanks. At first they showed disorientation; they bumped into their species-mates and occasionally swam away from the school. At the end of four hours, however, these fry could not be distinguished in behavior from the others. What this experiment showed is that fishes reared in isolation will soon join a school. It did not answer the question of whether or not schooling behavior would appear in fishes so reared.

With a more adequate supply of fry reared in isolation and in semi-isolation during the summer of 1960, we found that they would indeed form schools. The fry that had never had any contact with species-mates schooled within 10 minutes after being placed together in the test chamber. Those that had spent the first week after hatching in the company of species-mates also formed schools, but it took some of them at least 150 minutes to do so. What is more, we found that the shorter the time they had spent in isolation, the longer it took them to form a school. This suggests that their early experience with species-mates—at the period when the fry are still approaching one another at odd angles and darting away—may have set up some inhibitory process.

Although these experiments indicate that isolation in infancy does not keep these fishes from forming schools, the role of experience deserves further study. In this connection it should be added that schooling behavior was established

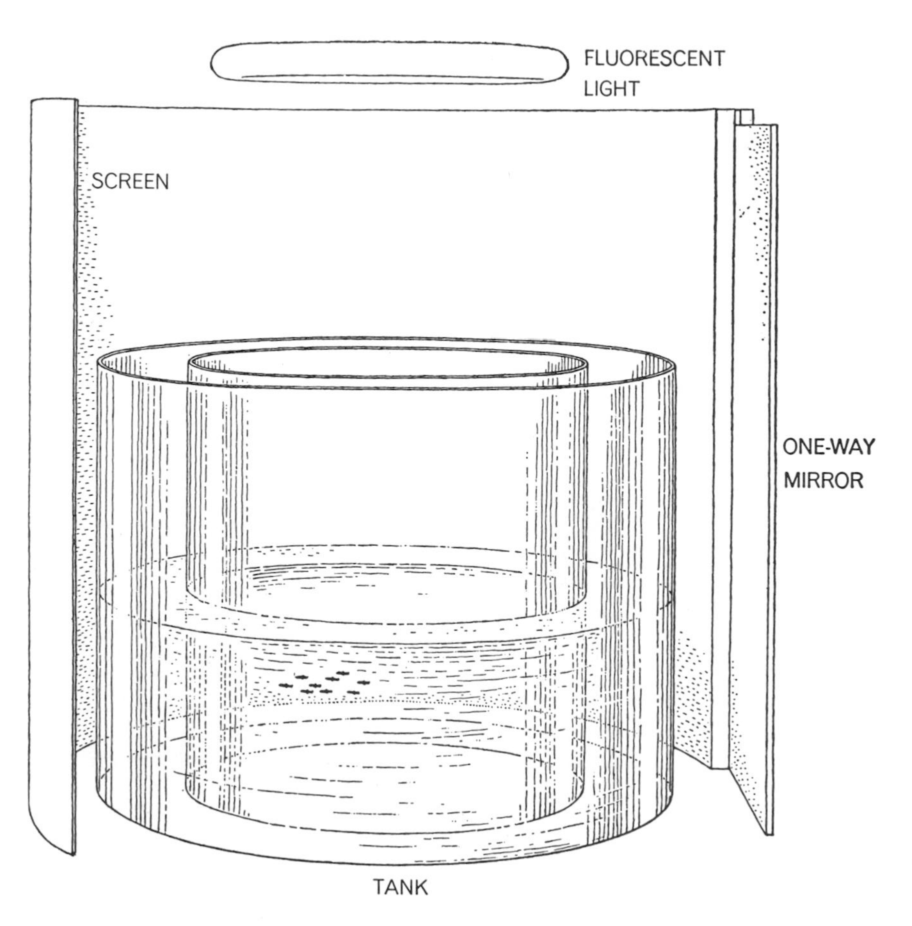

FISH TANK used by the author to study the development of schooling behavior in *Menidia* fry is doughnut-shaped and has a three-inch-wide channel in which the fry can swim continuously without reversing direction. The tank is completely encircled by a screen (*here cut away*). The fry are observed either from above or through the one-way mirror.

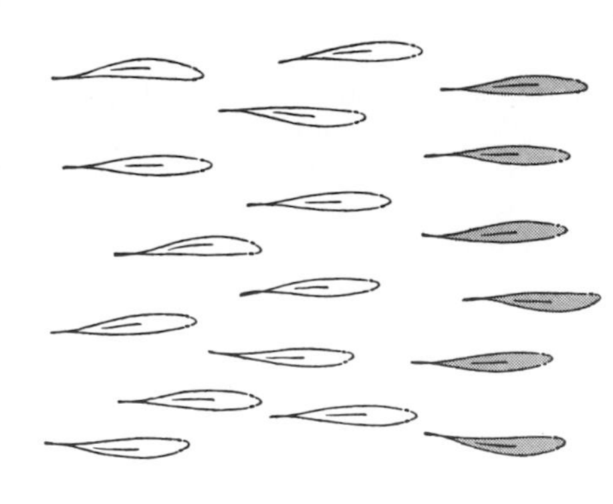

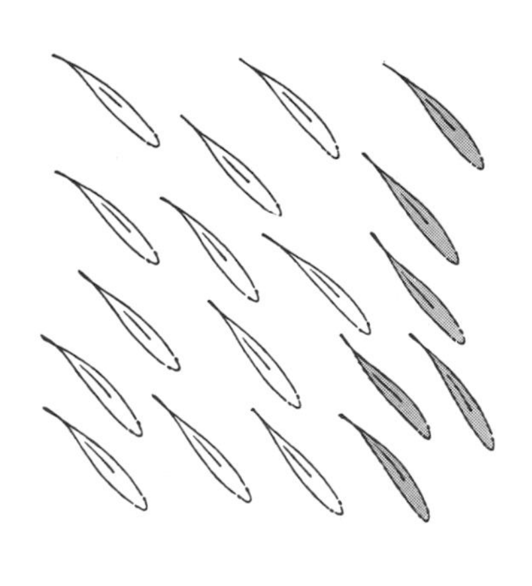

TURNING OF SCHOOL makes the relative position of the leading fish with respect to the school change. These fish (*gray*), which are originally at the leading edge (*left*), gradually shift around to the flank as the school turns (*middle and right*).

in our control communities when the fry were still a good deal smaller than the size at which we exposed our few precious isolates and semi-isolates to one another's company.

Another set of experiments with our laboratory fry produced evidence that the visual attraction of one fry for another develops in parallel with the emergence of schooling behavior. Very young fry showed no response at all to another fry swimming on the other side of a glass barrier. As the fry approached schooling age and size, however, they responded more and more actively to the visual image of the other fry. Finally they began to orient themselves in parallel to the fry on the other side of the barrier and were even observed to vibrate their bodies as they did so.

In a similar experiment with adult schooling fishes the visual attraction of one for another becomes readily apparent. Placed on each side of a glass partition, they swim toward each other immediately. In fact, fishes that cannot see cannot school. A fish blinded in one eye approaches and lines up with another fish on the side of the intact eye; a pair of fish blinded in different eyes swim at random when their sightless eyes are turned toward each other and school normally when they approach on the side with sight.

Just what visual cues are decisive in the mutual attraction of schooling fishes remains to be determined. Various experiments have shown that movement is important and that movement outweighs color and species, especially in attracting the initial approach. Albert E. Parr, now at the American Museum of Natural History, proposed in 1927 that fish-to-fish distances in schools might be explained by a balance of visual attraction and repulsion. According to Parr, the fish are repelled when they come too close together and attracted when they swim too far apart; the typical spacing in the school would thus represent the equilibrium of these two forces.

In a study of the schooling species around Cape Cod, Edward R. Baylor of the Woods Hole Oceanographic Institution and I found that many of these fishes are farsighted and that their retinas are therefore presented with a somewhat fuzzy image. The distribution of rod and cone cells in their retinas indicates, on the other hand, that their eyes may be well adapted for enhanced perception of contrast and so of motion against the hazy underwater background. This kind of vision would be highly adaptive in schooling behavior. Baylor and I also tried to modify the fish-to-fish schooling distance in pairs of fish by placing contact lenses over their eyes, but we observed no conclusive effects.

Although it appears that the visual apparatus is dominant in schooling behavior, there is also evidence that it does not serve as the exclusive channel of mutual attraction among fish. M. H. A. Keenleyside of the Fisheries Research Board of Canada has observed, for example, that *Pristella,* a species that sometimes schools, would respond to fish on the other side of a glass barrier by swimming back and forth along the barrier but would gradually lose interest, wandering away from the barrier more and more frequently and finally not returning at all. Sensory cues other than visual ones are most likely involved in establishing the parallel orientation and the fish-to-fish distances that give the school its ordered structure. It is difficult to determine which cues, because the experimenter cannot control for vision—a fish deprived of sight cannot make the initial approach so essential to the rest of the process.

Hearing, taste and smell have all been implicated, although inconclusively. James M. Moulton of Bowdoin College found that different schooling species produce different sounds, mostly of hydrodynamic origin, as the fish stream and veer through the water. Such sounds, in Moulton's opinion, may help to maintain the total school. There is no evidence, however, that sound helps to keep an individual fish oriented in position in the school. There is even less to be said for taste and smell, particularly in the case of oceanic fishes. Such odors as the fishes might produce would be diluted in the sea; although they might act on individuals at the trailing edge of a school, they could play little role in the behavior of those in the vanguard.

The one sensory system that would seem to be designed to play a role in the orientation of fish to fish is that associated with the lateral line—the nerve and its associated branches that are distributed over the head and run from head to tail along each side. It is thought that this organ is responsive to vibrations and water movements. Willem A. van Bergeijk and G. G. Harris at the Bell Telephone Laboratories have reported evidence indicating that the lateral line is sensitive particularly to "near field" motion of the water produced by propagated sound waves. Parallel orientation may well be facilitated by information about the movements of nearby fish picked up by the lateral line. The approach of one fish to another induced by visual attraction might also be checked by the increasing force of lateral-line perceptions of the movements of the same companion at closer range.

That schooling is a successful way of life can be judged from the fact that so many fishes have adopted it. Some 2,000 marine species school, and there is a major group—the Cypriniformes, consisting mainly of fresh-water fishes—that contains 2,000 more schooling species; among them are the common fresh-water minnows, or shiners, and the familiar characins of the tabletop aquarium. It is evident that these fishes must have converged on the schooling way of life by diverse evolutionary pathways. Of the marine fishes the best-known schooling orders are three that rank among the most numerous in the sea and constitute a vast portion of the world's fish supply. They are the Clupeiformes, the well-known herrings; the Mugiliformes, which include, in addition to the schooling mullets and our laboratory silversides, the solitary barracuda; the Perciformes, comprising the schooling jacks, pompanos, bluefishes, mackerels and tuna and the occasionally schooling snappers and grunts as well as numerous families of nonschoolers. Anatomically the Clupeiformes and the Mugiliformes are rather primitive fishes, whereas the Perciformes are advanced. Although unrelated, these fishes do have significant features in common. Like many other schooling fishes, they are generally sleek and silvery. Significantly, they also have the same small and flattened pectoral fins actuated by musculature that does not permit much mobility. As C. M. Breder, Jr., of the American Museum of Natural History was the first to observe, these fishes cannot swim backward. When they make a pass at a bit of food and happen to miss it, they must come around on a wide turn for another attempt. This limitation on their maneuverability must nonetheless be an advantage in the maintenance of a school, because it tends to keep them all moving forward.

Since the schooling families include anatomically primitive as well as advanced forms, the evidence from living species does not show whether schooling is a primitive or an advanced adaptation. The fossil record is equally inconclusive on this score. Herrings are found in great number in Eocene deposits and one may reasonably speculate that they were schooling then. But fishes were evolving long before the Eocene, and it is impossible to determine one way or the other whether the fishes of those times schooled.

In spite of all the indications that schooling is an effective adaptation, no student of the subject has been able to show why it is so effective. Many advantages can be cited in favor of the behavior, but none seems critical to survival. It is said, for example, that the school creates for its predators, as it does for human observers, the illusion that it is a huge and formidable animal of some kind and so frightens off the predator. No real evidence supports this idea, and one can more plausibly see the school as providing easy prey. If the predator misses one fish, there is always another. In an experiment with goldfish, on the other hand, Carl Welty of Beloit College found that the fish consumed fewer *Daphnia* when they

SCHOOL OF ATLANTIC MENHADEN in Long Island Sound was photographed from the air by Jan Hahn of the Woods Hole Oceanographic Institution. The menhaden, which is a species of herring, forms schools containing as many as a million members.

were fed too many than they did if they were allowed a smaller number. Welty suggested that large numbers of prey might "confuse" the predator. This idea finds support in a mathematical analysis by Vernon E. Brock and Robert H. Riffenburgh of the University of Hawaii, which shows that a school cannot be decimated by attackers once it exceeds a certain number. But one must then ask: Why do some predators school?

Another rationalization for schooling holds that it facilitates the finding of food. When it comes to the search itself, however, only the fish on the school's periphery will be in a position to locate the food; the talents of those in the center of the school are wasted. Of course, once the food is sighted, all may partake. The young of many fishes travel in schools, and their social feeding seemingly promotes more rapid growth. As our efforts to raise fry in isolation would indicate, the sight (or taste and smell) of other fish feeding induces fish to feed. Again, one must doubt that this advantage could account for the evolution of schooling behavior in so many different species.

Another advantage, often cited, has to do with the reproduction of the schooling species. When it is time to reproduce, there is no courtship behavior, no mate selection; as Parr observed some years ago, the males and females of schooling species are usually indistinguishable on casual inspection. The fishes simply shed their eggs and sperm in almost countless numbers into the plankton and leave the spawning site. This certainly enhances the probability of successful fertilization. In some of my collecting, however, I have found schools that were either all male or all female!

To the list of potential adaptive advantages I would like to add another one. Hydrodynamic considerations argue that schooling provides a more efficient way to move through the water. The exertion of each fish may be lessened because it can utilize the turbulence produced by the surrounding fish. Although the fish at the leading edge of the school may have to expend no less energy than solitary fish, the followers may receive enough assistance to help reduce their expenditure of energy. The attainment of maximum efficiency may dictate an optimum fish-to-fish distance in the school.

Study of the schooling of fishes has asked more questions than it has answered. But the questions have now begun to suggest fruitful programs of observation and experiment.

25

LEARNING IN THE OCTOPUS

BRIAN B. BOYCOTT
March 1965

In recent years a number of British students of animal behavior, of whom I am one, have done much of their experimental work at the Stazione Zoologica in Naples. The reason why these investigations have been pursued in Naples rather than in Britain is that our chosen experimental animal—*Octopus vulgaris*, or the common European octopus—is found in considerable numbers along the shores of the Mediterranean. *Octopus vulgaris* is a cooperative experimental subject. If it is provided with a shelter of bricks at one end of a tank of running seawater, it takes up residence in the shelter. When a crab or some other food object is placed at the other end of the tank, the octopus swims or walks the length of the tank, catches the prey with its arms and carries it home to be poisoned and eaten. Since it responds so consistently to the presence of prey, the animal is readily trained. It is also tolerant of surgery and survives the removal of the greater part of its brain. This makes the octopus an ideal animal with which to test directly the relation between the various parts of the brain and the various kinds of perception and learning.

There are many unanswered questions about such relations. We now know a great deal about conduction in nerve fibers, transmission from nerve fiber to nerve fiber at the synapses and the integrative action of nerve fibers in such aggregations of nerve cells as the spinal cord; we are almost wholly ignorant, however, of the levels of neural integration involved in such long-term activities as memory. We can still quote with sympathy the remark of the late Karl S. Lashley of Harvard University: "I sometimes feel, in reviewing the evidence on the localization of the memory trace, that the necessary conclusion is that learning is just not possible!"

It was J. Z. Young, then at the University of Oxford, who first began to exploit the possibility of using for memory studies various marine mollusks of the class Cephalopoda. Shortly before World War II he undertook to work with the cuttlefish *Sepia officinalis*. In a simple experiment he and F. K. Sanders removed from a cuttlefish that part of the brain known as the vertical lobe. They found that a cuttlefish so deprived would respond normally—that is, attack—when it was shown a prawn. If the prawn was pulled out of sight around a corner after the attack began, however, the cuttlefish could not pursue it. The animal might advance to where the prawn had first been presented, but it was apparently unable to make whatever associations were necessary to follow the prawn around the corner. One might say it could not remember to hunt when the prey was no longer in sight. Young and Sanders found that surgical lesions in certain other parts of the cuttlefish's brain did not affect this hunting behavior.

In 1947 I had the privilege of joining Young in his studies. Financed by the Nuffield Foundation, we began

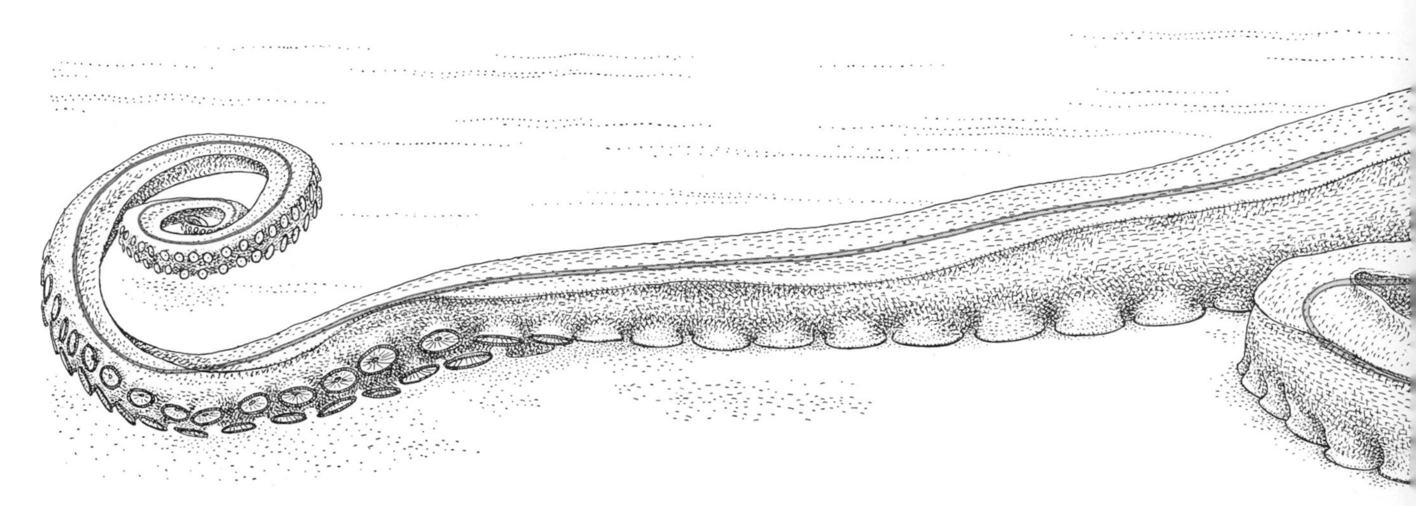

COMMON EUROPEAN OCTOPUS (*Octopus vulgaris*) is the experimental animal the author and his fellow-workers in Naples use for their investigations of perception and learning. The animal's brain (*in color between the eyes*) is about two cubic

work at the Stazione Zoologica, where both seawater aquariums and *Octopus vulgaris* were in abundant supply. The octopus was chosen in preference to the other common laboratory cephalopods—cuttlefishes and squids—because they do not survive so well in tanks and are less tolerant of surgery. At Naples today, in addition to Young's associates from University College London, there are investigators from the University of Oxford led by Stuart Sutherland and from the University of Cambridge led by Martin J. Wells, all going their various ways toward using the brains of octopuses for the analysis of perception and memory. At present most of the work is financed by the Office of Aerospace Research of the U.S. Air Force.

In our early experiments we attempted to train octopuses to do a variety of things, such as taking crabs out of one kind of pot but not out of another, to run a maze and so on. Our most successful experiment was to put a crab in the tank together with some kind of geometric figure—say a Plexiglas square five centimeters on a side—and give the octopus an electric shock when it made the normal attacking response. With this simple method we found that octopuses could learn not to attack a crab shown with a square but to go on attacking a crab shown without one [*see bottom illustrations on pages 248 and 249*]. Or we could train the animals to stop taking crabs but to go on eating sardines or vice versa. The purpose of these experiments was to elucidate the anatomy and connections of the animal's brain and relate them to its learning behavior.

Like the brains of most other invertebrates, the brain of the octopus surrounds its esophagus [*see illustrations on next page*]. The lobes of the brain under the esophagus contain nerve fibers that stimulate peripheral nerve centers, for example the ganglia in the arms and the mantle. These peripheral ganglia contain the nerve cells whose fibers in turn stimulate the muscles and other effectors of the body; through them local reflexes can occur. When all of the brain except the lobes under the esophagus is removed, the octopus remains alive but lies at the bottom of the tank; it breathes regularly but maintains no definite posture. If it is sufficiently stimulated, it responds with stereotyped behavior.

A greater variety of behavior can be obtained if some of the brain lobes above the esophagus are left intact. For instance, the upper brain's median basal lobe and anterior basal lobe send their fibers down to the lower lobes and through them evoke the patterns of nerve activity involved in walking and swimming. Above these two lobes are the vertical lobe, the superior frontal lobe and the inferior frontal lobe; their surgical removal does not result in any defects of behavior that are immediately obvious.

It is with these three lobes and the two optic lobes—which lie on each side of the central mass of the brain—that this article is mostly concerned. Using the electric-shock method of training

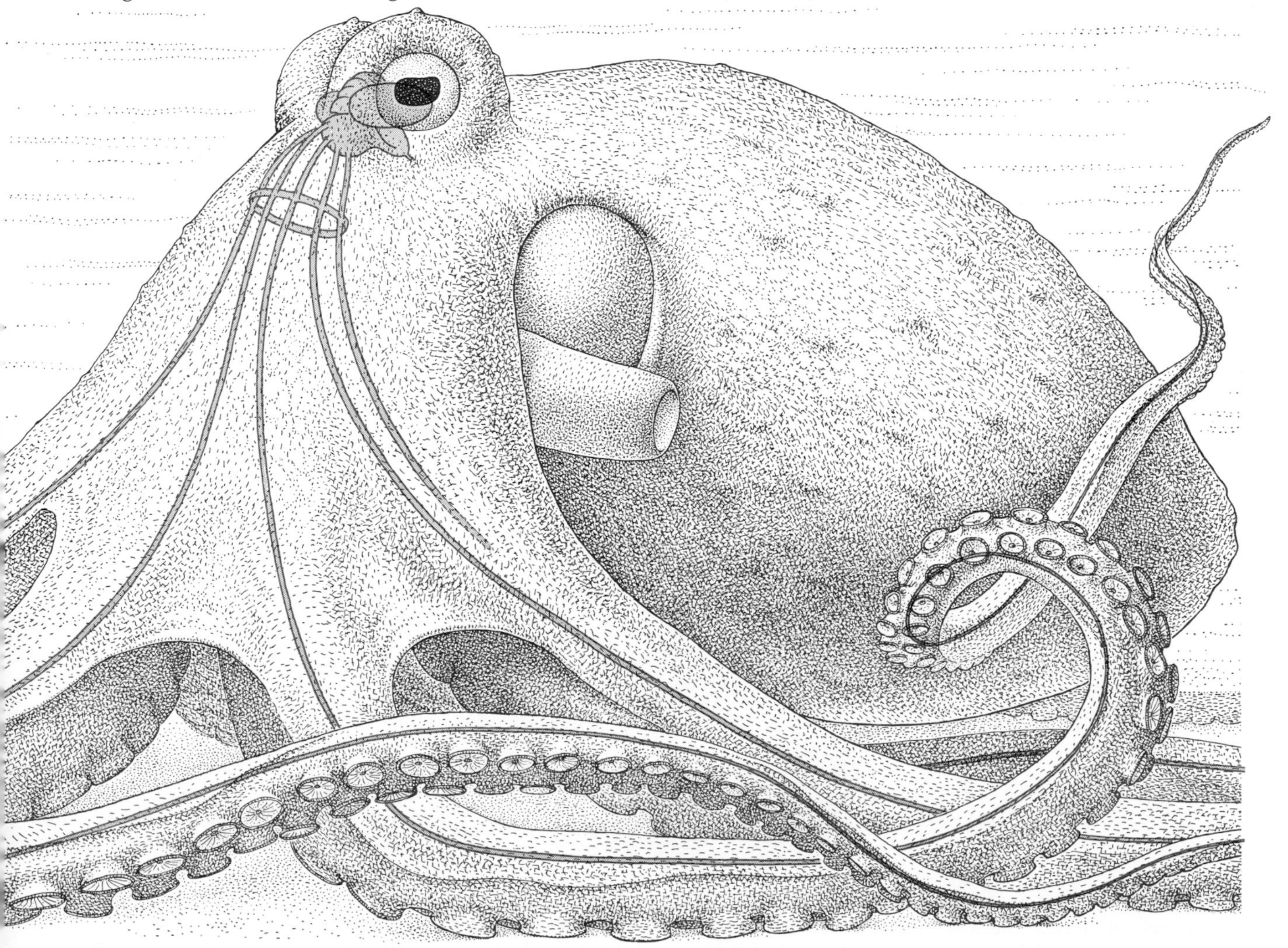

centimeters in size; the basket-like structure below it is composed of the eight major nerves of the arms, some of which are also outlined in color. The octopus adapts readily to life in a tank of seawater and can be trained easily through reward and punishment.

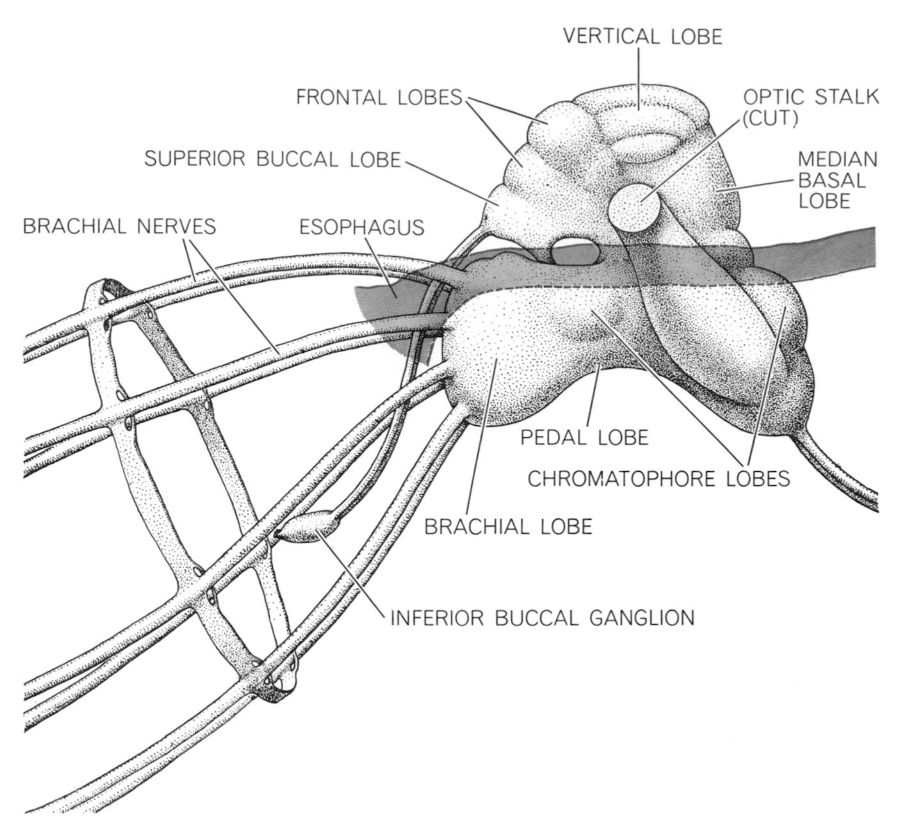

OCTOPUS BRAIN is shown in a side view with the left optic lobe removed (*see top view of brain below*). The labels identify external anatomical features of the brain and its nerve connections. As is the case with many other invertebrates, the brain of the octopus completely surrounds the animal's esophagus. Excision of the entire upper part of the brain is not fatal, but the octopus's behavior then exhibits neither learning nor memory.

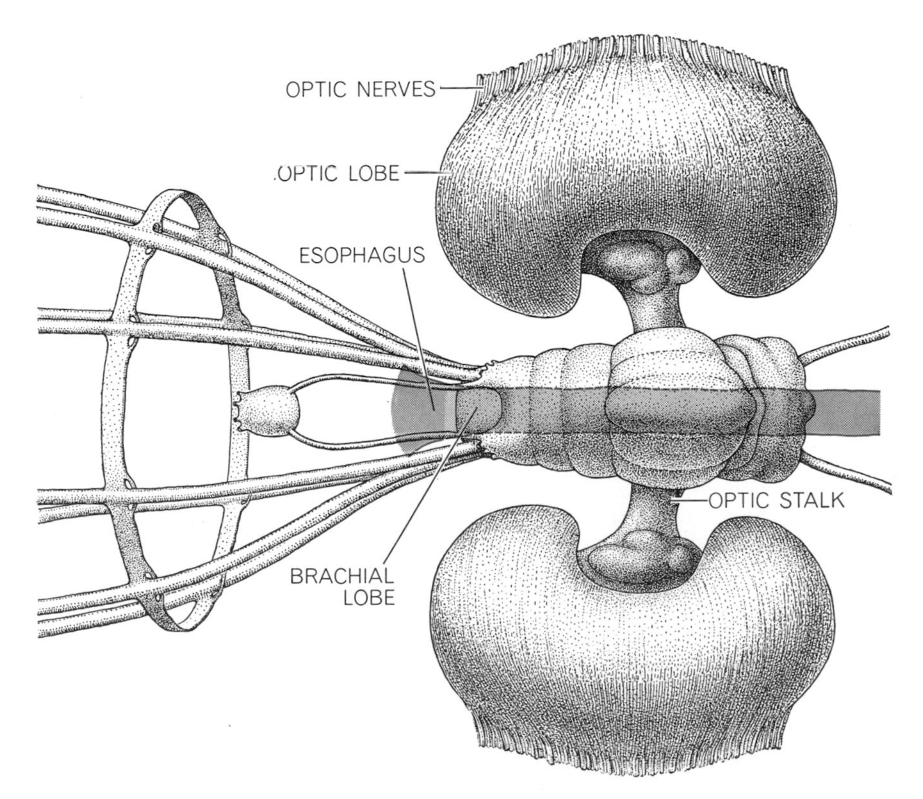

TOP VIEW OF BRAIN relates the two large optic lobes and their stalks to the central brain structure situated above and below the octopus's esophagus (*color*). Combined, the mass of the two optic lobes roughly equals that of the brain's central structure; the fringe of nerves at each lobe's outer edge connects to the retinal structures of the octopus's eyes.

we soon found that, as far as visual learning goes, removing either the vertical lobe, the superior frontal lobe or both, or cutting the nerve tracts between these two lobes, left the octopus unable to learn the required discriminations (or, if they had already been learned, unable to retain them). Since operations on other parts of the brain—performed on control animals—had no effect either on learning or on previously learned behavior, we seemed to have demonstrated that the vertical lobe and superior frontal lobe of the octopus brain are memory centers. In a sense they are, but this is an unduly simple view; in a recent summary of findings Young has listed no fewer than six different effects caused by the removal of or damage to the vertical-lobe system alone.

Karl Lashley, who studied the cerebral cortex of mammals, concluded that, in the organization of a memory, the involvement of specific groups of nerve cells is not as important as the total number of nerve cells available for organization. A similar situation appears to hold true in the functioning of the vertical lobe of the octopus brain; there is a definite relation between the amount of vertical lobe left intact and the accuracy with which a learned response is performed [*see top illustration on page 250*]. This seems to suggest that, at least in the octopus's vertical lobe and the mammalian cerebral cortex, memory is both everywhere and nowhere in particular.

Some of the difficulties such a conclusion presents may be due to a failure to distinguish experimentally between the two constituents of a memory. Whatever its nature, a memory must consist not only of a representation, in neural terms, of the learned situation but also of a mechanism that enables that representation to persist. A distinction must be made between the topology of what persists (the coding and spatial relations involved in the memory of a particular animal) and the mechanisms of persistence (the neural change that is presumably the same in the memory of any animal). Indeed, it may be that some of the theoretical confusion in the study of memory arises from the fact that experiments showing a quantitative relation between memory and nerve tissue tell us something about how the neural representation of memory is organized but nothing about how the representation is kept going.

In our experiments demonstrating that an octopus deprived of its vertical lobe could not be trained to discriminate between a crab alone (that is, reward) and a crab accompanied by a geometric figure and a shock (that is, punishment) our groups of trials were separated by intervals of approximately two hours. When we spaced the trials so that they were only five minutes apart, however, we found such animals were capable of learning [*see bottom illustration on page 250*]. Using the number of trials required as a criterion of learning, we found that these animals attained a level of performance as good as that of normal animals trained with longer time intervals between trials.

One significant difference remained: a normal octopus has a learning-retention period of two weeks or longer, but animals without a vertical lobe had retention periods of only 30 minutes to two hours. These observations suggest that the establishment of a memory involves two mechanisms. There is first a short-term, or transitory, memory that, by its continuing activity between intervals of training, leads to a long-term change in the brain. If there were no reinforcement, the short-term memory would wane; with reinforcement it keeps going and so induces the long-term—and by implication slower—changes that enable a brain to retain memories for long periods.

In 1957 Eliot Stellar of the University of Pennsylvania School of Medicine pointed out the parallels between our results with invertebrates and the unexpected discovery of a similar effect in man by Wilder Penfield, Brenda Milner and W. B. Scoville of the Montreal Neurological Institute. Epileptic patients who have been treated by surgical removal of the temporal lobes of the brain score as well in I.Q. tests after the operation as they do in tests before the onset of epilepsy. They remember their past, their profession and their relatives. They cannot, however, retain new information for more than short periods. Articles can be read and understood, but they are not remembered once they are finished and another topic is taken up. A relative may die but his death goes unremembered after an hour or so. This surgery involves the hippocampal system of the human brain; its effects seem to suggest that, although man's cerebral cortex incorporates a long-term memory system, the hippocampal system is essential to the establishment of new long-term memories.

Today a considerable body of behavioral and psychological evidence favors the separation of memory into short-term and long-term systems. At the neurological level this distinction has brought about a reaffirmation of the role in memory of what are called self-reexciting chains. A few years ago the concept of such chains had gone out of fashion because it had been found that neither convulsive shocks nor cooling the brain to a temperature so low that all activity ceased would abolish learned responses. It is now known that if such treatments are given during the early stages of learning—that is, before a memory is fully established—they have an effect; supposedly this is because they have interfered with the more active part of the process. As the surgical operations for epilepsy indicate, a long-term memory system is intact after removal of the temporal lobes. A short-

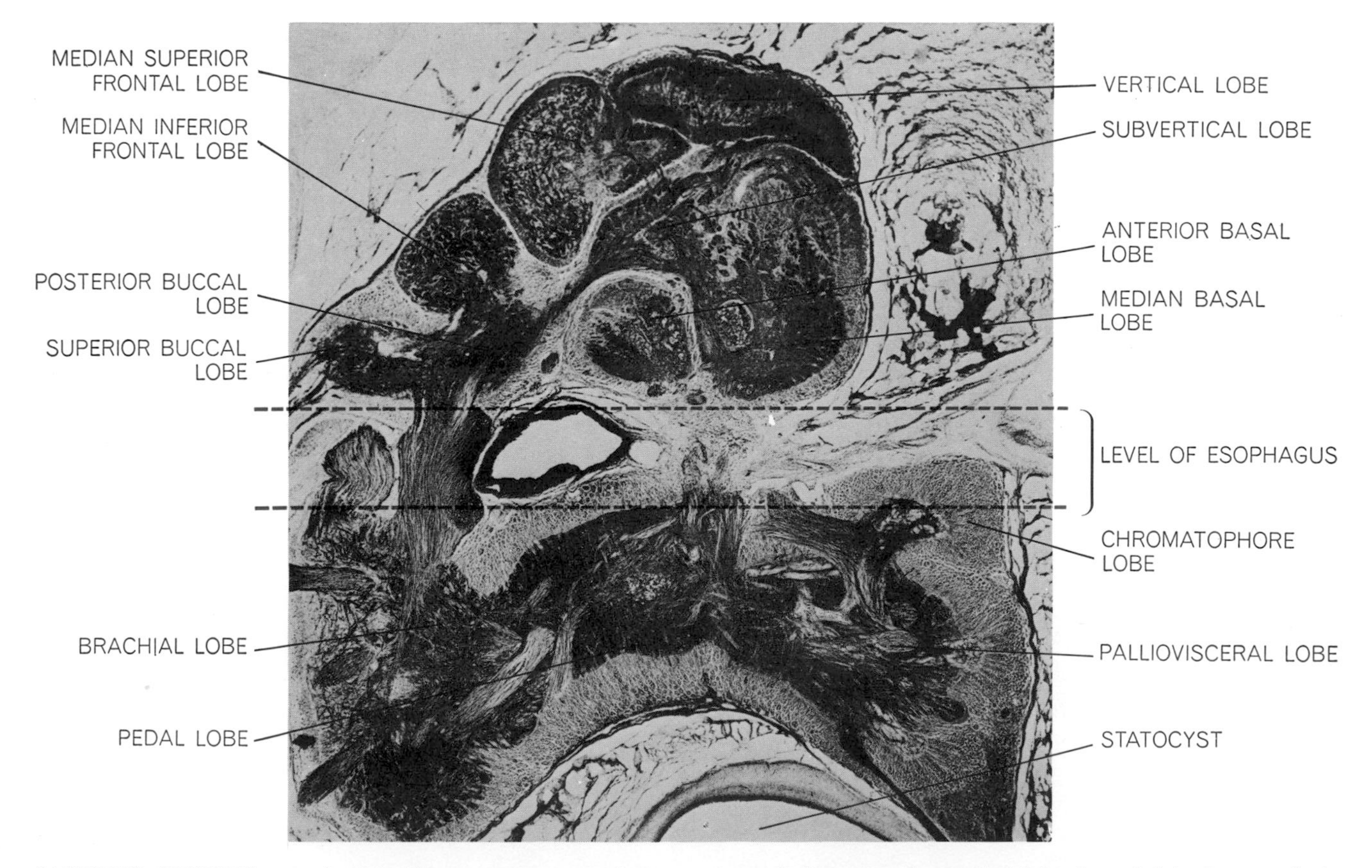

SAGITTAL SECTION stained with silver reveals some of the structures in the octopus brain. Broken lines (*color*) show the route of the esophagus, the boundary between the upper and lower parts of the brain. Labels identify eight lobes in the upper brain and four in the lower; experiments before and after surgical removal show that the vertical lobe (*top right*) plays a role in visual learning and that the inferior frontal lobe (*top left*) is one of two involved in tactile learning. The statocyst (*bottom right*) is not a part of the brain; it is one of the twin organs responsible for the octopus's sense of balance. Magnification is 15 diameters.

UNSCHOOLED OCTOPUS leaves the shelter at one end of its tank (*first photograph*) and walks toward the bait at the opposite end. The advancing animal uses only one of its eyes to guide it. When the bait, a crab, is in range, the octopus throws its leading

term memory system must also remain, however, because the patients can remember new information for short periods, particularly when they use mnemonic devices. On the basis of this interpretation it would appear that the hippocampal system may have the role of linking the two memory mechanisms—whatever that may mean.

For octopuses in our training situation it seems at first that when the vertical lobe of the brain is removed, the long-term memory system of the animal is completely abolished, leaving only the short-term system. We obtain a different result, however, if instead of showing such an animal a crab with or without a geometric figure we present it with figures only, rewarding it with a crab for an attack on one figure and punishing it with a shock for an attack on another. Under such conditions an octopus without its vertical lobe can learn the required discriminations and retain them. At least two conclusions can be drawn from this kind of result. The first is that the vertical lobe is essential to the memory system if the learned response involves a change in what might be termed innate behavior toward an object as familiar to an octopus as a crab. The second is that a long-term memory system for some responses

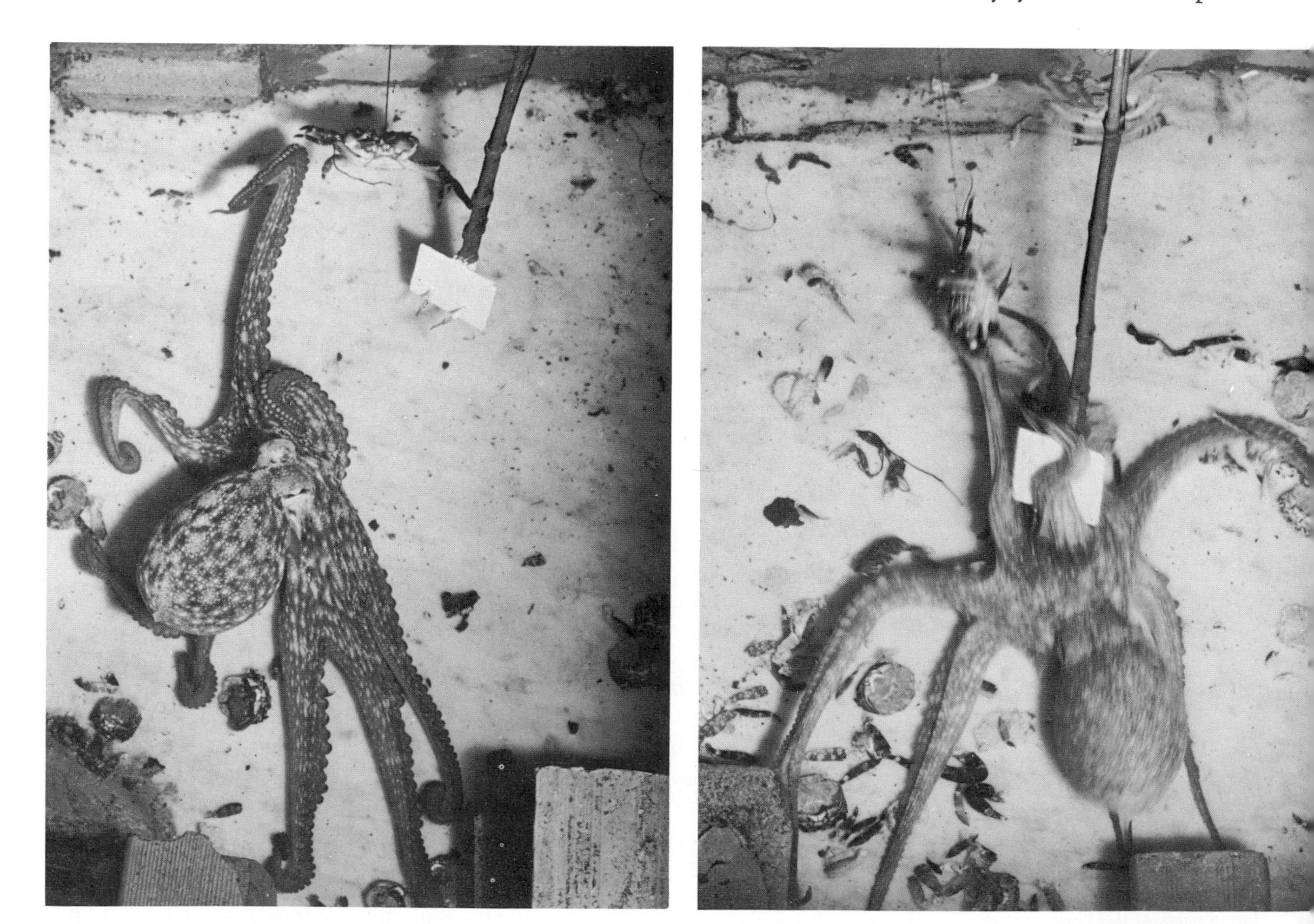

TRAINED OCTOPUS is cautious in its approach when a crab and a geometric figure are presented together (*first photograph*). If the animal seizes the crab, it receives an electric shock (*note darkened region at the base of the arm in second photograph*). As

arms forward to seize it (*second photograph*). Next it tucks the crab up toward its mouth (*third photograph*). The octopus then returns to its shelter (*fourth photograph*), where it kills the crab with a poisonous secretion from its salivary glands and eats it.

can be maintained in the absence of the vertical lobe.

Since we do not know (and probably never will know because it is so difficult to rear *Octopus vulgaris* from its larval stage) whether the octopus's response to a crab is learned or innate, our studies over the past eight years have involved experiments in which reward or punishment is given only after the animal has responded to an artificial situation, that is, the presentation of a figure of a given size, shape or color. It has been shown that animals without a vertical lobe can learn to attack unfamiliar figures for a reward, although they do so more slowly than normal animals. Once they have learned to attack such figures these octopuses retain their response for as long a time as normal animals do. If octopuses without a vertical lobe are required to reverse a learned visual response, however, they find it particularly difficult. When a shock is received for attacking a figure that formerly brought a reward, the animals can still learn to discriminate, but they make between four and five times as many mistakes as normal animals; moreover, their period of retention is shorter.

In addition to its large visual system

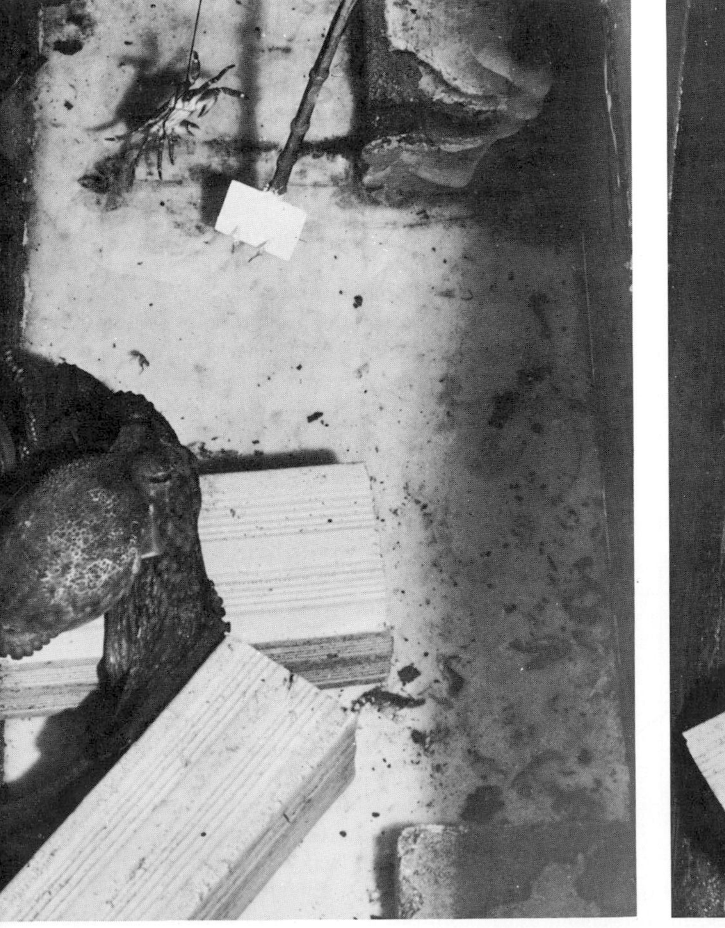

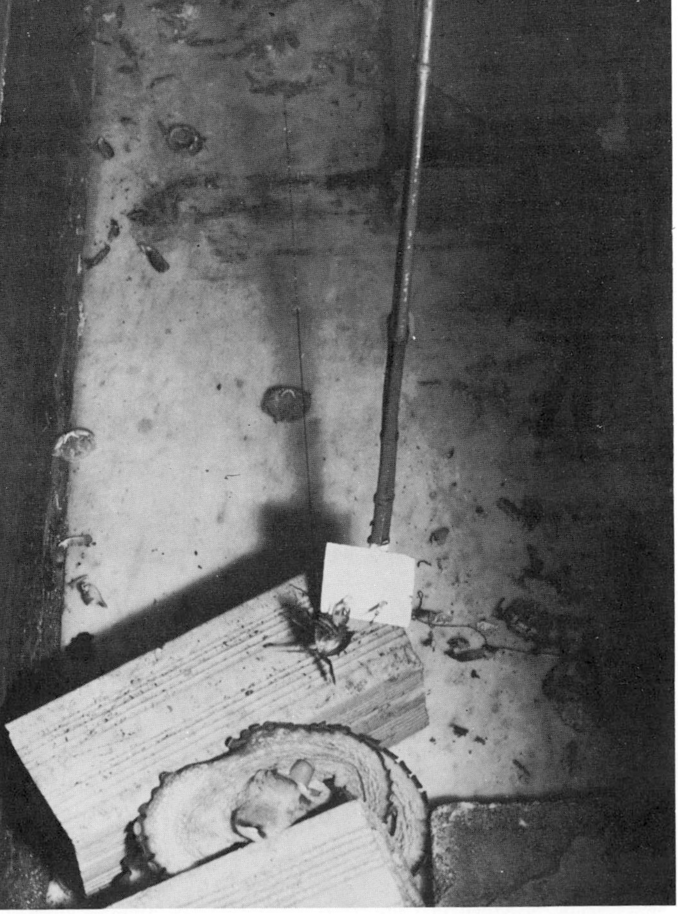

training continues, the octopus will often not even leave its shelter when crab and figure are presented (*third photograph*). If crab and figure are brought near a fully trained animal, it pales and pumps a jet of water at them (*fourth photograph*).

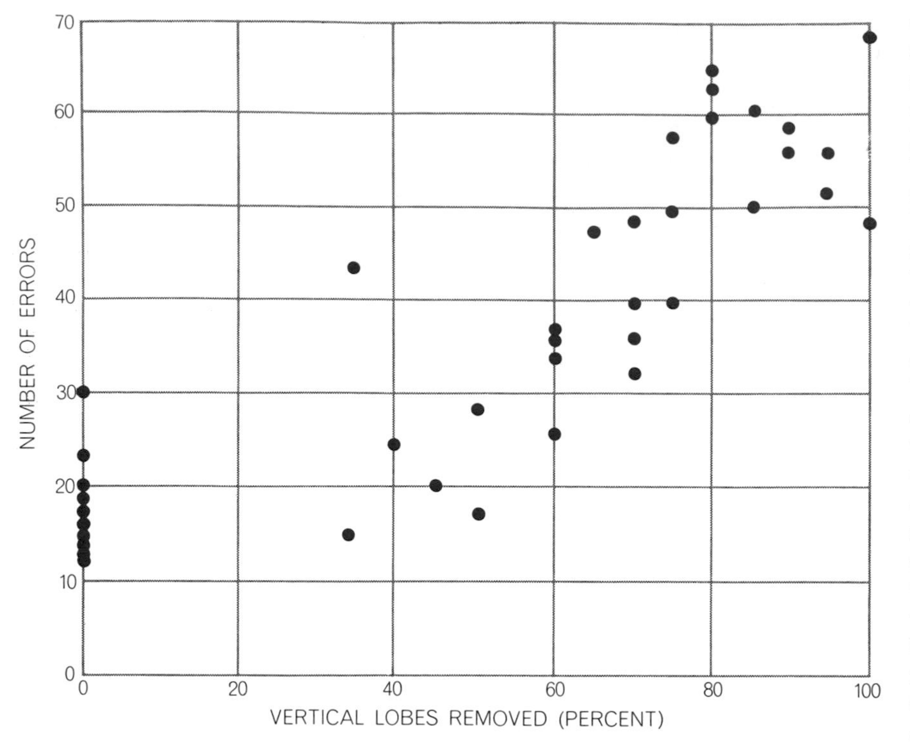

CONTRAST IN PERFORMANCE of normal (*far left*) and surgically altered octopuses shows that the number of errors increased more than threefold as larger and larger portions of the brain's vertical lobe were excised. This finding supports the conclusion that the organization of memory depends primarily on the number of brain cells available.

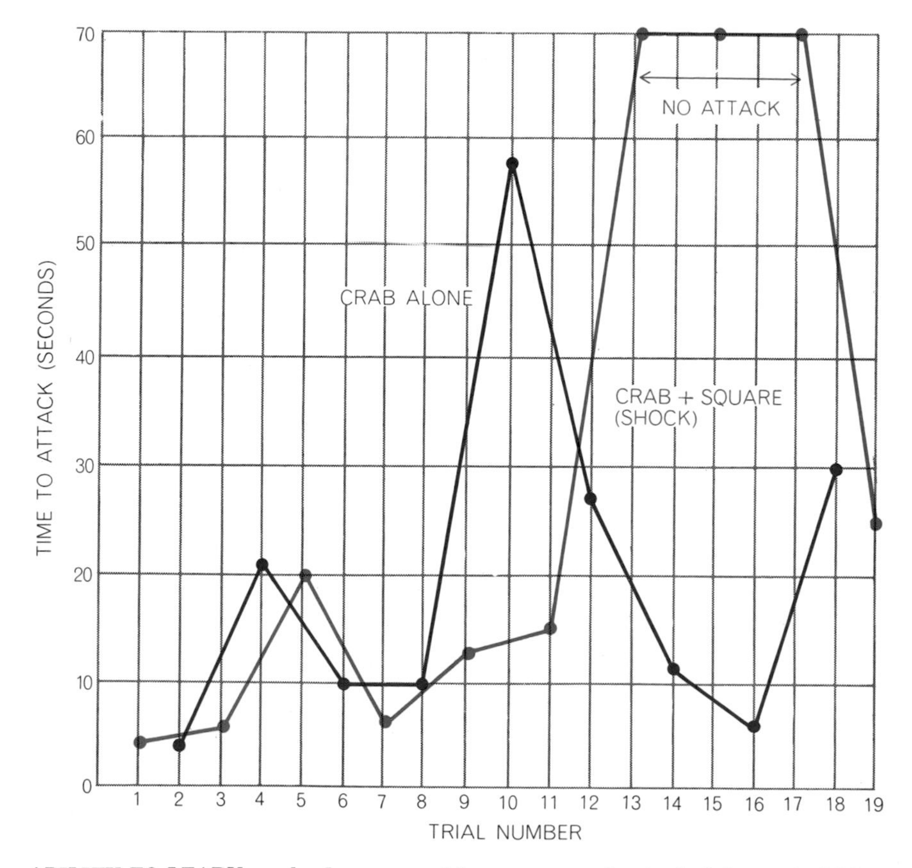

ABILITY TO LEARN can be demonstrated by an octopus deprived of the vertical lobes of its brain, provided that the trials are only a few minutes apart. In the example illustrated, little learning was apparent during 12 alternating exposures to negative and positive stimuli. Thereafter three successive negative stimuli were avoided by the octopus.

the octopus has a complex chemotactile sensory system. Most of the investigation of this system has been done by Martin Wells and his wife Joyce. By applying methods similar to those used for training the animals to make visual discriminations, they have been able to show that tactile learning in the octopus is about as rapid as visual learning. Octopuses have been trained to discriminate between a live bivalve and a counterfeit one consisting of shells of the same species that have been cleaned and filled with wax. They can discriminate between a bivalve with a ribbed shell and another species of comparable size but with a smooth shell. Just recently Wells has found that octopuses can detect hydrochloric acid, sucrose or quinine dissolved in seawater at concentrations 100 times less than those the human tongue can detect in distilled water. Presented with artificial objects, they can distinguish between grooved cylinders and smooth ones, although they cannot distinguish between two grooved objects that differ only in the direction in which the grooves run [*see illustration on opposite page*]. After intensive training they can discriminate a cube from a sphere about 75 percent of the time.

Through each arm of an octopus, which is studded with two rows of suckers, runs a cord of nerve fibers and ganglia. In these ganglia occur local reflexes along the arm and between the rows of suckers. It is supposed that, when the octopus makes a tactile discrimination, the state of excitation in the ganglia above each sucker is determined by the proportion of sense organs excited, and the degree to which these sense organs are stimulated determines the frequency with which nerve impulses are discharged in the fibers running from the ganglia to the brain. Learning in the isolated arm ganglia is probably not possible. Wells has found that for tactile learning to occur the upper brain's median inferior frontal lobe and subfrontal lobe are necessary. Damage to these regions of the brain does not affect visual learning, and for that reason the two lobes have often been used as the sites for control lesions in the investigation of visual learning.

The role of the median inferior frontal lobe seems to be to interrelate the information received from each of the octopus's eight arms; if the lobe is removed and one arm is trained to reject an object, then the other arms continue to accept the object. Without the subfrontal lobe the animals cannot even learn to reject objects by touch. As in the case of the vertical lobe in visual learning, the retention of small portions of the subfrontal lobe allows adequate learned performance. Wells believes that as few as 13,000 of the five million subfrontal-lobe cells may be sufficient for some learning to occur. The subfrontal lobe is structurally very similar to the vertical lobe; it must be considered the vertical lobe's counterpart in the chemotactile system. Removal of the vertical lobe nonetheless has an effect on chemotactile discrimination, mainly in the direction of slowing the rate at which learning occurs.

This account has discussed the main lines of work on memory systems that have been carried out with octopuses as experimental animals, together with some comparisons with human memory. Recently Young has summarized all the work on the cephalopod brain of the past 17 years and has devised a scheme of how such brains may work in the formation, storage and translation of memory into effective action.

Young proposes that in the course of evolution chemotactile and visual centers developed out of a primitive taste-and-bite reflex mechanism. As these "distance receptor" systems evolved, providing information as to where food might be obtained other than that received from direct contact, there came to be a more indirect relation between a change in the environment and the responses that such a change produced in the animal. As this happened, signal systems of greater duration than are provided by simple reflex mechanisms also had to evolve; learning had to become possible so that the animal could assess the significance of each distant environmental change.

Suppose, for example, a crab appears at a distance in the visual field of an octopus; as a result of what can be called "cue signals" there arises in the octopus brain a system for producing "graduated commands to attack." This command system will be weak at first but will grow stronger with reinforcement. The actual strengthening process will vary according to the reward or punishment met at each attack, because the outcome of each attack gives rise to a "result" signal. Such signals condition the distance-receptor systems that initially cued the attack—in the present example, the visual-receptor system. These result signals become distributed throughout the nervous tissue that carries a record or a particular event.

There is, of course, a delay between the moment the cue signals are received in the brain and the moment the result signals arrive. If the result signals are to produce the appropriate conditioning of memory elements, the address of these elements, so to speak, has to be held to allow correct delivery of the information of, say, taste or pain. In the brain of the octopus each optic lobe contains "classifying" cells, among them vertically and horizontally oriented sets of nerve fibers that are presumably related to the vertical and horizontal arrangement of elements in the retina of the octopus's eye. These classifying cells form synapses with "memory" cells in the optic lobes that in their turn activate the cells that signal either attack or retreat. According to Young's hypothesis, each of the memory cells at first has a pair of alternative pathways; the actual neural change during learning consists in closing one of the two pathways. This closing may be accomplished by small cells that are abundant in these learning centers and that can perhaps be switched on so as to produce a substance that inhibits transmission.

Suppose an attack has been evoked by means of this system; the memory cells activate not only an attack circuit but also a circuit reaching the vertical lobe of the upper brain. The signals indicating the results of the attack, such as taste or pain, arrive back and further reinforce the memory cells in the optic lobes, which have been under the influence of the appropriate pathways set up in the vertical lobe during the time interval between the cue signal and the result signal [*see illustration on next page*].

The hypothesis that the actual change represented by memory is produced by the small cells agrees with the fact that these cells are also present in the part of the brain that was shown by the Wellses to be the minimum necessary for tactile memory. Young suggests that the small cells were originally part of the primitive taste-and-bite reflex system, serving the function of

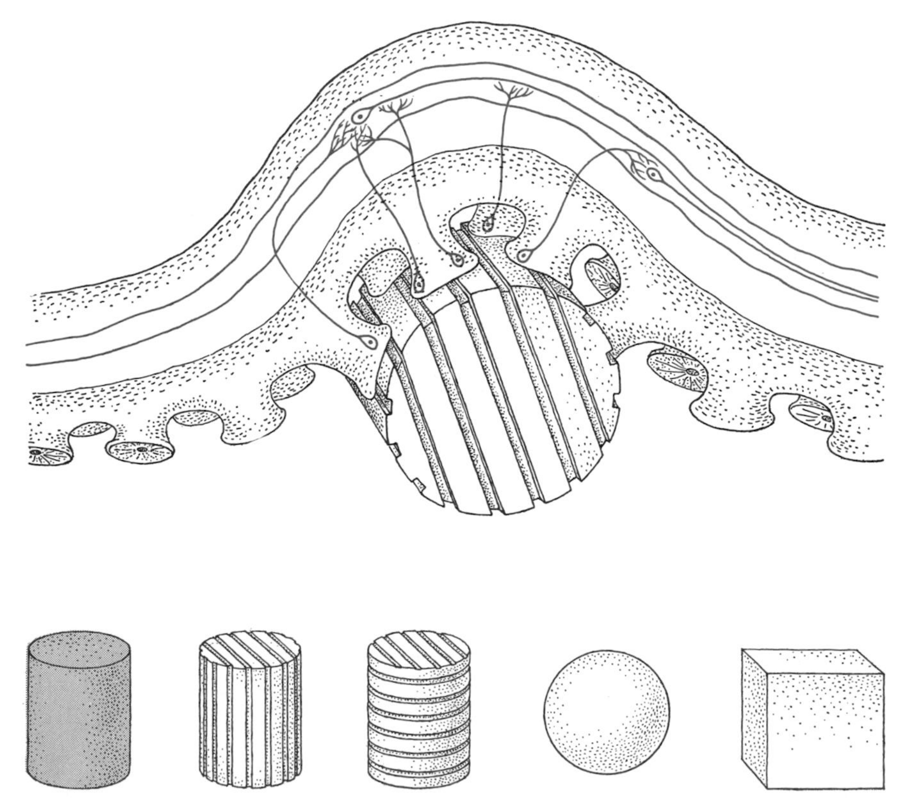

LEARNING BY TOUCH in the octopus was investigated by presenting objects with a variety of shapes and textures. In the case of a grooved cylinder (*top*) only the sense organs in contact with the surface are excited; those resting over the grooves remain inactive. Thus the octopus can learn to discriminate between a smooth cylinder (*gray*) and a grooved one (*color*), and even between a cube and a sphere; it cannot, however, discriminate between two cylinders that differ only in the orientation of the grooves.

temporary inhibition. The evolution of the memory consisted in making the inhibition last longer. The sets of auxiliary lobes associated with the memory system arose to allow for various combinations of inputs to be set up, to be combined with the signals that report the results of actions and finally to be "delivered to the correct address" in the memory.

There is much that is speculative about this description, but the fact remains that both the visual and the tactile memory systems of the octopus embrace sets of brain lobes arranged in similar circuits. This organization provides opportunities for study of the memory process that are made more challenging by Young's conviction that comparable circuits exist in the brains of mammals, including man.

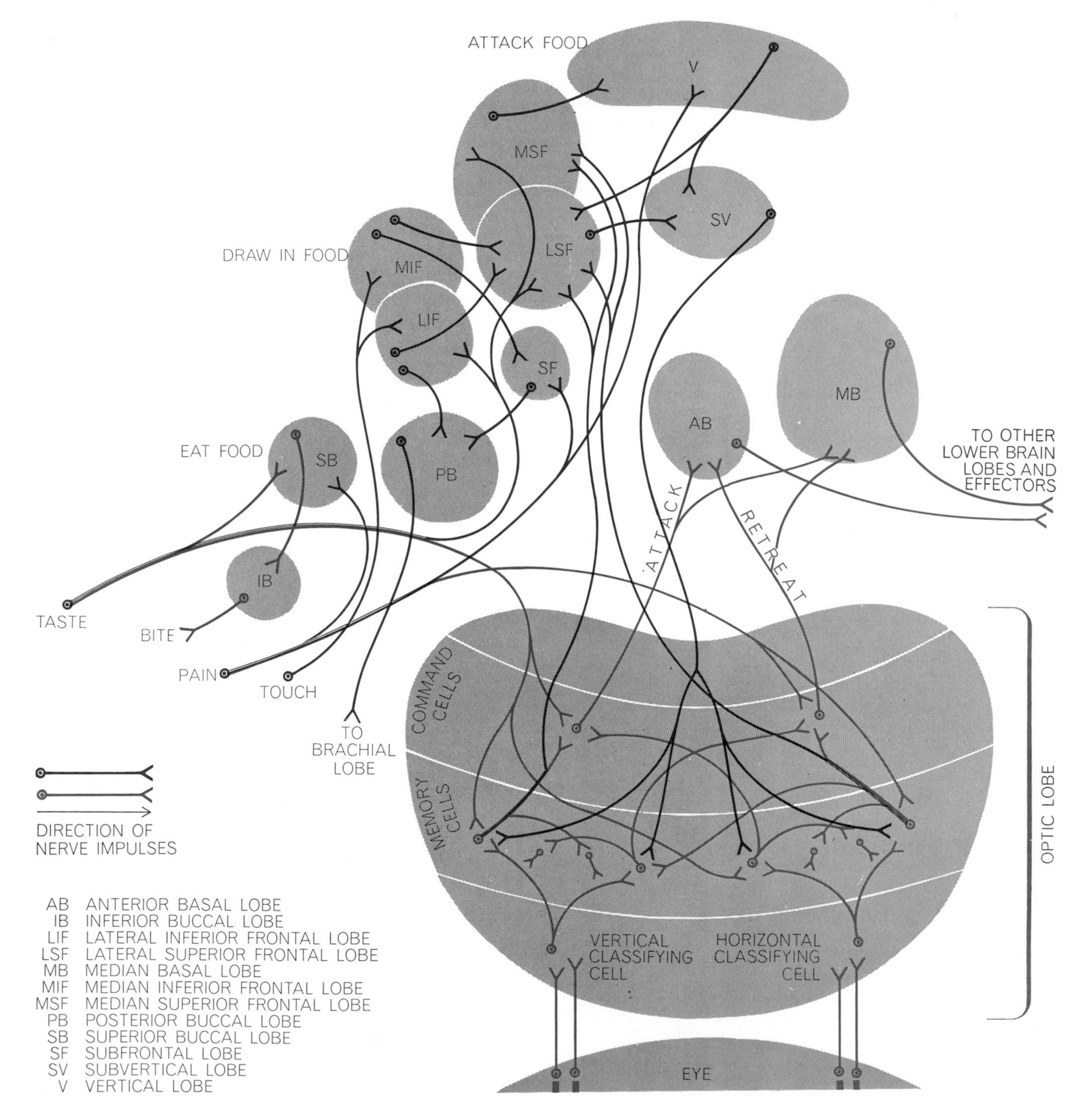

DUAL NATURE OF MEMORY can be traced out on an exploded view of the octopus brain. Circuits leading from the optic lobe (*color*) are the first to be activated on receipt of a visual cue by the lobe's classifying cells. The cue is then recorded in the memory cells and relayed to the command cells; the latter induce the octopus to attack or retreat. If an attack is rewarded, the returning "result" signal will reinforce a memory that registers the initiating cue favorably. If, instead, the attack brings pain, the reinforced memory will register the cue unfavorably and any similar cues encountered in the future will be channeled to the circuits governing retreat rather than attack. Additional circuits (*black*) connect the memory and command regions of the optic lobe to various lobes of the upper brain; thus each event and its outcome are also recorded and reinforced in these nervous tissues. In due course what appear to be the long-term components of the memory system become localized in individual upper brain lobes.

THE HOMING SALMON

ARTHUR D. HASLER AND JAMES A. LARSEN

August 1955

A learned naturalist once remarked that among the many riddles of nature, not the least mysterious is the migration of fishes. The homing of salmon is a particularly dramatic example. The Chinook salmon of the U. S. Northwest is born in a small stream, migrates downriver to the Pacific Ocean as a young smolt and, after living in the sea for as long as five years, swims back unerringly to the stream of its birth to spawn. Its determination to return to its birthplace is legendary. No one who has seen a 100-pound Chinook salmon fling itself into the air again and again until it is exhausted in a vain effort to surmount a waterfall can fail to marvel at the strength of the instinct that draws the salmon upriver to the stream where it was born.

How do salmon remember their birthplace, and how do they find their way back, sometimes from 800 or 900 miles away? This enigma, which has fascinated naturalists for many years, is the subject of the research to be reported here. The question has an economic as well as a scientific interest, because new dams which stand in the salmon's way have cut heavily into salmon fishing along the Pacific Coast. Before long nearly every stream of any appreciable size in the West will be blocked by dams. It is true that the dams have fish lifts and ladders designed to help salmon to hurdle them. Unfortunately, and for reasons which are different for nearly every dam so far designed, salmon are lost in tremendous numbers.

There are six common species of salmon. One, called the Atlantic salmon, is of the same genus as the steelhead trout. These two fish go to sea and come back upstream to spawn year after year. The other five salmon species, all on the Pacific Coast, are the Chinook (also called the king salmon), the sockeye, the silver, the humpback and the chum. The

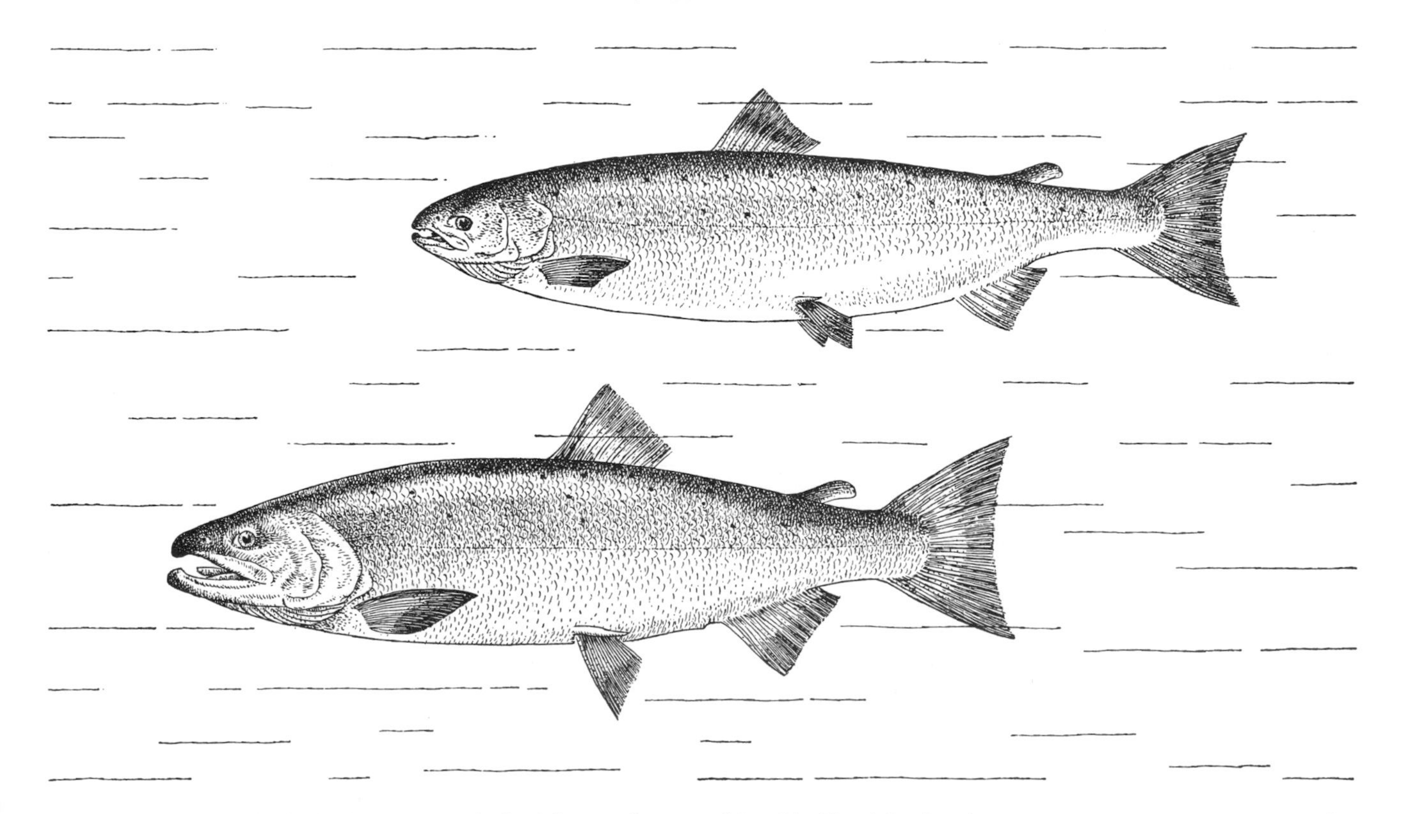

TWO COMMON SPECIES of salmon are (*top*) the Atlantic salmon (*Salmo salar*) and (*bottom*) the silver salmon (*Oncorhynchus kisutch*). The Atlantic salmon goes upstream to spawn year after year; the silver salmon, like other Pacific species, spawns only once.

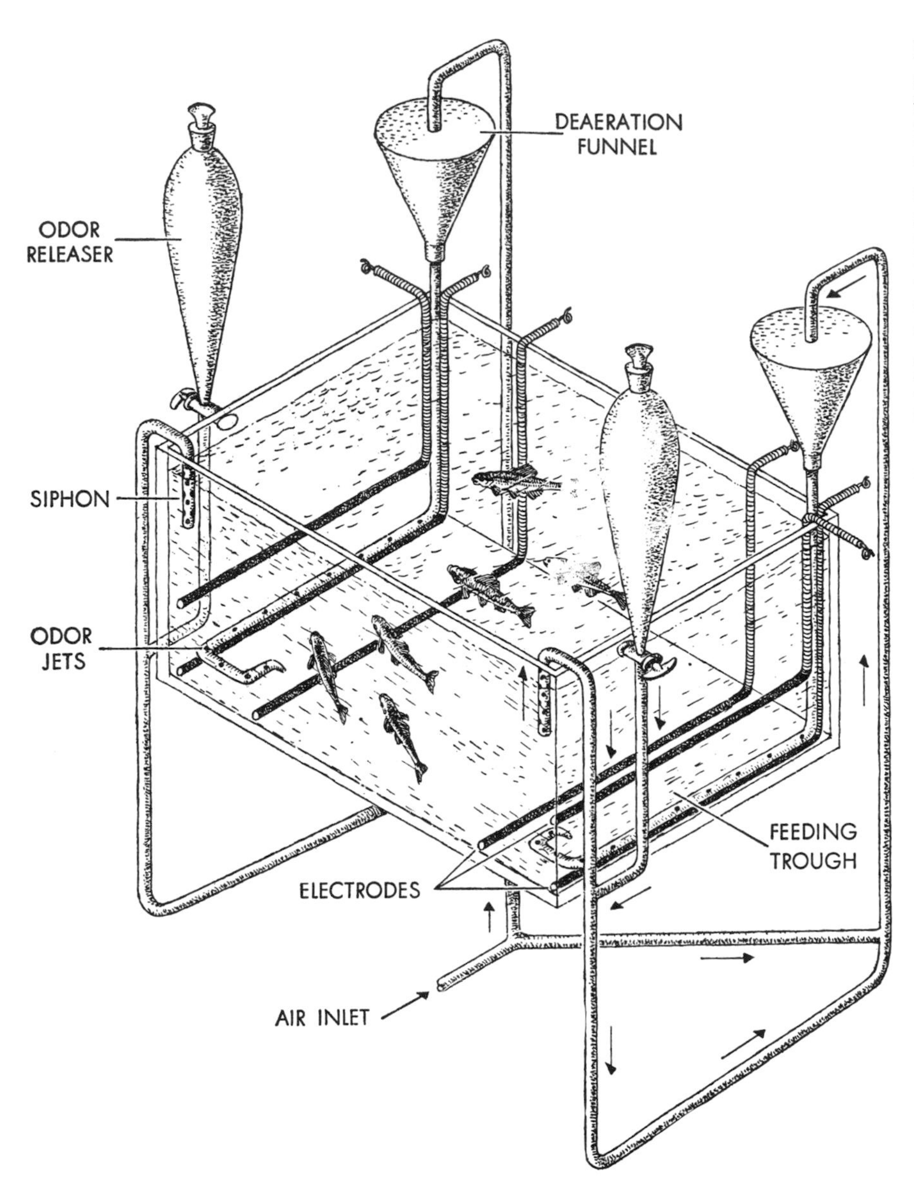

EXPERIMENTAL TANK was built in the Wisconsin Lake Laboratory to train fish to discriminate between two odors. In this isometric drawing the vessel at the left above the tank contains water of one odor. The vessel at the right contains water of another odor. When the valve below one of the vessels was opened, the water in it was mixed with water siphoned out of the tank. The mixed water was then pumped into the tank by air. When the fish (minnows or salmon) moved toward one of the odors, they were rewarded with food. When they moved toward the other odor, they were punished with a mild electric shock from the electrodes mounted inside the tank. Each of the fish was blinded to make sure that it would not associate reward and punishment with the movements of the experimenters.

Pacific salmon home only once: after spawning they die.

A young salmon first sees the light of day when it hatches and wriggles up through the pebbles of the stream where the egg was laid and fertilized. For a few weeks the fingerling feeds on insects and small aquatic animals. Then it answers its first migratory call and swims downstream to the sea. It must survive many hazards to mature: an estimated 15 per cent of the young salmon are lost at every large dam, such as Bonneville, on the downstream trip; others die in polluted streams; many are swallowed up by bigger fish in the ocean. When, after several years in the sea, the salmon is ready to spawn, it responds to the second great migratory call. It finds the mouth of the river by which it entered the ocean and then swims steadily upstream, unerringly choosing the correct turn at each tributary fork, until it arrives at the stream where it was hatched. Generation after generation, families of salmon return to the same rivulet so consistently that populations in streams not far apart follow distinctly separate lines of evolution.

The homing behavior of the salmon has been convincingly documented by many studies since the turn of the century. One of the most elaborate was made by Andrew L. Pritchard, Wilbert A. Clemens and Russell E. Foerster in Canada. They marked 469,326 young sockeye salmon born in a tributary of the Fraser River, and they recovered nearly 11,000 of these in the same parent stream after the fishes' migration to the ocean and back. What is more, not one of the marked fish was ever found to have strayed to another stream. This remarkable demonstration of the salmon's precision in homing has presented an exciting challenge to investigators.

At the Wisconsin Lake Laboratory during the past decade we have been studying the sense of smell in fish, beginning with minnows and going on to salmon. Our findings suggest that the salmon identifies the stream of its birth by odor and literally smells its way home from the sea.

Fish have an extremely sensitive sense of smell. This has often been observed by students of fish behavior. Karl von Frisch showed that odors from the injured skin of a fish produce a fright reaction among its schoolmates. He once noticed that when a bird dropped an injured fish in the water, the school of fish from which it had been seized quickly dispersed and later avoided the area. It is well known that sharks and tuna are drawn to a vessel by the odor of bait in the water. Indeed, the time-honored custom of spitting on bait may be founded on something more than superstition; laboratory studies have proved that human saliva is quite stimulating to the taste buds of a bullhead. The sense of taste of course is closely allied to the sense of smell. The bullhead has taste buds all over the surface of its body; they are especially numerous on its whiskers. It will quickly grab for a piece of meat that touches any part of its skin. But it becomes insensitive to taste and will not respond in this way if a nerve serving the skin buds is cut.

The smelling organs of fish have evolved in a great variety of forms. In the bony fishes the nose pits have two separate openings. The fish takes water into the front opening as it swims or breathes (sometimes assisting the intake with cilia), and then the water passes out through the second opening, which may be opened and closed rhythmically by the fish's breathing. Any odorous substances in the water stimulate the nasal receptors chemically, perhaps by an effect on enzyme reactions, and the re-

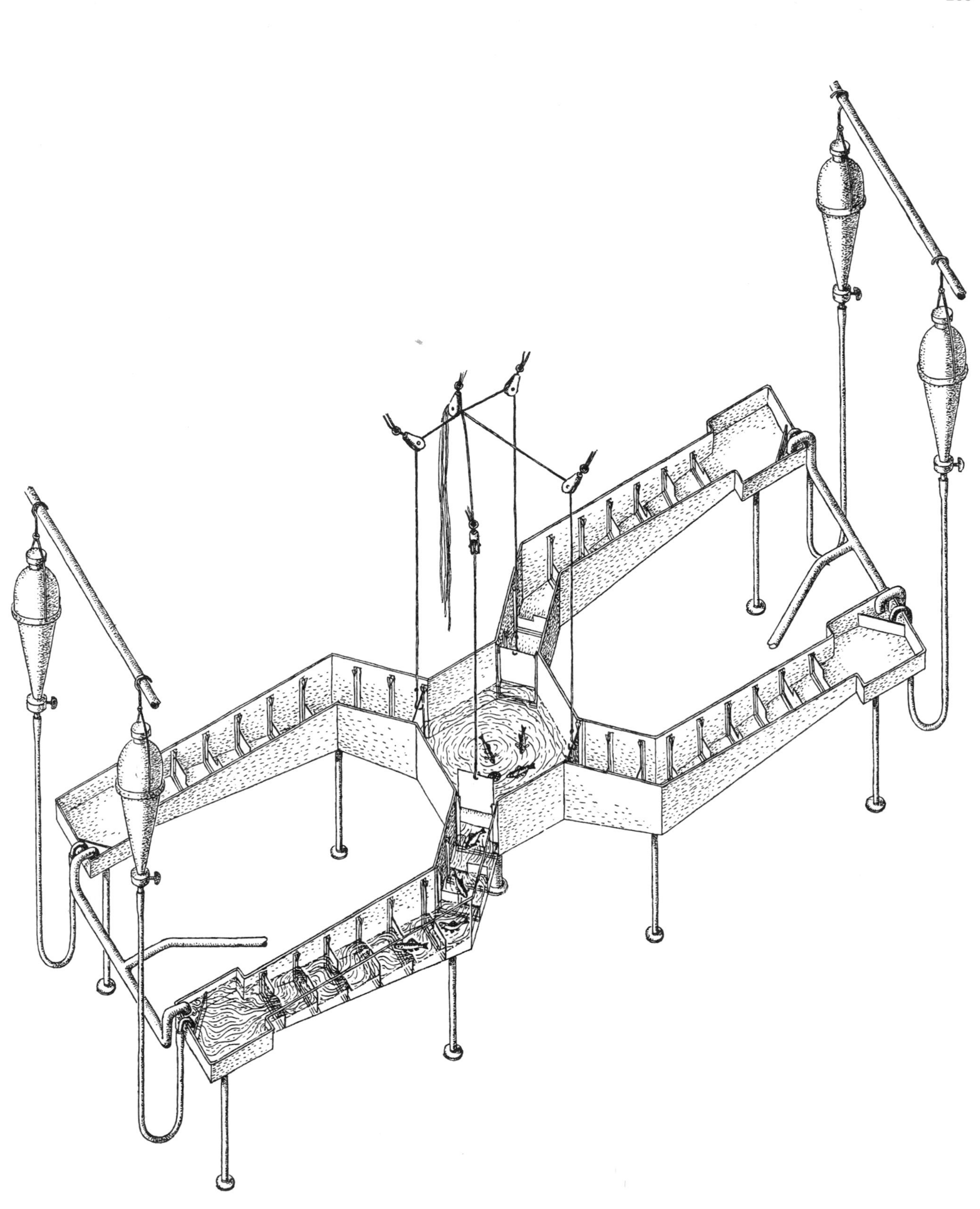

FOUR RUNWAYS are used to test the reaction of untrained salmon fingerlings to various odors. Water is introduced at the outer end of each runway and flows down a series of steps into a central compartment, where it drains. In the runway at the lower left the water cascades down to the central compartment in a series of miniature waterfalls; in the other runways the water is omitted to show the construction of the apparatus. Odors may be introduced into the apparatus from the vessels suspended above the runways. In an experiment salmon fingerlings are placed in the central compartment and an odor is introduced into one of the runways. When the four doors to the central compartment are opened, the fingerlings tend to enter the arms, proceeding upstream by jumping the waterfalls. Whether an odor attracts them, repels or has no effect is judged by the observed distribution of the fish in the runways.

sulting electrical impulses are relayed to the central nervous system by the olfactory nerve.

The human nose, and that of other land vertebrates, can smell a substance only if it is volatile and soluble in fat solvents. But in the final analysis smell is always aquatic, for a substance is not smelled until it passes into solution in the mucous film of the nasal passages. For fishes, of course, the odors are already in solution in their watery environment. Like any other animal, they can follow an odor to its source, as a hunting dog follows the scent of an animal. The quality or effect of a scent changes as the concentration changes; everyone knows that an odor may be pleasant at one concentration and unpleasant at another.

When we began our experiments, we first undertook to find out whether fish could distinguish the odors of different water plants. We used a specially developed aquarium with jets which could inject odors into the water. For responding to one odor (by moving toward the jet), the fish were rewarded with food; for responding to another odor, they were punished with a mild electric shock. After the fish were trained to make choices between odors, they were tested on dilute rinses from 14 different aquatic plants. They proved able to distinguish the odors of all these plants from one another.

Plants must play an important role in the life of many freshwater fish. Their odors may guide fish to feeding grounds when visibility is poor, as in muddy water or at night, and they may hold young fish from straying from protective cover. Odors may also warn fish away from poisons. In fact, we discovered that fish could be put to use to assay industrial pollutants: our trained minnows were able to detect phenol, a common pollutant, at concentrations far below those detectable by man.

All this suggested a clear-cut working hypothesis for investigating the mystery of the homing of salmon. We can suppose that every little stream has its own characteristic odor, which stays the same year after year; that young salmon become conditioned to this odor before they go to sea; that they remember the odor as they grow to maturity, and that they are able to find it and follow it to its source when they come back upstream to spawn.

Plainly there are quite a few ifs in this theory. The first one we tested was the question: Does each stream have its own odor? We took water from two creeks in Wisconsin and investigated whether fish could learn to discriminate between them. Our subjects, first minnows and then salmon, were indeed able to detect a difference. If, however, we destroyed a fish's nose tissue, it was no longer able to distinguish between the two water samples.

Chemical analysis indicated that the only major difference between the two waters lay in the organic material. By testing the fish with various fractions of the water separated by distillation, we confirmed that the identifying material was some volatile organic substance.

The idea that fish are guided by odors in their migrations was further supported by a field test. From each of two different branches of the Issaquah River in the State of Washington we took a number of sexually ripe silver salmon which had come home to spawn. We then plugged with cotton the noses of half the fish in each group and placed all the salmon in the river below the fork to make the upstream run again. Most of the fish with unplugged noses swam back to the stream they had selected the first time. But the "odor-blinded" fish migrated back in random fashion, picking the wrong stream as often as the right one.

In 1949 eggs from salmon of the Horsefly River in British Columbia were hatched and reared in a hatchery in a tributary called the Little Horsefly. Then they were flown a considerable distance and released in the main Horsefly River, from which they migrated to the sea. Three years later 13 of them had returned to their rearing place in the Little Horsefly, according to the report of the Canadian experimenters.

In our own laboratory experiments we tested the memory of fish for odors and found that they retained the ability to differentiate between odors for a long period after their training. Young fish remembered odors better than the old. That animals "remember" conditioning to which they have been exposed in their youth, and act accordingly, has been demonstrated in other fields. For instance, there is a fly which normally lays its eggs on the larvae of the flour moth, where the fly larvae then hatch and develop. But if larvae of this fly are raised on another host, the beeswax moth, when the flies mature they will seek out beeswax moth larvae on which to lay their eggs, in preference to the traditional host.

With respect to the homing of salmon we have shown, then, that different streams have different odors, that salmon respond to these odors and that they remember odors to which they have been conditioned. The next question is: Is a salmon's homeward migration guided solely by its sense of smell? If we could decoy homing salmon to a stream other than their birthplace, by means of an odor to which they were conditioned artificially, we might have not only a solution to the riddle that has puzzled scientists but also a practical means of saving the salmon—guiding them to breeding streams not obstructed by dams.

We set out to find a suitable substance to which salmon could be conditioned. A student, W. J. Wisby, and I [Arthur Hasler] designed an apparatus to test the reactions of salmon to various organic odors. It consists of a compartment from which radiate four runways, each with several steps which the fish must jump to climb the runway. Water cascades down each of the arms. An odorous substance is introduced into one of the arms, and its effect on the fish is judged by whether the odor appears to attract fish into that arm, to repel them or to be indifferent to them.

We needed a substance which initially would not be either attractive or repellent to salmon but to which they could be conditioned so that it would attract them. After testing several score organic odors, we found that dilute solutions of morpholine neither attracted nor repelled salmon but were detectable by them in extremely low concentrations—as low as one part per million. It appears that morpholine fits the requirements for the substance needed: it is soluble in water; it is detectable in extremely low concentrations; it is chemically stable under stream conditions. It is neither an attractant nor a repellent to unconditioned salmon, and would have meaning only to those conditioned to it.

Federal collaborators of ours are now conducting field tests on the Pacific Coast to learn whether salmon fry and fingerlings which have been conditioned to morpholine can be decoyed to a stream other than that of their birth when they return from the sea to spawn. Unfortunately this type of experiment may not be decisive. If the salmon are not decoyed to the new stream, it may simply mean that they cannot be drawn by a single substance but will react only to a combination of subtle odors in their parent stream. Perhaps adding morpholine to the water is like adding the whistle of a freight train to the quiet strains of a violin, cello and flute. The salmon may still seek out the subtle harmonies of an odor combination to which they have been reacting by instinct for centuries. But there is still hope that they may respond to the call of the whistle.

THE NAVIGATION OF THE GREEN TURTLE

ARCHIE CARR
May 1965

One of the stubborn puzzles of animal behavior is the ability of some animals to travel regularly to remote oceanic islands. The best-known of the blue-water navigators are birds; recently, however, the green turtle (*Chelonia mydas*) has given evidence of being as keen an island-finder as the sooty tern or the albatross. Because green turtles swim slowly, and do so at the surface of the water or a little below it, they are potentially easier to follow in their journeys than either birds or migratory fishes, seals and whales. The green turtle may therefore prove to be an important experimental subject for students of animal navigation.

The evidence that green turtles can find their way to remote oceanic islands is provided by female green turtles that normally inhabit feeding grounds along the coast of Brazil. It appears that once every two or three years these turtles swim all the way to Ascension Island—a target five miles wide and 1,400 miles away in the South Atlantic—to lay their eggs. By the processes of natural selection this population seems to have evolved the capacity to hold a true course across hundreds of miles of sea, using only animal senses as instruments of navigation. The difficulties facing such a voyage would seem insurmountable if it were not so clear that the turtles are somehow surmounting them.

The green turtle is one of the five kinds of sea turtle found throughout the warmer oceans of the world [*see illustration on next two pages*]. Adult green turtles, which may weigh more than 500 pounds, are herbivorous; they feed on the so-called turtle grass that grows abundantly in sheltered tropical shallows. Although the green turtles of the world are separated into reproductively isolated breeding colonies, they show little tendency to evolve into recognizable local races and species. The green turtles of the Pacific, for example, show only minor differences in form and color from those of the Atlantic. The one area in which what appears to be a well-differentiated species has evolved is the northern coast of Australia, where the form *Chelonia depressa* is found. Green turtles nest only in places where the average temperature of the surface water during the coldest month of the year is above 68 degrees Fahrenheit. In the Atlantic the northern limit of their nesting range seems to have been Bermuda; early voyagers to the New World destroyed the colony there. The most northerly nesting site known in the Pacific is French Frigate Shoal, an outlier of the Hawaiian Islands.

Until a few years ago what was known about the green turtle consisted mainly of cooking recipes and a sea of folklore from which rose only a few islands of fact. Among these were studies by Edward Banks in the Turtle Islands of Sarawak, by James Hornell in the Seychelles Islands off the east coast of Africa, by P. E. P. Deraniyagala in Ceylon and by F. W. Moorhouse along the Great Barrier Reef of Australia. More recently in Sarawak, Tom Harrisson and John R. Hendrickson have independently uncovered a great deal of information on the nesting ecology and reproductive cycles of the huge breeding colony there. Since 1955 my colleagues at the University of Florida and I have been working, with the aid of a series of National Science Foundation grants, at the only green turtle nesting beach that remains in the western Caribbean: Tortuguero in Costa Rica. What is now known about *Chelonia* from all these studies permits the piecing together of a coherent—albeit still somewhat fragmentary—account of its life.

Although green turtles are primarily sea animals, in a few places in the Pacific they sometimes go ashore to bask. They have been seen lying in the sun along with albatrosses and basking monk seals on such small Pacific islands as Pearl and Hermes Reef, Lisianski and Kure Atoll. The females of the Pacific populations sometimes even nest during the day. Neither nesting during the day nor basking appears to be the habit of green turtles in any Atlantic population. Once the Atlantic hatchlings leave the beach the males remain at sea for the rest of their lives. The females come ashore only to nest, and they do so after dark. They always return to the same general nesting area and often to the same narrow sector of beach.

The green turtle is one of the few reptiles that are known to reproduce at intervals of more than a year. In Sarawak the entire turtle population nests in a three-year cycle. At Tortuguero about a third of the colony returns to nest every two years and the other two-thirds follows a three-year schedule. In our eight years of tagging turtles at this site no turtle has returned after an absence of only one year. When the Tortuguero turtles finish nesting, they travel back to their home pastures of turtle grass and evidently remain there feeding until their reproductive rhythm directs them to return to the nesting beach. Such a feeding ground may be within a few dozen miles of the nesting beach or many hundreds of miles away.

The time of year at which the nesting season begins and the duration of the season vary from one region to another. In some places nesting is restricted to three or four months of the year; in others it may run through the entire year, with a peak during two or three months or even with two separate peaks. Mating takes place at the nesting ground and apparently nowhere else. Either the males accompany the females during their journey from the home pasture or they make an exactly timed rendezvous with them at the nesting beach. In any case, as soon as the first tracks of nesting turtles appear on the shore the animals can be seen at sea just beyond the surf, courting, fighting and mating. Because these activities are mostly out of sight little is known about them. From a low-flying airplane one sometimes gets a glimpse of two males splashing around a single female. Many females come ashore scratched and lacerated, evidently because of the violent attentions of the males.

Some females apparently mate just before their first nesting trip ashore; others do so later. At Tortuguero mating activity appears to end after the third or fourth week of the nesting season. Even when mating precedes the first landing, it must take place after at least some of the female's eggs have formed shells. It therefore seems unlikely that any of the eggs of the current season are fertilized as a result of current mating. The encounter probably serves to fertilize eggs for the next nesting season, two or three years ahead.

Once at the nesting beach the female goes ashore from three to seven times to deposit clutches of eggs in the sand. The interval between nesting trips is about 12 days, and the number of eggs laid on each occasion is 100 or so. Each egg is about two inches in diameter. The shell is flexible, and when the egg is first laid it has a curious dent in it that no amount of pressing will smooth out. The incubation period is about 57 days.

On hatching and emerging from the nest the young green turtles nearly always set off on a direct course for the sea, even when the water itself is completely hidden by dunes or other obstacles. Experiments with a number of different kinds of marine and freshwater turtles indicate that the mechanism of this orientation is an innate response to some quality of the light over open water. It is clear that no compass-like sense is involved: in a series of tests we flew little turtles from Caribbean nests across Costa Rica and allowed them to emerge from artificial nests on a Pacific beach. Even though the sea now lay in the opposite direction the young turtles reached the water as easily as their siblings on the home beach did.

After the hatchlings enter the water, either their sea-finding drive gives way to other orienting mechanisms or they simply keep swimming until whatever

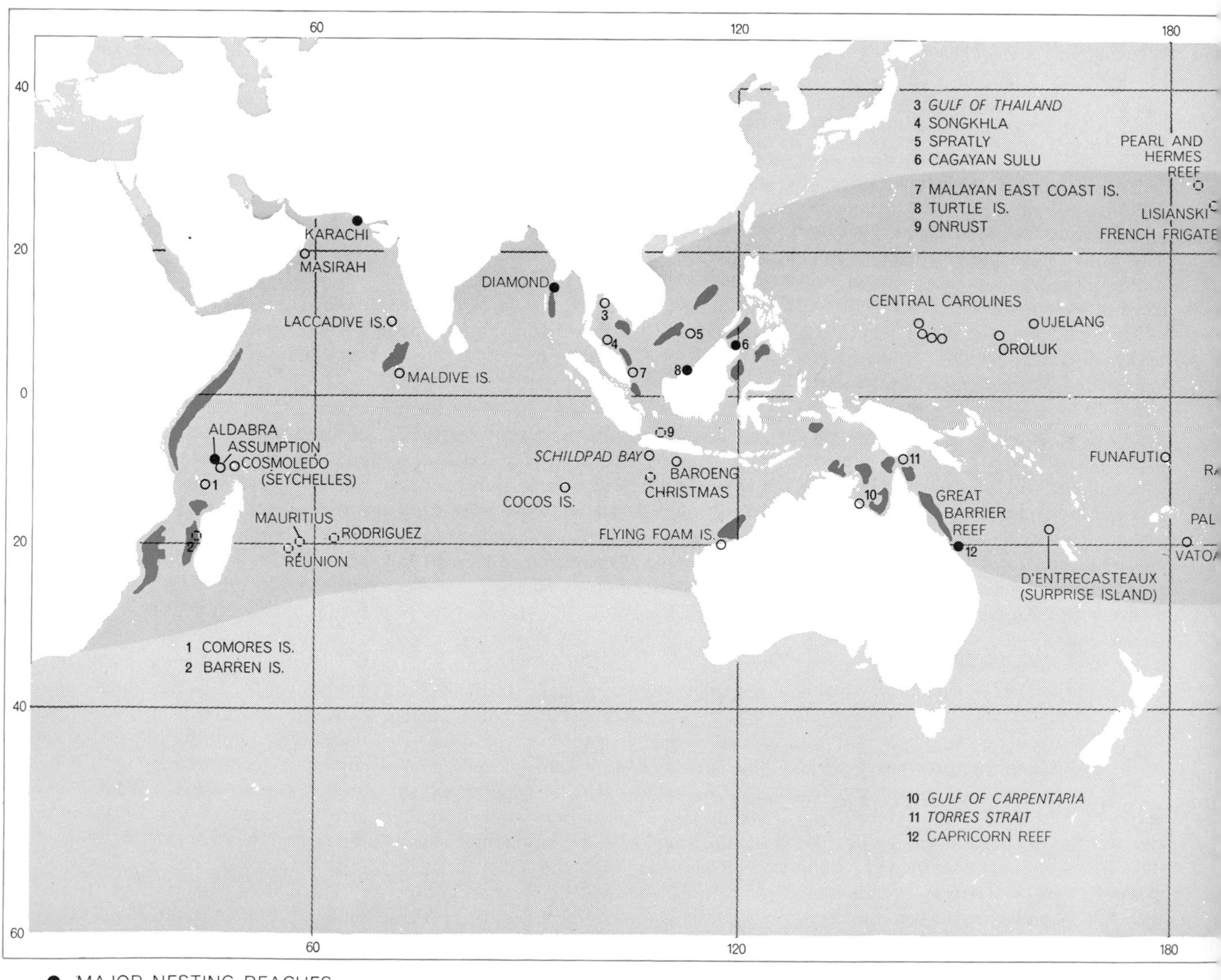

PAST AND PRESENT RANGE of the green turtle was compiled from the accounts of early explorers and later sources by James J. Parsons of the University of California at

difference they perceive between sea light and land light becomes too slight to provide a guiding stimulus. By this time they will probably have been picked up by a longshore current. Since the adults navigate over long distances, they must have some kind of compass sense. This sense may be latent in the hatchlings: recent laboratory experiments at Duke University by Klaus Fischer seem to show that this is the case. In nature, however, this sense evidently does not come into operation until the light-beacon sense fails. In any case, tests have shown that the sea-finding drive is not lost even when turtles are kept away from water for as long as a year after hatching. It is also evident that both capacities are present in the mature female. The compass sense must guide her from feeding ground to nesting beach, and the sea-finding drive directs her back to the ocean after she has nested—even though it is out of sight of her nesting place.

A recent experiment demonstrates the kind of ambiguity that can arise in the study of orientation among young green turtles. Several hundred 20-day-old hatchlings from Tortuguero were placed in a circular tank at the Lerner Marine Laboratory on the island of Bimini in the Bahamas. The nearest water was on the bay side of the island, some 40 yards away. The ocean was some 200 yards away in the opposite direction. From turtle's-eye level the rim of the tank blocked any view of either bay or ocean. The skyline was broken by trees and buildings.

The distribution of the young turtles in the tank was recorded over a three-day period, at 9:00 A.M., 4:00 P.M. and 11:00 P.M. each day. During most of that time the wind blew steadily from the bay side of the tank. At night, while the sleeping turtles floated on the surface, the steady breeze piled them up on the ocean side of the tank. Once awake in daylight, however, the turtles showed a marked orientation in the opposite direction. There was active—at times frantic—swimming toward and crowding along the wall nearest the bay [*see illustration on page 262*]. This bias was not simply a tendency to swim upwind; the bay side of the tank was equally favored during a few windless periods. The response could have been an innate direction preference based on a compass sense, but it was more probably the same light-seeking urge that guides hatchlings from the nest to the water. Two months later, however, when the same test was repeated with the same animals, they showed no seaward orientation at all.

Almost nothing is known about the movements and habits of green turtles

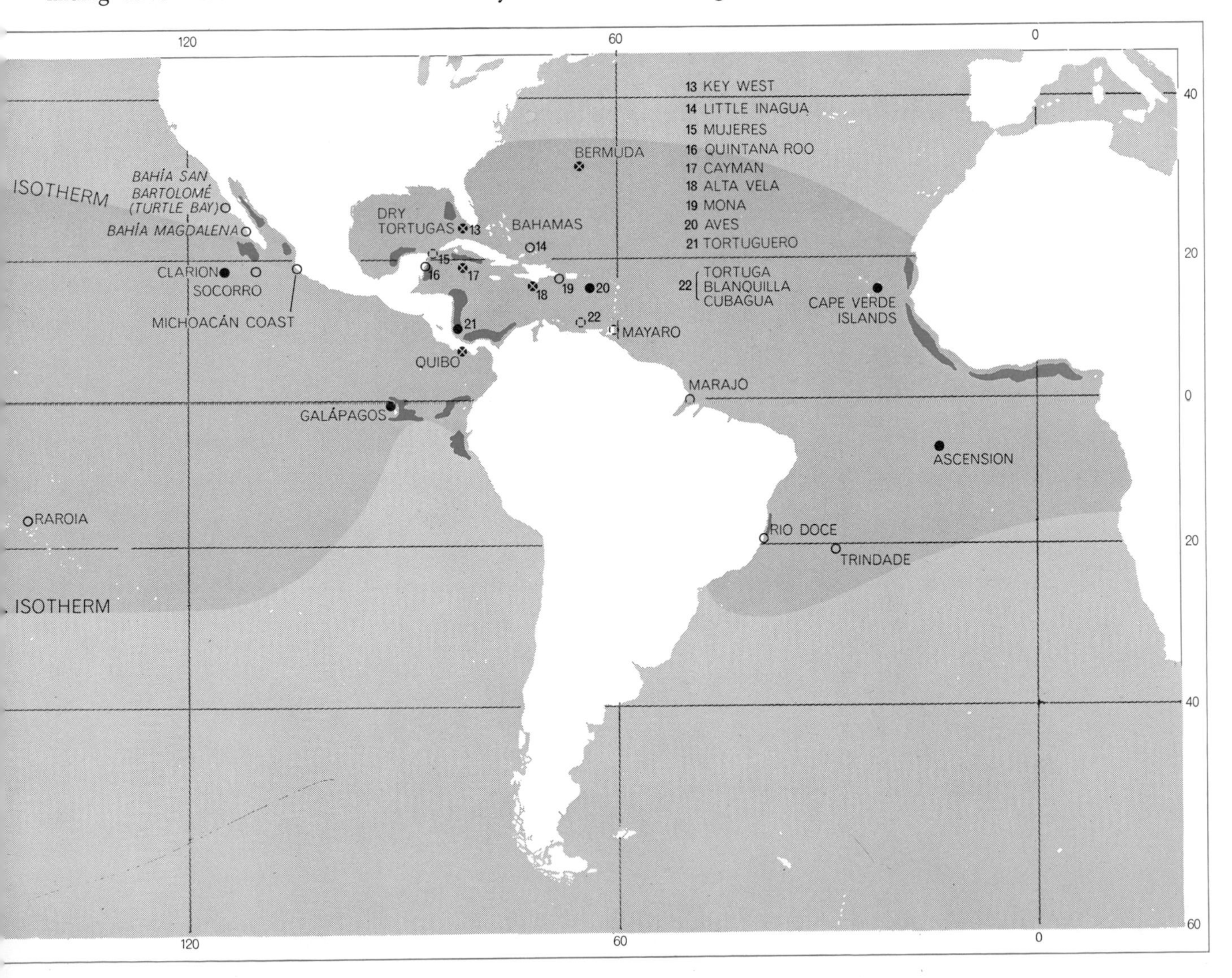

Berkeley. His task was complicated by the fact that many animals called green turtles in the literature are one of the four other kinds of sea turtle. The author has amended Parsons' 1962 data on the basis of fieldwork, including his own, done more recently.

GREEN TURTLE HATCHLINGS, with colored tags attached to their left hind flippers, scramble across the beach at Ascension Island, headed for the waters of the South Atlantic. In order to trace turtle migrations the author and his colleagues have tagged young and adult green turtles both at Ascension and in Costa Rica at the only nesting beach still frequented in the western Caribbean.

during the first year of life. They are mainly carnivorous at this age, but they are able to feed only on small, weak marine invertebrates. Such prey is scarce both at the nesting beaches and at the turtle-grass pastures that feed the grown turtles, and very young turtles are almost never seen in these places. It seems likely that the hatchlings must spend their first months moving from location to location at sea as growth qualifies them to feed on invertebrates of increasing size. All we really know is that the hatchlings disappear.

The only place in the Atlantic and Caribbean regions where we have been able to study green turtles of an age between hatching and maturity is off the west coast of Florida. There a migrating population of young turtles, ranging in weight from 10 to 90 pounds, shows up each April. At this stage most of them have become herbivorous, and they spend the summer months browsing in the turtle-grass flats between Tarpon Springs and the mouth of the Suwannee River. In November they move away to an unknown destination. These Florida visitors may have come from the Costa Rican nesting ground. We have no proof of this, however, because of the difficulty of devising a marking system that will survive the changes in size and proportion of the growing animals.

As recently as 10 years ago it was not definitely known that green turtles migrated long distances from home range to nesting beach. The first clear evidence of periodic long-range migrations came from the tagging program conducted at Tortuguero. During the past eight years 3,205 adult turtles have been tagged; 129 of these have later been recovered. Most of the tags have come to us from professional turtle fishermen operating off the coast of Nicaragua, but other recovery sites are distributed over an area more than 1,500 miles across at its widest point. The eight most distant recoveries are as follows: one from the Marquesas Keys off the tip of Florida, one from the northern coast of Cuba, four from the Gulf of Mexico off the Yucatán Peninsula and two from the Gulf of Maracaibo in Venezuela [*see illustration below*].

These returns furnish grounds for some generalizations that support the reality of migratory travel by *Chelonia*. One such generalization is that no turtle tagged at Tortuguero has ever been recovered there after the end of the nesting season. Another is that no turtle tagged at Tortuguero has ever been found nesting anywhere else. A third is that there is very little correlation between the time elapsed after tagging and the distance the tagged turtle traveled from Tortuguero. This strongly suggests that the turtles are not random wanderers but migrants following a fixed travel schedule between the nesting beach and their restricted home range. Otherwise the animals would tend to cover the same distance in the same time.

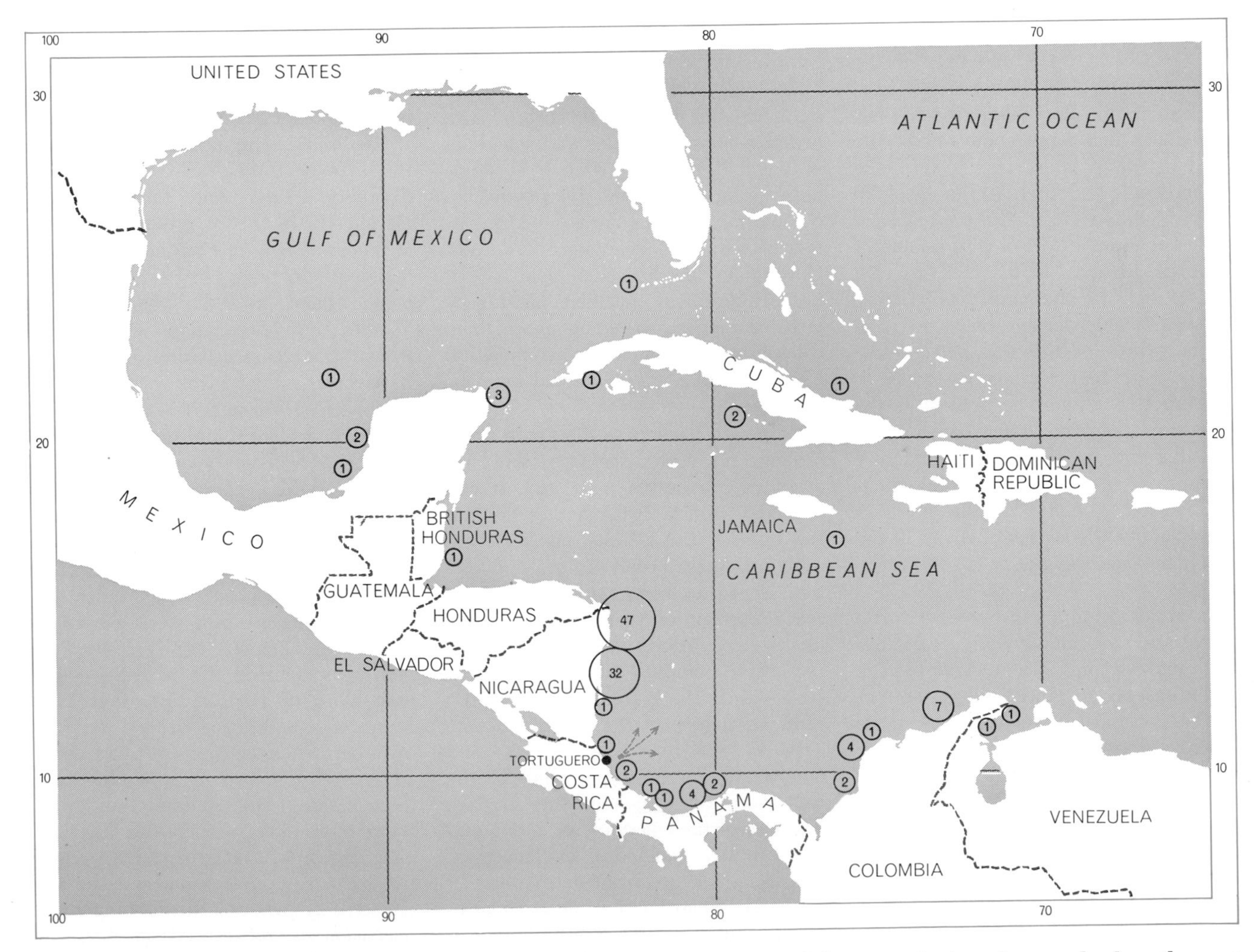

TAGGING AT TORTUGUERO, the nesting beach in Costa Rica, has marked 3,205 adult green turtles in the past eight years. So far 129 tags have been recovered. As the relative size of the circles indicates, most of the recoveries have been made along the coast of Nicaragua, but turtles from Tortuguero have also appeared as far away as Venezuela, the Gulf coast of Yucatán and Florida.

Although such findings make a strong case for periodic migration and homing by the green turtles that nest at Tortuguero, they do not prove a capacity for true navigation, that is, oriented travel involving something more than piloting by landmarks and an ability to keep headed in a fixed direction by using the sun or stars for a compass; such piloting is accomplished by many different kinds of animals. The beach at Tortuguero is part of a mainland shore. The green turtles could simply leave their distant pastures on an initially correct compass heading and, on making landfall, follow the coast until things look, smell or taste in ways that mean the ancestral breeding ground has been reached. In their successive nestings during a single season females often return to the same 200-yard stretch of nesting shore they had used earlier.

Indeed, it appears certain that once the turtles have reached the Tortuguero area, they do search the shore for cues to guide them to a nesting site. At the start of the nesting season distinctive "half-moon" turtle tracks appear, usually toward one end of the 22-mile length of nesting beach. These are semicircular or U-shaped trails left by females that have come out of the water, made a short trip toward the upper beach and then turned back to the surf without nesting. Such behavior implies that some sort of discriminatory process is involved in the selection of a nesting site. In the course of coming ashore a female nearly always stops in the backwash of the surf and presses her snout deliberately against the sand, sometimes repeating the process as she moves up the wet lower beach. This behavior appears to be an olfactory assessment of the shore, although it could also be tactile. Little more is known about the senses involved in the selection of nesting sites.

OCEAN-FINDING ABILITY is demonstrated by free-swimming green turtle hatchlings in a circular tank. Although the rim blocks any direct view, most of the turtles have gathered along the side nearest the water, guided by some difference in the light from that direction.

In searching for an instance of turtle migration that clearly involves an ability to make a long, oriented sea voyage in the absence of landmarks, I thought of the green turtle colony that nests on Ascension Island. The capacity for open-sea orientation is the ultimate puzzle in the study of animal navigation. Even human navigators, with their ability to measure the position of the sun and the stars, were unable to calculate positions at sea accurately until the development of precise chronometers in the 18th century. Work with various animals, notably homing pigeons and migratory birds, suggests that there are three inherent aids to navigation: a clock sense, a map sense and a compass sense. On the basis of these senses the navigation feats of animals that migrate overland can be explained, at least in theory. On the featureless open sea, however, the situation is quite different.

One difficulty in the study of open-sea navigation by animals is the dearth of information on routes and schedules of travel. I am not aware of a single instance in which the journey of an oceanic migrant has been traced in detail, with data on headings and speeds. Only when this has been done will it be possible to say that here piloting by visual guideposts is taking place, there celestial navigation can be assumed and elsewhere some cryptic signal must be involved.

Before we could use the green turtle nesting colony on Ascension Island for a study of open-sea navigation it was necessary to establish whether the turtles traveled there from some distance away or were merely a local population. That they were not local residents seemed certain on two grounds: (1) the turtles arrive to nest in substantial numbers each February but disappear by June and (2) there are no beds of turtle grass anywhere in the vicinity of Ascension. It was then necessary to find out where the feeding grounds of the Ascension nesting colony were located. A survey of the coasts of Argentina and Brazil in 1957 yielded no evidence that the abundant green turtle population of the Brazilian coast came from nesting grounds on the South American mainland. Turtles were known to nest on the island of Trindade, some 700 miles off the Brazilian coast, but their numbers were small. It therefore seemed likely that the green turtles that nest on Ascension feed along the coast of Brazil. The next step was to test the reality of this apparent instance of open-sea migration by means of a tagging program.

The green turtles at Ascension, like those at Tortuguero, breed in either two-year or three-year cycles. During February, March and April of 1960 Harold Hirth, then a graduate student at the University of Florida, tagged 206 female turtles at the six nesting beaches on the island. In 1963, when that part of the 1960 population which nests every three years was due to return to Ascension, a tag patrol was set up at the beaches. Three of the turtles Hirth had tagged were found again. In 1964, when the two-year group was due to return for the second time, two more of Hirth's turtles showed up [*see lower illustration on next page*]. Four of these five females had landed on the same short section of beach where Hirth had tagged them; the fifth went ashore on an adjacent beach recently formed by high seas. During the past five years nine more of Hirth's 1960 tags have been recovered from turtles captured by fishermen along the Brazilian coast.

A skeptical statistician might attribute these findings to random wandering. Short of tracking a turtle all the way, clinching proof that green turtles travel from Brazil to Ascension will be obtained only when a turtle tagged on Ascension is recaptured in Brazilian waters, released there and then captured again on Ascension. This is not likely to happen; by the time a tag reaches the University of Florida from Brazil the turtle that carried it has usually been eaten. Thus the evidence remains circumstantial: turtles tagged on Ascension have been captured off Brazil; others have disappeared for three or four years only to return to the same Ascension beaches on which they were first encountered. This does not prove the reality of the migratory pattern beyond all possible doubt, but I think it does so beyond reasonable doubt.

There are three basic questions to be asked about this particular migratory journey: How was it originally established as a behavioral adaptation? What route does it follow? What guides the turtles? With respect to the first question, the establishment of a nesting colony on a tiny mid-ocean island such as Ascension seems an evolutionary venture most unlikely to succeed. The problem is to visualize the selective process and the survival values in the earliest stages of the evolution of such a migratory pattern. If Ascension had once been a much more extensive area of land, the navigational equipment with which the green turtles now find the island could have been slowly refined by natural selection as the area shrank to a small island. I have found no one, however, who believes the area of Ascension can have become appreciably smaller during the past 50 million years or so. Certainly the water around it today is so deep that such a change seems unlikely.

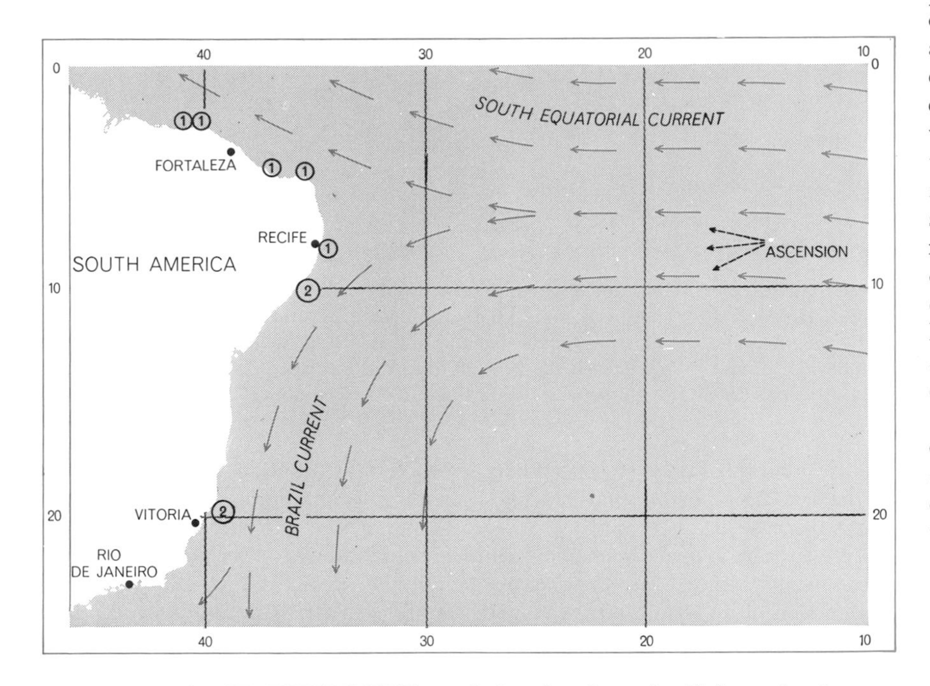

ASCENSION ISLAND EXPERIMENT was designed to determine if the turtles that nest there are the same animals that feed along the coast of Brazil, 1,400 miles or more away. In 1960, 206 turtles were tagged at Ascension; so far nine have been captured at the coastal points noted, proving that at least some Ascension turtles do travel to Brazil.

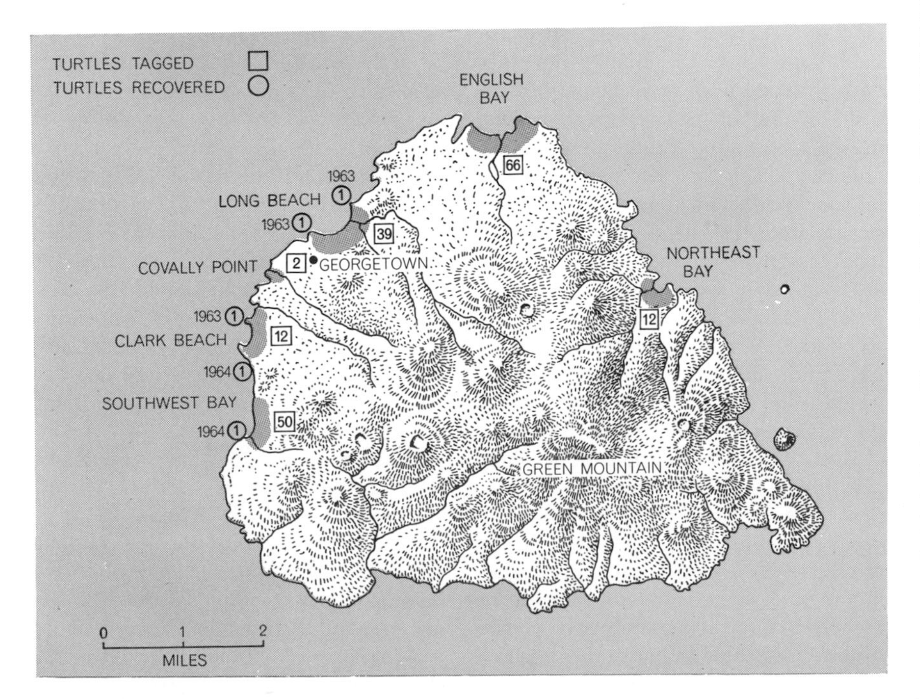

REPEATED NESTING AT ASCENSION has also been proved by the tagging experiment. In 1963 three of the turtles tagged in 1960 returned to the island and dug their nests in the same beaches they had used before. In 1964 two more tagged turtles appeared; one of them missed its home beach by a few hundred yards. All had presumably come from Brazil.

There is one possibility that may decrease the theoretical difficulty to some degree. Perhaps the island was originally colonized by green turtles that had been accidentally carried from West Africa by the South Equatorial Current, which flows from east to west. Another kind of sea turtle—the West African ridley (*Lepidochelys olivacea*)—has colonized the coast of the Guianas, north of Brazil, in just this fashion. If egg-bearing green turtles had landed by chance on Ascension and nested there, the process of selection would have had good material to work on. The hatchlings leaving the island would have been carried to the coast of Brazil by the South Equatorial Current and at the same time could have borne with them "imprinted" information that would help them to retrace their journey at maturity. Such a hypothesis of course does no more than get the nesting colony established on Ascension. It makes no attempt to solve the navigation puzzle.

The next question concerns the route followed from Ascension to Brazil and back. Apart from inference, nothing is known about this. The South Equatorial Current presumably carries the hatchlings to Brazil; the shortest path for mature turtles returning to the island would be directly eastward from Brazil against the thrust of the same current. Such a route would conform to a classic pattern for aquatic migrations: upstream movement for the strong adult animals and downstream travel for the weak and inexperienced young.

There are two other routes that would allow the entire round trip to be

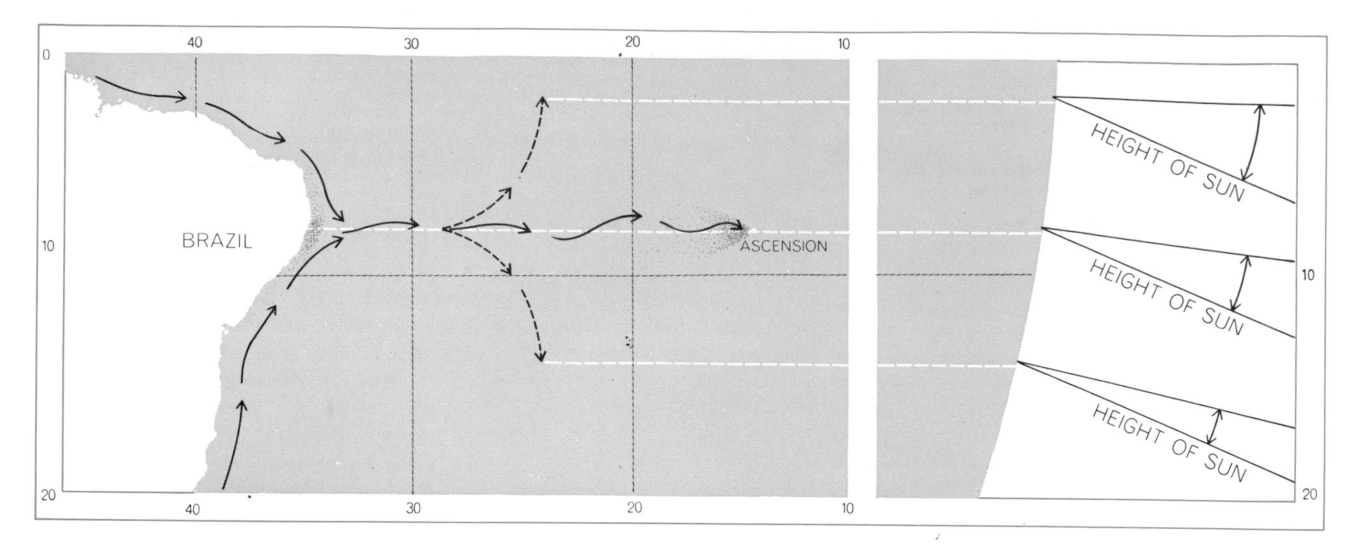

MEANS OF NAVIGATION available to the green turtles for their ocean voyage to Ascension remain conjectural. Distinctive chemical cues from the island, carried westward by the currents, may help the turtles to make final landfall; chemical or visual cues unique to the Brazilian coast may also guide them to the latitude of Ascension before they set out to sea. In the open ocean, however, some kind of guidance beyond a simple compass sense must be required if the turtles are to be able to correct for drift and reach the zone of chemical guidance. In this illustration the hypothetical cue is the sun's height above the horizon at noon.

made with the aid of favorable currents. If the Ascension-bound migrants swam northward from Brazil, they would enter the part of the Equatorial Current destined to become the Gulf Stream and could stay in this current until a full circle brought them westward to Ascension from the coast of West Africa. If instead they rode the Brazil Current southward, they could then travel the West Wind Drift to South Africa, catch the northward-flowing Benguela Current and rejoin the South Equatorial Current on its way to Ascension and beyond. The time required for these journeys seems prohibitive. Both would have to be negotiated without food, and the temperature of the West Wind Drift waters can drop to a chilling 40 degrees F. The direct easterly course, or some modification of it, appears to be the most logical route from Brazil to Ascension.

As for the final question, in attempting to judge how the Brazil-to-Ascension migration is guided the first step will be to determine how and where some sort of contact—direct or indirect—is established with the island. Even with a pinpoint target such as Ascension, migrating birds would be able to correct a fairly gross navigational error by visual means. To a turtle, however, an island is out of sight a few miles away. Green Mountain on Ascension stands 5,000 feet high and often has a corona of cloud that rises much higher. The sight of birds converging on the island from miles away at sea might provide an approaching turtle with still another visual guide. Such signals, however, would probably be picked up only in the last 20 miles or so. What cues might there be at longer range?

Very little is known about nonvisual phenomena associated with the presence or direction of an island, nor is anyone sure about what a green turtle can taste, smell or hear. It seems a point in favor of the upstream hypothesis of travel to Ascension, however, that such a route would allow the approaching turtles to detect an olfactory gradient—if indeed such a chemical cue is given off by the island. Perhaps when the hatchlings leave the island they take with them an imprinted memory of the taste or smell of Ascension water. Coming back as mature adults they may be able to detect this Ascension effusion in the westbound current far downstream from the island and to follow it until they make a visual landfall. There are two basic weaknesses in this proposal: no one knows how far downstream a green turtle can taste or smell an island, and there is no information to indicate the direction from which the migrants approach Ascension.

Even if we assume that far-reaching cues guide the final homing of the turtles to Ascension, there remain hundreds of miles of open ocean to be crossed—and crossed with precision—before such signs could conceivably be detected. For the greater part of this distance the animals must be navigating, and the most logical assumption appears to be that the navigating is done with information from celestial guideposts. Colin Pennycuick of the University of Cambridge has recently called my attention to an old island-finding technique used by human navigators before the accurate calculation of longitude was possible. Knowing the latitude of his target, the navigator would sail north or south by compass until the noon position of the sun showed that he had reached the desired latitude; he would then simply sail due east or west (depending on the location of his target) until he made landfall.

This human technique seems particularly worth considering in connection with the migration of green turtles to Ascension. Possibly the turtles go north or south along the Brazilian coast until they reach the vicinity of their first landfall as hatchlings. Visual or olfactory impressions from this first contact with the mainland would identify that coastal point, which should be close to the latitude of Ascension. Thereafter compass sense alone might conceivably guide the turtles on a journey due east to a point where cues from the island could be detected.

This proposal too has weaknesses. Any drifting due to wind or current during the hundreds of miles of open-sea swimming could put the migrant hopelessly off course in spite of a constant compass heading. Unless the turtle were able to make corrections by one means or another the landfall at Ascension would never take place. For human navigators today the ability to make such corrections requires finding not only latitude but longitude as well. Latitude can be judged from the height

RADIO-EQUIPPED TURTLE was used by the author to test the feasibility of tracking movements at sea from shore. To track the green turtles' long journey from Brazil to Ascension Island might require such advanced techniques as satellite-relayed telemetry.

of the sun at noon, but reckoning longitude requires the measurement not only of some celestial body's altitude but also of its azimuth. It is hard to see how this can be done effectively by an animal in the open sea; the azimuth measurement—the horizontal component of the movement of the sun and the stars—could not be made against a featureless horizon. Pennycuick's suggestion nonetheless has hypothetical merit in regard to the Ascension migration. Eastbound turtles that were displaced north or south during the journey could conceivably correct for drift on the basis of latitude reckoning alone. Even this, of course, would be an astounding animal adaptation.

It is clear that an adequate analysis of the travel orientation of the green turtle will require tracking the animals throughout entire migratory journeys. At the University of Florida we have made preliminary tracking tests in which a turtle tows a float from which a helium-filled balloon rises to mark the position of the migrant. This seems promising for short runs. With the support of the Office of Naval Research we are trying to work out procedures for tracking longer trips by mounting a radio transmitter on the back of a turtle [*see illustration at left*], on the floats or preferably on the balloons. The length of the journey from Brazil to Ascension, however, has so far made the task of keeping in touch with these migrants a formidable one.

This may soon no longer be the case. The National Aeronautics and Space Administration intends to load apparatus for a number of scientific experiments aboard space vehicles connected with the Apollo program. Tracking the Ascension migrants by satellite could easily prove to be the most efficient method of learning the route they follow. The green turtle could without inconvenience tow a raft, bearing a radio transmitter and power source, for long distances. Each time the satellite passed within range of the towed transmitter a signal would be received; these signals, rebroadcast to a control station, would allow a precise plotting of the position of the turtle.

One key experiment making use of such facilities comes quickly to mind. Several radio-equipped turtles could be released a few hundred miles east of Ascension, where because of the prevailing current no chemical cues from the island could possibly be present. If the turtles nonetheless made their way to the Ascension nesting beaches, this would prove that their feat of navigation is not chemically guided. If satellite-tracking can pin down facts such as that, what may be discovered about animal navigation in general should abundantly repay the effort.

28

THE BEHAVIOR OF SHARKS

PERRY W. GILBERT
July 1962

Among the many "crash" programs of applied research sponsored by the U.S. armed forces during World War II was the effort to develop a shark repellent. The aim, of course, was to protect military personnel who might be cast away at sea from one of the more unpleasant hazards of that experience. Investigators soon had encouraging results to report. They were able to show that decomposed shark flesh or, alternatively, copper acetate would inhibit the feeding activities of smooth dogfish, a harmless species, and certain other sharks. Whether or not the repellents that incorporated these findings have ever protected anyone from a shark attack is uncertain. During the past five years, however, it has been demonstrated that the odor of decomposed shark flesh or of copper acetate has no inhibitory effect whatever on the behavior of the species native to the Caribbean Sea, nor on the behavior of those Pacific Ocean sharks on which they have been tested.

A year or two ago people responsible for the safety of bathers at beaches found reason to hope that the device called the bubble curtain might keep sharks away. The curtain, set up offshore by leaking compressed air from perforations in a pipe or hose, was supposed to present some sort of sensory or psychological barrier to approach by sharks. Unfortunately, when it was tried out on sharks in a laboratory pen, it proved to be no barrier at all [*see top illustration on page* 273].

These two stories may serve to remind bathers, skin divers, small-boat sailors and others who venture into the ocean that there is as yet no sure protection from sharks in open water. They serve also to emphasize the practical motivation of a field of investigation that has been attracting increasing interest in this country and abroad in recent years. It has long been suspected that sharks possess a remarkable ability to locate their prey, often at a considerable distance. Study has accordingly been focused on the sensory organs that direct their predatory behavior. Although all sensory systems undoubtedly come into play, it appears that three are particularly involved. They are the familiar senses of smell and vision and the "vibration sense," peculiar to fishes and aquatic amphibians, that is embodied in the

LEMON SHARK PREPARES TO ATTACK a 150-pound chunk of blue marlin by circling it. The shark, which is about nine feet long, was photographed by the author from an underwater cage at the Lerner Marine Laboratory on the island of Bimini in the Bahamas.

lateralis system, located on the head and along each side of the body. These organs in the shark show a high degree of elaboration and specialization. Investigation of the feeding behavior of sharks, progressing in parallel with study of their anatomy, has now begun to yield the understanding that must come before control.

Although the anatomical work can be done in the laboratory, observation of a large shark requires facilities in scale with the formidable dimensions of the animal. At the Lerner Marine Laboratory of the American Museum of Natural History on the island of Bimini in the Bahamas the Office of Naval Research has provided two spacious pens, each measuring 40 by 80 feet; these make it possible to confine sharks up to 15 feet in length under conditions that reasonably approximate those of the open sea. In a small central pen a shark selected for study can be trapped and brought alongside the dock with the help of a net and an electric hoist. After the shark has been anesthetized by having its gills sprayed with an anesthetic known as M.S. 222, it can be lifted out of the water for whatever preparation the experiment requires. For about 20 minutes, without danger to either the investigator or the shark, the animal can be laid out on the dock for such a purpose as fixing plastic shields over its eyes to deprive it temporarily of sight.

In our work at the Lerner Laboratory my colleagues and I have found that it is unsafe to generalize from observation of a single shark or even a single species of shark. When we tested the effectiveness of the bubble curtain, for example, we found that it did stop one of a group of 12 large tiger sharks, whereas the other 11 swam heedlessly through it. We have repeatedly noted that six or seven lemon sharks in a group exhibit greater interest in the bait presented and attack it more vigorously than does a solitary animal. Our experience shows that it is important to learn as much as possible about the normal feeding behavior of a species before attempting to manipulate it experimentally. The knowledge that lemon sharks feed more actively during the evening and at night, and that they may cease feeding altogether for days at a time when the temperature of the water drops below 65 degrees Fahrenheit, can be crucial to the evaluation of a whole series of experiments. For these reasons we work not only with solitary sharks but also with groups of sharks of the same species and at different times of the day and year before we attempt to draw conclusions about their behavior.

Of the shark's three principal sensory systems, the sense of smell has long been regarded as the most acute; indeed, the shark has been described as a "swimming nose." For this characterization there is considerable anatomical support. The nostrils, located on the underside of the flat snout just ahead of the mouth, open into capacious cups, or sacs, lined with folds of tissue. Since the tissue con-

SHARK ATTACKS its prey by braking its rapid forward motion with its large pectoral fins, tilting its head and body upward and opening its jaws wide. Next it drops both upper and lower jaws to secure a firm grip on its prey and shakes its head and body violently from side to side until it has torn off a 10-to-15-pound chunk of tissue. Then, still shaking its head violently, it swims away quickly.

tains the olfactory cells, the folds greatly enlarge the sensory surface. When a shark takes water into its mouth to aerate its gills, suction causes some water to flow in and out of each olfactory sac. In addition the forward motion of the shark brings water through the funnel-shaped nostrils into the sacs. A fleshy flap extending across the portal of each sac separates the inflowing stream from the outflowing. Thus the olfactory system of a shark is constantly bathed by a current of water whether the shark is at rest or in motion. In one of the oddest of all sharks, the hammerhead, the nostrils (as well as the eyes) are located far apart at the ends of the "hammer." This anatomical fact, coupled with the hammerhead's habit of swinging its head from side to side through a considerable arc as it swims, may lend enhanced directionality to the species' sense of smell.

Ralph E. Sheldon, working at the Marine Biological Laboratory in Woods Hole, Mass., was the first to demonstrate scientifically that the sense of smell is important in guiding sharks to a meal. He observed that smooth dogfish had no difficulty distinguishing a cheesecloth packet that contained crushed crabmeat from identical packets containing stones. When Sheldon plugged the sharks' nostrils with cotton so that a current of water no longer washed the olfactory sacs, they no longer "homed in" on the food packets, although they swam quite close to them. To obviate the possibility that the cotton plugs merely rendered the sharks uncomfortable and so discouraged them from eating, Sheldon plugged the nostril on one side only of several dogfish; after a brief period of adjustment all but one of the animals readily located the food packet.

We have performed similar experiments at Bimini. To healthy adult and subadult lemon sharks, five to nine feet in length, we presented four identical perforated cans, only one of which contained chunks of fresh bonito or tuna. The sharks approached and circled the baited container five or six times oftener than they did all three unbaited ones together. When we plugged their nostrils with cotton dipped in a dilute anesthetic, they no longer showed any such preference.

AUTHOR SPRAYS mouth and gills of a lemon shark to prepare it for experiment. It takes about a minute to anesthetize a 500-pound shark. This photograph and the one at top of opposite page were made by Peter Stackpole and are reproduced through the courtesy of *Life*.

Obviously the distance at which a shark can smell an odorous substance depends on conditions in the water as well as on the shark's acuity of smell. If a strong current prevails, there is evidence that a shark may detect an odor a quarter of a mile from its source. It is a question of the degree to which the odorous substance is diluted in the water. In the case of salmon, it has been shown that they can detect a substance in dilutions down to one part in several million. We have seen lemon sharks detect small fluid samples of freshly caught tuna (one of the strongest shark attractants we have found) placed upcurrent at a distance of 75 feet. Given the speed of the current and the probable dimensions of the "olfactory corridor," as indicated by dyes released in the water, we estimated that the tuna juice must have been diluted to one part in 1.5 million.

Recently Albert L. Tester of the University of Hawaii reported that blacktip sharks and gray sharks, two Pacific species, showed a mild response to water siphoned into their tank from another in which fish were swimming. When these fish were agitated, the sharks showed a marked attraction to the siphoned water. This suggests that a frightened or distressed fish gives off some substance in sufficient quantity to attract sharks to it. As in Ernest Hemingway's *The Old Man and the Sea,* big-game fishermen frequently experience the frustration of losing a prize fish to attacking sharks. In these cases, however, sight and the vibration sense may play an equal role.

It has been postulated, in fact, that it is the lateralis system that enables sharks to locate such disturbances as the torpedoing of a ship from a long way off. The mass attacks by sharks that have occurred in connection with sea disasters suggest that the sharks in a wide area of the surrounding sea may somehow be attracted to the site. In sharks the lateralis system is large and well developed. It consists of fine canals, filled with watery solution, that are just beneath the skin of the head and along both sides of the body. (In the primitive frilled shark the lateral lines appear as open grooves.) The canals are connected at intervals to the surface through tiny

AUTHOR EXAMINES the eye of an anesthetized mako shark with an ophthalmoscope. The fish can safely be kept under anesthetic for as long as 20 minutes. It has been lifted horizontally from the water by an electric hoist and placed on the dock for examination.

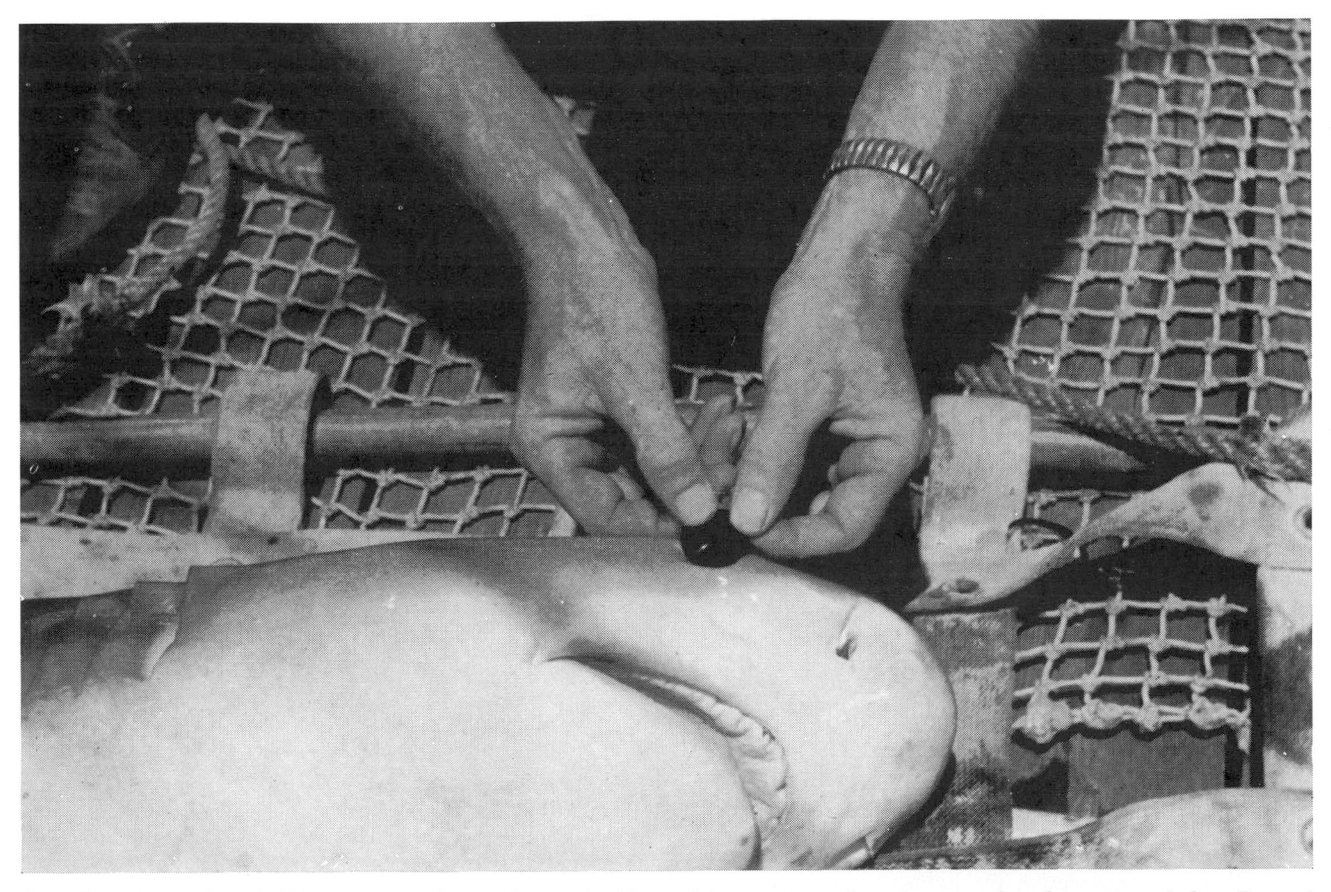

BLACK PLASTIC EYE SHIELD is inserted into a lemon shark's eye by the author. After the other eye had also been occluded, the fish was returned to the water for observation of the effects of blindness on its ability to locate chunks of bait placed in its pen.

tubules and pores. Clusters of sensory cells called neuromasts are arrayed along the inner surface of the canals. From the neuromasts bunches of tiny hairlike processes reach into the fluid that fills the canal. This structure plainly suggests that any movement of the fluid must cause the neuromast processes to move ever so slightly and thereby trigger the firing of a nerve impulse. Movement of the fluid could well be induced by a mechanical disturbance in the water such as the erratic movements of a wounded fish or the splashing of a human swimmer.

Nearly half a century ago George H. Parker of Harvard University found that a shark deprived of the senses of sight and of hearing would continue to respond to a source of disturbance in the water as long as its lateralis system remained intact. When Parker severed the nerve-trunk connections of the lateralis system, however, the shark ceased to respond.

Recently Otto E. Lowenstein of the University of Birmingham suggested that the system may serve to "echo-locate" objects by measuring the time relations of reflected vibrations set up by the swimming movements of the fish itself. This would be a particularly useful faculty in water that is too turbid or dark for the fish to see in. Just how sensitive the lateralis organs are and at what distances they can detect disturbances are unanswered questions that confront the ingenuity of the investigator.

Against the argument still heard from some quarters that vision plays a minor role in the feeding behavior of sharks, the anatomy of the shark's eye presents conclusive evidence. In its basic design this eye is a somewhat flattened version of the standard vertebrate eye, with iris, lens and retina and three fluid-filled chambers contained within the tough cartilaginous envelope of the sclera. Since the retina of the shark has no cone cells (except in one species), it can be concluded that the shark sees no color and has a vision of low acuity. On the other hand, the shark's retina is abundantly supplied with rods, which give it high sensitivity to contrasts of light and shadow and to motion. This sensitivity is greatly amplified by an extraordinary structure, the *tapetum lucidum:* a mirror-like reflecting layer that underlies the retina. Made up of tiny plates silvered with guanine crystals, the tapetum reflects incoming light back through the retina, thereby restimulating the light-sensitive rods. The tapetum thus helps

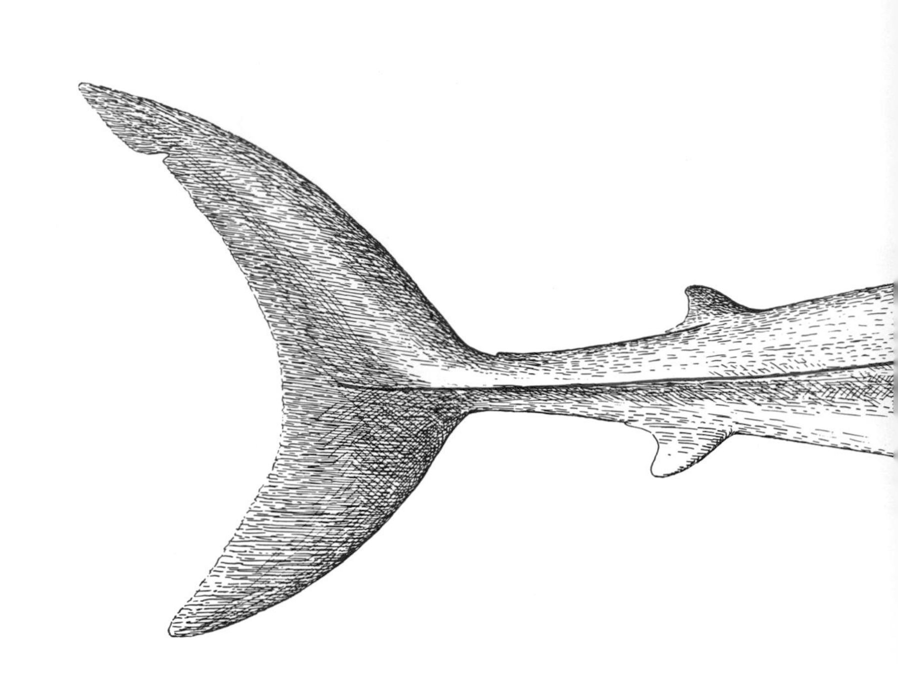

MAJOR SENSORY SYSTEMS guiding the predatory activities of the shark are the visual, the olfactory and the lateralis system, or vibration sense. These three are indicated in color

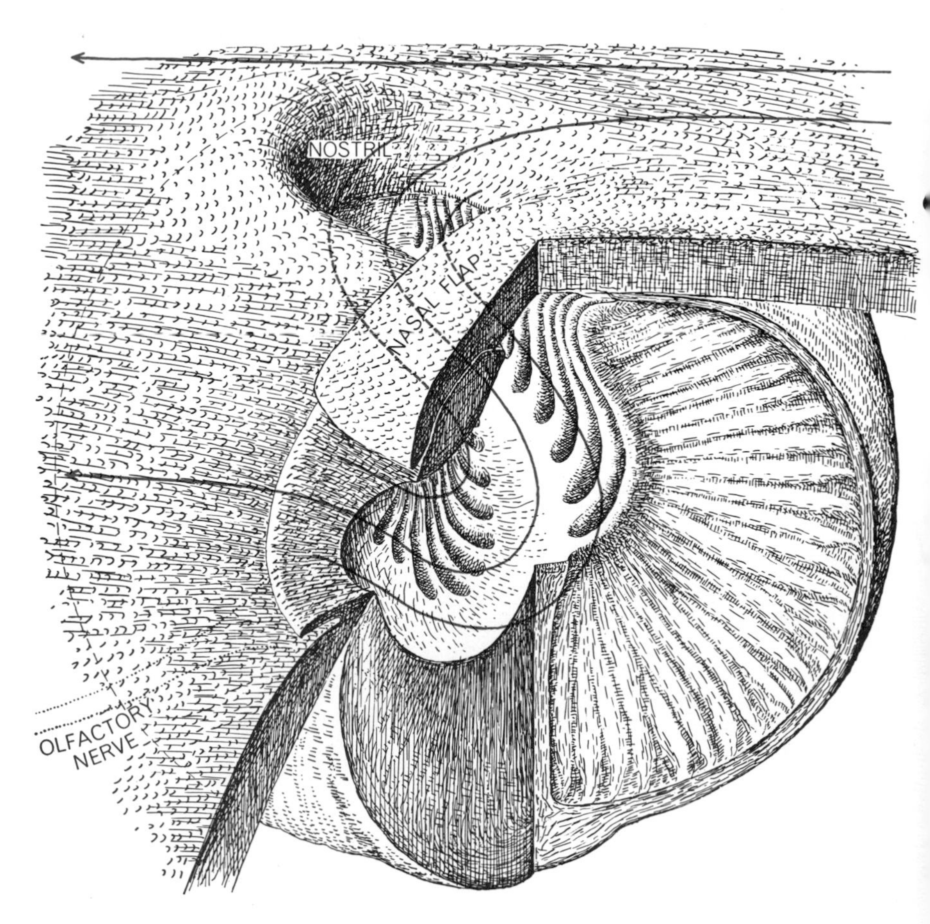

OLFACTORY AND VISUAL SYSTEMS of the shark are outlined. Each nostril opens into an olfactory sac (*a*), which is connected with the olfactory nerve. The sac is lined with folds of tissue containing the olfactory cells. These folds increase the olfactory surface and thus enhance the shark's sensitivity to the odors carried on the currents of water that bathe the sac. A fleshy flap (*b*) across the entrance to the sac divides the nostril, separating the

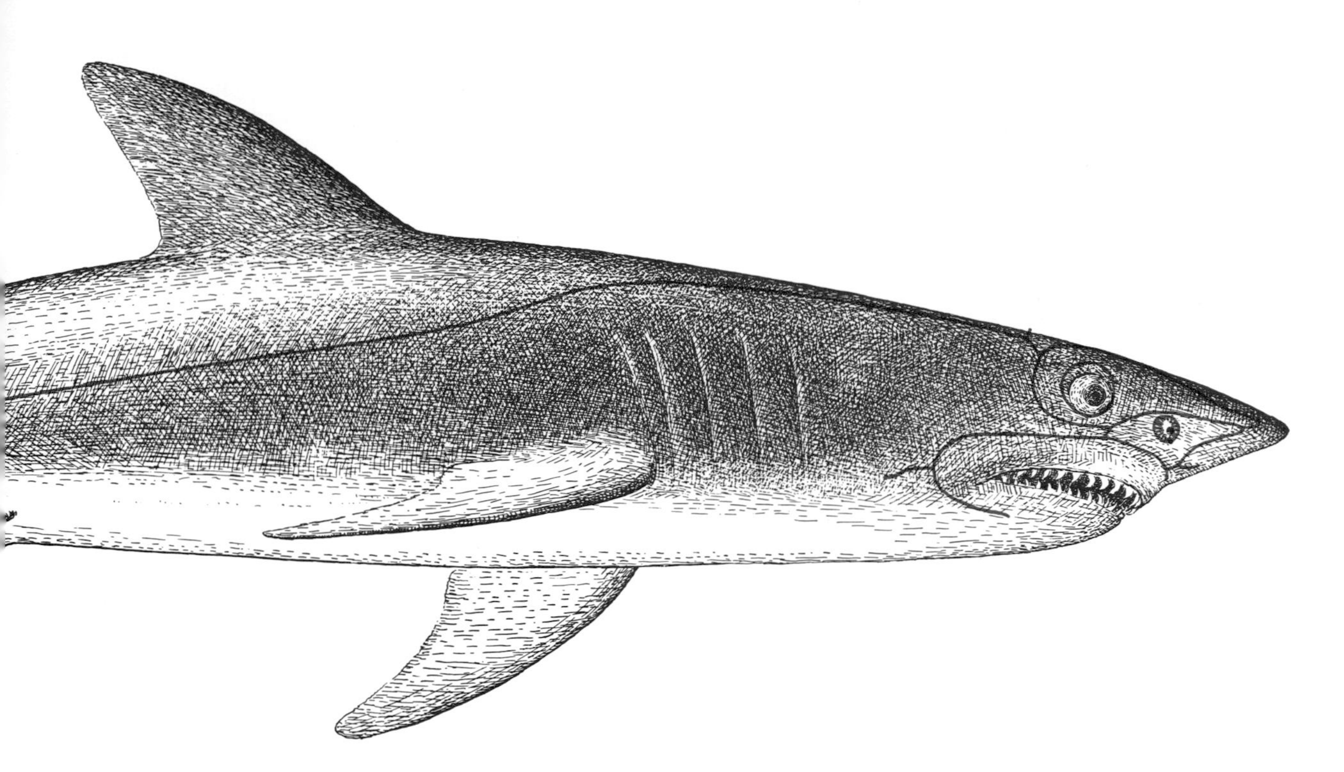

in this drawing of a mako shark. The visual and olfactory systems are localized in the head. The lateralis system has branches both in the head and along each side of the body. All of the shark's sensory systems are connected with appropriate centers in the brain.

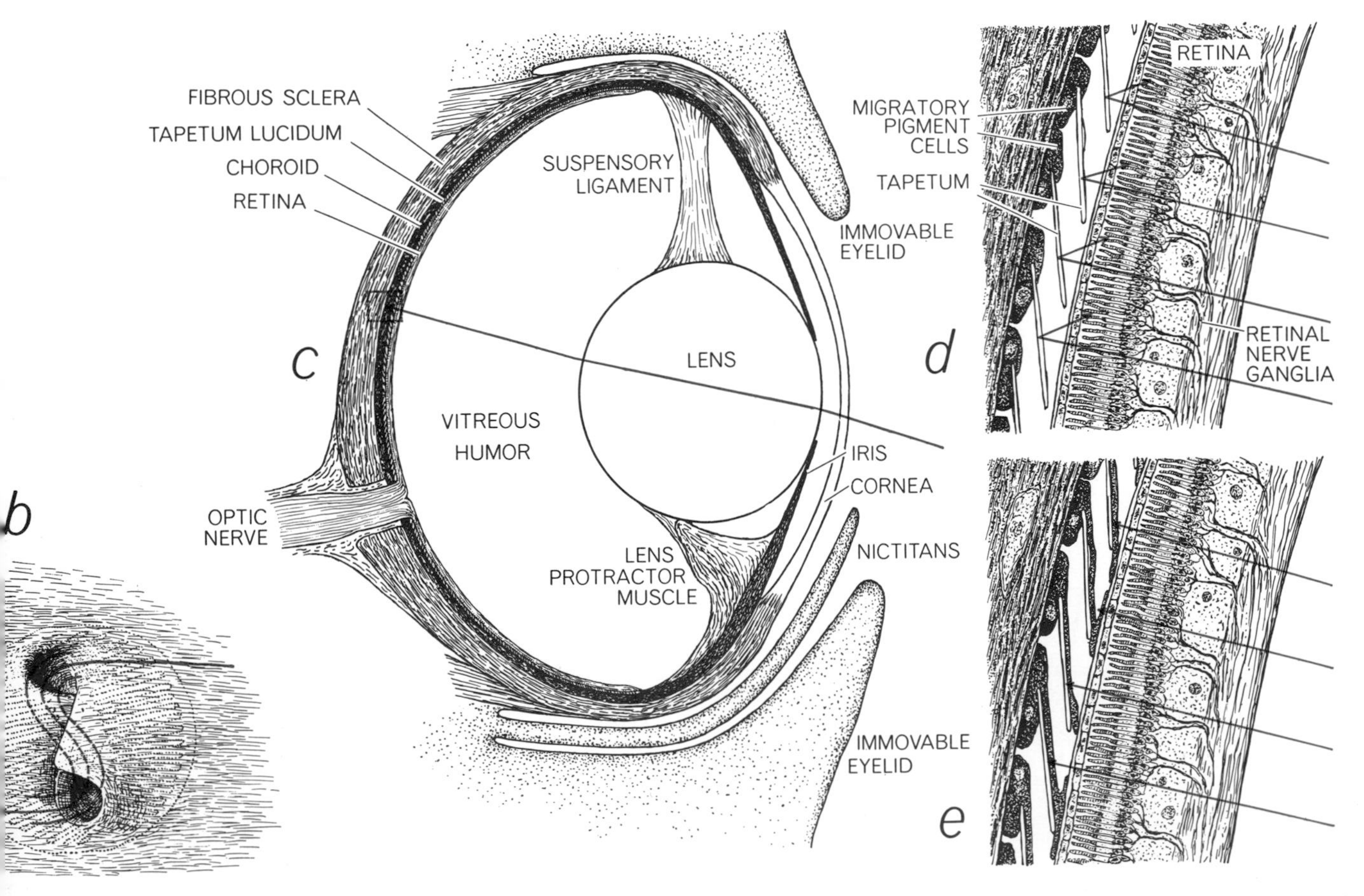

inflowing and outflowing water. The colored lines show this current flow. In design the shark's eye (*c*) is similar to the standard vertebrate eye. Most species have a movable eyelid, the *nictitans*, which protects the eyeball. The retina of only one species has cones, but all have many rods. Sharks therefore have high visual sensitivity. Behind the retina is the *tapetum lucidum* (*d*), a series of plates silvered with guanine crystals, which reflects light back through the retina. Many species of sharks have pigment cells in front of the tapetum that expand to cover the plates when it is light (*e*) and contract when the shark becomes dark-adapted.

shark species that feed at night or in very deep water make the most of the scant amount of light entering the eye.

In the species that feed during daylight hours this remarkable anatomical adaptation is compensated by another one: a curtain that temporarily occludes the tapetum. Pigment-containing cells expand by reflex over each tapetal plate as the shark moves into bright light. Conversely, as the shark becomes dark-adapted, the pigment cells contract, thereby uncovering the reflecting surface. The movement of the pigment curtain can be seen by looking into the dark-adapted eye of an anesthetized shark with an ophthalmoscope. In some shallow-water species the lower half of the tapetum, which faces upward toward the light, is permanently occluded. Only the upper portion of the tapetum, aimed toward the bottom over which the shark is swimming, reflects light back to the visual cells of the retina.

Eugenie Clark of the Cape Haze Marine Laboratory has recently tested the ability of an adult lemon shark to distinguish visually between targets of various shapes. She presented the fish with a diamond and a square, rewarding it with food when it pressed its snout against one and punishing it by bumping its snout when it pressed the other. Although both targets were white, the shark had little difficulty in learning to discriminate between them. It did have difficulty, however, when presented with a circular and a square target, both white; apparently it could not clearly distinguish these differences in shape. On the other hand, responding to difference in brightness, the shark learned to distinguish between red and white targets of the same shape fairly easily.

Our experiments at Bimini have clearly demonstrated the importance of vision in the feeding behavior of the sharks. A shark that can see has no trouble locating bait. On the other hand, sharks temporarily blinded by opaque plastic eye shields have great difficulty finding their food. Vision may not be a major factor in guiding sharks that swim in turbid waters. In very clear water such as prevails at Bimini, however, vision becomes important as soon as the shark is within 50 feet of a stationary or slowly moving lure. Thereafter, as the distance from the lure diminishes, the importance of vision increases. By the time a shark is 10 feet from its prey its sense of sight is probably its principal guide.

The sense of sight, in fact, is the target of what appears to be the only effective, although not universally practicable, shark repellent, the discovery of which was an inadvertent by-product of the wartime experiments with decomposed shark flesh and copper acetate. In some of these experiments nigrosine dye was added to the other ingredients. The repellent incorporating the dye was found to be effective against many species of shark. Indeed, the dye is effective entirely by itself. If an open bottle of nigrosine dye is placed in a circular tank with a free-swimming shark, the dye will gradually diffuse through the water, coloring it black. As the dye spreads from the center, the shark alters its pattern of

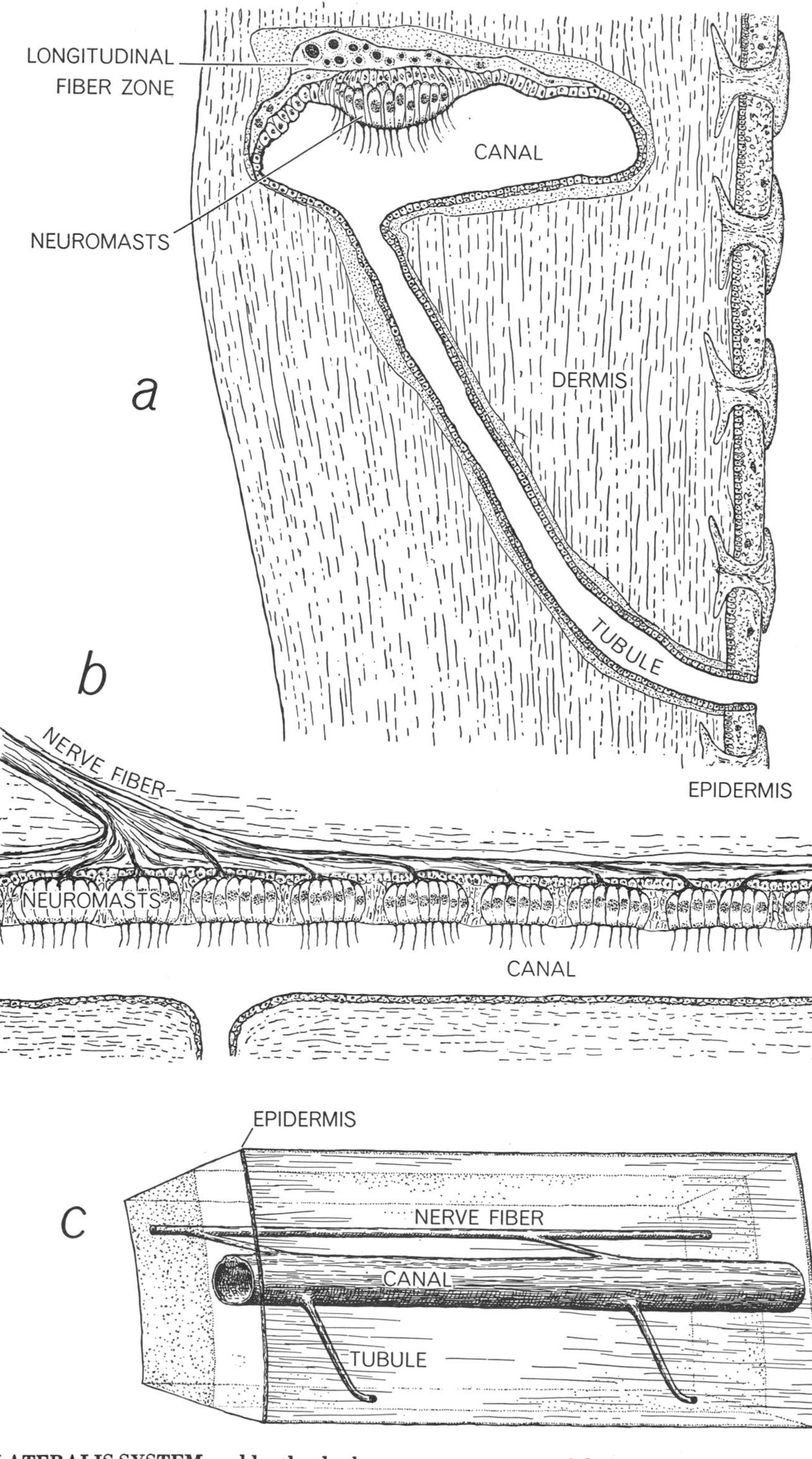

LATERALIS SYSTEM enables the shark to sense movements of the surrounding water. The system comprises a series of fluid-filled canals in the head and along each side of the trunk. The canals, seen here in three views, lie beneath the skin and open to it through small tubes. Neuromasts, hairlike processes connected to the nervous system, extend into the canals from their inner walls. Movement of the fluid in the canals, produced by disturbances in the water outside, causes the neuromasts to move and thereby triggers release of a nerve impulse.

BUBBLE CURTAIN, proposed as a means of protecting swimmers against shark attacks, is produced by pumping compressed air through a perforated hose. Experiments with tiger sharks at the Lerner Laboratory showed the bubble curtain to be ineffective.

GROUP OF LEMON SHARKS preparing to attack a 450-pound chunk of marlin was observed from an underwater cage. One fish moves in for the first bite. Then the others follow, each becoming more agitated until what is known as a "feeding frenzy" develops.

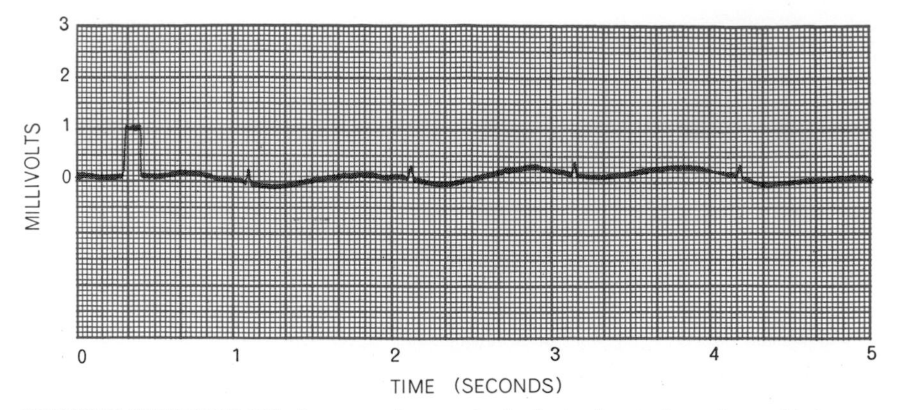

ELECTROCARDIOGRAM of a young lemon shark shows that its heart beats 50 times a minute. The first upward deflection standardizes the machine's response to one millivolt of current. Each subsequent deflection represents a ventricular excitation of the shark's heart.

SHARK swims freely in a circular pool while its electrocardiogram is made. These experiments represent the first time such measurements have been made of free-swimming sharks.

swimming to avoid the dark area, restricting its movement to the spots that remain clear. By the time the dye bottle is completely empty, the shark has been penned in a small, crescent-shaped segment of the tank, near one side. This effect is not produced by the smell of the dye; a shark with plugged nostrils will also avoid the dyed areas of the tank. If, however, a shark is fitted with opaque eye shields, it will swim straight through the dyed water.

In the large pens at Bimini we are now studying, with the help of underwater motion-picture photography, the behavior of sharks at the moment they attack and consume their prey. When a large dead fish such as a 400-pound blue marlin is offered, lemon sharks first slowly circle it at a distance of six to 10 feet. Then, as they swim faster, the circle tightens and presently one shark moves in for the first bite. Contrary to popular belief, the shark seldom rolls on its side. Braking its forward motion with its large pectoral fins, the shark points upward slightly as its mouth makes contact with the bait. It opens its jaws wide, the lower jaw dropping downward and the upper jaw protruding markedly from beneath the thin upper lip. If the first bite does not give the shark an adequate hold, it bites a second and third time until it anchors its teeth deeply. Then it closes its jaws and shakes the entire forward part of its body violently from side to side until it has torn 10 to 15 pounds of tissue from the marlin. Still shaking its head vigorously from side to side, the shark swims quickly away.

As the blood and body juices of the marlin flow from the wound, the other sharks in the pack become more and more agitated and move in rapidly for their share of the meal. Frequently three or four sharks will attack the marlin simultaneously. A wild scene, sometimes called a "feeding frenzy," now ensues. The behavior of the animals appears to be determined entirely by the visual sense. An observer can substitute tin cans and wooden boxes for the marlin, and the sharks will indiscriminately attack and consume them. It is clear that any effective shark repellent will have to take effect long before the animals go into a feeding frenzy. Research on living sharks is, however, still relatively young. Although there is much to be learned about these remarkable fish, the work now under way in many parts of the world holds out hope that new and better methods can be found to protect man from attack by sharks.

WHALES, PLANKTON, AND MAN

WILLIS E. PEQUEGNAT
January 1958

Among the "whale statements" in the prefatory pages of *Moby Dick* we find a quotation from Obed Macy's *History of Nantucket* which reads: "In the year 1690 some persons were on a high hill observing the whales spouting and sporting with each other, when one observed; there—pointing to the sea—is a green pasture where our children's grandchildren will go for bread."

Many people who are actively concerned with mankind's food problem think it is high time we made the Nantucketer's prophecy good. Of course to bake literal loaves of bread from seaweed would take some doing. But the whale feeds upon the plants of the sea only indirectly (by eating small herbivorous animals), and what is food for whales could also be food for man. The whale bears compelling testimony to the abundance of the ocean pasture. The sea is estimated to be as productive of organic food, acre for acre, as the land. With 70 per cent of the globe covered by oceans, in the aggregate the sea must produce five to 10 times as much living matter as the land. Yet we take only 1 per cent of our food from this source. Considering that half the world population lives on the edge of starvation and that we must feed 100,000 additional mouths every day, the sea's food potential cannot be neglected much longer.

We could, to begin with, increase our harvest from the sea by more intensive fishing for the conventional game. But I want to propose here a new kind of fishing which could be far more productive. My proposal admittedly is based on some purely theoretical deductions. But even if these deductions are only partly borne out, we may be able to open up at once a vast source of palatable high-protein food.

The whale's pasture is the upper, sunlit layer of the open ocean. The plant life of this zone consists mainly of microscopic floating organisms—the plants of the plankton. Principal among these plant organisms are the geometrically exquisite single-celled diatoms, housed in snowflakes of silica, and the whirling dinoflagellates, whose perpetual motion keeps them from sinking out of reach of the sunlight. These simple algae conduct the major portion of the photosynthesis that goes on in the ocean. They are the "grasses" of the sea. They synthesize carbohydrates, fats and proteins from carbon dioxide, sea water and certain nutrients. Under optimal conditions of light, temperature and nutrient supply, they may multiply to hundreds of thousands of cells per liter of water, clouding the water with a soupy "bloom"—yellow, green, red or brown, depending upon which species is dominant. At such times their prodigious capacity for multiplication becomes self-limiting, as they use up the supply of some critical nutrient or reduce the penetration of sunlight. More commonly the multiplication of the plant plankton is limited by the predation of the plankton animals that live upon it.

The smallest of these animals are single-celled protozoa which under the microscope look like tiny mollusks, encased in mineral shells of exotic design. The largest are wriggling crustaceans resembling small shrimp. Between these extremes of size is an endless variety of creatures, including the larval stages of many species of fish and of bottom-dwelling mollusks. Not all of these animals feed directly on the plants of the plankton; many of them prey on one another, the larger consuming the smaller. Finally the largest of the plankton animals are devoured by fishes, birds, seals and certain whales.

At each step in this food chain we must reckon with a considerable loss of organic matter. Only a small fraction of what an animal consumes goes to make up its substance. The rest is dissipated as energy in various forms, a large part of it being expended as mechanical energy in the creature's pursuit of its food. In general, 80 to 90 per cent of the organic matter is lost at each step. In the sea, as on the land, the hierarchy of predators forms a sharply narrowing pyramid. With 1,000 pounds of plant life at the base it shrinks to 100 pounds of animal plankton, then diminishes to 10 pounds of fish and is capped at its peak by only one pound of, say, sea lion.

Such calculation at once suggests that man might gain a great deal by short-cutting the food chain, perhaps going so far as to harvest the plant plankton directly. Except in a few regions, however, the plant plankton is too diffusely dispersed in the water. As a practical matter we need some animal to concentrate it for us. The ideal would be a plankton animal of appreciable size with a high rate of reproduction.

Now the whale makes exactly these requirements of its food supply. To achieve and maintain its huge bulk it must find its food close to the base of the pyramid. It is no accident that the most abundant species of whales, the baleen group, feed almost exclusively on a plankton animal. The baleen whales, among which is the blue or sulfur-bottom whale, the largest animal that has ever existed, are equipped with a mouth strainer, in the form of ingenious horny plates with fringed edges. When they scoop up their food in great mouthfuls of sea water, the strainer lets the plant plankton and small plankton animals escape but holds in the largest

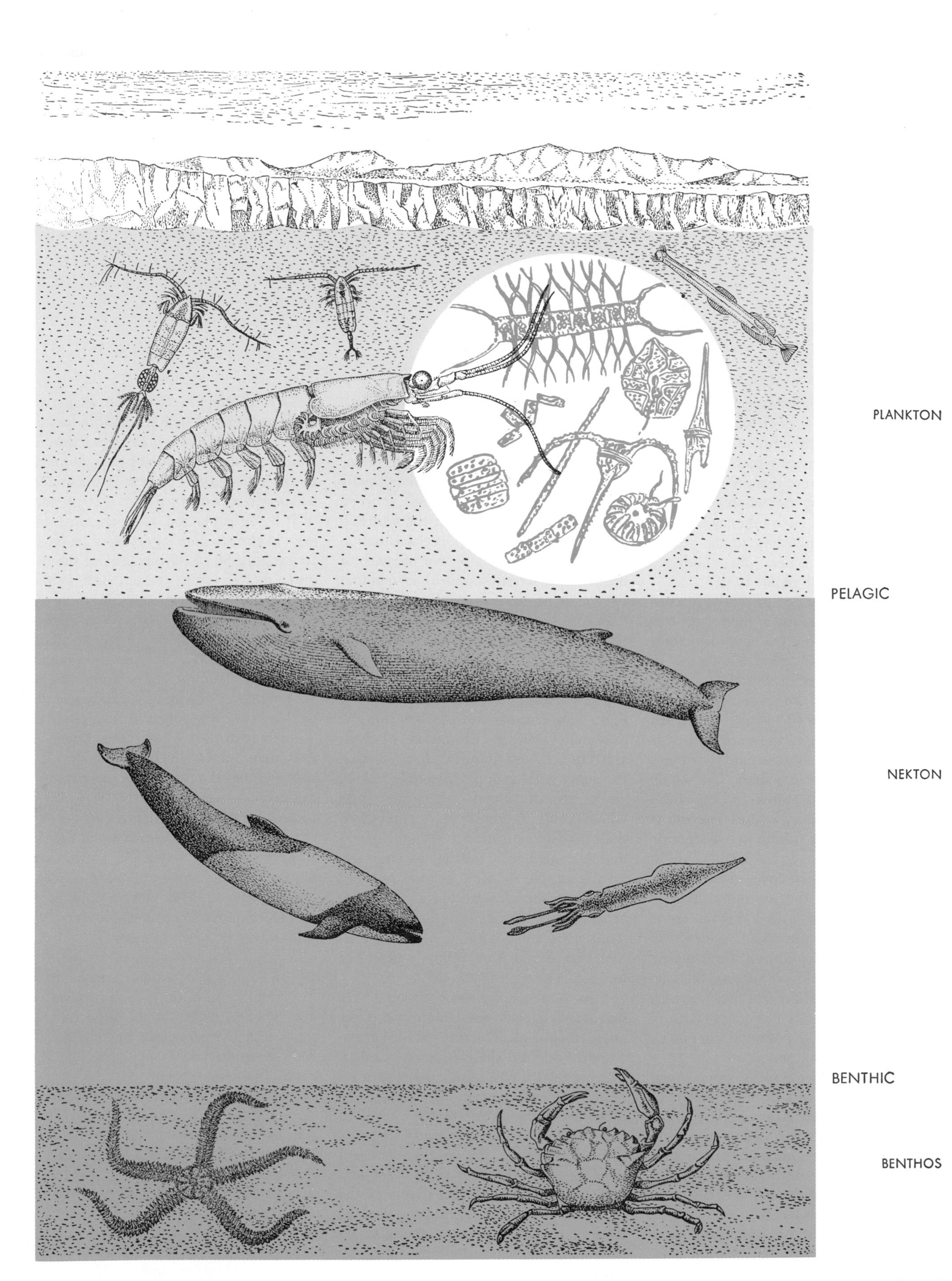

FOOD RELATIONSHIPS among typical Antarctic ocean organisms are shown in this drawing. Plant plankton (*circle*) and the animal plankton which they nourish inhabit the sunlit upper layer. Swimming animals (nekton) include whales, porpoises and squid. Whales feed on krill (*top center*), porpoises on krill and squid. Benthic animals (crabs and starfish) inhabit the bottom (benthos); they feed on animal remains from the upper layers. Organic material is ultimately swept upward by currents to renew the cycle.

of them—a bright red, shrimplike crustacean named *Euphausia superba* and known to whalers as krill or whale food.

Krill is an admirable answer to the whale's food needs. In the first place, the animal is a herbivore, bringing the whale to within one step of the plant plankton. The crustacean has efficient equipment for collecting the tiny plants that serve as its food: its hind limbs sweep currents of water toward a net of hairlike bristles under the forepart of its body, where the plants are entrapped and then swept into the mouth. In the second place, the animal is large enough (up to two and a half inches long) and abundant enough to feed the whale's gigantic appetite.

In the Antarctic waters, where many baleen whales feed, krill grows in enormous quantity. Since it takes a good many krill to make a meal for a whale, it is fortunate that they live close to the surface and are easy to catch. Cousins of this crustacean in other oceans are found only at depths of several hundred feet. But krill confine themselves to a thin zone within about 30 feet of the surface. They swarm in shoals and windrows from a few square yards to half an acre in size. Sometimes they are so densely packed that they give a reddish hue to the water. Aggregations of such swarms may extend for hundreds of square miles. The whales browse in their midst, singly and in herds, consuming vast numbers of them. Yet despite the fact that krill is the principal or exclusive food not only of whales but also of some species of seals, penguins, many other oceanic birds and hordes of fishes, the creature maintains a huge population, breeding larvae all summer long.

This plenteous animal surely deserves man's consideration as a possible food. To see whether it could fill the bill, we must answer several questions. Would it supply our nutritional requirements? Would people want to eat it? Is it abundant enough to give significant relief to the world food shortage? I have looked into each of these matters.

In connection with an entirely different line of study I had occasion to do a biochemical assay of krill. The animal proved to be fairly rich in protein and fats. It can be calculated that a pound of krill will yield at least 460 calories—about the same as other shellfish. The eyes of krill are unusually rich in vitamin A, a feature which might have by-product interest if the animal were fished on a large scale.

How palatable is krill? I happen to know of at least one occasion when people ate some krill (taking them from

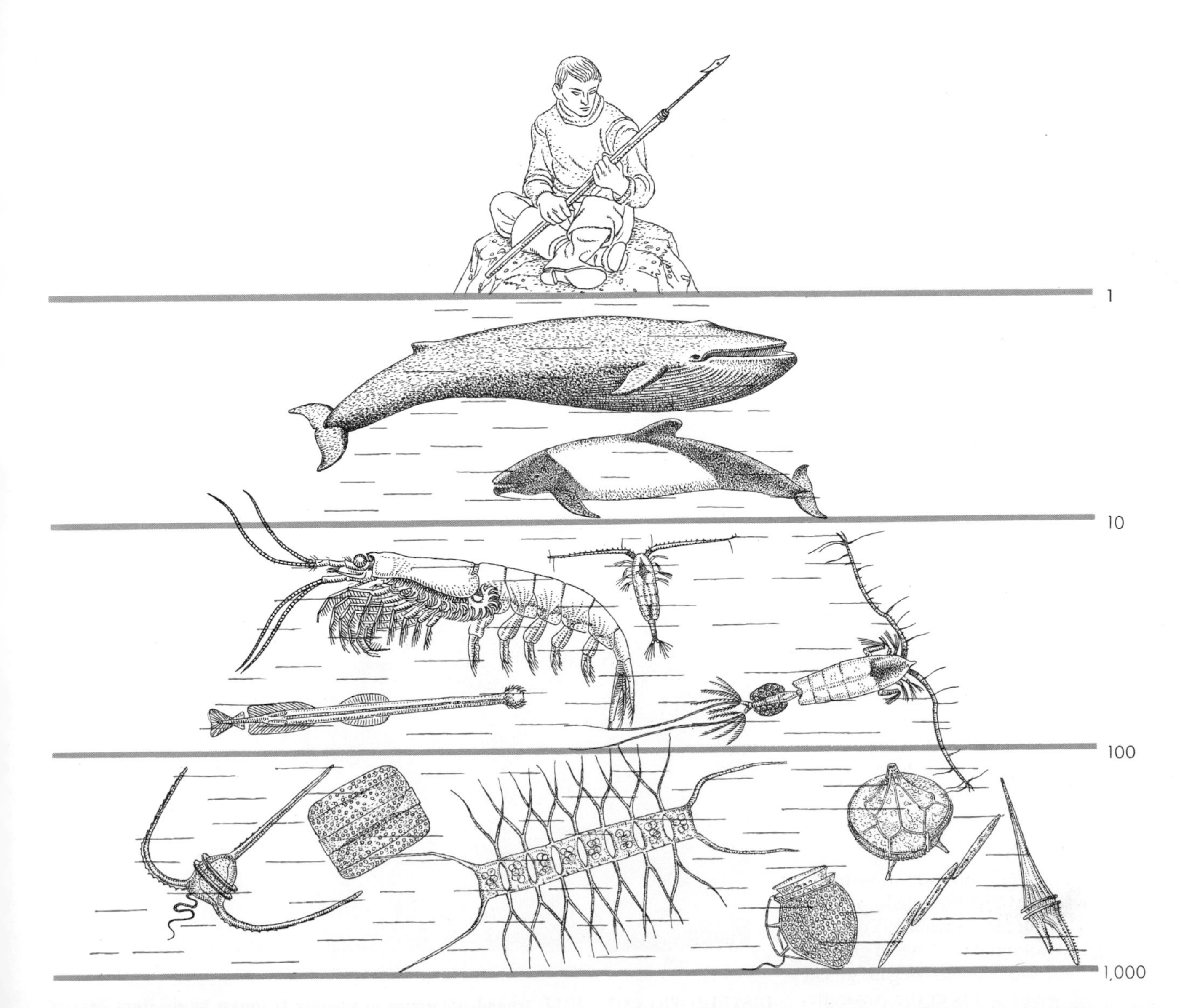

SIMPLIFIED PYRAMID of Antarctic life begins with plant plankton (*bottom*) and ends with man. Energy loss at each step is about 90 per cent; 1,000 pounds of plant plankton produce 100 pounds of animal plankton, 10 pounds of whale and one pound of man.

the freezer in my laboratory under the mistaken impression that they were shrimps). The consumers naturally were most uncommunicative, but I have reason to believe that the crustaceans were eaten with relish, and no untoward aftereffects were experienced. Their removal from the deep freezer can be taken also as a tribute to the appetizing appearance of these bright red animals.

As to their abundance in the sea, we must take an indirect census. Let us consult the whale, estimating its krill consumption from its energy requirements. I estimate, with advice from authorities on fluid dynamics, that an average-sized adult blue whale, weighing some 90 tons, develops about 10 horsepower when it swims at the moderate pace of four knots. (Whales have been observed to reach 12 knots easily.) Assuming that the muscles propelling the animal are about 20 per cent efficient in converting food into power, we can calculate that a blue whale needs about 780,000 calories per day for propulsion. To this we must add an energy allowance for maintenance of its body temperature and processes such as digestion and respiration. A rough estimate based on the whale's surface area puts this figure at 230,000 calories per day. The total so far, then, comes to more than a million calories per day. At about 460 calories per pound of krill, this means an average daily consumption of about 2,200 pounds—more than a ton of food.

We are not yet finished, however. During its growth period (*i.e.*, the first five years of its life) a blue whale gains at the average rate of 90 pounds per day. So a growing whale needs another 600 to 800 pounds of krill per day, raising the total to a ton and a half. Now we have to double this figure to arrive at the actual daily consumption, because we have reason to believe the blue whale feeds only half the year. For the winter it leaves its feeding grounds in the Antarctic and migrates north to more temperate waters. Whale authorities believe that the whale does not feed to any appreciable extent outside the krill range in the Antarctic Ocean but lives through the half year on stored energy. The grand total of a growing blue whale's food intake while it feeds is, then, about three tons of krill per day!

Now we are in a position to estimate how much krill the whale population as a whole may consume. To judge the highest potential we have to go back to 1910, before whalers began to cut down the whale herds in the Antarctic. The population at that time is estimated to have been half a million. Half a mil-

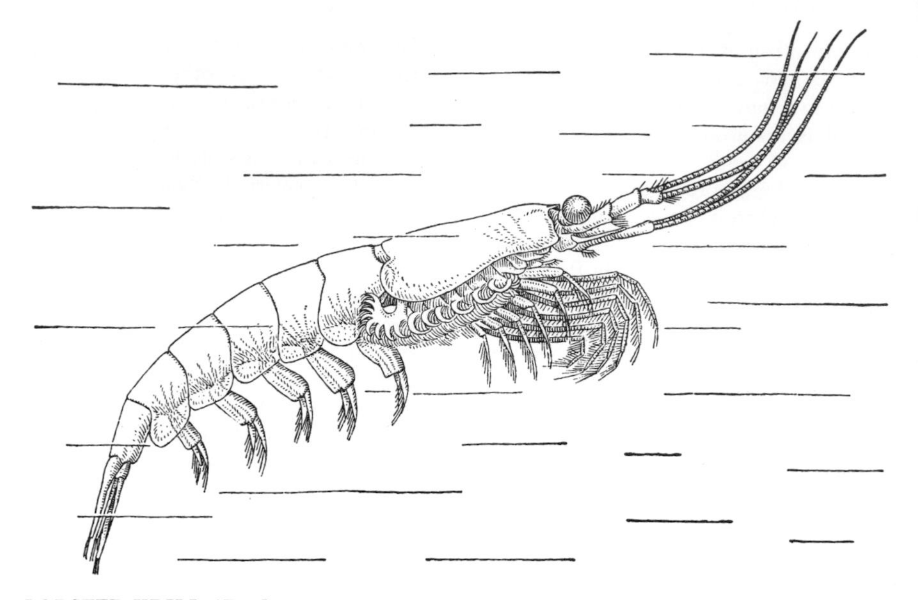

LOBSTER KRILL (*Euphausia superba*) is the largest of several related species consumed in enormous quantities by whales. Its forelimbs trap the plant plankton on which it feeds.

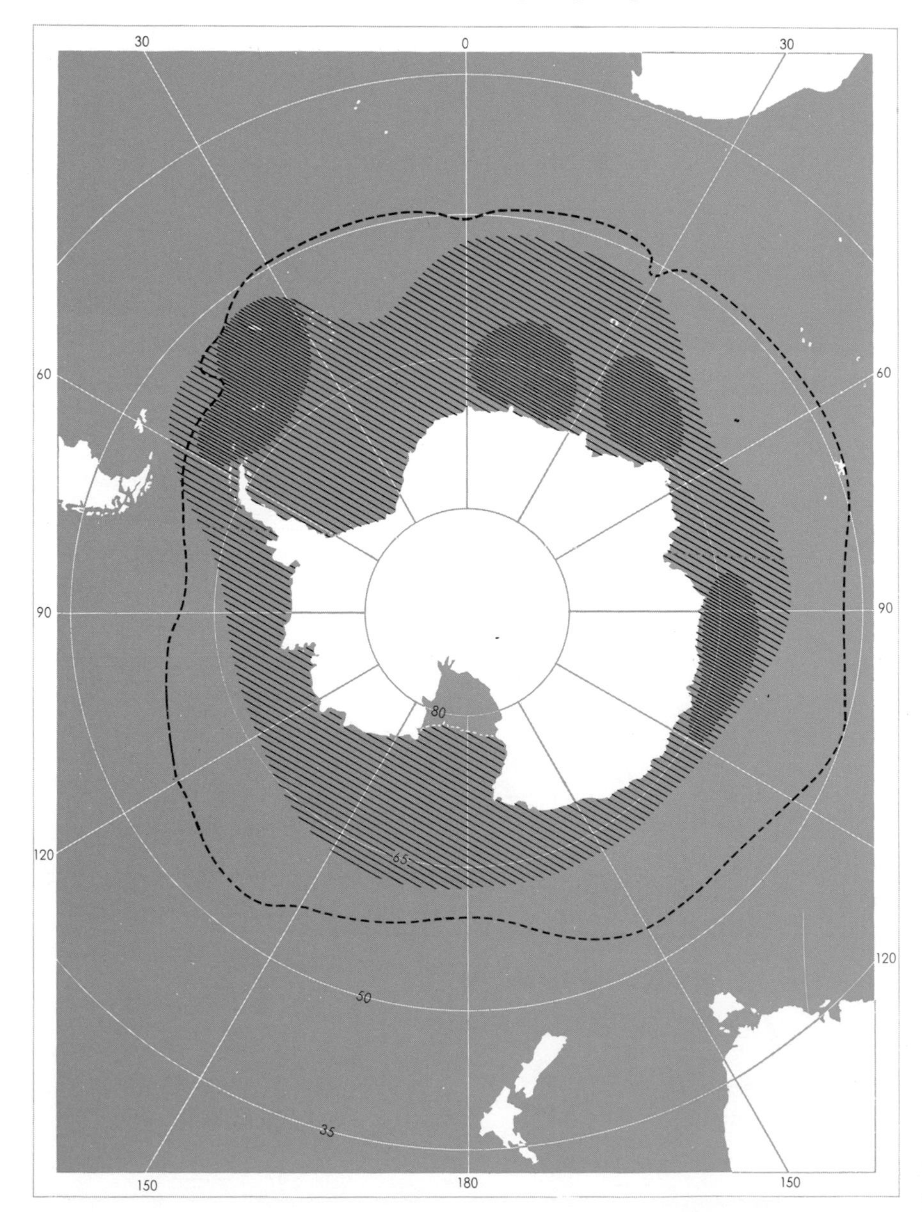

DISTRIBUTION OF KRILL around Antarctica in summer is shown by shading; heavier shading denotes denser concentrations. The broken line surrounds the "Antarctic convergence," within which upwelling nutrients support a luxuriant growth of plant plankton.

lion whales consuming three tons of krill per day for six months gives a total of 270 million tons of krill per year. Since it is highly unlikely that whales harvest more than 20 per cent of the krill in any given area, the total annual production of krill must have been at least 1,350 million tons. On the 3.5 million square miles of feeding grounds this averages to about 1,000 pounds per acre.

At that rate the krill crop in the ocean bests the animal yield on land, for rich pasture land gives about 700 pounds of cattle and sheep per year. The comparison is apt, because these animals, like krill, are herbivorous, converting plants to meat. Krill is, by this token, a far more efficient producer of food than carnivorous fishes, which under cultivation in ponds yield only 100 to 250 pounds per acre per year.

The whale's example shows that we can reap vastly increased returns from the biological economy of the oceans. The 270 million tons of krill on which the Antarctic whales fed in their heyday would be more than enough to supply the annual requirements of the entire U. S. population.

There are growing indications that krill may be a timely subject. Some 250 ships and 16,000 men—the largest whaling fleet in history—are now operating in the Antarctic Ocean. They depend mainly on the fin whale. If and when this species follows the blue and humpback whales into near-extinction, whalers may find it worth while to turn their attention to krill. My own calculations, based on operating costs of ships, processing costs and a likely market price for krill, indicate that even today krill-trawling might be more profitable than whaling.

We ought to begin a scientific exploration of this promising possibility in the Antarctic Ocean. A pilot project in krill-fishing would enable scientists, engineers and businessmen to join hands in a project which would not only advance knowledge but might also develop a valuable source of food.

30

THE COELACANTH

JACQUES MILLOT
December 1955

In 1938, off the east coast of South Africa near the mouth of the small Chalumna River, a fisherman brought up a strange fish. It came to the attention of the zoologist J. L. B. Smith, at Rhodes University College in Grahamstown, and he recognized the fish as an authentic coelacanth. The discovery revolutionized the zoological world. For the coelacanth was a member of a very ancient class of fishes which was supposed to have disappeared some 70 million years ago. This great group of fishes, called crossopterygians, flourished during that decisive era in the history of the earth when the first land animals evolved—when the fish, taking on legs and lungs, went forth to conquer the continents. The crossopterygians were distinguished by lobed fins which were the forerunners of the limbs of higher animals, and that fact alone makes them extraordinarily interesting.

The news that a representative of this supposedly extinct group was still in existence naturally created a great sensation everywhere. The coelacanth acquired a world-wide celebrity such as few animals have ever achieved. Zoologists became extremely eager to find other specimens in better condition, in order to investigate the anatomy and physiology of the fish. It was an unparalleled opportunity to study at first hand a living link from the ancient evolution of vertebrates.

For 14 years Smith and others searched without success in the Indian Ocean along the east coast of Africa. At last, in 1952, word came that a fisherman had pulled up a second coelacanth in the waters off Anjouan, an island in the French Comoro archipelago north of the Mozambique Channel. The South African Government provided Smith with an airplane to fetch it. Unfortunately this specimen was, like the preceding one, in such sad shape—mutilated and half decomposed—that no really useful study could be made of it.

The interested scientists decided that they must organize a methodical search, mobilizing all possible facilities to insure success. A system for catching, preserving and transporting the fish by air was therefore set up in the Comoro Islands by the Madagascar Institute of Scientific Research, aided by the local government. This organization has been rewarded by the capture of nine new coelacanths, all in good condition.

Thorough study of these fish is now in progress. Specialists of various countries

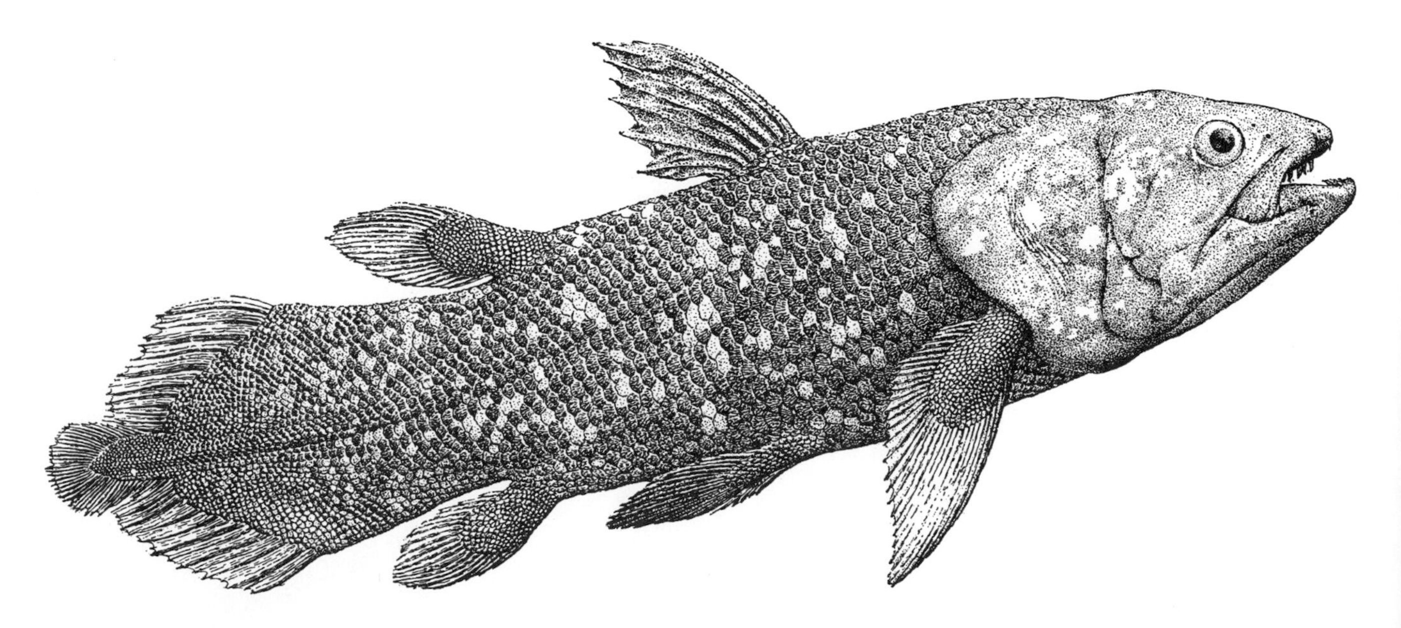

DRAWING of a coelacanth is based on specimens caught in the Comoro Islands. Its fins, with the exception of one on its back, are characteristically paddle-shaped. An adult coelacanth is between four and five-and-a-half feet long and weighs from 70 to 180 pounds.

are collaborating in a program which is centered in the laboratory of comparative anatomy of the National Museum of Natural History in Paris.

When I was a high-school student, we were taught that there were five major classes of vertebrates: the fishes, the amphibians, the reptiles, the birds and the mammals. This view is now out of date. The label "fish" covers a range of animals as diverse as those that live on land. For a zoologist the difference between, say, a lamprey and a carp is greater than between a frog and a man. It is more useful, therefore, to divide the vertebrates into two great groups: the fishes on the one hand and the "tetrapod" (four-limbed) animals of the land and air on the other. The fishes can be subdivided into four principal categories: the agnatha (without jaws), the placoderms (armored), the chondrichthyes (cartilaginous) and the osteichthyes (bony).

The agnatha were not only the earliest fishes but also the first vertebrates we know of. They sucked in food particles through round mouths. Lacking not only jaws but also paired fins, they probably swam clumsily, tadpole-fashion. In our time they are represented only by two degenerate descendants, the lampreys and the hagfishes, which lead a near-parasitic life [see the article "The Sea Lamprey," by Vernon C. Applegate and James W. Moffett, beginning on page 391]. At their apogee the agnatha apparently evolved into fish armored with bony plates. During the Devonian Period (some 400 million years ago), they gradually disappeared and were replaced by the placoderms, which retained the armor and acquired primitive jaws and paired fins. Then the placoderms in turn vanished as two new lines of fishes with jaws came into being—the cartilaginous and the bony fishes.

One of the great problems of evolution has been to find anatomical links between the fishes and their land-invading descendants, which emerged at the end of the Devonian epoch. Comparative anatomists have speculated for half a century on how the fin of the fish evolved into the forelimb of the frog, the forerunner of our own arm. It is hopeless to trace a connection if we look only at a modern fish. In the fish we know well—the goldfish of the aquarium, the trout of the dinner table—the fins are altogether different in structure and orientation from the limbs of a frog. For a long time evolutionists were troubled by this major gap between the fishes and the

X-RAY PHOTOGRAPH represents a vertical section through the tail of a coelacanth. At the top and bottom are the fin rays. In the middle is a section of the fish's primitive spine.

amphibians. But the gap has now been bridged by studies of ancient fishes, and this is where the coelacanth comes in.

The fins of the crossopterygians are strikingly different from those of all other fishes. Instead of being attached to the main body directly, they are borne on a scaly stalk protruding from the body. The fin articulates through a single structure, just as the limbs of a tetrapod—toad, eagle, dog or man—hinge on a single bone, *e.g.*, the humerus of the arm and the femur of the leg. The fin of the crossopterygians can be thought of as a true little limb—the missing link between the typical fin of fishes and the limb of the other vertebrates. Furthermore, the skull of the crossopterygians is constructed along the same general lines as that of primitive amphibians. These striking resemblances have quite naturally led paleontologists to consider the crossopterygians as the pivotal group that brought forth amphibians and the land and air animals.

The crossopterygians are very ancient bony fishes which appeared some 400 million years ago at the start of the Devonian. They include two distinct lines of descent. One line is the coelacanths (whose name, from the Greek, means literally "hollow spine"). The other is called the rhipidistians.

It sometimes happens in a family that two brothers, while bearing a close resemblance to each other physically, have contrasting temperaments and aptitudes and lead quite divergent lives. The same sort of thing can be noted in the history of animal descents. The coelacanths and the rhipidistians offer a striking example. The rhipidistians were the bright boys of the crossopterygian family, resolutely turned toward progress. They foreshadowed the later vertebrates not only in their fins and skull structure but also in possessing internal nostrils—a develop-

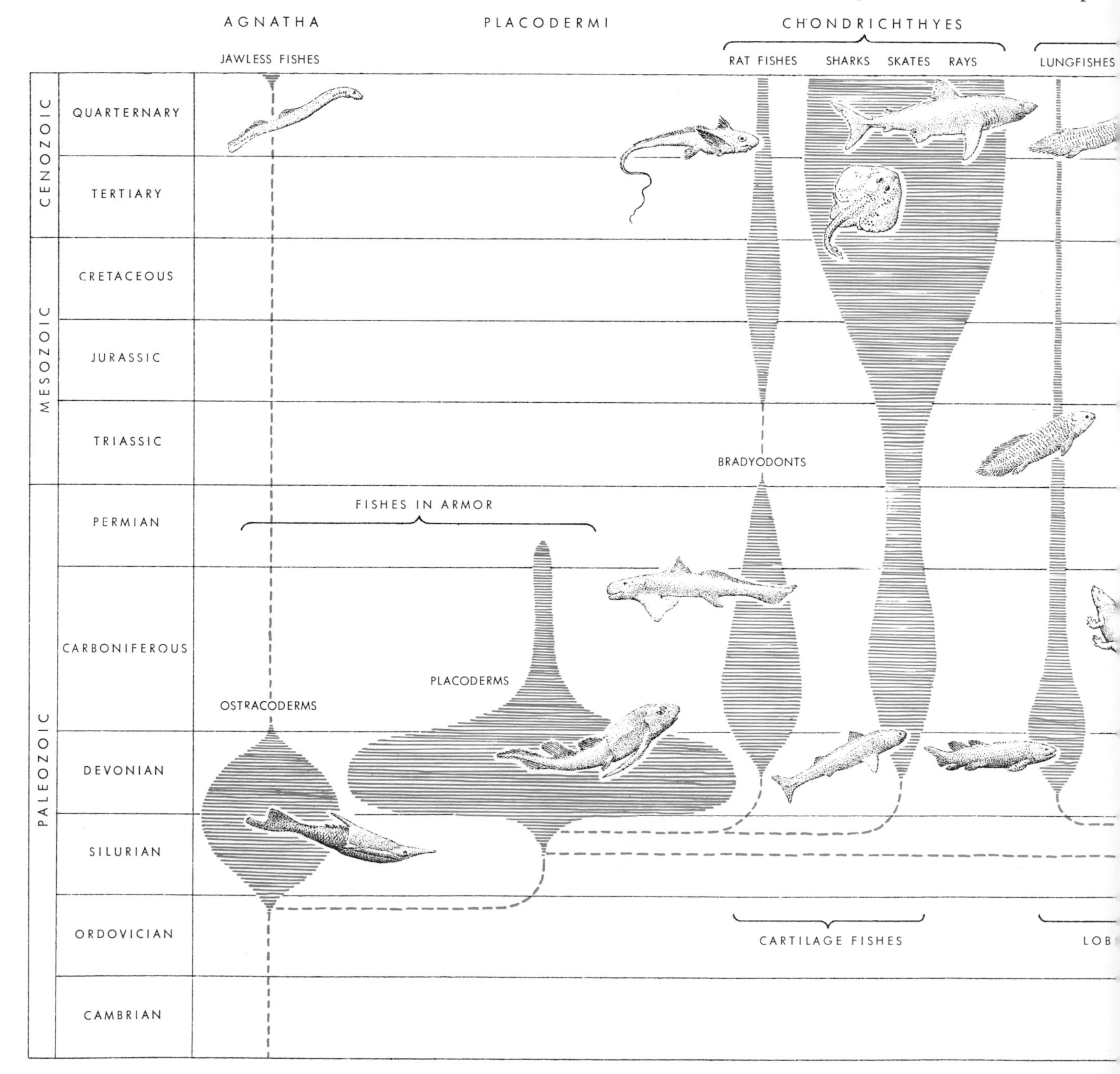

EVOLUTION OF FISHES is outlined in this chart. The vertical dimension of the chart is time, with the present at the top. The horizontal dimension of the colored areas indicates the relative abundance of each group at any one time. The dotted lines represent the

ment which seems to indicate that they possessed functional lungs and led an amphibian existence. Indeed, fossils of primitive amphibians are found beside them in many a geological deposit. The rhipidistians show a remarkable structural kinship to early amphibians found in deposits of late Devonian times in Greenland. They form a solid bridge between the fishes and the amphibians.

On the other hand, the coelacanths have shown themselves to be as stable and as obstinately conservative as their rhipidistian brothers were progressive. Throughout the hundreds of millions of years the coelacanths have kept the same form and structure. Here is one of the great mysteries of evolution—that of the unequal plasticity of living things. Why are certain organisms so labile while others apparently very like them retain their identity through the ages, no matter what vicissitudes they have to undergo? Whatever the reasons or the moral, the fact is that the unchanging coelacanths far outlived their rhipidistian brothers, for the latter disappeared by the end of the Devonian Period.

All of the coelacanths now captured belong to one species, named *Latimeria chalumnae* by Smith. They are good-sized fishes, between four and five-and-a-half feet long as adults and weighing from 70 to 180 pounds. The living coelacanths are a good deal larger than those previously found as fossils. Their body build calls to mind some of the big modern rock fishes, and their mode of life cannot be far different, though they live at greater depths. The coelacanth's flat, powerful tail is not distinctly differentiated from the rest of its body. Its body is entirely clothed in large, circular scales, whose size, exact shape and ornamentation vary with the region of the body. Alive the fish has a steely blue-gray color, flecked with

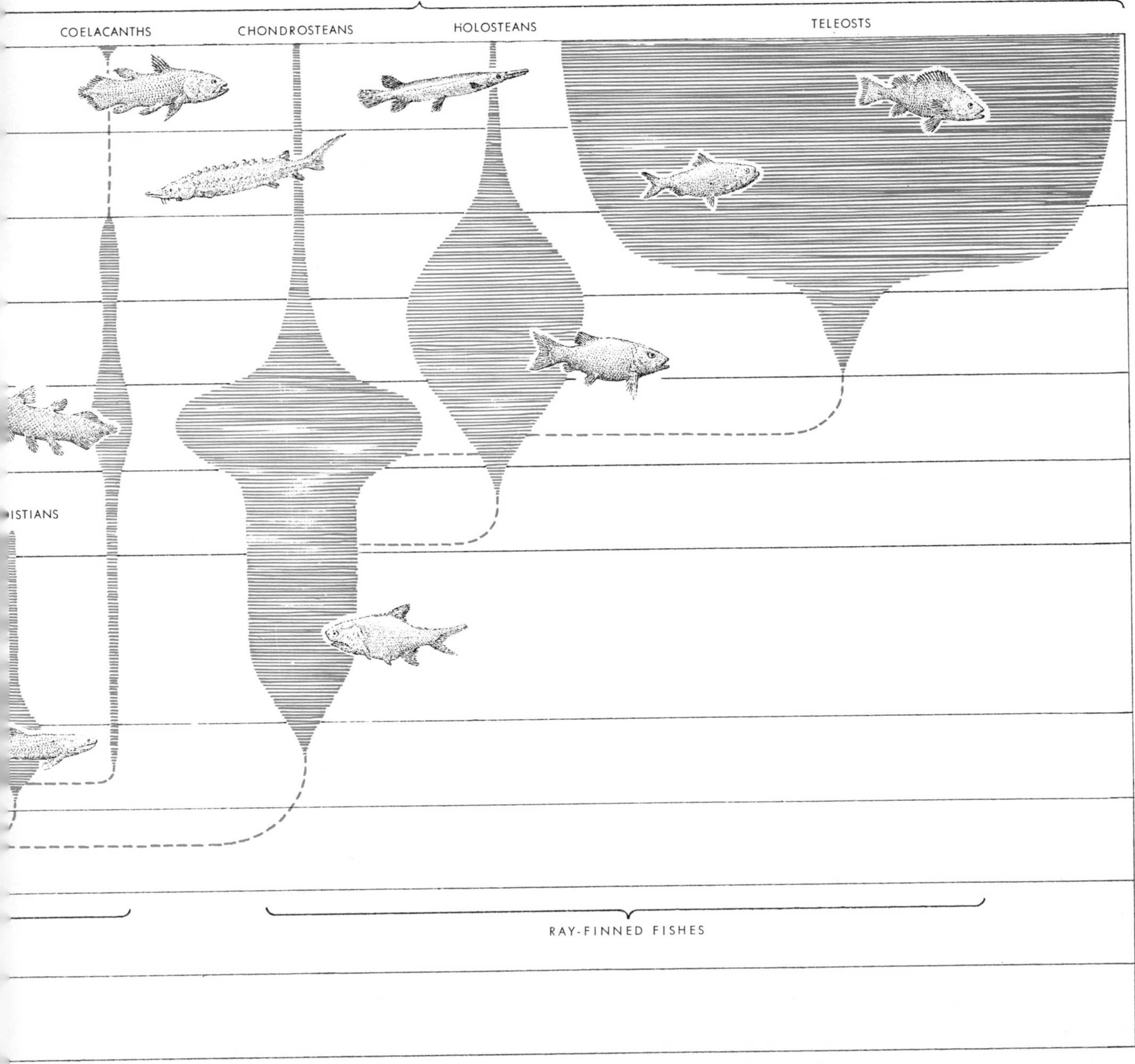

supposed evolutionary relationships of the groups. The typical species that illustrate the characteristics of each group are not drawn to the same scale. The chart was prepared with the assistance of Bobb Schaeffer of the American Museum of Natural History.

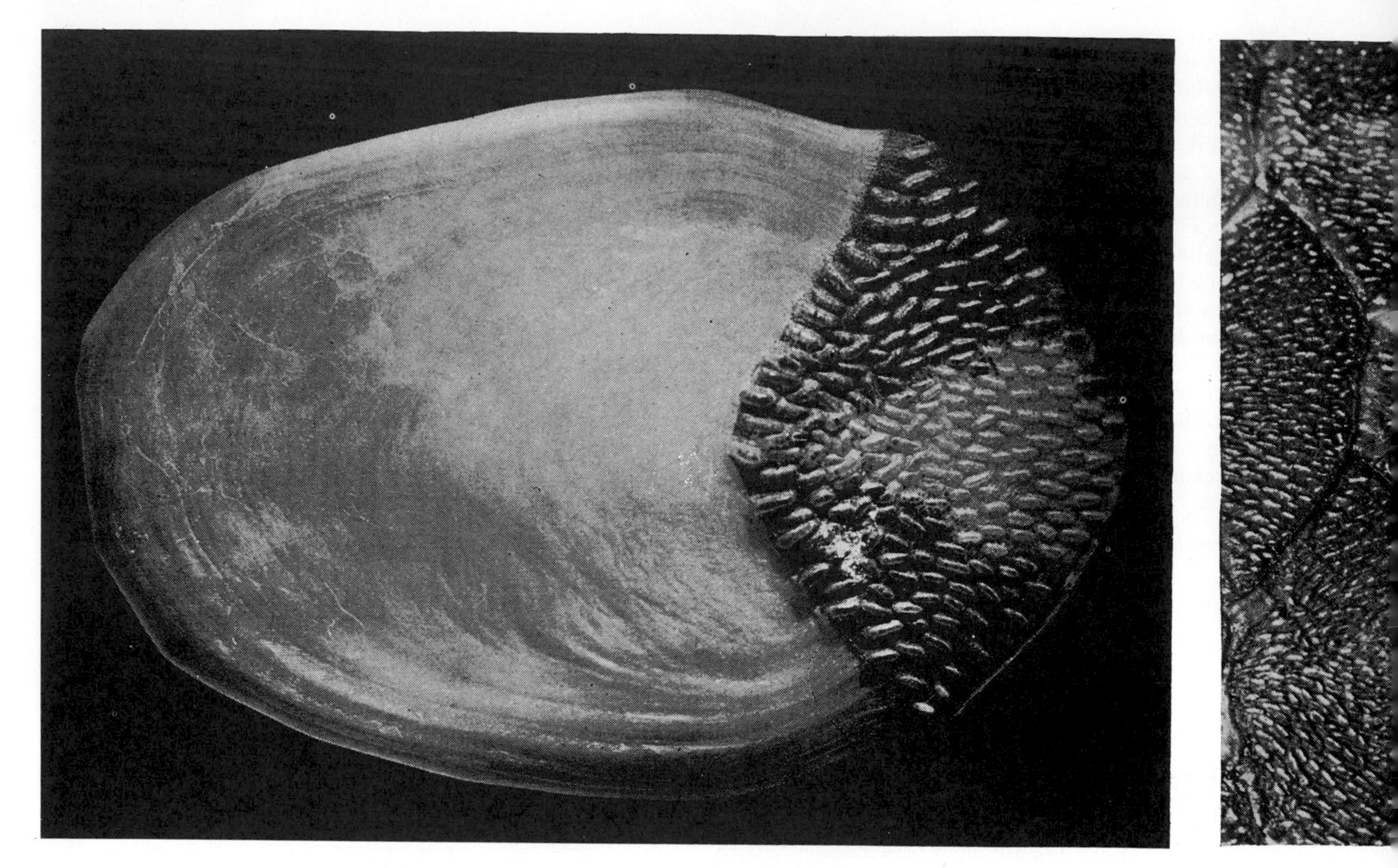

SINGE SCALE of a coelacanth is enlarged three times in the photograph at the left. At the left side of this photograph is the part of the scale that is overlaid with other scales. At the right side of the photograph is the exposed part of the scale, ornamented

light spots; after death its color changes rapidly, usually to chocolate brown.

Its fins at once identify the coelacanth as a crossopterygian. Six of its seven fins have a characteristic paddle shape and are borne on well-developed, scaly stalks [*see illustration on page 280*]. Each has about 30 fin rays. The seventh fin, on the forward part of its back, is fan-shaped and is attached directly to the back.

Coelacanth fossils have been found in many parts of the world—from Brazil to Spitsbergen, from Great Britain to Madagascar. But today the fish seems to be restricted to the waters washing the little Comoro archipelago, situated halfway between Madagascar and east Africa at the north entry to the Mozambique Channel. The first specimen, found off South Africa, must have been a wanderer from the Comoro area.

Although the local Comoro fishermen have been mobilized almost to a man and have been spurred on by sizable rewards, the catch of coelacanths is only three or four per year. The fishing is rather difficult: coelacanths live on the deep, rugged bottom and cannot be bagged with nets. Except for the first stray specimen, all have been caught at from 80 to 200 fathoms—the limit of the fishing lines. In all likelihood they go down to about 400 fathoms. Their eyes are phosphorescent, and they shy away sharply from the light. They survive only a few hours after being brought to the surface. One specimen, taken in perfect condition, was immediately placed in a submerged boat as an aquarium, but in spite of all the care lavished on it it died within a day, plainly through the combined effect of the decompression and the rise in temperature. Samples of water taken precisely where the fish were caught have shown temperatures of the order of 54 degrees Fahrenheit, as against the 79 degrees of surface waters.

The coelacanths are probably no great swimmers. They seem to be rather sedentary animals which live amidst basaltic rocks where they find shelter and lie in wait for their prey. Their powerful tail should enable them to hurl themselves at their victims with an irresistible pounce. Perhaps their stalked fins permit them to creep along the rocks like seals. But for the time being this is mere speculation. The only fish that was observed alive for any length of time was slow-moving (at the surface at least), but exhibited exceptional mobility of its fins. The pectoral fins, notably, can turn in just about all directions—nearly 180 degrees fore and aft as well as up and down. At the time of death the two fins may set themselves in opposite ways; the fish then presents an asymmetry highly disconcerting to an uninitiated observer.

The Mozambique coelacanths are carnivores feeding only on small fishes. They usually swallow their prey whole, and some of the little fishes have been found intact in the predator's stomach. For the most part the coelacanth's food fishes are believed to live at depths from 300 to 500 fathoms.

The discovery of the living coelacanths magnificently confirms, and amplifies, what paleontologists had deduced about the early vertebrates from their fossils. It gives ringing testimony to the remarkable powers for reading the past that mankind has developed through sophisticated modern paleontology. When we read archaeologists' conclusions about ancient life on our planet, pieced together from the most fragmentary evidence, their assertions must sometimes seem rash, and we wonder how much reliance can be placed on their picture. But now that we have a "living fossil," the coelacanth, to check their deductions, we can see that the paleontologists' reconstructions of the crossopterygian fishes are masterpieces of skill and insight.

The bringing to light of living coelacanths adds some 70 million years to

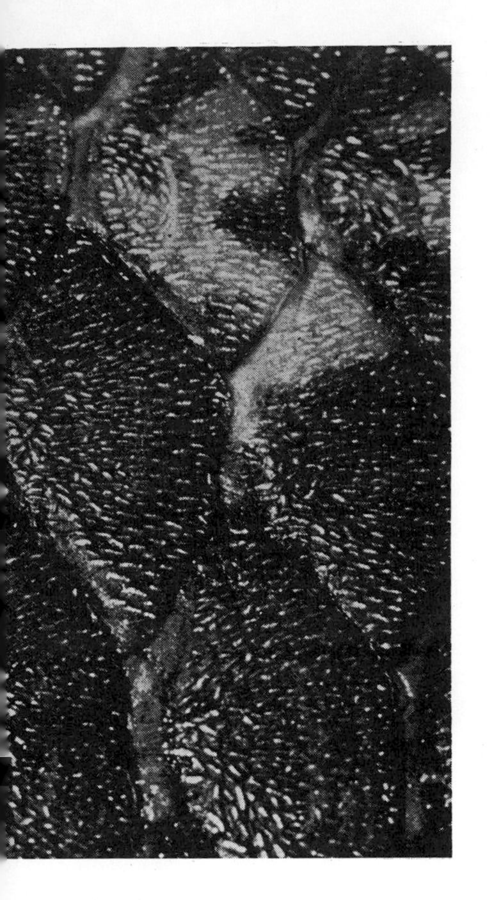

with tubercles. In the photograph at the right several scales are enlarged two times.

their estimated longevity. They are beyond compare the oldest "higher animals" on the earth. To appreciate their age, we must bear in mind that when they arrived the only other vertebrates in the world were a few strange fishes now long extinct; that in their heyday there was not yet the least sign of a reptile or an amphibian animal, to say nothing of a mammal or a bird. Compared with the coelacanths, the ancient and extinct dinosaurs lived only yesterday; the coelacanths are nearly 200 million years more ancient. And yet these astonishing fishes turn out to be still with us and still robust—by no means degenerate or tired of living.

What is even more remarkable is that in spite of drastic changes in the world environment, the coelacanths are still much the same organically as their ancestors. Their living organs yield many secrets of the past. As Smith has said, the coelacanths are incomparable "machines for reading time backward."

It has long been realized that the age-old evolution of animal organs is more or less reflected in the development of the embryo of a vertebrate animal today; biology students of my generation were suckled on the formula: "Ontogeny follows phylogeny"—the development of the individual repeats the development of the race. As Thomas Huxley used to say: "The individual must climb back to the top of his genealogical tree." The classic view is that in the course of a few weeks or months the young embryo retravels the entire evolutionary path traveled by his forefathers in the course of the ages. Whether this is strictly true has been much debated, but it is indisputable that the study of embryonic stages gives invaluable aid in reconstituting the ancestral stages.

Let us take the heart, for instance. In the very young embryo of a vertebrate animal, including man, the heart begins as a simple enlargement of the principal blood vessel. The enlarged section partitions itself into four chambers in a row. As the organ develops, the vessel coils back on itself S-fashion, and the two rear chambers fold over above or ahead of the forward ones. As a result the heart acquires a globular shape, with the auricle (consisting of the atrium and sinus) folded over the ventricle.

Now if this development truly repeats the evolution of the vertebrate heart, we should expect the coelacanth to have a more or less linear heart, with a sinus and atrium behind the ventricle. Dissections of the fish have verified the correctness of this prediction. The heart is not quite linear but has the shape of a flattened V. The atrium and venous sinus are behind the ventricle, with only the merest suggestion of an overlap. Altogether the form of the coelacanth's heart neatly matches a diagram, picturing what a primitive vertebrate's heart should be like, which was drawn some 30 years ago by the British zoologist Edwin S. Goodrich! Such confirmations are the compensation of men of science.

An equally striking finding emerges from examination of the coelacanth's pituitary gland. This tiny but important organ, situated just beneath the brain, originates in a curious manner in the embryo of a modern vertebrate. It develops from two separate bulbs: one coming from the floor of the cerebrum, the other from the back wall of the pharynx. In other words, one bulb develops from nerve tissue, the other from digestive tissue. In the course of the fetus' development, the pharyngeal bulb detaches itself completely from the digestive tube and joins the nerve bulb to form the pituitary, so that no indication of its origin in the pharynx remains. But in the coelacanth that connection persists: throughout the fish's life its pituitary gland is connected to the roof of the palate by a long tube, richly supplied with blood vessels and functioning in hormone production. Here again we have a telling illustration of an ancient stage in animal evolution.

Study of the coelacanth's nervous centers is especially interesting. Its brain is tiny and simple in construction; it occupies only a minute part of the fish's cranium, and is extremely small in proportion to the whole body. The brain of a 90-pound coelacanth weighs less than 50 grains—that is, no more than one 15,000th of the body weight. No present-day vertebrate that we know of has so small a brain in relation to its size. And yet the coelacanth's marked microcephalia has not prevented it from outclassing all other vertebrates in life's competition across the ages. Here, then, is food for thought: smallness of the nerve centers is in no way a hindrance to triumph in the struggle for life. Indeed, it would seem that other biological factors, such as fecundity and physiological adaptability, play a role far greater than intelligence in the survival of species. But it must be said that the coelacanthian brain, though tiny and simple, seems perfectly proportioned within itself. Its various elements seem to have remained in balance, none developing out of proportion to the others.

There are still a great many things to be learned from this uniquely instructive anatomy—from the fish's hollow spine extending far forward into the head; from its double respiratory system (both gills and lungs—now degenerate), which permitted the animal to evolve toward either an aquatic or a terrestrial life; from its unusually complex fins, which foreshadow our own limbs, and from its exceptionally rich musculature, which allows the fish to make precise and varied movements.

Detailed studies now in preparation will soon apprise the scientific world on all these points. In the meantime, research is continuing. It will throw light on the mode of life of these remarkable crossopterygians and will try to penetrate the secret of the adaptability which has enabled them to live through many geological eras under widely differing conditions without modifying their constitution. It will also try to penetrate their mode of reproduction and their stages of embryonic development, in which we can expect some real revelations about evolution.

Let us thank fortune for permitting the coelacanths to live on into our time. They open a window onto the past of the higher animals, including ourselves, and help us to reconstruct and to understand our history.

THE OCEANIC LIFE OF THE ANTARCTIC

ROBERT CUSHMAN MURPHY

September 1962

A marine biologist with his townet can locate the Antarctic Convergence, that outer oceanic frontier of the Antarctic, as readily as an oceanographer can with his instruments for measuring the salinity, temperature and flow of the waters. When *Euphausia superba*, the red shrimplike crustacean commonly called krill, shows up in the net, all hands can be certain that their ship has passed southward across the Convergence. The krill symbolizes life in the Antarctic more aptly than any penguin does. It is the key organism in the shortest food chain of one of the most abounding provinces of life on earth. Feeding directly on the one-celled plants of the sea, the krill in turn supports not only fish but also penguins and vast populations of winged sea birds, seals and whales. Thus in Antarctic waters the building of the body of the blue whale, the largest animal that has ever lived, goes on at only one remove from the organic fixation of the radiant energy of the sun by microscopic plants. Thanks also to the immense fertility of the water, maintained by the upwelling of mineral nutrients from below, the 12 million square miles of Antarctic ocean are richer in life than any other comparable oceanic area.

Throughout the region the abundance of life at sea contrasts with the poverty of life ashore. Especially on the Antarctic continent proper, the community of terrestrial life presents a study in adaptation to the extremes of cold and desiccation [see "The Terrestrial Life of the Antarctic," by George A. Llano, Offprint #865]. Hence the vast bulk of Antarctic life is marine. The land provides little more than a breeding place for birds and mammals that have otherwise forsaken it for the sea. Under the steady circumpolar drive of the prevailing westerly winds and the eastward-moving current at the Antarctic Convergence, the whole vast region tends toward considerable uniformity in the distribution of its living forms. For this reason the Antarctic offers an admirable field for elucidation of the broad principles of marine ecology.

ANTARCTIC SPECIES thrive in a world of ice and water. The crabeater seals and emperor penguins in the aerial photograph on the opposite page are on an ice floe off Cape Crozier on Ross Island. Leopard seals prey on penguins, but only in the water.

The unparalleled lushness of the Antarctic Ocean arises from the turnover of the waters of the Atlantic, Pacific and Indian oceans set in motion by the bottom current of chilled water that runs outward from the continental shelf of Antarctica. The surface water is rich in nitrogen, bound in nitrate and nitrite salts; nitrogen-liberating bacteria, plentiful in warm seas, are scarce or inhibited in Antarctic water. The phosphate content is so high that it is never fully utilized by the microscopic plants of the plankton, as is common in summer in northern temperate latitudes. Whereas lime is scarcer than it is in ocean waters that are supplied by continental runoff in the north, silica is abundant. Silica is preponderant, therefore, in the hard parts of the plants and invertebrate animals of the plankton. The one-celled algae, for example, appear in the siliceous, snowflake diatom forms rather than in the limy dinoflagellate forms that prevail elsewhere in the world ocean. Since the capacity of water for dissolved gases varies inversely with temperature, the amount of oxygen in Antarctic water is of the order of 95 per cent of saturation in winter, with frequent supersaturation. Oxygen content is lowest in late summer, when temperatures are highest, but it is ample even then.

With a constant supply of nutrients for diatoms and other one-celled plants, which are the pasture of the sea, the stage is set for maintaining the pyramid of oceanic life. In addition to the food chain centered on the krill and other euphausians, there are chains in which squids or small schooling fishes supply the staple food of larger fishes, petrels, penguins, seals and whales; these cycles are less well known. No matter what links form the chain, the primary foodstuff is ultimately restored to the water in the form of excreta and dead bodies, broken down by bacterial decay. Upwelling then returns it to the surface layer, where photosynthesis takes place, and the whole process is repeated. In the Antarctic, sea water is not to be regarded as merely saline H_2O. It is also a broth that has been physiologically conditioned by the metabolism of organisms, to the general benefit of their populations.

In a surprising way the very coldness of the water accounts in part for the unique wealth of Antarctic marine life. As long ago as 1908 Jacques Loeb, then at the University of California, demonstrated that the duration of life and the rate of development respond differently to temperature. Working with sea urchin eggs, he found that reduction of temperature by 10 degrees centigrade theoretically increased the length of life 1,000 times, whereas the corresponding period of development was increased only about three times. From this he concluded that the chemical processes controlling development are altogether different from those causing old age and death. The sense of Loeb's discovery is demonstrated in the Antarctic water. There, at a temperature of 0 degrees C., individual organisms have longer life spans. As a result many more successive generations of each species of marine organism live contemporaneously than exist in warmer waters. On the other hand, the number of species in each great family of plants and animals native to the Antarctic is small in comparison with that of the tropics. Therefore in general it can be said of Antarctic waters that they abound in larger numbers of fewer kinds of plants and animals than milder oceans do. The same is also true of the higher vertebrates—the birds and mammals—that surmount the pyramid of Antarctic marine life.

Echo-sounding gear has shown that the animals of the plankton in the Antarctic, as in other oceans, congregate in layers at various depths and in sufficient density to return a blurred echo. These "deep scattering" layers migrate vertically in the course of the day, usually sinking during daylight hours and rising toward the surface at night. In some Antarctic species an annual cycle of horizontal migration is superimposed on this diurnal oscillation; certain crustaceans tend to drift northward during the summer and descend at the Antarctic Convergence into deeper southward-moving water masses. They thereby

MARINE ANIMALS of the Antarctic are seen in their normal surroundings on this page and the next three pages. Their approximate distribution on the mainland, shelf and pack ice and Antarctic islands and in the open sea is indicated by the schematic cross sec-

maintain a favorable latitudinal placement by means of a vertical circular overturn. Francis C. Fraser of the British Museum has observed that in addition to such seasonal migration the krill carries out migratory movements in correlation with successive stages in its life history. Larval forms appear to congregate in southward-flowing waters between the surface and the bottom. Upward movement then brings a constant replacement of adolescents at the edge of the pack ice, after which the maturing organisms are carried northward at the surface as far as the Antarctic Convergence.

Since the upper Antarctic water masses flow eastward under the influence of the prevailing westerly winds, the movement of the zooplankton also has an eastward component that promotes high uniformity in the distribution of species around the polar continent. As a result the distribution of animals that feed on the plankton shows a corresponding uniformity. Under particular circumstances, given peculiarities in oceanic circulation or the availability of especially favorable breeding places, some species cluster in more or less constant nodes of concentration. A few species, such as the chin-strap penguin, seem to be only now at the point of extending their ranges "all the way around."

Although even less is known about

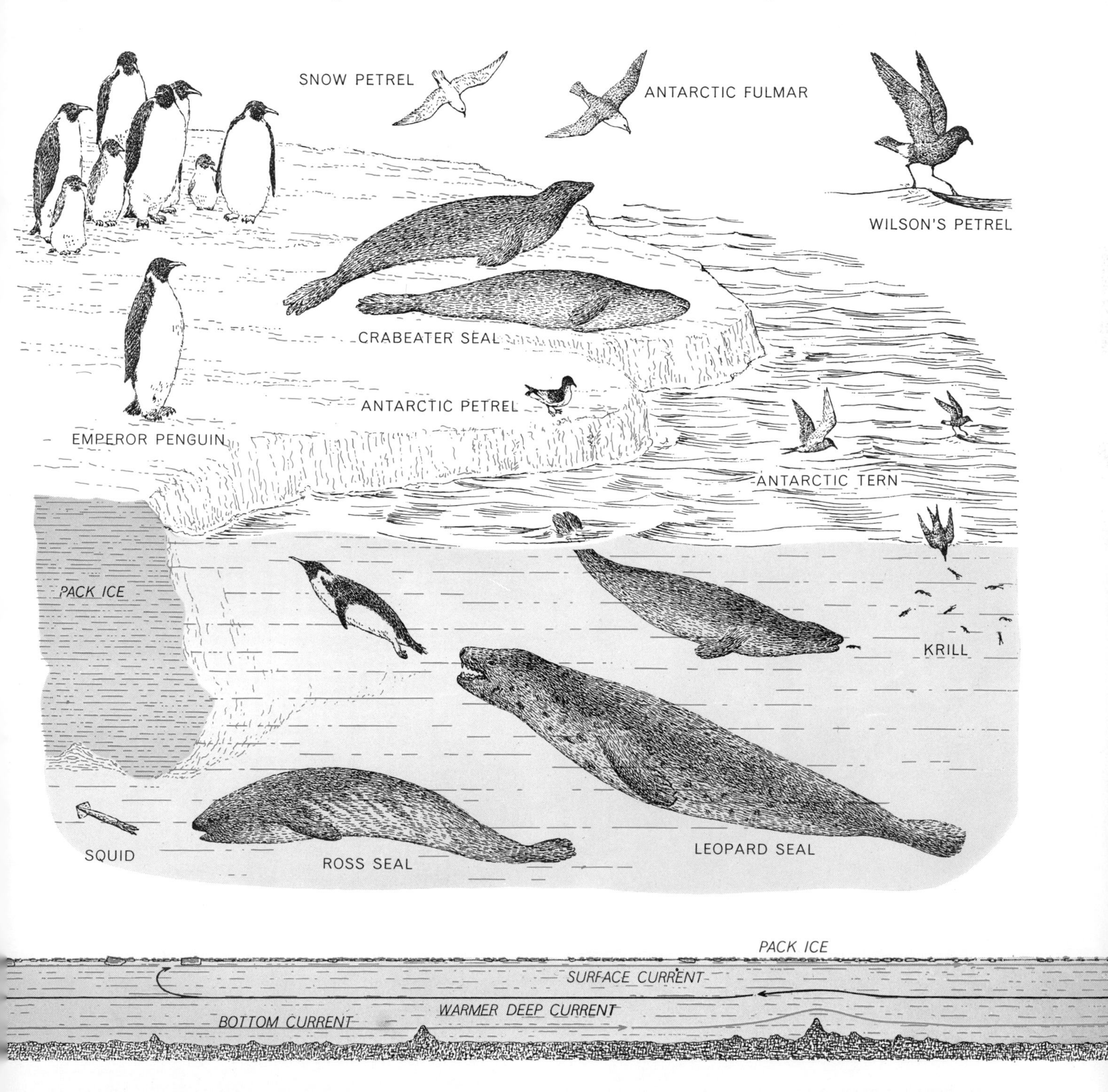

tion across the bottom of the four pages. The leopard seal, normally a creature of the outer pack ice, comes in close to shore to prey on penguins. Of the winged birds, only the Antarctic skua, the Antarctic tern and a few of the petrels actually breed on the mainland.

the bottom of the Antarctic Ocean than about its overlying waters, it appears that the community of life in the depths reflects the same circumpolar pattern, with a more or less well-defined northern boundary at the Convergence. Since the Convergence is generally thought of as a surface or near-surface phenomenon, the reason for the sharp transformation at the bottom below it is not immediately clear. It may be related to differences in the rain of organic detritus from waters near the surface. The bottom world of mollusks, brachiopods, pycnogonids, echinoderms, corals, tunicates, hydroids, holothurians and marine worms largely remains to be explored. Of particular interest are the sponges, which seem to be represented here in greater profusion than they are in tropical oceans. Sponges are usually held to have gained their ascendancy in the Cretaceous period of from 140 million to about 70 million years ago, but some zoologists believe that their true climax is today and in the Antarctic.

In contrast to the rich life of the deep waters, the coast line of Antarctica and the shores and submerged ledges of the Antarctic islands are surprisingly barren of bottom life and fish. Large attached algae are scarce in the intertidal band on the coasts. Perhaps only because rising and falling ice scrapes the rock clean, the absence of wrack of any sort in the dark but clear water below

ANTARCTIC OCEAN, rich in plant and animal plankton, is a favorite feeding ground for whales. The blue whale, the largest creature on earth, feeds on swarms of the crustacean *Euphausia superba*, or krill, as do the humpback and finback whales and many

the cliffs of the continent and its neighboring islands is one of the striking distinctions of the area. In addition, the thickness and density of the fast ice and close-in pack ice filter out sunlight. Photosynthesis is feeble or gives out altogether, and the ecological system is vitiated. Most of the animals seem to be wandering carnivorous types.

As for the vast regions of water that underlie the great ice shelves of the Antarctic continent, such as those of the Ross and Weddell seas, it has long been held that these are quite deficient in life. This supposition has been upset recently by the finding of large fishes together with bottom invertebrates frozen *in situ* and exposed well above sea level on the wind-scoured surface of the Ross Ice Shelf near the U.S. base at McMurdo Sound. These remains, on top of ice more than 100 feet thick, had apparently been trapped by freezing at the bottom of the shelf when ice touched the sea floor. Thereafter they were brought up slowly as the wind ablated the upper surface of the shelf and new ice nourished the bottom. Preliminary carbon-14 dating indicates that it may have required about 1,100 years for these specimens to work their way up through the ice. If the explanation of how they did so is correct—and there seems to be no other explanation—it confirms a glaciological hypothesis advanced 40 years ago. It also serves as a stimulus for the chal-

fishes, seals and birds of the Antarctic. The killer whale preys on seals and other whales. The drawing at the right illustrates the multiple use of the small areas available for nesting on most of the Antarctic islands, with several species of birds crowded together.

lenging task of exploring the underside of the ice shelves, surely one of the strangest of environments.

The size of some of the fishes, ranging from 40 to 150 pounds, was another surprise. Nearly all the previously recorded members of the endemic Nototheniid group to which they belong were relatively small.

The hagfishes, rays and eelpouts are among the few fishes of northern affinities that have representatives in the Antarctic. The order of Nototheniiformes embraces 90 per cent of all the fishes found in the pack ice waters; few, if any, of these fishes range beyond the border of the Antarctic Zone. Certain species bear chin barbels, with or without terminal "baits." Some of them look like sculpins; others might be called crocodile-headed or dragon-headed, as reflected in such generic names as *Bathydraco* ("depth monster"). Those of less forbidding appearance are commonly called Antarctic cod, although another Antarctic family, the Muraenolepidae, is more closely related to the true cods. One curious family in this order, the ice fishes (Chaenichthyidae), includes species that have no circulating red blood; the recent discovery by the Soviet worker L. D. Martsinkevitch of

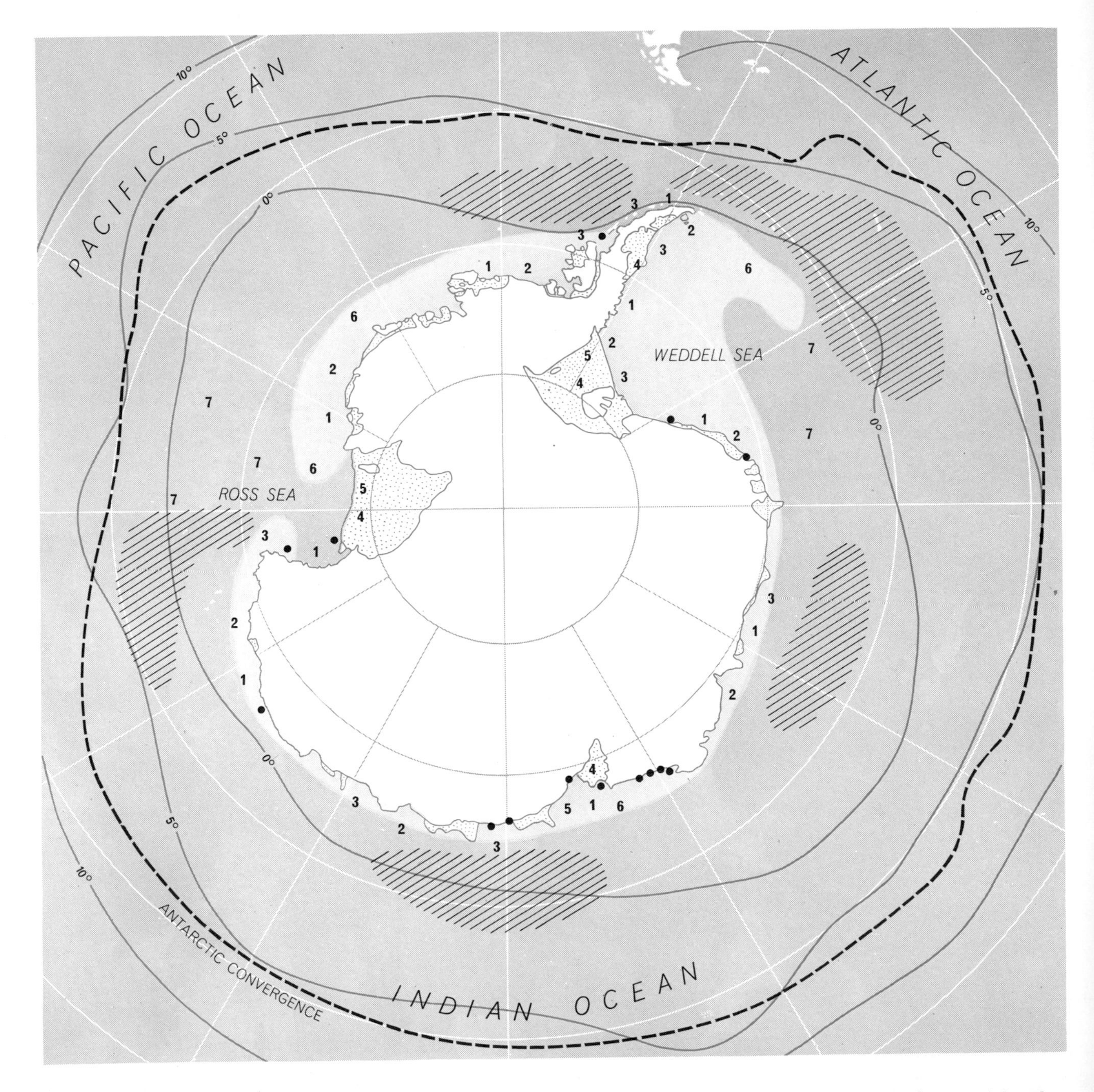

HABITATS of the most important marine animals of the Antarctic continent and its surrounding waters are shown on this map. In the case of emperor penguins the known rookeries are marked, and major concentrations of whales are shown. In the case of the other animals the numbers are spotted to indicate their approximate sequence as one approaches the Pole rather than specific sites at

hemoglobin-bearing corpuscles in certain visceral organs of these fishes only partly clears up the respiratory mystery. As among other groups of Antarctic animals, the southernmost Nototheniids tend to be completely circumpolar, whereas kinds that are characteristic of less polar waters are likely to be limited to different peripheral areas. From all of these observations ichthyologists conclude that the Antarctic fish fauna must have evolved during a period of cold isolation, dating perhaps from the earlier half of the Tertiary period, which is reckoned as running from about 70 million years ago to the beginning of the Pleistocene, one million years ago.

A complete chapter in an early work on the natural history of Iceland consists of the single sentence: "There are no reptiles in Iceland." For the same reason this discussion must now skip from fish to the warm-blooded vertebrates.

Like the submarine denizens of the Antarctic waters, the birds exhibit a degree of endemism that suggests long isolation. Penguins, of course, come first to mind. Actually penguins are Southern Hemisphere birds, one of several disparate types of wingless bird that evolved on islands and subcontinents below the Equator in the absence of four-footed predators. The ranges of penguins are by no means restricted to the Antarctic; some kinds are of temperate or subtropical distribution. Yet there are four or five substantially Antarctic types. The point is made more forcefully by the petrels: here is an avian order as worldwide as salt water, but it has a sizable number of exclusively Antarctic forms.

PERMANENT PACK ICE
WHALES
● EMPEROR PENGUIN ROOKERIES
1 ADÉLIE PENGUIN
2 SKUA
3 ANTARCTIC TERN AND PETRELS
4 WEDDELL SEAL
5 ROSS AND CRABEATER SEALS
6 LEOPARD SEAL
7 NOTOTHENIIFORM FISHES

which they congregate. The domain of the key crustacean *Euphausia superba* extends from near the coast to the Convergence.

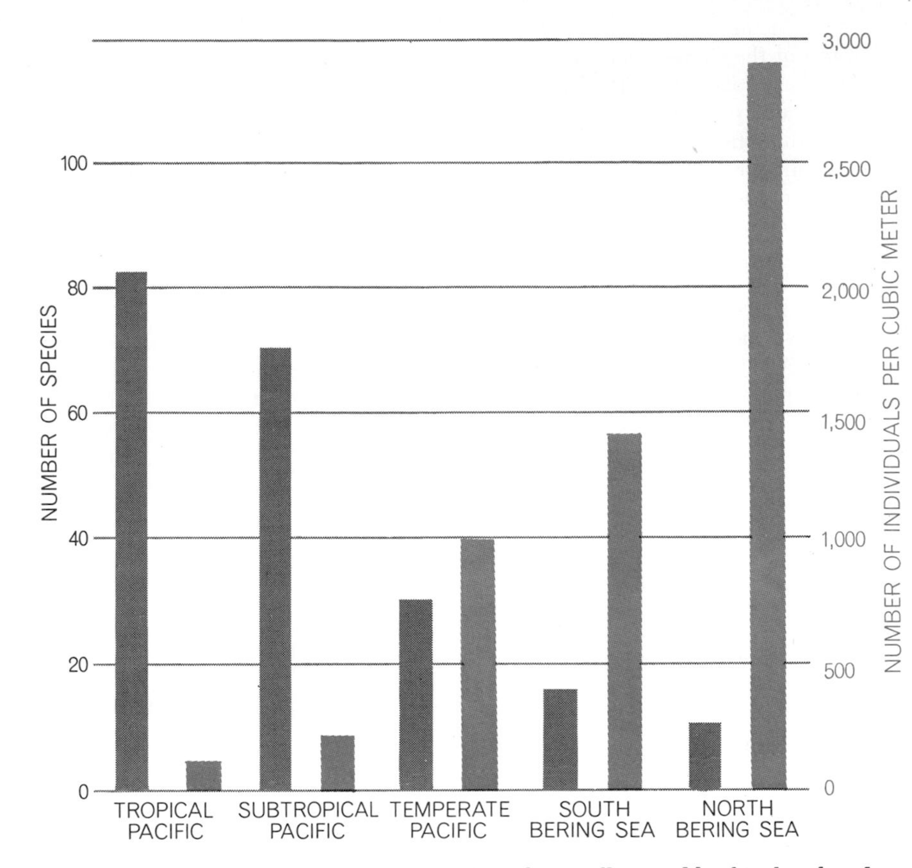

TEMPERATURE DEPENDENCE of species and numbers is illustrated by this chart based on a study by A. K. Brodskij of *Calanus*, a crustacean, in the Northern Hemisphere. As the water becomes colder the number of species found in the upper 50 meters decreases (*gray bars*), but the number of individuals in a given volume increases (*colored bars*).

Ignoring the trifling representation of land birds in the outer sub-Antarctic islands—including the pipit of South Georgia, the southernmost of all terrestrial birds—the typically Antarctic species are sea birds and ought to be of as much interest to the oceanographer as they are to the ornithologist. There are about 30 species that can be reckoned as Antarctic; of these some 15 nest on the Antarctic continent itself. This is not a large number, but the size of many of the populations of each species is exceedingly large.

The discrepancy between the number of species and the number of individuals conforms to the general pattern of life in the Antarctic region. It can also be correlated with a similar discrepancy between breeding space and food resources. The supply of food is enormous, but suitable nesting places are relatively scanty. This leads to great concentrations of birds in the available territory ashore as well as to an interesting multiple use of the territory. Giant petrels, for example, are to be seen nesting on open flat surfaces, with smaller burrowing petrels in the ground around and beneath them, storm petrels in the intervening hummocks of lichens or mosses and still other birds occupying adjacent ledges and rocky niches or rock piles at the foot of talus slopes.

Prominent among the members of certain communities is the sheathbill (*Chionis*), an aberrant relative of the snipes, which is a shore bird in heritage but a sea bird in habitus. I have motion pictures of the sheathbill showing that it excretes drops of brine through its nostrils, a faculty supposedly reserved to "sea birds" and dependent on special salt glands in their nostrils. The sheathbill is the only bird without webbed toes that breeds in the Antarctic. As a scavenger and a hanger-on around the breeding sites of penguins, petrels and cormorants, it gets its food from the sea but, so to speak, secondhand. The same is true of the Antarctic skua, a relative of the gulls, which preys on the eggs and chicks of penguins. Since the International Geophysical Year the skua has been accommodating itself increasingly to leavings found around scientific stations.

In climatic conditioning and adaptation there are, of course, degrees of "Antarcticity" among the Antarctic birds. The wandering albatross is Antarctic in the sense that it nests on islands south of the Convergence, but only one member of its family, the light-mantled sooty albatross, penetrates deeply into the Antarctic Zone. Among penguins, the emperor and the Adélie breed on the shores of the Antarctic continent, whereas the chin-strap, gentoo and macaroni penguins occupy outer but still Antarctic belts. The emperor penguin is beyond doubt the most polar of all birds. Its range extends no farther south than that of the Adélie and less so than that of the snow petrel. But its imperial title is clinched by the fact that it breeds in the midwinter dark on the fast ice of the continent, carrying and incubating its single egg on top of its feet. By this regime, unique among all birds, the emperor incidentally escapes the predation of the skua, which winters on the northward oceans.

The 11 petrels that breed in the Antarctic Zone represent 11 different genera, which in turn include every natural subdivision of the order, from great albatrosses to tiny storm petrels. The zonal ties of these birds, however, vary widely. Only the snow petrel (*Pagodroma*) is unequivocally restricted to Antarctica. Other species nest there but may migrate to milder zones or even beyond to the tropics. Such diversity has obvious evolutionary significance: the species that make the farthest seasonal departures, Wilson's petrel for instance, have temperate and subtropical relatives. The more exclusively Antarctic petrels belong to the group known as the fulmars, which in both north and south polar regions are associated with water of minimum temperature.

For food both the endemic forms and the seasonal invaders prey primarily on krill, squids and a scattering of other invertebrates; the snow petrel, however,

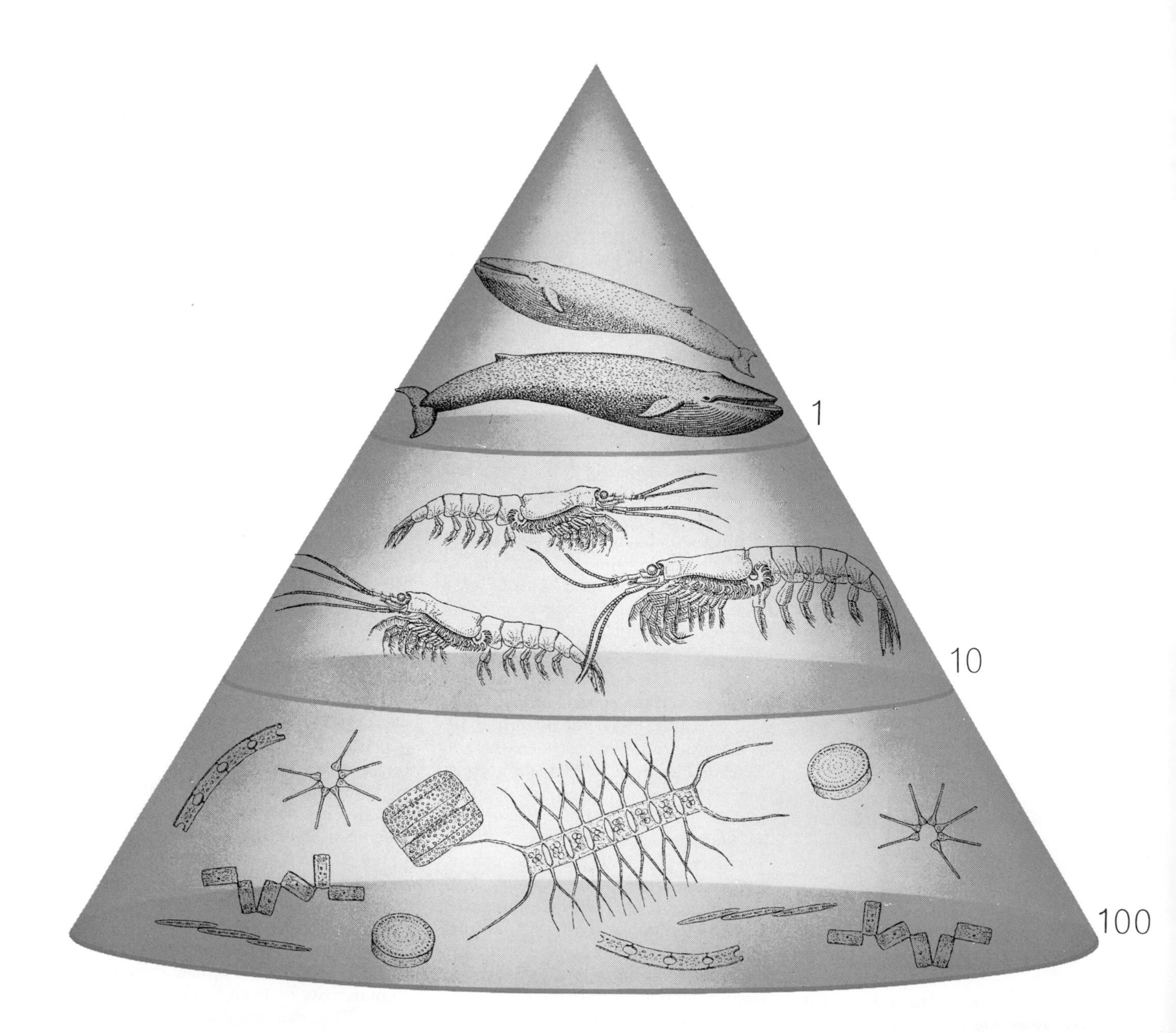

SIMPLIFIED FOOD CHAIN shows how each step in the process involves a "diminishing return." That is, it takes 100 units of phytoplankton, such as diatoms, to grow 10 units of krill, which in turn is enough to grow only one unit of its predator, the whale.

regularly captures many small surface fish in the pack ice. Although the rigorous environment exacts a high egg and chick mortality among all of these birds, adult viability seems to be exceptionally favorable. Recoveries made after intervals of one or two years have many times shown banded petrels still linked with their former mates and nest sites. The hazards of predators, such as the skua, and the severe physical milieu are alike weighted heavily against immature stages of all Antarctic species. But a single egg, produced only once a year, is sufficient to maintain populations estimated to be among the largest in the class Aves.

Of the three Antarctic terns, one (*Sterna virgata*) barely merits the designation, having a limited range on islands south of the Indian Ocean. The second, the Antarctic tern (*Sterna vitata*), is truly Antarctic, nesting on many islands in the circumpolar ring and even on parts of the continent; it is a coastal and relatively landbound bird. The third is the Arctic tern (*Sterna paradisea*), which migrates southward when winter approaches in the high Northern Hemisphere latitudes and probably enjoys more annual daylight than any other animal. During the Antarctic summer it waxes fat on krill in the pack ice.

The occurrence of four endemic species of seal again suggests the isolation of the Antarctic through a prolonged period of evolutionary history. The Antarctic seals appear to be genetically remote from the northern seals and still more so from their neighbors just beyond the Convergence: the southern fur seals and sea lions. All four of the typically Antarctic seals belong to the family (Phocidae) of the common harbor seal. All the species have undergone the same evolutionary divergence committing them more completely to life in the sea, a divergence that is beautifully expressed in both structure and function. They cannot "gallop" on all fours when ashore; their hind limbs, their principal means of propulsion while swimming, are trailed when they are out of water, forming merely a tail end to the body. In forward movement on land or ice, however, the hitching of the forelimbs is augmented by eellike sinuations. One of the species, the leopard seal, even manages to wriggle along on a flat surface with its foreflippers appressed against its sides. All the species, incidentally, mate in the water.

As oceanic rather than land animals, the Antarctic seals are circumpolar in distribution. But their ecological niches are so distinct that they completely

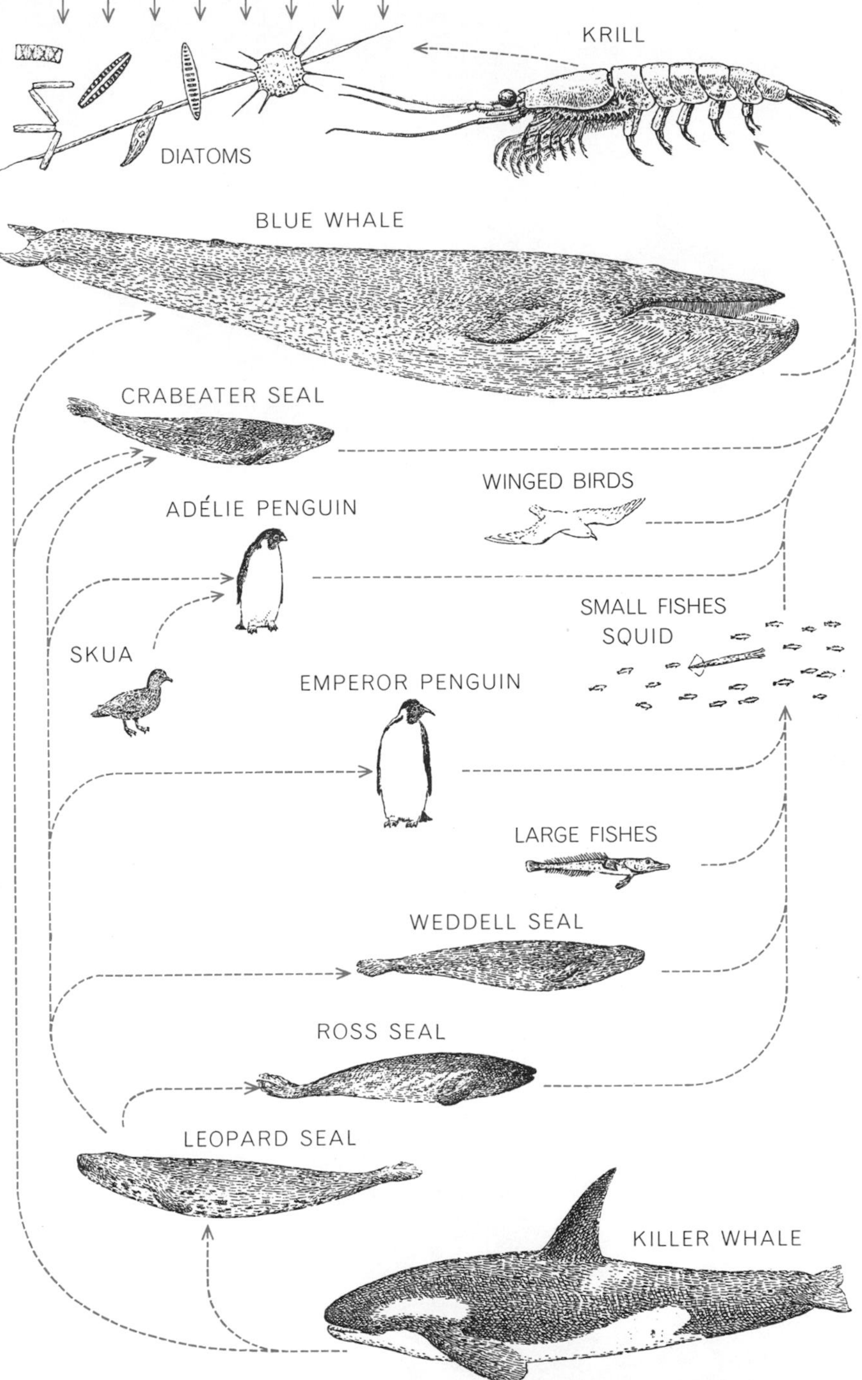

FOOD CYCLE of the marine animals is based on the krill *Euphausia superba*. This crustacean feeds on diatoms: microscopic marine plants that utilize the energy of the sun (*short arrows*) to transform nutrients in the water into living tissue. The krill is in turn the food of whales, penguins and other birds, the crabeater seal, squids and fishes. In addition to this basic cycle there is predation by seals, penguins and large fishes on squids and small fishes; leopard seal preys on penguins and other seals and the skua eats penguin eggs and chicks.

escape interspecific competition. Only one of them, the crabeater seal, has a relatively large population. The rarest, the Ross seal, may total fewer than 50,000 animals.

The first seal to be expected on a southward voyage would be the leopard seal, named for its color pattern, its ferocity or both. This animal is a solitary, large-headed predator, its jaws armed behind the sharp canines with rows of magnificent tridentated teeth. It also has anatomical specializations for swallowing large gobbets of food, one of these being a flat, ribbon-like windpipe that takes up no room in its throat except when the animal is actually drawing breath. Primarily a devourer of penguins

ADÉLIE PENGUINS congregate in large rookeries at the beginning of the Antarctic summer, hatch their eggs in December and remain ashore until the young can swim, in February. Then the Adélies move out to feed in the pack ice as the emperors come south.

(I have taken 160 pounds of penguin remains from the stomach of a female leopard seal on South Georgia), the leopard seal has also been seen dismembering the young of other seals. It is chiefly an inhabitant of the outer pack ice but it follows the migratory Adélie penguins down to the ice foot of the continental coast. Females, larger than males, attain a length of about 13 feet.

Next on a southerly course would be the crabeater, more properly the krill-eater, seal. The bulk of its food is the same reddish euphausian that is the mainstay of Antarctic whales and birds. Its cheek teeth are imitation baleen: the cusps interlock, forming a strainer functioning like that in the mouth of a whalebone whale. It can leap out of water like a penguin and it has a long conical head. Often met in big groups in the ice floes and usually reluctant to be stampeded from its basking place, the crabeater is the pack ice seal. Strangely, however, numerous bodies of crabeater seals have been found high up in the dry valleys on the continent of Antarctica, mummified by freezing and desiccation. Carbon-14 dating has recently shown some such examples to be more than 2,000 years old. A recent report by John H. Dearborn of Stanford University supports the opinion long held by George A. Llano, that such carcasses chiefly represent young seals that had remained too late in the autumn on the surface of the ice. When the leads froze over, they had no choice but to scatter aimlessly, gradually losing weight and starving. Those that died on the fast ice were rafted out to sea at the spring breakup; those that wandered inland left the famous and formerly puzzling mummies.

The Ross, or singing, seal (*Ommatophoca rossi*), named in honor of the Antarctic explorer James Clark Ross, is the rarest and least known of the Antarctic species. It is a creature of dense and tight pack ice and has almost never been found anywhere else. The smallest of the several south polar seals, attaining a length of about eight feet, it has a stout, turtle-necked appearance and its short head can be drawn backward more or less within the skin folds. Only a few score Ross seals have ever been observed and little has been learned directly about their life history. But the exceptionally large eye, the heavy, pressure-resistant construction of neck and thorax, the great size of both fore and hind flippers, and the reduction of the dentition to a series of curved, delicate and extremely sharp spikes enable one to surmise that

ADÉLIE PARENT shelters its chick beneath its warm body. Adélies prefer to build pebble nests on exposed or lightly snow-covered ground. Emperors incubate and hatch eggs on ice.

SKUA, called the eagle of the Antarctic, screams across a penguin rookery in search of stray chicks. Skuas start to breed in the Antarctic just as the Adélie chicks are being hatched.

LEOPARD SEAL, a voracious enemy of penguins, cruises in the water near an ice floe (*top picture*). Adélies escape (*bottom picture*) by propelling themselves out of the water to the ice, where the seal cannot overtake them. Adélies can shoot as high as seven feet.

it is a deep diver for the squids on which it mainly subsists. Squids, by the way, are the most numerous of all higher invertebrates in the ocean and several species are known only through examples recovered from the stomachs of sperm whales or seals.

The most polar seal of all is the Weddell seal, a denizen of the fast ice all around the continent. It is nonmigratory, remaining at high latitudes, except for stragglers, throughout the winter night. Its food consists of fish, together with squids and bottom invertebrates, which it must capture in the season of complete darkness as well as in the summer daylight. Edward Wilson, who died with Scott on the return from the South Pole, aptly matched the habits of the Weddell seal with those of the emperor penguin, whereas the crabeater could be likened to the Adélie penguin.

The canine and caniniform incisor teeth of this seal function as an extremely efficient saw for cutting through thick and flinty ice. It swings its head in a semicircle and splinters and pulverizes the ice below its snout. The water temperature in winter, only a degree or two below the freezing point of fresh water, is much higher than that of the atmosphere. The Weddell seals are able to keep "warm," therefore, by remaining submerged throughout much of the colder season. Their snorts and calls can be heard through the ice from air-filled chambers kept open with their circular saws. In common with most other seals they carry their unborn pups to a late stage of development; weaning and independence come at an earlier age than they do for most large mammals. The pups first enter the water when they are only about three weeks old.

Recently Carl R. Eklund of the Polar Branch of the Army Research Office and Earl L. Atwood, Jr., of the U.S. Fish and Wildlife Service have applied the technique of visual sampling and statistical analysis in an effort to estimate the populations of three of the Antarctic seals, the leopard, Ross and crabeater. They conclude that the Ross seal population may be larger than hitherto suspected. The species clings, however, to areas of very dense pack ice, which, before the advent of modern naval icebreakers, had never been penetrated by ships. Eklund and Atwood tentatively find that Ross seals may make up .8 per cent of the total population of these three Antarctic species, leopard seals 2.2 per cent and crabeaters 97 per cent. Converted into demographic numbers, this would mean about 50,000 Ross seals, 150,000 leopard seals and between five million and eight million crabeaters.

Although whales range all the oceans of the world, the Antarctic is the region in which they are found in greatest abundance. It is also the home of a few kinds that are restricted or adapted to the peculiarly severe environment. Today whales constitute the chief commercial resource of the Antarctic. Overfishing has been rampant, in spite of earnest efforts toward international regulation. In a single season of the peak period, about 1937, more than 45,000 whales were killed south of the Antarctic Convergence.

Any doubt about the thoroughness of modern techniques of exploitation is dispelled by figures showing the successive reduction of one species after another, beginning with the humpback whale. The latter, a relatively small, fat and easily handled whale, at first made up nearly 100 per cent of the catch. It now constitutes less than 2 per cent. Today whales are taken on quota and during a limited season in "blue whale units," according to which each of the smaller species is assigned a ratio to the size of the biggest animal of all. The right whale, formerly the most valuable of the whalebone whales, has had to be given special international protection, and harpoon guns, or the still newer and more efficient electrocution, now bring the other major baleen species—the humpback, the sei, the finback and the blue—into the hold of the modern whaling ship. Female whales accompanied by calves are legally protected at all times.

All the species of the genus *Balaenoptera* filter crustaceans out of the water through their baleen. The sei whale, and to a certain extent the finback, also capture good-sized fish.

The entire lives of the southern whales are dominated by the seasonal breeding and feeding migrations. It is the presence of the richest of pastures that brings these giant animals to Antarctic waters, where they reach their peak numbers about February and are at a minimum in July and August. The Antarctic also nurtures a host of lesser cetaceans about which little is yet known. They include such baleen species as the little piked whale (*Balaenoptera acutorostrata*), beaked or toothed whales such as *Berardius bairdii*, the pygmy right whale (*Neobalaena*) and the ferocious killer whale (*Grampus orca*), the whale that is the predator of whales and other mammals of the sea.

FROZEN REMAINS of large fish were found on the exposed ice of the Ross Ice Shelf in 1960. Carbon-14 dating established that the carcasses were about 1,100 years old. Apparently the fish had been trapped when ice touched the sea floor. As the bottom ice melted and the top was eroded by wind, the remains slowly worked their way up through 100 feet of ice.

LARGE FISH, a Nototheniid 52 inches long and weighing 58 pounds, was caught with the aid of a seal. The men were preparing to pull up a fish trap when a seal burst through a hole in the ice holding the fish in its mouth. The men, less surprised than the seal, got the fish.

THE ICE FISH

JOHAN T. RUUD
November 1965

Shortly after I received my undergraduate degree in zoology I spent a season in the Antarctic aboard the whaling factory ship *Vikingen*. One day in October, 1929, as the ship lay at anchor in a fjord of the South Atlantic island of South Georgia, I walked along the deck and stopped to watch some members of the crew get ready for a fishing trip in the fjord. One of them, an old whale-flenser who had spent several seasons around South Georgia, said to me: "Do you know there are fishes here that have no blood?" Certain that he was trying to pull a greenhorn's leg, I answered: "Oh, yes? Please bring some back with you."

Knowing my animal physiology, I was perfectly sure that such a fish could not exist; in fact, many textbooks firmly state that all vertebrates possess red blood corpuscles containing the respiratory pigment hemoglobin. There are a few trivial exceptions to this rule. The larvae of the common eel do not develop red blood corpuscles until they metamorphose into young adults. The lamprey, the primitive eellike fish that feeds by sucking the blood of other fishes, has a respiratory pigment that is slightly different from hemoglobin.

When the flenser and his friends returned from their day's fishing, I asked them: "Where are your bloodless fish?" As I expected, they replied that they had not caught any. Then, after my return home in 1930, I happened to mention the episode to a fellow student, Ditlef Rustad. Two years earlier Rustad had spent a season in the Antarctic as the biologist on a Norwegian expedition, and much to my surprise he said: "I have seen such a fish." He produced photographs of what he called a "white crocodile fish," which he had examined at Bouvet Island in the South Atlantic.

The seemingly bloodless fish proved to be a member of the chaenichthyid family; the family is one of many in the order Perciformes, or perchlike fishes, the largest and most diverse group of bony fishes. Within this enormous order the chaenichthyids belong to the suborder Notothenioidei, a group of fishes found exclusively in the Antarctic or waters adjacent to it. On the basis of Rustad's photographs the Swedish ichthyologist Orvar Nybelin later declared that the crocodile fish from Bouvet Island was a new species, which he named *Chaenocephalus bouvetensis*. This species was distinct from another species of the same genus: *C. aceratus*, which had been known from South Georgia and other islands in the region. In his description of the new species, published in 1947, Nybelin quoted Rustad to the effect that the blood of the fish was colorless and noted that *C. aceratus* was also said to have colorless blood. So it was that 20 years after Rustad's visit to Bouvet Island the relevant ichthyological literature mentioned for the first time a fact of which whalers in the Antarctic had been aware for half a century. Only one earlier reference to an antarctic fish with colorless blood is known: in 1931 L. Harrison Matthews, now scientific director of the Zoological Society of London, mentioned in his book *South Georgia* that such fishes were found there and also that their gills were white instead of the usual red. The whalers had called the fishes crocodile fish because of the shape of their jaws or ice fish because of their transparency.

In 1948 my attention was directed to the ice fish for a second time. One of my students, Fredrik Beyer, returned that year from the Antarctic, where he had served as the biologist accompanying the "Brattegg" expedition. He brought back with him some specimens of a third chaenichthyid species—*Champsocephalus esox;* they had been caught with hook and line in the harbor of Punta Arenas, the subantarctic port in the Straits of Magellan. Beyer reported that when the fish were alive they had had creamy white gills and colorless blood.

A further opportunity to look into the riddle of the ice fish came in 1950, when a whaling company operating around South Georgia asked our group at the University of Oslo to undertake

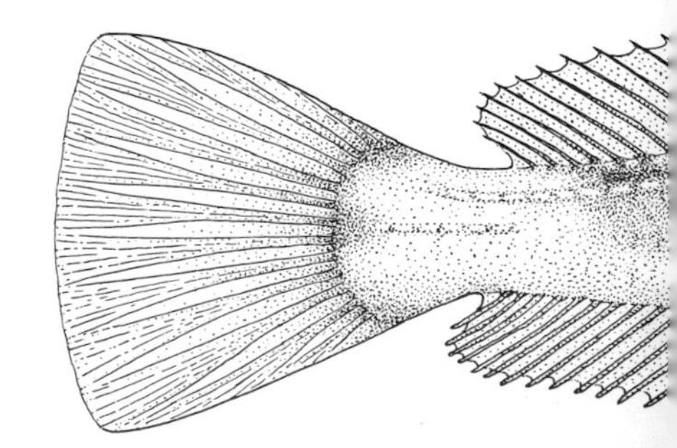

ANTARCTIC ICE FISH caught off the island of South Georgia is a member of the

a survey of the South Georgia "cod"—actually a species of the nototheniid family, which, like the chaenichthyids, belongs to the Notothenioidei. This fish was so common in those waters that the development of a commercial fishery was being considered. Another of my students, Steinar Olsen, undertook the survey; he also promised to keep his eyes open for any species of ice fish he might run into. In 1951 Olsen returned with preserved specimens of three separate species of ice fish—and samples of their blood. The blood was nearly transparent, with a slightly yellowish cast. Under the microscope a few cells could be recognized as white corpuscles, but neither red corpuscles nor any blood pigment could be found.

So far, so good: we had some facts to go on. The ice fish evidently lacked both red blood corpuscles and hemoglobin. But vertebrates ought to have some means of storing oxygen in their blood. On what kind of mechanism did the ice fish rely? The first step toward answering the question would be to determine the oxygen-storage capacity of a freshly drawn sample of blood, and to do this it would be necessary to go to the Antarctic equipped with the appropriate apparatus. My chance came in 1953, when I volunteered as leader of an international whale-marking expedition.

Our plan was to cruise in antarctic waters for six or seven weeks before the 1953–1954 whaling season began, marking as many whales as possible; our marking party would then be landed at Husvik Harbor on South Georgia just before Christmas to await a vessel due to sail back to Europe sometime in January. This schedule would give me two or three weeks on South Georgia during which I might have an opportunity to catch and study some ice fish.

Everything went as planned; we landed at South Georgia a few days before Christmas and three ice fish—all of them belonging to the local species *C. aceratus*—were brought to me on the first and second day of my visit. This was before I had been able to set up my little laboratory and test my apparatus. The obvious thing to do was to draw the blood from the specimens and store it in a refrigerator until I had everything under control; at the same time I assumed that additional ice fish could be obtained more or less at will.

The gills of *C. aceratus* were indeed creamy white, and the freshly drawn blood was nearly transparent. I found that, if the blood was left undisturbed, it clotted to the consistency of jelly in two or three minutes. When the blood was spun in a centrifuge, a sediment of white corpuscles settled out of it. This sediment constituted less than 1 percent of the blood by volume; the fluid above it was as clear as water.

After the blood samples had been adequately aerated, I determined their oxygen content; it was .67 percent by volume. This was a surprisingly low value. Ice fish are not small; some are two feet long or longer. It is generally assumed that a good-sized animal must have some reasonably efficient means of storing oxygen in its blood, so that its circulatory system can carry this vital element from the respiratory membranes to the tissues in the body most distant from them. Therefore my first conclusion was that something was wrong either with my apparatus or with the stored blood samples.

To check the apparatus I ran an analysis of the oxygen content of my own blood and the blood of some red-blooded nototheniid fishes; everything came out correct. Since there was no experimental error, the next check required another ice fish. Then my wait for more specimens of ice fish began. They seemed to have disappeared completely from the vicinity of Husvik Harbor; no ice fish entered our fish traps.

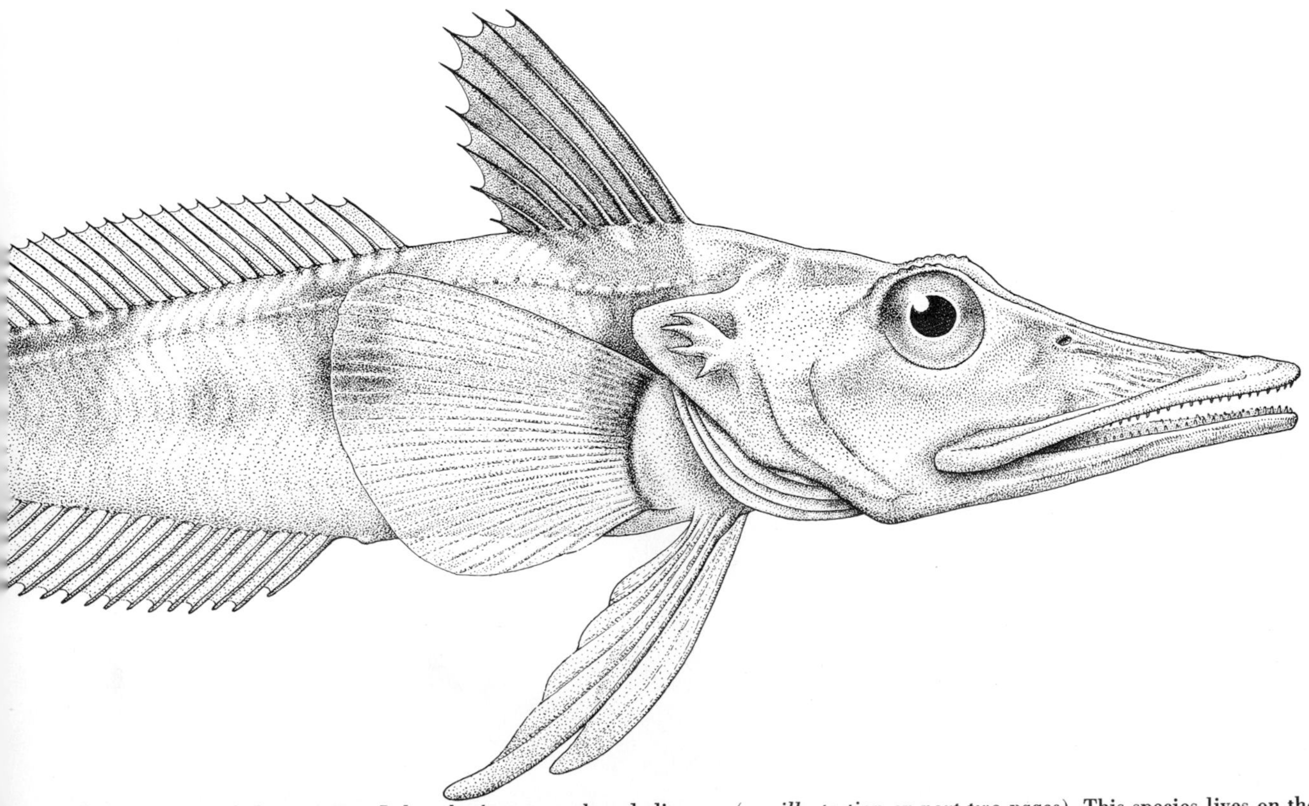

species *Chaenocephalus aceratus*. It has the large mouth and slim body characteristic of all 16 species in the family Chaenichthyidae (*see illustration on next two pages*). This species lives on the bottom. It is not a small fish; it grows to a length of about two feet.

and none took my baited hooks. As the day of our marking party's departure for Europe drew near I sent out an appeal to all the other whaling stations on South Georgia. Success! Felix Richards, the physician at the Leith Harbor station, appeared one afternoon in a motorboat carrying a live ice fish in a barrel of seawater.

I went to work immediately with some freshly drawn blood. The average of eight analyses was .72 percent of oxygen by volume—a result essentially the same as the earlier one. It was evident that the oxygen in saturated ice fish blood is no more than the amount the blood plasma can take up in solution; no other mechanism for the storage of oxygen could possibly be present.

This conclusion raised two related questions. First, how is it possible for a vertebrate of this size to survive in spite of what amounted to a complete anemia? Second, how did this lack of red blood corpuscles and hemoglobin evolve? The questions can be asked about all 16 species of chaenichthyid fishes. So far, however, the only species for which the oxygen capacity of the blood has been determined is *C. aceratus.* Let us examine what is known of the ecology of this species to see what light it may cast on our two questions.

First, Olsen's studies of 1950 and 1951 on South Georgia showed that although the adults of two other local chaenichthyid species live in the open ocean, feeding on the same shoals of

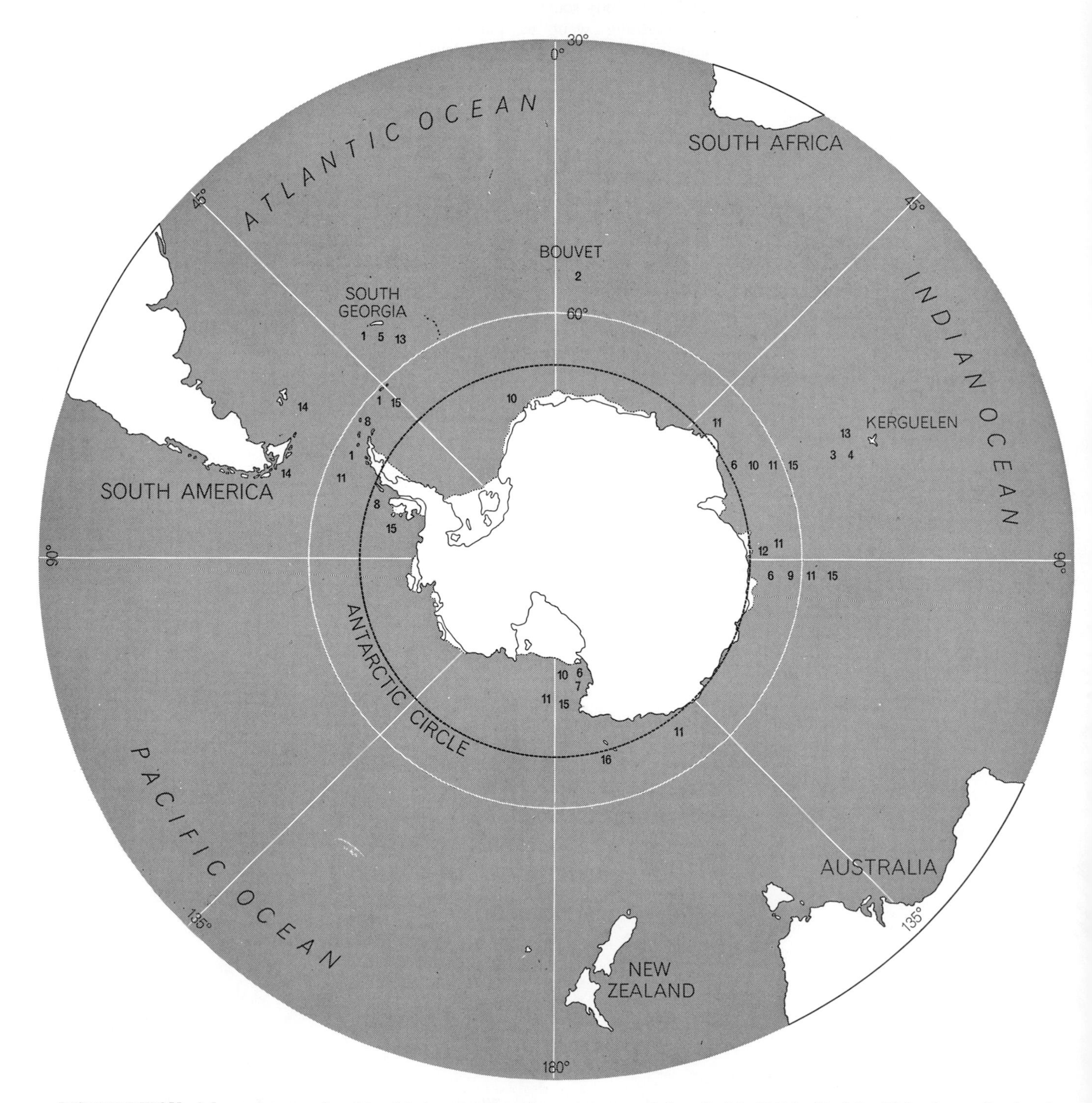

DISTRIBUTION of the various species of ice fish in antarctic and adjacent waters is plotted on this map. The first specimen of this family was taken near Kerguelen Island early in the 19th century and described in 1844 by Sir John Richardson, who placed it in a new genus, *Chaenichthys*. The newest family member is *Chionodraco myersi*, described in 1960 by Hugh H. DeWitt and

shrimplike krill that are the staple food of the baleen whales in antarctic waters, the young of these species and both the young and the adults of *C. aceratus* live near the bottom in deep fjords and along the continental shelf. This means that these particular ice fishes have a quite stable environment; these deep waters are well aerated, rich in food and vary in temperature only a few degrees throughout the year (from 2 or 3 degrees above zero centigrade to 1.7 degrees below zero). This steady low temperature is significant because in cold-blooded animals such as fishes the consumption of oxygen roughly doubles with a rise in temperature of 10 degrees C. It appears that even during the cool antarctic summer *C. aceratus* moves from offshore waters to deeper ones, and so avoids the increase in metabolic rate that would result from an increase in the water temperature.

Next, the shape of the ice fish's body —the big head and mouth and the rather narrow body and tail—suggests a sluggish way of life. These antarctic fishes resemble the angler fish, the sculpin or the sea robin of warmer waters; can we assume that their existence is somewhat similar? Imagine an ice fish resting on the bottom waiting for prey—mostly smaller fishes—to come within reach of a quick, snapping movement. The lurking ice fish is nearly invisible; except for a few dark spots arranged in vertical bands its body is transparent, merging almost perfectly with the stems of the surrounding kelp.

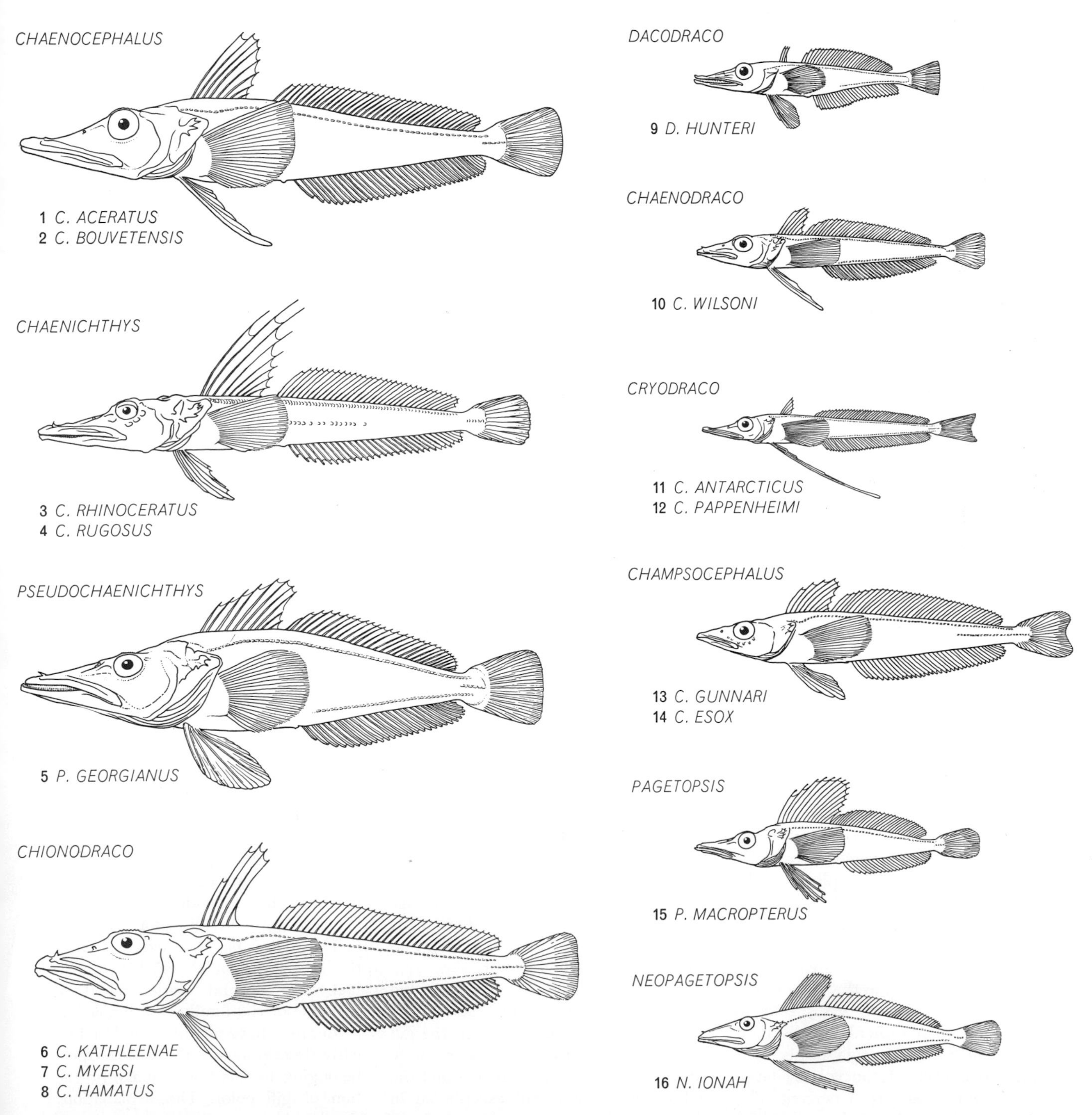

James C. Tyler. This species and *Cryodraco pappenheimi* are the only two ice fish that are not surely known to have white gills or colorless blood. The outline drawings show the conformation and the relative size of representative species among the 10 genera that comprise the ice fish family; each is shown at one-sixth of natural size. Where several species are named only the first is illustrated.

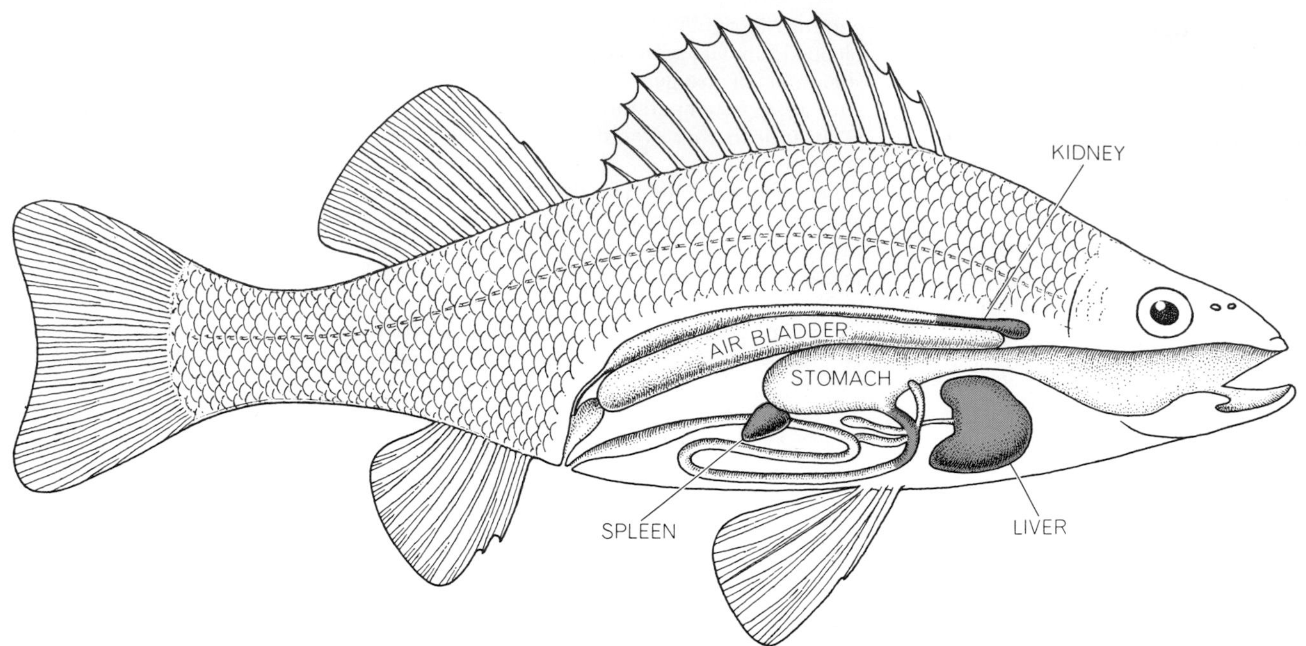

BLOOD-PRODUCING ORGANS of the spiny-rayed fishes are shown in this drawing of a sectioned yellow perch. They are the front portion of the kidney, the front portion of the alimentary tract, the spleen and, to a lesser extent, the liver. A microscopic examination of 18 spleens taken from specimens of *C. aceratus* revealed no evidence of the development of red blood corpuscles.

The size of its mouth means that quite large prey can be taken with a single bite; thereafter the ice fish can take a long rest while it digests the food that came its way.

What, in terms of oxygen demand, must the ice fish accomplish in the course of a quick meal of this kind? For the first stage of muscular activity no oxygen is required; the dominant chemical event is the production of the waste product lactic acid. It is only later that oxygen is needed, partly for the combustion of the lactic acid and partly for the regeneration of other substances involved in the first stage. This oxygen debt is paid off by drawing on the store of oxygen held in solution in the blood plasma. Thus far in the sequence, at least, the anemic ice fish is no worse off than any other vertebrate.

In vertebrates that have red blood corpuscles the oxygen removed from the blood plasma is readily replenished by drawing on the oxygenated hemoglobin in the red cells. Moreover, the amount of oxygen stored in the oxygenated hemoglobin is usually much greater than the amount dissolved in the plasma. For example, the blood of two specimens of red-blooded nototheniid fishes I examined on South Georgia respectively contained oxygen at the volume levels of 5.99 percent and 6.24 percent. This is an oxygen capacity of the same order of magnitude as that possessed by fishes that live under similar circumstances in arctic waters; in both cases about 90 percent of the available oxygen is bound to the hemoglobin molecules in the red cells.

What can the anemic ice fish substitute for this second store of oxygen on which most vertebrates draw? What alternative mechanism exists for paying the oxygen debt built up in muscles and other tissues? The oxygen removed from the blood plasma can only be replaced by means of a quicker intake of oxygen from the environment, together with a quicker or greater transport of oxygen from the site of intake to the site of consumption. These two generalized mechanisms are exactly the ones on which I believe the ice fish relies.

To consider oxygen transport first, I am fairly confident that compared with other fishes the ice fish has a larger volume of body fluid and blood. The heart of the ice fish is unusually well developed, and the only red muscle to be found in the fish is the dilated part of the aorta called the bulbus arteriosus. This suggests that the circulation of the blood through the ice fish's gills and body is strong and persistent.

As for the intake of oxygen from the surrounding medium, two anatomical facts must be considered. First, the gill chamber of the ice fish appears quite large, but measurements reveal that the gill surface area is not exceptional in size. Moreover, studies soon to be published by Jon Steen and Trom Berg disclose that the epidermal membrane covering the gills is fairly thick; this does not favor a quick exchange of gases.

What may be significant is the second fact: the ice fish is almost completely without scales, which means that respiration can occur through the naked skin over most of its body. A former student, Finn Walvig, has examined the skin's microscopic structure and found the skin richly supplied with blood vessels. The space between these vessels and the surrounding water is so large, however, that the ice fish's skin is also not particularly well suited to the exchange of gases.

We must therefore assume that the ice fish's respiratory exchange of gases is not unusually effective. It remains possible that the ice fish overcomes its oxygen handicap to some extent by circulating an unusually large volume of body fluid; this is known to be the case in other fishes under emergency oxygen conditions. What remains to be determined is to what extent, if any, the ice fish may be dependent on anaerobic (oxygenless) metabolism for its survival.

Thus far I had learned that five of the 16 species of chaenichthyid fishes had white gills and colorless blood. Was this true of the family as a whole? The reader may have already wondered why early descriptions of the various species belonging to this family make no mention of gill color. The explanation is simple. Almost without exception the fish collections brought back from the Antarctic by explorers were preserved in alcohol; this bleaches the specimen's gills and sometimes further discolors them because the alcohol extracts pigments from the body of the specimen. The ichthyologist whose eventual task it is to describe such preserved fishes cannot possibly give any verdict about the original coloration of the gills; indeed, he can be expected to assume that in life they were the usual red color. Only if the man who actually collected the fish and saw it alive has made a note of the gill color—as Rustad did for *C. bouvetensis*—will the describer have the benefit of this information.

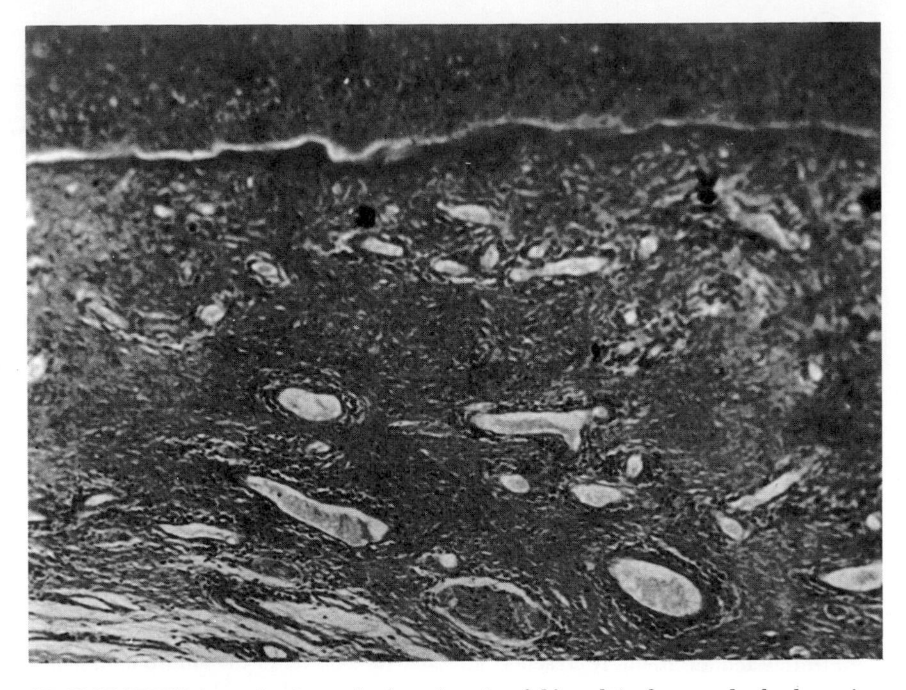

SKIN SECTION from the forward edge of an ice fish's pelvic fin reveals the layer just below the epidermis (*top*) that is richly supplied with capillaries and larger blood vessels. Although the number of blood vessels elsewhere in the skin of *C. aceratus* varies, the overall vascular system is well developed. The author suggests that the blood vessels of the skin constitute a second respiratory surface supplementing the ice fish's inefficient gills.

Fortunately there are other ways to overcome this absence of data, and I believe I have done so for 11 of the 16 chaenichthyid species. In the course of my work on *C. aceratus* I observed that the clot of blood remaining in the bulbus arteriosus of the fish after death had a consistency and appearance quite different from clots found in the bulbus of red-blooded fishes. The clot in a red-blooded fish is soft and opaque; under the pressure of a needle it disintegrates easily into individual corpuscles, the majority of which can be recognized as red blood cells. A clot removed from the bulbus of an ice fish, on the other hand, is nearly transparent and has a rubber-like consistency; it resists any attempt to break it down into smaller particles. This is exactly what one would expect of clotted blood plasma containing no cells apart from a minute number of white corpuscles.

Through the courtesy of F. C. Fraser, keeper of zoology at the British Museum (Natural History), I was able to remove and examine the clots from the bulbus of 11 separate species of chaenichthyid fishes in the museum's collection. All the clots appeared to be much the same as those I had found in *C. aceratus*. Although such an observation is not conclusive, I felt it was safe to assume that 12 species of chaenichthyid fishes, and probably all the other chaenichthyid species, have in common an almost complete lack of red corpuscles.

Our second basic question—how this anemia arose—must now be examined. Are the chaenichthyid fishes completely unable to produce red corpuscles? Or is the formation of red cells only suppressed and are the few cells that form nonetheless then destroyed so rapidly that they are extremely scarce or totally missing? In a normal bony fish the blood-forming tissues are the kidney, the spleen, the upper end of the alimentary tract and, to a lesser extent, the liver. Because the spleen is easy both to dissect out and to preserve, Walvig and I decided to enlist the cooperation of the shore party at Husvig Harbor on South Georgia in collecting a number of ice fish spleens. Walvig analyzed the collection and published his results in 1958; in summary, no cells could be found that appeared to be red corpuscles in the making. This finding reinforced my opinion that the red-cell deficiency is total, at least in the case of *C. aceratus*.

Other findings cast doubt on whether this is true of all chaenichthyid species. The Soviet biologist L. D. Marcinkevic examined seven chaenichthyid species during an Antarctic expedition in 1957. His conclusions were published in 1958; he states that he has found red corpuscles in limited numbers in the fishes' blood vessels and also that he has found evidence of red-cell formation in the liver. Although the scarcity of red cells in Marcinkevic's specimens—they are about half as abundant as white cells—means that they could play no significant part in oxygen transport, the possibility remains that what I have assumed to be a total deficiency may instead be only partial. If such is the case, further investigation may reveal varying degrees of reduction in the number of red corpuscles and the amount of hemoglobin in the separate species of chaenichthyid fishes.

With this theoretical consideration in mind, James C. Tyler of the Academy of Natural Sciences in Philadelphia has proposed that there may exist "white-blooded," or hemoglobin-deficient, species of fish among several other antarctic fish families. In fact, two nototheniid species he collected in 1958 and 1959 proved to possess fewer red cells and lower amounts of hemoglobin than is normal among bony fishes. The deficiencies, however, were far from being in a class with those of *C. aceratus* and the other species of the chaenichthyid family I have examined.

Without attempting to guess what future studies will show, I believe that the answer to the question of how these anemic fishes have managed to evolve is to be found in their unique ecological setting. In the case of the ice fish, for example, its major physiological asset—an unusually large volume of blood circulation—by itself would probably not be enough to keep the animal alive in temperate waters. Only when this characteristic is combined with the Antarctic's low and stable water temperature and that environment's abundant supply of food and oxygen is the survival of these peculiar animals possible. In fact, it is hard to imagine any other marine or freshwater environment that would offer a similar chance of survival should another family of anemic fishes begin to evolve elsewhere in the world.

33

THE SEA'S DEEP SCATTERING LAYERS

ROBERT S. DIETZ
August 1962

Nautical charts display hundreds of shoals rising from the deep sea and marked "ED"—existence doubtful. Each of them represents an echo sounding made by some ship passing through the area. Lacking the time to make a careful survey and fearful of running aground, the captain simply reports the reading to a hydrographic office, where it is duly recorded. More likely than not the sounding is spurious—a reflection not from the true bottom but from a "phantom bottom" now known to exist throughout most of the seas. A ship later passing one of the supposed shoals may find blue water to depths of two or three miles. But hydrographers are naturally reluctant to erase any possible hazard to navigation, so the charts remain cluttered with fictitious banks.

The existence of a phantom sound-reflecting layer was not recognized until 1942. At that time physicists were experimenting off San Diego with underwater sound for detecting submarines. Beyond the continental shelf, over water several thousand feet deep, their transmitted sound pulses, or pings, regularly and annoyingly returned an echo from about 900 feet. Unlike the sharp echo from a submarine, it was a diffuse, soft reverberation. On pen tracings of echoes from various depths it appeared as a layer of heavy shadowing. The zone confounded experiments only during the day; at sundown it would rise nearly to the surface and diffuse. With the first light it would re-form and descend to its normal depth. It never reflected all the sound energy striking it; the echo from the ocean bottom could always be detected through it, although sometimes only faintly. The source of the unexplained reverberation was named the deep scattering layer, or DSL (soon amended to DSL's because three and sometimes as many as five layers are often found).

At first it was supposed that the DSL's, like the D layer of the ionosphere, which fades away at night, had a physical cause, such as a temperature discontinuity. No one, however, could suggest a physical effect that would account for the diurnal migration. Martin W. Johnson, a zoologist at the Scripps Institution of Oceanography, surmised almost at once that the echo must come from marine animals that rise to the surface at night and return to the depths in the morning.

This interpretation is now universally accepted. What animals they might be, how they survived the enormous changes of pressure and temperature during their migrations, what physiological mechanisms were involved in the process—all these were mysteries. Even now the answers are far from complete. Nevertheless, this nuisance to the physicist has presented the biologist with a powerful new ecological tool for understanding the mass distribution of life in the sea.

Among the first animals suggested as a deep sound-scatterer was the squid, which lives throughout the oceans. Some investigators have wishfully supposed that the DSL's consist of vast schools of large, commercially valuable fish. It has been found, however, that such large fishes, which are sometimes present in the scattering bands, return hard echoes that stand out against the soft reverberations of the DSL's. Those of us who study echograms call the animals "tent fish" or "blob fish," depending on whether their echo is traced as an inverted V or an irregular bulge [*see illustration on pages 308 and 309*]. Almost always these larger fishes rise to the surface at night. Sometimes they can be "seen" on the echogram working their way toward the surface in the evening along with the DSL's. Curiously they rarely show up in the descending scattering layers of early morning. Instead they suddenly appear at full depth.

Today it is clear that the scattering layers consist of small, nocturnal marine organisms. They cannot be too small; at the frequencies used by echo

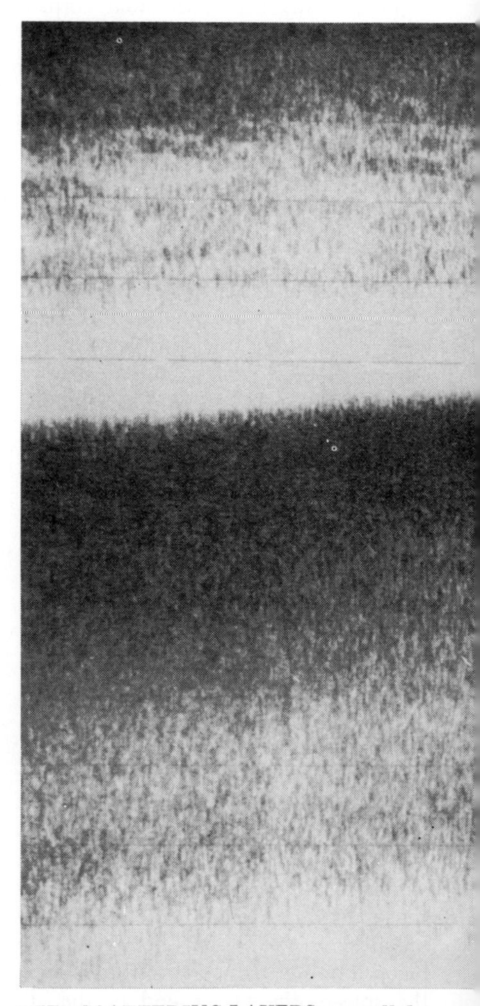

DEEP SCATTERING LAYERS are well developed in this echogram, made in the deep

sounders nothing shorter than about an inch will scatter sound effectively. Among habitats only the sky offers as little refuge as the sunlit upper regions of the sea. An animal sighted by a predator in either place is almost as helpless as grass in a meadow. In the open sea many small organisms conceal themselves by assuming the same transparency and refractive properties as the water itself. Others contain poisons or have hard shells or sharp spines. The animals of the deep scattering layers hide in the dark. In the daytime they seek the deep water, where sunlight hardly penetrates; they rise only at night to browse in the plankton-rich surface waters.

During the day DSL's lie at depths roughly between 700 and 2,400 feet. At night they rise almost to the surface and diffuse, or they may merge into a broad band extending down to 500 feet. The nature and the complexity of the bands vary with time and place. Off California they generally lie at about 950 feet, 1,400 feet and 1,700 feet. Most places in the ocean usually have three layers, the deepest at an average of 1,900 feet. In studying echograms covering about 200,000 miles of ship tracks, the deepest one I found was at 2,350 feet south of the Aleutian Islands. This must mark the approximate boundary of the twilight zone even in the clearest waters. From bathyscaph dives we know that the last glimmer of light fades into blackness at about 2,000 feet. The twilight organisms have more sensitive eyes than ours and their threshold of vision extends somewhat deeper. Even if their eyes can respond to single photons of light, as has been suggested, the greatest depth at which they can possibly see anything is no more than 3,000 feet.

Occasionally discrete echoes have been observed from depths below 3,000 feet. They may be caused by animals that sometimes descend this deep, but no one really knows. Scattering bands also appear in the waters above the continental shelves, which are no more than 600 feet deep. None of these compare in importance or stability with the "true" DSL's, which may be defined as the layers between 700 and 2,400 feet, most of which rise to the surface at night.

Sometimes the echograms show very diffuse layers that stay at the same depth day and night. They may in part represent stay-at-homes among normally migrating species—immature forms or individuals that are resting or breeding. Some of the layers must consist of non-migrating organisms that always stay in the twilight zone, probably salps, ctenophores, medusae, pteropods, pelagic worms and others. In any case they too are less important than the migrating DSL's.

Although we are far from knowing all the animals that make up DSL's in all parts of the world, it is clear that two groups predominate: (1) myctophids, or lantern fish, so named because of the luminous spots on their bodies, and similar small bathypelagic fishes; (2) shrimp-like crustaceans called euphausiids and sergestids. Nets towed at night near the surface commonly pick up from one to five of these individuals per cubic yard; exceptionally rich bands bring in as many as 20 per cubic yard.

Whether the fish or the crustaceans

Pacific off the coast of Peru. The three layers can be seen descending to depth (*starting at right*) with the coming of dawn. Echogram covers about one hour. This echogram and the ones that follow were made by vessels of the Scripps Institution of Oceanography.

play the most important role in sound-scattering is an open question. Edward Brinton of the Scripps Institution of Oceanography has found the euphausiids to be 10 times more abundant than fish off San Diego. In their effect on sound pulses the lantern fish may offset the abundance of the euphausiids by their larger size—several inches compared with one inch for the crustaceans. Furthermore, many lantern fish have swim bladders, each of which contains a minute bubble of air that can resonate with the sound waves to make a highly effective scatterer.

The striking changes in animal population with depth in the sea were known long before the discovery of the scattering layers. Most biologists have tried to explain the phenomenon in terms of temperature. The late Danish oceanographer Anton F. Bruun divided the oceans into two temperature levels separated by the 10-degree-centigrade isotherm. The upper, warmer zone he termed the thermosphere; the lower, colder region, the psychrosphere. But the DSL's suggest that light exerts a considerably more important control than temperature does. From the evidence of scattering layers the sea can be divided into three light zones: the sunlit zone, inhabited by both plants and animals, extending down to about 500 feet; the twilight zone, populated only by animals, from 500 feet down to 2,400 feet; and the black abyssal region with its few and highly specialized animals. The lower boundary of Bruun's thermosphere rises with distance from the Equator, eventually reaching the surface in the higher latitudes. The scattering layers show no such warping. Of course toward higher latitudes tropical species fade out and boreal species become predominant, but all the animals move in response to light, not to water temperature.

Both the crustaceans and the lantern fish of the DSL's display acute sensitivity to light. On the darkest nights dip nets pick them up at the very surface, but even moonlight sends them down many feet. Echograms occasionally show scattering layers descending to moderate depth as the full moon rises. With the first glimmer of daylight, often an hour before actual sunrise, the various organisms begin their descent. Off San Diego the myctophids, the most sensitive to light, generally start down first, because they must dive to the greatest and darkest depth. A short time later the second layer, probably consisting of euphausiids, takes form and settles, and it is soon followed by the third layer, probably composed of sergestid shrimps. (The order cannot be exactly specified because it depends on the species present in a given population at a given time.) The layers never cross one another; each appears to be precisely adjusted to some particular level of twilight. Diving at speeds of as much as 25 feet a minute, the scatterers are at least halfway to their ultimate levels at sunrise. Within one hour after sunrise they attain their preferred depth. As the sun approaches the zenith, however, they sink a bit lower to their maximum depth to avoid the penetrating light rays. When passing clouds darken the sky, the scatterers react by rising to somewhat higher levels than normal for the daytime.

While aboard the U.S.S. *Cacapon* in 1947 I traced the DSL's all the way from California to the Antarctic. By day the ever present layers, sometimes dense and at other times thin, hung like decks of stratus clouds between 900 and 2,100 feet. Each night the scatterers rose toward the surface and diffused. As the days grew longer the diurnal migrations remained precisely synchronized to sunrise and sunset. But near Antarctica, where the nights were reduced to a mere four hours, the migrations seemed to break up in confusion. The organisms of the shadows apparently could not cope with a 20-hour day.

The DSL's are almost, but not quite, universal in the deep ocean. In the nearly lifeless central South Pacific the bands become extremely faint and may even disappear completely. They do not exist in the Arctic Ocean, where the permanent cover of pack ice cuts off so much sunlight that diatoms cannot flourish. The Arctic Ocean is the most sterile of all the seas.

Most attempts to photograph the scattering layers on film or by television have failed. Deep trawling with nets has also met with indifferent success in catch-

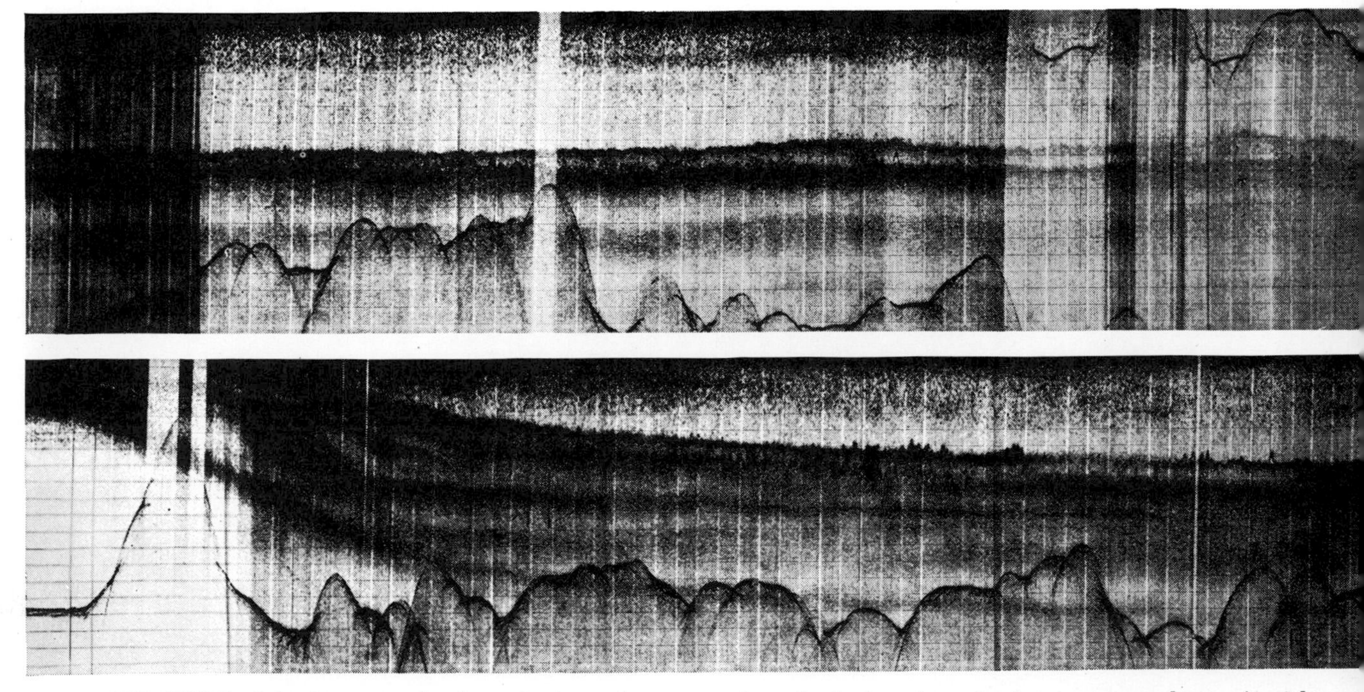

FULL DAY'S CYCLE of the deep scattering layers begins at far upper right with descent from surface at daylight. The organisms remain at depth throughout the day, in separate layers. At night they rise again to feed (*far lower left*). "Tent fish" and "blob fish"

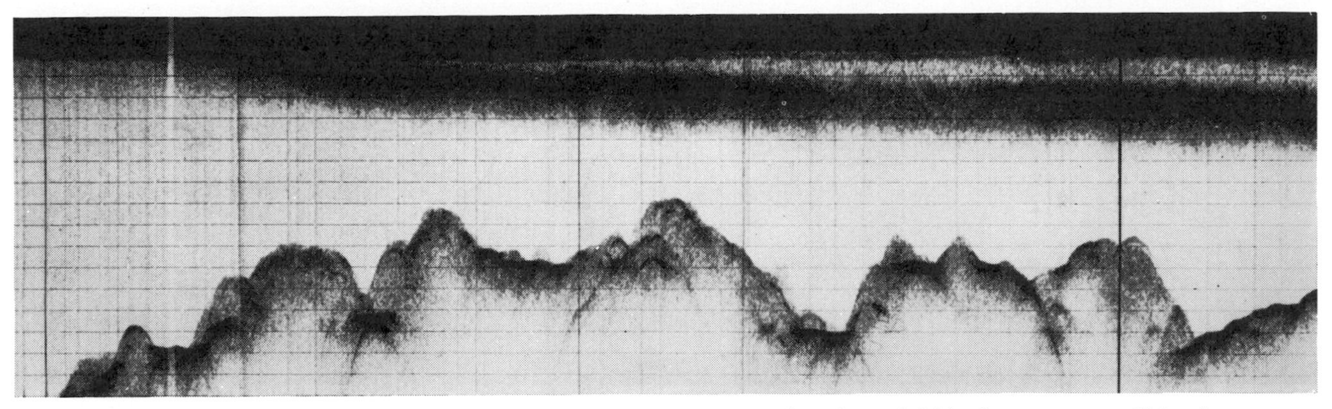

EFFECT OF BRIGHT MOON on deep scattering organisms is to hold them at shallow depth. Deep scattering layer can be seen rising to surface (*toward left*) after moon sets. This echogram was recorded between 11:45 p.m. and 1:15 a.m. in the eastern Pacific.

ing and identifying the DSL organisms. They are too small and too widely dispersed. Bathyscaph divers looking through portholes have fared somewhat better, at least when they did not expect to see great schools of fish. Crustaceans and lantern fish have appeared occasionally at the depth of the scattering layers but seldom in real abundance. On a bathyscaph dive to 3,600 feet in the Mediterranean I saw no organisms big enough to be effective sound-scatterers. Down to 2,100 feet there was an increasing abundance of minute suspended detritus from living plants and animals called sea snow; it scattered our underwater light beam as dust motes scatter a shaft of sunlight. Just below 2,100 feet the water abruptly became crystal clear; apparently this is the boundary between the ocean's twilight zone —the realm of the scattering layers and their light-sensitive organisms—and the almost lifeless, eternally dark waters of the abyss.

In bathyscaph dives off San Diego, Eric G. Barham of the U.S. Navy Electronics Laboratory has had better luck than I have had. He has repeatedly seen concentrations of four-to-six-inch fish between 650 and 1,000 feet; they were probably young hake. Hard echoes (tent fish) often come from these depths. Euphausiids have been found in the stomachs of these fish, so it seems likely that the hake feed on the scattering layer organisms.

In a recent bathyscaph dive into the San Diego Trough, Barham reported that he saw much sea snow but no large organisms between 850 and 1,200 feet. Then he entered a zone inhabited by deep-sea prawns. From 1,200 to 1,500 feet he saw so many of these sergestids that he could not count them. In the next 200 feet he encountered a large number of lantern fish. Below 1,700 feet the bathyscaph entered a region relatively free of large organisms. Then from 2,150 to 2,300 feet it sank through a zone containing the greatest concentration of fish he had seen on any dive. Again these appeared to be lantern fish, as many as eight in view at a time until he could not keep count of the sightings. Most tended to avoid the lighted area, and Barham saw them only at the edge of the cone of light. Within the abyssal zone, from 2,300 feet to the bottom at 3,900 feet, he saw only an occasional red

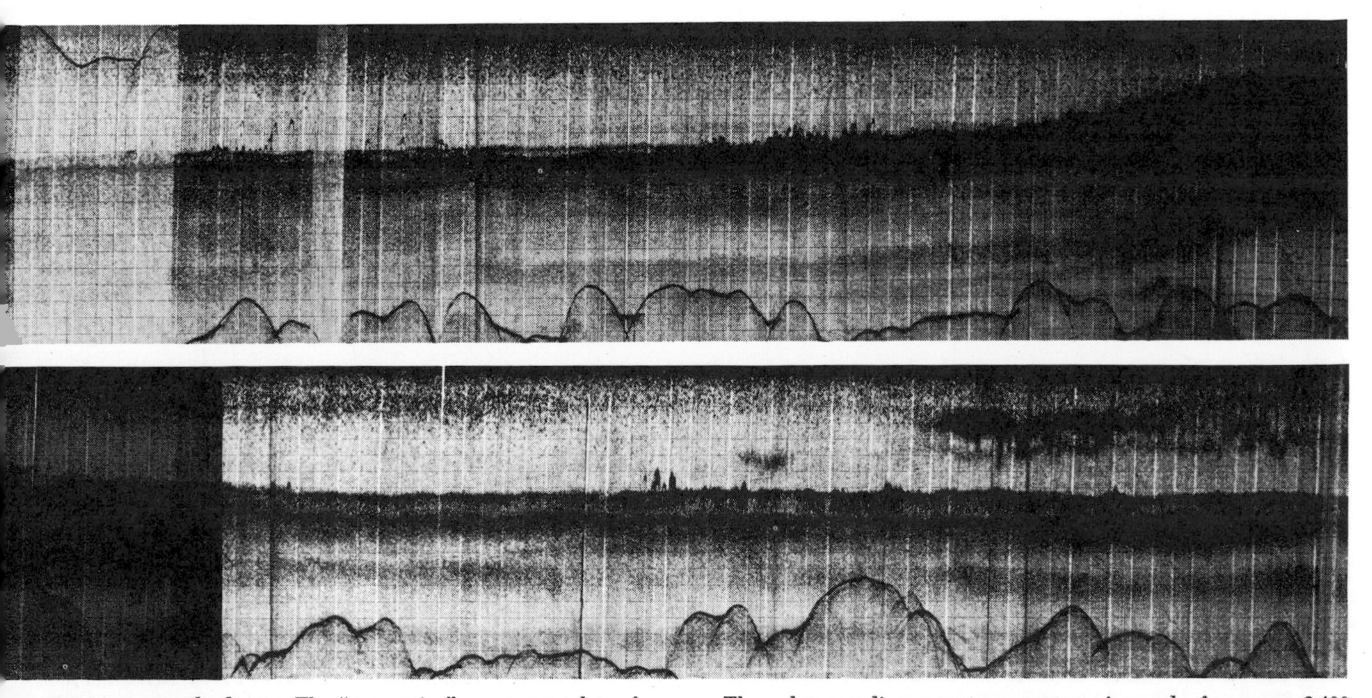

appear among the layers. The "mountains" represent echoes from the sea bottom, exaggerated 20 times in the vertical dimension. The echo-recording apparatus repeats its cycle for every 2,400 feet of depth; the bottom echo is actually on the fifth cycle.

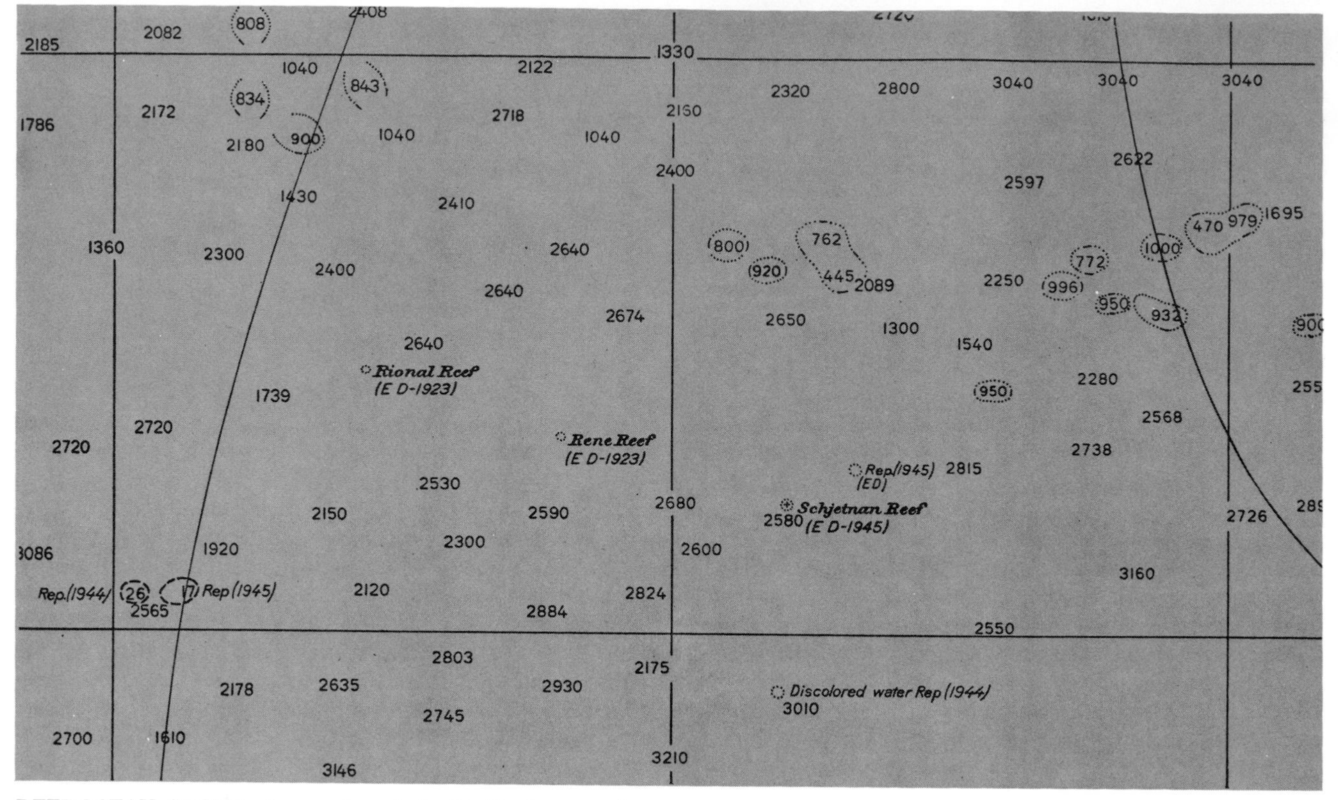

DEEP-OCEAN CHART of a portion of the Pacific between the Hawaiian Islands and the Marshall Islands displays four "reefs" marked "ED" (existence doubtful). Those reported in 1945 were found by echo sounders and could represent scattering phenomena.

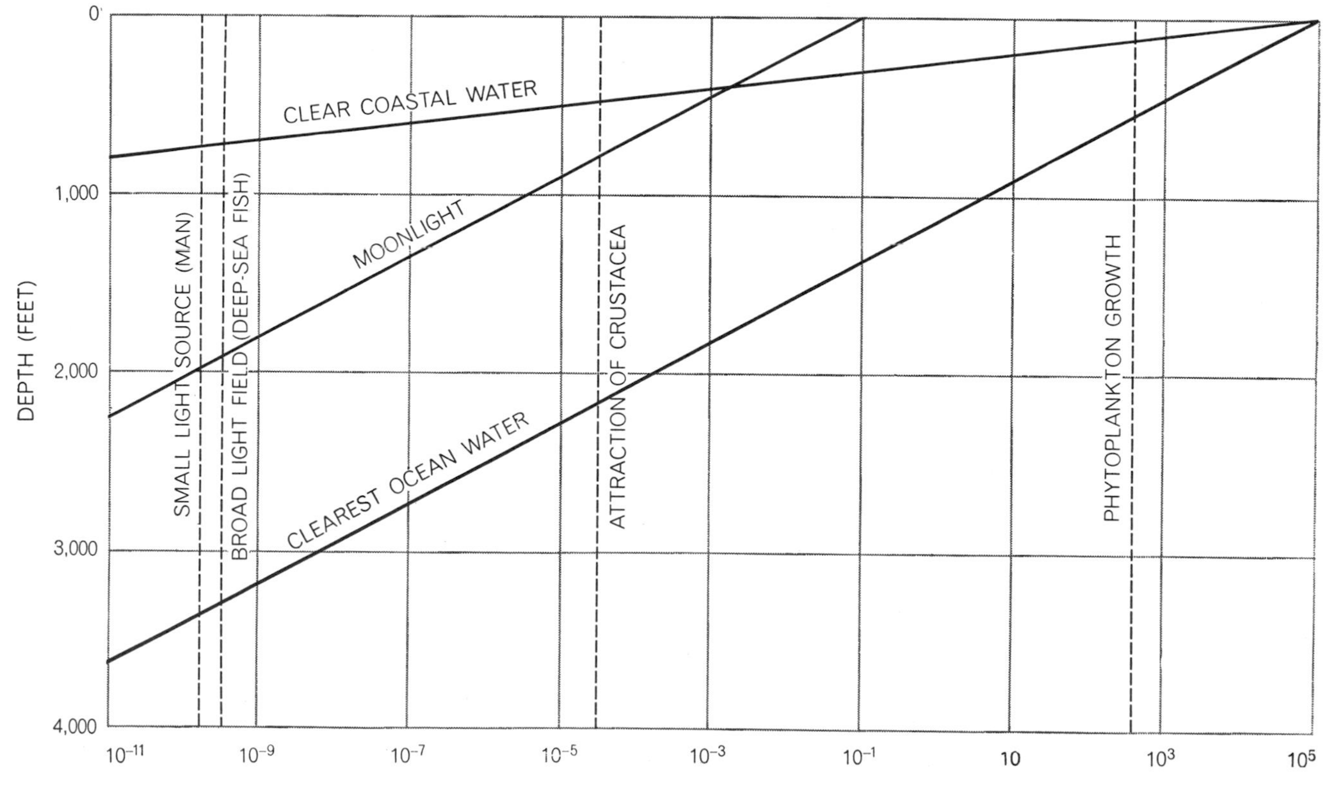

INTENSITY OF LIGHT decreases with depth. The lines "Clear coastal water" and "Clearest ocean water" refer to sunlight. Coastal water may appear clear to the eye but is always turbid. Black vertical lines denote amount of light necessary for each function. "Small light source (man)" is the faintest light source a man can detect. The broad light field for deep-sea fishes indicates minimum quantity of general daylight these fishes can see. More light is needed to attract crustaceans, and phytoplankton require light many orders of magnitude more intense in order to grow. The diagram is adapted from a paper by G. L. Clarke of Harvard University and the Woods Hole Oceanographic Institution and by Eric J. Denton of the Marine Biological Association laboratory in Plymouth, England.

prawn, some shrimplike mysids, medusae and two types of worm.

In spite of their distribution, the DSL's can never be a sea-food cornucopia for man. The echo effects indicate a concentration no greater than a few animals per cubic yard. Nevertheless, the animals of the scattering layer must constitute an important source of food for commercially important fish; surely they are an important link in the food chain of the oceans.

The DSL forms themselves feed on tiny surface organisms: euphausiids and sergestids are herbivores, grazing on diatoms; lantern fish eat other crustaceans that are nourished by diatoms. The DSL population in its turn falls prey not only to larger deep-sea fishes but also to surface swimmers. One reason oceanic banks and shelf margins make good fishing grounds may be that they are often richly populated by animals from the DSL's. During the night, when they are near the surface, many of the little creatures must drift over the banks and shores. In the morning they are trapped in the shallow lighted waters and are quickly devoured by predatory fishes. Above the submerged tops of sea mounts the DSL's sometimes practically disappear, presumably as a result of such devastating grazing.

As human beings we sometimes tend to forget that we are not the only end of nature's branching food chain. Until quite recently the source of food of the fur seal was a mystery. Every year during the breeding season some three and a half million of these voracious mammals migrate from their deep-water habitat far out in the northern Pacific and come ashore on the Pribilof Islands. Experience with fur seals in zoos indicates that it must take three and a half billion pounds of fish a year to feed the enormous wild population. Yet the fur seals never seem to compete with commercial fishing. Clearly they are tapping some vast reservoir of noncommercial sea food. Recently we have learned that in the deep ocean they feed largely on lantern fish, probably catching them at night when they rise to the surface.

Another eventual consumer of the twilight animals is the cachalot, or sperm whale. Unlike the seal, it goes down to hunt for its food. Although it is a mammal, forever tied to the surface, it sometimes takes a single gulp of air and dives as much as two-thirds of a mile to forage. The cachalot does not feed on the scattering-layer organisms but eats the squids and larger fishes—the echogram

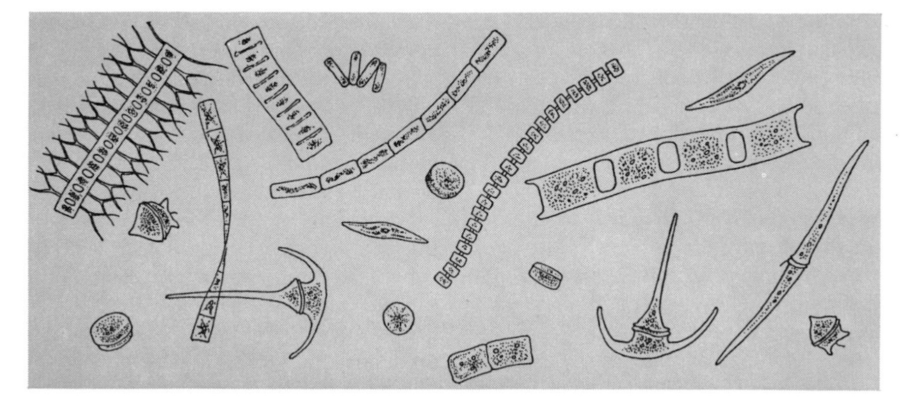

PHYTOPLANKTON, tiny drifting plants, are the "grass" of the sea: they form the basis of the food chain. These are diatoms and dinoflagellates, enlarged approximately 200 diameters.

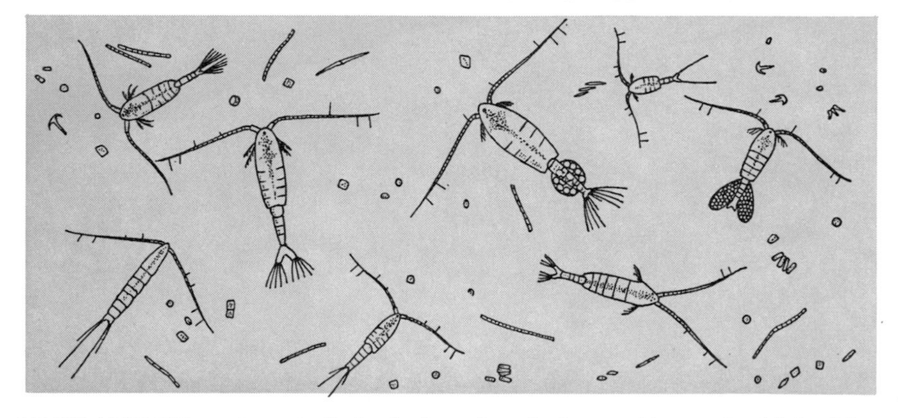

ZOOPLANKTON are tiny animals that feed on phytoplankton and in turn furnish food for larger animals. These copepods, enlarged about 10 diameters, drift near the sea's surface.

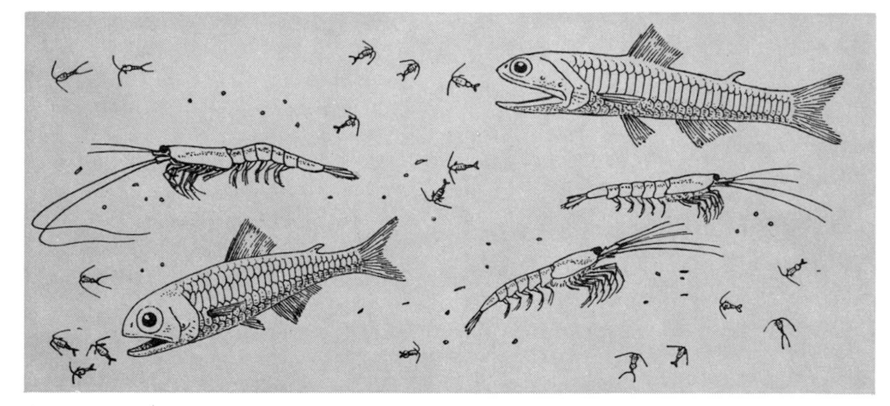

"DEEP SCATTERERS" rise from depths at night to feed in photic zone. The myctophids (lantern fish) eat copepods; the euphausiids (crustaceans) consume the phytoplankton.

LARGER ANIMALS, such as this squid and herring, may well eat the organisms of the deep scattering layers at night near the surface. Fur seals also feed extensively on myctophids.

blob fish and tent fish—that do feed on the scatterers. These constitute its whole diet. Many rare species of deep-sea fishes are known only from the stomach contents of the sperm whales that abound around the Azores and are still harpooned from open boats as they were by the New Bedford whalers of old. Apparently the animals of the deep scattering layers are safe during the day from foraging by the plankton-sieving baleen whales. According to the whale expert Raymond M. Gilmore of the Museum of Natural History in San Diego, Calif., these whales limit their dives to the upper 200 to 300 feet of the sea, or considerably above the DSL's.

A basic tenet of oceanography is that parcels of water throughout the ocean, although they look exactly the same, in reality differ from one another. We do not need precise chemical analyses to prove the point: the organisms of the sea are remarkably sensitive indicators. Nothing illustrates variation in the sea more dramatically than echograms and their DSL's. For example, off Peru the echograms are heavily blackened by the teeming life of the Humboldt Current, but a little farther out the life dwindles to nothingness. The scattering is an index of organic productivity and in turn of those chemical factors that control the distribution of living things.

It is unfortunate that the world's expanding population cannot look to the deep scattering layers as a direct source of food. Nevertheless, the organisms of the DSL's are well up in the pyramid that requires 1,000 pounds of diatom fodder to support the growth of a pound of commercial fish. The deep scattering layers play a major role in the biological economy of the seas.

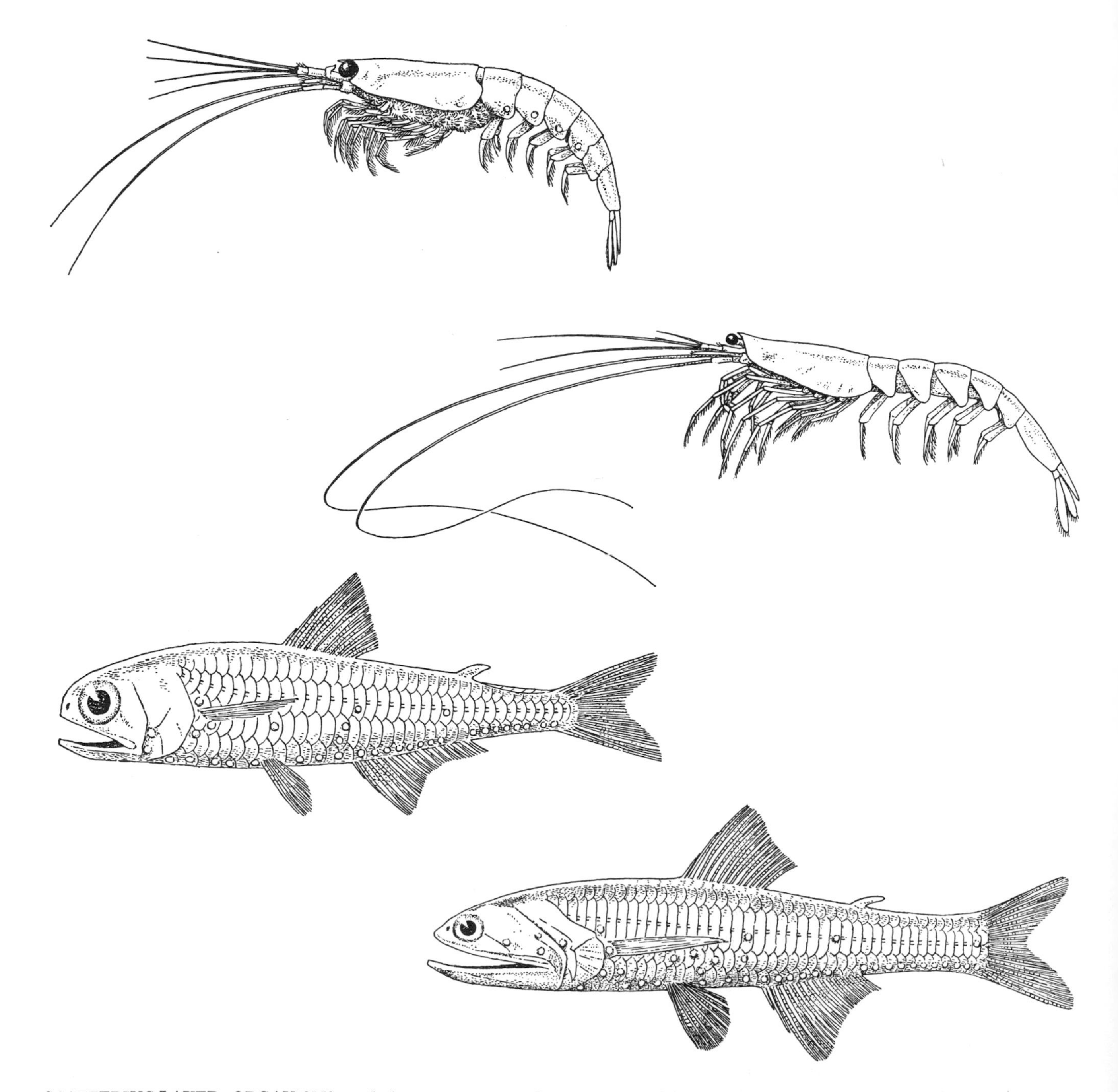

SCATTERING-LAYER ORGANISMS include crustaceans and fishes. Among them are *(top to bottom)* a euphausiid, a sergestid and two forms of myctophid, or lantern fish. The animals range in size from about an inch for the euphausiid up to three inches for the lantern fish. The four spots on the euphausiid and the similar spots on the myctophids are photophores, or light-producing organs.

THE LAST OF THE GREAT WHALES

SCOTT MCVAY
August 1966

The order Cetacea—the whales—consists of more than 100 species. The 12 largest either were in the past or are now commercially important to man. At one time their oil was valued as a lamp fuel and a high-grade lubricant. Today most whale oil is made into margarine and soap, some whale meat is eaten and the rest of the animal is utilized as feed for domestic animals and fertilizer. Whale hunting is profitable, and as a result a majority of the 12 commercially hunted species have been all but exterminated. In the 18th and 19th centuries the whaling vessels of a dozen nations sailed all the oceans in pursuit of five species of whales. In that period four of the five were hunted almost to the point of extinction; the complete disappearance of these whales during the age of sail was probably prevented only by a slump in the demand for whale oil late in the 19th century. With the age of steam another seven species (which were unsuitable as quarry in the previous era) became the hunted. These species are now also in danger of extinction. It is the purpose of this article to trace the circumstances that have allowed this sorry episode to be repeated in so short a time.

The ultimate fate of the great whales has been a question for more than a century. Herman Melville included such a query among the observations that make his *Moby Dick* an encyclopedia of whales and whaling: "Owing to the almost omniscient look-outs at the mastheads of the whale-ships, now penetrating even through Behring's straits, and into the remotest secret drawers and lockers of the world; and the thousand harpoons and lances darting along all the continental coasts; the moot point is, whether Leviathan can long endure so wide a chase, and so remorseless a havoc; whether he must not at last be exterminated from the waters." Melville's conclusion was that Leviathan could endure. This was not unreasonable at the time (1851), when whalers pursued whales in open boats and killed them with lances. The five species of whales that were hunted then included only one from the suborder Odontoceti, or toothed whales; this was the sperm whale (*Physeter catodon*). The other four were baleen whales, of the suborder Mysticeti. These were the bowhead (*Balaena mysticetus*), the two right whales (*Eubalaena glacialis* and *E. australis*) and a lesser cousin, the gray whale (*Eschrichtius glaucus*). All five were hunted because they do not swim too fast to be overtaken by oarsmen and because they float when they are dead.

The bowhead whale and the right whales suffered near extinction. Very few of these once abundant animals have been seen in the past decade. In May, 1963, for example, the Norwegian ship *Rossfjord* was steaming west of the Russian island Novaya Zemlya when a whale with "jaws [that] were extremely curved" was sighted. The event was duly noted in the *Norwegian Whaling Gazette* with the observation that the whale—clearly a right whale—belonged to "a species that the crew had not previously seen."

Today all four of these baleen whales (the gray, the bowhead and both right whales) are nominally protected by international agreement, although three right whales that were sighted in the Antarctic in the early 1960's were promptly killed and processed. Perhaps because of its gregarious way of life, the gray whale has managed a slow recovery from the pressure of hunting. An estimated 6,000 gray whales now migrate annually from the Arctic Ocean to their breeding grounds off the coast of Lower California [see "The Return of the Gray Whale," by Raymond M. Gilmore; SCIENTIFIC AMERICAN, January, 1955]. No one knows today how many (or how few) bowhead and right whales are still alive.

Of the eight great whales that are the quarry of modern whaling fleets, seven belong to the suborder of baleen whales and six of these to the genus *Balaenoptera* [*see illustration on next two pages*]. The largest of the six is the largest animal known to evolutionary history: the blue whale (*Balaenoptera musculus*). Weighing as much as 25 elephants and attaining a length of as much as 85 feet, this is the true Leviathan. It was known to Melville, who remarked that blue whales (he called them sulfur-bottoms) are seldom seen except in the remoter southern seas and are never chased because they can "run away with rope-walks of line." The modern catcher ship, armed with cannon-launched explosive harpoons, proved to be the blue whale's nemesis. Over the past 60 years antarctic waters have yielded more than 325,000 blue whales, with an aggregate weight in excess of 26 million tons [see "The Blue Whale," by Johan T. Ruud; SCIENTIFIC AMERICAN, December, 1956]. Even though the blue whale is now a rare animal, many of the statistics concerning baleen whales caught in the Antarctic continue to be reckoned in terms of "blue-whale units," as are the whaling nations' annual antarctic quotas.

Five other baleen whales of commercial significance are the finback whale (*Balaenoptera physalus*), the sei whale (*B. borealis*) and three smaller whales: Bryde's whale (*B. edeni*) and the two minke whales (*B. acutorostrata* and *B. bonaerensis*). The finback averages little more than 65 feet in length and yields only 10 tons of oil, compared with the blue whale's 20 tons. Accordingly in terms of blue-whale units it takes two finbacks to equal one blue. The sei is slighter, averaging 55 feet in length. It is comparatively blubber-poor and meat-rich; six sei whales equal one blue-whale unit. Bryde's whale and the minkes, the first averaging 45 feet and the other two 30 feet, are not separately identified in whaling statistics. When taken, they are probably counted as sei whales; they will receive no further mention here.

The remaining baleen whale to be taken commercially is the small but oil-rich humpback whale (*Megaptera novaeangliae*). This animal is shorter than the sei, averaging about 45 feet, but it is so stocky that it yields some eight tons of oil. Two and a half humpbacks thus equal one blue-whale unit. The humpback has evidently never existed in large enough numbers to constitute a major whaling resource. Nonetheless, until recently some 1,000 humpbacks were taken each year.

The eighth and last whale that is hunted today, surprisingly enough, is one that somehow escaped the near extermination that was the lot of the gray, bowhead and right whales in the days of sail. This is the sperm whale, which dives deep to hunt for squid along the ocean floor (its deepest-known dive is about 3,500 feet) and can remain sub-

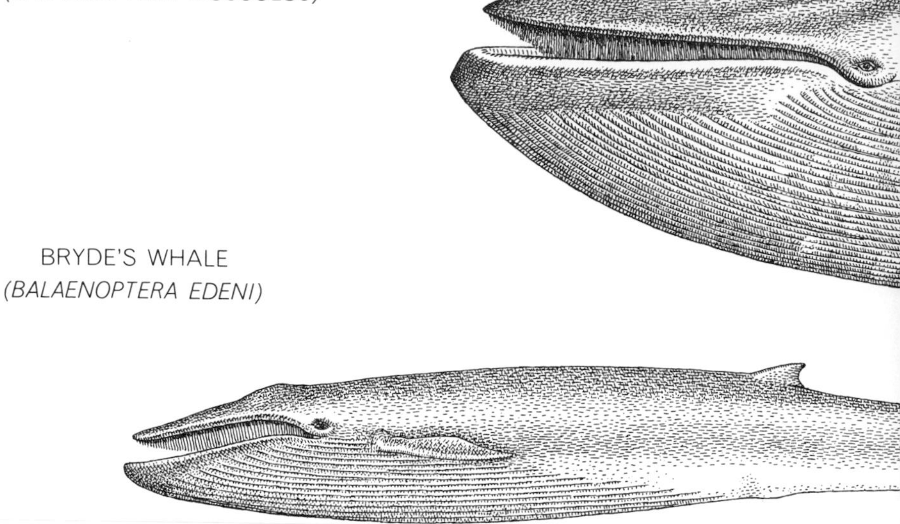

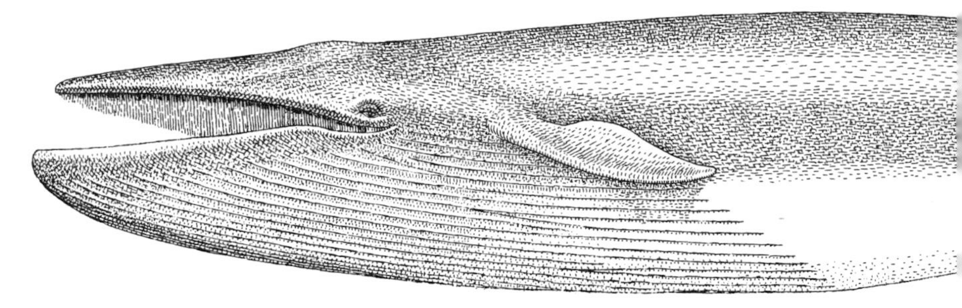

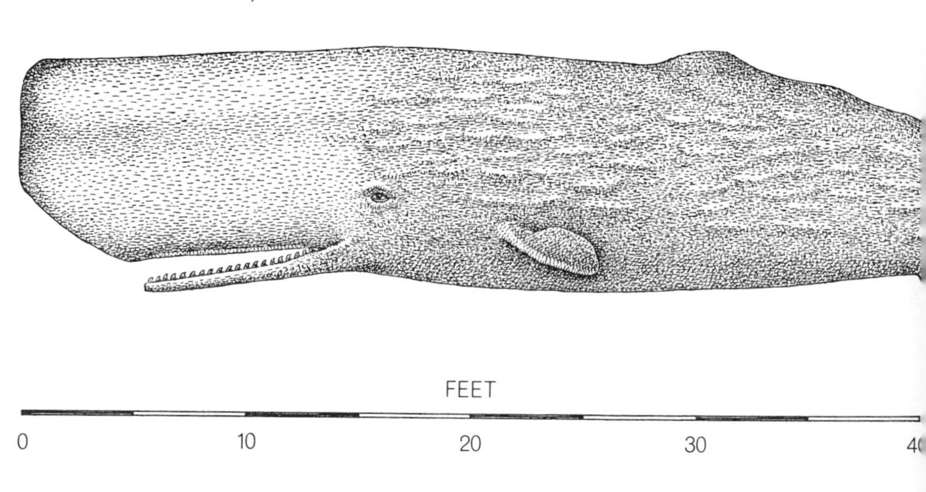

WHALERS' QUARRY TODAY consists mainly of animals of the baleen suborder; six of the seven species are illustrated above. The blue whale, the world's largest animal, is on the verge of extinction, as is the oil-rich humpback. The finback is seriously overhunted and the same fate is befalling the sei whale. Kills of Bryde's whale and two minke whale

pecies (one of which is not illustrated) are not reported by name n the whaling statistics; they are probably counted as sei. As a esult their present numbers are unknown. The toothed suborder f whales includes such familiar animals as the porpoises, the dolphins, the grampus, pothead and the true and false killer whales. Three of the toothed whales are illustrated. Only the largest, the sperm whale, is commercially valued and endangered by overhunting, although the narwhal and white whale are hunted occasionally.

CATCHER VESSEL pitches in a heavy antarctic swell. The stubby device seen silhouetted on the bow platform is the cannon that fires harpoons loaded with explosives. A dead whale, probably a finback, has had its flukes bobbed and is chained by its tail to the ship's side.

merged as long as 90 minutes. The sperm's huge square head contains the largest brain in the animal kingdom: it weighs more than 20 pounds. Sperm whales roam the mid-latitude oceans in groups that whalers call pods, including bulls, cows and calves. Once a year the mature bulls leave the pack and travel to antarctic waters for the summer months. As a result the whaling fleets have an opportunity to kill sperm bulls in the Antarctic and bulls and cows alike during the voyage to and from the fishery.

As the population of baleen whales in antarctic waters has dwindled, the whaling fleets have begun to spend more time in the North Pacific, where the primary quarry is the sperm whale. Every year since 1962 the industry has killed more sperm whales than whales of any other single species; the peak catch (in 1964) was more than 29,000. Sperm whales are not counted in terms of blue-whale units, and as yet there is no limit on the number that may be taken in any year. The only limitation is on the minimum size of sperm cow that may be killed. The bowhead and the right whales, on the other hand, are supposedly protected everywhere in the world, as are the humpback and the blue whales throughout the Pacific and the humpback south of the Equator around the world.

Before World War II whaling was a *laissez faire* enterprise. Then in the postwar epoch of international cooperation 17 interested nations entered into a convention designed to regulate the whaling industry. In December, 1946, the International Whaling Commission was established as an executive body to oversee the conservation and sensible utilization of the world's whale resources. The participating nations in the Western Hemisphere were Argentina, Brazil, Canada, Mexico, Panama and the U.S., although only Argentina and Panama were then active in the industry. (Both have since abandoned whaling.) Among the nations of Europe, Denmark, France, Britain, the Netherlands, Norway and the U.S.S.R. were signatories. So were Iceland, South Africa, Australia, New Zealand and Japan. At that time five nations operated factory ships and catcher fleets; today only Japan, Norway and the U.S.S.R. are major whaling nations. The Netherlands and Britain have abandoned their fleets, although the British continue their shore-based antarctic whaling enterprise (which is now conducted jointly with the Japanese) on the subantarctic island of South Georgia.

Since its activation in the fall of 1948 the International Whaling Commission has been charged with such tasks as protecting overexploited whale species, setting minimum-size limits below which various species may not be taken, setting maximum annual catch quotas for the antarctic fishery and designating areas closed to hunting. Although each commissioner is in principle responsible for his nation's observance of the commission's regulations, the commission itself unfortunately has neither inspection nor enforcement powers; any member nation can repudiate or simply ignore the commission's actions. Nonmember nations, of course, are equally unrestricted in their whaling activities. It calls for little political insight to forecast that recommendations made by the nonwhaling members of such a body will be ignored by the whaling members. What comes as a surprise is the fact that both the whaling and the nonwhaling nations on the commission were unresponsive to the significance of the whaling statistics that were presented to them each year during the 1950's.

In the years after World War II the waters of the Antarctic constituted the world's last great whaling ground. Record catches such as the one of 1930–1931 were never repeated, but up to 1950–1951 the whalers killed some 7,000 blue whales each season. The commission's annual quota for the antarctic fishery was 16,000 blue-whale units; the catch of finbacks (at the rate of two for each blue) helped to fulfill the quota. During this period about 18,000 finbacks were killed each season [*see bottom illustration on page 318*].

As the 1950's progressed, however, an ominous trend was evident. The blue whales were becoming scarce. The blue-whale kill, which totaled only about 5,000 in 1951–1952, fell below 2,000 in 1955–1956 and was down to 1,200 by 1958–1959. To counterbalance the declining catch of blues, the whalers pursued the finbacks more vigorously. By the end of the 1950's, 25,000 or more finbacks were being taken each season. Those familiar with patterns of predation could see that the blue whales were being fished out and were in need of immediate full protection. The finbacks in turn probably could not survive another decade like the preceding one, during which some 240,000 animals had been subtracted from the stock.

By 1960, in spite of continued indifference on the part of the whaling nations, the commission finally decided to undertake some fact-finding. A special three-man committee was assigned the task of assessing the antarctic whale populations, even though an expert as well-regarded as the Dutch cetologist E. J. Slijper declared that the danger of their extinction was "surely remote." The committee was also to recommend any actions necessary to maintain the fishery as a continuing resource. It was agreed that, to avoid bias, the three men should be neither citizens of any nation active in antarctic whaling nor experts on whales. Three specialists in the field of population dynamics were chosen: K. Radway Allen of New Zealand, Douglas G. Chapman of the U.S. and Sidney J. Holt. Although Holt is a British subject, he has the status of an international civil servant by virtue of his employment with the Food and Agriculture Organization of the United Nations. The three men were asked to report their findings to the commission at its annual meeting in 1963.

As the special committee set about its three-year job, the world's whaling fleets continued to kill whales indiscriminately in the antarctic fishery. For the first season following the appointment of the committee the International Whaling Commission failed to set any quota for the antarctic catch. The whaling industry took more than 16,000 units that season, as well as some 4,500 nonquota sperm whales. Within the 16,000 units the catch consisted of 1,740 blue whales (510 more than the previous season) and 27,374 finbacks (a record number for kills of that species). The industry also took 4,310 sei whales, 718 humpbacks and even two protected right whales that surfaced within range of the gunners.

During the second year of the committee's work the whaling industry did less well in the antarctic fishery. Once again the commission set no quota; the industry only managed to process 15,229 units. A few more sei and sperm whales were killed than during the previous season and another right whale was illegally shot. Among the whales that counted most—the blues and the finbacks—a larger number were immature and the numbers of both were diminishing.

The final season before the committee's report was due (1962–1963) produced a similar record. The commission set a quota of 15,000 blue-whale units. There was a modest increase in the sei and sperm kill but a sharp decline in blue and finback kills. For the first time since World War II the kill of blues fell below 1,000.

In July, 1963, the commission met in London and the special committee presented its report. The committee stated that in the antarctic fishery both the blue whale and the humpback were in serious danger of extermination. It was estimated that no more than 1,950 blue whales—possibly as few as 650—still survived in antarctic waters. The committee also noted that overfishing had reduced the stock of finbacks to approximately 40,000, far below the population level required for a maximum yield. It was recommended that the taking of blues and humpbacks be immediately prohibited and that the annual kill of finbacks be limited to 5,000 or fewer. Elimination of the blue-whale-unit system of accounting and substitution of separate quotas for each whale species was also strongly recommended.

Finally the committee's three members gave the whaling industry a prediction with a clear practical meaning. They forecast that, if unrestricted whaling were permitted in the 1963–1964 season, the industry would not be able

FINBACK WHALE is winched from the sea up the ramp of a factory ship. Its tongue is grotesquely expanded by the action of air that was pumped into the corpse to keep it afloat. On the flensing deck *(rear)* another whale is having its blubber stripped off for trying out.

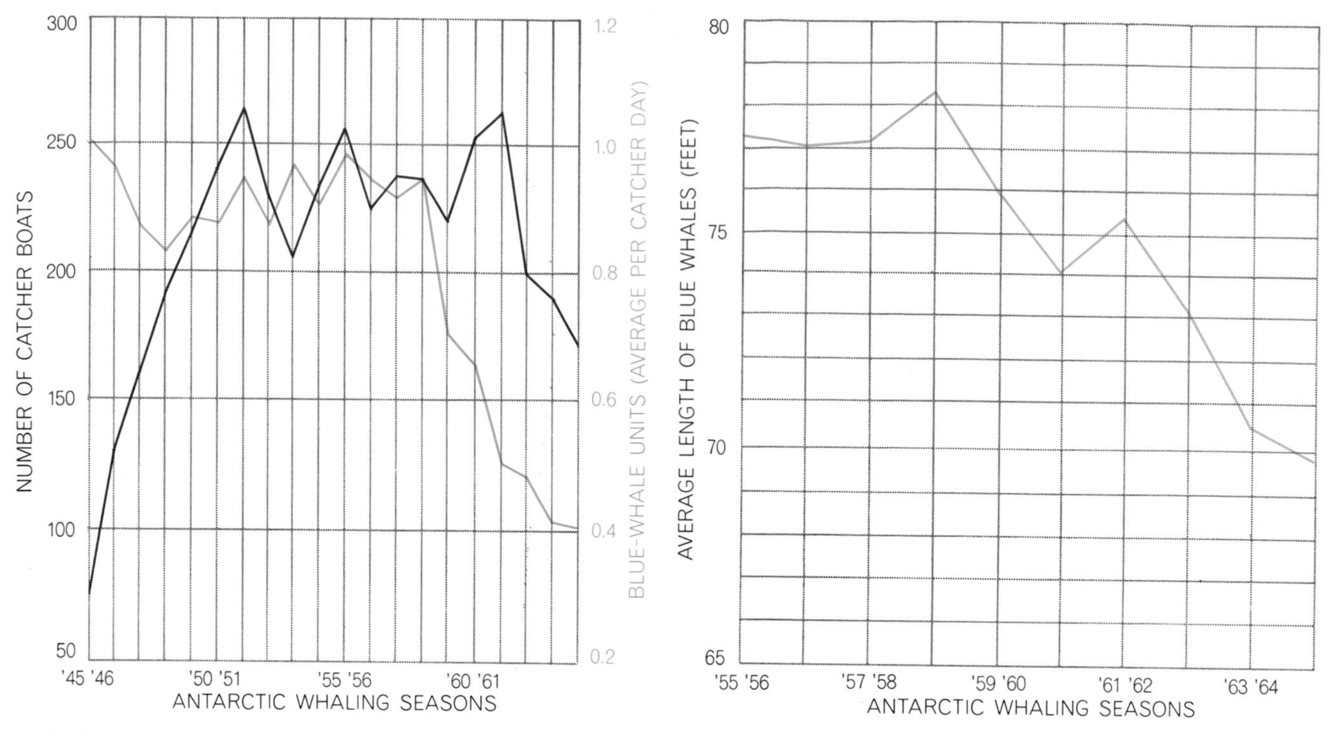

PRODUCTIVITY of the antarctic fishery began a sharp decline in the late 1950's (*color*). The decline, measured in terms of the catch per catcher-day, continued unchecked into the early 1960's although the number of catcher vessels (*black*) remained well above 200.

NINE-SEASON RECORD of the average length of blue whales killed in the Antarctic shows a decline that began in the 1959–1960 season. The average mature female is 77 feet long and a mature male 74 feet. Evidently recent catches have included immature whales.

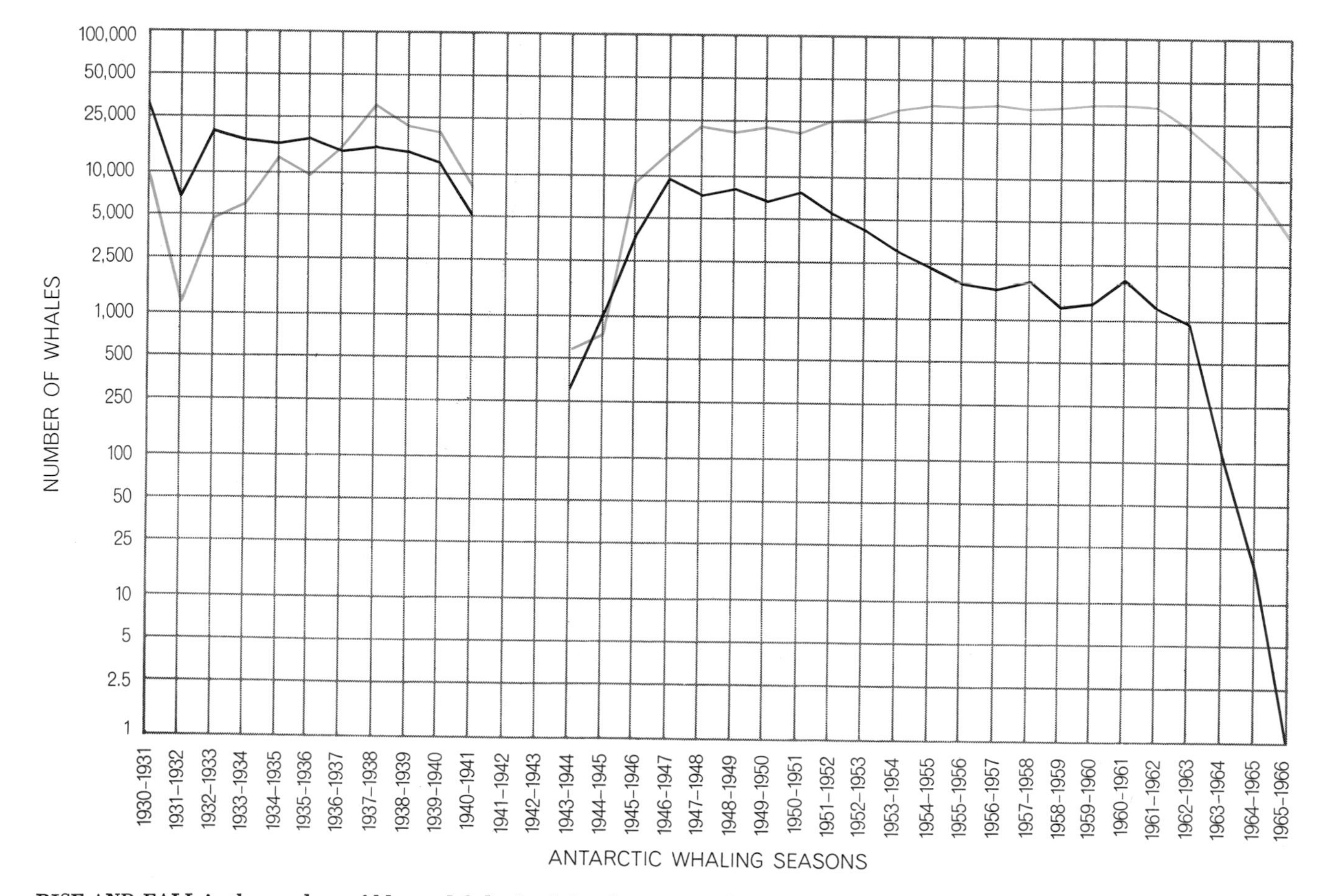

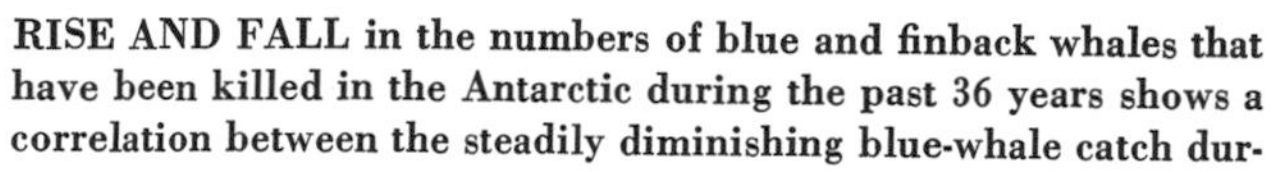
RISE AND FALL in the numbers of blue and finback whales that have been killed in the Antarctic during the past 36 years shows a correlation between the steadily diminishing blue-whale catch during the 1950's and an increase in the catch of finbacks (*color*). Just as the blue-whale stock dwindled away under the pressure of overhunting, the finback stock is now showing a severe decline.

to harvest more than 8,500 blue-whale units and would slaughter 14,000 finbacks—nearly three times the recommended number—in the process.

The prediction was disregarded. The commission voted a 1963–1964 quota of 10,000 blue-whale units for the antarctic fishery on a motion by the Japanese commissioner that was seconded by the Russian commissioner. In view of the committee's predicted maximum catch this was in effect no quota at all.

As for the committee's other findings, the commission failed to act on some and was lukewarm toward others. In the case of the now economically insignificant humpback the commission could afford to be forthright; that whale was declared protected anywhere in the world south of the Equator. In the case of the blue whale a partial sanctuary was established in all waters south of 40 degrees south latitude except some 3.3 million square miles from 40 to 55 degrees south latitude and from the Greenwich meridian to 80 degrees east longitude. Japanese whalers had taken some 700 small blue whales in this zone during the 1962–1963 season; they considered it too good a hunting ground to be put out of bounds. The suggested elimination of the blue-whale unit and the establishment of species quotas were ignored. The commission added a fourth man (John A. Gulland of Great Britain) to the special committee and asked that a further report be presented in 1964. At that time it was intended to set the antarctic quota at a level in line with the committee's findings.

In 1963 a few members of the commission may have viewed with skepticism the ability of the three committee members to make accurate forecasts of a phenomenon as full of variables and unknowns as the effects of harvesting the antarctic whale stock. If so, the results of the 1963–1964 season settled their doubts. Sixteen factory ships and their catcher fleets worked the fishery; the industry's statistics, weighing such factors as the number of days each catcher was able to spend in hunting, showed that the total effort of the catchers was 91 percent of that during the season of 1962–1963. The number of blue whales killed, however, was the lowest of any season in the industry's history: a mere 112. The committee had predicted a total catch of 8,500 blue-whale units; the fleet managed to process 8,429 units. The committee had predicted that 14,000 finbacks would be killed; the finback toll came to 13,870. The committee's forecasts thus proved to be highly accurate.

At the commission's 1964 meeting the enlarged committee added the sei whale, formerly the least prized of any in the antarctic fishery, to the list of the overhunted. The committee pointed out that the total sei population had probably never exceeded 60,000, yet more than 20,000 sei had been killed in the course of the four previous seasons. With a whale stock of reasonable size it is a rule of thumb that 10 to 15 percent of the population can be harvested annually without causing a decline. With the expectation that more and more sei whales would be killed each season as the finback population thinned out, the committee anticipated sei kills above the sustainable level in the immediate future. In line with the commission's declared intention of setting quotas according to the committee's findings, it was proposed that the antarctic quota be drastically reduced in three annual steps. A limit of 4,000 units was sought for the 1964–1965 season, a limit of 3,000 units for 1965–1966 and one of 2,000 units for 1966–1967. This degree of restraint, the population experts declared, was necessary merely to hold the number of whales in the Antarctic at the present level. They once more ap-

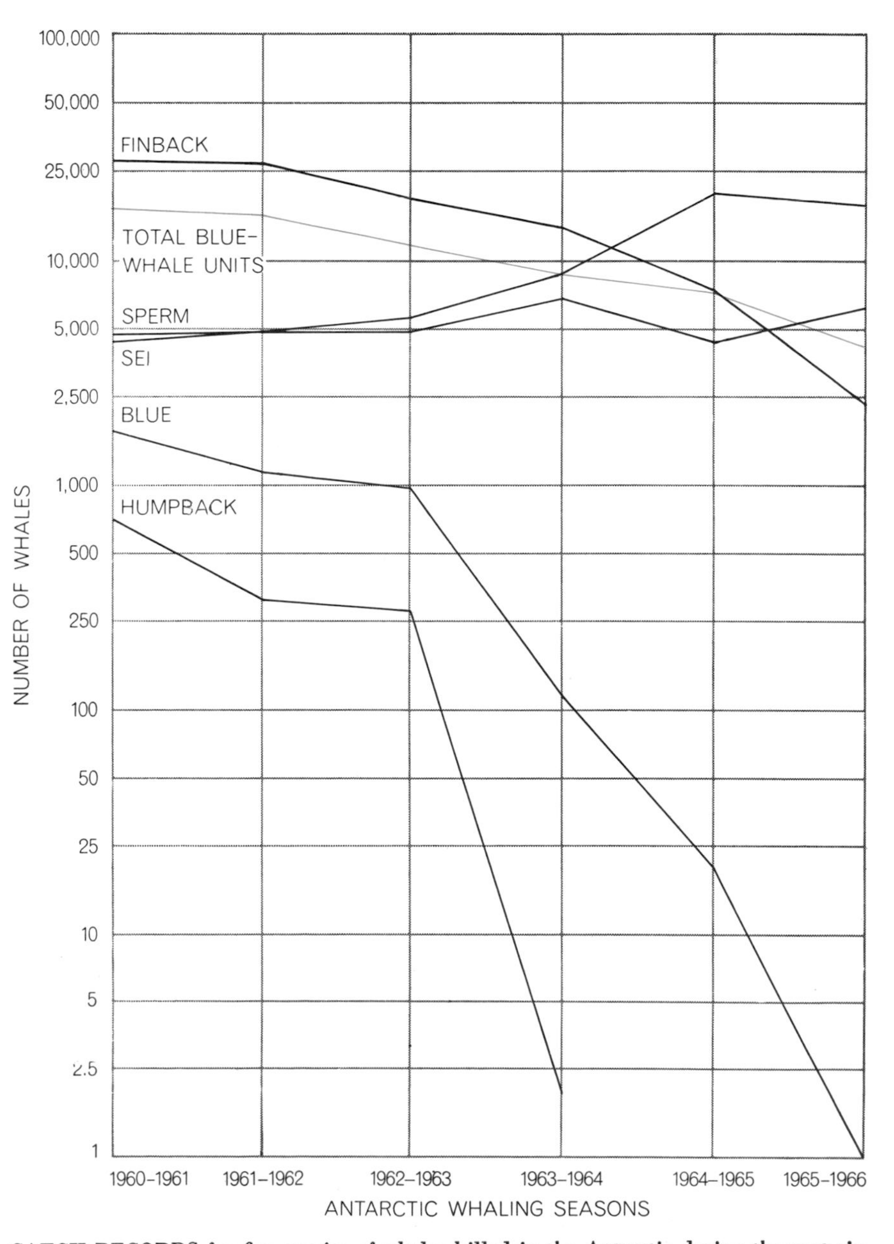

CATCH RECORDS for five species of whales killed in the Antarctic during the past six seasons show that increasing catches of sperm and sei whales have failed to counterbalance the decline in the fishery's productivity. The yield, calculated in blue-whale units (*color*), has dropped from 16,375 units in 1960–1961 to one-quarter of that amount in 1965–1966.

pealed—once more unsuccessfully—for the establishment of species quotas in place of the blue-whale-unit system.

Finally, a new gloomy statistic came before the commission: in 1963, for the first time since modern whaling had begun in the Antarctic, the larger part of the world catch had been taken in other waters. The most heavily fished area had been the North Pacific; most of the whales taken there had been sperm, which continued to be free of quota restrictions. The committee pointed out that this diversion of the industry's effort from the dwindling antarctic whale resource to the North Pacific could have ominous consequences for the sperm-whale stock.

The proposal for a reduced antarctic quota came to a vote in the commission. Japan, the U.S.S.R., Norway and the Netherlands all voted against it; in spite of the commission's declared intent to give substance to the committee's findings, the 1964 meeting adjourned with no commission quota set for the next antarctic season. The four whaling nations subsequently agreed in private that they would limit themselves to 8,000 blue-whale units in the 1964–1965 season, a figure twice the one recommended by the committee.

The 1964–1965 season was disastrous. The industry processed only 7,052 units, more than 10 percent short of its self-established quota. Only 20 blue whales were killed. The finback kill, declining for the fourth successive year, was 7,308 animals, or only a quarter of the 1960–1961 peak. As the committee had anticipated, the industry made up the difference by overkilling sei whales —almost 20,000 of them. In spite of this slaughter and a fairly large nonquota sperm-whale catch, the antarctic fishery for a second year supplied the industry with fewer whales than were taken in other waters.

For the first time since the beginning of the crisis the International Whaling Commission convened a special meeting. At this meeting (in May, 1965) the commission established a quota for the 1965–1966 antarctic season that reflected, at least in part, the committee's concern. The catch, even the whaling nations agreed, should not exceed 4,500 blue-whale units. At the regular June meeting that followed, the reduced quota was approved, and the commission agreed on a plan for further successive reductions of the antarctic fishery's quotas in the seasons to come.

The commission also attempted a first step toward partial protection of the world's sperm-whale stock. Up to that time the only restriction governing sperm kills was that cows less than 38 feet in length should not be taken. In 1964 the worldwide sperm catch had risen to a high of more than 29,000. The fleets en route to and from the Antarctic had killed 4,316 sperm whales. Once they were in the antarctic fishery they had taken 4,211 more; most of the rest had been killed in the North Pacific. In the hope of protecting sperm cows in Temperate Zone waters the commission ordered a worldwide hunting ban in the area between

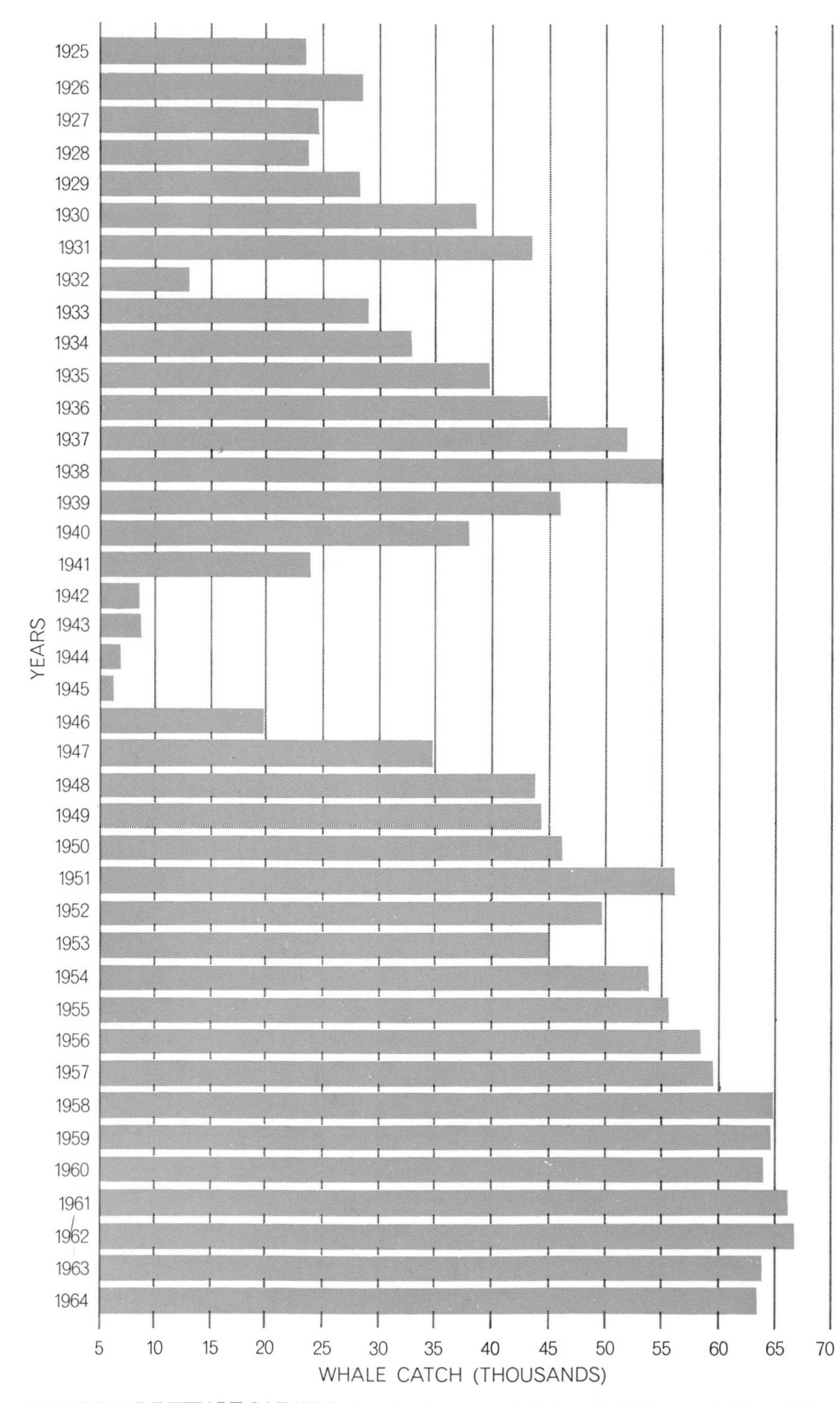

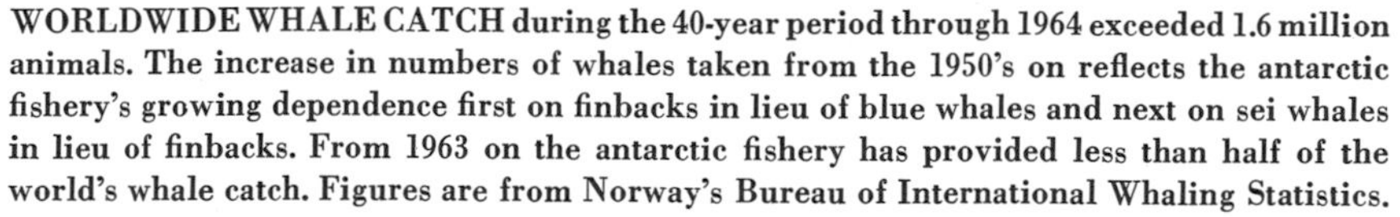

WORLDWIDE WHALE CATCH during the 40-year period through 1964 exceeded 1.6 million animals. The increase in numbers of whales taken from the 1950's on reflects the antarctic fishery's growing dependence first on finbacks in lieu of blue whales and next on sei whales in lieu of finbacks. From 1963 on the antarctic fishery has provided less than half of the world's whale catch. Figures are from Norway's Bureau of International Whaling Statistics.

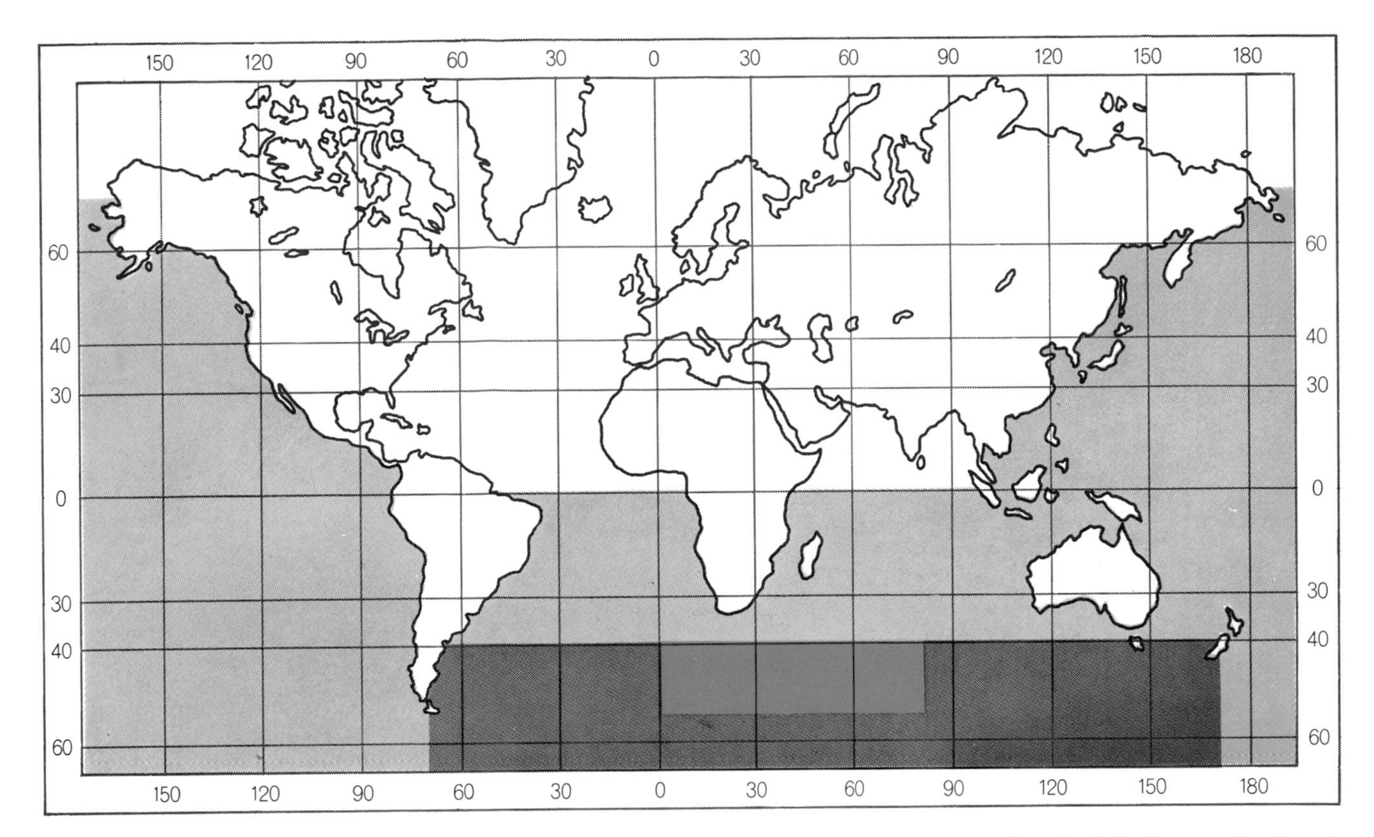

REFUGE AREAS within which designated species of whales are safe from hunting by member nations of the International Whaling Commission have been expanded in recent years. The nearly extinct bowhead and right whales enjoy worldwide protection. The blue and humpback whales are protected throughout the Pacific. The humpback is also protected south of the Equator elsewhere in the world, but the blue whale's protection elsewhere is limited to a zone south of 40 degrees south latitude (*gray area*). Even within this zone of protection 3.3 million square miles of ocean were kept open for blue-whale hunting at the request of Japan until a year ago.

40 degrees north latitude and 40 degrees south latitude. All three whaling nations, however, objected to the commission's order; the ban has simply been ignored. The worldwide sperm catch in 1965 was somewhat lower than in 1964: future catch records will reveal whether or not this drop reflects overkilling.

The imperative need for a reduced antarctic quota was demonstrated clearly enough by the results of the 1965–1966 season. With a quota of 4,500 blue-whale units the antarctic fleets could process only 4,089 units. The finback kill reached a new low, less than 10 percent of the 1960–1961 peak; in spite of protection, one blue whale and one humpback were taken. Again it was the sei whales that bore the brunt of the slaughter, and even their numbers were less than in the previous season. The conclusion was inescapable: The antarctic fishery, with its seasonal yield down to less than 100,000 tons of oil, was no longer economically significant.

Can the antarctic whale fishery ever be restored? Even with the progressively reduced quotas now envisioned it may take 100 years before this resource recovers to the level of the middle 1950's. Recovery to the levels before World War II, which depended almost exclusively on the blue whale, is improbable. This giant mammal—one of the most remarkable ever to appear on the earth—has probably been reduced in numbers below the level that allows a species to survive; its worldwide population today may well be less than 1,000. Certainly there are 594 fewer blue whales this year than last. Whaling from land stations in Peru and Chile—two nations that are not parties to the international convention—accounted for 449 of these casualties in 1965.

The crucial question today does not concern the antarctic fishery but rather the future of all whaling. Continued overhunting of the finback whale can be expected to do to that species what has already been done to the blue whale. Last year, in addition to the antarctic season's kill of 2,314 finbacks, roughly 4,500 more of these animals were taken in other waters. Throughout the world's oceans today the only great whale that survives in economically significant numbers is the sperm. Yet a measure of the industry's lack of concern for its dwindling resources may be gained from recent efforts to establish quotas in the North Pacific, where the absence of quotas or restrictions allows hunting all year long. In spite of efforts by Canada and the U.S. the two whaling nations with fleets in the North Pacific—the U.S.S.R. and Japan—could not even agree to restrict the finback kill, let alone that of either sei or sperm.

Each of two past eras of whaling has virtually eradicated its own most highly prized whale species. The bowhead whale and the right whales are monuments to man's thoughtlessness in the days of sail. The blue whale and humpback—and possibly the finback and sei as well—are monuments to an industry's lack of foresight in the days of steam. The whaling nations today face a third and almost certainly a final decision. If essentially unrestricted whaling continues, the only surviving stock of any economic importance—the sperm whale, of whose numbers more than 250,000 have been killed in the past 12 years—is doomed to become a monument to international folly. Only sharply reduced annual harvests and protective regulations that are both enforceable and enforced offer the possibility that the last of the great whales will survive.

V

MAN AND THE SEA

V

MAN AND THE SEA

INTRODUCTION

For the pragmatist, the most important question about the exploitation of the sea's resources for the benefit of mankind, particularly in the decade ahead, may be stated simply: Is it a good way to make a profit?

Although many government officials, international agencies, and the news media may proclaim that we are on the threshhold of reaping the harvest of the sea—a fish in every pot, so to speak—most oceanographers are wary of the promoters' golden rainbow. The truth is that development of any marine resource requires massive amounts of capital and that sustained development and reward require the profit-motivated efforts of those in the free-enterprise system.

Although such a thesis may not be popular, it does have the positive attribute of ensuring broad commitment on the part of the industrial community; this is not only desirable—it is necessary. When we assess the possibilities for exploiting (in the finest sense of the term) the ocean for man's benefit and material well-being, there appear to be four things—minerals, food, recreation, and waste dumping—that hold promise of early reward. In anticipating the fruits of expansion of these four realms of ocean-oriented enterprise, we must surely consider the problems attendant to each.

By the end of the present century, the demand for petroleum products will be between four and five times what it is today. To meet such a demand, the resources of the continental shelf must be exploited. Searches on the land areas of the world, principally by the major petroleum companies, have already extended from the northern tip of Alaska to the southern tip of Tierra del Fuego. Likewise, during the 1920s and 1930s and in the decade immediately after World War II, intensive exploration was conducted throughout the United States, Canada, and the vast sedimentary basins of South America, Africa, the Middle East, and the Southeast Pacific region. Although many giant fields were found as a result of these surveys, large oil fields were becoming increasingly more difficult and more costly to find by the beginning of the 1960s. Moreover, because the total volume of potential oil-bearing rocks is clearly finite, it became apparent that most of the giant oil fields on land had been found. This does not mean that giant oil fields will not be found on land in the future, but the decline in number of major discoveries each year over the past decade suggests that few such fields remain to be found. This realization has forced the petroleum industry to expand ocean-floor exploration, and it is inevitable that the accompanying increases in seismic surveying, drilling, platform construction, and tanker transport will lead to additional conflict between the petroleum companies and others who use or relate to the sea—commercial fishermen, sportsmen, scientists, conservationists, hard-mineral producers, and those who reside in homes along the coast. Nevertheless, no one interest group can be ethically restricted (if not legally restricted) from pursuing its interests in the common domain of the sea while others are not restricted from doing so. This attitude is based on a simple concept—simple in theory, if not in practice—that man's exploitation of the sea must be carried out under the principle of multiple use. In effect, this requires co-operation between the users, sensible governmental regulations, and concern shown by all parties in adhering to rigid rules of conservation. There is the further requirement that each user must not degrade the environment in conducting his individual activity, although he may be *legally* within his rights to do so. This last requirement, essentially a moral commitment, is no less important than the others.

Another problem encountered in the search for oil on the continental shelf is the requirement that large sums of money be paid to governmental agencies, both state and national, as lease bonuses. Certainly, all developers of offshore resources expect to pay a fair share of the cost incurred by government in its administration of submerged lands, but to render large sums of money for lease bonuses, in excess of actual administrative needs of the regulatory agencies, greatly restricts the industry in how effectively and how quickly it can search for and find new petroleum resources offshore. It is the opinion of many scientists and engineers that it would be of far greater benefit to the government and to the whole of mankind were the offshore-oil industry permitted instead to use the money now required as lease bonuses to finance additional exploration and development. This is a sound argument, if for no other reason than to ensure a strong, viable, and bold offshore-oil industry for this nation. There is no doubt that offshore-oil exploration is costly. Richard J. Howe, of Esso Production Research Company, writing in *Ocean Industry* in November of 1968, listed the cumulative expenditures for domestic offshore-oil production as follows:

Lease bonus and rental payments	\$ 4.00 billion
Royalty payments	\$ 1.85 billion
Geophysical surveys	\$ 1.10 billion
Drilling and completing wells	\$ 3.10 billion
Platforms and pipelines	\$ 1.85 billion
Operating costs	\$ 0.85 billion
Total (through 1968)	\$12.75 billion

From examination of this list, it is apparent that approximately one-third of the total cost has been to meet lease and rental payments to government. Surely, the same four billion dollars could have been more wisely spent, for both government and industry, by expanding offshore exploration, the results of which would have produced more than that sum in royalties and taxes alone, considering the present life expectancy of the average offshore oil field.

The problems mentioned above are not the only ones to be considered. There is a growing concern by conservationists, land owners, and municipalities that, in some instances, the exploration for and exploitation of offshore oil is a threat to wildlife, property values, and public recreation sites. It cannot be denied that such concern is justified.

When we consider the exploitation of sea-floor deposits of hard minerals—gold, titanium, copper, chromium, nickel, and others—we must first consider the projected world demand for such minerals. It is expected that, by the year 1985, the demand for the principal industrial minerals will have doubled that of today, and that by the year 2000—only fifteen years beyond—the demand will have tripled. These are conservative estimates, and if the industrialization of some emerging nations proceeds more rapidly than we now predict, these estimates will be far too low. Furthermore, the changing winds of international politics may place the United States in jeopardy of losing access to important industrial and strategic metals mined in foreign countries. We have recently seen evidence of this as a result of the rise of nationalism in at least two South American countries, where American-owned firms that have produced significant amounts of copper and associated metals for decades have now been taken over by their host countries. A similar situation can develop in any foreign country where the United States presently obtains ores or refined metals. The solution to such a problem need not be armed conflict, international embargo, or time-consuming debates over conference tables. We simply need to find sufficient alternative sources of important minerals within territory that this nation now holds and will continue to hold.

Inasmuch as the mineral resources beneath the American continental shelf, the American portion of the Great Lakes, and the American insular possessions are politically safe, this nation would be wise to start now to locate and evaluate such underwater mineral resources as can be readily found. Although some metal deposits on the ocean floor may be of a quality below that that we now consider economically feasible to exploit, their potential for mining should be greatly increased if, at some future date, this nation finds that there is a real need for them. Moreover, by locating new mineral deposits now and in the decade ahead, those who control our present foreign sources will undoubtedly think twice before capriciously placing a higher price on their raw material. In many ways, but most particularly by preventing a foreign government or supplier from placing an economic or political squeeze on the United States by threatening to withhold ore or to inflate its price, we can achieve less conflict with our neighbors. All too often, we have no alternative but to comply. To assure ourselves of adequate known supplies of important metals, we should, in this decade, initiate a broad survey to locate America's underwater mineral resources.

Such an endeavor requires the cooperative enterprise of industry, government, and the academic community. We need to take immediate steps to ensure proper legal protection by passing modern exploration-permit and lease laws for all submerged lands, to ensure an adequate financial incentive—possibly through partial tax relief on marginal production—and to ensure an available corps of highly trained ocean-mineral explorationists. The National Sea Grant Program has already aided several universities through large funding grants that are designated for training students in underwater minerals exploration, in devising new ocean mining systems, and in preparing model legislation to control industrial activities in this field.

The technological problems encountered in deep-ocean mining of manganese nodules have not been completely solved, but the team of researchers of Deepsea Ventures, Inc., headed by the brilliant ocean engineer John Flipse, has certainly proven that hydraulic mining of nodules from the sea floor is possible. Tests made in the Atlantic during 1970 were highly successful, and a viable system for commercial recovery in the Pacific is now being planned. Of equal importance is the need for an efficient, profitable process for separating copper, nickel, cobalt, and manganese from the nodules. Again, Deepsea Ventures has developed a pilot-plant model of its proposed refining system. Early results appear promising indeed, but only after a full scale refining plant is built and in operation will we know if a truly profitable process has been achieved.

We have not reached the end of the rainbow in deep-ocean mining, however—far from it. F. L. LaQue, a former vice president of International Nickel Company and now Senior Lecturer at Scripps Institution of Oceanography, has presented some timely observations on the subject.[1] He particularly cites the disparity between the quantity of metals available from the nodules and the world demand for them.

Such constraints as legal ownership, political pressures, competition with ores on land, costly research and development, remote locations for mining, unknown labor requirements, the vagaries of international trade, and the operational restrictions of a hostile sea are all very real. Only as these separate problems are overcome, are we to see deep ocean mining become a productive industry.

A decade or more from now, but certainly not beyond the end of the third decade hence, ocean mining on a large scale will become a reality. Already some of this nation's giant minerals corporations are surveying the sea floor for new resources. What happens, for example, when ocean-mining technology and metal-recovery processes become competitive with conventional sources on land? What will be the fate of those living in mining towns so dependent on a single commodity, when an ocean floor *resource* of the same metal becomes, as it might for copper, an *ore*? One solution to potential socioeconomic problems is early planning on the part of skilled sociologists. Perhaps given adequate lead time, alternative measures can be introduced to promote an easy transition. These and many other challenging "marine-affairs" problems will face the young lawyer, sociologist, and engineer in the years ahead.

Let us turn our attention to another and perhaps more immediately critical problem involving the sea's resources, that of food. S. J. Holt, in his provocative article "The Food Resources of the Ocean," has summarized the several problems facing the commercial fisherman and, in a broader sense, mankind. We have watched the world fish catch triple in the past thirty years. It is surprising that, of the 50 million metric tons of fish caught in 1968, for example, only about one-half was consumed directly by man as food. The remainder was used to feed animals!

1. F. L. LaQue, "Prospects for and from Deep Ocean Mining." *Marine Technology Society Journal*, Vol. 5, No. 2, pp. 5–15, 1971.

This does not mean that all of that consumed by animals was lost to man. Indeed, much of the chicken one buys in the neighborhood supermarket, in some communities, has been fed on fish products, usually a pelletized meal. In the fish-to-chicken conversion to make the protein more palatable, there is a built-in inefficiency, similar to the krill-whale-man problem discussed earlier.

Neglecting other priorities, we might best concentrate on two aspects of the food-from-the-sea picture: protein-poor diet and the population spiral. Nutritionists have discovered that foods rich in protein are absolutely necessary for proper mental development in young people. In affluent nations, the consumption of meat ensures an adequate protein level in the diet. In many parts of the world, however, particularly in countries where meat is rarely eaten, and a staple diet of grain, roots, and vegetables is common, the growing child suffers from a lack of necessary protein. Although a food intake sufficient to ward off hunger may be had through starches and other basics, the lack of protein causes irreparable damage. In adulthood, such a handicapped person appears indolent and shows limited mental abilities. Because these are undesirable qualities, which for people of an emerging nation can be depressing, an early priority in man's use of food from the sea should be the introduction of fish protein into the diets of those peoples not now receiving adequate protein. To meet this need on a global basis in the decade ahead will require the concerted effort of governments, the commercial fishing industry, and the compassionate concern of all people. Although there is no hope to help those adults who lacked protein in their diet as youths, we can do much to help the next generation through the use of protein from the sea.

The second most important consideration regarding food from the sea is the problem of feeding a spiraling population. If we view fish and the edible invertebrates of the sea as a means for meeting the *total* food requirements for keeping the hungry masses of this earth from starvation, we grossly overestimate the global food potential of the sea. Vast stocks of edible fish have already been removed from the sea by over-fishing, and many fisheries are on the brink of extinction. Moreover, large areas of the ocean are virtually barren or have insufficient fish to warrant exploitation. These two limitations must be balanced against a steadily climbing world birth rate, which, unfortunately, is largely confined to the poorer nations of Africa, the Far East, and Central and South America. We cannot feed *all* hungry people of the world with food from the sea; our only hope lies in some effective means of controlling population growth. Once a reasonably stable zero birth rate is achieved for underprivileged peoples, we *may* be able to use the additional food resources of the sea to bridge the gap between starvation and an adequate diet. This does not mean that we cannot feed many millions from the sea. We can, especially if new fishing methods are introduced and if we utilize such marine life as the lowly krill, but there is a limit to the food-production capacity of the sea. In order to meet the demand in this century, we must begin today to limit, through population control, the number of people who must be fed.

Mariculture, the farming of marine life, holds considerable promise for providing new food resources. Although fish farming has been practiced for centuries on certain Pacific islands, it was never developed to the level at which it was possible to produce enough fish to feed anyone but the local villagers. Recent research by Kenneth Norris and his staff at the Oceanic Research Institute in Hawaii, however, has shown that it is possible to produce vast quantities of mullet, a common food fish of Hawaii, by scientifically inducing the birth of thousands of fish from previously collected eggs. Again, the basic research was done in the laboratory. By utilizing such research results, many protein-deficient peoples may find it possible to work fish farms throughout the tropical Pacific islands and elsewhere. The Japanese have developed mariculture to its highest level of productivity by cultivating exportable amounts of oysters, fish, shrimp, and other marine life. Although such success promises hope in meeting the food requirements of many, the practice of mariculture is restricted to areas where the oceanic variables are favorable. Unfortunately, such factors as water temperature, dissolved oxygen, basin closure, and a supply of particulate food for the farmed organisms are not everywhere favorable.

We have commented on the problems attendant to man's pursuit of new mineral and food resources from the sea in the decade ahead. Although these are important, even critical, there are many other problems, soon to be encountered, that will also require solution. As our own nation expands in population, we must find new sites for recreation, new sites for industrial plants and facilities, and new places to store raw materials awaiting consumption. The sea and, to a lesser extent, the Great Lakes provide the most immediate solution.

Space now occupied on land by petroleum-tank farms, reserve coal piles, parking lots, and many process plants is space not available for playgrounds or parks or new homes. There is, then, a conflict between the separate users and would-be users. As a solution to this, particularly in congested urban areas along the coast, my friend Carl Austin of the U.S. Navy's research center at China Lake, California, has suggested that we move most of the space-taking tanks, coal piles, and similar effects of man from the land to a place on the nearby ocean or Great Lakes floor. Austin and his colleagues have designed structures for the bottom of the sea, and have developed the ocean engineering to implement such a move. The decision makers in this nation, from the taxpayer to the government administrator, may soon have the opportunity to convert the congested megalopolis to an urban community of aesthetic appeal. In pragmatic assessment, we may save our cities by hiding parts of them under water.

Although most closely related to urban growth, the need for expanded recreational sites is common to all parts of the country, and because the major population centers are located on our East and West coasts and around the Great Lakes, it is logical to look to the coastal zone as the focal point for recreational development. Unfortunately, as we mentioned previously in our comments on beaches, there are heavy demands from many quarters—industry, the public, real estate developers, state and national government, park agencies, conservationists, and the military—for their share of the coastal zone. The problem is one of priorities, and it cannot be solved by political decree alone. Space within the shore zone is limited, and there is simply not enough of it to meet all demands. What might we do to enhance our shores for recreation?

The President's Commission on Marine Sciences, Engineering and Resources, in its 1969 report, suggested that we (1) require provisions for public access to the waters in private development areas; (2) require public approval of land fills along our coasts, particularly where such fills might degrade the natural environment; (3) require private developers to build docks or picnic areas that will be open to all; (4) develop artificial islands near urban areas in order to expand shore recreational areas; (5) cordon off and chemically treat segments of polluted waters, while planning to correct the surrounding pollution, in congested Great Lakes areas; (6) open the beaches on military bases to the public, at least on a part-time basis; and (7) convert part of old ports into recreation centers.

Such recommendations as these are sure to cause conflict. The right of an individual to own and control waterfront property in many areas is legally protected. Further, those who seek the solitude of shore homes are not inclined to welcome mass invasion of "their" beaches by swarms of pleasure-seekers, nor are real-estate developers likely to be enthusiastic about building public picnic areas in the midst of an exclusive development of private homes. Nevertheless, compromise must be achieved if we, as a nation, expect to meet minimum demands for new coastal recreation centers by 1980.

The development of the seashore, as the development of any resource of the sea, proceeds more quickly if, as we said before, it proves to be a good way to make a profit. Accordingly, the President's Commission has also recommended that "seasteads," similar to homesteads, be granted to entrepreneurs who wish to build fishing piers, underwater-diving parks, amusement centers, and similar money-making ventures. The holders of seastead rights would have rights and access to the waters and to the bottom, but not to mineral resources below the bottom.

Of the four most important ocean resources—minerals, food, recreation, and dumping sites—the last is perhaps of more vital interest now than the other three. The reason I say this is that, although we will have future opportunities to reconsider and alter our schemes for minerals, food, and recreational developments, in dumping toxic waste at sea (including poison gas, radioactive chemicals, and munitions, we will have only *one* opportunity to do it right.

The use of the sea as a dumping site for such wastes as dredge spoil, garbage, chemical, solid waste, and sewage is, in many ways, desirable. The chief concern is *where* such debris should be dumped. When dumping outdated munitions, for example, we might consider some basic oceanographic relationships. In their day-to-day use of nautical charts, mariners and oceanographers alike have become familiar with the many designated areas for dumping explosives. Although the original purpose in specifying precise limits for dumping explosives was to avoid the possibilities of mariners dropping anchor on submerged explosives and fishermen hauling them up with their nets, the selection was not always made with proper consideration of currents and water depth. Indeed, many of the designated sites are on the continental shelf, and several are in the paths of known currents. If we apply only those oceanographic principles that one may learn from reading this book, it is obvious that the desirable sites are in deep submarine canyons, beyond the edge of the continental shelf. Where it is absolutely necessary to dump on the shelf, we should do so in localized depressions or in other sites where sediments will eventually blanket the discarded munitions. This is also desirable for other dangerous debris.

Regardless of the specific material to be dumped, the site should be selected to conform with the most desirable oceanographic conditions. In short, these are deep water, preferably a soft muddy bottom, a lack of currents, and a site where no future engineering activities or explorations for mineral resources are likely to be conducted. Although this last requirement is difficult to predict, some reasonable judgment can be made regarding it.

It may also be necessary to continue to use parts of the Great Lakes as dumping sites. The important consideration here is that such debris as may be dumped in the deeper parts of the Great Lakes should be inert. This means that organic waste matter and industrial chemicals that consume an excessive amount of oxygen in their degradation process should not be dumped. Dumping in the Great Lakes should be carried on only after a careful scientific survey has conclusively shown that dumping the same debris on land causes or will cause greater harm to the environment there than dumping it in the lakes.

Inasmuch as these conflicts must be resolved if we are to preserve coastal and Great Lakes waters, an enlightened public must demand that remedial measures be initiated at the earliest possible date. Beyond the need for these efforts, there must also be statutory penalties for those who violate the water-quality control regulations already in existence on both the state and national levels. Furthermore, a strong enforcement agency should be responsible for seeing that dumping and other potentially degrading uses of the sea are conducted in a prescribed manner. It might be that the U.S. Coast Guard is the most likely and most able candidate for enforcing dumping regulations.

The use of the sea floor and selected parts of the Great Lakes for restricted dumping grounds is also related to economics. Because of the high cost of removing man's wastes, particularly from urban communities, many tax dollars that might have been used for constructing playgrounds, building libraries, and promoting welfare are not available for those purposes. Should the feasibility studies currently being conducted by several ocean-engineering corporations and some municipalities show that a significant savings can be made by using the sea as a dumping site, and doing so with proper safeguards, we should then make a concerted effort to implement such a plan.

Man must provide new technological innovations to carry out the various proposals we have discussed so far. We will require an entirely new approach to training engineers for ocean-oriented assignments. Willard Bascom ("Technology and the Ocean") shows us some of the challenges facing man as he begins to work over and beneath the sea. His comments on new metals that will resist marine corrosion, the design of new mining vessels, semisubmersible platforms, and research vehicles suggest that ocean engineering has hardly begun to scratch the surface of its potential. For many who enter this new field of endeavor, a significant portion of their activities will require that they live for extended periods in pressurized chambers moored at the bottom of the ocean. As an air-breathing animal, man has much to learn about the physiological and psychological limitations of his own body and mind under submarine conditions. The hazards pointed out by Joseph B. MacInnis in "Living Under the Sea" are many and complex, and to overcome them will require cooperative research between physicians, engineers, and scientists. Because many of the problems encountered by ocean engineers working below the sea, such as hypoxia and hyperoxia, are also encountered by pleasure-seeking scuba divers, the results of research in professional diving will be of immediate value to those who simply dive for fun.

The hazards of scuba diving, which are well known to experienced divers, are still taken lightly by the public. I am always amazed at how easy it is for a novice to purchase the equipment used in scuba diving and by how blandly he can proceed to the nearest body of water and, through ignorance of gas mixtures, body intolerance, and the limitations of the apparatus itself, place himself in immediate danger. Although the various scuba-diving clubs have fought for rigorous safety training and certification of skilled divers, it is still possible in many communities to purchase scuba equipment and use it without prior instruction or licensing.

Should scuba divers be required by law to pass an examination that ensures that they are aware of, and able to cope with, the hazards of diving? Again, we see a conflict developing that revolves around an abridgment of personal freedom and the concern for human safety. As our population and our leisure time increase, we are sure to face legal and moral questions about such personal uses of the sea.

Throughout this and the other introductory commentaries, we have considered problems that relate mainly to the ocean proper. We might, however, conclude with a discussion of those problems peculiar to the Great Lakes. These large "inland seas" may be considered oceans, in terms of physical, chemical, and geological processes, although they do indeed differ from their saltwater counterparts in salinity and in the kinds of life that they support. Within the next decade, as least one out of every five persons in the United States will live, work, or vacation in the Great Lakes region. Such human demands on these lakes will increase in geometric proportions, and already we see a serious deterioration of these magnificent bodies of water.

Man has abused the Great Lakes since he first discovered them, and as the number of communities that depend upon these lakes for recreation, for water supply, and as a site for dumping has increased, there has been a correlative decrease in water quality.

With the construction of the Welland Ship Canal in 1829, which provided a passage around Niagara Falls to Lake Erie, man began his systematic and careless program of destroying the quality of the lake waters and the life within them. The Welland Canal allowed the devastating sea lamprey to enter the Great Lakes and, lake by lake, to destroy much of the vast lake-trout fishery, even though no lamprey was discovered in Lake Erie until 1921. This time-lag effect suggests that the peril caused by man's tampering with the environment may not be immediately obvious, and that due regard should be given to alternatives in any engineering development. The tragedy of the lamprey is in no way more vividly seen than by contrasting the lake-trout production of Lake Michigan for 1943 with that for 1952. In 1943, 6,860,000 pounds of lake trout were caught; in 1952, the catch was only 3000 pounds. This drastic decrease is sobering testimony to the havoc man can so easily wreak on the natural environment. Moreover, the loss of this income-producing fishery has been keenly felt in the many small fishing ports surrounding Lake Michigan. Other species of fish are in danger in the Great Lakes, as evidenced by the drastic decline in the catches of walleye and blue pike. Even the whitefish has experienced a devastating decline in number, as pointed out by Charles F. Powers and Andrew Robertson in "The Aging Great Lakes."

The question most frequently asked by some Great Lakes limnologists is not "Can we save the commercial-fishing stocks?" but rather "Do we want to save them?" Obviously, the cost of revitalizing the commercial-fishing industry of the Great Lakes would be tremendous, and in view of the small portion of the total economic base that commercial fishing has been, it may be more prudent to use our resources for improving sport fishing and water quality.

Introduction of the coho salmon into the waters of Lake Michigan in recent years has met with outstanding success. The increase in tourists and fishermen is well known to the several natural-resource agencies, and even more so to the operators of lakeshore motels, restaurants, and fishing camps and to bait-and-tackle dealers. This infusion of new money spent on recreation has led many state planners in the upper Great Lakes region to give tourism a far greater priority for support than the re-establishment of the commercial-fishing industry. Even the prestigious Great Lakes fishery laboratory at Ann Arbor, formerly concerned with commercial fisheries, has changed its research objectives in the past few years and is now concentrating on problems related to sport fishing in the Great Lakes.

The Great Lakes are now coming to the attention of the major mining companies, which for years have mined copper, iron, zinc, lead, nickel, cobalt, and noble metals from the mineralized rocks of the Great Lakes region, both in Canada and in the United States. Recent discoveries of manganese deposits on the floor of Green Bay and parts of Lake Michigan have encouraged further exploration of the submerged land beneath these lakes. If we examine the known distribution of ore bodies and mineral deposits around Lake Superior and consider their likely projection beneath the lake floor, there is good reason to believe that major metal deposits do exist beneath this lake. The same is true for parts of the other Great Lakes, chiefly where crystalline bedrock is exposed nearby. The University of Wisconsin has initiated a major program of exploration for underwater minerals in the upper Great Lakes. This research is supported primarily by the National Sea Grant Program, although industrial participation and support is also given. The search by Wisconsin scientists for copper beneath Lake Superior waters off the Keweenaw Peninsula is already in progress. Even sand and gravel deposits on the floor of Lake Erie are being mined for use in the surrounding metropolitan areas.

Remembering the cavalier attitudes behind many early mining operations, many conservationists and public-spirited citizens are already alarmed at the thought of mining beneath any of the Great Lakes. Their demands that mineral exploration be stopped and that mining not be allowed are in direct conflict with the extractive industries who sorely need new sources of ore to feed the smelters and refineries of the Great Lakes industrial community. It is suggested that this conflict may best be resolved by invoking the principle of multiple use of the environment. This would provide a base for negotiating a compromise that would protect the primary interests of both groups. Actually, the alarm sounded by environmental groups may be premature; it may be altogether unnecessary. Geologists have already come to believe that most of the copper mining, should such develop, will be done reasonably near the shore. Advances in ocean engineering and underwater technology during the past three years suggest that lake-bottom deposits can be mined by one or more of several "clean" ore-recovery systems. The use of sealed shafts, deeply bored tunnels, enclosed remote-control dredges, and other systems still being developed could provide the means for mining the lake bottoms without environmental damage. Various mining companies, through their own initiative, are now seeking ways to exploit mineral deposits without harming the environment. Conflict is costly in terms of litigation expense, operating losses, and, most particularly, public relations. For these reasons alone, the industrial sector can be expected to pursue underwater mining cautiously and with due regard for the environment.

As we approach the end of these introductory comments, I should like to mention a few examples of the kinds of marine research that are sure to be in the fore during the decade ahead.

Man's use of the sea in the development of new recreation sites, mariculture, mineral extraction, engineering, and dumping must be carefully monitored. Although the older and more conventional techniques of water sampling, current measurement, and bottom coring—all conducted from relatively slow-moving ships—will continue to be used in environmental assessment, we must initiate new surveys that are rapid, accurate, and not a great deal more costly. One approach, exemplified by the very able work of Robert A. Ragotzkie, a physical oceanographer, is the use of infrared-sensing systems mounted in aircraft that can fly many traverses over a large body of water in a very short time. Determination of surface-temperature variations of lakes and oceans provides for early recognition of thermal pollution, the introduction of anomalously warm water into the natural environment. Moreover, remote sensing using electronic and photographic systems can be so planned that information on sites of silting, changes in beaches, shoaling, algal blooms, and sources of chemical pollutants may be readily obtained, as may much information of great importance to underwater mining and engineering operations. We predict, then, that oceanographers and limnologists concerned with man's use of the environment will increase their airborne surveys tenfold by the end of the 1970s.

The industrial sector engaged in marine-resource ventures is already tailoring its research-and-development programs to meet problems expected ten years hence. This differs from the approach taken in the past by many firms, which have looked to a three- or four-year payoff in applied research. Such men as William D. Folta and Raymond M. Thompson of Inlet Corporation have encouraged a vigorous, cooperative research attack by joint industry-university teams on those problems of sea-floor exploration that they predict will arise in the late 1970s. They have, at the same time, worked hard to encourage academic, industrial, and agency investigators to focus their efforts on the contemporary technological, scientific, and legal problems common to all of them. As far as commercial enterprise is concerned, we should see a new trend of cooperation and mutual assistance in the decade ahead, quite unlike the rigid proprietary ways of the past.

The academic community is also concerned with the ocean-related problems of the decades ahead. The distinguished Paul Fye, director of the Woods Hole Oceanographic Institution, has only recently announced the establishment of an entirely new marine-affairs program at Woods Hole. The objective of this forward-looking endeavor is to provide, for example, specialized training for young lawyers who wish to pursue careers in ocean law, international politics related to the sea, and other associated problem areas. Similar programs are now found at several universities in this country, testifying to the academic community's concern for this important aspect of man's use of the global ocean. In short, the inevitable increase in the exploitation of the sea in the decade ahead and the growing need for common solutions to common problems suggest that closer ties and more effective cooperation between industry, government, and academia are absolutely necessary. Likewise, the maximum environment must now be monitored more carefully and more often; a major increase in the number of research aircraft would help ensure this.

In conclusion, we must recognize that man, in his relation to the ocean and to the Great Lakes, is faced with more problems than answers, and that, although the sciences of the sea are mysterious to most people, this nation's role in the use and control of the ocean is decided by the wishes of its citizens. Perhaps we are asking too much of our people when we seek their opinions on such complex national and international problems as those relating to the ocean. I think not. Decision makers in government do listen to those who, from a base of knowledge, voice constructive comments and recommendations regarding marine affairs.

I end this commentary in agreement with a thought-provoking statement made by the President's Commission on Marine Science, Engineering and Resources:

> The Commission shares the conviction that marine scientific inquiry and resource development, as well as meteorological prediction, offer many real opportunities to emphasize the common interests of all nations and to benefit mankind. The gap between the living standards of the rich and poor nations is ever widening. The world cannot be stable if a handful of nations enjoy most of the planet's riches while the majority exists at or below subsistence levels, and many of the efforts to aid the less fortunate nations will involve uses of the sea.[2]

2. Commission on Marine Science, Engineering and Resources, *Our Nation and the Sea*, (Washington, D.C.: Government Printing Office, 1969), p. 3.

FRESH WATER FROM SALT

DAVID S. JENKINS
March 1957

In the face of our increasing control over nature, it is ironic that we have steadily been losing ground with regard to one of mankind's most vital needs—water. Over much of the civilized world, water shortage is a grave and growing problem.

To be sure, water has always been a major concern of man. The children of Israel recovered their faith in God only when Moses smote the rock and produced water. Egypt rose and fell with the flow of the Nile and even today is placing its hopes for the future on plans to develop the resources of that great river. Few things have more powerfully influenced the course of the human race than the perennial search for fresh water.

But in today's world the need for water has become acute in many areas. In some arid countries the per capita consumption of water, thanks to improved sanitation, has suddenly risen from two or three quarts per day to 20 or 30 or more. Underdeveloped countries seeking to raise their standard of living by industrialization and irrigation find themselves with huge new needs for water supplies. Even our own water-favored country is beginning to be concerned, with many communities already facing shortages. The problem is widespread, not confined to localities such as the drought-stricken Southwest. Since 1900 the U. S. has increased its consumption of water almost sevenfold. By 1975 our water requirement will have nearly doubled again. We shall then be using about 27 per cent of the total supply of natural fresh water in our rivers, lakes, springs and wells. Many areas will have reached the limit of their local resources. The remaining 73 per cent of the total supply, largely stream water, will probably be prohibitively expensive to collect, store and distribute to the places where it is needed.

There are two major steps we can take. The first, and most important for the years immediately ahead, is to reduce our lavish waste of water. Among other things, we can reduce the pollution of our streams, recover used water by purification, capture floodwaters and manage the industrial use of water more carefully. Some industries have already shown what can be accomplished along this line. For instance, the Kaiser steel mill at Fontana, Calif., has reduced the consumption of water in manufacturing steel from the average of 65,000 gallons per ton to only 1,400 gallons per ton, by recirculating the water it uses.

But in the last analysis we must also increase the water supply itself. There is one way this can be achieved on a very large scale: by converting salt water. There is plenty of water in the oceans, and man's ingenuity is certainly equal to the task of converting sea water and other saline waters into fresh water at a reasonable cost.

Five years ago the U. S. Congress, recognizing the gravity of the situation, passed a Saline Water Conversion Act which authorized a program of research and development. Under the administration of the Department of the Interior a coordinated campaign of investigation is now under way in the U. S., with cooperation from abroad. Government agencies, private industries and other institutions are engaged in more than a score of laboratory investigations, a few of which have progressed to pilot-plant tests. The salt-water conversion methods under investigation include a number of well-known processes and several completely new ideas.

Fresh water is commonly defined as water containing less than 1,000 parts per million of dissolved salts. But how fresh the water needs to be depends on the use to be made of it. Drinking water, according to the U. S. Public Health Service standards, should have no more than 1,000 and preferably less than 500 parts per million. In general salinity of water for agricultural irrigation should be no more than 1,200 parts per million, the allowable concentration depending on the specific salts it contains. For some industrial purposes, such as cooling and flushing, unrefined sea water will do; on the other hand, in high-pressure boilers it may be necessary to have almost pure water containing not more than two or three parts per million of salt. Thus the economic feasibility of sea-water conversion depends on the use to which the water is to be put: for some purposes the cost may be reasonable, for others not. For the guidance of the research program a survey of the various industries' water requirements is urgently needed.

The salinity of the waters available for conversion varies greatly. The oceans are

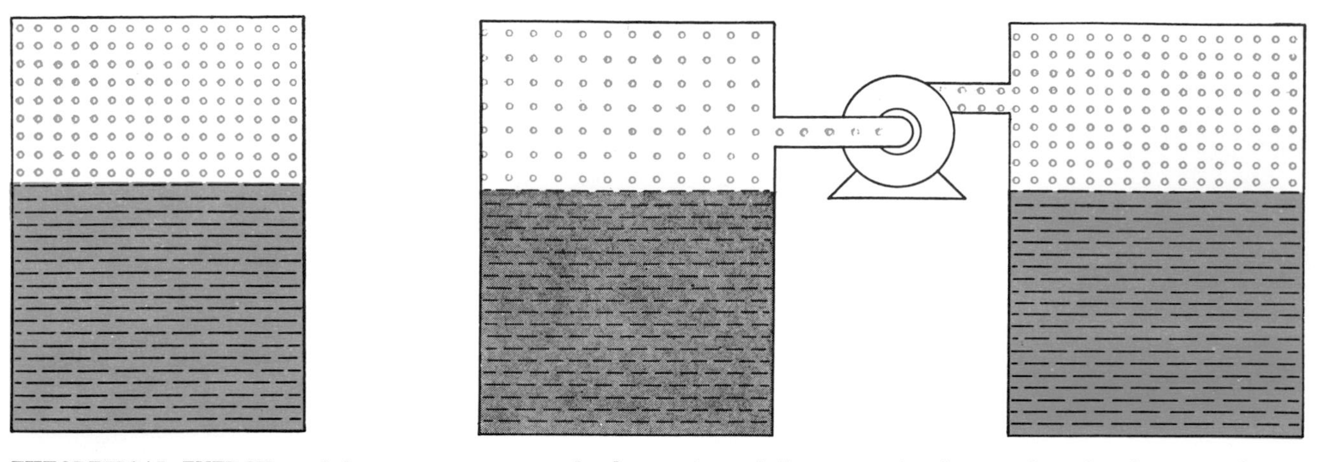

THEORETICAL ENERGY needed to separate water molecules from salt ions can be calculated from experiment shown here. At the same pressure and temperature, more water molecules will go into the vapor phase from fresh water (*left*) than from salt water (*center*). By compressing the vapor from the salt water to the same density or vapor pressure as that of the fresh water vapor, it can be just made to condense to fresh water (*right*). The energy needed is equal to the energy that binds the water molecules to salt ions.

fairly uniform, averaging about 35,000 parts per million of dissolved salt. But in the Persian Gulf it is nearly 40,000 parts per million; in Chesapeake Bay about 15,000; in the Baltic Sea only 7,000. Any water less salty than the oceans but with more than 1,000 parts per million is called brackish.

Common salt, sodium chloride, accounts for most of the saltiness of sea water. However, sea water contains small amounts of many other salts—some 44 dissolved elements in all [*see table on page 342*].

What is required to desalt water? The basic facts are simple enough. A salt dissolved in water is separated into ions —*e.g.*, in the case of sodium chloride: the positively charged sodium ion and the negatively charged chlorine ion. The ions are bound to water molecules by their electric charges. The problem, then, is to pull the water molecules and the ions apart.

We can make a calculation of how much energy this takes. Consider two sealed flasks, one partly filled with pure water, the other with sea water. At room temperature, say, a certain amount of the water in each flask evaporates into the unfilled part of the vessel, and this establishes a certain equilibrium vapor pressure. The vapor pressure in the container of sea water is lower than in the flask of pure water, because its water molecules, being bound to salt ions, do

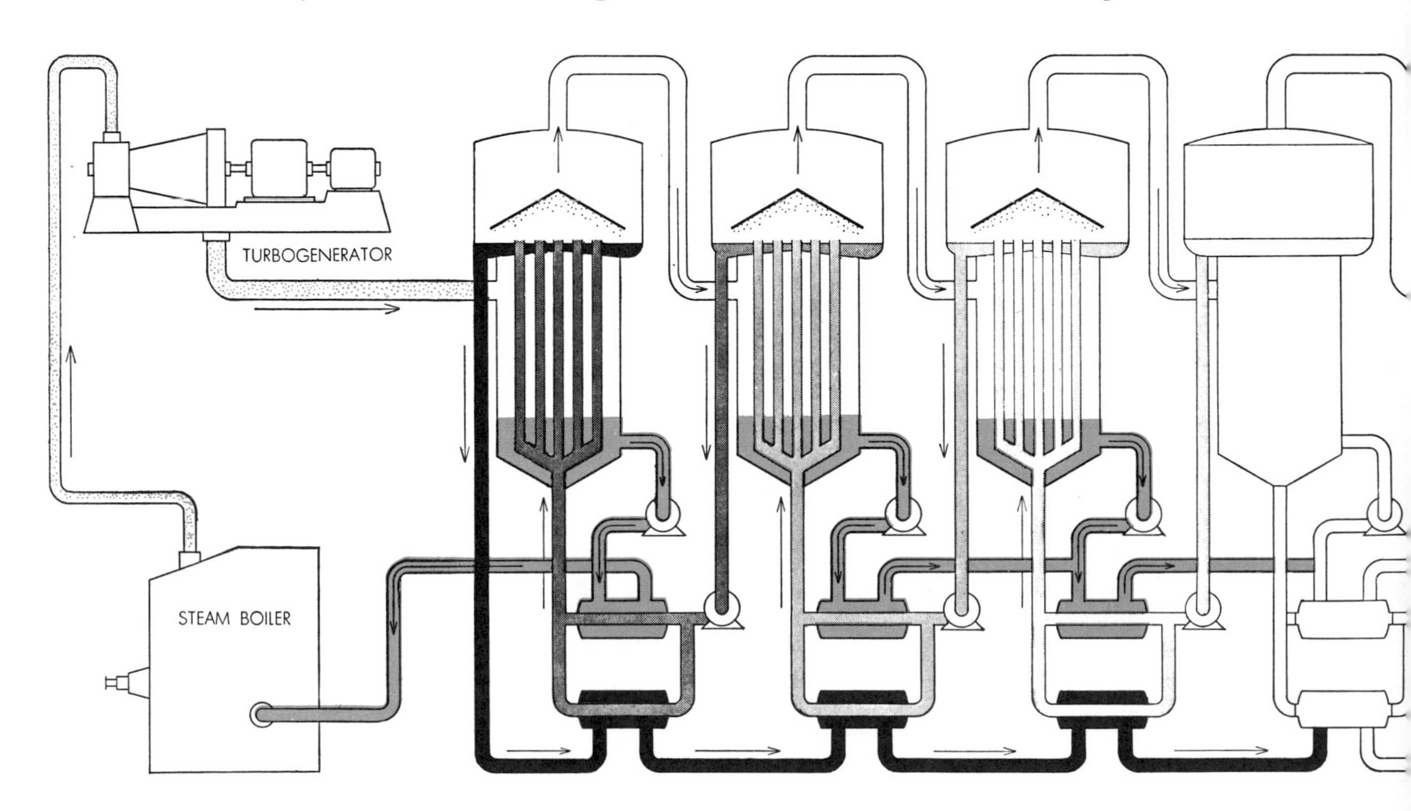

MULTI-STAGE STILL uses the latent heat released by condensation of water vapor at each stage to evaporate the brine flowing in from the next stage. The first stage (*left*) provides a condenser for the steam from a turbogenerator. The steam (*colored dots*) is condensed to water by the cooler salt water (*gray*) flowing in through the evaporator tubes. The salt water, heated by the condensation of the steam, separates into water vapor (*colored dots*) and droplets of brine (*black dots*) in the steam chest above the evaporator.

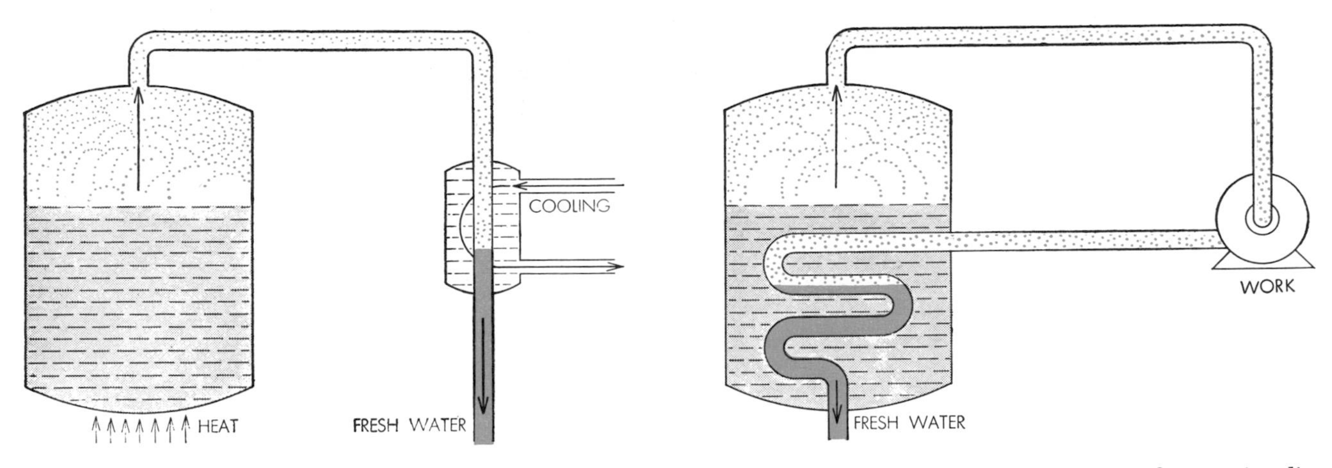

ACTUAL ENERGY needed to evaporate fresh water from salt water is necessarily much greater than the theoretical (*opposite page*) which assumes a barely perceptible rate of evaporation and 100 per cent efficiency in conversion of energy. Simple distillation (*left*) may require 1,000 times as much energy. Compression distillation (*right*) is more efficient. Compression of the vapor raises its temperature; the superheated vapor is piped through the boiler where it condenses, yielding its latent heat to evaporate more brine.

not evaporate as easily. Now the extra energy needed to separate water molecules from the ions can easily be measured: it is just equal to the energy we have to supply to compress the vapor from the sea-water bottle so that its pressure is the same as that of the vapor in the bottle of pure water [*see diagrams on opposite page*]. For sea water of average saltiness this energy amounts to 2.8 kilowatt hours per 1,000 gallons.

But this is merely the minimum amount needed to tip the scale of vapor-pressure equilibrium so that evaporation can proceed—at a barely perceptible rate. To raise the rate of evaporation to a useful level requires a great deal more energy. Furthermore, there are fundamental limitations on the efficiency of conversion of energy to useful work in any process or machine; consequently a considerable part of the energy we feed into the machine is unavoidably wasted. In practice, to separate water from salt by evaporation in a simple still takes 1,000 times the amount of energy given above as the theoretical thermodynamic requirement—that is, about 2,800 kilowatt hours of energy (in the form of heat) per 1,000 gallons of water. But there are, of course, much more efficient processes than simple, single-stage distillation. It is estimated that some of the processes now under study may reduce the energy requirement to about four or five times the thermodynamic minimum,

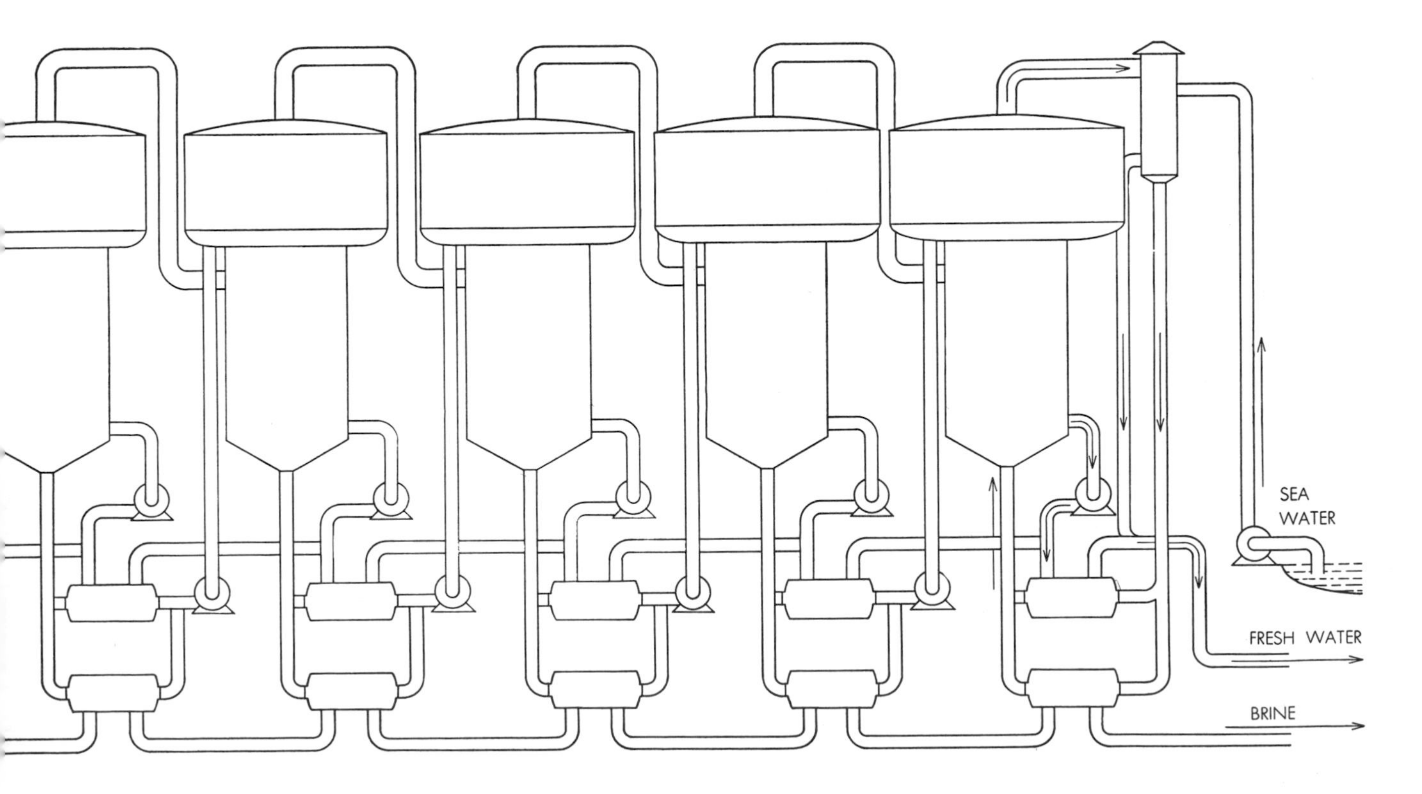

Here the brine droplets are stopped by the conical baffle plate and fall to the bottom of the steam chest, draining out at left. The water vapor rushes out at the top and to the right into the second stage. There it gives up its heat to evaporate the salt water flowing in from the third stage and condenses to fresh water. Because the temperature of the salt water at each stage is lower, the boiling point must be brought lower; this is accomplished by reducing the pressure at each stage, as indicated by increased size of steam chests from left to right.

EXPERIMENTAL SOLAR STILLS designed by Maria Telkes are shown on rooftop at New York University. In unit at left water vapor rising from shallow pool of sea water condenses on underside of inclined glass or plastic roof. Units at right are "sloping stills," in which sea water is flowed through a black wick tilted at right angles to sun's rays with considerable increase in output.

or about 10 to 15 kilowatt hours per 1,000 gallons.

Let us see what devices we can employ to enhance the efficiency of distillation. The simple still evaporates water to steam at atmospheric pressure. But now if we compress the steam to a few pounds per square inch above atmospheric pressure, the temperature of the steam will rise slightly, and we can use this added heat to evaporate more water. In other words, we have increased the yield of distilled water without feeding more heat to the system, merely spending a little energy to drive a mechanical compressor. This method, called "compression distillation," reduces the total energy requirement from 2,800 kilowatt hours to about 200 kilowatt hours per 1,000 gallons of water. The idea is more than 100 years old: it was first patented by a Frenchman, Pierre Pelletan, in 1840. The U. S. armed services used it extensively during World War II for supplying water to troops in areas lacking ready fresh water.

In the past three years interest in compression distillation has been heightened by an exciting new system. It was devised by Kenneth C. D. Hickman, a collaborator in the governmental research program. In essence what Hickman has added is a simple device for increasing phenomenally the rate of heat transfer to the water: namely, spreading it out in a thin film. The salient feature of his device is a rotating drum, shaped something like a child's musical top [*see diagram on page 344*]. Salt water at a temperature of 125 degrees Fahrenheit is sprayed on the inside surface of the drum. The centrifugal force of the drum's rotation spreads the water over this surface as a very thin, turbulent film. Some of the water evaporates (the unevaporated brine is constantly drawn off through a scoop). The water vapor leaves the drum via a pipe where a blower compresses it slightly, raising its temperature. The warmed vapor then circulates to the outside surface of the drum; there it condenses and gives up its latent heat; the drum shell transmits this heat to the film of water on its inside surface, speeding evaporation. The condensed vapor is collected as distilled water.

This system is recommended not only by its simplicity and low power requirement but also by another great advantage: the low operating temperatures (125 to 150 degrees F.). In distillation processes using much higher temperatures, the sea-water salts are deposited on metal surfaces as scale. Scale formation, which impedes the transfer of heat to the water, is the greatest single enemy of efforts to bring down the cost of distillation. In Hickman's apparatus little scale forms, because of the low operating temperatures. The main limitation of the rotary compression still is that such a still obviously must be limited in size.

There are other highly promising attacks on the problem of improving the efficiency of distillation. One of the most hopeful is the multi-stage still. In this system the latent heat released by the condensation of the evaporated water at each stage is used for the next stage, providing a chain effect. For example, in the first stage sea water is evaporated to steam at atmospheric pressure or higher; the steam passes to coils in a second evaporator, condenses there and is collected as distilled water; in condensing it releases its latent heat to evaporate sea water in that container, and so on through a number of stages [*see diagram on pages 338 and 339*]. Vacuum pumps keep each successive evaporator under lower pressure, so that its water boils at a lower temperature. Such a system could operate on the exhaust (waste) steam from an electricity-generating plant. The main problem is to prevent the formation of scale, and the method is promising enough to justify the considerable research being conducted on that aspect. If scale can be eliminated, a 20-stage still may be feasible, and it might produce fresh water at between 30 and 40 cents per 1,000 gallons.

There is a comparatively new distillation method which uses a sudden reduction of pressure instead of heat to evaporate water. At a given temperature, the amount of water vapor that air can hold depends on the air pressure. If salt water is fed into a closed chamber in which the pressure is lower than outside, part of the water will "flash" to steam. This method is being used extensively in multi-stage systems. In French West Africa engineers are attempting to develop a flash evaporator which will operate on the temperature differences between upper and lower levels of the ocean, using the colder water to chill and condense the vapor from the flash chamber. The U. S. Department of the Interior and the University of California have been working on similar systems.

Various other distillation processes have been studied, including distillation at "supercritical" temperatures and pressures. It is not possible to discuss here all the distillation ideas that are under study. But even this brief description of the work in progress must make clear that the prospects for distilling salt water to fresh at a reduced cost are bright.

The sun, which showers us with a vast abundance of energy free of charge, is responsible for our natural supply of fresh water by its evaporation of the seas. Is there any way to harness solar energy to provide us with more? Some ingenious solar stills have been proposed.

The simplest form of solar still is a pan

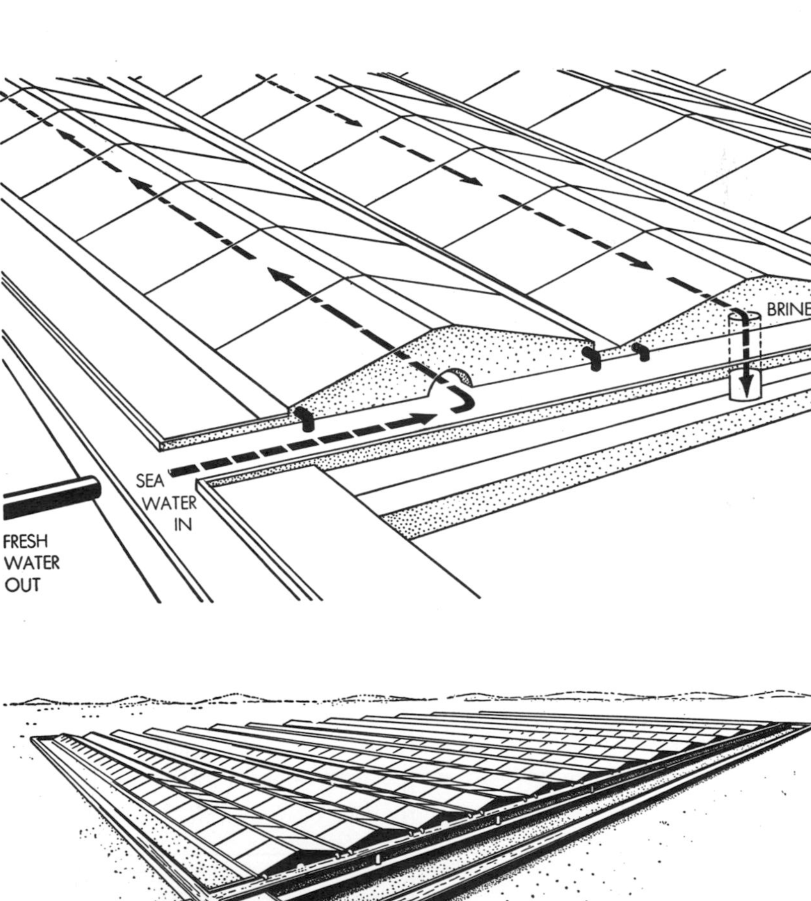

SOLAR STILL of type under development by Department of the Interior is diagrammed. Water vapor evaporating from salt water condenses on underside of glass or plastic plates and drains into gutter on either side of unit (*top diagram*). Incoming sea water (*middle diagram*) flows into first unit and returns through second. Outlet pipes for fresh water and brine are immersed in incoming sea water in order to preheat it by heat-exchange effect. Drawing at bottom shows the plan of a large installation which might cover many acres.

containing a shallow layer of salt water (say about an inch) and covered with a sloping glass plate. The glass is transparent to the sun's radiation but holds in the heat reradiated within the pan. Water evaporated from the bottom condenses on the glass, trickles down its sloping surface and is collected in a trough. This type of still, using only about half of the incoming solar energy, can produce little more than a pint of fresh water per day per square foot of area, even in the hot, clear climate of Arizona.

Some economy can be achieved by reducing the cost of the equipment. Several manufacturers have recently produced transparent plastic films which can replace glass at much less expense. One of them is a fluorocarbon called Teflon, reported to resist all forms of weathering. E. I. du Pont de Nemours and Company has designed an arrangement in which the Teflon canopy is supported by inflating it to slightly higher than atmospheric pressure, eliminating the need for a supporting frame.

Several radically new designs for solar stills are now under serious study. In Denver George O. G. Lof, a consulting engineer, is investigating for the Department of the Interior a still in which the ground acts as a storage bank for the sun's heat. A basin containing a foot of water is placed directly on the ground, so that solar heat absorbed by the water is transmitted to the ground. This heat reservoir then continues to evaporate water when the sun is not shining. If the loss of heat by radiation at night is not too great, it has been estimated that this type of still may produce up to a fifth of a gallon of fresh water per day per square foot at something like 50 cents per thousand gallons.

Maria Telkes of New York University has designed an interesting 10-stage still. It operates without machinery or any energy requirement except solar heat (or comparatively low-temperature heat from some other source). The apparatus is a sandwich-like arrangement of alternate absorbing and condensing layers. A black wick in sheet form, soaked with salt water, absorbs the sun's heat. The evaporated water condenses in the next layer, gives up heat to warm the next wick, and so on. This arrangement produces five or six times as much water as a single-stage solar still per square foot of area exposed to the sun.

Solar stills of various types are being developed in the U. S., North Africa, Australia, Spain, Italy and elsewhere.

Now let us turn from distillation to other methods of separating salt from water. In recent years the ion-exchange method has been used by industry for special purposes, such as refining brackish water. But it appears that ion-exchange systems will not become sufficiently economical for large-scale desalting of sea water.

In the ion-exchange process for treating water, the salt water is washed through resins or other material where its salt ions are replaced by unobjectionable ions. Ion exchange has been employed for softening water, for purifying water for special industrial purposes and for desalting brackish well water in the Sahara Desert and elsewhere.

Now the ion-exchange principle has been applied to form selective membranes which can separate ions from water. Ion-exchanges within the membrane make it impermeable either to positive or to negative ions. In the case of a membrane impermeable to positive ions but not to negative ones, an electric current will drive the negative ions through the membrane while it repels the positive ones. If a current is applied in a tank of salt water divided into compartments by a series of membranes, alternately permeable to positive and to negative ions, the salt ions collect in alternate compartments and the water in

EXPRESSED AS SALTS	PARTS PER MILLION PARTS SEA WATER (APPROXIMATE)
SODIUM CHLORIDE ($NaCl$)	27,213
MAGNESIUM CHLORIDE ($MgCl_2$)	3,807
MAGNESIUM SULFATE ($MgSO_4$)	1,658
CALCIUM SULFATE ($CaSO_4$)	1,260
POTASSIUM SULFATE (K_2SO_4)	863
CALCIUM CARBONATE ($CaCO_3$)	123
MAGNESIUM BROMIDE ($MgBr_2$)	76
TOTAL	35,000
EXPRESSED AS IONS	
CATIONS	
SODIUM (Na^+)	10,722
MAGNESIUM (Mg^{++})	1,297
CALCIUM (Ca^{++})	417
POTASSIUM (K^+)	382
TOTAL	12,818
ANIONS	
CHLORIDE (Cl^-)	19,337
SULFATE (SO_4^-)	2,705
BICARBONATE (HCO_3^-)	97
CARBONATE (CO_3^{--})	7
BROMIDE (Br^-)	66
TOTAL	22,212

SALTS IN SEA WATER are of many varieties, the principal being shown here. Sea water contains 44 principal elements, including gold in the amount of .000006 parts per million.

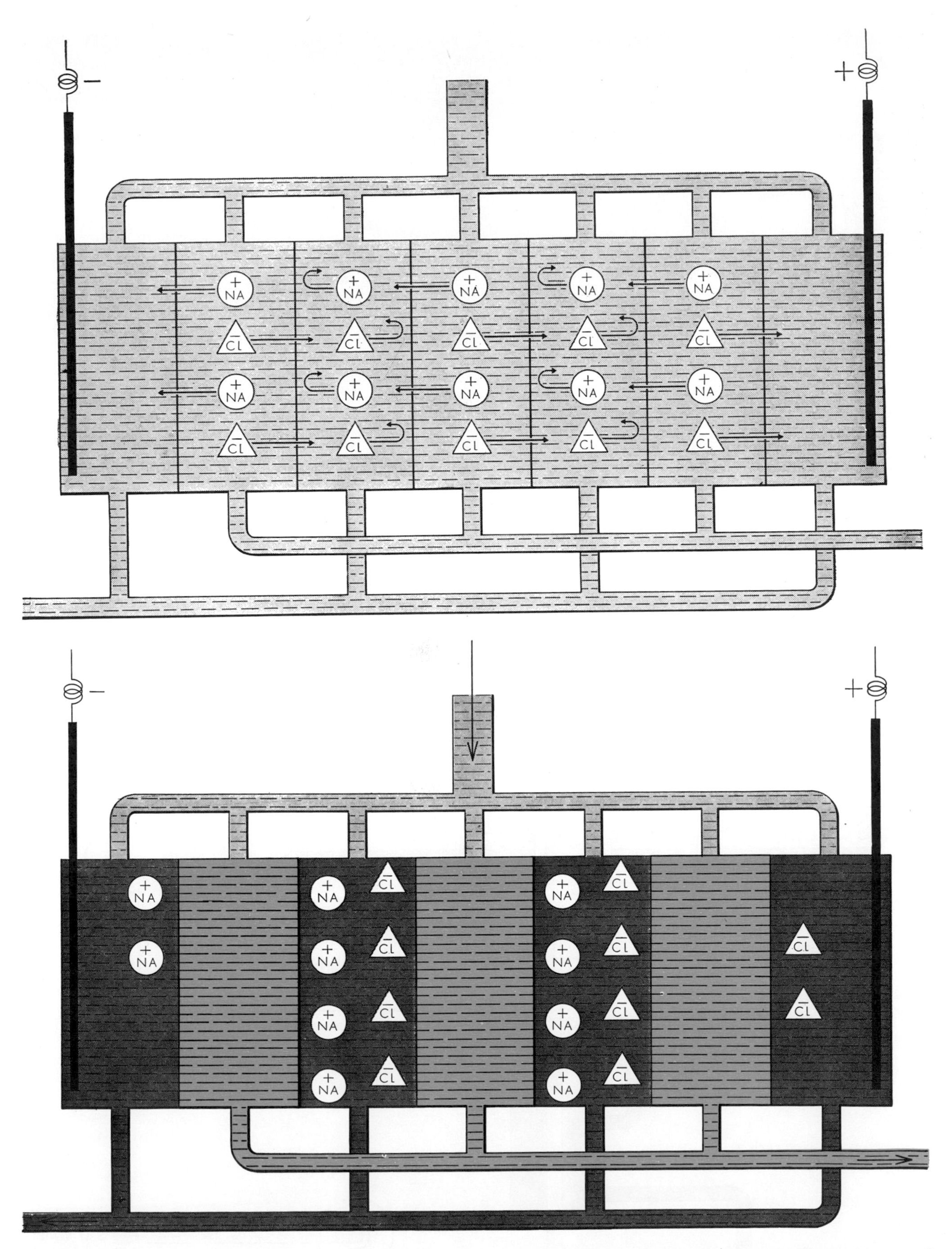

ION-EXCHANGE SEPARATION of fresh water from salt employs membranes which are alternately permeable to the sodium or chlorine ion and impermeable to the other. By applying an electric current across the system (*top diagram*), the sodium ions are attracted toward one end of the system and the chlorine ions toward the other. The ions are thus concentrated in alternate cells, leaving desalted water in the cells between (*bottom diagram*). The brine can then be drawn off via one pipe and desalted water via another.

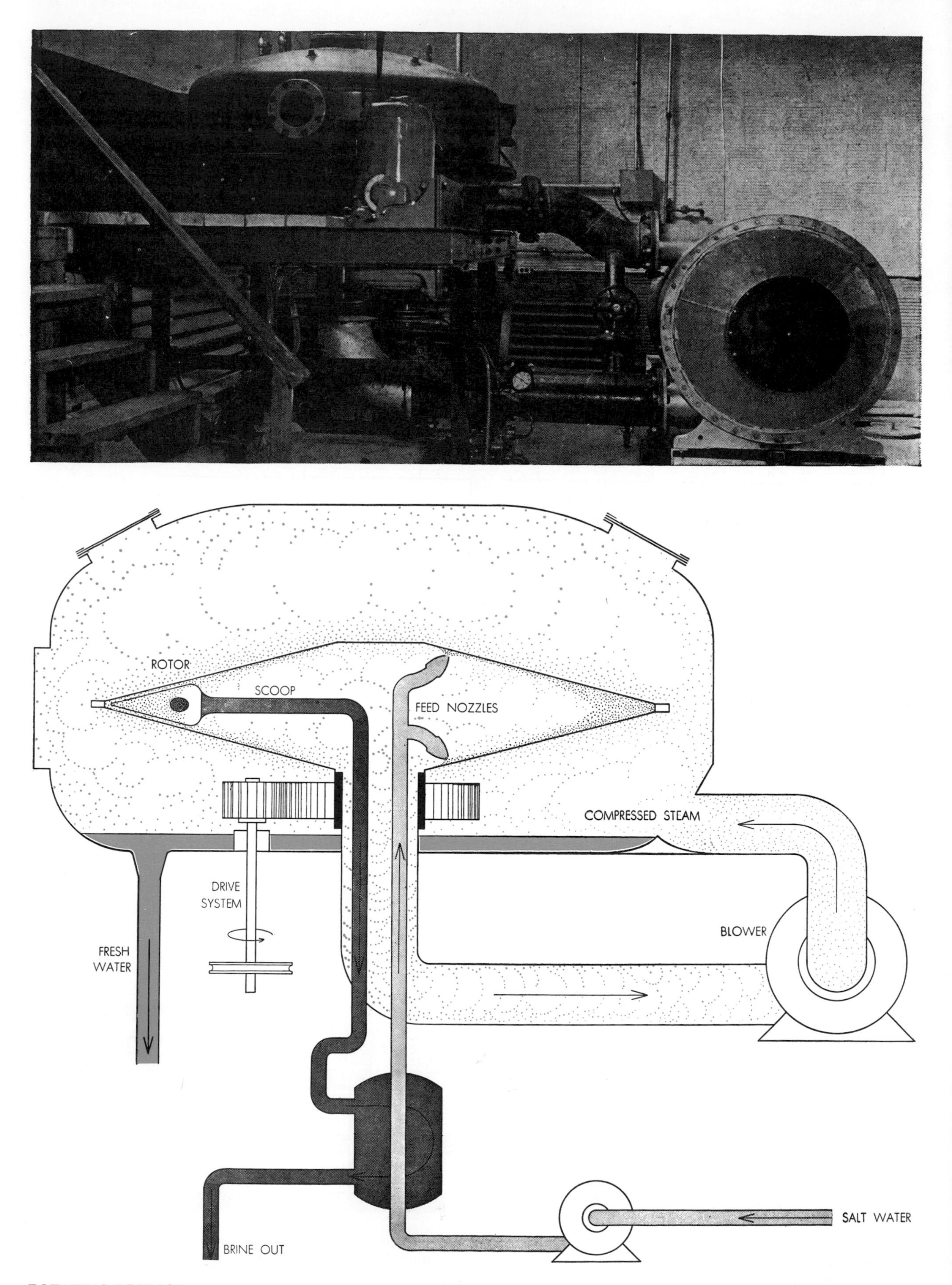

ROTATING-DRUM STILL incorporates principle of compression distillation (*see diagram on page 339*). Salt water is jetted against the hot inner surface of the drum and spread out in thin film by centrifugal force. Water vapor (*colored dots*) is sucked out and compressed by blower at right and then condenses on outer surface of drum. Waste brine is scooped into drain at left inside drum.

the intervening ones is desalted [*see diagram on page 343*].

This process, called electrodialysis, is being developed by research groups in the U. S., the Netherlands, England and the Union of South Africa. Because of the electric power requirement, it does not look economically promising for converting sea water, but it offers good prospects for desalting brackish waters.

George W. Murphy, now at the University of Oklahoma, has proposed using the electrical charge of ions from a strong brine, instead of an electric current, as the driving power to push ions through the membrane. Research on this possibility is being done at the Southern Research Institute in Birmingham.

Another promising membrane method is based on the phenomenon of osmosis. As every student of chemistry knows, if a salty solution is divided from a less salty one by an osmotic membrane, which is impermeable to salt but not to water, water passes through the membrane into the more salty solution, tending to equalize the salinity on both sides of the membrane. But this process can be reversed by applying to the more concentrated solution a mechanical pressure greater than the osmotic pressure acting on the water (which amounts to 350 pounds per square inch between fresh water and sea water). That is to say, the "reverse osmosis" forces water through the membrane out of the salty solution, while the membrane holds back the salt ions. Charles E. Reid of the University of Florida has demonstrated that with membranes made of cellulose acetate, 90 to 95 per cent of the salt can be removed from sea water in one pass.

Two other processes which offer promise are being developed. One is separation of salts from water by freezing. The Carrier Corporation, under contract with the Department of the Interior, is conducting research on a very attractive combination of freezing and evaporation, and similar developments have been reported by Israel and Yugoslavia. The other promising process is separation of water by dissolving it in organic solvents which do not dissolve salts.

We are already converting salt water to fresh for some purposes. In five to 15 years we should be able to convert it at reasonable cost for a much larger number of uses, with industrial uses first. Undoubtedly man will find not one but many solutions of the problem, and will begin manufacturing water by a number of devices and on a scale ranging from small household stills to large municipal and industrial plants.

MULTI-ROTOR STILL made up of eight rotating drums of type shown on opposite page, each eight feet in diameter, is under development at Badger Manufacturing Company in Cambridge, Mass. Rotary drum achieves high rate of heat transfer and thus high efficiency.

THE PHYSICAL RESOURCES OF THE OCEAN

EDWARD WENK, JR.
September 1969

Men have caught fish in the ocean and extracted salt from its brine for thousands of years, but only within the past decade have they begun to appreciate the full potential of the resources of the sea. Three converging influences have been responsible for today's intensive exploration and development of these resources. First, scientific oceanography is generating new knowledge of what is in and under the sea. Second, new technologies make it feasible to reach and extract or harvest resources that were once inaccessible. Third, the growth of population and the industrialization of society are creating new demands for every kind of raw material.

The ocean's resources include the vast waters themselves, as a processing plant to convert solar energy into protein [see the article "The Food Resources of the Ocean," by S. J. Holt, beginning on page 356], a storehouse of dissolved minerals and fresh water, a receptacle for wastes, a source of tidal energy and a medium for new kinds of transportation. They also include the sea floor, sediments and rocks below the waters as sites of fossil fuel and mineral deposits; the seacoast as a unique resource that is vulnerable to rapid, irrevocable degradation by man.

Because the oceans are so wide and so deep, statistics on their gross resource potential are impressive. It is important to understand, however, that the immediate significance of these resources and their long-term relevance to society involve both exploration and development, and development depends on economic, social, legal and political considerations. One special feature of marine resources that may at first retard development may in the long run promote it: the fact that almost without exception sea-floor resources are in areas not subject to private ownership (although the resources will be largely privately developed). More than 85 percent of the ocean bottom lies beyond the present boundaries of national jurisdictions, and in the areas that are subject to national control the resources are considered common property. This circumstance may uniquely invoke a balancing of public and private interests, disciplined resource management and enhanced international cooperation.

OFFSHORE OIL PLATFORM in the photograph on the opposite page is in Alaska's Cook Inlet, 60 miles southwest of Anchorage. Wells are drilled from derricks set over the massive legs, 14 feet in diameter. The plume of flame is burning natural gas, a waste product in this case. Oil and gas account for more than 90 percent of the value of minerals now being retrieved from the oceans.

The 350 million cubic miles of ocean water constitute the earth's largest continuous ore body. Dissolved solids amount to 35,000 parts per million, so that each cubic mile (4.7 billion tons of water) contains about 165 million tons of solids. Although most chemical elements have been detected (and probably all are present) in seawater, only common salt (sodium chloride), magnesium and bromine are now being extracted in significant amounts. The production of salt (which can be traced back to Neolithic times and resulted in the first U.S. patent) is currently valued at $175 million per year worldwide. Magnesium, the third most abundant element in the oceans, is by far the most valuable mineral extracted from seawater in this country, with annual production worth about $70 million. Although the ocean contains bromine in concentrations of only 65 parts per million, it is the source of 70 percent of the world's production of this element, which is used principally in antiknock compounds for gasoline. The economic recovery of other chemicals from seawater is questionable because of extraction costs. In a cubic mile of seawater the value of 17 critical metals (including cobalt, copper, gold, silver, uranium and zinc) is less than $1 million at current prices; a plant to handle a cubic mile of water per year would have to process 2.1 million gallons per minute every minute of the year, and operating it would cost significantly more than the value of all its products.

One of the potential resources of seawater that has been most difficult to extract economically is water itself—fresh water. As requirements for water for domestic use, agriculture and industry rise sharply, however, desalting the sea becomes increasingly attractive. More than 680 desalination plants with a capacity of more than 25,000 gallons of fresh water per day are now in operation or under construction around the globe, and the growth rate is projected at 25 percent per year over the next decade. The cost of desalting has been decreased by new technology to less than 85 cents per 1,000 gallons, but this is still generally prohibitive in the U.S., where the cost of 1,000 gallons of fresh water is about 35 cents. In water-deficient areas or where the local water supply is unfit for consumption, however, desalted water is competitive. This accounts for the presence of more than 50 plants in Kuwait, 22 on Ascension Island and the 2.6-million-gallon facility at Key West, the first U.S. city to obtain its water supply directly from the ocean. Considerably lower costs will be attained within the next decade where large-scale desalting operations are combined with nuclear-fueled power plants to take advantage of their output of waste heat.

Once upon a time man could safely utilize the waters of the sea as a recep-

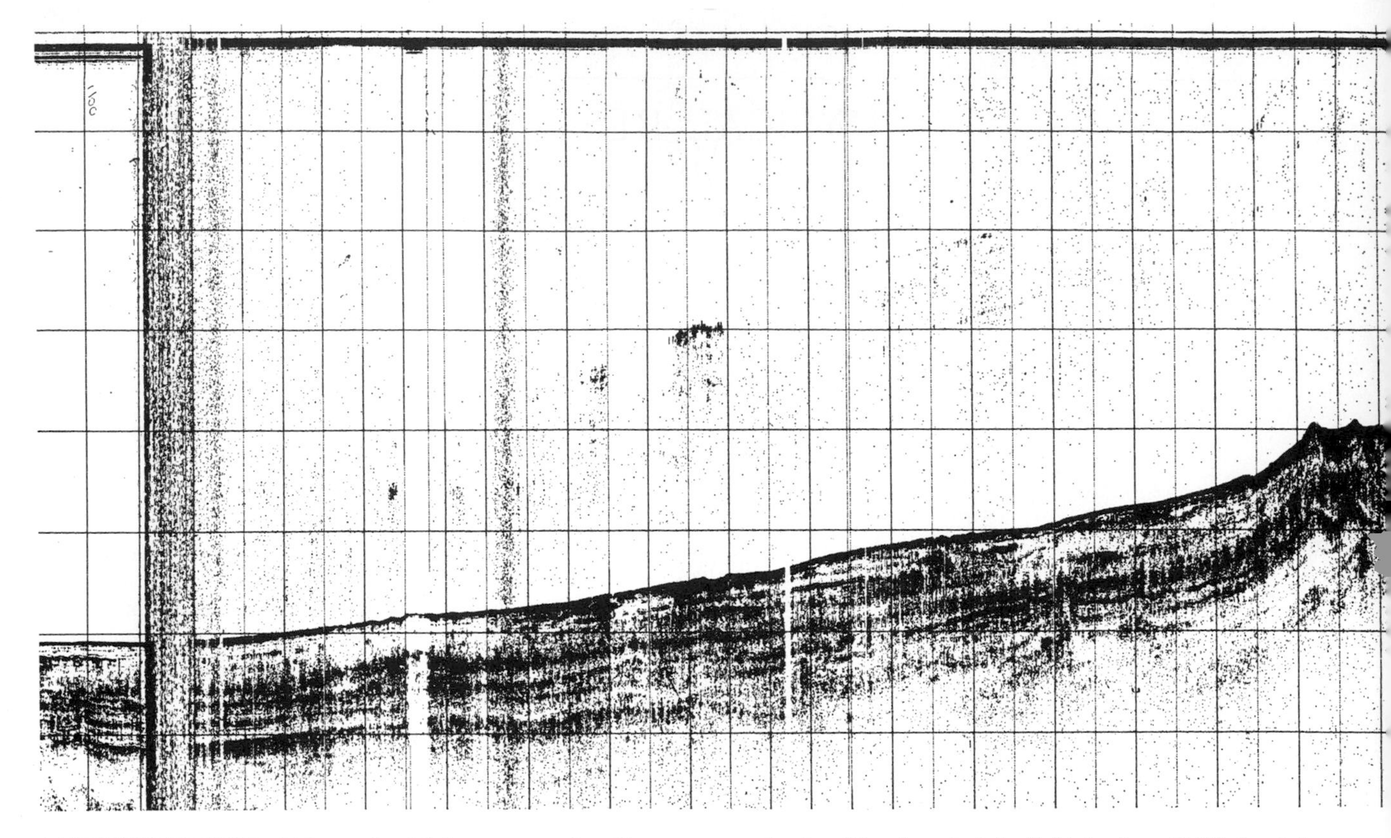

CONTINENTAL RISE, which may be rich in resources, is evident in this seismic profile made off Liberia by the Global Ocean Floor Analysis and Research Center of the U.S. Naval Oceanographic Office. The hard, straight line across the top of the record is a water reflection. The abyssal plain (*left*) is about 15,000 feet below sea level. From this plain a thick apron of land-derived layered sediments comprising the continental rise slants gently up to the toe of the continental slope. The continental slope, which is here

tacle for sewage and other effluents from municipalities and industries, confident that the wastes would rapidly be diluted, dispersed and degraded. With the growth of population and the concentration of coastal industry that is no longer possible. The sheer bulk of the material disposed of and the presence of new types of nondegradable waste products are a special threat to coastal waters—the same waters that, as we shall see, are subject to increasing demands from a wide range of competing activities. In addition pollutants are now beginning to concentrate at an alarming rate far from shore in the open ocean. Since tetraethyl lead was introduced into gasoline 45 years ago, lead concentrations in Pacific Ocean waters have jumped tenfold. Toxic DDT residues have been detected in the Bay of Bengal, having drifted with

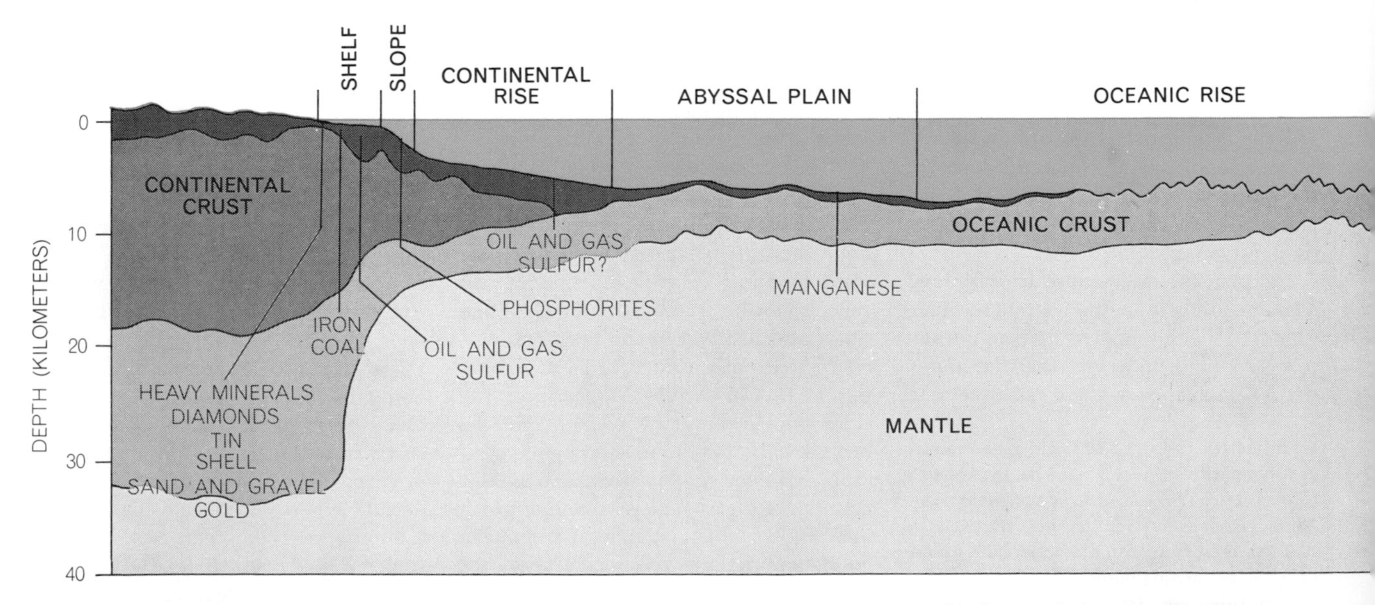

OCEAN-FLOOR RESOURCES that are known or believed to exist in the various physiographic provinces are indicated on a schematic cross section of a generalized ocean basin extending from a continent out to a mid-ocean ridge. Some of these resources are now

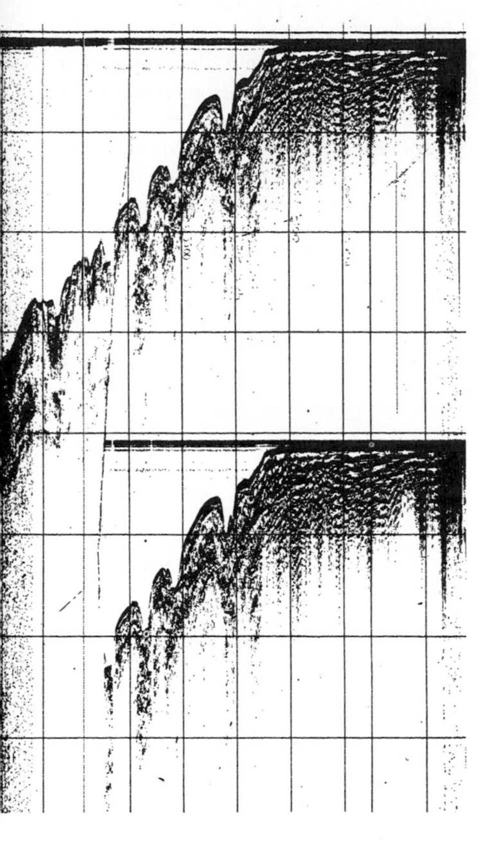

marked by large sedimentary ridges, ascends more steeply to the shallow continental shelf. The "multiple" (*right*) is in effect an echo of the structures shown above it.

the wind from as far away as Africa. And man-made radioactivity from nuclear fallout can be isolated in any 50-gallon water sample taken anywhere in the ocean.

The mineral resources of the seabed, unlike those of the essentially uniform overlying waters, occur primarily in scattered, highly localized deposits and structures on top of and within the sediments and rocks of the ocean floor. They include (1) fluids and soluble minerals, such as oil, gas, sulfur and potash, that can be extracted through boreholes; (2) consolidated subsurface deposits, such as coal, iron ore and other metals found in veins, which are so far mined only from tunnels originating on land, and (3) unconsolidated surface deposits that can be dredged, such as heavy metals in ancient beaches and stream beds, oyster shell, sand and gravel, diamonds, and "authigenic" minerals: nodules of manganese and phosphorite that have been formed by slow precipitation from seawater. Economic exploitation has so far been confined to the continental shelves in waters less than 350 feet deep and within 70 miles of the coastline.

Oil and gas represent more than 90 percent by value of all minerals obtained from the oceans and have the greatest potential for the near future. Offshore sources are responsible for 17 percent of the oil and 6 percent of the natural gas produced by non-Communist countries. Projections indicate that by 1980 a third of the oil production—four times the present output of 6.5 million barrels a day—will come from the ocean; the increase in gas production is expected to be comparable. Subsea oil and gas are now produced or are about to be produced by 28 countries; another 50 are engaged in exploratory surveys. Since 1946 more than 10,000 wells have been drilled off U.S. coasts and more than $13 billion has been invested in petroleum exploration and development. The promise of large oil reserves has stimulated industry to invest more than $1.7 billion since mid-1967 to obtain Federal leases off Louisiana, Texas and California that guarantee only the right to search for and develop unproved reserves. To date more than 6.5 million acres of the outer continental shelf off the U.S. have been leased, which is half of the acreage offered, resulting in lease income to the Federal Government of $3.4 billion. With more than 90 percent of the most favorable inland areas explored and less than 10 percent of the U.S. shelf areas surveyed in detail, the prospects are encouraging for additional large oil finds off U.S. coasts.

Sulfur, one of the world's prime industrial chemicals, is found in the cap rock of salt domes buried within continental and sea-floor sediments. The sulfur is recovered rather inexpensively by melting it with superheated water piped down from the surface and then forcing it up with compressed air. Only 5 percent of the explored salt domes contain commercial quantities of sulfur, and offshore production has been limited to two mines off Louisiana that supply two million tons, worth $37 million, a year. Now a critical shortage of sulfur and the recent discovery by the deep-drilling ship *Glomar Challenger* of sulfur-bearing domes in the deepest part of the Gulf of Mexico have stimulated an intensive search for offshore sulfur deposits.

Undersea subfloor mining can be traced back to 1620, when coal was ex-

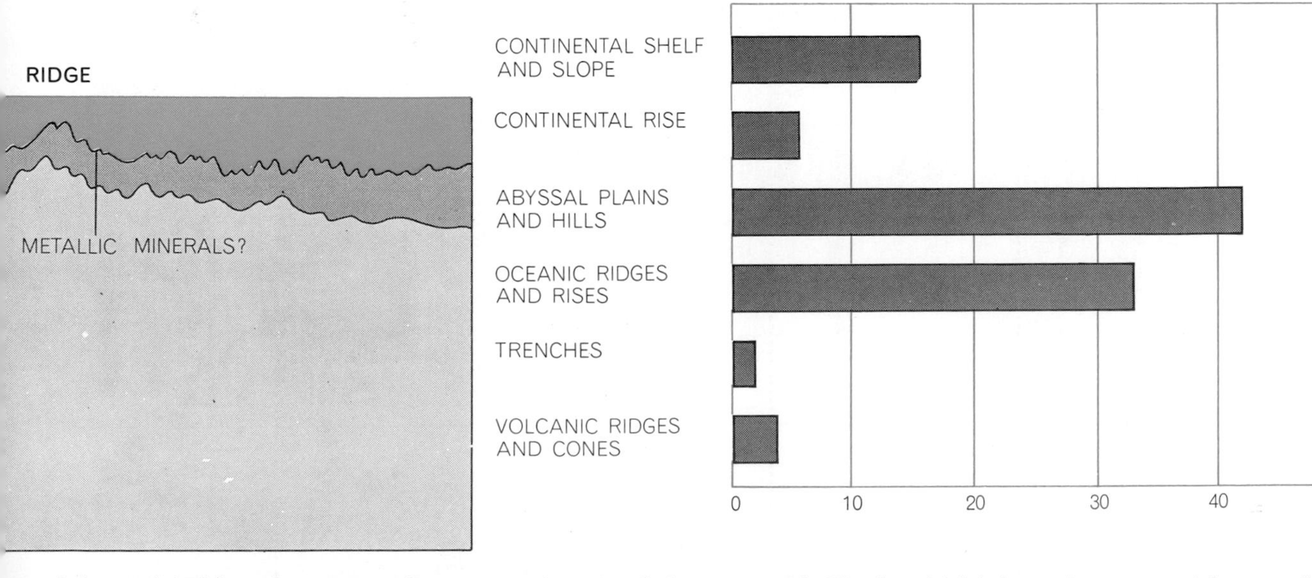

being exploited but others may not be economic for years. Sedimentary layers (*black*) are the most likely site of recoverable raw materials. The chart (*right*) shows what percent of the ocean floor's 140 million square miles of area is occupied by each province.

tracted in Scotland through shafts that were driven seaward from an offshore island. To date 100 subsea mines with shaft entries on land have recovered coal, iron ore, nickel-copper ores, tin and limestone off a number of countries in all parts of the world. Coal extracted from as deep as 8,000 feet below sea level accounts for almost 30 percent of Japan's total production and more than 10 percent of Britain's. With present technology subsea mining can be conducted economically as far as 15 miles offshore, given mineral deposits that are worth $10 to $15 per ton and occur in reserves of more than $100 million. The economically feasible distance should increase to 30 miles by 1980 with the development of new methods for rapid underground excavation. Eventually shafts may be driven directly from the seabed

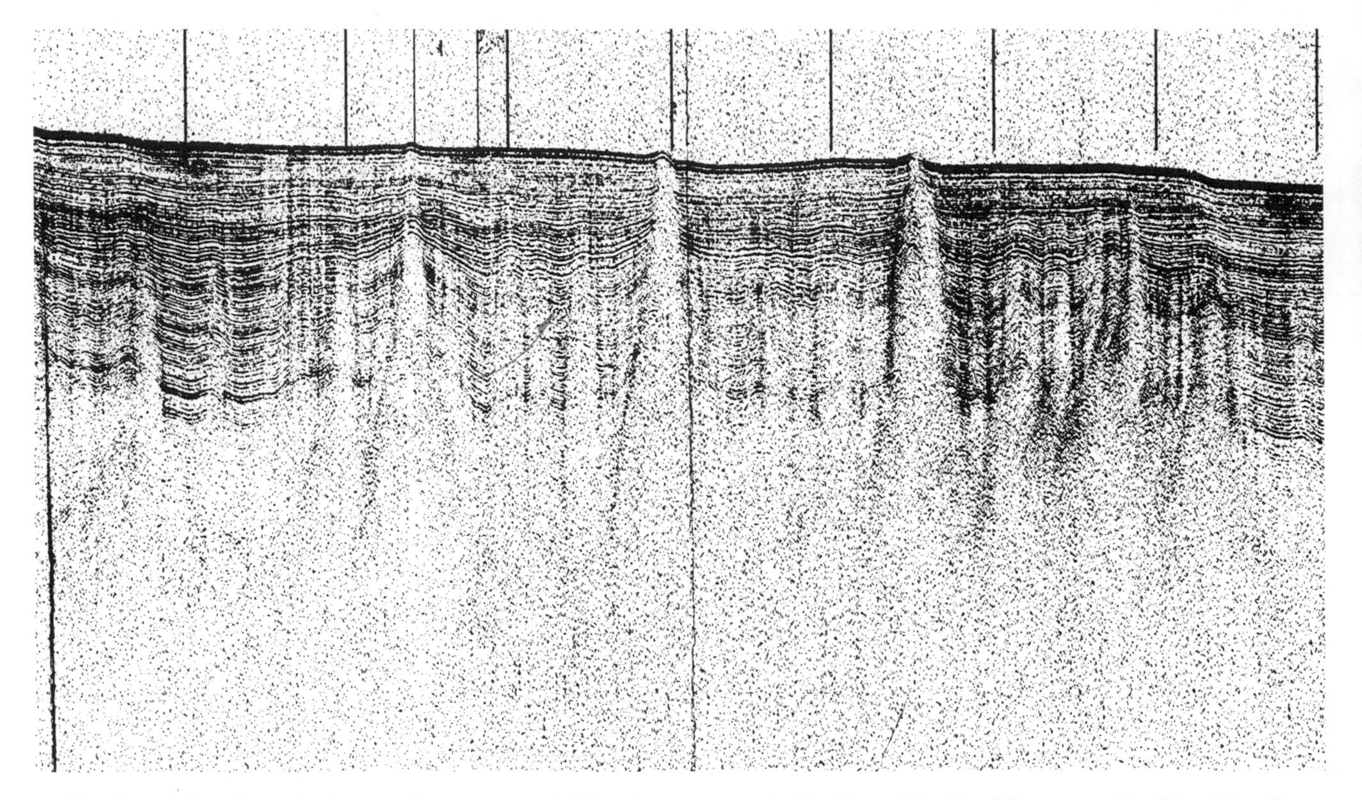

POSSIBILITY OF OIL in the deep-sea floor was revealed by this record from some 250 miles northwest of the Cape Verde Islands. The record, like the one on the preceding pages, was made by the research ship *Kane* of the Naval Oceanographic Office. The tall narrow structures appear to be salt domes, along the flanks of which surface-seeking oil is often trapped in tilted sedimentary layers.

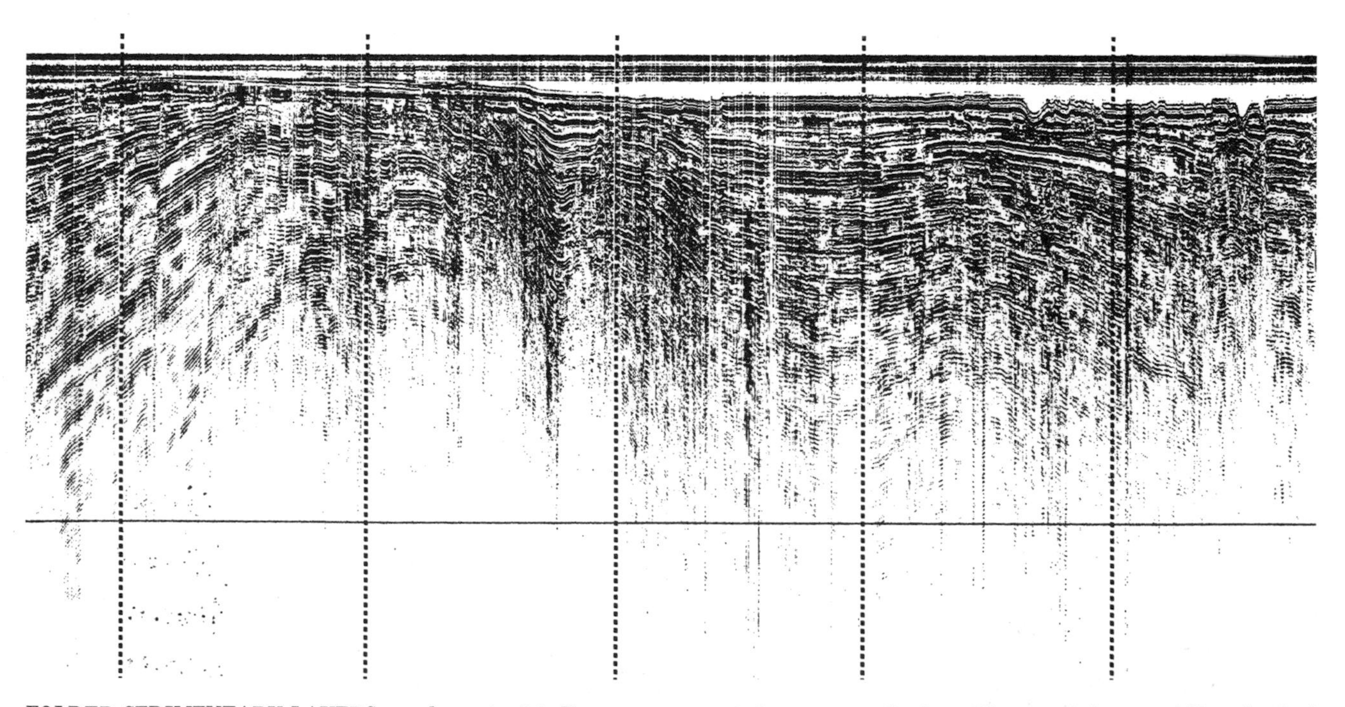

FOLDED SEDIMENTARY LAYERS are shown in this *Kane* record made on the continental shelf north of Trinidad. The band at top is from water reflections. The record shows anticlines (arches) and synclines (troughs); oil is often trapped in crests of anticlines.

if ore deposits are located in ocean-floor rock far from land.

Seventy percent of the world's continental shelves consist of ancient unconsolidated sediments from which are dredged such commodities as sand, gravel, oyster shell, tin, heavy-mineral sands and diamonds. Dredging is an attractive mining technique because of low capital investment, quick returns and high profits and the operational mobility offered by floating dredges. So far it has been limited to nearshore waters less than 235 feet deep and protected from severe weather effects. As knowledge of resources in deeper water increases, industry will undoubtedly upgrade its dredging technology.

Of the many potentially valuable surface deposits, sand and gravel are the most important in dollar terms, and only these and oyster shells are now mined off the U.S. coast. Some 20 million tons of oyster shells are extracted from U.S. continental shelves annually as a source of lime; sand and gravel run about 50 million cubic yards. As coastal metropolitan areas spread out, they cover dry-land deposits of the very construction materials required to sustain their expansion; in these circumstances sea-floor sources such as one recently found off New Jersey, which is thought to contain a billion tons of gravel, become commercially valuable.

In the deeper waters of the continental shelves, on the upper parts of the slopes and on submarine banks and ridges widespread deposits of marine phosphorite nodules are found at depths between 100 and 1,000 feet. The best-known large deposits are off southern California, where total reserves are estimated at 1.5 billion tons, and off northwestern Mexico, Peru and Chile, the southeastern U.S. and the Union of South Africa. The only major attempt at mining was made in 1961, when a company leased an area off California, but that lease was returned unexploited to the Federal Government four years later. With large land sources generally available to meet the demand for phosphates for fertilizer and other products, offshore exploitation of this resource is not likely to occur soon, except possibly in phosphate-poor countries.

The only known minerals on the floor of the deep ocean that appear to be of potential economic importance are the well-publicized manganese nodules, formed by the precipitation from seawater of manganese oxides and other mineral salts, usually on a small nucleus such as a bit of stone or a shark's tooth.

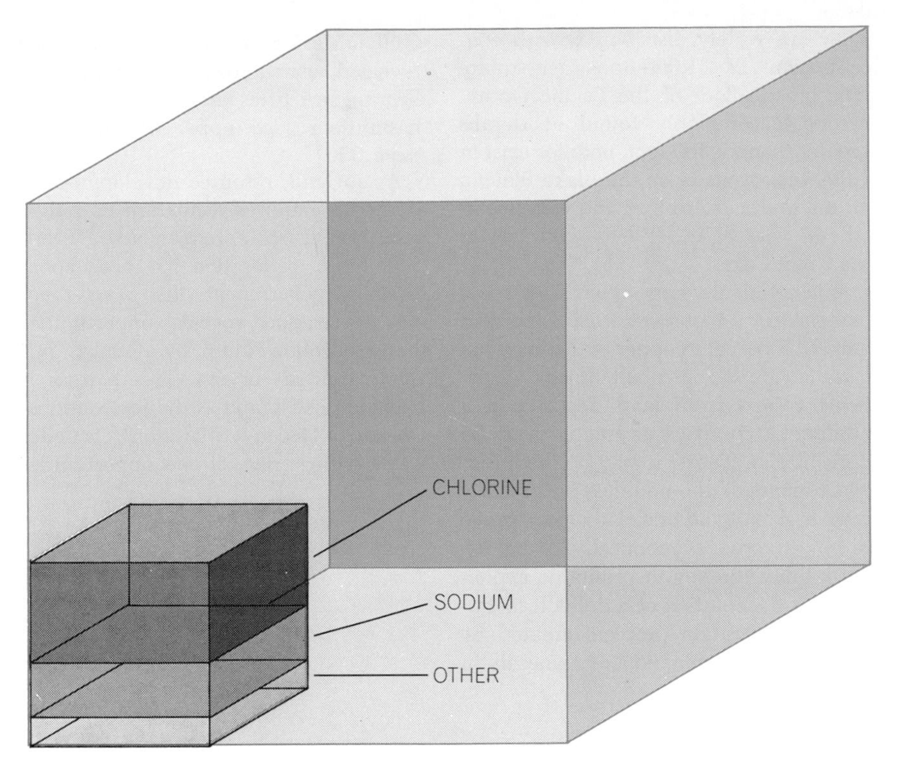

SEAWATER contains an average of 35,000 parts per million of dissolved solids. In a cubic mile of seawater, weighing 4.7 billion tons, there are therefore about 165 million tons of dissolved matter, mostly chlorine and sodium (*gray cube*). The volume of the ocean is about 350 million cubic miles, giving a theoretical mineral reserve of about 60 quadrillion tons.

ELEMENT	TONS PER CUBIC MILE	ELEMENT	TONS PER CUBIC MILE
CHLORINE	89,500,000	NICKEL	9
SODIUM	49,500,000	VANADIUM	9
MAGNESIUM	6,400,000	MANGANESE	9
SULFUR	4,200,000	TITANIUM	5
CALCIUM	1,900,000	ANTIMONY	2
POTASSIUM	1,800,000	COBALT	2
BROMINE	306,000	CESIUM	2
CARBON	132,000	CERIUM	2
STRONTIUM	38,000	YTTRIUM	1
BORON	23,000	SILVER	1
SILICON	14,000	LANTHANUM	1
FLUORINE	6,100	KRYPTON	1
ARGON	2,800	NEON	.5
NITROGEN	2,400	CADMIUM	.5
LITHIUM	800	TUNGSTEN	.5
RUBIDIUM	570	XENON	.5
PHOSPHORUS	330	GERMANIUM	.3
IODINE	280	CHROMIUM	.2
BARIUM	140	THORIUM	.2
INDIUM	94	SCANDIUM	.2
ZINC	47	LEAD	.1
IRON	47	MERCURY	.1
ALUMINUM	47	GALLIUM	.1
MOLYBDENUM	47	BISMUTH	.1
SELENIUM	19	NIOBIUM	.05
TIN	14	THALLIUM	.05
COPPER	14	HELIUM	.03
ARSENIC	14	GOLD	.02
URANIUM	14		

CONCENTRATION of 57 elements in seawater is given in this table. Only sodium chloride (common salt), magnesium and bromine are now being extracted in significant amounts.

They are widely distributed, with concentrations of 31,000 tons per square mile on the floor of the Pacific Ocean. Although commonly found at depths greater than 12,500 feet, nodules exist in 1,000 feet of water on the Blake Plateau off the southeastern U.S. and were located last year at a depth of 200 feet in the Great Lakes.

The nodules average about 24 percent manganese, 14 percent iron, 1 percent nickel, .5 percent copper and somewhat less than .5 percent cobalt. Since ore now being mined from land deposits in a number of countries averages 35 to 55 percent manganese, it may be the minor constituents of the nodules, particularly copper, cobalt and nickel, that first prove to be attractive economically. Many experts think the key to profitable exploitation is the solution of a difficult metallurgical separation problem created by the unique combination of minerals in the nodules.

Few discoveries have created more excitement among earth scientists than the location, by different expeditions in 1964, 1965 and 1966, of three undersea pools of hot, high-density brines in the middle of the Red Sea. The brines contain minerals in concentrations as high as 300,000 parts per million—nearly 10 times as much solid matter as is commonly dissolved in ocean water—and overlie sediments rich in such heavy metals as zinc, copper, lead, silver and gold. Similar deposits may be characteristic of other enclosed basins associated, as is the Red Sea, with rift valleys.

As this decade ends resource exploration is advancing on many fronts. Chromite has been found by Russian oceanographers in sea-floor rifts in the Indian Ocean, and zirconium, titanium and other heavy minerals have been detected in sediments from extensive areas off the Texas coast. Methane deposits sufficient to supply Italy's needs for at least six years have been confirmed in the Adriatic Sea. New oil fields of economic value have been discovered off Mexico, Trinidad, Brazil, Dahomey and Australia. Surveys of the Yellow Sea and the East China Sea indicate that the continental shelf between Taiwan and Japan may contain one of the richest oil reserves in the world. It is now becoming clear that the continental rises, which lie at depths ranging from about 5,000 to 18,000 feet and contain a far larger total volume of sediments than the shelves, may hold significant petroleum reserves. Within the past year the *Glomar Challenger* has drilled into oil-bearing sediments lying under 11,700 feet of water in the Gulf of Mexico, and seismic surveys have revealed what appear to be typical oil-bearing structures under the deep ocean-basin floor [*see upper illustration on page 350*].

As on land, resource development of a frontier requires a mixture of public and private entrepreneurship. Historically basic exploration has been sponsored by government; this broad-ranging exploration reveals opportunities that are followed up by detailed privately funded surveys. This pattern is likely to persist, and as the International Decade of Ocean Exploration gets under way a wide variety of new opportunities for marine resource development will surely come to light.

Limitations on the exploitation of the oceans stem partly from lack of knowledge about the distribution of resources and the state of the art of undersea technology. The major limits, however, are set by venture economics, the motivating factor for the profit sector. That factor is influenced by the availability of competing land deposits, by extraction technology and the legal situation and, most critically, by market demand. On the basis of projections of world population and gross national products to the year 2000, which indicate respective in-

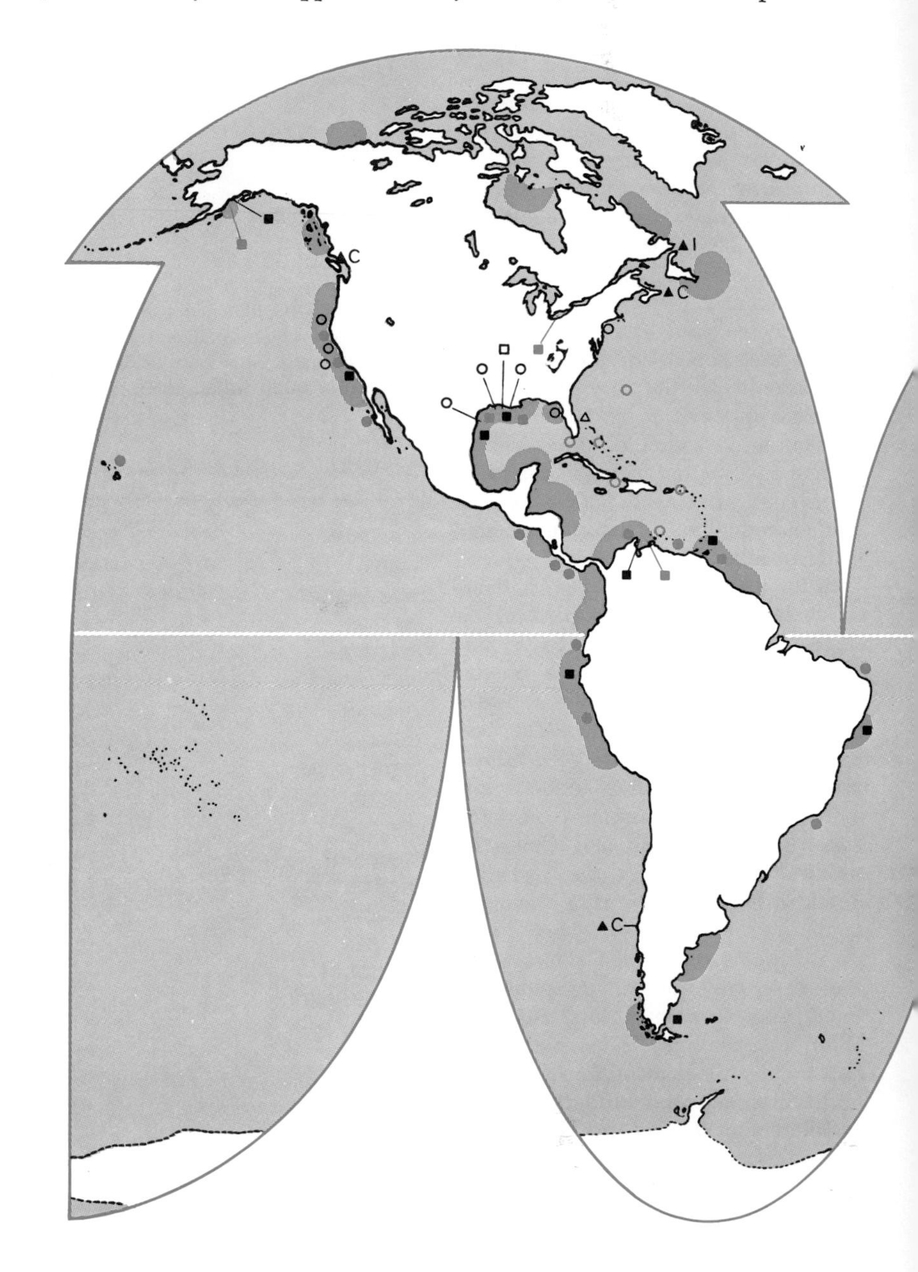

CURRENT PRODUCTION of major ocean resources (except sand, gravel and shell) is mapped with areas of oil and gas exploration. Data come from U.S. Geological Survey, *Oil &*

creases of almost 100 and 500 percent over 1965, a sharp rise in total resource demand can be anticipated, and with it a greater role for the sea.

Other major impediments to the rapid development of ocean resources arise from social and legal constraints. Damage to beaches and wildlife from oil leaks, as in the Santa Barbara Channel, and uncertainty about the effect of dredging on marine organisms have brought public awareness that offshore development may have detrimental consequences. The public, the owner of the resources, is demanding greater safeguards, questioning the wisdom of resource development in areas where it may threaten the environment. In deeper waters seabed development comes up against the potent issue of ownership. There are major questions about the boundaries of national jurisdictions and about the jurisdiction over the seabed beyond such boundaries [see "The Ocean and Man," by Warren S. Wooster; SCIENTIFIC AMERICAN Offprint 888].

The coastal margin—the ribbon of land and water where people and oceans meet and are profoundly influenced by each other—has only recently come to be recognized and treated as a valuable and perishable resource. It is actually a complex of unique physical resources: estuaries and lagoons, marshes, beaches and cliffs, bays and harbors, islands and spits and peninsulas.

In the year 2000 half of the estimated 312-million population of the U.S. will live on 5 percent of the land area in three coastal urban belts: the megalopolises of the Atlantic, the Pacific and the Great Lakes. Along with the people will come an intensification of competing demands for the limited resources of the narrow, fragile coastal zone. To make matters worse, the coastal resource is shrinking under the pressure of natural forces (hur-

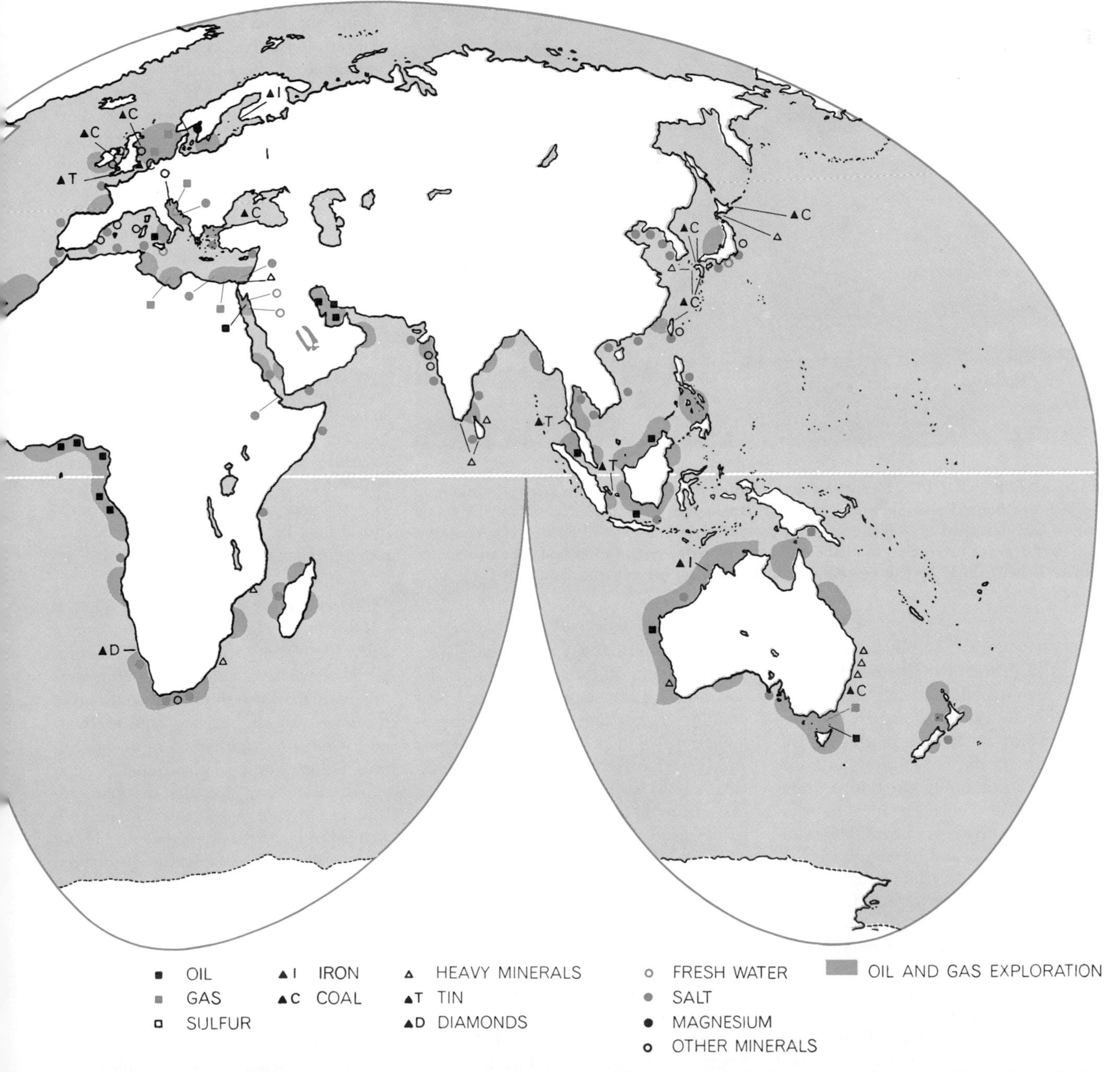

***Gas Journal*, the magazine *Offshore* and other sources. Oil, gas and sulfur are produced by drilling; coal and iron ore from mines driven from dry land; heavy minerals, tin and diamonds by dredging; fresh water, salt, magnesium and other minerals from seawater.**

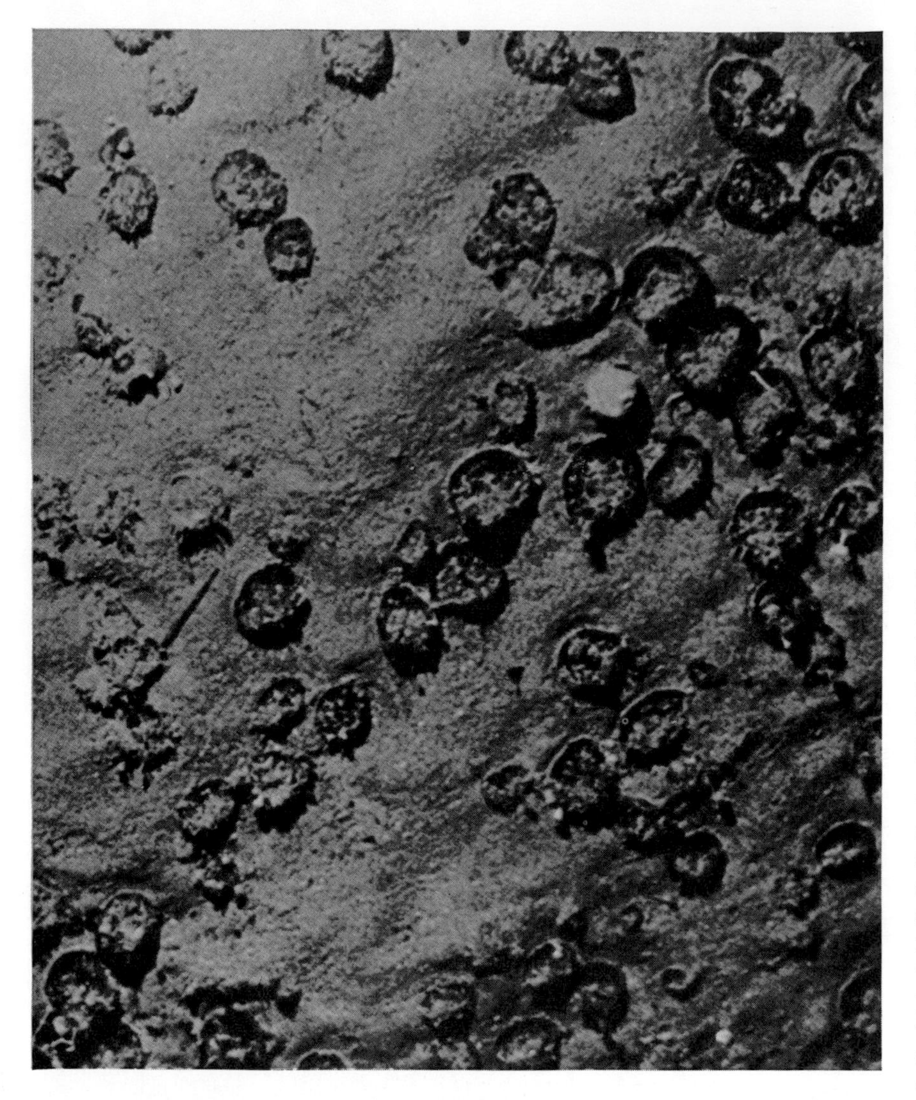

MANGANESE NODULES, formed by precipitation from seawater, are generally found on the deep-sea floor. These nodules were photographed on the Blake Plateau off the southeastern U.S., less than 3,000 feet deep, by a prospecting ship operated by Deepsea Ventures, Inc. They average two inches in diameter, about a quarter-pound in weight. The manganese content is between 15 and 30 percent, the nickel and copper content about 1 percent each.

ricanes have caused $5 billion in damage to the U.S. economy in the past 15 years) and human exploitation and neglect.

More than a tenth of the 10.7 million square miles of shellfish-producing waters bordering the U.S. is now unusable because of pollution. Dredging, drainage projects and even chemical mosquito-control programs are having devastating effects on fish and other aquatic life. The amount of industrial waste reaching the oceans will increase sevenfold within the decade. Whereas 14 nuclear-powered generating plants are operating in the U.S. today, more than 100 are scheduled by 1975, with nine planned for Long Island Sound alone. Thermal pollution from the discharge of hot water is therefore a potential threat to coastal waters as well as inland lakes and rivers.

In the competition for the zone's resources among different uses—industrial and housing development, ports, shipbuilding, recreation, commercial fisheries and waste disposal—natural wetlands and estuarine open spaces are losing out. Of the tidal wetlands along the Atlantic coast from Maine to Delaware, 45,000 acres were lost between 1955 and 1964. An inventory shows that 34 percent of that area was dried up by being used as a dumping ground for dredging operations; 27 percent was filled for housing developments; 15 percent went to recreational developments (parks, beaches and marinas) and 10 percent to bridges, roads, parking lots and airports; 7 percent was turned into industrial sites and 6 percent into garbage and trash dumps. (In Maryland 176 acres of submerged land in Chesapeake Bay were sold recently for $100 an acre and, after being filled with dredged bay-bottom muck, were subdivided into lots selling for between $4,000 and $8,000 each.)

With the demand for marine recreation growing with the coastal population, pressure is increasing on the one-third of the coastal zone that has recreational potential. Only about 6.5 percent of this is now in public ownership, yet in order to meet the projected demand it is considered essential that about 15 percent be accessible to the public. The mere fact that coastal land with recreational potential exists, moreover, is far from meaning that it will ever be put to recreational use. Swimming, boating and skin diving are often incompatible with competing alternative uses, many of which appear to have equally valid claims. In the face of conflicts between public and private, and long-term and short-term, benefits, how and by whom will the ultimate decisions be made on the proper utilization of coastal land?

Management of the coastal zone is unwieldy because the environment is almost hopelessly fragmented by political subdivisions: 24 states, more than 240 counties, some 600 coastal cities, townships, towns and villages and numerous regional authorities and special districts with their own regulatory powers. Superimposed on the many public jurisdictions there is another tapestry of private ownership. Because the states hold coastal resources in trust out to the three-mile limit, the Federal Government has a restricted role in resolving disputes, but it may be able to exert leadership in defining the issues.

Thoreau once admonished: "What is the use of a house if you haven't got a tolerable planet to put it on." Unless rational alternatives among competing uses are evaluated, the trend will continue to be toward single-purpose uses, motivated by short-term advantages to individuals, industry or local governments. Such exploitation may actually dissipate resources. Private beach development restricts public access; dredging and filling downgrade commercial fishing; offshore drilling rigs limit freedom of navigation. Each single-purpose use may seem justifiable on its own, but the overall effect of piecemeal development can be chaos.

In this technological age man can do many more of the things he wants to do. The oceans place before him a vast store of little-developed material resources; the tools of science and technology are at his disposal. This combination of a new frontier, new knowledge and new technical capability may be unique in the human experience. We are accumulating the basic information with which to define the ecological base from which we operate, to understand the natural forces at work and to predict the consequences of each insult to the environment. With this new comprehension it will soon be possible to develop the engineering with which to harvest mineral wealth, maintain water quality, inhibit beach erosion, create modern ports and harbors—and to establish the criteria for making necessary choices among courses of action and the law and institutions to effectuate them. In time we may even be able to correct mistakes that were made long ago in ignorance or that occur in the future because of man's stupidity, neglect or greed.

TIDAL WETLANDS, an important coastal resource, are disappearing rapidly. The top photograph shows Boca Ciega Bay, near St. Petersburg on the west coast of Florida, as it was in 1949. The bottom photograph shows the same area filled and developed, in 1969.

37

THE FOOD RESOURCES OF THE OCEAN

S. J. HOLT
September 1969

I suppose we shall never know what was man's first use of the ocean. It may have been as a medium of transport or as a source of food. It is certain, however, that from early times up to the present the most important human uses of the ocean have been these same two: shipping and fishing. Today, when so much is being said and written about our new interests in the ocean, it is particularly important to retain our perspective. The annual income to the world's fishermen from marine catches is now roughly $8 billion. The world ocean-freight bill is nearly twice that. In contrast, the wellhead value of oil and gas from the seabed is barely half the value of the fish catch, and all the other ocean mineral production adds little more than another $250 million.

Of course, the present pattern is likely to change, although how rapidly or dramatically we do not know. What is certain is that we shall use the ocean more intensively and in a greater variety of ways. Our greatest need is to use it wisely. This necessarily means that we use it in a regulated way, so that each ocean resource, according to its nature, is efficiently exploited but also conserved. Such regulation must be in large measure of an international kind, particularly insofar as living resources are concerned. This will be so whatever may be the eventual legal regime of the high seas and the underlying bed. The obvious fact about most of the ocean's living resources is their mobility. For the most part they are lively animals, caring nothing about the lines we draw on charts.

The general goal of ecological research, to which marine biology makes an important contribution, is to achieve an understanding of and to turn to our advantage all the biological processes that give our planet its special character. Marine biology is focused on the problems of biological production, which are closely related to the problems of production in the economic sense. Our most compelling interest is narrower. It lies in ocean life as a renewable resource: primarily of protein-rich foods and food supplements for ourselves and our domestic animals, and secondarily of materials and drugs. I hope to show how in this field science, industry and government need each other now and will do so even more in the future. First, however, let me establish some facts about present fishing industries, the state of the art governing them and the state of the relevant science.

The present ocean harvest is about 55 million metric tons per year. More than 90 percent of this harvest is finfish; the rest consists of whales, crustaceans and mollusks and some other invertebrates. Although significant catches are reported by virtually all coastal countries, three-quarters of the total harvest is taken by only 14 countries, each of which produces more than a million tons annually and some much more. In the century from 1850 to 1950 the world catch increased tenfold—an average rate of about 25 percent per decade. In the next decade it nearly doubled, and this rapid growth is continuing [*see illustration on page 361*]. It is now a commonplace that fish is one of the few major foodstuffs showing an increase in global production that continues to exceed the growth rate of the human population.

This increase has been accompanied by a changing pattern of use. Although some products of high unit value as luxury foods, such as shellfish, have maintained or even enhanced their relative economic importance, the trend has been for less of the catch to be used directly as human food and for more to be reduced to meal for animal feed. Just before World War II less than 10 percent of the world catch was turned into meal; by 1967 half of it was so used. Over the same period the proportion of the catch preserved by drying or smoking declined from 28 to 13 percent and the proportion sold fresh from 53 to 31 percent. The relative consumption of canned fish has hardly changed but that of frozen fish has grown from practically nothing to 12 percent.

While we are comparing the prewar or immediate postwar situation with the present, we might take a look at the composition of the catch by groups of species. In 1948 the clupeoid fishes (herrings, pilchards, anchovies and so on), which live mainly in the upper levels of the ocean, already dominated the scene (33 percent of the total by weight) and provided most of the material for fish meal. Today they bulk even larger (45 percent) in spite of the decline of several great stocks of them (in the North Sea and off California, for example). The next most important group, the gadoid fishes (cod, haddock, hake and so on), which live mainly on or near the bottom, comprised a quarter of the total in 1948. Although the catch of these fishes has continued to increase absolutely, the

SCHOOL OF FISH is spotted from the air at night by detecting the bioluminescent glow caused by the school's movement through the water. As the survey aircraft flew over the Gulf of Mexico at an altitude of 3,500 feet, the faint illumination in the water was amplified some 55,000 times by an image intensifier before appearing on the television screen seen in the photograph on the opposite page. The fish are Atlantic thread herring. Detection of fish from the air is one of several means of increasing fishery efficiency being tested at the Pascagoula, Miss., research base of the U.S. Bureau of Commercial Fisheries.

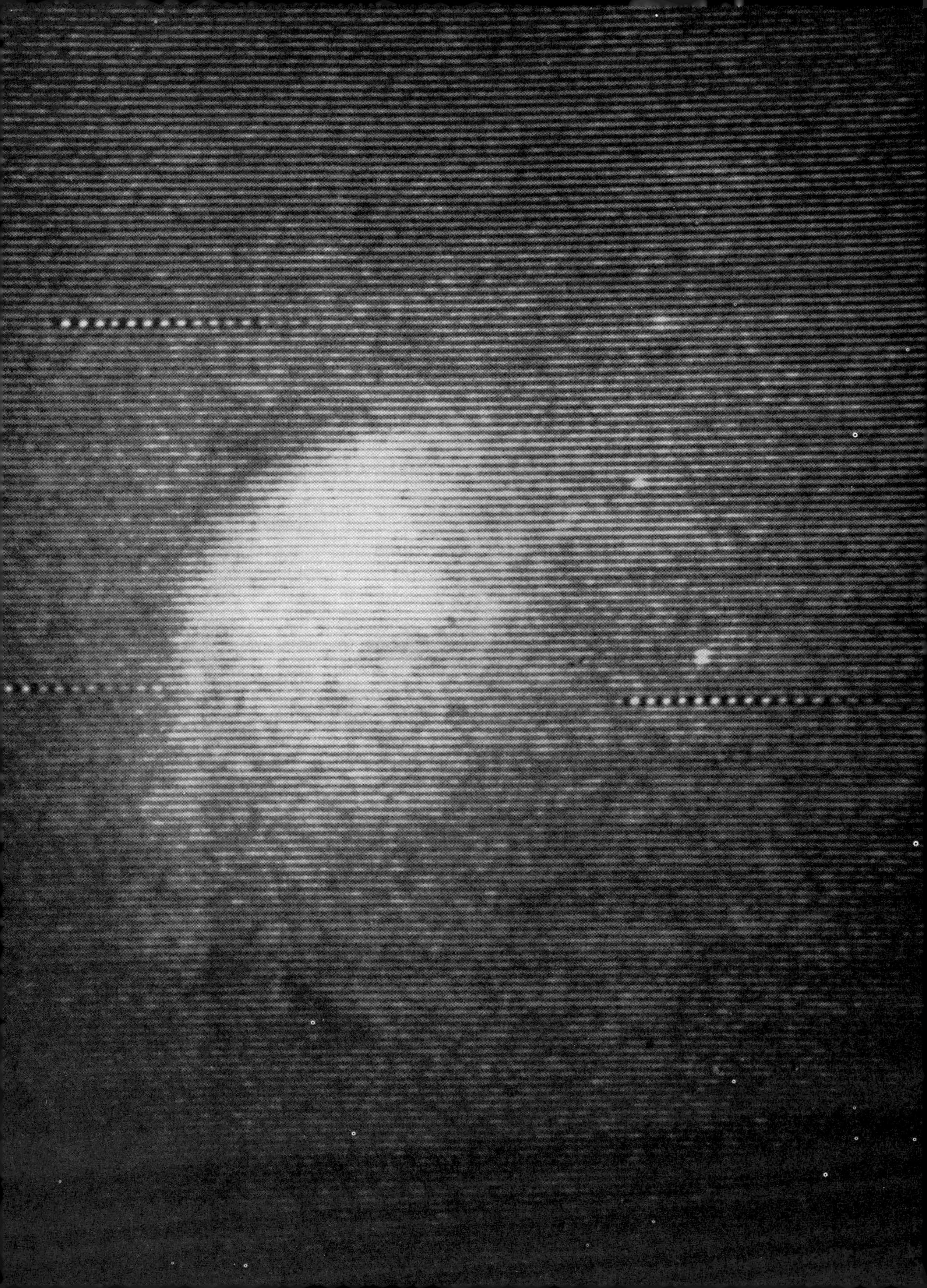

proportion is now reduced to 15 percent. The flounders and other flatfishes, the rosefish and other sea perches and the mullets and jacks have collectively stayed at about 15 percent; the tunas and mackerels, at 7 percent. Nearly a fifth of the total catch continues to be recorded in statistics as "Unsorted and other"—a vast number of species and groups, each contributing a small amount to a considerable whole.

The rise of shrimp and fish meal production together account for another major trend in the pattern of fisheries development. A fifth of the 1957 catch was sold in foreign markets; by 1967, two-fifths were entering international trade and export values totaled $2.5 billion. Furthermore, during this same period the participation of the less developed countries in the export trade grew from a sixth to well over 25 percent. Most of these shipments were destined for markets in the richer countries, particularly shrimp for North America and fish meal for North America, Europe and Japan. More recently several of the less developed countries have also become importers of fish meal, for example Mexico and Venezuela, South Korea and the Republic of China.

The U.S. catch has stayed for many years in the region of two million tons, a low figure considering the size of the country, the length of the coastline and the ready accessibility of large resources on the Atlantic, Gulf and Pacific seaboards. The high level of consumption in the U.S. (about 70 pounds per capita) has been achieved through a steady growth in imports of fish and fish meal: from 25 percent of the total in 1950 to more than 70 percent in 1967. In North America 6 percent of the world's human population uses 12 percent of the world's catch, yet fishermen other than Americans take nearly twice the amount of fish that Americans take from the waters most readily accessible to the U.S.

There has not been a marked change in the broad geography of fishing [*see illustration on these two pages*]. The Pacific Ocean provides the biggest

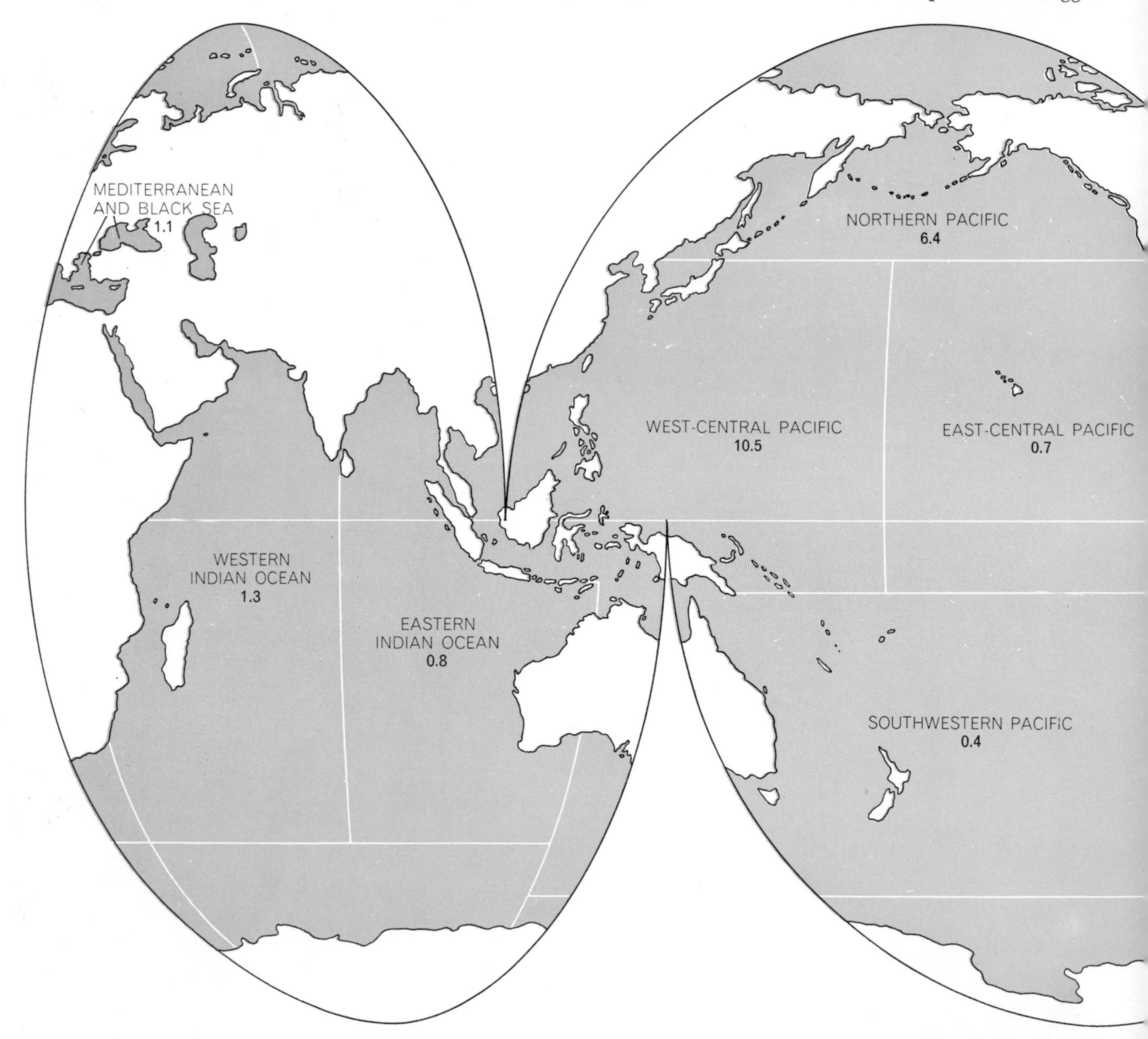

MAJOR MARINE FISHERY AREAS are 14 in number: two in the Indian Ocean (*left*), five in the Pacific Ocean (*center*) and six in the Atlantic (*right*). Due to the phenomenal expansion of the Peru fishery, the total Pacific yield is now a third larger than the Atlantic total. The bulk of Atlantic and Pacific catches, however, is still taken well north of the Equator. The Indian Ocean, with a

share (53 percent) but the Atlantic (40 percent, to which we may add 2 percent for the Mediterranean) is yielding considerably more per unit area. The Indian Ocean is still the source of less than 5 percent of the catch, and since it is not a biologically poor ocean it is an obvious target for future development. Within the major ocean areas, however, there have been significant changes. In the Pacific particular areas such as the waters off Peru and Chile and the Gulf of Thailand have rapidly acquired importance. The central and southern parts of the Atlantic, both east and west, are of growing interest to more nations. Although, with certain exceptions, the traditional fisheries in the colder waters of the Northern Hemisphere still dominate the statistics, the emergence of some of the less developed countries as modern fishing nations and the introduction of long-range fleets mean that tropical and subtropical waters are beginning to contribute significantly to world production.

Finally, in this brief review of the trends of the past decade or so we must mention the changing importance of countries as fishing powers. Peru has become the leading country in terms of sheer magnitude of catch (although not of value or diversity) through the development of the world's greatest one-species fishery: 10 million tons of anchovies per year, almost all of which is reduced to meal [*see illustration on page 363*]. The U.S.S.R. has also emerged as a fishing power of global dimension, fishing for a large variety of products throughout the oceans of the world, particularly with large factory ships and freezer-trawlers.

At this point it is time to inquire about the future expectations of the ocean as a major source of protein. In spite of the growth I have described, fisheries still contribute only a tenth of the animal protein in our diet, although this proportion varies considerably from one part of the world to another. Before such an inquiry can be pursued, however, it is necessary to say something about the problem of overfishing.

A stock of fish is, generally speaking, at its most abundant when it is not being exploited; in that virgin state it will include a relatively high proportion of the larger and older individuals of the species. Every year a number of young recruits enter the stock, and all the fish—but particularly the younger ones—put on weight. This overall growth is balanced by the natural death of fish of all ages from disease, predation and perhaps senility. When fishing begins, the large stock yields large catches to each fishing vessel, but because the pioneering vessels are few, the total catch is small.

Increased fishing tends to reduce the level of abundance of the stock progressively. At these reduced levels the losses accountable to "natural" death will be less than the gains accountable to recruitment and individual growth. If, then, the catch is less than the difference between natural gains and losses, the stock will tend to increase again; if the catch is more, the stock will decrease. When the stock neither decreases nor increases, we achieve a sustained yield. This sustained yield is small when the stock is large and also when the stock is small; it is at its greatest when the stock is at an intermediate level—somewhere between two-thirds and one-third of the virgin abundance. In this intermediate stage the average size of the individuals will be smaller and the age will be younger than in the unfished condition, and individual growth will be highest in relation to the natural mortality.

The largest catch that on the average can be taken year after year without causing a shift in abundance, either up or down, is called the maximum sustainable yield. It can best be obtained by leaving the younger fish alone and fishing the older ones heavily, but we can also get near to it by fishing moderately, taking fish of all sizes and ages. This

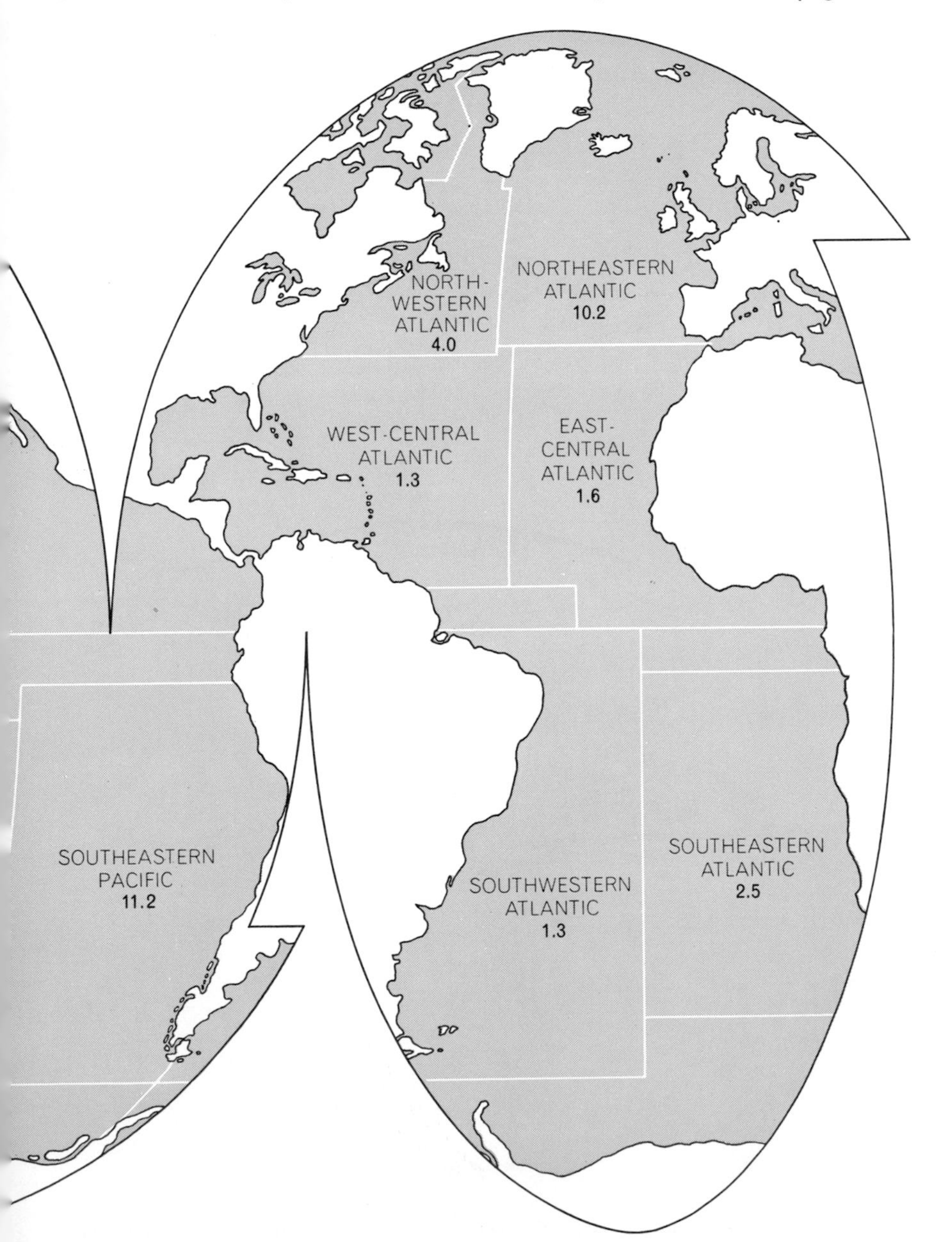

total catch of little more than two million metric tons, live weight, is the world's major underexploited region. The number below each area name shows the millions of metric tons landed during 1967, as reported by the UN Food and Agriculture Organization.

phenomenon—catches that first increase and then decrease as the intensity of fishing increases—does not depend on any correlation between the number of parent fish and the number of recruits they produce for the following generation. In fact, many kinds of fish lay so many eggs, and the factors governing survival of the eggs to the recruit stage are so many and so complex, that it is not easy to observe any dependence of the number of recruits on the number of their parents over a wide range of stock levels.

Only when fishing is intense, and the stock is accordingly reduced to a small fraction of its virgin size, do we see a decline in the number of recruits coming in each year. Even then there is often a wide annual fluctuation in this number. Indeed, such fluctuation, which causes the stock as a whole to vary greatly in abundance from year to year, is one of the most significant characteristics of living marine resources. Fluctuation in number, together with the considerable variation in "availability" (the change in the geographic location of the fish with respect to the normal fishing area), largely account for the notorious riskiness of fishing as an industry.

For some species the characteristics of growth, natural mortality and recruit-

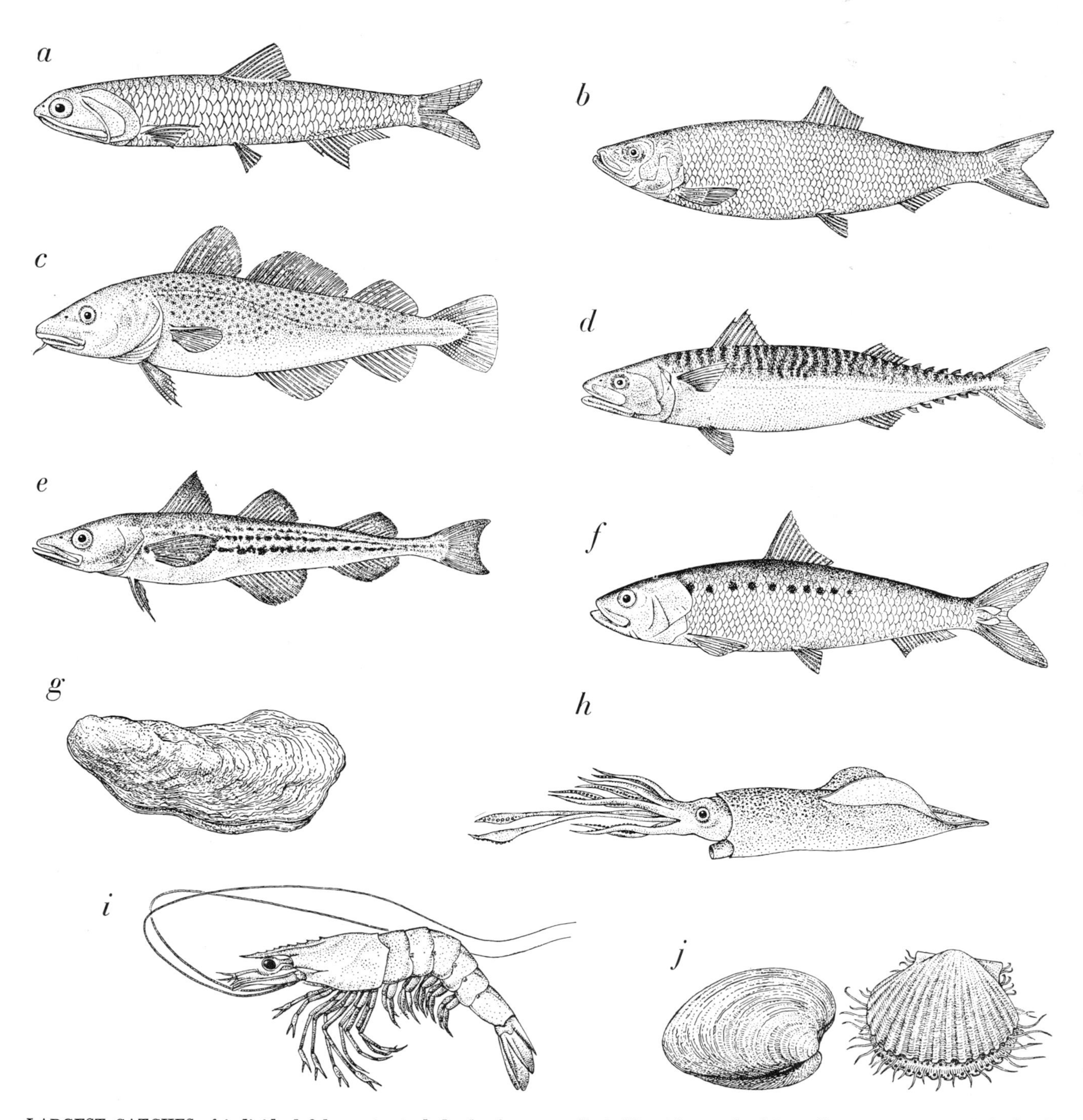

LARGEST CATCHES of individual fish species include the five fishes shown here (*left*). They are, according to the most recent detailed FAO fishery statistics (1967), the Peruvian anchoveta (*a*), with a catch of more than 10.5 million metric tons; the Atlantic herring (*b*), with a catch of more than 3.8 million tons; the Atlantic cod (*c*), with a catch of 3.1 million tons; the Alaska walleye pollack (*d*), with a catch of 1.7 million metric tons, and the South African pilchard (*e*), with a catch of 1.1 million tons. No single invertebrate species (*right*) is harvested in similar quantities. Taken as a group, however, various oyster species (*f*) totaled .83 million tons in 1967; squids (*g*), .75 million tons; shrimps and prawns (*h*), .69 million tons; clams and cockles (*i*), .48 million tons.

ment are such that the maximum sustainable yield is sharply defined. The catch will decline quite steeply with a change in the amount of fishing (measured in terms of the number of vessels, the tonnage of the fleet, the days spent at sea or other appropriate index) to either below or above an optimum. In other species the maximum is not so sharply defined; as fishing intensifies above an optimum level the sustained catch will not significantly decline, but it will not rise much either.

Such differences in the dynamics of different types of fish stock contribute to the differences in the historical development of various fisheries. If it is unregulated, however, each fishery tends to expand beyond its optimum point unless something such as inadequate demand hinders its expansion. The reason is painfully simple. It will usually still be profitable for an individual fisherman or ship to continue fishing after the *total* catch from the stock is no longer increasing or is declining, and even though his own rate of catch may also be declining. By the same token, it may continue to be profitable for the individual fisherman to use a small-meshed net and thereby catch young as well as older fish, but in doing so he will reduce both his own possible catch and that of others in future years. Naturally if the total catch is declining, or not increasing much, as the amount of fishing continues to increase, the net economic yield from the fishery—that is, the difference between the total costs of fishing and the value of the entire catch—will be well past its maximum. The well-known case of the decline of the Antarctic baleen whales provides a dramatic example of overfishing and, one would hope, a strong incentive for the more rational conduct of ocean fisheries in the future.

There is, then, a limit to the amount that can be taken year after year from each natural stock of fish. The extent to which we can expect to increase our fish catches in the future will depend on three considerations. First, how many as yet unfished stocks await exploitation, and how big are they in terms of potential sustainable yield? Second, how many of the stocks on which the existing fisheries are based are already reaching or have passed their limit of yield? Third, how successful will we be in managing our fisheries to ensure maximum sustainable yields from the stocks?

The first major conference to examine the state of marine fish stocks on a global basis was the United Nations Scientific Conference on the Conservation and Utilization of Resources, held in 1949 at Lake Success, N.Y. The small group of fishery scientists gathered there concluded that the only overfished stocks at that time were those of a few high-priced species in the North Atlantic and North Pacific, particularly plaice, halibut and salmon. They produced a map showing 30 other known major stocks they believed to be underfished. The situation was reexamined in 1968. Fishing on half of those 30 stocks is now close to or beyond that required for maximum yield. The fully fished or overfished stocks include some tunas in most ocean areas, the herring, the cod and ocean perch in the North Atlantic and the anchovy in the southeastern Pacific. The point is that the history of development of a fishery from small beginnings to the stage of full utilization or overutilization can, in the modern world, be compressed into a very few years. This happened with the anchovy off Peru, as a result of a massive local fishery growth, and it has happened to some demersal, or bottom-dwelling, fishes elsewhere through the large-scale redeployment of long-distance trawlers from one ocean area to another.

It is clear that the classical process of fleets moving from an overfished area to another area, usually more distant and less fished, cannot continue indefinitely. It is true that since the Lake Success

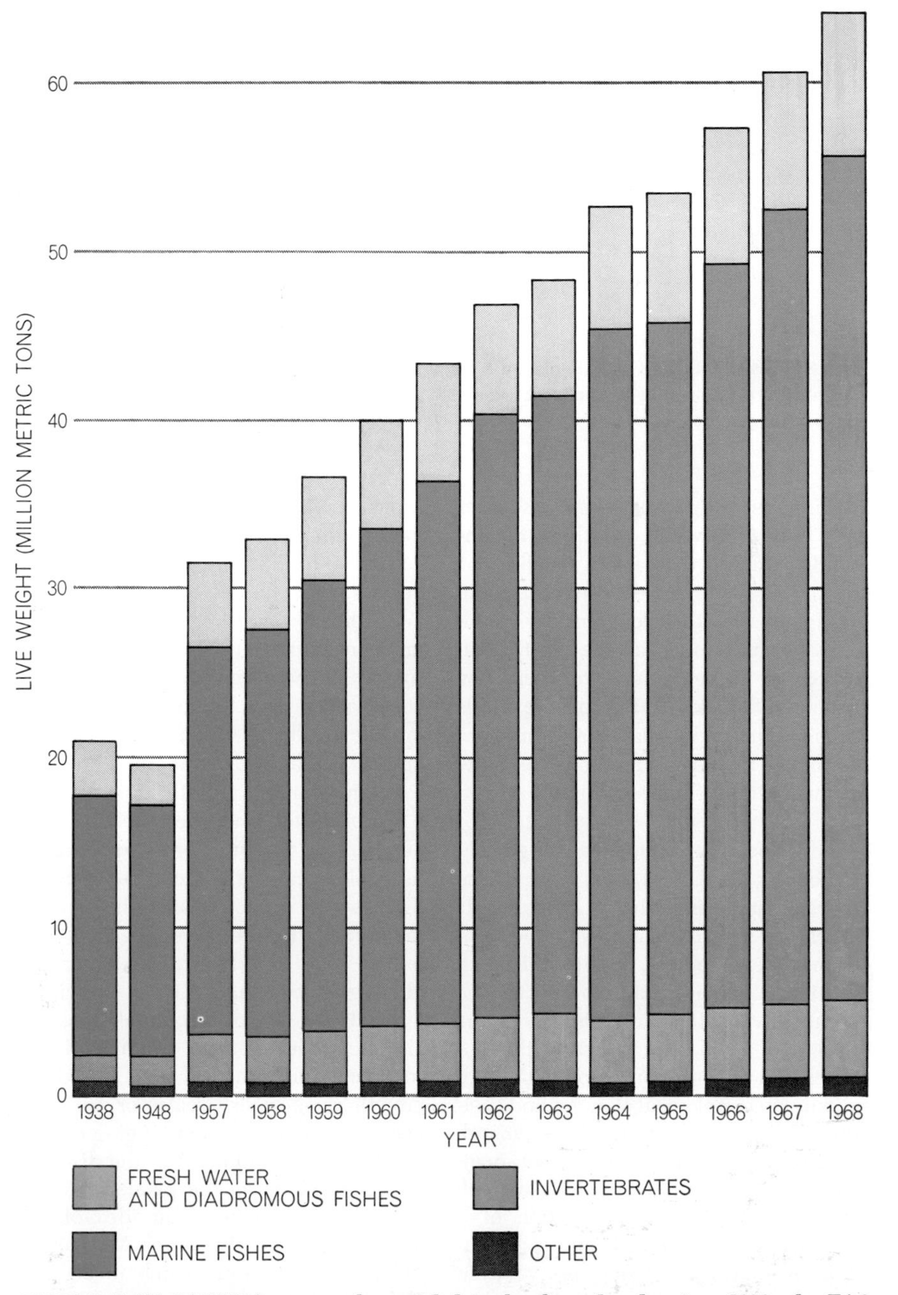

WORLD FISH CATCH has more than tripled in the three decades since 1938; the FAO estimate of the 1968 total is 64 million metric tons. The largest part consists of marine fishes. Humans directly consume only half of the catch; the rest becomes livestock feed.

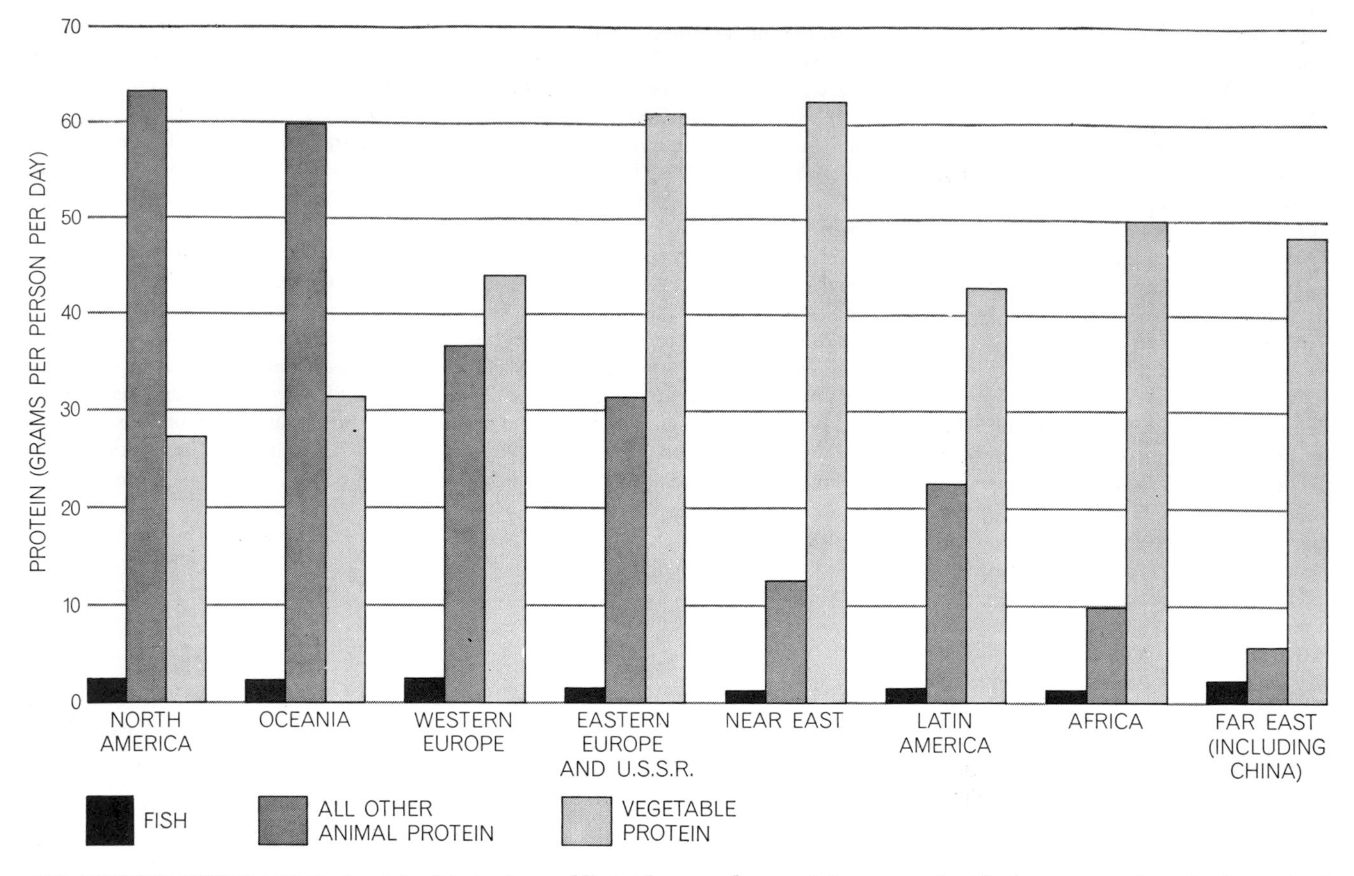

RELATIVELY MINOR ROLE played by fish in the world's total consumption of protein is apparent when the grams of fish eaten per person per day in various parts of the world (*left column in each group*) is compared with the consumption of other animal protein (*middle column*) and vegetable protein (*right column*). The supply is nonetheless growing more rapidly than world population.

meeting several other large resources have been discovered, mostly in the Indian Ocean and the eastern Pacific, and additional stocks have been utilized in fishing areas with a long history of intensive fishing, such as the North Sea. In another 20 years, however, very few substantial stocks of fish of the kinds and sizes of commercial interest and accessible to the fishing methods we know now will remain underexploited.

The Food and Agriculture Organization of the UN is now in the later stages of preparing what is known as its Indicative World Plan (IWP) for agricultural development. Under this plan an attempt is being made to forecast the production of foodstuffs in the years 1975 and 1985. For fisheries this involves appraising resource potential, envisioning technological changes and their consequences, and predicting demand. The latter predictions are not yet available, but the resource appraisals are well advanced. With the cooperation of a large number of scientists and organizations estimates are being prepared in great detail on an area basis. They deal with the potential of known stocks, both those fished actively at present and those exploited little or not at all. Some of these estimates are reliable; others are naturally little more than reasonable guesses. One fact is abundantly clear: We still have very scrappy knowledge, in quantitative terms, of the living resources of the ocean. We can, however, check orders of magnitude by comparing the results of different methods of appraisal. Thus where there is good information on the growth and mortality rates of fishes and measures of their numbers in absolute terms, quite good projections can be made. Most types of fish can now in fact virtually be counted individually by the use of specially calibrated echo sounders for area surveys, although this technique is not yet widely applied. The size of fish populations can also be deduced from catch statistics, from measurements of age based on growth rings in fish scales or bands in fish ear stones, and from tagging experiments. Counts and maps of the distribution of fish eggs in the plankton can in some cases give us a fair idea of fish abundance in relative terms. We can try to predict the future catch in an area little fished at present by comparing the present catch with the catch in another area that has similar oceanographic characteristics and basic biological productivity and that is already yielding near its maximum. Finally, we have estimates of the food supply available to the fish in a particular area, or of the primary production there, and from what we know about metabolic and ecological efficiency we can try to deduce fish production.

So far as the data permit these methods are being applied to major groups of fishes area by area. Although individual area and group predictions will not all be reliable, the global totals and subtotals may be. The best figure seems to be that the potential catch is about three times the present one; it might be as little as twice or as much as four times. A similar range has been given in estimates of the potential yield from waters adjacent to the U.S.: 20 million tons compared with the present catch of rather less than six million tons. This is more than enough to meet the U.S. demand, which is expected to reach 10 million tons by 1975 and 12 million by 1985.

Judging from the rate of fishery development in the recent past, it would be entirely reasonable to suppose that the maximum sustainable world catch of between 100 and 200 million tons could be reached by the second IWP target date, 1985, or at least by the end of the century. The real question is whether or not this will be economically worth the effort. Here any forecast is, in my view, on soft ground. First, to double the catch we have to more than double the amount of fishing, because the stocks decline in abundance as they are exploited. Moreover, as we approach the global maximum more of the stocks that are lightly fished at present will be brought down to intermediate levels. Second, fishing will become even more competitive and costly if the nations fail to agree, and agree soon, on regulations to cure overfishing situations. Third, it is quite uncertain what will happen in the long run to the costs of production and the price of protein of marine origin in relation to other protein sources, particularly from mineral or vegetable bases.

In putting forward these arguments I am not trying to damp enthusiasm for the sea as a major source of food for coming generations; quite the contrary. I do insist, however, that it would be dangerous for those of us who are interested in such development to assume that past growth will be maintained along familiar lines. We need to rationalize present types of fishing while preparing ourselves actively for a "great leap forward." Fishing as we now know it will need to be made even more efficient; we shall need to consider the direct use of the smaller organisms in the ocean that mostly constitute the diet of the fish we now catch; we shall need to try harder to improve on nature by breeding, rearing and husbanding useful marine animals and cultivating their pasture. To achieve this will require a much larger scale and range of scientific research, wedded to engineering progress; expansion by perhaps an order of magnitude in investment and in the employment of highly skilled labor, and a modified legal regime for the ocean and its bed not only to protect the investments but also to ensure orderly development and provide for the safety of men and their installations.

To many people the improvement of present fishing activities will mean increasing the efficiency of fishing gear and ships. There is surely much that could be done to this end. We are only just beginning to understand how trawls, traps, lines and seines really work. For example, every few years someone tries a new design or rigging for a deep-sea trawl, often based on sound engineering and hydrodynamic studies. Rarely do these "improved" rigs catch more than the old ones; sometimes they catch much less. The error has been in thinking that the trawl is simply a bag, collecting more or less passive fish, or at least predictably active ones. This is not so at all. We really have to deal with a complex, dynamic relation between the lively animals and their environment, which includes in addition to the physical and biological environment the fishing gear itself. We can expect success in understanding and exploiting this relation now that we can telemeter the fishing gear, study its hydrodynamics at full scale as well as with models in towing tanks, monitor it (and the fish) by means of underwater television, acoustic equipment and divers, and observe and experiment with fish behavior both in the sea and in large tanks. We also probably have something to learn from studying, before they become extinct, some kinds of traditional "primitive" fishing gear still used in Asia, South America and elsewhere—mainly traps that take advantage of subtleties of fish behavior observed over many centuries.

Successful fishing depends not so much on the size of fish stocks as on their concentration in space and time. All fishermen use knowledge of such concentrations; they catch fish where they have gathered to feed or to reproduce, or where they are on the move in streams or schools. Future fishing methods will surely involve a more active role for the fishermen in causing the fish to congregate. In many parts of the world lights or sound are already used to attract fish. We can expect more sophistication in the employment of these and other stimuli, alone and in combination.

Fishing operations as a whole also depend on locating areas of concentration and on the efficient prediction, or at least the prompt observation, of changes in these areas. The large stocks of pelagic, or open-sea, fishes are produced mainly in areas of "divergencies," where water is rising from deeper levels toward the surface and hence where surface waters are flowing outward. Many such areas are the "upwellings" off the western coasts of continental masses, for example off western and southwestern Africa, western India and western South America. Here seasonal winds, currents and continental configurations combine to cause a periodic enrichment of the surface waters.

Divergencies are also associated with

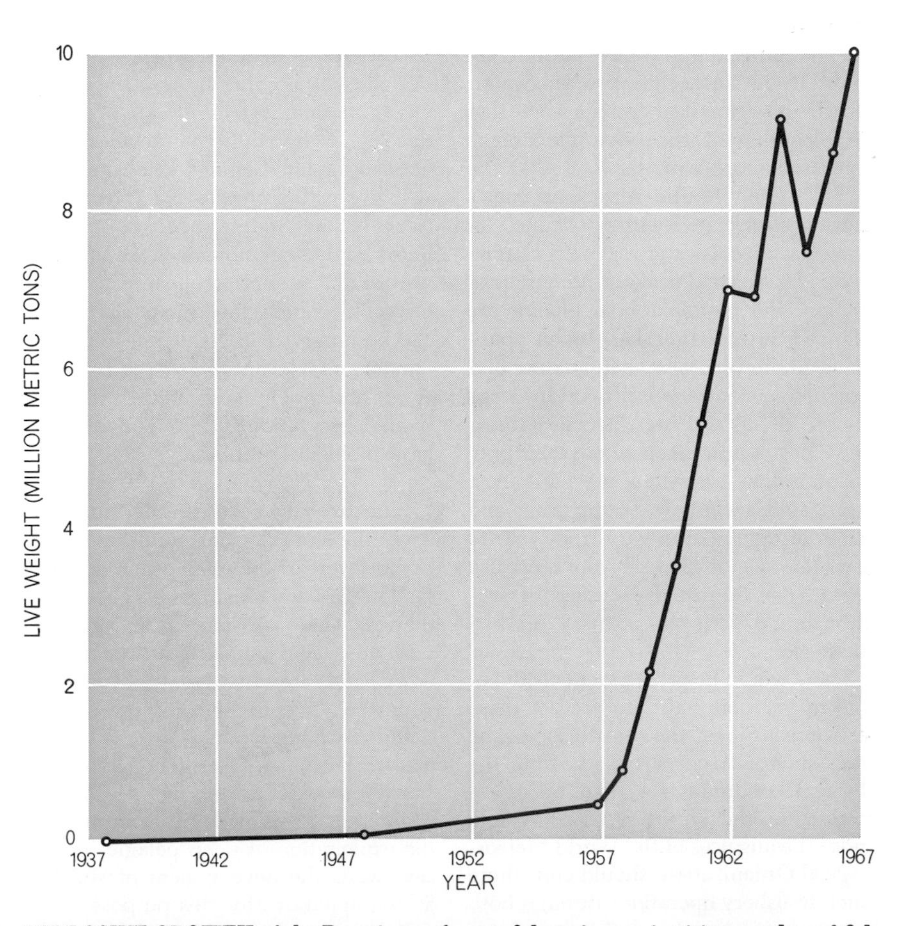

EXPLOSIVE GROWTH of the Peruvian anchoveta fishery is seen in rising number of fish taken between 1938 and 1967. Until 1958 the catch remained below half a million tons. By 1967, with more than 10.5 million tons taken, the fishery sorely needed management.

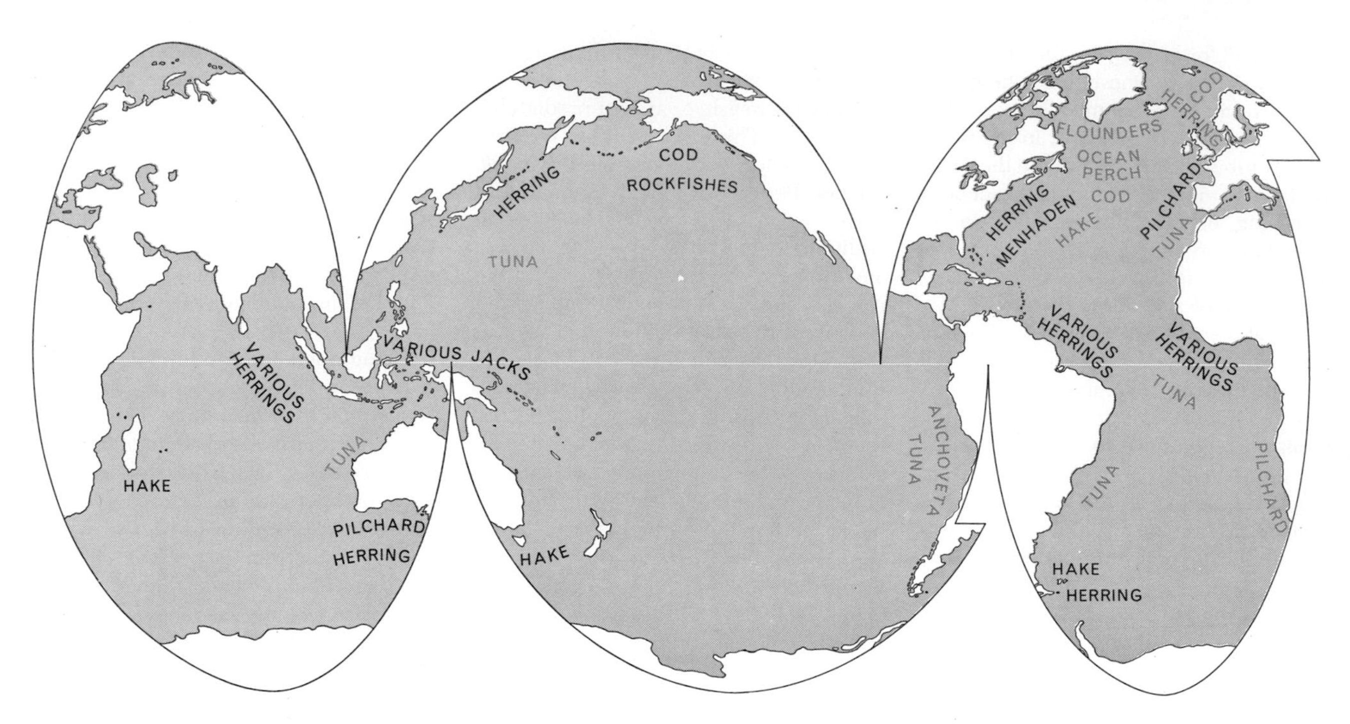

EXPLOITATION OF FISHERIES during the past 20 years is evident from this map, which locates 30 major fish stocks that were thought to be underfished in 1949. Today 14 of the stocks (*color*) are probably fully exploited or in danger of being overfished.

certain current systems in the open sea. The classical notion is that biological production is high in such areas because nutrient salts, needed for plant growth and in limited supply, are thereby renewed in the surface layers of the water. On the other hand, there is a view that the blooming of the phytoplankton is associated more with the fact that the water coming to the surface is cooler than it is associated with its richness in nutrients. A cool-water regime is characterized by seasonal peaks of primary production; the phytoplankton blooms are followed, after a time lag, by an abundance of herbivorous zooplankton that provides concentrations of food for large schools of fish. Fish, like fishermen, thrive best not so much where their prey are abundant as where they are most aggregated. In any event, the times and places of aggregation vary from year to year. The size of the herbivore crop also varies according to the success of synchronization with the primary production cycle.

There would be great practical advantage to our being able to predict these variations. Since the weather regime plays such a large part in creating the physical conditions for high biological production, the World Weather Watch, under the auspices of the World Meteorological Organization, should contribute much to fishery operations through both long-range forecasting and better short-term forecasting. Of course our interest is not merely in atmospheric forecasts, nor in the state of the sea surface, but in the deeper interaction of atmosphere and ocean. Thus, from the point of view of fisheries, an equal and complementary partner in the World Weather Watch will be the Integrated Global Ocean Station System (IGOSS) now being developed by the Intergovernmental Oceanographic Commission. The IGOSS will give us the physical data, from networks of satellite-interrogated automatic buoys and other advanced ocean data acquisition systems (collectively called ODAS), by which the ocean circulation can be observed in "real time" and the parameters relevant to fisheries forecast. A last and much more difficult link will be the observation and prediction of the basic biological processes.

So far we have been considering mainly the stocks of pelagic fishes in the upper layers of the open ocean and the shallower waters over the continental shelves. There are also large aggregations of pelagic animals that live farther down and are associated particularly with the "deep scattering layer," the sound-reflecting stratum observed in all oceans. The more widespread use of submersible research vessels will tell us more about the layer's biological nature, but the exploitation of deep pelagic resources awaits the development of suitable fishing apparatus for this purpose.

Important advances have been made in recent years in the design of pelagic trawls and in means of guiding them in three dimensions and "locking" them onto fish concentrations. We shall perhaps have such gear not only for fishing much more deeply than at present but also for automatically homing on deep-dwelling concentrations of fishes, squids and so on, using acoustic links for the purpose. The Indian Ocean might become the part of the world where such methods are first deployed on a large scale; certainly there is evidence of a great but scarcely utilized pelagic resource in that ocean, and around its edge are human populations sorely in need of protein. The Gulf of Guinea is another place where oceanographic knowledge and new fishing methods should make accessible more of the large sardine stock that is now effectively exploited only during the short season of upwelling off Ghana and nearby countries, when the schools come near the surface and can be taken by purse seines.

The bottom-living fishes and the shellfishes (both mollusks and crustaceans) are already more fully utilized than the smaller pelagic fishes. On the whole they are the species to which man attaches a particularly high value, but they cannot have as high a global abundance as the pelagic fishes. The reason is that they are living at the end of a longer food chain. All the rest of ocean life depends on an annual primary production of 150 billion tons of phytoplankton in the 2 to 3 percent of the water mass into which light penetrates and photosynthesis can occur. Below this "photic" zone dead and dying organisms sink as a continual rain of organic matter and are eaten or decompose. Out in the deep ocean little, if any, of this organic matter reaches the bottom, but nearer land a substantial quantity does; it nourishes an entire community of marine life, the benthos, which itself provides food for animals such as cod, ocean perch, flounder and shrimp that dwell or visit there.

Thus virtually everywhere on the bed of the continental shelf there is a thriving demersal resource, but it does not end there. Where the shelf is narrow but primary production above is high, as in the upwelling areas, or where the zone of high primary production stretches well away from the coast, we may find considerable demersal resources on the continental slopes beyond the shelf, far deeper than the 200 meters that is the average limiting depth of the shelf itself. Present bottom-trawling methods will work down to 1,000 meters or more, and it seems that, at least on some slopes, useful resources of shrimps and bottom-dwelling fishes will be found even down to 1,500 meters. We still know very little about the nature and abundance of these resources, and current techniques of acoustic surveying are not of much use in evaluating them. The total area of the continental slope from, say, 200 to 1,500 meters is roughly the same as that of the entire continental shelf, so that when we have extended our preliminary surveys there we might need to revise our IWP ceiling upward somewhat.

Another problem is posed for us by the way that, as fishing is intensified throughout the world, it becomes at the same time less selective. This may not apply to a particular type of fishing operation, which may be highly selective with regard to the species captured. Partly as a result of the developments in processing and trade, and partly because of the decline of some species, however, we are using more and more of the species that abound. This holds particularly for species in warmer waters, and also for some species previously neglected in cool waters, such as the sand eel in the North Sea. This means that it is no longer so reasonable to calculate the potential of each important species stock separately, as we used to do. Instead we need new theoretical models for that part of the marine ecosystem which consists of animals in the wide range of sizes we now utilize: from an inch or so up to several feet. As we move toward fuller utilization of all these animals we shall need to take proper account of the interactions among them. This will mean devising quantitative methods for evaluating the competition among them for a common food supply and also examining the dynamic relations between the predators and the prey among them.

These changes in the degree and quality of exploitation will add one more dimension to the problems we already face in creating an effective international system of management of fishing activities, particularly on the high seas. This system consists at present of a large number—more than 20—of regional or specialized intergovernmental organizations established under bilateral or multilateral treaties, or under the constitution of the FAO. The purpose of each is to conduct and coordinate research leading to resource assessments, or to promulgate regulations for the better conduct of the fisheries, or both. The organizations are supplemented by the 1958 Geneva Convention on Fishing and Conservation of the Living Resources of the High Seas. The oldest of them, the International Council for the Exploration of the Sea, based in Copenhagen and concerned particularly with fishery research in the northeastern Atlantic and the Arctic, has had more than half a century of activity. The youngest is the International Commission for the Conservation of Atlantic Tunas; the convention that establishes it comes into force this year.

For the past two decades many have hoped that such treaty bodies would ensure a smooth and reasonably rapid approach to an international regime for ocean fisheries. Indeed, a few of the organizations have fair successes to their credit. The fact is, however, that the fisheries have been changing faster than the international machinery to deal with them. National fishery research budgets and organizational arrangements for guiding research, collecting proper statistics and so on have been largely inadequate to the task of assessing resources. Nations have given, and continue to give, ludicrously low-level support to the bodies of which they are members, and the bodies themselves do not have the powers they need properly to manage the fisheries and conserve the resources. Add to this the trend to high mobility and range of today's fishing fleets, the problems of species interaction and the growing number of nations at various stages of economic development participating in international fisheries, and the regional bodies are indeed in trouble! There is some awareness of this, yet the FAO, having for years been unable to give adequate financial support to the fishery bodies it set up years ago in the Indo-Pacific area, the Mediterranean and the southwestern Atlantic, has been pushed, mainly through the enthusiasm of its new intergovernmental Committee on Fisheries, to establish still other bodies (in the Indian Ocean and in the east-central and southeastern Atlantic) that will be no better supported than the ex-

RUSSIAN FACTORY SHIP *Polar Star* lies hove to in the Barents Sea in June, 1968, as two vessels from its fleet of trawlers unload their catch for processing. The worldwide activities of the Russian fishing fleet have made the U.S.S.R. the third-largest fishing nation.

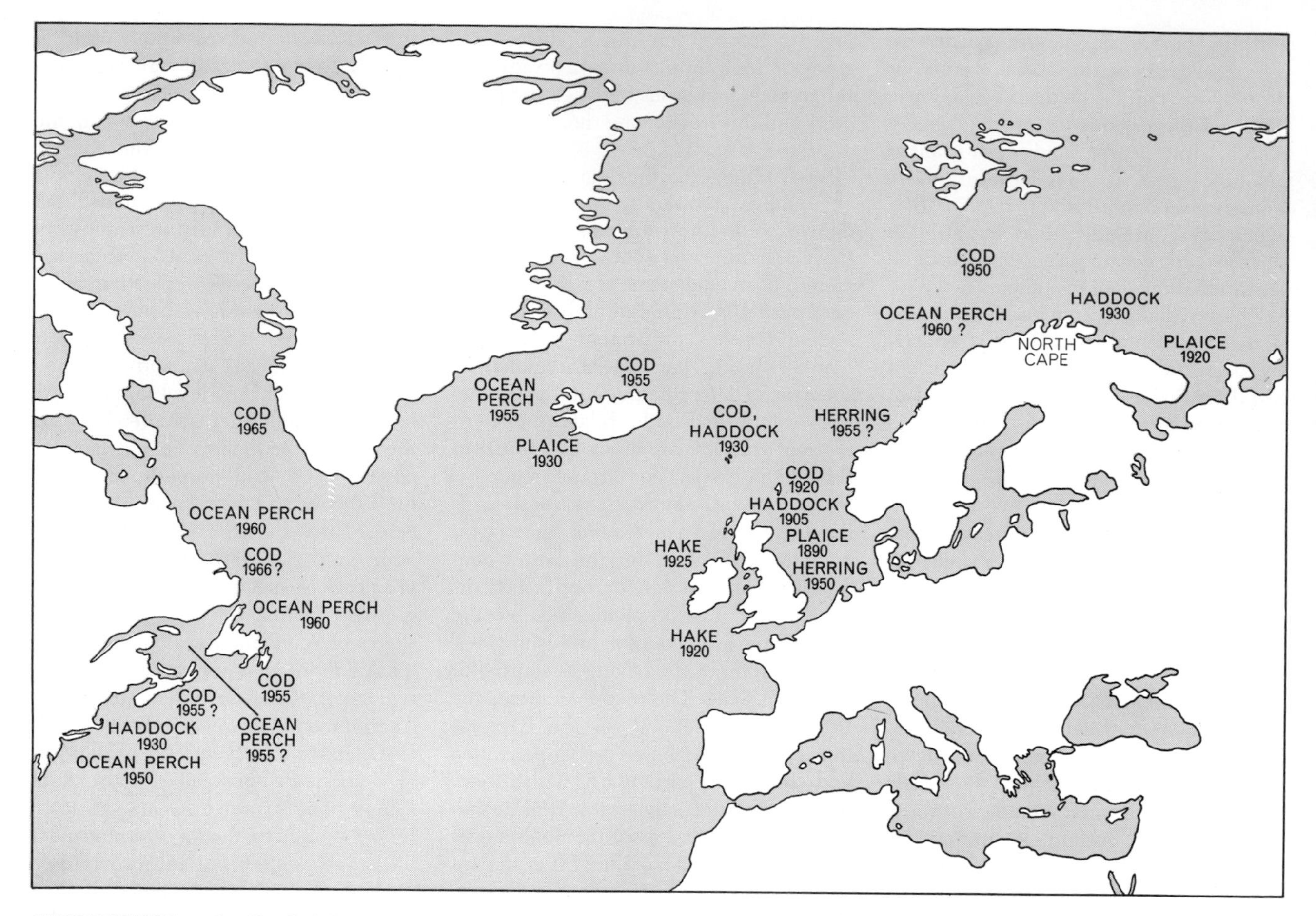

OVERFISHING in the North Atlantic and adjacent waters began some 80 years ago in the North Sea, when further increases in fishing the plaice stock no longer produced an increase in the catch of that fish. By 1950 the same was true of North Sea cod, haddock and herring, of cod, haddock and plaice off the North Cape and in the Barents Sea, of plaice, haddock and cod south and east of Iceland and of the ocean perch and haddock in the Gulf of Maine. In the period between 1956 and 1966 the same became true of ocean perch off Newfoundland and off Labrador and of cod west of Greenland. It may also be true of North Cape ocean perch and Labrador cod.

isting ones. A grand plan to double the finance and staff of the FAO's Department of Fisheries (including the secretariats and working budgets of the associated regional fishery bodies) over the six-year period 1966–1971, which member nations endorsed in principle in 1965, will be barely half-fulfilled in that time, and the various nations concerned are meanwhile being equally parsimonious in financing the other international fishery bodies.

Several of these bodies are now facing a crucial, and essentially political, problem: How are sustainable yields to be shared equitably among participating nations? It is now quite evident that there is really no escape from the paramount need, if high yields are to be sustained; this is to limit the fishing effort deployed in the intensive fisheries. This could be achieved by setting total quotas for each species in each type of fishery, but this only leads to an unseemly scramble by each nation for as large a share as possible of the quota. This can only be avoided by agreement on national allocations of the quotas. On what basis can such agreement be reached? On the historical trends of national participation? If so, over what period: the past two years, the past five, the past 20? On the need for protein, on the size or wealth of the population or on the proximity of coasts to fishing grounds? Might we try to devise a system for maximizing economic efficiency in securing an optimum net economic yield? How can this be measured in an international fishery? Would some form of license auction be equitable, or inevitably loaded in favor of wealthy nations? The total number or tonnage of fishing vessels might be fixed, as the United Kingdom suggested in 1946 should be done in the North Sea, but what flags should the ships fly and in what proportion? Might we even consider "internationalizing" the resources, granting fishing concessions and using at least a part of the economic yield from the concessions to finance marine research, develop fish-farming, police the seas and aid the participation of less developed nations?

Some of my scientific colleagues are optimistic about the outcome of current negotiations on these questions, and indeed when the countries participating are a handful of nations at a similar stage of economic and technical development, as was the case for Antarctic whaling, agreement can sometimes be reached by hard bargaining. What happens, however, when the participating countries are numerous, widely varying in their interests and ranging from the most powerful nations on earth to states most newly emerged to independence? I must confess that many of us were optimistic when 20 years ago we began proposing quite reasonable net-mesh regulations to conserve the young of certain fish stocks. Then we saw these simple—I suppose oversimple—ideas bog down in consideration of precisely how to measure a mesh of a particular kind of twine, and how to take account of the innumerable special situations that countries pleaded for, so that fishery research sometimes seemed to be becoming perverted from its earlier clarity and broad perspective.

Apprehension and doubt about the ultimate value of the concept of regulation through regional commissions of the present type have, I think, contributed to the interest in recent years in alternative regimes: either the "appropriation" of high-seas resources to some form of international "ownership" instead of today's condition of no ownership or, at the other extreme, the appropriation of increasingly wide ocean areas to national ownership by coastal states. As is well known, a similar dialectic is in progress in connection with the seabed and its mineral resources. Either solution would have both advantages and disadvantages, depending on one's viewpoint, on the time scale considered and on political philosophy. I do not propose to discuss these matters here, although personally I am increasingly firm in the conclusion that mankind has much more to gain in the long run from the "international" solution, with both seabed and fishery resources being considered as our common heritage. We now at least have a fair idea of what is economically at stake.

Here are some examples. The wasted effort in capture of cod alone in the northeastern Atlantic and salmon alone in the northern Pacific could, if rationally deployed elsewhere, increase the total world catch by 5 percent. The present catch of cod, valued at $350 million per year, could be taken with only half the effort currently expended, and the annual saving in fishing effort would amount to $150 million or more. The cost of harvesting salmon off the West Coast of North America could be reduced by three-quarters if management policy permitted use of the most efficient fishing gear; the introduction of such a policy would increase net economic returns by $750,000 annually.

The annual benefit that would accrue from the introduction and enforcement of mesh regulations in the demersal fishery—mainly the hake fishery—in the east-central Atlantic off West Africa is of the order of $1 million. Failure to regulate the Antarctic whaling industry effectively in earlier years, when stocks of blue whales and fin whales were near their optimum size, is now costing us tens of millions of dollars annually in loss of this valuable but only slowly renewable resource. Even under stringent regulation this loss will continue for the decades these stocks will need to recover. Yellowfin tuna in the eastern tropical Pacific are almost fully exploited. There is an annual catch quota, but it is not allocated to nations or ships, with the classic inevitable results: an increase in the catching capacity of fleets, their use in shorter and shorter "open" seasons and an annual waste of perhaps 30 percent of the net value of this important fishery.

Such regulations as exist are extremely difficult to enforce (or to be seen to be enforced, which is almost as important). The tighter the squeeze on the natural resources, the greater the suspicion of fishermen that "the others" are not abiding by the regulations, and the greater the incentive to flout the regulations oneself. There has been occasional provision in treaties, or in *ad hoc* arrangements, to place neutral inspectors or internationally accredited observers aboard fishing vessels and mother ships (as in Antarctic whaling, where arrangements were completed but never implemented!). Such arrangements are exceptional. In point of fact the effective supervision of a fishing fleet is an enormously difficult undertaking. Even to know where the vessels are going, let alone what they are catching, is quite a problem. Perhaps one day artificial satellites will monitor sealed transmitters compulsorily carried on each vessel. But how to ensure compliance with minimum landing-size regulations when increasing quantities of the catch are being processed at sea? With factory ships roaming the entire ocean, even the statistics reporting catches by species and area can become more rather than less difficult to obtain.

Some of these considerations and pessimism about their early solution have, I think, played their part in stimulating other approaches to harvesting the sea. One of these is the theory of "working back down the food chain." For every ton of fish we catch, the theory goes, we might instead catch say 10 tons of the organisms on which those fish feed. Thus by harvesting the smaller organisms we could move away from the fish ceiling of 100 million or 200 million tons and closer to the 150 billion tons of primary production. The snag is the question of concentration. The billion tons or so of "fish food" is neither in a form of direct interest to man nor is it so concentrated in space as the animals it nourishes. In fact, the 10-to-one ratio of fish food to fish represents a use of energy—perhaps a rather efficient use—by which biomass is concentrated; if the fish did not expend this energy in feeding, man might have to expend a similar amount of energy—in fuel, for example—in order to collect the dispersed fish food. I am sure the technological problems of our using fish food will be solved, but only careful analysis will reveal whether or not it is better to turn fish food, by way of fish meal, into chickens or rainbow trout than to harvest the marine fish instead.

There are a few situations, however, where the concentration, abundance and homogeneity of fish food are sufficient to be of interest in the near future. The best-known of these is the euphausiid "krill" in Antarctic waters: small shrimplike crustaceans that form the main food of the baleen whales. Russian investigators and some others are seriously charting krill distribution and production, relating them to the oceanographic features of the Southern Ocean, experiment-

JAPANESE MARICULTURE includes the raising of several kinds of marine algae. This array of posts and netting in the Inland Sea supports a crop of an edible seaweed, *Porphyra*.

AUSTRALIAN MARICULTURE includes the production of some 60 million oysters per year in the brackish estuaries of New South Wales. The long racks in the photograph have been exposed by low tide; they support thousands of sticks covered with maturing oysters.

ing with special gear for catching the krill (something between a mid-water trawl and a magnified plankton net) and developing methods for turning them into meal and acceptable pastes. The krill alone could produce a weight of yield, although surely not a value, at least as great as the present world fish catch, but we might have to forgo the whales. Similarly, the deep scattering layers in other oceans might provide very large quantities of smaller marine animals in harvestable concentration.

An approach opposite to working down the food chain is to look to the improvement of the natural fish resources, and particularly to the cultivation of highly valued species. Schemes for transplanting young fish to good high-seas feeding areas, or for increasing recruitment by rearing young fish to viable size, are hampered by the problem of protecting what would need to be quite large investments. What farmer would bother to breed domestic animals if he were not assured by the law of the land that others would not come and take them as soon as they were nicely fattened? Thus mariculture in the open sea awaits a regime of law there, and effective management as well as more research.

Meanwhile attention is increasingly given to the possibilities of raising more fish and shellfish in coastal waters, where the effort would at least have the protection of national law. Old traditions of shellfish culture are being reexamined, and one can be confident that scientific bases for further growth will be found. All such activities depend ultimately on what I call "productivity traps": the utilization of natural or artificially modified features of the marine environment to trap biological production originating in a wider area, and by such a biological route that more of the production is embodied in organisms of direct interest to man. In this way we open the immense possibilities of using mangrove swamps and productive estuarine areas, building artificial reefs, breeding even more efficient homing species such as the salmon, enhancing natural production with nutrients or warm water from coastal power stations, controlling predators and competitors, shortening food chains and so on. Progress in such endeavors will require a better predictive ecology than we now have, and also many pilot experiments with corresponding risks of failure as well as chances of success.

The greatest threat to mariculture is perhaps the growing pollution of the sea. This is becoming a real problem for fisheries generally, particularly coastal ones, and mariculture would thrive best in just those regions that are most threatened by pollution, namely the ones near large coastal populations and technological centers. We should not expect, I think, that the ocean would not be used at all as a receptacle for waste—it is in some ways so good for such a purpose: its large volume, its deep holes, the hydrolyzing, corrosive and biologically degrading properties of seawater and the microbes in it. We should expect, however, that this use will not be an indiscriminate one, that this use of the ocean would be internationally registered, controlled and monitored, and that there would be strict regulation of any dumping of noxious substances (obsolete weapons of chemical and biological warfare, for example), including the injection of such substances by pipelines extending from the coast. There are signs that nations are becoming ready to accept such responsibilities, and to act in concert to overcome the problems. Let us hope that progress in this respect will be faster than it has been in arranging for the management of some fisheries, or in a few decades there may be few coastal fisheries left worth managing.

I have stressed the need for scientific research to ensure the future use of the sea as a source of food. This need seems to me self-evident, but it is undervalued by many persons and organizations concerned with economic development. It is relatively easy to secure a million dollars of international development funds for the worthy purpose of assisting a country to participate in an international fishery or to set up a training school for its fishermen and explore the country's continental shelf for fish or shrimps. It is more difficult to justify a similar or lesser expenditure on the scientific assessment of the new fishery's resources and the investigation of its ocean environment. It is much more difficult to secure even quite limited support for international measures that might ensure the continued profitability of the new fishery for all participants.

Looking back a decade instead of forward, we recall that Lionel A. Walford of the U.S. Fish and Wildlife Service wrote, in a study he made for the Conservation Foundation: "The sea is a mysterious wilderness, full of secrets. It is inhabited only by wild animals and, with the exception of a few special situations, is uncultivated. Most of what we know about it we have had to learn indirectly with mechanical contrivances to probe, feel, sample, fish." There are presumably fewer wild animals now than there were then—at least fewer useful ones—but there seems to be a good chance that by the turn of the century the sea will be less a wilderness and more cultivated. Much remains for us and our children to do to make sure that by then it is not a contaminated wilderness or a battlefield for ever sharper clashes between nations and between the different users of its resources.

38

TECHNOLOGY AND THE OCEAN

WILLARD BASCOM
September 1969

Without technology, meaning knowledge fortified by machinery and tools, men would be ineffective against the sea. During the past decade the technology that can be brought to bear in the oceans has improved enormously and in many ways. The improvements have not only increased knowledge of the oceans but also speeded the flow of commerce while decreasing its cost, brought new mineral provinces within reach and made food from the sea more readily available.

With today's technology it is possible, given a sufficient investment of time and money, to design and build marine hardware that can do almost anything. The problem is to decide whether it is sensible to make a given investment. Industry decides on the basis of whether a proposed technological step will solve a specific problem and improve the firm's competitive position. Government has more latitude: it does not need to show a prompt return on investment, and it can better afford the high risk of developing expensive and exotic devices for which there may be no immediate or clearly defined need. The gains in ocean technology have resulted from the largely independent efforts of both industry and government.

This article will deal broadly with the progress in ocean technology over the past decade, concentrating on developments that seem to be the most important at present. I shall begin by making my own selection of the five most important advances. The main criterion in this selection is that each advance represents an order-of-magnitude improvement: in one way or another it is a tenfold step forward since the beginning of the decade. I have also given weight to the social and economic significance of these developments and to the degree of engineering imagination and perseverance that each one required.

The first development is the supership. Not long ago a "supertanker" carried 35,000 deadweight tons. Now a fleet of ships with nearly 10 times that capacity is coming into being. For these vessels the Panama Canal and the Suez Canal are obsolete. By the same token the ships are making large new demands on the technology that provides the terminal facilities.

Second is the deep-diving submarine. Man can now go to the deep-ocean bottom in an "underwater balloon" submersible such as the *Trieste*, which reached a depth of 36,000 feet in the Mariana Trench 200 miles southwest of Guam. Somewhat more conveniently he can go to a depth of about 6,000 feet in any of several small submarines. This rapidly developing technology still has a long way to go, but it has certainly improved by an order of magnitude in the past decade. Several techniques have been employed to solve the problem of how to make a submarine hull that is strong enough to resist great pressure and still light enough to return to the surface.

The third development is the ability to drill in deep water. This category includes both the drilling that is done in very deep water for scientific purposes and the use of full-scale drilling equipment on a floating platform to obtain oil from the continental shelf. The first deep-ocean drilling, which was carried out eight years ago by the National Academy of Sciences in water 12,000 feet deep near Guadalupe Island off the west coast of Mexico, improved on four previous records by an order of magnitude: the ship held its position at sea for a month without anchors, drilled in water 20 times deeper than that at earlier marine drilling sites, penetrated 600 feet of the deep-sea floor and lifted weights of 40 tons from the bottom. These records have since been improved on even more. In fact, virtually all floating drilling equipment, including semisubmersible platforms and self-propelled vessels, has been designed and built in the past decade.

Fourth is the ability to navigate precisely. A ship in mid-ocean has rarely known its position within a mile; indeed, five miles is probably closer to the truth, notwithstanding assertions to the contrary. Now a ship 1,000 miles from land can fix its position within .1 mile. If the vessel is within 500 miles of land, the position can be ascertained within .01 mile. The position of a ship within 10 miles of land can be fixed to an accuracy of 10 feet. The techniques for these determinations include orbiting satellites, inertial guidance systems and a number of electronic devices that compare phases of radio waves.

Finally I would cite the ability to examine the ocean bottom in detail from the surface by means of television and side-looking sonar. These techniques, together with their recording devices and the capacity for precise navigation, have made it possible to inspect the sea floor much as land areas have been examined by aerial photography. New television tubes that amplify light by a factor of 30,000 make it possible to eliminate artificial lighting, thereby eliminating also the backscatter of light by small particles in the water.

The supership and the improvement in drilling are mainly industrial developments. The evolution of navigation technology has resulted largely from government efforts. Both industry and government have figured prominently

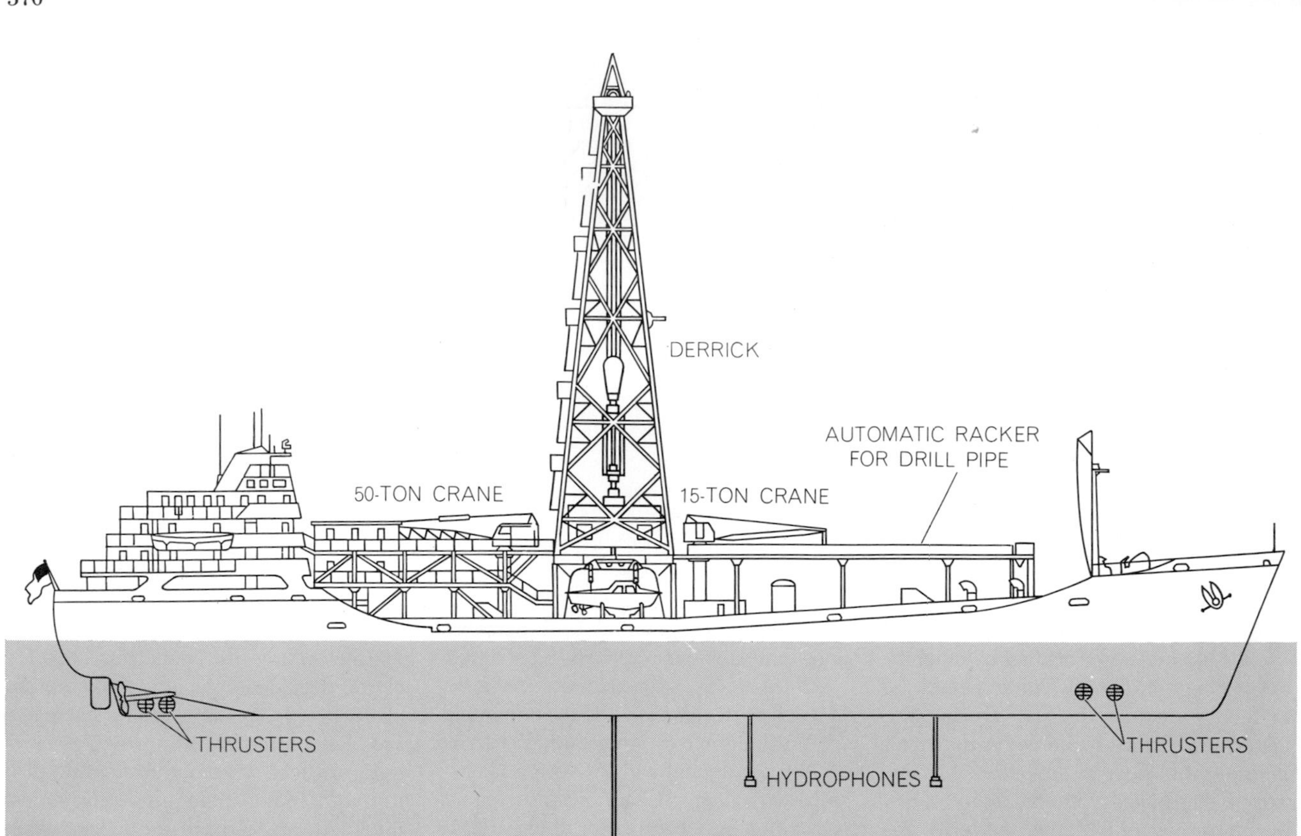

GLOMAR CHALLENGER has a 142-foot derrick as her most conspicuous feature. Her automatic pipe racker can hold 23,000 feet of drill pipe. Positioning equipment includes two tunnel thrusters at the bow and two near the stern to provide for sidewise maneuvers. When the ship is on station (*above*), four hydrophones are extended under the hull to receive signals from a sonar beacon on the ocean floor. The signals are fed into a computer that controls the thrusters to maintain the ship's position over the drill hole. At the sea bottom (*below*), as much as four miles under the ship, the drill penetrates as much as 2,500 feet of sediment and basement rock.

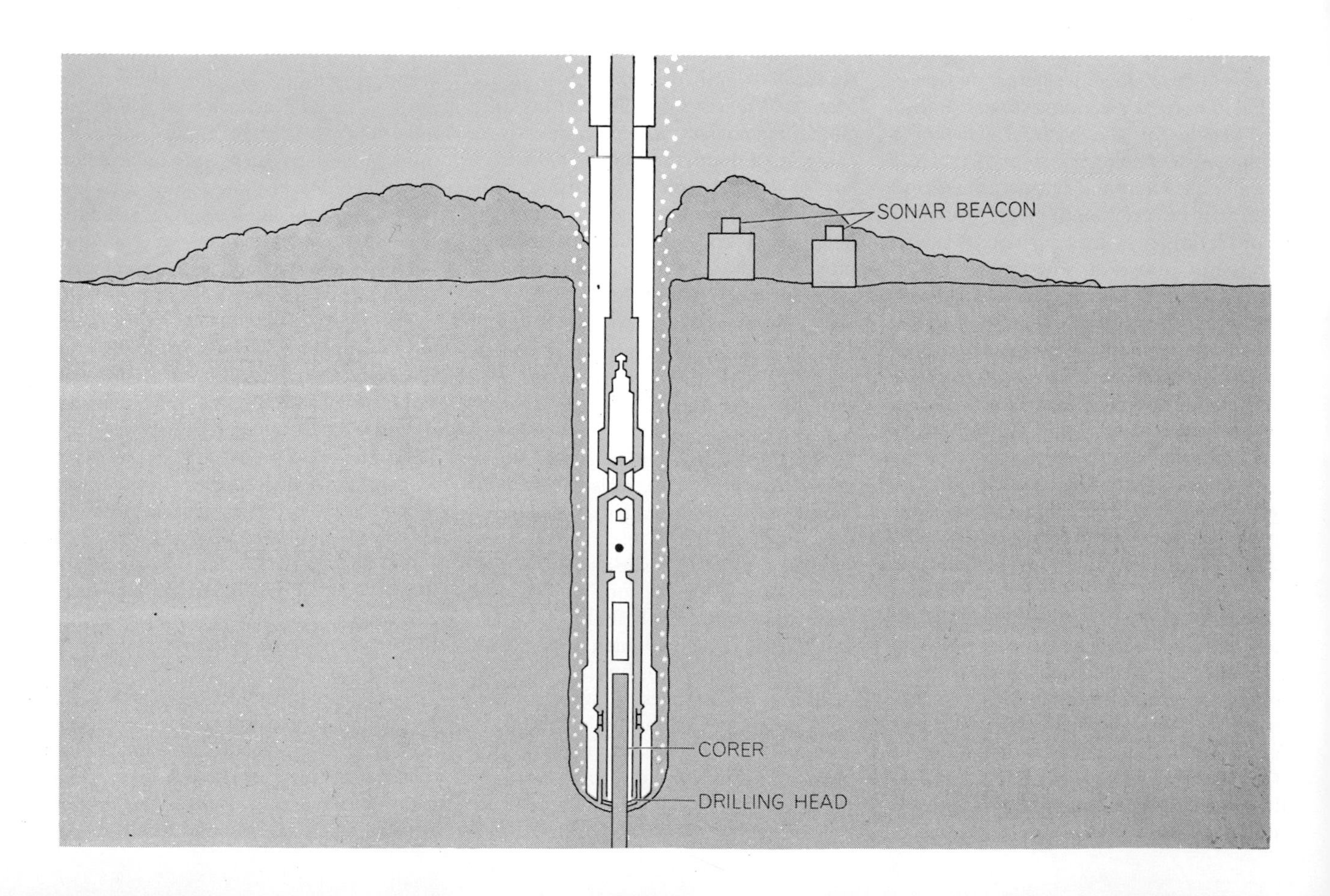

in the development of deep-diving submarines and techniques for examining the bottom with television and sonar.

In considering the application of these and other techniques one might classify them according to who uses them. For example, scientific investigators use research ships and submarines, instruments, buoys, samplers and computers. Industry constantly seeks better methods for mining, fishing, salvage and the production of oil. Waterborne commerce needs better ships, better cargo-handling methods and better port facilities. Exploration becomes more efficient as improved navigational systems, vehicles, geophysical tools and communications equipment become available. Adventure and recreation offer new toys such as air-cushion vehicles and scuba equipment.

The entire area of military technology, which is the most sophisticated of all, must be outside the scope of this article. The best of modern seaborne military technology is done in secrecy, with budgets far in excess of those spent for any of the other areas. Thus we shall not go into such matters as the duel between the submarine builder, who endlessly tries to make submarines go deeper, faster and quieter, and the antisubmarine expert, who tries to detect, identify and destroy the steadily improving submarines.

In any case, the classification of marine technology according to users is somewhat impractical because there is so much overlap. For example, certain kinds of diving and television equipment might be used by all the groups. Therefore I shall discuss marine technology in terms of materials, vehicles, instruments and systems.

What characteristics should a marine material have? It should be light, strong, easy to form and connect, rigid or flexible as desired and inexpensive. The difficulty ocean design engineers have in finding a material that meets most of the requirements for a given task has led them to speak whimsically of an ideal material called "nonobtainium." The problem is that characteristics such as lightness and strength are relative. Nonetheless, engineers and manufacturers recall that not many years ago fiber glass, Dacron and titanium were not obtainable, and so they are optimistic about the development of materials that come ever closer to the qualities of nonobtainium.

In the past decade the steel available for marine purposes has improved substantially under the spur of demands for submarines that can withstand the pressure of great depth, drill pipe (unsupported by the hole wall that pipe in a land well has) that must survive high bending stresses, and great lengths of oceanographic cable that must not twist. For example, a new kind of maraging steel with a high nickel content is tougher, more resistant to notching and less subject to corrosion fatigue than the steel formerly available. The minimum yield strength of conveniently available steel shell plate has risen from 80,000 pounds per square inch to 130,000 pounds per square inch and more for the shells of deep-diving submarines. Steel in wire form now attains a strength of 350,000 pounds per square inch. Steel is becoming more uniform and reliable as the processes of mixing and rolling are subjected to better quality control. Indeed, some metallurgists believe nearly any metal requirement can be met by properly alloyed steel.

Also available for marine purposes are new, high-strength aluminum alloys, such as 5456 (a designation indicating the mix of metals in the alloy), that have a strength of more than 30,000 pounds per square inch after welding and are resistant to corrosion. They can also be cut with ordinary power saws instead of torches and welded by a technique that is easily taught. With this material small boats, ships up to 2,000 tons and superstructures for much larger ships can be built, as can a number of other structures where lightness and flexibility are important.

Titanium is becoming more readily available. When special properties of lightness, strength (as high as 120,000 pounds per square inch) and good resistance to corrosion are required, its relatively high cost becomes acceptable.

Glass, fiber glass and plastics are the glamorous materials of oceanography. They are virtually free of the problems of corrosion and electrolysis that have afflicted most materials in a marine environment, and they are easily formed

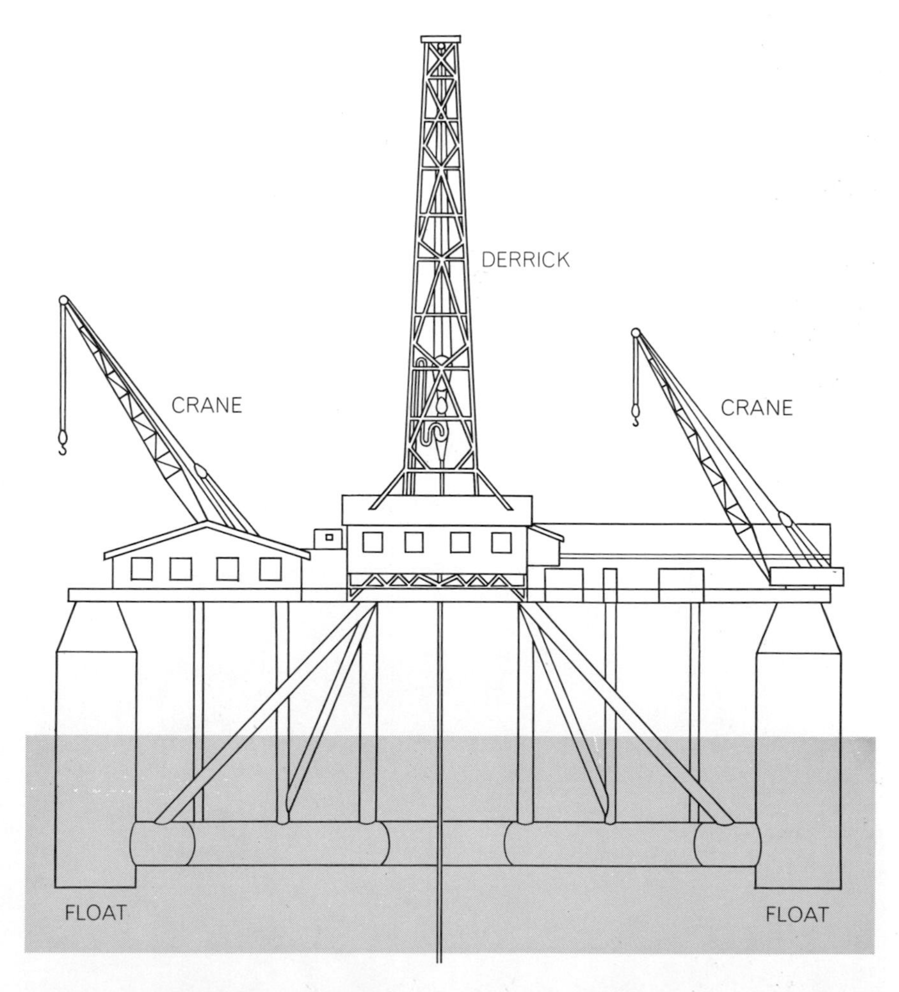

SEMISUBMERSIBLE PLATFORM, ***Blue Water 3,*** **is now drilling for oil off Trinidad. When the platform has been towed to its position, water is drawn into the four corner cylinders to make the structure submerge enough so that wave motions have little effect on it. The platform, which is 220 feet by 198 feet, was designed for work in the open ocean.**

into complex shapes. Constant research is improving the strength and versatility of these materials.

Glass is less fragile than most people think and has excellent properties in compression. It is finding increasing favor among the designers of small submarines, who want a glass-bubble pressure hull that is also a superwindow. Glass microspheres, which do the same thing as a submarine hull but on a microscopic scale, are packaged in blocks of epoxy and used to furnish incompressible flotation at depths of as much as 20,000 feet. The best such material to withstand the pressure at that depth so far weighs about 40 pounds per cubic foot; in seawater at 64 pounds per cubic foot the material therefore has a net buoyancy of 24 pounds per cubic foot.

The many remarkable characteristics

SUPERTANKER *UNIVERSE IRELAND* is seen at her loading berth in the Persian Gulf. The vessel, which is 1,135 feet in length, carries 312,000 deadweight tons of oil from the Persian Gulf to Ireland, going around Africa at an average speed of 15 knots.

of fiber glass are widely known. Its outstanding virtue as a marine material is that it enables precisely shaped hulls with complex lines to be reproduced easily. The result has been a revolution in the construction of small craft over the past decade. Fiber-glass hulls, which are light and strong without a rib structure, are a major contribution to marine technology and a boon to the small-boat owners, whose maintenance problems are reduced accordingly.

Among the plastics polyvinyl chloride has found use in marine pipelines subject to severe internal corrosion. It is light, inexpensive and easily joined. Nylon and polypropylene for cordage and fishing nets and Dacron for sails are appreciated by all sailors because the materials are light and elastic and do not rot. Ship-bottom paints, designed to reduce fouling by marine organisms, have been greatly improved. Inorganic zinc underpaints promise to decrease substantially the pitting of hulls and decks, which should increase the life of ships and the time between dry-dockings.

A remarkable collection of marine vehicles and equipment has made its appearance in the past 10 years. Ships now exist that can go up, down and sideways and can flip. They skim, fly and dive. Some of them are amphibious and some go through ice, over ice or under it.

This versatility is important; a ship cannot be efficient unless it has been designed to do exactly what the user wants. Widely varying requirements mean very different sea vehicles. A distinct place on the spectrum is occupied by the superships I mentioned earlier. They are bigger than anyone dreamed of only a few years ago; in fact, they are almost the largest man-made structures. The largest vessel now in service is the *Universe Ireland*, a tanker with a capacity of 312,000 deadweight tons. The ship is 1,135 feet long and delivers 37,400 shaft horsepower. It plies between the Persian Gulf and Ireland, going around Africa at 15 knots and pushing a 12-foot breaking wave ahead of it. Even the enormous vessels of the *Universe Ireland* class will soon be surpassed in size by ships being built in Japan and West Germany. They will be so large they will not be able to dry-dock in any yard except where they were built, and they will not be able to enter any ordinary harbor because they will draw up to 80 feet of water.

A variation in the tanker field is the conversion of the comparatively small (114,000 deadweight-ton capacity) *Man-*

ALUMINAUT, a mobile submersible capable of carrying two crew and four passengers to depths of 15,000 feet and of probing the bottom or moving heavy objects with manipulator arms attached to the hull, is photographed during a dive. The craft is 51 feet in length.

MANIPULATOR ARM of *Aluminaut* explores bottom off Bimini at a depth of nearly 1,800 feet. Numbered sample boxes are nearby. Thin layer of sand is rippled by a current moving it over a rock base; the dark areas are debris that are caught in filamentous organisms attached to rock. Photograph was made by A. Conrad Neumann of the University of Miami.

hattan to a supericebreaker. The purpose is to move the petroleum from the large new oil fields on the northern slope of Alaska to more moderate climates. A fleet of such ships may be able to keep open a northwest passage from the U.S. East Coast to Alaska. From ships of the *Manhattan* class it is only a small step conceptually to a ship five times larger that could cross the Arctic Ocean at will, treating the ice, which averages eight feet in thickness, as an annoying scum.

The ships that go up include both the ground-effect machine, which can rise a few feet above the surface of the sea on a cushion of air, and the hydrofoil, which has a hull that flies above the surface at high speed with the support of small, precisely shaped underwater foils. The newest versions of these "flying boats" represent substantial technical achievements, and yet neither vehicle seems likely to become a very important factor in marine affairs because each has basic problems, such as the danger to the hydrofoil of hitting heavy flotsam and the inability of the ground-effect machine to carry large loads or to operate in high waves. The ground-effect machine does have a potential, not much exploit-

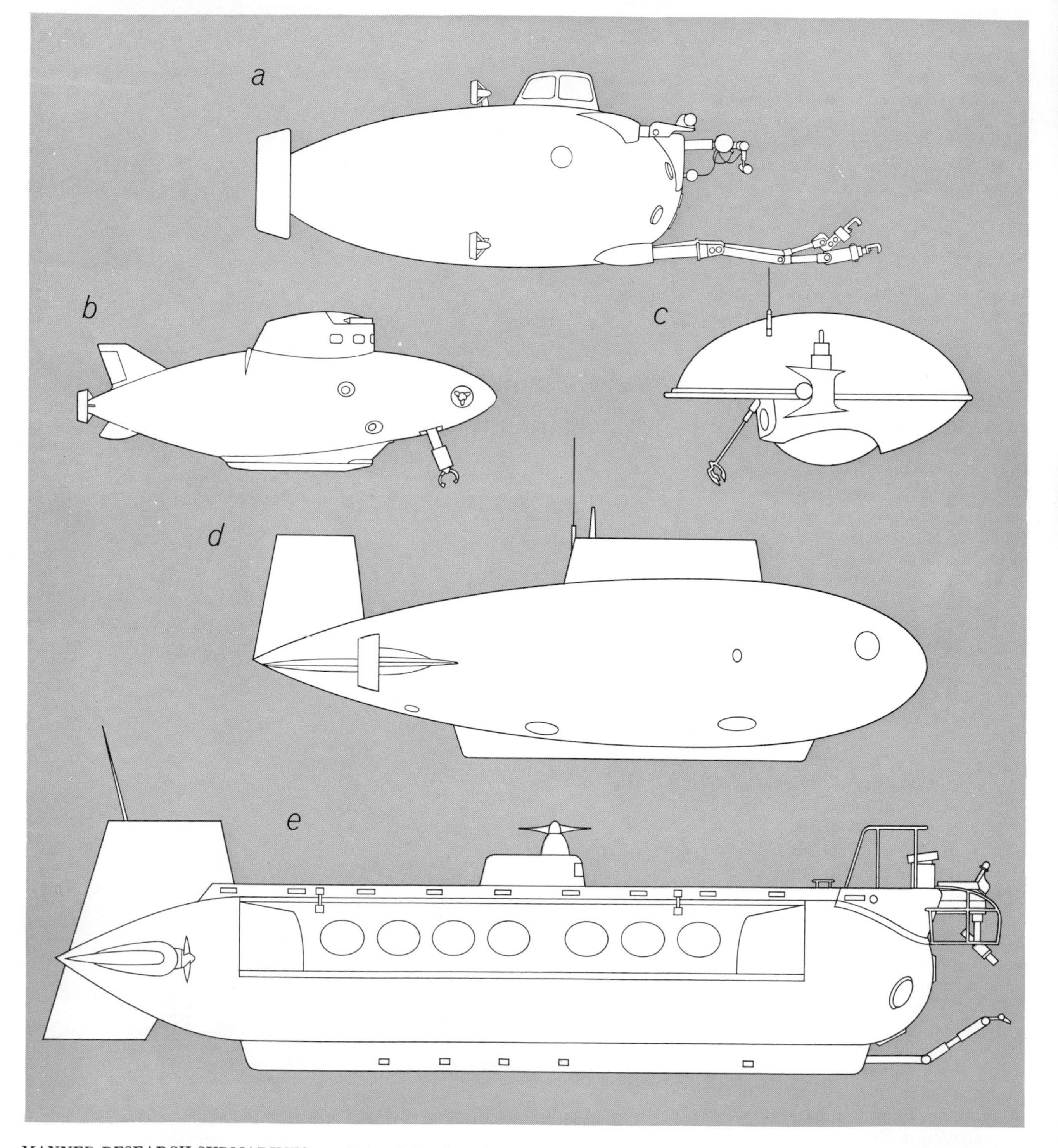

MANNED RESEARCH SUBMARINES are designed for deep diving. They include (*a*) *Beaver IV*, which can dive to 2,000 feet; (*b*) *Star III*, to 2,000 feet; (*c*) *Deepstar IV*, to 4,000 feet; (*d*) *Deep Quest*, to 8,000 feet, and (*e*) *Aluminaut*, which is designed to go to 15,000 feet with a staff of six. Vessels are drawn to scale. *Aluminaut* is made primarily of aluminum; the others are steel craft.

ed as yet, stemming from its ability to run up a beach and cross mud flats, ice and smooth land surfaces.

The ships that go down are of course submarines. Nuclear power in military submarines dates back further than the decade under discussion, but large advances in nuclear propulsion have been made during the decade. The circumnavigation of the earth without surfacing and the trip under ice to the North Pole were both made possible by nuclear power and highly developed life-support systems for keeping the crews alive and well on the long missions.

Quite a number of small, deep-diving submarines are in existence. Of them the *Aluminaut,* designed to go to 15,000 feet while carrying six people, has accomplished the most. (Because of the problem of obtaining life insurance for its crew its deepest dive has been about 6,000 feet.) Among the many other small submarines are the ones of the *Alvin* class, which can dive to 6,000 feet; *Deep Quest,* to 8,000 feet; *Deepstar IV,* to 4,000 feet; *Beaver IV* and *Star III,* to 2,000 feet, and *Deep Diver,* to 1,000 feet. There is therefore a considerable choice of vehicles, instruments and sup-

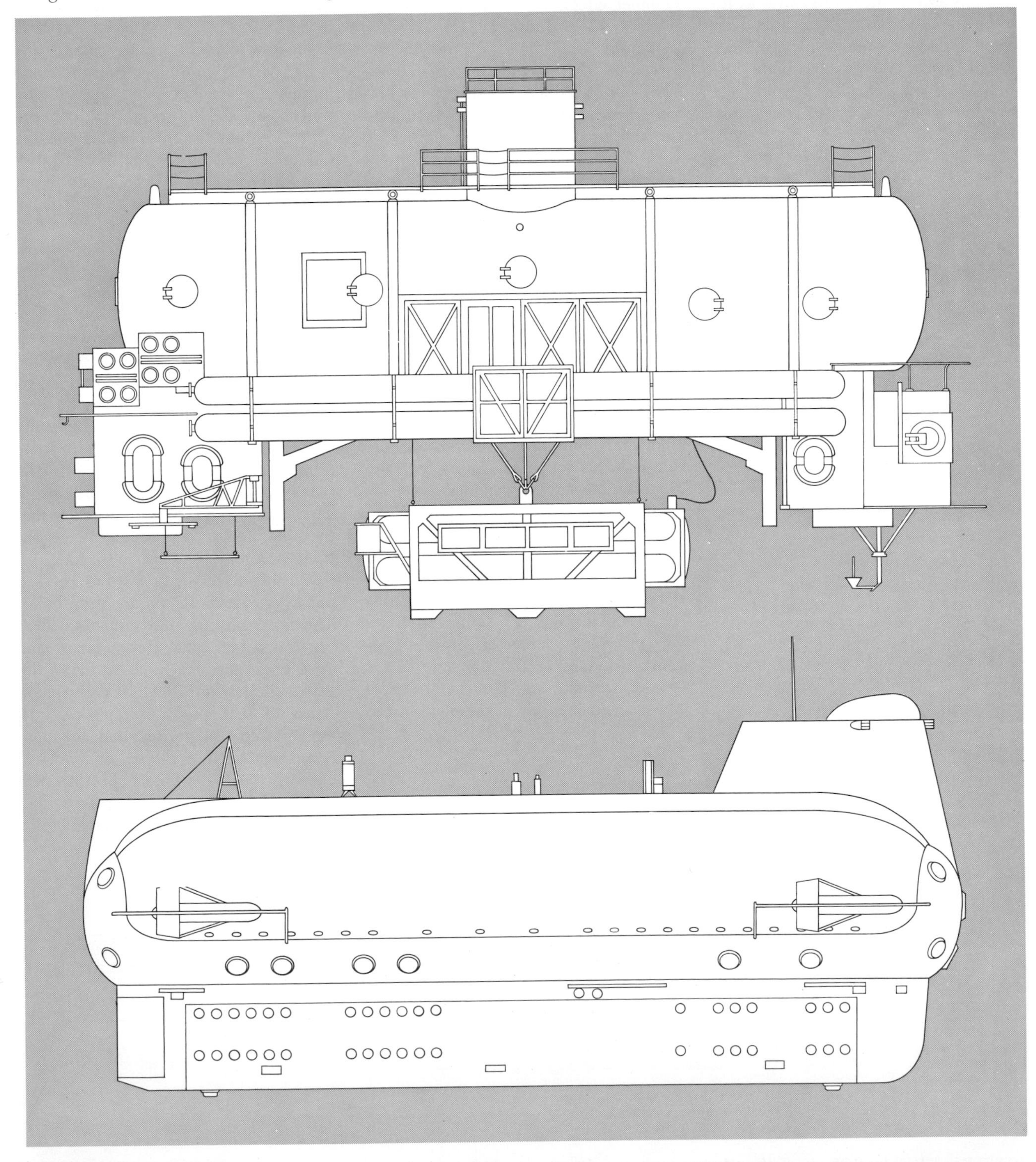

UNDERWATER LABORATORIES include the Navy's *Sealab III* (*top*) and the Grumman-Piccard submersible *Ben Franklin* (*bottom*). *Sealab* is designed to operate on the sea bottom, where it will provide living quarters for divers who will venture forth periodically in heated diving suits to explore the bottom. The first mission of *Ben Franklin* was a submerged drift up the Gulf Stream.

porting facilities. The problem is that there are few customers with the inclination to employ these vehicles at $1,000 or more per hour of diving.

The ships that move sideways are those with trainable propellers or vertical-axis propellers or tunnel thrusters. (A tunnel thruster enables a pilot to move the ship's bow sideways.) Vessels so equipped have found use in self-docking situations and in such waterways as the St. Lawrence Seaway. Dynamic positioning, which means holding position without anchors, is possible with ships that have precise local-navigation systems and central control of several maneuvering propellers.

The ship that flips is *FLIP*, operated by the Scripps Institution of Oceanography. It has two positions of stability. While it is under tow it lies on the surface and looks like a barge made from a big piece of pipe. On station it ballasts itself so as to float on end, much like a big, habitable buoy. In this position *FLIP*, because of its size, is detuned from the motion of the sea surface: it does not move vertically under the influence of ordinary waves and swell. As a result it is an excellent platform for making underwater sound measurements.

Among instruments and tools the now venerable sonar, the sound-ranging device, still figures prominently. It has been improved substantially. Frequencies have risen steadily, making it possible to narrow the beam width to searchlight dimensions, with the result that the distance to (or depth of) discrete areas of the ocean bottom can be measured more accurately. Sonars employing the Doppler effect, which is the change of pitch of a sound resulting from relative motion between the source and the observer, make possible the direct measurement of a ship's speed over the bottom—a measurement that is essential to the high-quality navigation required for such purposes as determining gravity at sea by means of a shipboard gravity meter. Frequency-scanning sonars are now available that better match the signal with the reflector.

Hydrophone arrays, sometimes a mile long, make it possible to use the low-frequency sound created by a series of gas explosions to examine the rocks under the sea bottom in great detail. The result is a continuous picture of a vertical geologic section. Such continuous-reflection seismic profiles have revealed folds and faults in sub-bottom rocks to depths of as much as 15,000 feet and have found many new undersea oil deposits.

Satellites are valuable ocean instruments. They are the essential elements in the system that makes it possible to determine a ship's position accurately wherever it may be. Other satellites transmit photographs of weather patterns, cloud cover and the state of the sea. By combining the picture with other weather data, meteorologists can produce accurate charts with reliable and up-to-date information. The information is useful in routing ships and forecasting waves.

Buoys moored in the deep ocean hold instruments for measuring, recording and transmitting sea and weather conditions. A number of buoys are already in use, producing an abundance of hitherto unavailable information at minimal expense. It seems likely that hundreds of additional buoys will be put to work in the next few years.

Shipboard computers are becoming an accepted convenience. Such a computer plots the ship's position continuously and matches it with accumulating data of other kinds so that investigators aboard the ship have an information system describing the pulse of the sea below them.

Occasionally a single device can revolutionize an industry. Such a device is the Puretic power block, which handles fishing nets. It is, like many good inventions, basically simple: it is a wide-mouthed, rubber-lined pulley driven by a small hydraulic motor. During the past decade it has been adopted by many fishing fleets and now accounts for some 40 percent of the world's catch. With the block it is possible to handle much larger nets with fewer men. One result is that the tuna industry has shifted almost entirely from line fishing to net fishing.

Another trend in fishing has been to put fish-processing equipment on boats. The equipment includes automatic filleting machines, quick-freeze boxes and even packaging machines so that a finished frozen product can be delivered at dockside. Scallops, for example, can now be shucked and eviscerated on shipboard, so that the scalloper can remain at sea for a week at a time and return with a cargo of ready-to-eat scallops.

Barge-mounted cranes capable of lift-

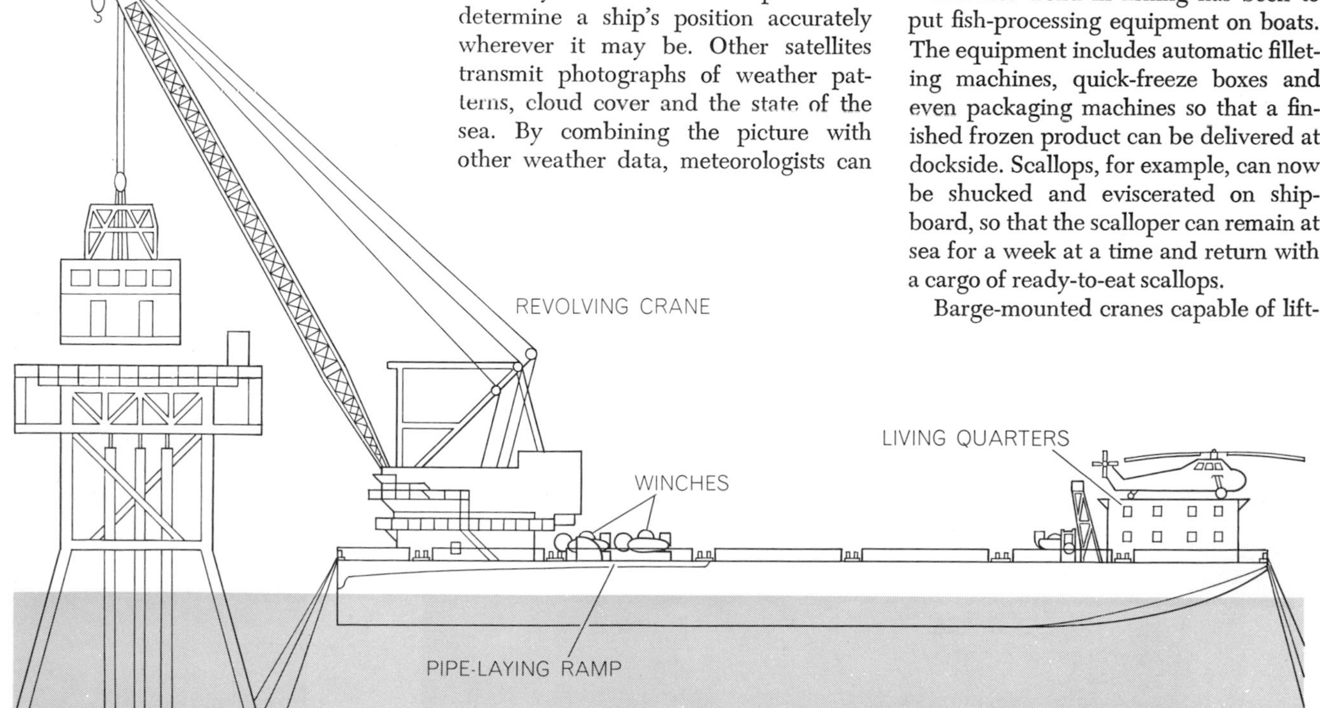

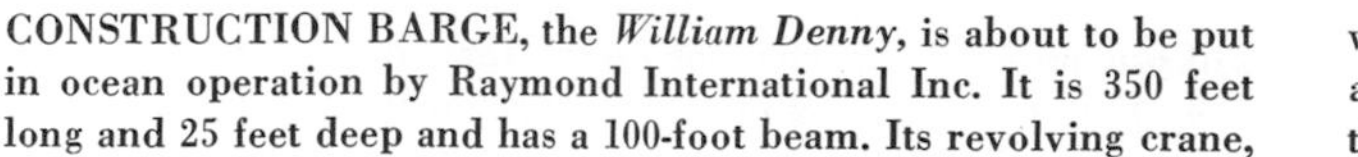
CONSTRUCTION BARGE, the *William Denny*, is about to be put in ocean operation by Raymond International Inc. It is 350 feet long and 25 feet deep and has a 100-foot beam. Its revolving crane, which has a 250-foot boom, can lift 500 tons at a 70-foot radius and 100 tons at a 215-foot radius. The craft can lift 750 tons over the stern. It can build structures, drive piles and lay pipelines.

ing 600-ton loads as much as 200 feet above the water are now available along many coasts. The result is a change in construction techniques. For example, a bridge can be built in large sections, which are then hoisted into place by the crane.

Shipyards are using elevators called Syncrolifts to lift and launch ships weighing as much as 6,000 tons. The machines are replacing dry docks and marine railways. The Syncrolift is simply a big platform that can be lowered below keel depth. A ship is then floated in, and a dozen or more synchronized winches hoist it up to the level of a transfer railway, which moves it to a position in the yard. In this way the yard can work on several ships simultaneously.

The first undersea dredge has just made its appearance. The machine moves along the bottom on crawler tracks in depths of as much as 200 feet. Hence it is not affected by the wave action that makes life on floating dredges hard. The machine was designed to replace eroding beaches with sand from offshore: the dredge has a 700-horsepower pump that moves the sand slurry a mile to shore.

It is fashionable now to speak of the "systems approach," which is a way of expressing the obvious idea that all the elements in the solution of a problem should fit together and be headed toward the same goal. All the ships and instru-

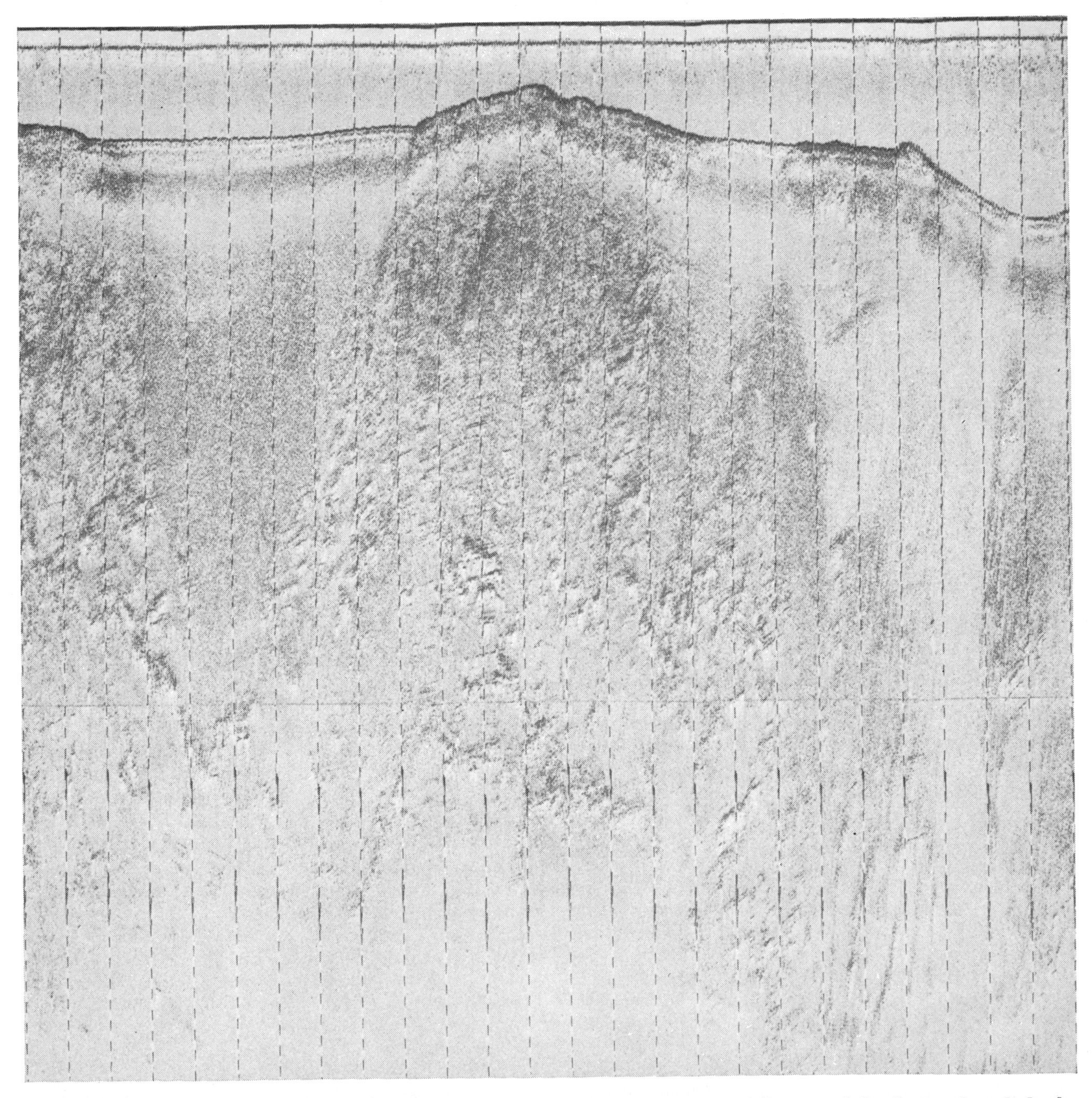

SIDE-LOOKING SONAR produced this view of the ocean bottom on the continental shelf northeast of Boston. From right to left the record covers a distance of about two kilometers along the ship's track. Broken lines show one-minute intervals. Irregular line near top is a profile view of the sea bottom as it appeared at the instrument's horizon. Irregular portions of the photograph are bedrock; darker flat areas are sand waves and gravel; lighter flat areas are smooth sand. Bottom was 60 to 140 meters below the ship. Record was made by John E. Sanders of Barnard College and K. O. Emery and Elazar Uchupi of Woods Hole Oceanographic Institution.

ments I have described are employed as parts of systems. There are, however, several integrated combinations of technology that are best described under the heading of systems. They include containerized cargo-handling, desalination of seawater, deep-ocean drilling and deep diving.

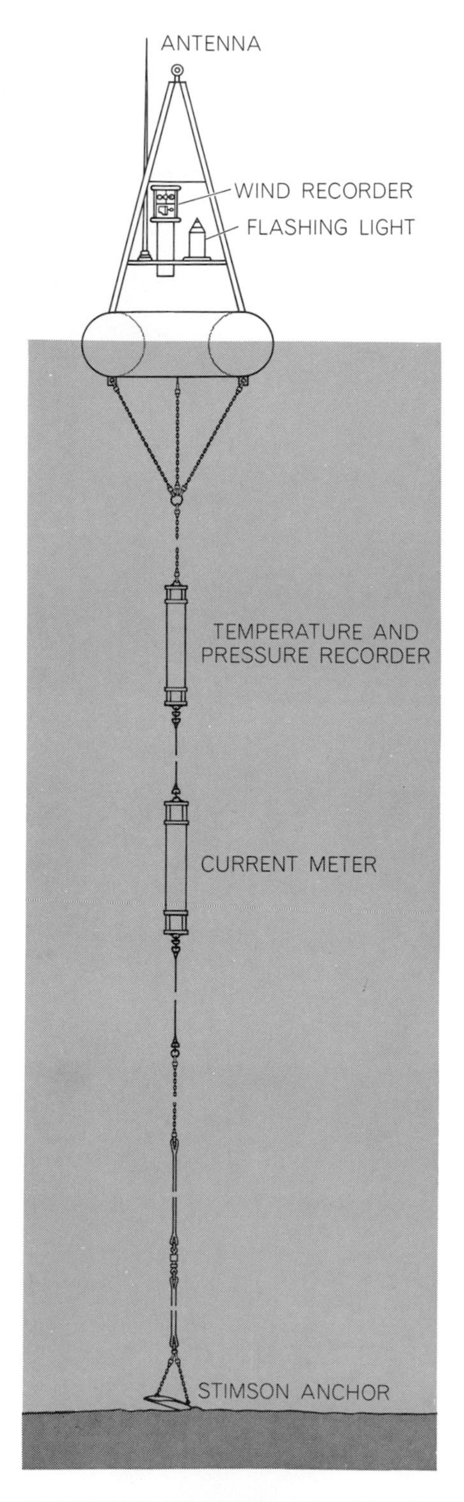

BIG RESEARCH BUOY employed by the Woods Hole Oceanographic Institution gathers and transmits data from the surface to the sea floor. At the surface it records wind speed and direction; below the surface it measures temperature, pressure and current.

Containerization has become a magic word on the waterfront. The basic idea is that a shipper can move his goods from his inland manufacturing point to an inland customer overseas in a private container. A container moves by train or truck to a marshaling yard on the waterfront. There it is picked up by a straddle truck and moved within reach of a gantry crane, which sets it in slots on a container ship. The contents are safe from pilferage, weather and damage. A harbor facility, dealing with containers of standardized size, can semiautomatically unload and reload a large cargo ship in less than 24 hours. Labor cost is lowered; the ship spends more time at sea and less alongside a dock, and the freight moves faster and more cheaply than on a break-bulk cargo ship [see "Cargo-handling," by Roger H. Gilman; SCIENTIFIC AMERICAN, October, 1968].

Methods for desalting water have been improved substantially. The worldwide use of desalted water from the sea is now almost 100 million gallons per day. Most of the water is obtained by various distillation processes; the average cost is estimated to be about 75 cents per 1,000 gallons. Other means of desalination, such as vacuum freezing and reverse osmosis, are being developed. Major nuclear plants that would produce both fresh water and electricity are under study. Most of the desalted water now obtained or in prospect is for household and industrial purposes. The day of cheap irrigation water in large quantities is still far away.

Offshore drilling from floating platforms is less than a decade old and has evolved rapidly. The self-propelled drilling ship and the semisubmersible platform, both of which drill while anchored, represent the two ends of the spectrum. The ship emphasizes speed of movement to the drilling site; the platform provides more steadiness and room for working. A semisubmersible platform is towed to its drilling site, where it takes on enough water ballast to submerge its lower portions. In that position it floats on large cylindrical columns. The arrangement is such that the platform is little affected by waves or other motions of the sea.

The unanchored deep-ocean drilling system, which consists of a drilling rig in a ship hull, has so far been used only for scientific work. It has improved substantially the ability of geologists and other investigators to explore the strata under the deep ocean. The technology dates from 1961 and includes dynamic positioning, the control of stress in a long and unsupported drill pipe, placement of conductor pipe (leading the drill through the soft sea bottom to bedrock) in deep water and the use of a seawater turbodrill to drill hard rock in more than 12,000 feet of water.

Later developments have led to the system employed on the *Glomar Challenger,* which is operated by the Scripps Institution of Oceanography in a National Science Foundation program involving the coring of deep-ocean sediments in water depths of up to 20,000 feet [*see illustration on page 370*]. The developments include acoustical position-sensing equipment and automatic control of the propulsion units, so that dynamic positioning is much more reliable. The ship has successfully drilled several dozen holes in water depths to 17,000 feet, penetrating as much as 2,500 feet of the sea bottom. The cores thus obtained have yielded much valuable information. Moreover, the discovery of hydrocarbons on Sigsbee Knolls deep in the Gulf of Mexico has done much to modify geological thinking about the possibility of oil in deep water.

Offshore oil production is moving steadily into deeper and rougher water and more remote areas. If it is to be profitable, several producing wells must be established in each cluster and the capital cost should not exceed the present cost of producing oil in 200 feet of water. Probably it is possible to build stationary platforms that would resemble existing ones for depths of up to 600 feet. The cost would be high, however, particularly for a system that involved completing the well atop the platform (installing the pipes and valves and related equipment needed to put the well into production after drilling has reached oil). The current trend is toward the use of floating drilling platforms such as the semisubmersible ones, with completion of the well being made on the sea floor. A system of this kind would include remotely controlled valves and flow lines to central collecting points. In depths of 1,000 feet or more submarine work chambers analogous to pressure-resistant elevators will lower workmen to the bottom; while they remain inside at normal surface pressure they will be able to remove and replace heavy components, make flow-line and electrical connections and inspect the machinery.

Deep diving is receiving increasing attention in ocean technology. Men go into increasingly deeper water by means of systems that grow ever more complicated, involving a variety of chambers, hoists, gases and instruments.

There are two competing methods: the bounce dive and the saturation dive. In a bounce dive the diver goes from atmospheric pressure to the required depth in a chamber, breathing gaseous mixtures that change in accordance with depth and physiological requirements. He works for a few minutes (perhaps 10) and then returns to the surface in a fully pressurized chamber for slow decompression on the deck of the mother ship. The saturation-dive system makes possible multiple dives. In it the diver's body is saturated with inert gases while he lives in a pressure chamber on shipboard. When it is time to dive, he moves into a similarly pressurized capsule that is lowered to the bottom. Since his body is already prepared for the pressure, he can immediately go to work. He can work much longer than the bounce diver

HOISTING DEVICE called the Syncrolift allows the Canadian submarine *Ojibwa* to be lifted out of the water (*above*) and pulled ashore on rails (*below*) for repair in a dockyard at Halifax, Nova Scotia. The platform is lowered under the keel of a vessel, the ship is floated in and the platform is raised by the array of winches visible in the photograph. The device has a capacity of 6,000 tons.

DIVING GEOLOGISTS of the Shell Oil Company probe the ocean floor at a depth of about 20 feet on the Bahama Banks. The pipe and hose slanting across the center are parts of an air-lift apparatus that they are using to obtain information helpful in the search for oil.

before he returns to the capsule and thence to the chamber on deck. Since he lives on the surface but at the pressure of the bottom, the procedure can be repeated for many days. Then the diver takes a slow decompression to atmospheric pressure. Divers using each of these systems have reached 1,000 feet.

Another scheme has divers living on the bottom in shallow water at ambient pressures in undersea chambers. Examples are the experiments that have been carried out in such chamber systems as the U.S. Navy's *Sealab II*, the University of Miami's *Tektite* and the French *Conshelf*.

Doubtless oceanographers can point to elements of ocean technology I have overlooked. The developments have been too rapid and profuse for one man to be familiar with them all. Sometimes one feels that the first question to be asked on hearing about a new oceanographic device is: "Is it obsolete yet?" The answer should be: "We're working on that problem and it soon will be."

LIVING UNDER THE SEA

JOSEPH B. MACINNIS

March 1966

It is one thing to glimpse a new world and quite another to establish permanent outposts in it, to explore it and to work and live in it. In recent years parts of the ocean floor have been studied in considerable detail, but almost entirely by surface-bound investigators. They have sounded the oceans with electronic devices, dangled instruments thousands of feet below the surface and secured samples of the bottom, and a few have undertaken brief expeditions in submersible vehicles to the greatest depths. Now, however, men are beginning to try to live underwater—to remain on the bottom exposed to the ocean's pressure for long periods and to move about and work there as free divers.

The submerged domain potentially available to man for firsthand investigation and eventual exploitation can be regarded as a new continent with an area of about 11,500,000 square miles—the size of Africa. It comprises the gently sloping shoulders of the continents, the continental shelves that rim the ocean basins. The shelves range up to several hundred miles in width and are generally covered by 600 feet of water or less. That they are submerged at all is an accident of this epoch's sea level: the ocean basins are filled to overflowing and the sea has spilled over, making ocean floor of what is really a seaward extension of the coastal topography. Geologically the shelf belongs more to the continents than to the oceans. Its basement rock is continental granite rather than oceanic basalt and is covered largely with continental sediments rather than abyssal ooze.

Not surprisingly, mineral deposits similar to those under dry land lie under the shelf. Oil and natural gas are the foremost examples. In 1964 alone the petroleum industry spent $5 billion to find and recover offshore oil; only recently the continental shelf in the North Sea has become the site of extensive exploration for oil and gas. Drilling and capping a well from the surface is not easy. The prospect of more efficient oil and gas operations in deeper water by men working on the floor of the shelf is one of the primary reasons for the surge of activity directed toward living under the sea. There are other reasons. One is the increasing interest in all aspects of oceanography, coupled with an awareness of the geological, biological and meteorological information to be gained in direct undersea investigations. Another is the advance in free-diving techniques that began with the invention of "self-contained underwater breathing apparatus" (SCUBA) in the 1940's. Finally, there is a need for improved methods of underwater salvage and submarine rescue.

The reasons for going underwater are balanced by an impressive list of potential hazards. Most of them stem from the effects of pressure, which increases at the rate of one atmosphere (14.7 pounds per square inch, or 760 millimeters of mercury) with every 33 feet of depth in seawater.

The best-known hazard and one of the most dangerous is decompression sickness—"the bends." Under pressure the inert gas in a breathing mixture (nitrogen or helium) diffuses into the blood and other tissues. If the pressure is relieved too quickly, bubbles form in the tissues much as they do in a bottle of carbonated water when it is opened. Sudden decompression from a long, deep dive can be fatal; even a slight miscalculation of decompression requirements can cause serious injury to the joints or the central nervous system. A diver must therefore be decompressed slowly, according to a careful schedule, so that the inert gas can be washed out of the tissues by the blood and then exhaled by the lungs. Whereas the demands of decompression become more stringent with depth, with time they increase only up to a point. After about 24 hours at a given depth the tissues become essentially saturated with inert gas

UNDERWATER DWELLING called the SPID (for "submerged, portable, inflatable dwelling") was designed by Edwin A. Link as a base of operations for long dives to the continental shelf. In the photograph on the following page the SPID is undergoing a pressure test at 70 feet. In the summer of 1964 two divers occupied the SPID for two days at 432 feet below the surface.

at a pressure equivalent to the depth; they do not take up significantly more gas no matter how long the diver stays at that level. Therefore if a diver must descend to a certain depth to accomplish a time-consuming underwater task, it is far more efficient for him to stay there than to return to the surface repeatedly, spending hours in decompression each time. Although this "saturation diving" is efficient, it imposes an extra technical burden, because the schedules for the ultimate decompression must be calculated and controlled with particular care.

Pressure also has significant effects on a diver's breathing requirements. For one thing, hyperoxia (too much oxygen) becomes almost as dangerous as hypoxia (too little). Acute hyperoxia can affect the central nervous system, causing localized muscular twitching and convulsions; chronic hyperoxia impairs the process of gas exchange in the alveoli, or air sacs, of the lung. Optimum oxygen levels are still under investigation; they vary with the duration, depth and phase of the dive and the muscular effort required of the diver. It is clear, however, that the "partial pressure" of oxygen should be kept between about 150 and 400 millimeters of mercury during the at-depth phase of a long saturation dive. The partial pressure of oxygen in the air we breathe at sea level is 160 millimeters of mercury (21 percent of 760). If oxygen is kept at 21 percent of the mixture, however, its partial pressure increases with depth—rising to 1,127 millimeters 200 feet down, for example. As a result the proportion of oxygen in the air or other breathing mixture must be cut back sharply from 21 percent. The band of permissible percentages narrows rapidly with depth [*see illustration on page 387*], calling for increasing accuracy in the systems that analyze and control the gas mixture.

Nitrogen, which is physiologically inert at sea level, has an anesthetic effect under pressure. At depths greater than 100 feet it begins to produce "nitrogen narcosis," an impairment in judgment and motor ability that can render a diver completely unable to cope with emergencies. Helium has been found to be much less narcotic and is currently used instead of nitrogen in almost all deep-sea dives. Being less dense, it also offers less breathing resistance under pressure; this can be important to a working diver. Helium has two disadvantages, however. Because its thermal conductivity is almost six times as great as nitrogen's, it accelerates the loss of body heat and makes a diver uncomfortably cold even at temperatures of 70 or 80 degrees. Helium also distorts the resonance of a diver's voice, making his speech almost unintelligible and thus giving rise to a serious communication problem.

In any confined environment the buildup of exhaled carbon dioxide must be monitored carefully. In our diving experiments for Ocean Systems, Inc., we try to keep the partial pressure of this gas below seven millimeters of mercury (compared with the sea-level pressure in fresh air of .3 millimeter), but at the U.S. Naval Medical Research Laboratory in New London, Conn., Karl E. Schaefer has found that at sea level slightly higher levels are tolerable for several weeks. In any case, carbon dioxide accumulates rapidly in a small space and soon reaches a toxic level, causing dizziness, headache and an increase in the rate of breathing. It must therefore be continuously "scrubbed" out of the diver's atmosphere, usually by being passed through some chemical with which it will react. Other gases, such as carbon monoxide and certain volatile hydrocarbons, can also reach toxic levels quickly if they are allowed to concentrate in the diver's breathing mixture.

There are sometimes other obstacles to casual access to the ocean floor: a demoralizing lack of visibility, strong currents, uncertain bottom profiles. There are also dangerous marine animals, ranging in size from a unicellular infective fungus to the widely feared great white shark. Finally, the water of the continental shelf is cold. Temperatures average between 40 and 60 degrees, and without protective clothing a diver soon becomes totally ineffective.

Faced with these difficulties commercial divers and undersea investigators found it impossible to spend time and do useful work on the continental shelf. Those who went down in pressurized suits and thick-hulled submersible vehicles were held prisoner by their protective armor. Free divers, on the other hand, could not go very deep or stay very long. In 1956 Edwin A. Link, the inventor of the Link Trainer for simulated flight training, was engaged in undersea archaeological investigations. He recognized that a diver could work more effectively at substantial depths if he could live there for prolonged periods instead of having to be decompressed to the surface after each day's work. Link set out to build a vehicle that could operate as an underwater elevator, a diving bell and a decompression chamber. The "submersible decompression chamber" (SDC) he designed is an aluminum cylinder 11 feet long and three feet in diameter [*see illustration on page 386*]. With its outer hatches closed it is a sealed capsule in which a diver can be lowered to the bottom. On the bottom, with the internal gas pressure equal to ambient water pressure and the hatches open, the SDC serves as a dry refuge from which the occupant can operate as a free diver. Then, with the hatches again closed, it becomes a sealed chamber in which the diver can be decompressed safely and efficiently on shipboard or during his ascent to the surface. An inner hatch provides an air lock through which someone else can enter the chamber (or pass food and other supplies into it) during the decompression phase.

Before open-sea experiments with the SDC were possible some preliminary research was necessary. How deep could a man go as a free diver? How long could he stay down? What would be the acute and the long-term medical effects of the pressure itself and of the synthetic atmosphere? What would be the response to the cold, the confinement and the psychological hazards of deep submergence? Some early and significant answers were provided by Captain George F. Bond, a U.S. Navy physician who in 1957 conceived and carried out a series of simulated dives in a compression chamber on land at the Naval Medical Research Laboratory. Bond's group first exposed small animals, including some primates, to a pressure equivalent to a depth of 200 feet. Volunteer Navy divers then lived in the chamber under precisely controlled conditions of pressure, temperature and humidity. These experiments showed, among other things, that men could breathe helium instead of nitrogen for long periods without ill effects and encouraged Link to move ahead.

Early in September, 1962, the SDC underwent its critical test in the Mediterranean Sea off Villefranche on the French Riviera. A young Belgian diver, Robert Sténuit, descended in it to 200 feet and lived there for 24 hours, swimming out into the water to work and returning to rest in the warm safety of the pressurized chamber. When the time came to return to the surface, Sténuit did not have to face hours of dangling on a lifeline or perching on a platform, decompressing slowly in the cold water. Instead he sealed himself into the chamber, was hoisted to the deck of Link's research vessel, the *Sea Diver*, and there was decompressed in

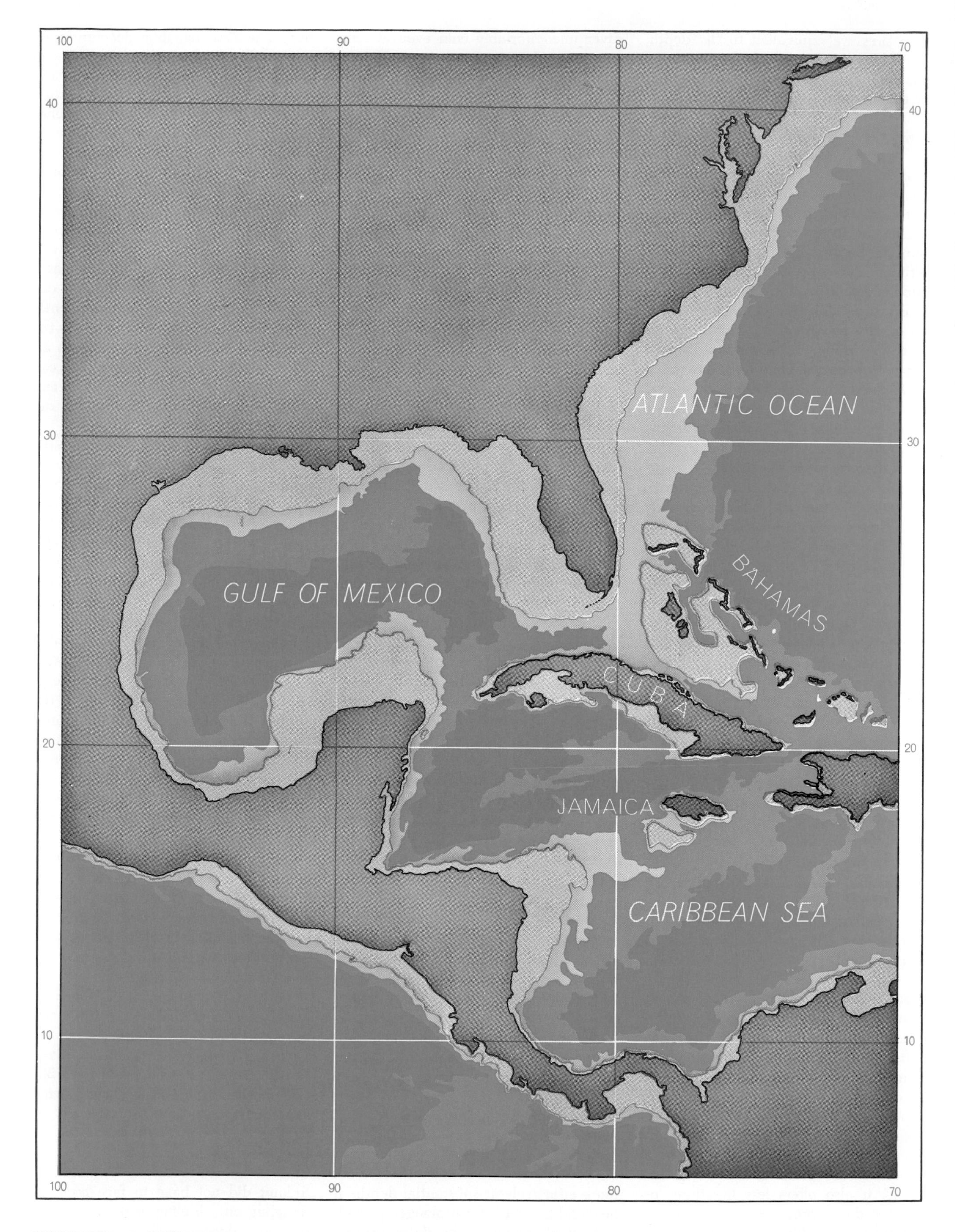

CONTINENTAL SHELF (*lightest areas*) off part of North America is shown. It is less a part of the ocean basin than it is an extension of the continental land mass. As in most parts of the world, the shelf slopes gently to about 600 feet below sea level; then the continental slope plunges toward the floor of the ocean basin. On this map, based on charts of the International Hydrographic Bureau, the contour intervals are in meters rather than feet. The lightest tone shows the bottom from sea level down to 200 meters (655 feet); successively darker blacks indicate bottom from 200 to 1,000, 1,000 to 3,000 and deeper than 3,000 meters.

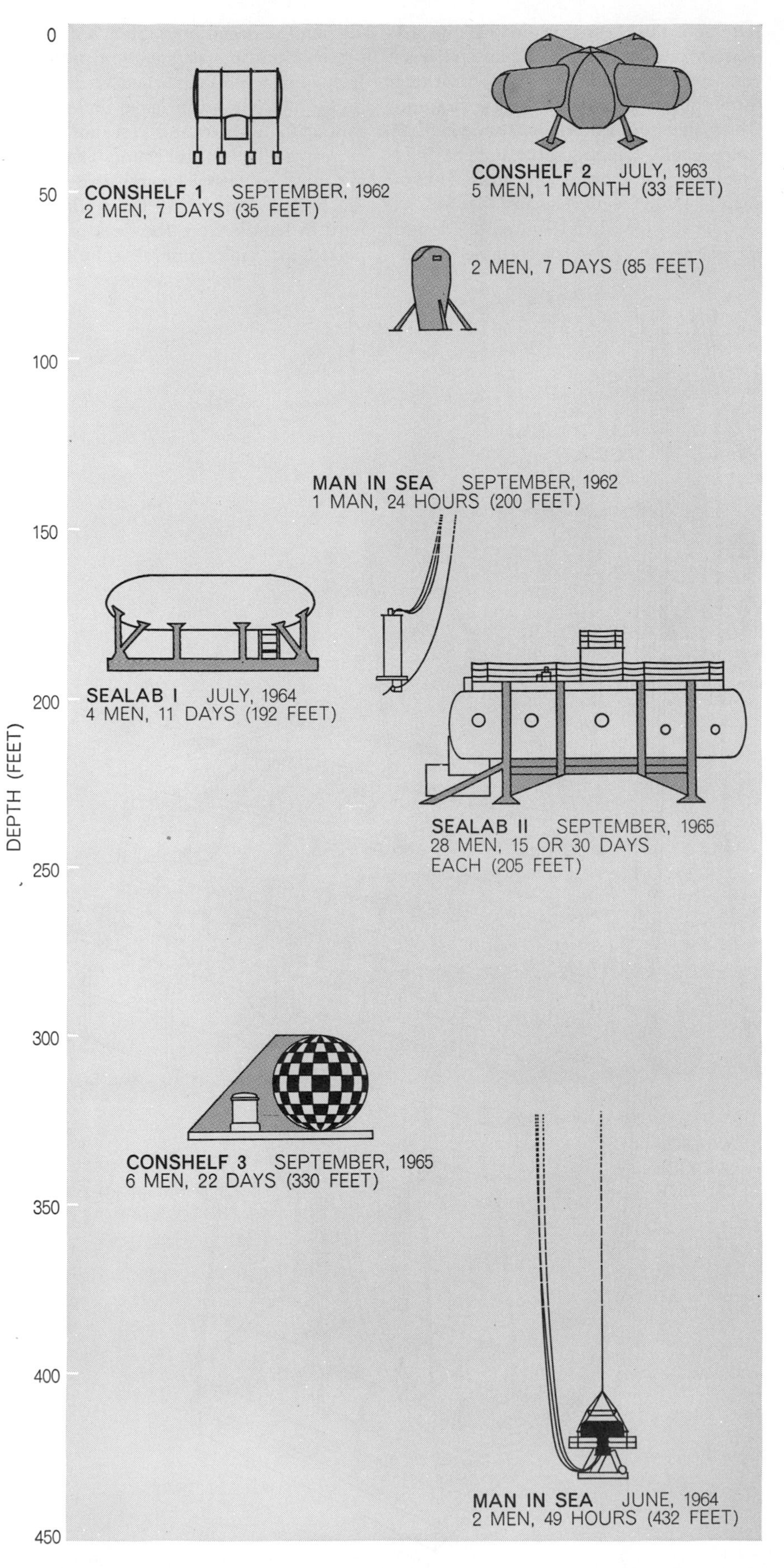

SATURATION DIVING, in which the divers stay down for prolonged periods, is made possible by underwater shelters. The chart gives data for seven such dives. "Man in Sea" is the Link project, "Conshelf" is Jacques-Yves Cousteau's and "Sealab" is the U.S. Navy's.

safety and relative (although somewhat cramped) comfort.

Meanwhile, moving ahead independently, the French undersea investigator and inventor of the aqualung, Jacques-Yves Cousteau, had undertaken experiments aimed at establishing manned undersea stations on the continental shelf. The first experiment in his "Conshelf" program was carried out near Marseilles in mid-September, 1962, when two men lived under 35 feet of water for a week. The divers worked in the sea several hours a day, returning to a cylindrical cabin to eat and sleep.

Cousteau's group went on, in the summer of 1963, to establish a complex underwater settlement at the remote Sha'ab Rumi Reef in the Red Sea. The hub of the settlement was "Starfish House," an assembly of cylindrical chambers in 33 feet of water. It housed five men and their eating, sleeping and laboratory facilities for a month. Nearby was a submarine "hangar" from which a two-man "diving saucer," with its pilot and passenger protected from the sea pressure, made a number of trips as deep as 1,000 feet to collect samples and make observations at the edge of the reef. Down the coral slope from Starfish House, at a depth of 85 feet, two men lived in the controlled oxygen-helium-nitrogen environment of a "deep cabin" for seven days, making short excursion dives as deep as 360 feet. One of the most interesting things about Conshelf II was its demonstration that—at least for relatively shallow depths—the participants did not have to be experienced divers or even young men in particularly good physical condition. Instead they were picked for their vocational ability as mechanics, scientific workers, cooks and so on. The experiment showed that biological investigations and submarine operations could be carried out from submerged stations.

In the U.S. meanwhile Link and Bond were designing pressure experiments and engineering diving systems that would enable free divers to reach greater depths safely. From late 1963 until March, 1964, a series of simulated saturation dives—the first such dives deeper than 200 feet—were carried out under the technical direction of Captain R. D. Workman at the Navy's Experimental Diving Unit in Washington. The tests showed that divers suffered no harmful effects when exposed to depths of 300 and 400 feet for 24 hours and that they could be decompressed successfully on a linear decompression schedule.

Link had decided that the second

phase of his "Man in Sea" project would attempt to demonstrate that men could work effectively at 400 feet for several days. He established a "life-support" team under the direction of Christian J. Lambertsen of the University of Pennsylvania School of Medicine to undertake preliminary research and supervise the medical aspects of the dive. Under Lambertsen's direction James G. Dickson and I first evaluated the accuracy and reliability of gas analyzers that would monitor the divers' breathing atmosphere. In addition to proving out the system, our experiments showed that mice could tolerate saturation at (and decompression from) pressures equivalent to 4,000 feet of seawater.

The 400-foot dive required the design of a larger and more comfortable "dwelling" on the ocean floor. Such a dwelling presents unusual engineering problems. It must provide shelter and warmth and be easy to enter and leave underwater, simple to operate and resistant to the corrosive effects of seawater. The dwelling must be heavy enough to settle on the bottom but not so heavy that it is hard to handle from the deck of a support ship. Link's unique solution was

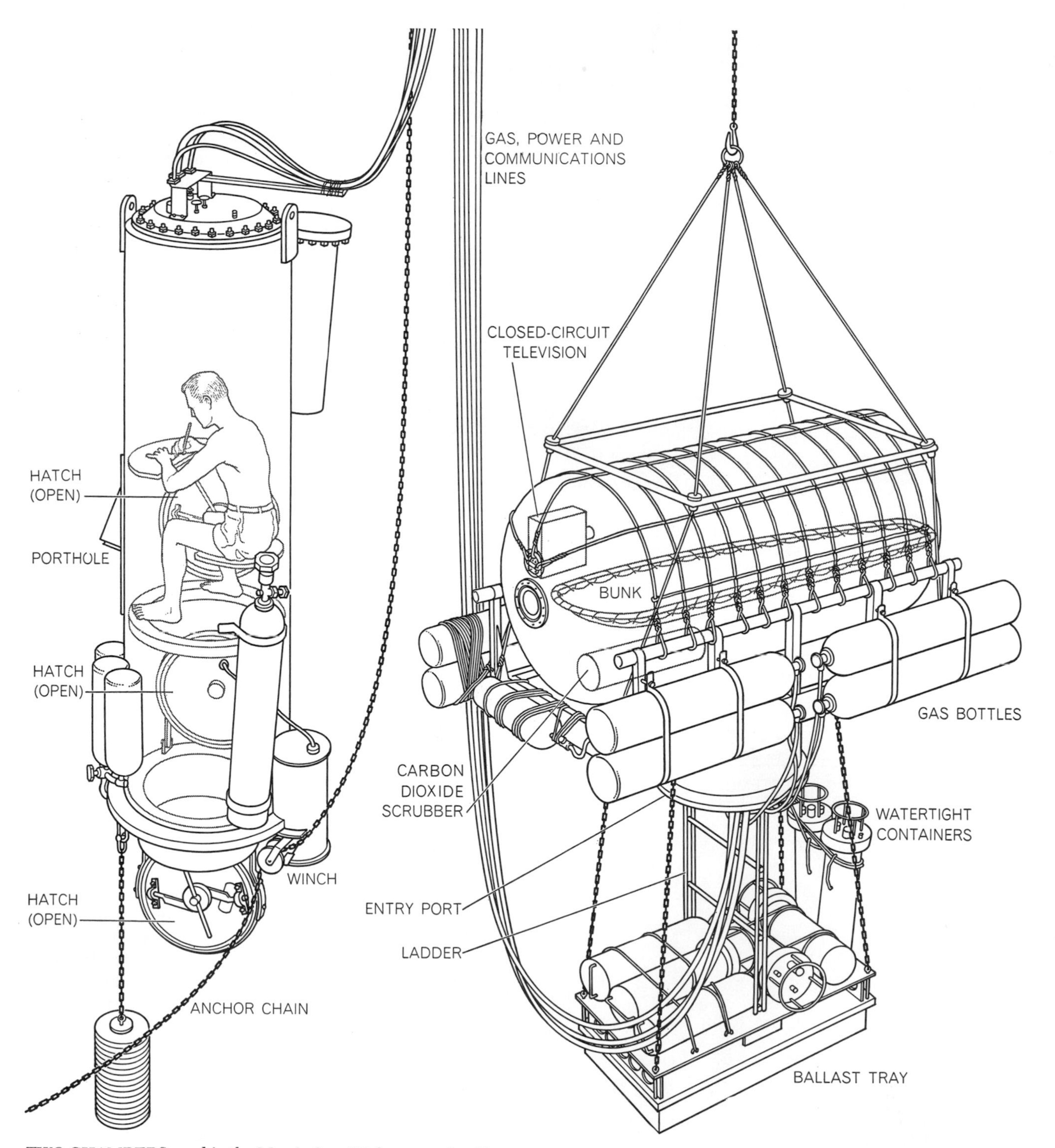

TWO CHAMBERS used in the Man in Sea 432-foot, two-day dive are diagrammed. The "submersible decompression chamber," or SDC (*left*), is an aluminum cylinder 11 feet long and three feet in diameter. With the hatches open and the inside gas pressure equal to the external water pressure, the SDC serves as a diving bell. The SPID (*right and photograph on page 382*) is an eight-by-four-foot inflatable rubber dwelling with a steel frame and ballast tray. Access to it is through an open entry port at the bottom.

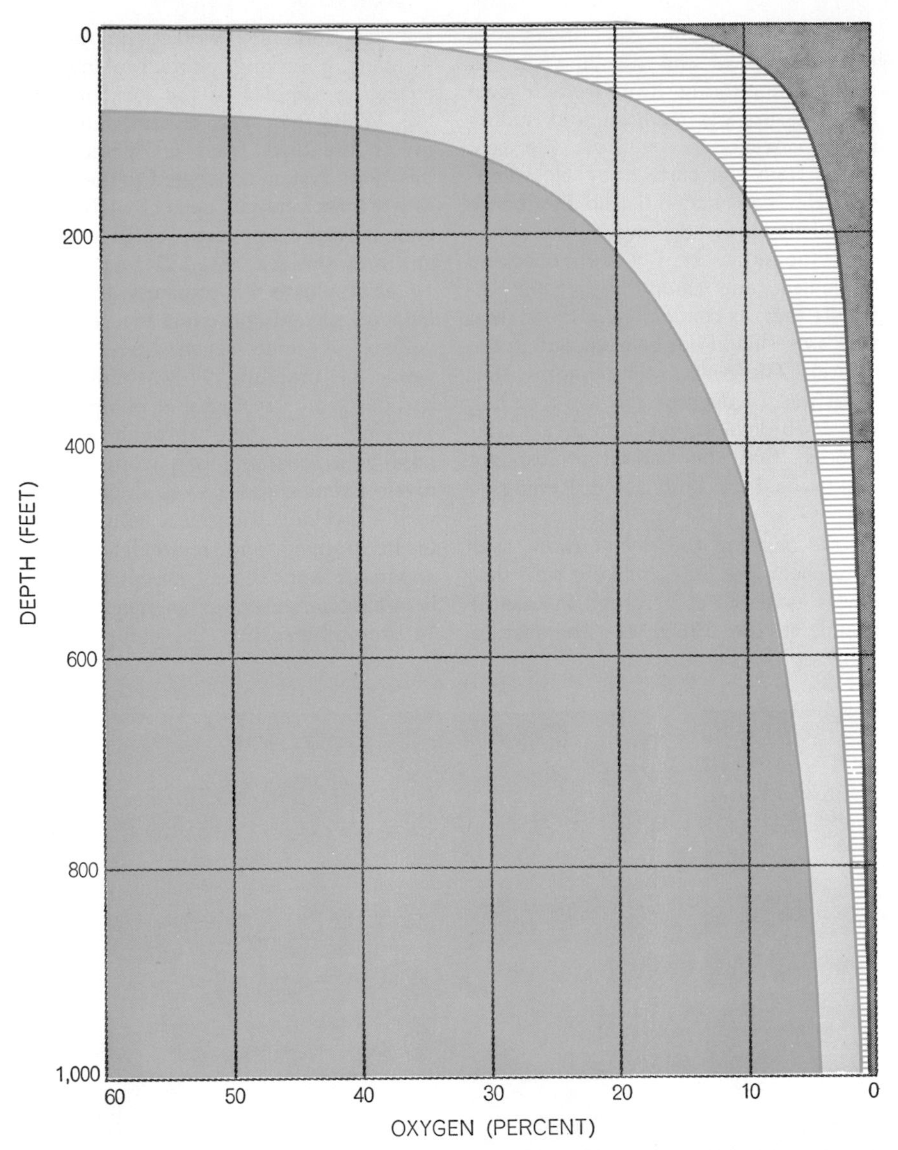

PERMISSIBLE RANGE (*barred white band*) of oxygen content in the breathing gas supplied to a diver narrows sharply with depth, requiring close control in order to avoid zones in which there is a danger of hypoxia (*gray*) or hyperoxia damaging to the lungs (*light color*) and the central nervous system (*dark color*). For several days' exposure at sea level a safe range is from 20 to 60 percent. With depth the "partial pressure" of oxygen increases, however; at 300 feet, or 10 atmospheres, 2 percent of the breathing mixture represents as much oxygen as 20 percent would represent under sea-level conditions.

in effect an underwater tent: a fat rubber sausage eight feet long and four feet in diameter, mounted on a rigid steel frame [*see illustration on preceding page*]. Deflated at the surface, the "submerged, portable, inflatable dwelling" (SPID) is remarkably easy to handle—an important advantage when undersea habitations are established in remote locations. As it is submerged, the tent is inflated so that its internal gas pressure is equal to the ambient water pressure. There are no hatches; an open, cufflike entry port in the floor of the SPID allows easy access and provides the necessary vertical latitude for variations in the pressure differential. Inside the SPID and in watertight containers on the frame and on the ballast tray below it are stored supplies and equipment: gas cylinders and the gas-circulating system, a closed-circuit television camera, communications equipment, food, water, tools and underwater breathing gear.

In the 400-foot dive the SPID was to be one of three major pressure chambers. The second was the proved SDC and the third was a new deck decompression chamber. This time the SDC was to serve as an elevator and also as a backup refuge on the bottom but not as the main decompression chamber. After a long, deep dive decompression takes several days, and it is important that the divers be as comfortable as possible. An eight-by-five-foot decompression chamber with a four-foot air lock was therefore secured to the deck of the *Sea Diver*. The SDC could be mated to it so that the divers could be transferred to it under deep-sea pressure. Decompression could then proceed under the direct supervision of life-support personnel.

Early in June, 1964, Link and his research group sailed to the Bahamas to test the three-chamber diving concept. We checked the chambers with dives to 40 and 70 feet, spending several weeks refining techniques for handling the SDC and the SPID and coping with potential emergencies. The exact site for the dive, chosen with the cooperation of Navy personnel using sonar and underwater television, was a gentle coral sand slope 432 feet deep, about three miles northwest of Great Stirrup Cay.

On June 28 the underwater dwelling, with its vital gear carefully stowed aboard, was lowered slowly to the ocean floor. When it had settled on the shelf, the oxygen level inside was adjusted to 3.8 percent, the equivalent of a sea-level partial pressure of 400 millimeters of mercury. The inert gas was helium with a trace of nitrogen (because there had been air in the tent to start with). Then the SPID was left, a habitable outpost autonomous except for communications, power and gas lines, ready for its occupants.

The next step was to transport Sténuit and Jon Lindbergh, another experienced diver, to the shelf. As usual, the SDC was placed in the water at the surface so that the divers could enter it from below. At 10:15 A.M. on June 30 Sténuit and Lindbergh went over the side and swam up into the chamber, closed the outer hatches and checked their instruments. At 10:45, still at the surface, the SDC was pressurized to the equivalent of 150 feet with oxygen and helium to check for leaks; one minor leak was discovered and repaired. At noon the chamber started down, slipping through the clear purple water toward the deep shelf. When it reached 300 feet, Lindbergh reported the bottom in sight. At 1:00 P.M. the anchor weights touched bottom and the chamber came to a stop five feet above the sand. It was just 15 feet from the waiting SPID. During the descent the SDC's internal pressure had been brought to 200 feet; now pressurization was completed. At 1:15 the bottom hatches were opened and Sténuit swam over and entered the dwelling. Lindbergh joined him and they began to arrange the SPID for their stay.

At that point Lindbergh reported that the carbon dioxide scrubber had been flooded and was not functioning. The divers found the backup scrubber in its watertight container and prepared to set it up as the carbon dioxide level rose to almost 20 millimeters of mercury. Then they found they could not get at the reserve scrubber: the pressure-equalizing valve that would make it possible to open the container was missing. With the carbon dioxide level rising rapidly as a result of their muscular exertion, they had to leave the dwelling and return to the SDC. We had hoped to maintain the diving team on the shelf with a minimum of support from the surface, but it now became necessary to send a spare scrubber down on a line from the *Sea Diver*. The divers installed it in the SPID and the dwelling was soon habitable.

Later that evening the divers took over control of the dwelling's atmosphere, monitoring it with their own high-pressure gas analyzer and adding makeup oxygen as required. We kept watch from the surface by closed-circuit television as Sténuit and Lindbergh settled down for the night. While one slept the other kept watch, checking instruments and communications (a procedure that, as confidence in the system increases, should not be necessary in the future). The water temperature that night was 72 degrees and the dwelling was at 76 degrees, yet both divers later reported that the helium atmosphere was too cold for comfortable sleeping.

In the morning the divers swam over to check the SDC, making sure that it was available as a refuge in case of trouble in the SPID. For the rest of the day both men worked out of the dwelling, observing, photographing and collecting samples of the local marine life. While they were in the water the divers breathed from a "closed" rebreathing system connected to the SPID rather than from an "open" SCUBA system. An open apparatus spills exhaled gas into the sea. At 432 feet, under 14 atmospheres of pressure, each exhalation expends gas equal to a sea-level volume of some seven liters, which would be prohibitively wasteful. Link had designed a system that pumped the dwelling atmosphere through a long hose to a breathing bag worn by the diver. Exhaled gases were drawn back to the dwelling through a second hose to be purified and recirculated. The apparatus worked well except that the breathing mixture was so dense under 14 atmospheres that the pumps could

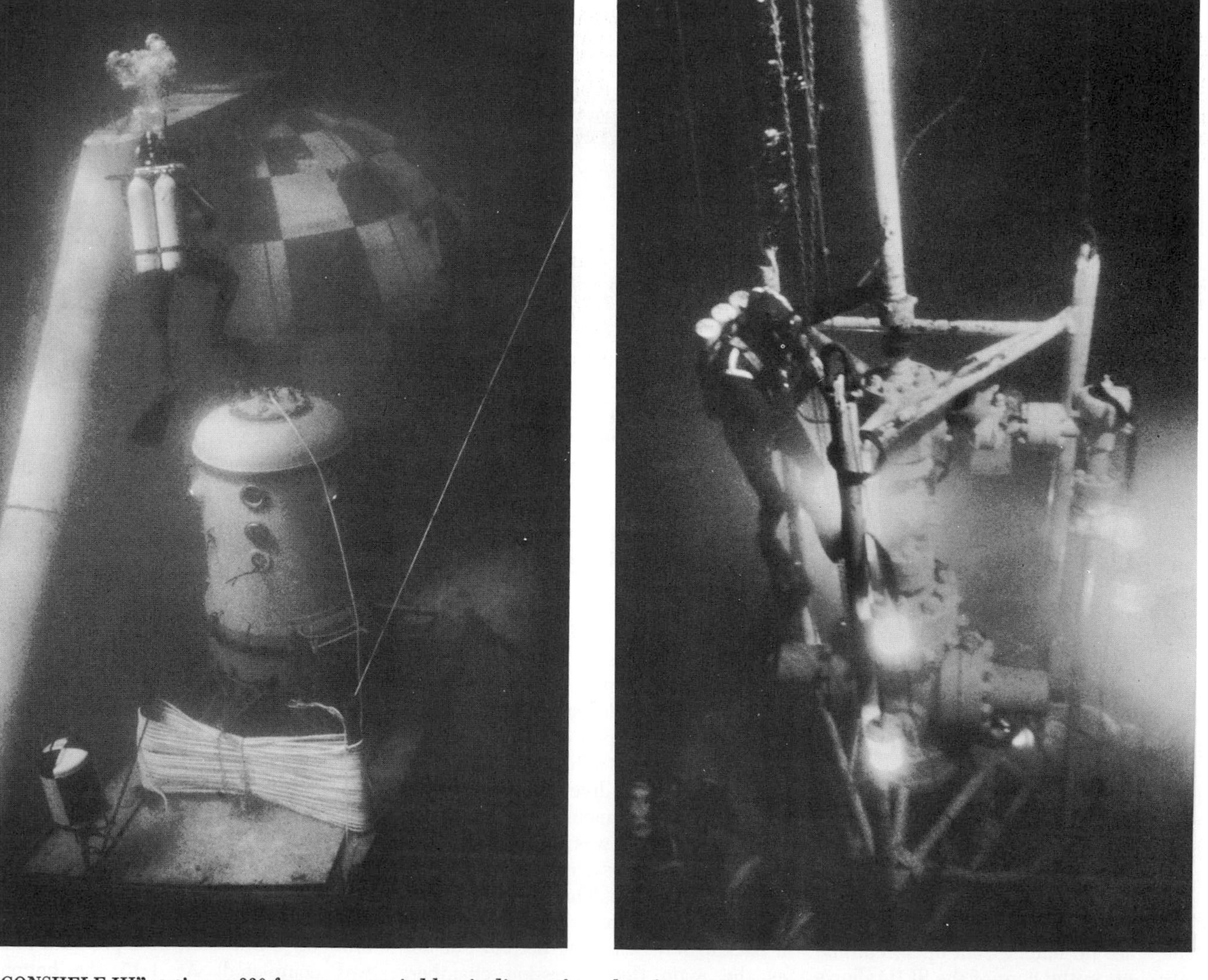

"CONSHELF III" station at 330 feet was occupied by six divers of Cousteau's group last fall. The spherical dwelling in which they lived is shown at the left in a photograph made from Cousteau's diving saucer. The elongated shape (*far left*) is a fin for stability under tow; the turret-shaped structure is a compression chamber for emergency escape to the surface. The major task accomplished by the divers was the installation and repair of an oil-well head (*right*). They were able to manipulate repair tools to handle emergency breakdown situations met in actual production. In the photograph a diver is guiding a tool pipe into the wellhead.

DEFLATED SPID is hoisted over the side of Link's vessel *Sea Diver*. One end of the SDC is visible in the left foreground, with part of the deck decompression chamber beyond it.

not move quite enough of it to meet the divers' maximum respiratory demand.

During the second evening we carried out and recorded voice-communication tests with the divers breathing either pure air or a mixture of 75 percent air and 25 percent helium. Voice quality was considerably better than in a helium atmosphere, but even 30 seconds of breathing air caused a noticeable degree of nitrogen narcosis. At 11:00 P.M. the two men bedded down for the second night. They were disturbed from time to time by heavy thumps against the outside of the dwelling. It developed that large groupers, attracted by the small fishes that swarmed into the shaft of light spilling from the open port of the SPID, were charging the swarm and hitting the dark bulk of the dwelling.

The next day the divers measured the visibility in the remarkably clear water; they could see almost 150 feet in the horizontal plane and 200 feet vertically. Then they took more photographs and collected animal and plant specimens. At 1:30 P.M. on July 2 both men were back in the SDC with the hatches secured. At 2:20, after 49 hours on the deep shelf, the SDC began its ascent. The internal pressure was maintained at 432 feet; although the divers were being lifted toward the surface, they were not yet being decompressed. At 3:15 the dripping SDC was hoisted onto its cradle aboard the *Sea Diver*. Now the internal pressure was decreased to 400 feet to establish a one-atmosphere differential between the divers' tissues and the chamber environment and make it possible for helium to begin escaping effectively from their tissues. Then, at 4:00, the SDC was mated to the deck decompression chamber, which was also at a pressure of 400 feet. Sténuit and Lindbergh, transferred to the deck chamber and we began to advance them to surface pressure at the rate of five feet, or about .15 atmosphere, per hour. With the divers safe in their chamber another advantage of deck decompression became evident: mobility. While decompression proceeded the *Sea Diver* weighed anchor and steamed for Florida. By the time it moored in Miami on the afternoon of July 5 the pressure had been reduced to 35 feet.

During the shallow stages of decompression, breathing pure oxygen establishes a larger outward pressure gradient in the lung for the inert gas one is anxious to flush out of the diver's tissues and thus helps to prevent the bends. Since breathing pure oxygen under pressure for a sustained period can cause lung damage, Lambertsen had in the past suggested alternating between pure oxygen and compressed air. We instituted this interrupted oxygen-breathing schedule when the divers reached 30 feet. Still, we had one period of concern about decompression sickness. At about 20 feet Sténuit reported a vague "sawdust feeling" in his fingers that seemed to progress to the wrists. I examined him under pressure in the chamber. There were no abnormal neurological findings, but decompression sickness is so diverse in its manifestations that almost any symptom has to be taken seriously. Dickson and I therefore recompressed the chamber one atmosphere and then resumed decompression at the slower rate of four feet per hour. Finally, at noon on July 6, Sténuit and Lindbergh emerged from the chamber in excellent condition after 92 hours of decompression. The important point about saturation diving is that their decompression time would have been the same if they had stayed down 49 days instead of 49 hours.

Their dive had shown that men could live and work effectively more than 400 feet below the surface for a substantial period, protected by an almost autonomous undersea dwelling, and be successfully recovered from such depths and decompressed on the surface at sea. More specifically, it demonstrated the flexibility and mobility of the three-chamber concept. It also emphasized some problems, including the voice distortion caused by helium and the need for a larger breathing-gas supply to support muscular exertion. It showed that the control of humidity in an atmosphere in direct contact with the sea is extraordinarily difficult. The relative humidity in the chamber was close to 100 percent and both divers complained of softened skin and rashes. Temperature was a problem too. Both men preferred having the chamber temperature between 82 and 85 degrees. In the water, we realized, heated suits are required to keep divers comfortable even in the Caribbean Sea.

There have been a number of other recent saturation diving experiments, two of them conducted by Bond's Navy group. The first, "Sealab I," took place off Bermuda later in July, 1964. Four men lived for 10 days in a large cylindrical chamber 192 feet below the surface. Last summer the Navy conducted "Sealab II," a massive 45-day effort involving three teams of 10 men, each of which spent 15 days underwater. (One man went down for two nonconsecutive 15-day periods and one, the astronaut Scott Carpenter, stayed for 30 consecutive days.) The base of operations was a cabin 57 by 12 feet in size submerged in 205 feet of water near the Scripps Institution of Oceanography at La Jolla, Calif. The Sealab "aquanauts" salvaged an airplane hulk, did biological and oceanographic research and conducted psychological and physiological tests. Electrically heated suits made it possible for them to work comfortably in the 55-degree water.

In the Mediterranean off Cap Ferrat, Cousteau's group last fall made another significant advance in underwater living. Six men lived in a spherical dwelling 330 feet below the surface for almost 22 days, linked to the surface only by an electrical and communications cable. Cousteau's "oceanauts" concentrated on difficult underwater work, including the successful emplacement and operation at 370 feet of a five-ton oil-well head in which oil under pressure was simulated by compressed air.

As men go deeper and stay longer the hazards increase and safety margins narrow. New questions arise. At what depth will even helium become too narcotic or too dense to breathe? Can hydrogen serve as an acceptable substitute? At what depth will pressure effects cause unacceptable changes in tissue structure? What will be the decompression obligation after saturation at 1,000 feet or more? And what are the residual effects of repeated exposure to great depths?

Again, the answers are beginning to come from dry-land experimentation. Last fall two Ocean Systems divers simulated a dive to 650 feet in our test chamber. They stayed at that pressure for 48 hours, becoming completely saturated with 20 atmospheres of helium. Our results indicated that helium is safe—at least at the depth and for the length of time involved in the test—and suggested that it may be possible to continue with helium as the inert gas even beyond 1,000 feet. We found that breathing an oxygen-neon mixture for 30 minutes at 650 feet caused no measurable narcotic or other detrimental effects and that it markedly improved voice quality. Heart and lung function, exercise tolerance, psychomotor performance and blood and urine characteristics were all within normal limits. I think the most significant result of this longest deep-pressure experiment to date was our impression that divers will be able to perform physical and mental work almost as effectively at 650 feet as at the surface.

There do not, then, seem to be any physiological or psychological barriers that will prevent the occupation of any part of the continental shelf. Nonetheless, it is important to recognize that so far all efforts to live under the sea have been investigations or demonstrations of man's ability to do so. In the last analysis men will live underwater only when specific tasks, with economic or other motivations, present themselves. At this point, however, the gates of the deep shelf have been opened.

"SEALAB II" CHAMBER housed 10 Navy divers at a time during a 45-day test at 205 feet last summer. The chamber is a 57-by-12-foot cylinder with several interior compartments. Entry is through the port at the lower left, which is protected by a wire shark barrier.

THE SEA LAMPREY

VERNON C. APPLEGATE AND JAMES W. MOFFETT
April 1955

For more than 80 years fishing in the Great Lakes has been a sizable industry and a popular recreation for fishermen of the U. S. and Canada. Each year it yields a commercial catch of more than 100 million pounds of choice food, to say nothing of the millions of pounds caught by sportsmen. The most prized fish, and the backbone of the fishing industry, has been the lake trout. In good years the trout catch amounted to more than 15 million pounds, worth nearly $8 million. But in the past 15 years the Great Lakes trout has suffered a disaster. The U. S., preoccupied with more spectacular troubles on a global scale, has not paid a great deal of attention to this calamity in its own backyard, though it threatens to destroy an important industry and relaxation.

The trout catastrophe began in Lake Huron in 1939. The fish suddenly began to decline in numbers, and within 14 years it had all but disappeared from that lake; the catch dropped from more than five million pounds a year to 344,000 pounds in 1953. The same fate began to overtake Lake Michigan's trout in 1946, and the catch there fell from more than five and a half million pounds to a mere 402 pounds in 1953. Now the slaughter has started in Lake Superior and has begun to cut sharply into its annual trout catch of four and a half million pounds.

Neither overfishing nor weather nor disease is responsible for the annihilation of the trout. The culprit is an eel-like fish known as the sea lamprey. It is a murderous animal efficiently equipped with tools for destroying fish much larger than itself. The lamprey has a sucker-like mouth, sharp teeth and a tongue as rough as a file. Attaching itself to its victim with its mouth, it rasps a hole in the fish's body and sucks the blood and body juices; it is assisted in this by a substance in its saliva, called lamphredin, which prevents coagulation of the blood and dissolves the torn flesh. The victim thrashes about violently but cannot shake off its parasite. The lamprey, a swift swimmer with excellent vision, makes easy prey of fishes, because they are not alarmed by it and tend to ignore it until it strikes. Once it has gained a hold, the lamprey hangs on until it is satiated or the victim dies. A full-grown lamprey may kill a delicate fish such as the trout in as little as four hours. When the victim is more hardy, or the lamprey small, the parasite may cling and feed on the fish for days or even weeks. In the laboratory large lampreys stick to their victims for an average of about 40 hours if the fish survive that long.

The sea lamprey is a newcomer to the upper Great Lakes. It is a marine species which, like certain salmon, hatches in a fresh-water stream, migrates to the ocean to spend its adult life, and then comes back to fresh water to spawn. In some places it has adjusted

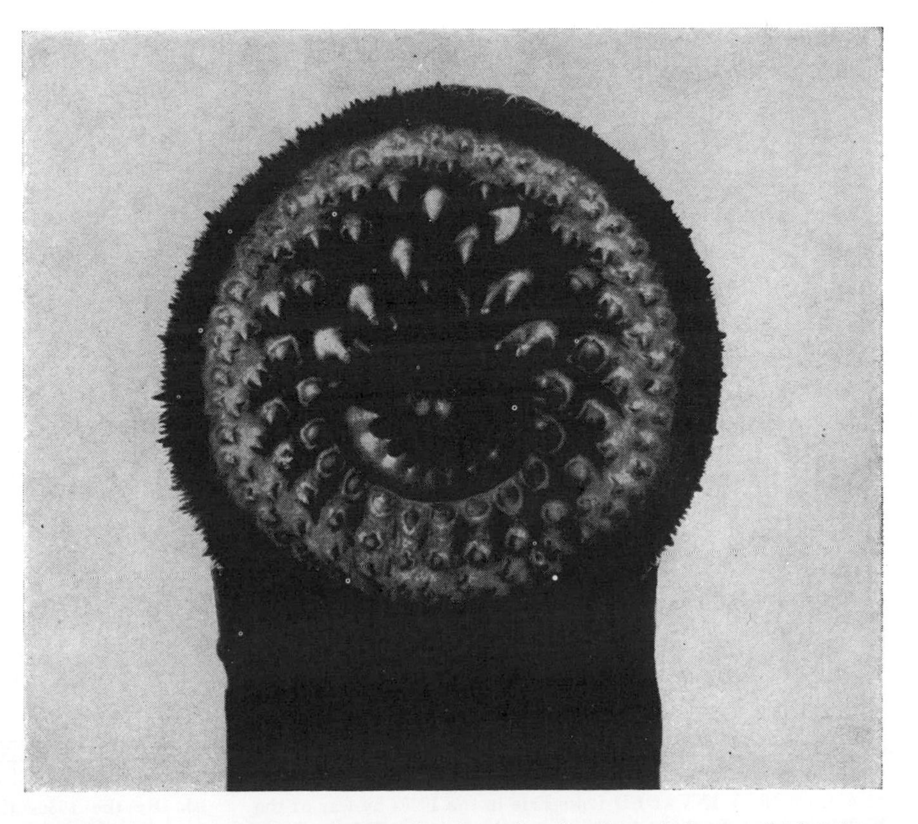

MOUTH of the sea lamprey is photographed through a flat piece of glass to which it is attached. The mouth is lined with horny teeth. In the center of the mouth is the rasped tongue.

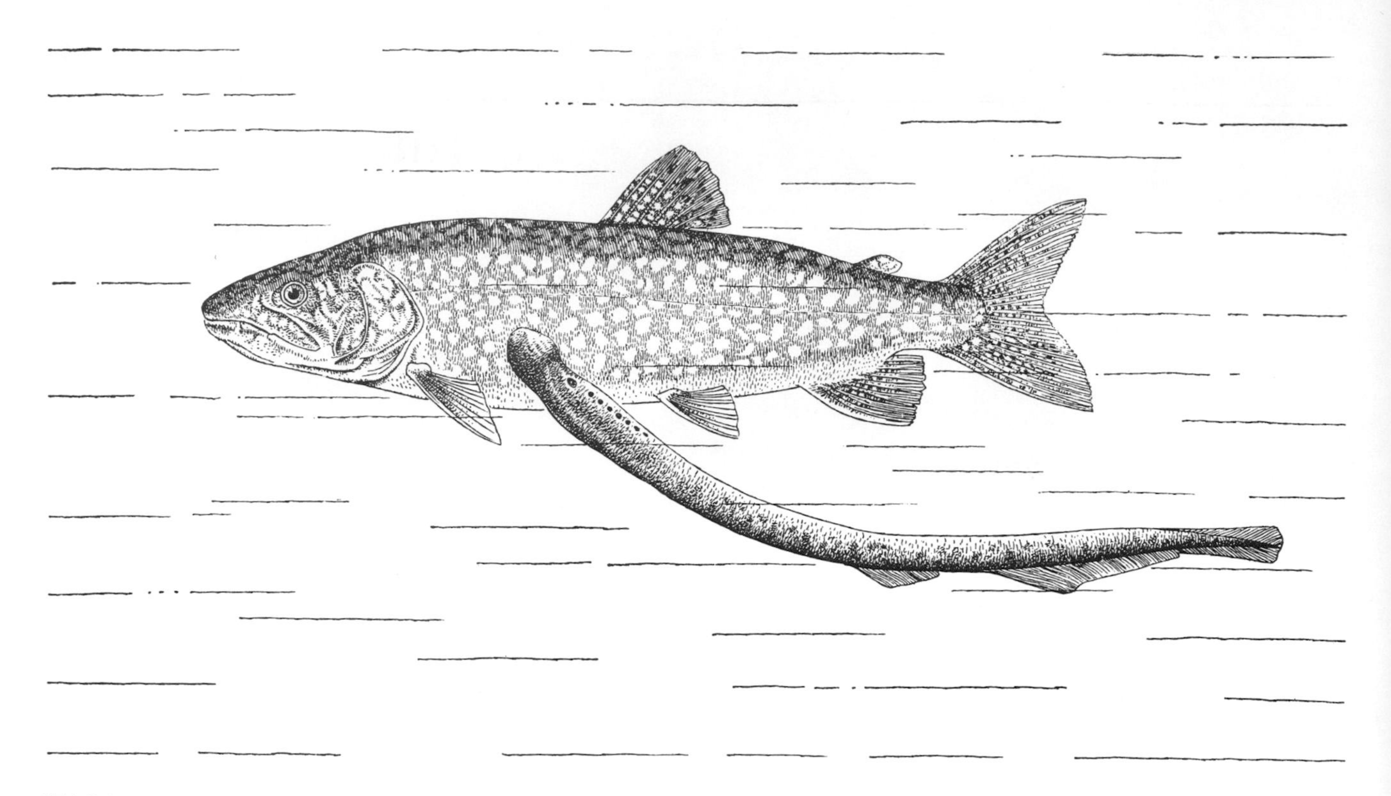

SEA LAMPREY FEEDS on a lake trout. The lamprey may remain attached to a fish for weeks. Some fish, however, die in as little as four hours. The adult lamprey ranges from 12 to 24 inches in length. Its back is dark blue and its belly silvery white.

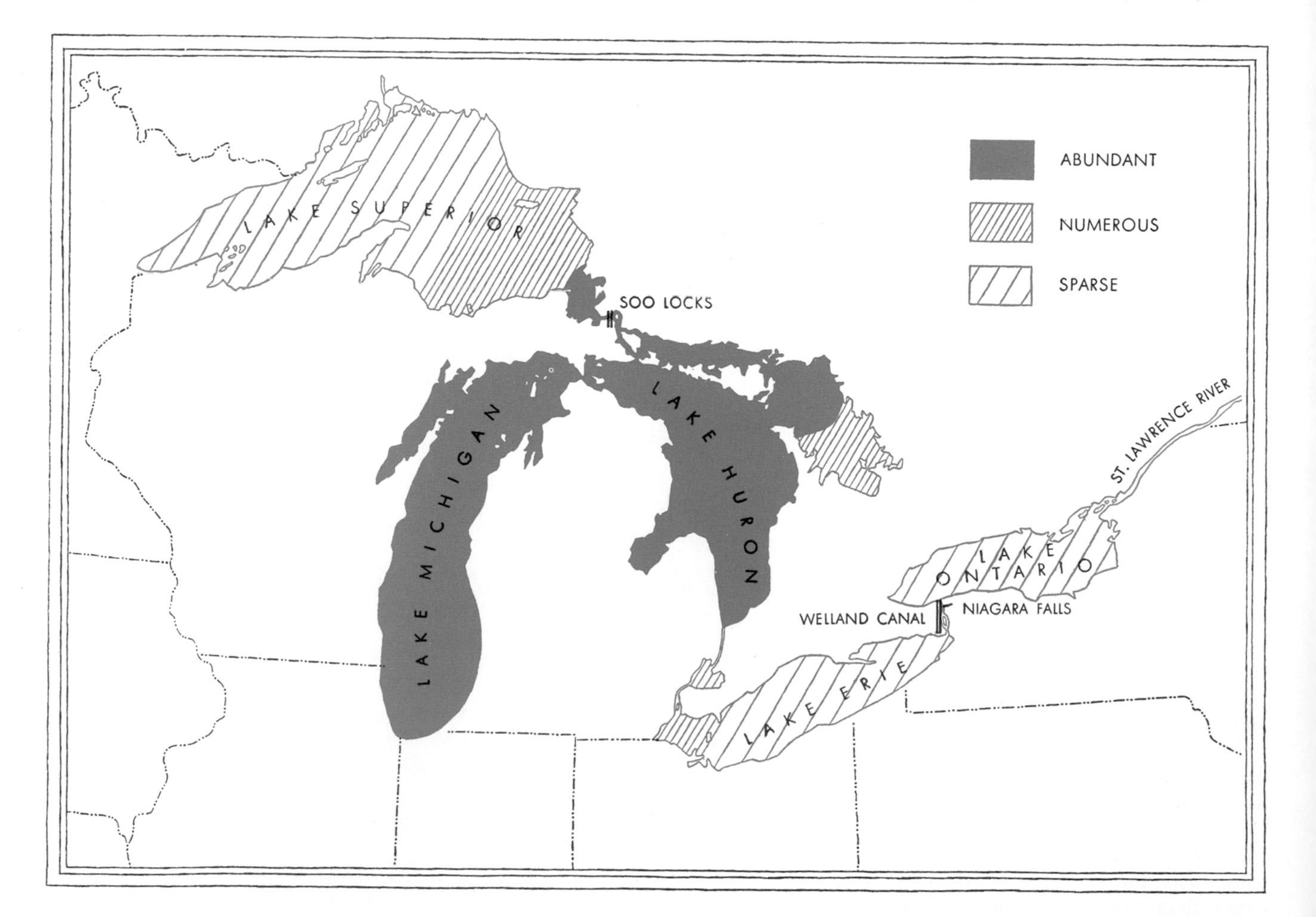

SEA LAMPREY INVADED Lake Erie in the 1920s by way of the Welland Canal. In this lake, however, the lamprey did not flourish. By the 1930s the lamprey had reached Lake Huron and Lake Michigan. It is now becoming established in Lake Superior.

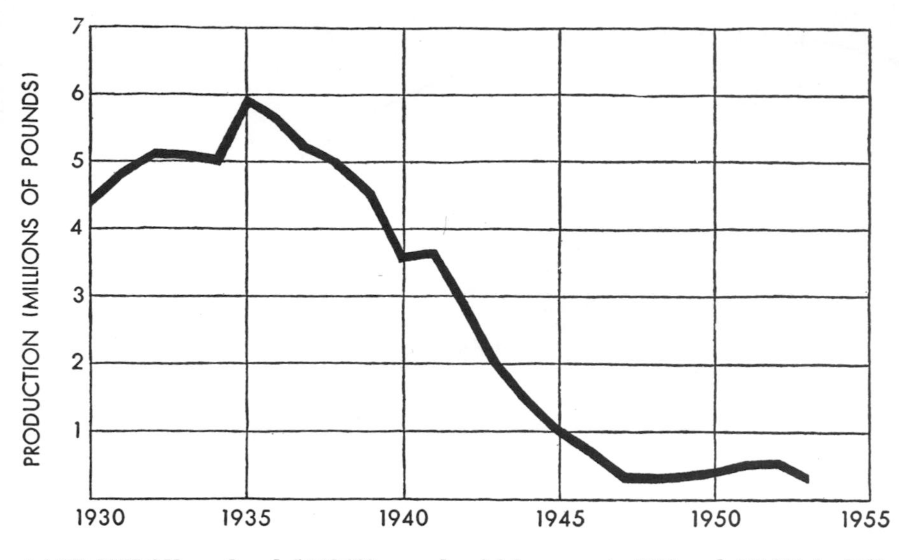

LAKE HURON produced 5,998,000 pounds of lake trout in 1935 and 344,000 in 1953. At 1950 dockside prices the 1935 catch was worth $2,999,000; the 1953 catch, $172,000.

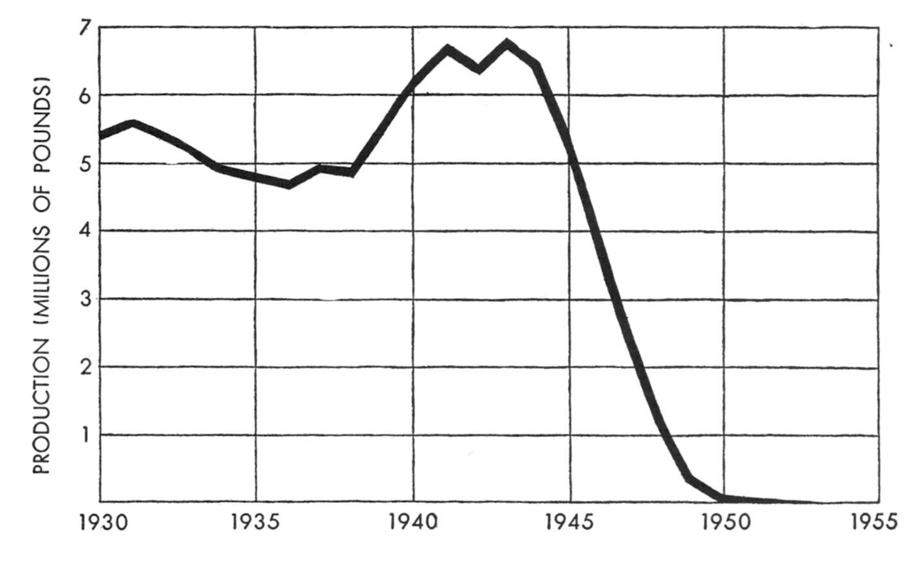

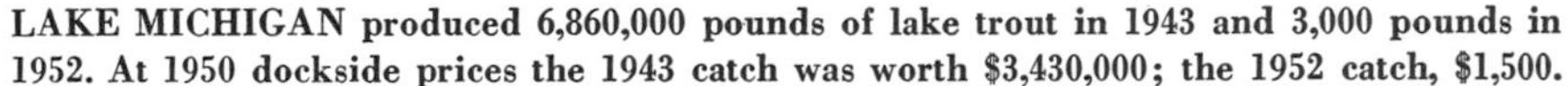

LAKE MICHIGAN produced 6,860,000 pounds of lake trout in 1943 and 3,000 pounds in 1952. At 1950 dockside prices the 1943 catch was worth $3,430,000; the 1952 catch, $1,500.

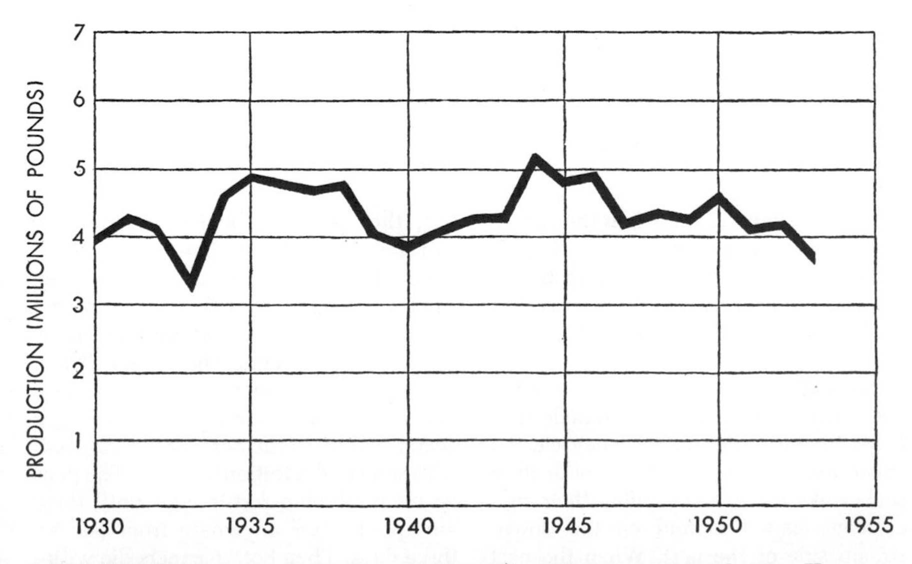

LAKE SUPERIOR does not yet show a catastrophic decline in trout production. However, production fell off from a high of 5,293,000 pounds in 1944 to 3,784,000 pounds in 1953.

itself to spend its entire life cycle in fresh water, passing its adulthood in lakes instead of in the ocean. It is an old inhabitant, for example, of the St. Lawrence River and Lake Ontario. Until 1829 the Niagara Falls blocked it from migrating into the other Great Lakes. Then the building of the Welland Ship Canal provided a passage around the Falls to Lake Erie, but the lamprey seems to have been slow to take advantage of the route. No lamprey was seen in Lake Erie until 1921.

In Lake Erie the lamprey did not flourish; the waters were too warm and the spawning conditions poor. But by the late 1930s the destroyer had penetrated into the next of the Great Lakes, Lake Huron. Fishermen's nets began to bring up trout and other fish with ugly wounds on their bodies. Sometimes the fish had lampreys still clinging to them. Lake Huron was a particularly favorable environment for the lampreys; they multiplied rapidly and made great inroads into the fish of that lake. Meanwhile they also spread through the Straits of Mackinac into Lake Michigan and increased meteorically there. Apparently further migration into Lake Superior was slowed by the locks and dams at the head of Saint Marys River, but the lampreys finally cleared that hurdle and are now well established in Superior.

The kill of trout by the lampreys was prodigious. Experiments in laboratory aquaria have shown that during its period of active feeding a lamprey kills a minimum of 20 pounds of fish. As many as 25,000 spawning lampreys have been trapped in a single northern Lake Huron stream in a year; simple arithmetic shows that this one group must have destroyed 500,000 pounds of fish.

Commercial fishing for trout in Lakes Huron and Michigan came to an end several years ago. As the trout gave out, the lampreys turned more and more to other fish—whitefish, suckers, walleyes and so forth. Today much of the fishing industry in the Great Lakes is in serious economic difficulty. If the Lake Superior trout go, the industry there probably will collapse. To try to save the trout and other fish, the U. S. Fish and Wildlife Service, the Great Lakes states and Canada have been carrying on research and testing measures against the lampreys. A treaty for joint action by the U. S. and Canada was signed on September 10, 1954, and awaits ratification.

As in any pest-control problem, we must find the vulnerable points in the life cycle of the animal to attack it effectively. The life cycle of the sea lam-

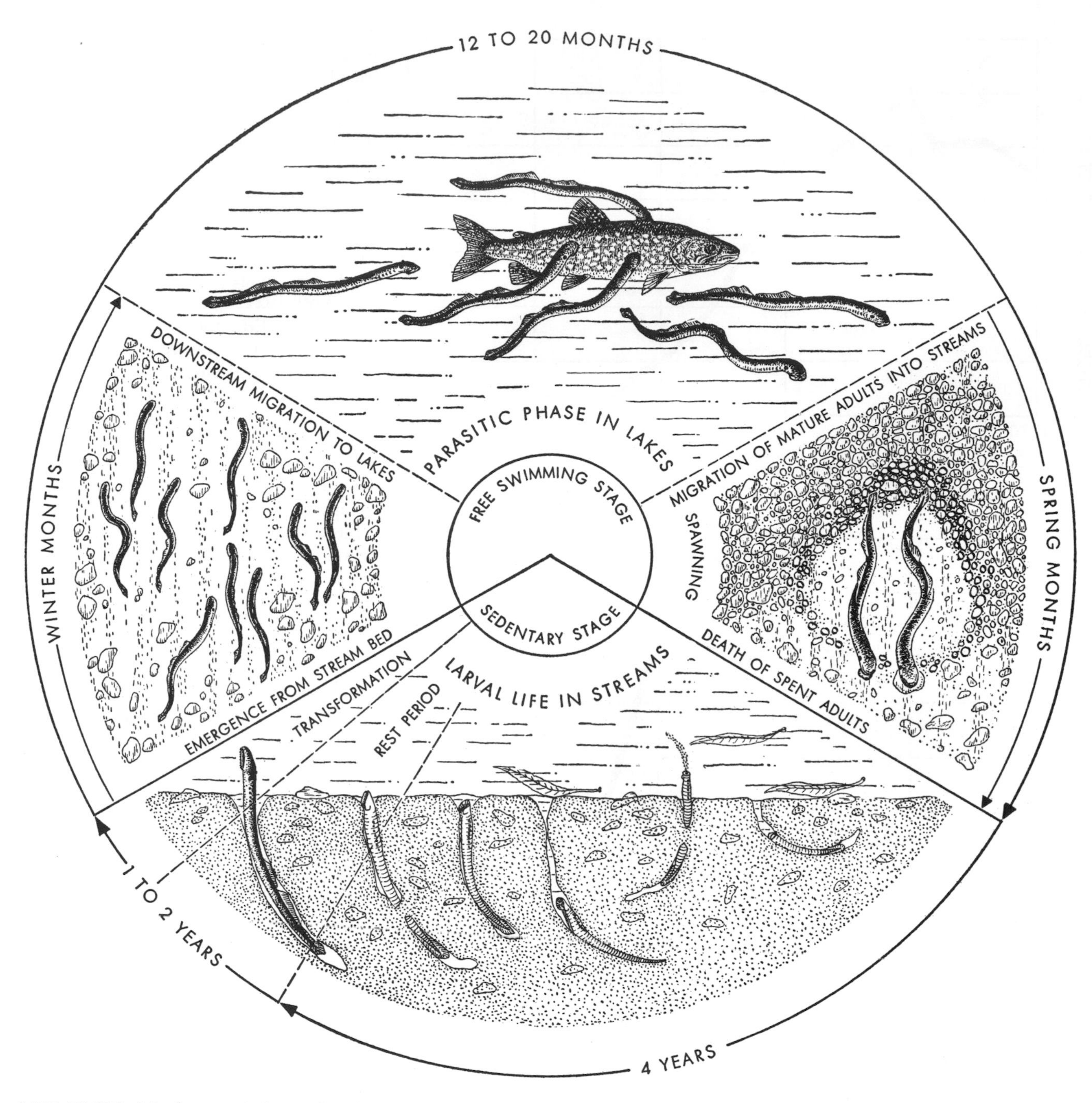

LIFE CYCLE of the lamprey is depicted in this chart. The lamprey lives 6½ to 7½ years. The size of the segments in the chart do not correspond to the length of time the lamprey spends in each stage. The lamprey spends most of its life as a larva.

prey begins in the shallow riffles of a stream. Here it passes the major portion of its life as a blind, harmless larva. Of its approximately seven-year life span, a lamprey spends only the last 18 months in the lakes as a parasite. At the end of that time it goes back to the stream to spawn and die. Let us follow the cycle in some detail from the spawning stage.

The old sea lampreys begin their migration up the tributaries of the Great Lakes to spawn in early spring, the time of migration in each lake depending on the water temperature. They congregate in bays and in the estuaries of rivers during late winter, and when the stream temperature rises above 40 degrees Fahrenheit they start moving upstream. They seek out streams with a gravel or sand bottom and a moderately strong current.

Normally the male starts building the nest; then he is joined by a female who helps in the construction. They clear a small area, picking up stones with their sucker-like mouths and piling them in a crescent-shaped mound on the downstream side of the nest. When the nest is finished and the temperature of the stream is warm enough (over 50 degrees F.), the spawning begins. The female extrudes a small number of eggs; the male at once fertilizes them, and the eggs are carried by the current to the gravel rim of the nest, where they lodge in the spaces among the stones. Then the female lays another batch of eggs and the process is repeated. The eggs accumulating in the nest rim are covered with sand and additional stones. The pair go on producing fertile eggs until they are spent—after anywhere from one to three days. Then both partners die within a matter of hours.

The female has deposited from 24,000

to 107,000 eggs; the average is about 61,500. Fortunately less than 1 per cent of these eggs will hatch out into larvae.

Hatching takes 10 to 12 days. The hatched larvae remain buried in the sand and gravel until about the 20th day. Then the tiny creatures, only about a quarter of an inch long and hardly thicker than a fine needle, emerge from the nest and drift downstream until they reach quiet waters. Here they dive for the bottom and each digs an individual burrow. This will be its home for about five years, unless erosion washes it away. Throughout its larval life the young lamprey is blind and harmless. It sucks food, mainly microscopic organisms, from the water passing the mouth of its burrow. A filtering apparatus in its throat keeps out debris and passes food organisms to its digestive tract.

After four years the larva undergoes a striking metamorphosis. It develops large, prominent eyes, a round mouth lined with horny teeth, a filelike tongue and enlarged fins. Its slim body becomes dark blue above and silvery white beneath. The new young lamprey, some four to seven inches long, may emerge from its mud flat when late fall rains raise the river level, but usually it waits until the spring breakup and flood. It drifts downstream to the big lake and begins its parasitic existence, living on the blood of fish. Feeding upon a succession of hosts, it grows very rapidly, attaining a final length of 12 to 24 inches.

The adult lamprey apparently has a great range of movement. Marked lampreys, released in the autumn at the northern tip of Lake Huron, were recovered throughout the length and breadth of the lake by the following spring. Several individuals had traveled nearly the entire length of the lake, a distance of over 200 miles. But details of the lampreys' movements in the lakes are scanty. There is some evidence that they migrate first to deep water, where they attack lake trout, chubs and other deep-water species. As the lampreys grow larger, they move shoreward and in the fall are found in relatively shallow water. It is at this time that attacks on whitefish, suckers and other shallow-water fish reach their maximum.

Toward the end of winter sexually maturing lampreys begin to assemble off the mouths of streams. During this waiting period tremendous internal changes occur. The sex glands expand enormously, while the digestive tract shrinks and the lamprey becomes incapable of feeding. From now on it will live only on its own tissues. Even its muscles, skin and eyes deteriorate. If the lamprey is delayed in reaching its spawning grounds, death may overtake it before it can spawn.

Plainly the most vulnerable times in the lamprey's life are its periods in the stream—as a larva or young migrant and later when it goes back to spawn. The vulnerability is enhanced by the fact that only about 200 streams tributary to Lakes Huron, Michigan and Superior are suitable for spawning.

One attack on the lampreys has been to build mechanical weirs and traps and barrier dams to block their spawning runs. Although effective, these devices have numerous drawbacks. They are generally expensive to install and maintain; the weirs and traps must be cleared regularly. They may break down under flood conditions. The reproductive potential of the sea lamprey is so great that even a few escaping individuals can "seed" a stream sufficiently to maintain the population.

When the shortcomings of the me-

MECHANICAL WEIR crosses Carp Creek, a Michigan tributary of Lake Huron. It is designed to trap spawning lampreys as they go upstream. This kind of weir has been supplanted by electromechanical weirs such as the one depicted on the following page.

chanical barriers became evident, we turned to electricity. Linear arrays of electrodes were set up in the water to create electrical fields just strong enough to stop the movement of lampreys upstream to their spawning grounds. Using regular 110-volt alternating current, the electrical devices are more economical to construct and operate than purely mechanical structures. Unfortunately, however, the prevention of spawning, even if completely successful, will not show results for at least seven years. It will not kill off the generations of larvae already in the streams.

If we could destroy the larvae or the young downstream migrants, we might reduce the population substantially in less than two years. But so far no practicable means of achieving either objective has been found. Traps for capturing the downstream migrants do not work during the flood stages, when most migration occurs. A dam with an inclined-screen trap is effective, but it is expensive to build and requires continual attention. Furthermore, in many streams the topography precludes the use of this type of structure. Attempts were made to electrocute the young migrants with a simple system of electrodes in the stream, but these experiments were discontinued when it was discovered that young lampreys are extraordinarily resistant to electrical currents and the power required would be prohibitively costly. As for destroying the larvae, the problems become even more difficult. It was found that young American eels, a notably voracious species, would destroy lamprey larvae, but they kill desirable fish as well. So do most poisons. Recent investigations have, however, encouraged the hope that we may find chemicals which are toxic to lampreys and relatively harmless to other fish. Thousands of chemicals are being tested in an effort to discover a specific larvicide.

Notwithstanding the defects of the available control methods, the urgency of the Great Lakes fishing industry's plight has prompted us to apply some of them while better methods are being sought. Electromechanical barriers have been installed in 44 tributaries of Lake Superior on the U. S. side and 24 more are under construction in Canada. Practically all the spawning streams in the U. S. part of the Lake Superior basin have been blocked. In Lakes Michigan and Huron of course it is too late to save the trout. But if the trout can be protected in Lake Superior, they will provide a supply of eggs for restocking Michigan and Huron when the sea lamprey has been brought under control.

Fortunately no other fish has usurped the environmental niche of the trout in the Great Lakes. The small fishes on which trout feed have increased to the point of overcrowding, and there will be an abundance of food for trout when they can return. Another encouraging factor is that lampreys apparently do not single out trout if there are larger fish around. Thus when effective control of lamprey spawning comes into sight, we can begin to plant lake trout with the hope that the lamprey can be exterminated before the young trout grow large enough to be attacked. This may reduce the time required to develop a breeding population of trout from 11 years to seven.

Complete eradication of sea lampreys from the Great Lakes above Niagara Falls is our objective. It may prove to be a long operation, as difficult as the campaign sometimes necessary to stamp out an agricultural or forest pest, but we are confident that we shall ultimately succeed.

ELECTROMECHANICAL WEIR crosses the Ocqueoc River, also a Michigan tributary of Lake Huron. In the spring of 1954 10,183 lampreys on their way upstream to spawn were killed or captured at this site. No lampreys were found on the upstream side.

THE AGING GREAT LAKES

CHARLES F. POWERS AND ANDREW ROBERTSON
November 1966

The five Great Lakes in the heartland of North America constitute the greatest reservoir of fresh water on the surface of the earth. Lake Superior, with an area of 31,820 square miles (nearly half the area of New England), is the world's largest freshwater lake; Lake Huron ranks fourth in the world, Lake Michigan fifth, Lake Erie 11th and Lake Ontario 13th. Together the five lakes cover 95,200 square miles and contain 5,457 cubic miles of water. They provide a continuous waterway into the heart of the continent that reaches nearly 2,000 miles from the mouth of the St. Lawrence River to Duluth at the western tip of Lake Superior.

The Great Lakes are obviously an inestimable natural resource for the development of the U.S. and Canada. They supply vast amounts of water for various needs: drinking, industrial uses and so forth. They serve as a transportation system linking many large inland cities to one another and to the sea. Their falls and rapids generate huge supplies of hydroelectric power. Their fish life is a large potential source of food. And finally, they serve as an immense playground for human relaxation, through boating, swimming and fishing.

The settlements and industries that have grown up around this attractive resource are already very substantial. Although less than 3.5 percent of the total U.S. land area lies in the Great Lakes basin, it is the home of more than 13.5 percent of the nation's population (and about a third of Canada's population). In the southern part of the basin, from Milwaukee on the west to Quebec on the east, is a string of cities that is approaching the nature and dimensions of a megalopolis. Many economists believe the Great Lakes region is likely to become the fastest-growing area in the U.S. Their forecast is based mainly on the fact that whereas most other regions of the country are experiencing increasing shortages of water, the Great Lakes area enjoys a seemingly inexhaustible supply.

Unfortunately the forecast is now troubled by a large question mark. The viability of this great water resource is by no means assured. Even under natural conditions the life of an inland lake is limited. It is subject to aging processes that in the course of time foul its waters and eventually exhaust them. The Great Lakes are comparatively young, and their natural aging would not be a cause for present concern, since the natural processes proceed at the slow pace of the geological time scale. The aging of these lakes is now being accelerated tremendously, however, by man's activities. Basically the destructive agent is pollution. The ill effect of pollution is not limited to the circumstance that it renders the waters unclean. Pollution also hastens the degeneration and eventual extinction of the lakes as bodies of water.

These conclusions are based on recent extensive studies of the Great Lakes by a number of universities and governmental agencies in the U.S. and Canada. Employing various research techniques, including those of oceanography, the studies have produced new basic knowledge about the natural history and ecology of the Great Lakes and recent major changes that have occurred in them.

The natural aging of a lake results from a process called "eutrophication," which means biological enrichment of its water. A newly formed lake begins as a body of cold, clear, nearly sterile water. Gradually streams from its drainage basin bring in nutrient substances, such as phosphorus and nitrogen, and the lake water's increasing fertility gives rise to an accumulating growth of aquatic organisms, both plant and animal. As the living matter increases and organic deposits pile up on the lake bottom, the lake becomes smaller and shallower, its waters become warmer, plants take root in the bottom and gradually take over more and more of the space, and their remains accelerate the filling of the basin. Eventually the lake becomes a marsh, is overrun by vegetation from the surrounding area and thus disappears.

As a lake ages, its animal and plant life changes. Its fish life shifts from forms that prefer cold water to those that do better in a warmer, shallower environment; for example, trout and whitefish give way to bass, sunfish and perch. These in turn are succeeded by frogs, mud minnows and other animals that thrive in a marshy environment.

The natural processes are so slow that

PROCESS OF EXTINCTION that is the destiny of all lakes is seen in action in the aerial photograph on the following page. Cattaraugus Creek, a stream that forms the boundary between Erie and Chautauqua counties in New York, enters Lake Erie at this point southwest of Buffalo. (North is to the right.) The stream is not polluted but it carries silt and nutrients. The silt acts to fill the lake; the nutrients feed various forms of plant life that encroach on the shallows and add to the accumulation of bottom deposits. The aging process, which eventually converts every lake into dry land, is greatly accelerated when, as in the case of America's Great Lakes, human and industrial wastes are added to the normal runoff load.

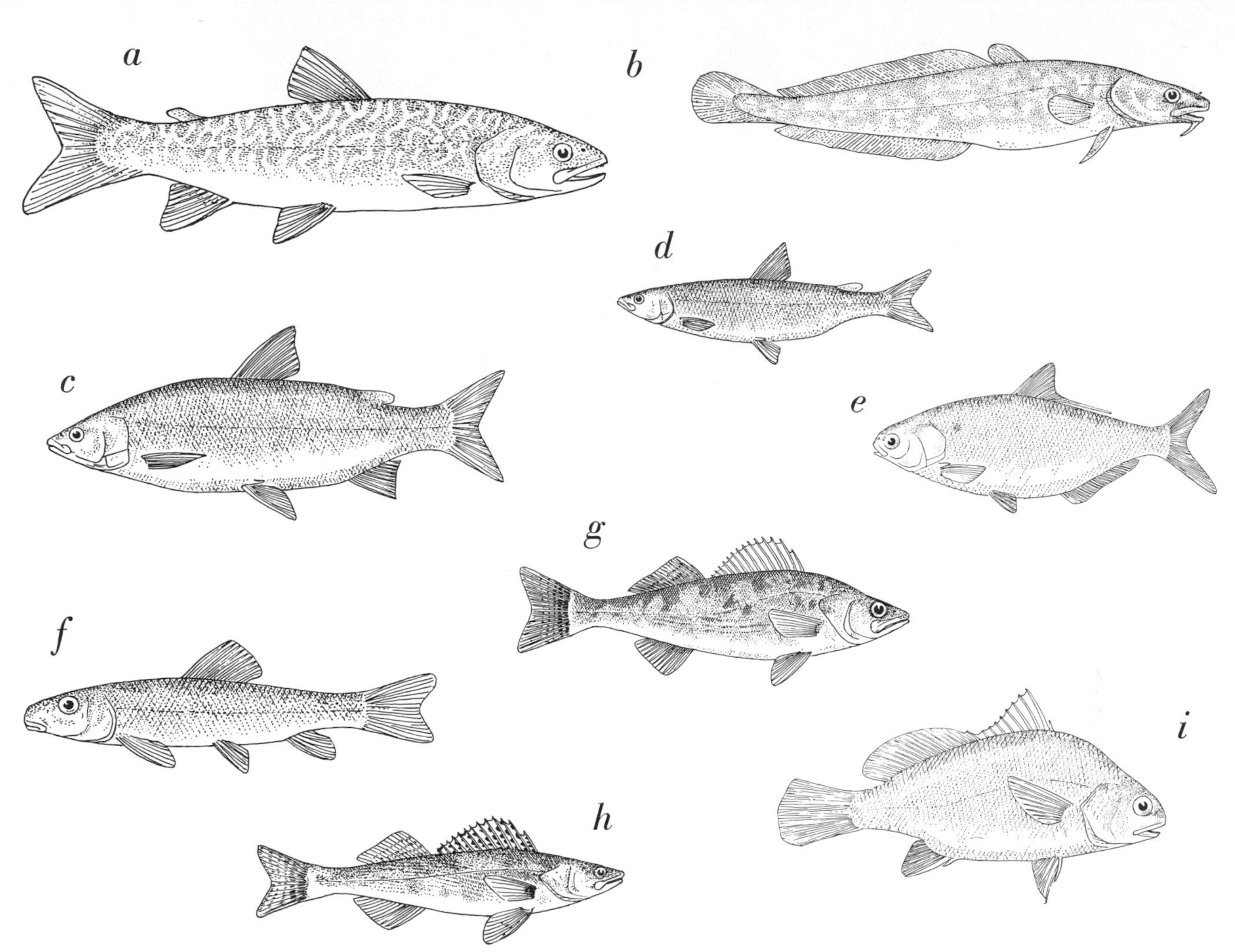

OLDER FISH POPULATION of the Great Lakes includes seven fishes that have nearly disappeared in the past two decades. Among them are the lakes' two largest species, the lake trout (*a*) and the burbot (*b*), and four smaller but economically important fishes, the whitefish (*c*), its close relative the chub, or lake herring (*d*), the walleye (*g*) and its close relative the blue pike (*not illustrated*). All six, as well as the sucker (*f*), have been victims of the parasitic sea lamprey, a fish that was confined to Lake Ontario until completion of the Welland Canal in 1932. Other indigenous fishes illustrated are the gizzard shad (*e*), the sauger (*h*) and the sheepshead (*i*).

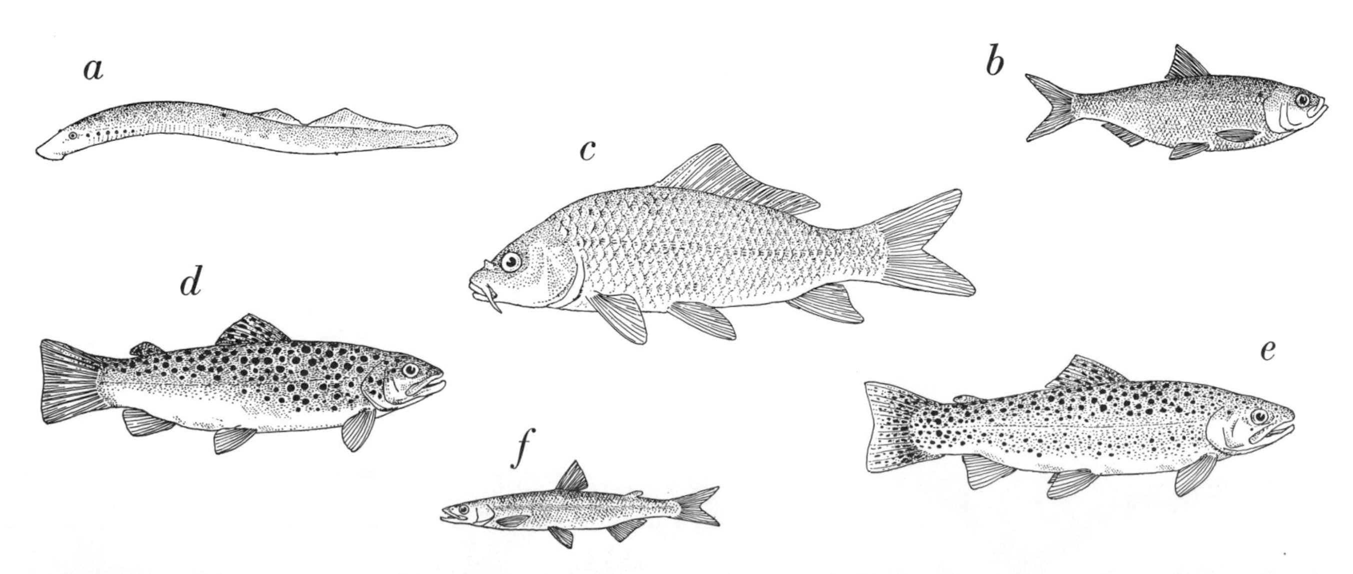

INTRUSIVE FISH POPULATION, responsible for disrupting the previous ecological balance of the four upper Great Lakes, are of two classes: those introduced by man and those that have entered on their own. The lamprey (*a*) is one of the voluntary intruders, as is the alewife (*b*), which also entered the upper lakes from Lake Ontario via the Welland Canal. Not a predator of adult fishes, the alewife nonetheless threatens the indigenous fish population. It feeds on these fishes' eggs and also consumes much of the other available food. Species introduced by man are the European carp (*c*) and brown trout (*d*), the rainbow trout (*e*) and the smelt (*f*).

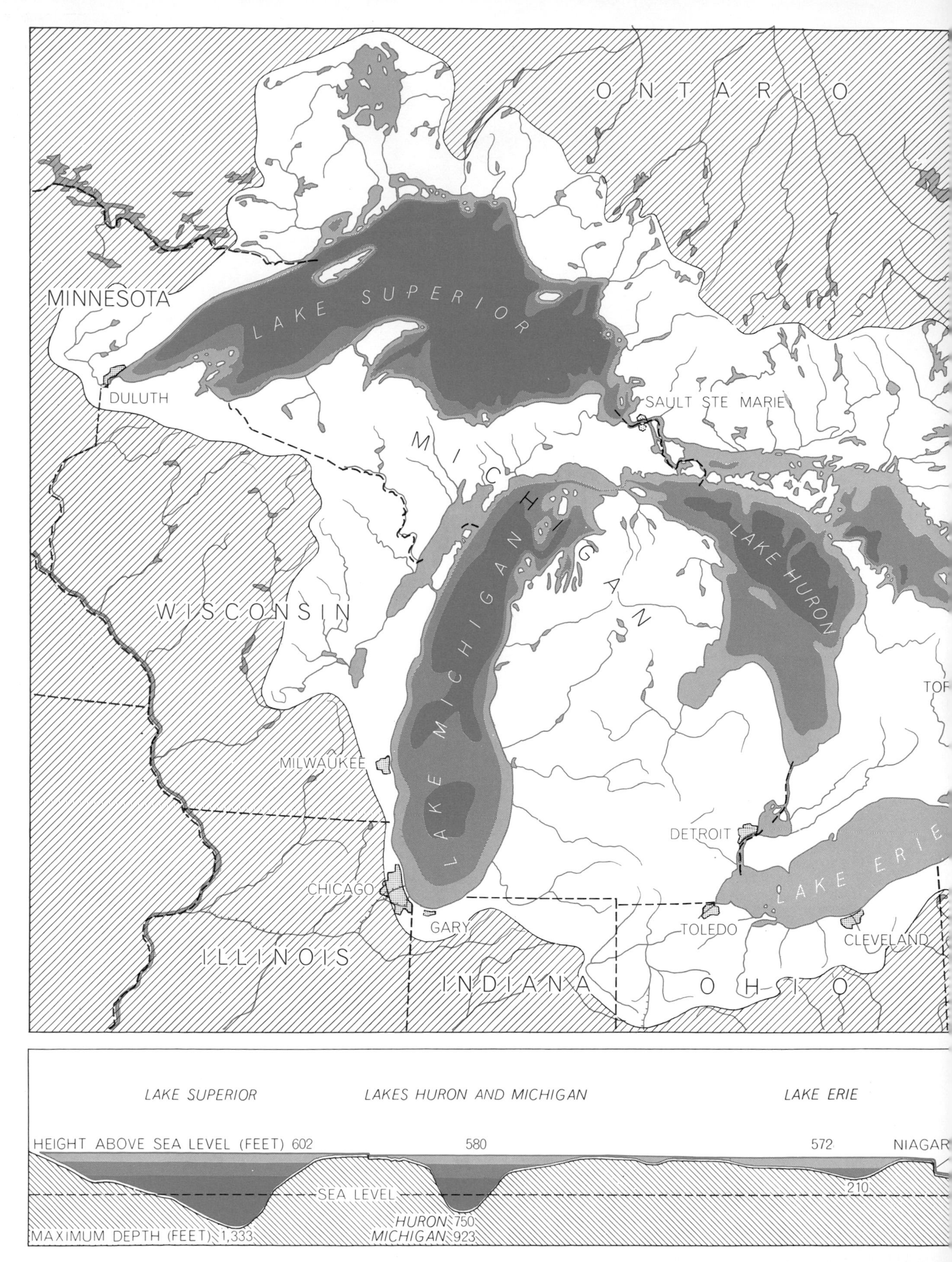

AMERICAN GREAT LAKES comprise the world's largest fresh-water reservoir. Their drainage area (*white*) is not big enough to counteract the loss of lake water through discharge and evaporation but the lakes' level is kept stable by the inflow of groundwater and capture of the rain and snow that fall on 95,200 square miles of water surface. Superior (*left*) is the deepest of the five and Erie the shallowest (*vertical scale of profile is exaggerated*). Erosion will have destroyed the escarpment forming Niagara Falls 25,000

years from now; Lake Erie will then empty, leaving little more than a marshy stream to channel water from the upper lakes into Lake Ontario and on to the Atlantic Ocean.

the lifetime of a lake may span geological eras. Its rate of aging will depend on physical and geographic factors such as the initial size of the lake, the mineral content of the basin and the climate of the region. The activities of man can greatly accelerate this process. Over the past 50 years it has become clear that the large-scale human use of certain lakes has speeded up their aging by a considerable factor. A particularly dramatic example is Lake Zurich in Switzerland: the lower basin of that lake, which receives large amounts of human pollution, has gone from youth to old age in less than a century. In the U.S. similarly rapid aging has been noted in Lake Washington at Seattle and the Yahara lake chain in Wisconsin.

When the European explorers of North America first saw the Great Lakes, the lakes were in a quite youthful stage: cold, clear, deep and extremely pure. In the geological sense they are indeed young—born of the most recent ice age. Before the Pleistocene their present sites were only river valleys. The advancing glaciers deepened and enlarged these valleys; after the glaciers began to retreat some 20,000 years ago the scoured-out basins filled with the melting water. The succeeding advances and retreats of the ice further deepened and reshaped the lakes until the last melting of the ice sheet left them in their present form.

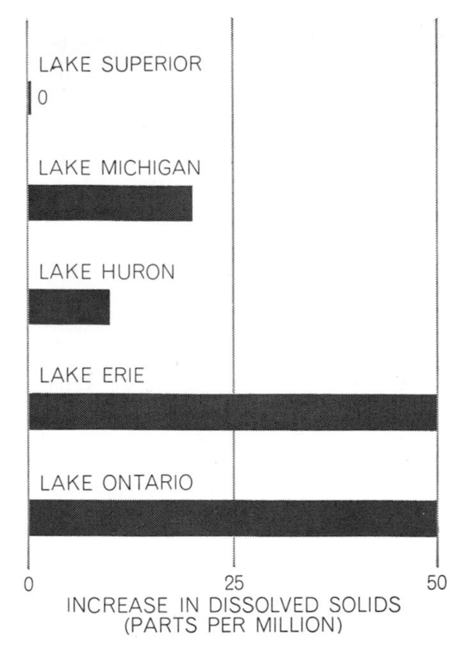

INDEX OF POLLUTION is provided by the extent to which the lakes' content of dissolved solids has increased in the past 50 years. Lake Superior shows no increase, and the modest increase for Lake Huron is attributable to its receipt of Lake Michigan water, which is more heavily polluted. Lake Erie's major cities and its small volume of water account for its rising solids content. Lake Ontario's pollution is a combination of what is received from Lake Erie and what the cities along its shores contribute.

The land area drained by the Great Lakes (194,039 square miles) is relatively small: it is only about twice the area of the lakes themselves, whereas the ratio for most other large lakes is at least six to one. The drainage alone is not sufficient to replace the water lost from the Great Lakes by evaporation and discharge into the ocean by way of the St. Lawrence. Thanks to their immense surface area, however, their capture of rainfall and snowfall, supplemented by inflow of groundwater, maintains the lakes at a fairly stable level. The level varies somewhat, of course, with the seasons (it is a foot to a foot and a half higher in summer than in winter) and with longer-range fluctuations in rainfall. Prolonged spells of abnormal precipitation or drought have raised or lowered the level by as much as 10 feet, thereby causing serious flooding along the lake shores or leaving boat moorings high and dry.

The five lakes differ considerably from one another, not only in surface area but also in the depth and quality of their waters. Lake Superior averages 487 feet in depth, whereas shallow Lake Erie averages only 58 feet. There is also a large difference in the lakes' altitude: Lake Superior, at the western end, stands 356 feet higher above sea level than Lake Ontario at the eastern extreme. Most of the drop in elevation occurs in the Niagara River between Lake Erie and Lake Ontario. At Niagara Falls, where the river plunges over the edge of an escarpment, the drop is 167 feet. This escarpment, forming a dam across the eastern end of Lake Erie, is continuously being eroded away, and it is estimated that in 25,000 years it will be so worn down that Lake Erie will be drained and become little more than a marshy stream.

The lakes are all linked together by a system of natural rivers and straits. To this system man has added navigable canals that today make it possible for large ocean-going ships to travel from the Atlantic to the western end of Lake Superior. Hundreds of millions of tons of goods travel up and down the Great Lakes each year, and on the U.S. side alone there are more than 60 commercial ports. The Sault Ste Marie Canal (the "Soo"), which connects Lake Superior and Lake Huron, carries a greater annual tonnage of shipping than the Panama Canal. Other major man-made links in the system are the Welland Canal, which bypasses the Niagara River's falls and rapids to connect Lake Erie and Lake Ontario, and the recently completed St. Lawrence Seaway, which makes the St. Lawrence River fully navigable from Lake Ontario to the Atlantic Ocean.

One of the first signs that man's activities might have catastrophic effects on the natural resources of the Great Lakes came as an inadvertent result of the building of the Welland Canal. The new channel allowed the sea lamprey of the Atlantic, which had previously been unable to penetrate any farther than Lake Ontario, to make its way into the other lakes. The lamprey is a parasite that preys on other fishes, rasping a hole in their skin and sucking out their blood and other body fluids. It usually attacks the largest fish available. By the 1950's it had killed off nearly all the lake trout and burbot (a relative of the cod that is also called the eelpout) in Lake Huron, Lake Michigan and Lake Superior. The lamprey then turned its attention to smaller species such as the whitefish, the chub (a smaller relative of the whitefish), the blue pike, the walleye and the sucker. Its depredations not only destroyed a large part of the fishing industry of the Great Lakes but also brought radical changes in the ecology of these lakes.

Since the late 1950's U.S. and Canadian agencies have been carrying on a determined campaign to eradicate the lamprey, using a specific larvicide to kill immature lampreys in streams where the species spawns [see the article "The Sea Lamprey," by Vernon C. Applegate and James W. Moffett, beginning on page 391]. The program has succeeded in cutting back greatly the lamprey population in Lake Superior; it is now being applied to the streams feeding into Lake Michigan and will be extended next to Lake Huron. Efforts have already been started to reestablish a growing lake-trout population in Lake Superior.

Meanwhile a second invader that also penetrated the lakes through the Welland Canal has become prominent. This fish is the alewife, a small member of the herring family. The alewife, which ranges up to about nine inches in length, does not attack adult fishes, but it feeds on their eggs and competes with their young for food. In the past decade it has multiplied so rapidly that it is now the dominant fish species in Lake Huron and Lake Michigan and seems to be on the way to taking over Lake Superior.

Recently attempts have been made to convert the alewives from a liability to an asset. The Pacific coho, or silver salmon, has been introduced into Lake Superior and Lake Michigan on an experimental basis. This fish should thrive feeding on the alewife and yet be protected from its depredations, because the eggs and young of the coho are found in tributary streams the alewives do not frequent. Other fishes such as the Atlantic striped bass are being con-

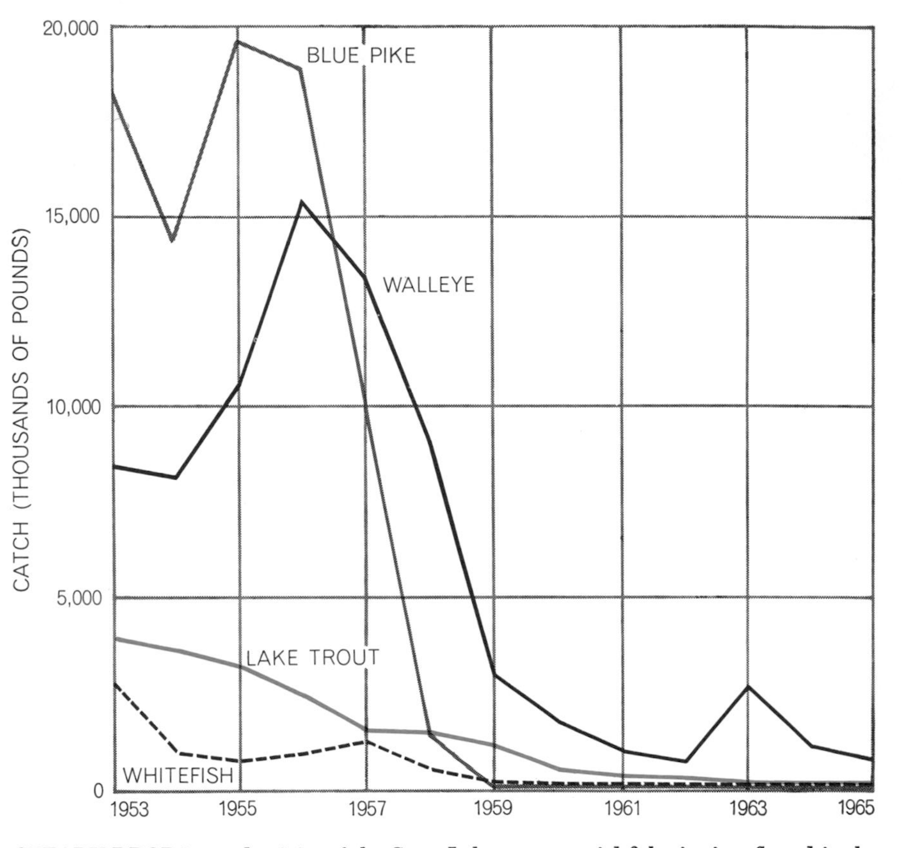

STEADY DROP in productivity of the Great Lakes commercial fisheries is reflected in the numbers of three species taken in Lake Erie (*black*) and one taken in Lake Superior (*color*) between 1953 and 1965. The lake-trout catch in Lake Michigan once rivaled Lake Superior's; from 1941 through 1946 it averaged more than six million pounds. The decline then began to be significant. Within a decade the Lake Michigan fishery ceased to yield lake trout.

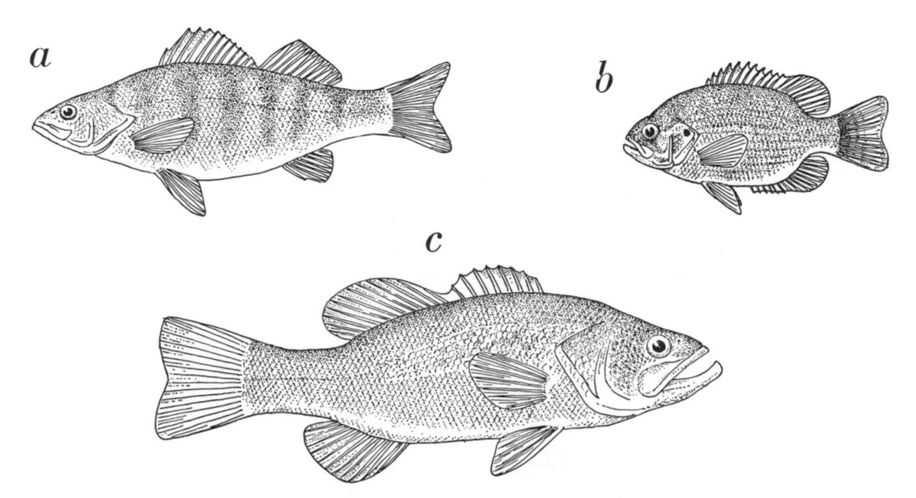

SUCCESSOR POPULATION of Great Lakes fishes when the lakes reach old age will probably include species that inhabit the lakes' shallow waters today. Among these are the yellow perch (*a*), the kind of sunfish known as rock bass (*b*) and the largemouth bass (*c*).

sidered for introduction to supplement the coho.

The introduction of a new fish into a lake is always an unpredictable matter. It may, as in the accidental admission of the lamprey and the alewife, disrupt the ecological balance with disastrous results. Even when the introduction is made intentionally with a favorable prognosis, it frequently does not work out according to expectations. The carp, prized as a food fish in many countries of Europe, was stocked in the Great Lakes many years ago and has established itself in all the lakes except Lake Superior. Commercial and sport fishermen in these lakes, however, have come to regard the carp as a nuisance. North Americans generally consider it inedible, chiefly because they have not learned how to prepare and cook it prop-

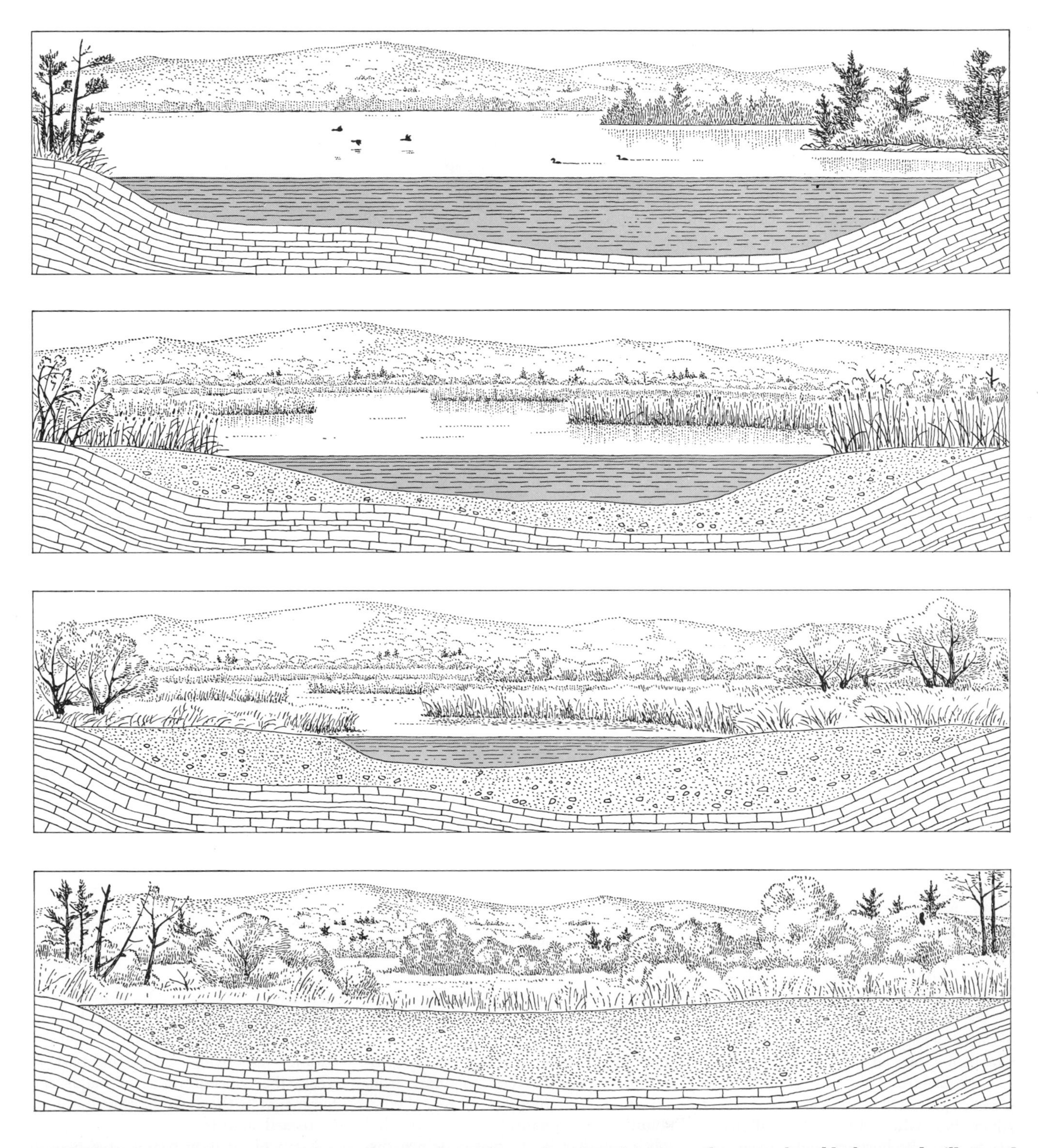

OBLITERATION of a lake is a process that starts at the edge of the water (*top*): a few bog-adapted conifers rise in a forest of hardwoods. Next the debris of shallow-water plants turns the lake margin into marsh that is gradually invaded by mosses and bog plants, bog-adapted bushes and trees such as blueberry and willow, and additional conifers. Eventually the lake, however deep, is entirely filled with silt from its tributaries and with plant debris. In the final stage (*bottom*) the last central bog soon grows up into forest.

erly. On the other hand, the smelt, introduced into the upper lakes from Lake Ontario early in this century, has become prized by fishermen and is taken in large numbers in the Great Lakes today. What effect it will eventually have on the ecology of the lakes remains to be seen.

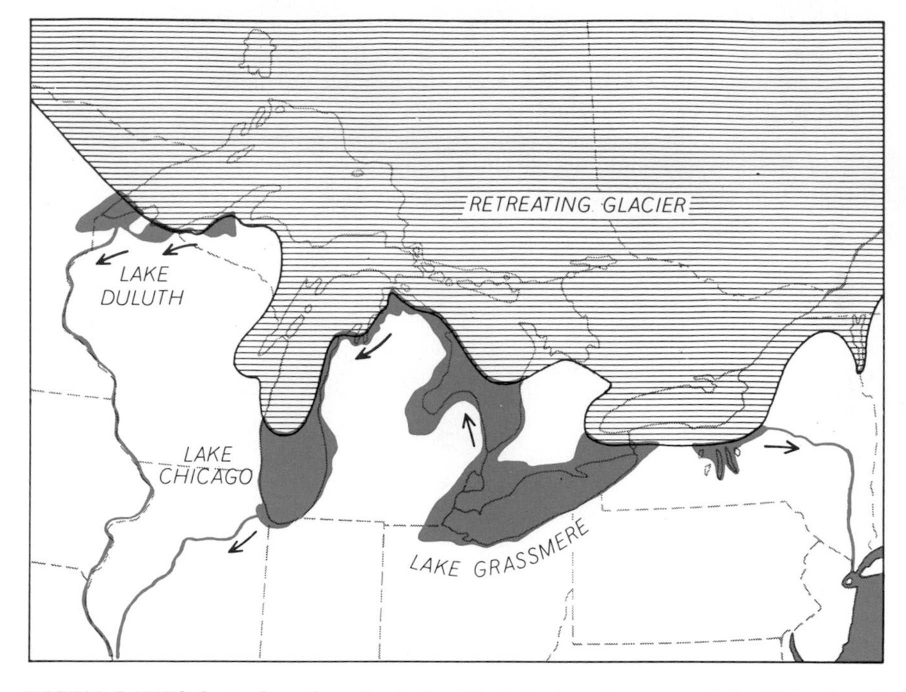

YOUNG LAKES drained southward, via the Mississippi to the west and the Mohawk and Hudson to the east, during the last glacial retreat of the Pleistocene some 10,000 years ago.

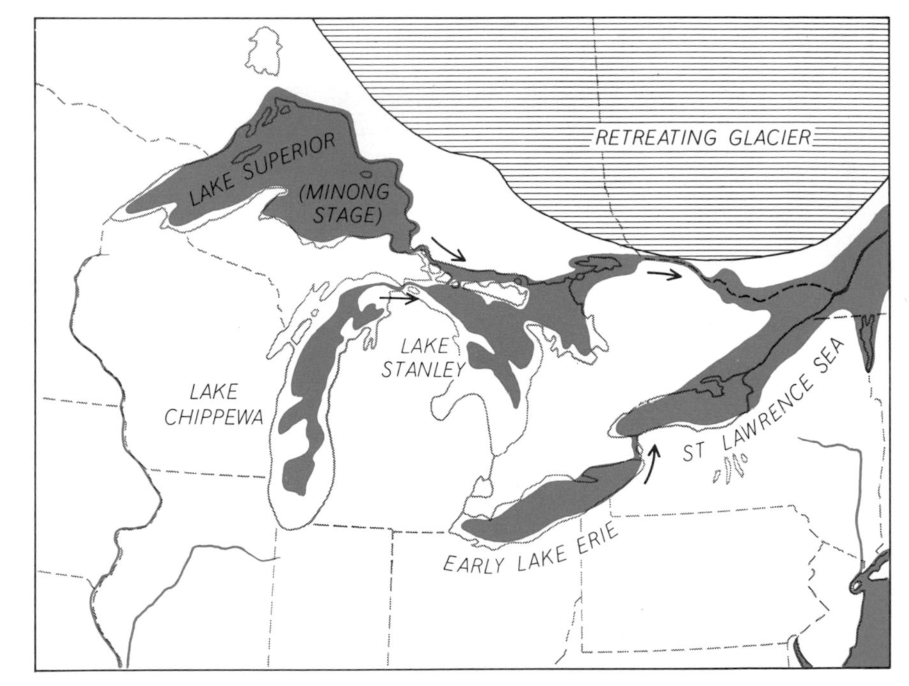

MATURING LAKES altered the southward drainage pattern and began to channel their waters eastward into the prehistoric St. Lawrence Sea as the glacial retreat continued.

The Great Lakes are so young that, biologically speaking, they must be considered in a formative stage. So far only a few species of fishes have been able to invade them and adapt to their specialized environment, particularly in their deep waters. As time goes on, more species will arrive in the lakes and evolve into forms specially adapted to the environmental conditions. Lake Baikal in Siberia, a very large and ancient body of fresh water, offers a good illustration of such a history: it has developed a well-diversified and distinctive population of aquatic animals including a freshwater seal. As diversity in the Great Lakes increases, it will become less and less likely that the arrival or disappearance of one or two species (such as the lake trout and burbot) will result in any profound alteration of the ecological balance.

Pollution, however, is a decidedly different factor. Its effects are always drastic—and generally for the worse. This is clearly evident in Lake Erie, the most polluted of the Great Lakes. The catch of blue pike from this lake dropped from 18,857,000 pounds in 1956 to less than 500 pounds in 1965, and that of the walleye fell from 15,405,000 pounds in 1956 to 790,000 pounds in 1965. There was also a sharp decline in lake herring, whitefish and sauger (a small relative of the walleye). While these most desirable fishes decreased, there were rises in the catch of sheepshead (the freshwater drum), carp, yellow perch and smelt. Other signs in the lake gave evidence of an environment increasingly unfavorable for desirable fish; among these were the severe depletion of oxygen in the bottom waters, the disappearance of mayfly larvae (a fish food), which used to be extremely abundant in the shallow western end of the lake, and spectacular growths of floating algae—a certain sign of advanced age in a lake.

Lake Erie receives, to begin with, the grossly polluted water of the Detroit River, into which 1.6 billion gallons of waste are discharged daily from the cities and industries along the riverbanks. To this pollution an enormous amount is added by the great urban and industrial complex around the lake itself. A recent study of the Detroit River by the U.S. Public Health Service showed that its waters contain large quantities of sewage bacteria, phenols, iron, oil, ammonia, chlorides, nitrogen compounds, phosphates and suspended solids. Similar waste materials are discharged into the lake by the steel, chemical, refining and manufacturing plants along the lake. Pollution is particularly serious in Lake Erie because of the lake's shallowness; its volume of water is too small to dilute the pollutants effectively. Over the past 50 years the concentrations of major contaminants in the Lake Erie waters have increased sharply.

Many of the industrial wastes, notably phenols and ammonia, act as poisons to the fish and other animal life in the lake. Solid material settles to the bottom and smothers bottom-dwelling organisms. Moreover, some of the solids decompose and in doing so deplete the water of one of its most vital constituents: dissolved oxygen. Algae, on the other hand, thrive in the polluted waters, particularly since the sewage wastes contain considerable amounts of the plant-fertilizing elements nitrogen and phosphorus. The algae contribute to the depletion of oxygen (when they die and decay), give the lake water disagreeable tastes and odors and frustrate the attempts of water-purifying plants to filter the water.

In addition to Lake Erie, the southern end of Lake Michigan has also become seriously polluted. Interestingly the city of Chicago, the dominant metropolis of this area, apparently does not contribute substantially to the lake pollution; it discharges its sewage into the Mississippi River system instead of the lake. The main discharge into Lake Michigan comes from the large industrial concentration—steel mills, refineries and other establishments—clustered along its southern shores. The Public Health Service has found that the lake water in this area contains high concentrations of inorganic nitrogen, phosphate, phenols and ammonia.

Apart from the southern end, most of the water of Lake Michigan is still of reasonably good quality. In Lake Ontario, although it receives a considerable discharge of wastes, the situation is not yet as serious as in Lake Erie because Ontario's much larger volume of water provides a higher dilution factor. Lake Huron, bordered by a comparatively small population, so far shows only minor pollution effects, and Lake Superior almost none. Nevertheless, the growth of the entire region and the spreading pollution of the lakes and their tributary waters make the long-range outlook disquieting. Already the quality of the waters over a considerable portion of the lake system has greatly deteriorated, and many bathing beaches must be closed periodically because of pollution.

It is clear that in less than 150 years man has brought about changes in the Great Lakes that probably would have taken many centuries under natural conditions. These changes, shortening the usable life of the lakes, seem to be accumulating at an ever increasing rate. We still know far too little about the complicated processes that are under way or about what measures are necessary to conserve this great continental resource. Obviously the problem calls for much more study and for action that will not be too little and too late. No doubt the Great Lakes will be there for a long time to come; they are not likely to dry up in the foreseeable future. But it will be tragic irony if one day we have to look out over their vast waters and reflect bitterly, with the Ancient Mariner, that there is not a drop to drink. To realize that this is not an unthinkable eventuality we need only remind ourselves of the water crisis in New York City, where water last year had to be drastically rationed while billions of gallons in the grossly polluted Hudson River flowed uselessly by the city.

FOR FURTHER READING

The following list of books and *Scientific American* Offprints has been prepared to assist the student who desires to further his knowledge of oceanography. These books and offprints are currently in print. Only a representative coverage has been provided; additional references, including scientific journals, are cited in titles listed. Those books marked with an asterisk (*) are available in paperback.

General References

*THE SEA AROUND US. Rachel Carson. The New American Library, 1961.

THIS GREAT AND WIDE SEA. R. E. Coker. The University of North Carolina Press, 1947.

THE PHYSICAL GEOGRAPHY OF THE OCEANS. Charles H. Cotter. American Elsevier Publishing Company, Inc., 1965.

FRONTIERS OF THE SEA—THE STORY OF OCEANOGRAPHIC EXPLORATION. Robert C. Cowen. Doubleday & Company, Inc., 1969.

*THE SILENT WORLD. Jacques Y. Cousteau with Frederic Dumas. Pocket Books, Inc., 1953.

EXPLORING THE SECRETS OF THE SEA. William J. Cromie. Prentice-Hall, Inc., 1962.

DOWN TO THE SEA—A CENTURY OF OCEANOGRAPHY. J. R. Dean. Brown, Son & Ferguson, Ltd., 1966.

GENERAL OCEANOGRAPHY—AN INTRODUCTION. Günter Dietrich. Interscience Publishers, 1963.

THE EARTH AND ITS OCEANS. Alyn C. Duxbury. Addison-Wesley Publishing Company, 1971.

ENCYCLOPEDIA OF OCEANOGRAPHY. Edited by Rhodes W. Fairbridge. Reinhold Company, 1966.

LIMNOLOGY IN NORTH AMERICA. Edited by David G. Frey. The University of Wisconsin Press, 1963.

*OCEANOGRAPHY. M. Grant Gross. Charles E. Merrill Publishing Company, 1971.

*A GLOSSARY OF OCEAN SCIENCE & UNDERSEA TECHNOLOGY TERMS. Edited by Lee M. Hunt and Donald G. Groves. Compass Publications, Inc., 1965.

OCEANOGRAPHY FOR GEOGRAPHERS. Cuchlaine A. M. King. Edward Arnold Publishers Ltd., 1962.

ESTUARIES. Edited by George H. Lauff. Publication No. 83. American Association for the Advancement of Science, 1967.

*NEW WORLDS OF OCEANOGRAPHY. E. John Long. Pyramid Publications, 1965.

OCEAN SCIENCES. Edited by E. John Long. United States Naval Institute, 1964.

*ANATOMY OF AN EXPEDITION. Henry W. Menard. McGraw-Hill Book Company, 1969.

WESTWARD HO WITH THE ALBATROSS. Hans Pettersson. E. P. Dutton & Co., Inc., 1953.

MARINE GEOTECHNIQUE. Edited by Adrian F. Richards. University of Illinois Press, 1967.

OCEANOGRAPHY. Edited by Mary Sears. American Association for the Advancement of Science, 1961.

*THE OCEAN LABORATORY. Athelstan Spilhaus. Creative Educational Society, Inc., 1967.

*OCEANS. Karl K. Turekian. Prentice-Hall, Inc., 1968.

OCEANOGRAPHY—AN INTRODUCTION TO THE MARINE SCIENCES. Jerome Williams. Little, Brown and Company, Inc., 1962.

SEA AND AIR—THE NAVAL ENVIRONMENT. Jerome Williams, John J. Higginson, and John D. Rohrbough. United States Naval Institute, 1968.

AN OCEANIC QUEST: THE INTERNATIONAL DECADE OF OCEAN EXPLORATION. National Academy of Sciences Publication No. 1709, 1969.

PAPERS IN MARINE BIOLOGY AND OCEANOGRAPHY. Supplement to Vol. 3, DEEP-SEA RESEARCH. Pergamon Press, 1955.

Physical and Chemical Oceanography

AN INTRODUCTION TO PHYSICAL OCEANOGRAPHY. William S. von Arx. Addison-Wesley Publishing Company, Inc., 1962.

*WAVES AND BEACHES—THE DYNAMICS OF THE OCEAN SURFACE. Willard Bascom. Doubleday & Company, Inc., 1964.

EBB AND FLOW—THE TIDES OF EARTH, AIR, AND WATER. Albert Defant. The University of Michigan Press, 1958.

MARINE CHEMISTRY. R. A. Horne. Wiley-Interscience, 1969.

ELEMENTS OF PHYSICAL OCEANOGRAPHY. Hugh J. McLellan. Pergamon Press, 1965.

*DESCRIPTIVE PHYSICAL OCEANOGRAPHY—AN INTRODUCTION. George L. Pickard. Pergamon Press, 1963.

CHEMICAL OCEANOGRAPHY. 2 Vols. Edited by J. P. Riley and G. Skirrow. Academic Press, 1965.

THE GULF STREAM—A PHYSICAL AND DYNAMICAL DESCRIPTION. Henry Stommel. University of California Press, 1960.

OCEANOGRAPHY FOR METEOROLOGISTS. Harald U. Sverdrup. Prentice-Hall, Inc., 1942.

THE OCEANS: THEIR PHYSICS, CHEMISTRY AND GENERAL BIOLOGY. Harald U. Sverdrup, Martin W. Johnson and Richard H. Fleming. Prentice-Hall, Inc., 1942.

Geological Oceanography

HOT BRINES AND RECENT HEAVY METAL DEPOSITS IN THE RED SEA. Edited by Egon T. Degens and David A. Ross. Springer-Verlag New York Inc., 1969.

THE DEEP AND THE PAST. David B. Ericson and Goesta Wollin. Alfred A. Knopf, 1964.

ELASTIC WAVES IN LAYERED MEDIA. W. Maurice Ewing, Wencelas S. Jardetzky and Frank Press. McGraw-Hill Book Co., 1957.

GEOLOGY OF THE GREAT LAKES. Jack L. Hough. University of Illinois Press, 1958.

*AN INTRODUCTION TO MARINE GEOLOGY. M. J. Keen. Pergamon Press, 1968.

BEACHES AND COASTS. Cuchlaine A. M. King. Edward Arnold Publishers Ltd., 1959.

THE OCEAN FLOOR. Hans Pettersson. Yale University Press, 1954.

*THE EARTH BENEATH THE SEA. Francis P. Shepard. Atheneum, 1968.

SUBMARINE CANYONS AND OTHER SEA VALLEYS. Francis P. Shepard and Robert F. Dill. Rand McNally & Company, 1966.

SUBMARINE GEOLOGY. 2nd Edition. Francis P. Shepard. Harper and Row, Publishers, 1963.

PROCESSES OF COASTAL DEVELOPMENT. V. P. Zenkovich. Interscience Publishers, 1967.

INITIAL REPORTS OF THE DEEP SEA DRILLING PROJECT. (JOIDES). National Science Foundation, 1969 onward, continuing series of volumes.

Biological Oceanography

OCEANOGRAPHY AND MARINE BIOLOGY—A BOOK OF TECHNIQUES. H. Barnes. George Allen & Unwin Ltd., 1959.

THE PHYSIOLOGY OF FISHES. Vol. II: Behavior. Edited by Margaret E. Brown. Academic Press Inc., 1957.

*THE EDGE OF THE SEA. Rachel Carson. The New American Library, 1955.

*THE LIVING SEA. Jacques Y. Cousteau with James Dugan. Pocket Books, Inc., 1964.

FISHERIES BIOLOGY: A STUDY IN POPULATION DYNAMICS. D. H. Cushing. University of Wisconsin Press, 1968.

THE OPEN SEA: FISH AND FISHERIES. Alister C. Hardy. Houghton Mifflin Company, 1959.

THE OPEN SEA. ITS NATURAL HISTORY: THE WORLD OF PLANKTON. Alister C. Hardy. Collins, 1956.

TREATISE ON MARINE ECOLOGY AND PALEOECOLOGY. Vol. 1: Ecology. Edited by Joel W. Hedgpeth. The Geological Society of America, 1957.

ASPECTS OF DEEP SEA BIOLOGY. Norman B. Marshall. Philosophical Library, Inc., 1954.

BIRD NAVIGATION. G. V. T. Mathews. Cambridge University Press, 1955.

MARINE ECOLOGY. Hilary B. Moore. John Wiley & Sons, Inc., 1958.

FUNDAMENTALS OF ECOLOGY. Eugene P. Odum. W. B. Saunders Company, 1953.

*BIOLOGY OF THE OCEANS. Donald J. Reish. Dickenson Publishing Company, 1969.

THE YEAR OF THE WHALE. Victor B. Scheffer. Charles Scribner's Sons, 1969.

Man and the Ocean

RADIOACTIVE CONTAMINATION OF THE SEA. Edited by V. I. Baranov and L. M. Khitrov. Israel Program for Scientific Translations, 1966.

*THE FOREST AND THE SEA—A LOOK AT THE ECONOMY OF NATURE AND THE ECOLOGY OF MAN. Marston Bates. The New American Library, 1960.

THE NEW WORLD OF THE OCEANS—MEN AND OCEANOGRAPHY. Daniel Behrman. Little, Brown and Company, 1969.

WORLD WITHOUT SUN. Jacques-Yves Cousteau. Edited by James Dugan. Harper and Row, Publishers, 1965.

*MAN UNDER THE SEA. James Dugan. Collier Books, 1956.

CABLESHIPS AND SUBMARINE CABLES. K. R. Haigh. Adlard Coles Ltd., 1968.

THE MINERAL RESOURCES OF THE SEA. John L. Mero. Elsevier Publishing Company, 1965.

UNDERWATER MEDICINE. Stanley Miles. J. B. Lippincott Company, 1966.

SEVEN MILES DOWN: THE STORY OF THE BATHYSCAPH TRIESTE. Jacques Piccard and Robert S. Dietz. G. P. Putnam's Sons, 1961.

*MANNED SUBMERSIBLES AND UNDERWATER SURVEYING. U.S. Naval Oceanographic Office SP-153. U.S. Government Printing Office, 1970.

PETROLEUM RESOURCES UNDER THE OCEAN FLOOR. National Petroleum Council, 1969.

POSSIBILITIES OF INCREASING WORLD FOOD PRODUCTION: BASIC STUDY NO. 10. Food and Agriculture Organization, United Nations, 1963.

SHORE PROTECTION, PLANNING AND DESIGN. U.S. Army Coastal Engineering Research Center. U.S. Government Printing Office, 1966.

Marine Affairs

*OCEAN SCIENCES, TECHNOLOGY, AND THE FUTURE INTERNATIONAL LAW OF THE SEA. William T. Burke. Ohio State University Press, 1966.

THE POLAR REGIONS IN THEIR RELATION TO HUMAN AFFAIRS. Laurence M. Gould. American Geophysical Society, 1958.

USES OF THE SEAS. Edited by Edmund A. Gullion. Prentice-Hall, Inc., 1968.

PETROLEUM ECONOMICS AND OFFSHORE MINING LEGISLATION. Anton Pedro Henrik Van Meurs. Elsevier Publishing Company, 1971.

THE LAW OF THE SEA: INTERNATIONAL RULES AND ORGANIZATION FOR THE SEA. PROCEEDINGS OF THE THIRD ANNUAL CONFERENCE, 1968. Law of the Sea Institute, University of Rhode Island, 1969.

MARINE SCIENCE AFFAIRS—A YEAR OF PLANS AND PROGRESS. A Report to the President from the National Council on Marine Resources and Engineering Development. U.S. Government Printing Office, 1968.

University Curricula in Oceanography

UNIVERSITY CURRICULA IN THE MARINE SCIENCES AND RELATED FIELDS. National Council on Marine Resources and Engineering Development. (Information on courses and degree programs in oceanography and marine affairs.) U.S. Government Printing Office, 1970.

Scientific American Offprints

262. WATER. Arthur M. Buswell and Worth H. Rodebush. April 1956.

803. SAND. Ph. H. Kuenen. April 1960.

805. THE CHANGING LEVEL OF THE SEA. Rhodes W. Fairbridge. May 1960.

808. THE CONTINENTAL SHELF. Henry C. Stetson. March 1955.

810. ANATOMY OF THE ATLANTIC. Henry Stommel. January 1955.

857. THE ANTARCTIC. A. P. Crary. September 1962.

871. CORALS AS PALEONTOLOGICAL CLOCKS. S. K. Runcorn. October 1966.

875. SEA-FLOOR SPREADING. J. R. Heirtzler. December 1968.

879. MAN AND THE SEA. Roger Revelle. September 1969.

888. THE OCEAN AND MAN. Warren S. Wooster. September 1969.

890. MODELS OF OCEANIC CIRCULATION. D. James Baker, Jr. January 1970.

1019. THE SWIMMING ENERGETICS OF SALMON. J. R. Brett. August 1965.

1035. THE HAGFISH. David Jensen. February 1966.

1054. THE NAVIGATION OF PENGUINS. John T. Emlen and Richard L. Penney. October 1966.

1068. ORTHODOX AND UNORTHODOX METHODS OF MEETING WORLD FOOD NEEDS. N. W. Pirie. February 1967.

1072. THE DIVING WOMEN OF KOREA AND JAPAN. Suk Ki Hong and Hermann Rahn. May 1967.

1118. SALT GLANDS. Knut Schmidt-Nielsen. January 1959.

1123. EXPERIMENTS IN WATER-BREATHING. Johannes A. Klystra. August 1968.

1135. THERMAL POLLUTION AND AQUATIC LIFE. John R. Clark. March 1969.

1156. THE WEDDELL SEAL. Gerald L. Kooyman. August 1969.

1205. MARINE FARMING. Gifford M. Pinchot. December 1970.

1209. REFLECTORS IN FISHES. Eric Denton. January 1971.

1212. GIANT BRAIN CELLS IN MOLLUSKS. A. O. D. Willows. February 1971.

1222. THE CHEMICAL LANGUAGE OF FISHES. John H. Todd. May 1971.

INDEX